INTRODUCTION TO SURFACE CHEMISTRY AND CATALYSIS

INTRODUCTION TO SURFACE CHEMISTRY AND CATALYSIS

GABOR A. SOMORJAI
Department of Chemistry
University of California
Berkeley, California

A Wiley-Interscience Publication
JOHN WILEY & SONS, INC.
New York • Chichester • Brisbane • Toronto • Singapore

This text is printed on acid-free paper.

Copyright © 1994 by John Wiley & Sons, Inc.

All rights reserved. Published simultaneously in Canada.

Library of Congress Cataloging in Publication Data:
Somorjai, Gabor A.
    Introduction to surface chemistry and catalysis/Gabor A.
  Somorjai.
      p. cm.
      "A Wiley-Interscience publication."
      Includes bibliographical references and index.
      ISBN 0-471-03192-5 (acid-free)
      1. Surface chemistry. 2. Catalysis. I. Title.
  QD506.S589 1993
    541.3'3—dc20   93-12436

10 9

To my wife Judith and our children Nicole and John

CONTENTS

PREFACE

The study of surfaces and interfaces has occupied me fully over the last 30 years and provided the challenge and the fulfillment in my professional career. I am fascinated by macroscopic surface phenomena and the challenge to understand them on a molecular level. The wonder of catalysis that produces molecules (including those essential for life) at high rates and nearly 100% selectivity has intrigued me since my early days as a university student. Our ability to walk requires the conscious application of controlled friction; the understanding of this phenonemon on the molecular scale still defies my comprehension. How does the brain work? This complex and useful surface device will continue to be a frontier field of science for decades.

Since the early 1960s, techniques have become available in ever-increasing numbers for the study of surfaces on the molecular level. Thus from the early days as a young scientist I could fully participate in the development of molecular surface science. Our research on atomic surface structure and composition, on chemical bonding in molecular monolayers, and on heterogeneous catalysis uncovered new concepts and helped the development of new techniques. Applications of the newly discovered molecular concepts to adsorption, to heterogeneous catalysis, and most recently to coatings and tribology in my laboratory and in others also led to the increasing use of modern surface science and its techniques in the technology.

I have had the good fortune to work with outstanding students and postdoctoral fellows. Their research accomplishments appear on virtually every page of this book. The Department of Chemistry, the Berkeley campus of the University of California, and the Lawrence Berkeley Laboratory provide an ideal environment to work because of their intense but friendly atmosphere and constant pursuit of excellence. I learn much of what I know from my colleagues and students in Berkeley. The Department of Energy, through its Basic Energy Sciences Materials Science Division, provides most of the funding (for which I am most grateful) for my research through the Lawrence Berkeley Laboratory.

I would like to acknowledge the contribution to this book of many of my colleagues working in the field of surface science and catalysis throughout the world. Their results are displayed in many of the figures and tables and in the text. I would like to acknowledge the competent help of Mrs. Gloria Osterloh and Mr. Robert Kehr and would like to thank Dr. Alexis Schach von Wittenau for his help with the references and final editing.

The results of frontier research, after publication in scientific journals, appear in review papers and monographs. All this material, after suitable sifting to preserve the blocks of new knowledge that have advanced the discipline, must find its way into textbooks that are used by students. I wrote my first textbook, *Principles of Surface Chemistry*, in 1972 in order to aid the incorporation of new results obtained by modern surface science in the chemical curriculum. This was followed by two monographs, one based on the Baker Lectures I delivered at Cornell University in 1977, entitled *Chemistry in Two Dimensions: Surfaces* (published in 1981), and another, *Monolayers on Solid Surfaces*, that I wrote with Dr. M.A. Van Hove and published in 1979. The present book attempts to distill the new results of surface science and present new concepts and, by necessity, to build on many of the topics that were presented in the books I wrote earlier. I hope that the experience I gained by writing those books will benefit the students who attempt to learn modern surface science and catalysis from this book.

GABOR A. SOMORJAI

Berkeley, California
January 1994

GENERAL INTRODUCTION

The purpose of this book is to describe the present state of development of modern surface science at an introductory level to students of physical sciences and engineering. Junior standing in chemistry, physics, engineering, or the life sciences would qualify the student to take a course that would make use of this text. Teachers of introductory general chemistry courses usually given during the first year of university or college enrollment could use certain chapters (with deletions of some of the derivations) to supplement discussions of thermodynamics or catalysis, for example. We have used some of the chapters as supplementary material in our freshman and our core physical chemistry courses at Berkeley. The book should also be useful as a reference for professionals in need of data and concepts related to the properties of surfaces and interfaces.

The introductory chapter reviews the nature of various surfaces and interfaces that we encounter in everyday life. It also reviews the concept of adsorption and the surface-science techniques used to obtain much of the available information on the properties of surfaces. The second chapter discusses the structure of clean and adsorbate-covered surfaces. Chapter 3 focuses on the equilibrium thermodynamic properties of surfaces and interfaces, including the properties of curved surfaces. Special emphasis is placed upon surface segregation, an important phenomenon in surface chemistry. The motion of surface atoms about their equilibrium positions and along the surface—that is, diffusion—is reviewed in Chapter 4. Energy transfer during gas–surface interactions and discussion of elementary surface reaction steps, adsorption, surface diffusion, and desorption are part of this chapter. The electrical properties of surfaces are the subject of Chapter 5. Properties of the surface space charge and how surface properties and adsorption modify the work function are discussed. Surface ionization, the excitation of valence and inner-shell electrons, and the effects of large electric fields at small tips (field emission, ionization, and tunneling) are reviewed. Chapter 6 focuses on the nature of the surface chemical bond, whose unique character is revealed by recent surface-science studies. Both adsorbate–substrate and adsorbate–adsorbate interactions are emphasized. Chapter

7 is devoted to surface catalysis. Its concepts—the nature of bonding that makes surface reaction turnovers possible—are reviewed. Case histories of ammonia synthesis, carbon monoxide hydrogenation, and platinum-catalyzed hydrocarbon conversion are presented to show both (a) the present state of understanding of the surface science of catalysis and (b) the complexity and importance of the field. In Chapter 8 the mechanical properties of surfaces are reviewed to bring into focus the special importance of the buried interface. Adhesion and tribological properties, friction, crack formation, and lubrication are discussed.

Molecular-level understanding of surface phenomena is exposed throughout the text where it is available, and its relation to macroscopic surface properties is described whenever possible.

At the end of each chapter there is a review of the concepts introduced in that chapter and a set of problems making use of those concepts. The problems are of three types: those that can be solved after careful reading of the chapter; those denoted by one star, which refer the reader to important papers in the literature; and problems denoted by two stars, which require an extended study of the subject. Solutions are provided for the first two classes of problems. This way the reader can explore important areas of surfaces and interfaces that are outside the scope of the book.

Much of the discussion focuses on the properties of the solid–gas and solid–vacuum interfaces because most of the results of modern surface-science studies on the molecular level come from the scrutiny of these interfaces. Investigation of the solid–liquid and solid–solid interfaces are frontier topics of surface science at present, and I hope that their properties can be reviewed in more depth in future editions of this book.

Most of the chapters discuss local properties of surface atoms and molecules: atomic structure, chemical bonding, adsorption, catalysis, and mechanical properties. Transport properties, electron transport, surface magnetism, and optical properties are important subjects of surface science but are not treated here.

The study of surfaces is an exciting science that is at the intellectual frontier of physical sciences, life sciences, and engineering. It is also eminently useful in many applications in technology. It is my hope that some of the readers will share my excitement and love of this subject and discipline.

LISTS OF CONSTANTS

Fundamental Constants

Constant	Symbol	Value
Speed of light	c	2.998×10^{10} cm/sec $= 2.998 \times 10^8$ m/sec
Planck's constant	h	6.626×10^{-27} erg $\cdot$ sec $= 6.626 \times 10^{-34}$ J $\cdot$ sec
Avogadro's number	N_A	6.022×10^{23} molecules/mole
Electron charge	e	1.602×10^{-21} coulombs $= 4.803 \times 10^{-10}$ esu
Gas constant	R	1.987 cal/deg/mole $= 8.315$ J/deg/mole
Boltzmann's constant	k_B	1.381×10^{-16} erg/deg $= 1.381 \times 10^{-23}$ J/deg $= R/N_A$
Gravitational constant	g	9.807 m/sec^2
Permittivity of vacuum	ϵ_0	8.854×10^{-12} C^2/J/m

Other Conversion Factors

1 atm	$=$	1.013×10^5 kg/m/sec^2
	$=$	1.013×10^5 N/m^2
	$=$	1.013×10^5 Pa
1 torr	$=$	133.3 N/m^2
1 debye	$=$	3.336×10^{-30} C $\cdot$ m

Energy Conversion Table[a]

	erg	joule	cal	eV	cm^{-1}
1 erg	1	10^{-7}	2.389×10^{-8}	6.242×10^{11}	5.034×10^{15}
1 joule	10^7	1	0.2389	6.242×10^{18}	5.034×10^{22}
1 cal	4.184×10^7	4.184	1	2.612×10^{19}	2.106×10^{23}
1 eV	1.602×10^{-12}	1.602×10^{-19}	3.829×10^{-20}	1	8066.0
1 cm^{-1}	1.986×10^{-16}	1.986×10^{-23}	4.747×10^{-24}	1.240×10^{-4}	1

[a]For example, 1 erg $= 2.389 \times 10^{-8}$ cal.

LIST OF SYMBOLS

Symbol		First Used	Meaning
A	$\mathcal{A}$	Chapter 3	Surface area of a solid
	A	Chapter 4	Preexponential factor
	A_d^α	Chapter 4	Preexponential factor for desorption
	A	Chapter 8	Area of contact
	A^s	Chapter 3	Surface work content (energy per unit area)
	A	Chapter 4	Profile amplitude
	A_0	Chapter 4	Initial profile amplitude
	$|a|$	Chapter 2	Magnitude of X-ray unit-cell vector
	a	Chapter 3	Surface area covered by 1 mole of a component in a binary system
	a	Chapter 4	Amplitude of vibration
	a_i	Chapter 3	Surface area covered by 1 mole of component i in a binary system
	a_i	Chapter 7	Activity of species i
	a_s	Chapter 4	Area per reaction site
	α	Chapter 4	Rate of temperature rise
	α	Chapter 5	Polarizability
	α	Chapter 7	Arbitrary constant
	α_E	Chapter 4	Energy accommodation coefficent
	α_R	Chapter 4	Rotational-energy accommodation coefficient
	α_T	Chapter 4	Translational-energy accommodation coefficient
	α_V	Chapter 4	Vibrational-energy accommodation coefficient
C	C_2	Chapter 3	Two-dimensional sound velocity
	C_3	Chapter 3	Three-dimensional sound velocity
	C_P^s	Chapter 3	Specific surface heat capacity
	C_V	Chapter 3	Vibrational heat capacity

	Symbol	First Used	Meaning
	c	Chapter 6	Velocity of light
	χ	Chapter 6	Diamagnetic susceptibility
D	$\mathfrak{D}$	Chapter 1	Dispersion
	D	Chapter 4	Diffusion constant
	D_0	Chapter 4	Diffusion constant prefactor
	D_s	Chapter 4	Surface diffusion
	d	Chapter 3	Layer thickness
	d	Chapter 4	Profile periodicity
E	E	Chapter 2	Particle energy
	E	Chapter 3	Total energy of a solid
	E	Chapter 4	Average energy of a harmonic oscillator
	E	Chapter 8	Bond energy
	E^0	Chapter 3	Energy per atom in a solid
	E^s	Chapter 3	Energy per unit surface area
	E_s	Chapter 3	Strain energy induced into a solvent lattice by a solute atom
	ΔE_D^*	Chapter 4	Activation energy for diffusion
	ΔE^*	Chapter 4	Activation energy
	E_B	Chapter 5	Binding energy
	ΔE_{X_2}	Chapter 3	Dissociation energy of X_2 gas molecule
	E_d	Chapter 4	Activation energy for desorption
	E_F	Chapter 4	Fermi Energy
	$E_{i,j}$	Chapter 3	Bond energy between atoms i and j
	E_{kin}	Chapter 5	Kinetic energy
	ΔE	Chapter 7	Dissociation energy
	ΔE^*	Chapter 7	Activation energy
	ΔE^*	Chapter 4	Potential energy barrier height
	ΔE_{bond}	Chapter 3	Surface chemical bond energy
	ΔE_{des}	Chapter 5	Desorption energy
	ΔE_{des}^+	Chapter 5	Desorption energy of positive ion
	e	Chapter 5	Charge of an electron
	ϵ	Chapter 5	Dielectric constant
	ϵ_0	Chapter 5	Permittivity of free space
F	F	Chapter 1	Flux
	$\mathfrak{F}$	Chapter 8	Force
	F_{hkl}	Chapter 4	(Complex) scattering amplitude at $T = 0$
	f	Chapter 4	Restoring force
	$f(\nu)$	Chapter 3	Frequency distribution
	f_i	Chapter 7	Partial fugacity of species i
	$f(E)$	Chapter 5	Boltzmann distribution
G	G	Chapter 3	Total free energy of a system
	G_{sm}	Chapter 3	Shear modulus of a solvent in its pure form
	G^s	Chapter 3	Surface Gibbs free energy (per unit area)

Symbol	First Used	Meaning
$\overline{G}$	Chapter 3	Molar Gibbs free energy
G^0	Chapter 3	Free energy per atom
ΔG^0_{298}	Chapter 7	Standard free energy
g	Chapter 3	Gravitational acceleration
g_+/g_0	Chapter 5	Ratio of statistical weights of ionic and atomic states
g_-/g_0	Chapter 5	Ratio of statistical weights of ionic and atomic states
Γ_A	Chapter 3	Surface excess concentration of component A
γ	Chapter 3	Surface tension or interfacial energy
γ_i	Chapter 3	Surface tension or interfacial energy of component i
γ_i	Chapter 7	Activity coefficient of species i
γ^0	Chapter 3	Surface tension at absolute zero

H

Symbol	First Used	Meaning
H	Chapter 3	Total enthalpy of a system
H^s	Chapter 3	Specific surface enthalpy
H_0	Chapter 7	Hammett acidity function
ΔH	Chapter 1	Enthalpy
ΔH^0_{298}	Chapter 7	Standard enthalpy
ΔH_{ads}	Chapter 1	Heat of adsorption
ΔH_{ads}^{diff}	Chapter 3	Differential heat of adsorption
ΔH_m	Chapter 3	Heat of mixing
ΔH_{segr}	Chapter 1	Heat of segregation
ΔH_{subl}	Chapter 3	Heat of sublimation
h	Chapter 3	Height of a capillary column
h	Chapter 2	Planck's constant
$\hbar$	Chapter 2	Planck's constant divided by 2π

I

Symbol	First Used	Meaning
I	Chapter 5	Tunneling current
I_{hkl}	Chapter 4	Intensity of a diffracted beam

J

Symbol	First Used	Meaning
$\mathcal{J}$	Chapter 7	Reaction turnover frequency
j	Chapter 5	Current density
j_+	Chapter 5	Ion flux
j_-	Chapter 5	Ion flux
j_0	Chapter 5	Neutral flux

K

Symbol	First Used	Meaning
K	Chapter 4	Ratio of adsorption rate to desorption rate
K_{sm}	Chapter 3	Bulk shear modulus of a solute in its pure state
K_a	Chapter 7	Dissociation equilibrium constant
K_{HA}	Chapter 7	Dissociation constant of acid
k	Chapter 3	Force constant
k	Chapter 7	Rate constant
$\vec{k}$	Chapter 4	Electron wavevector
k_B	Chapter 3	Boltzmann's constant
k_a^*	Chapter 4	Adsorption rate constant
k_d^*	Chapter 4	Desorption rate constant

Symbol		First Used	Meaning
	$k_d^{(\alpha)}$	Chapter 4	Rate constant of desorption
	k_r	Chapter 4	Reaction rate constant
	k_w	Chapter 3	Wetting coefficient
	$\Delta \vec{k}$	Chapter 4	Scattering vector
L	L	Chapter 1	Langmuir (unit of gas exposure)
	l	Chapter 3	Ratio of the number of nearest neighbors in the plane of the atom in question to the bulk coordination number
	λ	Chapter 2	deBroglie wavelength
M	M	Chapter 1	Molecular weight
	M	Chapter 1	Atomic weight
	M	Chapter 3	Symbol for a metal atom in a chemical reaction
	$M(P)$	Chapter 7	Weight fraction of hydrocarbon with P carbon atoms
	m	Chapter 2	Particle mass
	m	Chapter 3	Ratio of the number of nearest neighbors in plane below the atom in question to the bulk coordination number
	m	Chapter 5	Electron mass
	μ	Chapter 5	Induced dipole moment
	μ	Chapter 6	Permanent dipole moment
	μ	Chapter 8	Coefficient of friction
	μ_i	Chapter 3	Chemical potential of component i
	μ_i^0	Chapter 3	Chemical potential of a pure component i
N	N_A	Chapter 1	Avogadro's number
	N_0	Chapter 5	Electron density
	N	Chapter 3	Number of atoms in a solid
	n	Chapter 4	Number of frequencies
	n	Chapter 6	Number of electrons
	N_D^+	Chapter 5	Concentration of ionized donors
	N_S	Chapter 5	Step density
	n_A	Chapter 3	Surface excess (moles) of component A
	n_e^{bulk}	Chapter 5	Bulk charge carrier concentration
	n_s	Chapter 5	Number of surface charges
	ν	Chapter 3	Vibration frequency
	ν	Chapter 4	Escape frequency
	ν_0	Chapter 4	Vibration frequency
	ν_D	Chapter 3	Maximum frequency
O	Ω	Chapter 3	Regular solution parameter
	ω	Chapter 4	Angular frequency
P	P	Chapter 1	Pressure
	P_0	Chapter 3	Saturation pressure

	Symbol	First Used	Meaning
	P_i	Chapter 4	Partial pressure of reactant i
	pK_a	Chapter 7	Negative logarithm of K_a
	p	Chapter 8	Contact pressure
	ϕ	Chapter 5	Work function
	ψ	Chapter 3	Contact angle
	ψ	Chapter 4	Angle of incidence
Q	q	Chapter 5	Unit of charge
	q	Chapter 8	Unit of heat
	q_{ads}	Chapter 3	Heat of adsorption per molecule
R	R	Chapter 1	Gas constant
	R	Chapter 4	Rotational energy
	$\Re$	Chapter 7	Specific turnover rate
	RP	Chapter 7	Reaction probability
	R_P	Chapter 7	Rate of propagation
	R_P	Chapter 7	Rate of termination
	r	Chapter 3	Radius of curvature of a curved surface
	r	Chapter 4	Relaxed length of a spring
	r	Chapter 6	Distance
	r_a	Chapter 4	Rate of adsorption for gas species a
	r_i	Chapter 3	Radius of atom i in a binary solution
	$r_{critical}$	Chapter 3	Critical radius of a droplet
	ρ	Chapter 1	Bulk density
	ρ_e	Chapter 5	Charge density
	$\Delta\rho$	Chapter 3	Density difference between gas and liquid phase
S	$\mathcal{S}$	Chapter 8	Shearing force
	S	Chapter 3	Total entropy of a solid
	S_e	Chapter 5	Electron affinity
	S^0	Chapter 3	Entropy of a solid, per atom
	$S^{(\alpha)}$	Chapter 4	Sticking coefficient
	$S_0^{(\alpha)}$	Chapter 4	Sticking coefficient at 0 coverage
	S_0	Chapter 7	Initial sticking coefficient
	S_0	Chapter 4	Reactive sticking coefficient
	S_j	Chapter 7	Catalytic selectivity
	S	Chapter 7	Sticking coefficient
	s	Chapter 8	Shearing strength
	σ	Chapter 1	Surface atom concentration
	σ	Chapter 4	Initial surface concentration
	σ_i	Chapter 7	Concentration of species i
	σ_0	Chapter 1	Surface concentration of adsorption sites
T	T	Chapter 1	Temperature
	T	Chapter 4	Translational energy
	T_c	Chapter 3	Critical temperature
	T_m	Chapter 4	Melting temperature

Symbol		First Used	Meaning
	T_P	Chapter 4	Desorption-peak maximum temperature
	T_s	Chapter 4	Surface temperature
	t	Chapter 4	Time
	τ, τ_0	Chapter 1	Residence time
	Θ_D	Chapter 3	Debye temperature
	Θ_{bulk}	Chapter 3	Bulk Debye temperature
	Θ_{surface}	Chapter 3	Surface Debye temperature
	θ	Chapter 3	Degree of coverage
V	V	Chapter 3	Volume
	V	Chapter 4	Potential energy
	V	Chapter 4	Vibrational energy
	V^s	Chapter 5	Electrostatic energy in surface region
	V_{ion}	Chapter 5	Ionization potential
	V	Chapter 3	Molar volume
	V_s	Chapter 4	Vibrational mode
	v	Chapter 8	Velocity
W	W	Chapter 3	Work
	$\mathcal{W}$	Chapter 8	Load
	W^r	Chapter 8	Work of restructuring
	W^s	Chapter 3	Work needed to increase total surface area
	W^s	Chapter 8	Work of separation
	W^s	Chapter 3	Work needed to increase total surface area, per unit surface area
	W^s_A	Chapter 3	Work of adhesion, per unit surface area
	w	Chapter 5	Tunnel barrier height
X	X	Chapter 3	Symbol for an arbitrary gas species in a chemical reaction
	ξ	Chapter 3	Ratio of surface heat capacity to bulk heat capacity
	ξ	Chapter 4	Trapping probability
	x	Chapter 4	Displacement
	$\sqrt{\langle x_\perp^2 \rangle}$	Chapter 4	Root-mean-square displacement perpendicular to the surface
	x_i	Chapter 3	Mole fraction of component i
	x_i^b	Chapter 3	Mole fraction of component i in bulk
	x_i^s	Chapter 3	Mole fraction of component i on surface
Z	z	Chapter 3	Number of nearest neighbors

INTRODUCTION TO SURFACE CHEMISTRY AND CATALYSIS

1

SURFACES—AN INTRODUCTION

1.1 HISTORICAL PERSPECTIVE

Surface science in general and surface chemistry in particular have a long and distinguished history. The spontaneous spreading of oil on water was described in ancient times and was studied by Benjamin Franklin. A timeline of the historical development of surface chemistry since then is shown in Figure 1.1. The application of catalysis started in the early 1800s, with the discovery of the platinum-surface-catalyzed reaction of H_2 and O_2 in 1823 by Döbereiner. He used this reaction in his "lighter" (i.e., a portable flame source), of which he sold a large number. By 1835 the discovery of heterogeneous catalysis was complete thanks to the studies of Kirchhoff, Davy, Henry, Philips, Faraday, and Berzelius [1]. It was at about this time that the Daguerre process was introduced for photography. The study of tribology, or friction, also started around this time, coinciding with the industrial revolution, although some level of understanding of friction appears in the work of Leonardo da Vinci. Surface-catalyzed-chemistry-based technologies first appeared in the period of 1860 to 1912, starting with the Deacon process ($2HCl + \frac{1}{2}O_2 \rightarrow H_2O + Cl_2$), SO_2 oxidation to SO_3 (Messel, 1875), the reaction of methane with

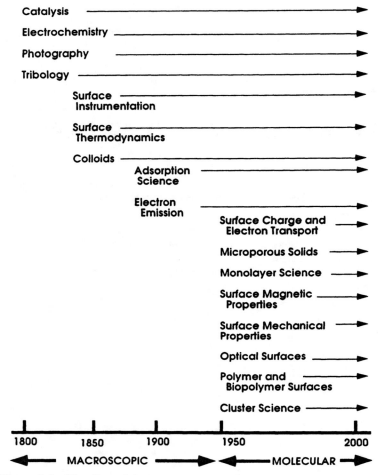

Catalysis

Electrochemistry

Photography

Tribology

Surface
Instrumentation

Surface
Thermodynamics

Colloids

Adsorption
Science

Electron
Emission

Surface Charge and
Electron Transport

Microporous Solids

Monolayer Science

Surface Magnetic
Properties

Surface Mechanical
Properties

Optical Surfaces

Polymer and
Biopolymer Surfaces

Cluster Science

1800   1850   1900   1950   2000

◀— MACROSCOPIC —▶ ◀— MOLECULAR —▶

Figure 1.1. Timeline of the historical development of surface chemistry.

steam to produce CO and H_2 (Mond, 1888), ammonia oxidation (Ostwald, 1901), ethylene hydrogenation (Sabatier, 1902), and ammonia synthesis (Haber, Mittasch, 1905–1912). Surface tension measurements and recognition of equilibrium constraints on surface chemical processes led to the development of the thermodynamics of surface phases by Gibbs (1877). The existence of polyatomic or polymolecular aggregates that lack crystallinity and diffuse slowly (gelatine and albumin, for example) was described in 1861 by Graham, who called these systems "colloids." Polymolecular aggregates that exhibit internal structure were called "micelles" by Nageli, and stable metal colloids were prepared by Faraday. However, the colloid subfield of surface chemistry gained prominence in the beginning of the 20th century with the rise of the paint industry and the preparation of artificial rubbers. Studies of light bulb filament lifetimes, high-surface-area gas absorbers in the gas mask, and gas separation technologies in other forms led to investigations of atomic and molecular adsorption (Langmuir, 1915). The properties of chemisorbed and physisorbed monolayers, adsorption isotherms, dissociative adsorption, energy exchange, and sticking upon gas–surface collisions were studied. Studies of electrode surfaces in electrochemistry led to the detection of the surface space charge [2] (for a review of electrochemistry in the 19th century, see reference [3]). The surface diffraction of electrons was discovered by Davisson and Germer (1927). Major academic and industrial laboratories focusing on surface studies have been formed in Germany (Haber, Polanyi, Farkas, Bonhoefer), the United Kingdom (Rideal, Roberts, Bowden), the United States (Langmuir, Emmett, Harkins, Taylor, Ipatief, Adams), and many other countries. They have helped to bring surface chemistry into the center of development of chemistry—both because of the intellectual challenge to understand the rich diversity of surface phenomena and because of its importance in chemical and energy conversion technologies.

In the early 1950s, focus in chemistry research shifted to studies of gas-phase molecular processes as many new techniques were developed to study gas-phase species on the molecular level. This was not the case in surface and interface chemistry, although the newly developed field-ion and electron microscopies did provide atomic level information on surface structure. The development of surface-chemistry-based technologies continued at a very high rate, however, especially in areas of petroleum refining and the production of commodity chemicals. Then, in the late 1950s, the rise of the solid-state-device-based electronics industry and the availability of economical ultrahigh vacuum systems—developed by research in space sciences—provided surface chemistry with new challenges and opportunities, resulting in an explosive growth of the discipline. Clean surfaces of single crystals could be studied for the first time, and the preparation of surfaces and interfaces with known atomic structure and controlled composition was driving the development of microelectronics and computer technologies. New surface instrumentation and techniques have been developed that permit the study of surface properties on the atomic scale. In Table 1.1 (p. 19) many of the most frequently used surface characterization techniques are listed. Most of these have been developed since the 1960s.

As a result of this sudden availability of surface characterization techniques, macroscopic surface phenomena (adsorption, bonding, catalysis, oxidation and other surface reactions, diffusion, desorption, melting and other phase transformation, growth, nucleation, charge transport, atom, ion and electron scattering, friction, hardness, lubrication) are being reexamined on the molecular scale. This has led to a remarkable growth of surface chemistry that has continued uninterrupted up to the present. The discipline has again become one of the frontier areas of chemistry. The newly gained knowledge of the molecular ingredients of surface phenomena has given birth to a steady stream of high technology products, including: new hard coatings that passivate surfaces; chemically treated glass, semiconductor, metal, and polymer surfaces where the treatment imparts unique surface properties; newly designed catalysts, chemical sensors, and carbon fiber composites; surface-space-charge-based copying; and new methods of electrical, magnetic, and optical signal processing and storage. Molecular surface chemistry is being utilized increasingly in biological sciences.

1.2 SURFACES AND INTERFACES—CLASSIFICATION OF PROPERTIES

Condensed phases—solids and liquids—must have surfaces or interfaces. The suit of an astronaut maneuvering in outer space represents a solid–vacuum interface (Fig-

Figure 1.2. Interfaces are ever-present in our lives. (**a**) An astronaut representing the solid–vacuum interface; (**b**) a jumping basketball player representing the solid–gas interface; (**c**) a sailboat representing the solid–liquid interface; and (**d**) a tire representing the solid–solid interface.

ure 1.2a); a basketball player jumping to score is a moving solid–gas interface (Figure 1.2b); a sailboat moving over the waves is a solid–liquid interface (Figure 1.2c); a tire slides at the solid–solid interface (Figure 1.2d). The surface of a lake is a liquid–gas interface. Olive oil poured on top of an open bottle of wine to prevent air oxidation forms a liquid–liquid interface. These interfaces exhibit some remarkable physical and chemical properties. The chemical behavior of surfaces is responsible for heterogeneous catalysis (ammonia synthesis, for example) and gas separations (as in the extraction of oxygen and nitrogen from air) by selective adsorption. Mechanical surface properties give rise to adhesion, friction, or slide. Magnetic surfaces are used for information storage (e.g., magnetic tape or computer disk drive). Optical surface phenomena are responsible for color and texture perception, total internal reflection needed for transmission through glass fibers, and the generation of second and higher harmonic frequencies in nonlinear laser optics. The electrical behavior of surfaces often gives rise to surface charge build-up, which is used for image transfer in xerography and for electron transport in integrated circuitry (Figure 1.3).

Surfaces and interfaces are the favorite media of evolution. Both photosynthetic and biological systems—the brain (Figure 1.4) and the leaf (Figure 1.5)—evolve and improve by ever increasing their interface area or their interface-to-volume ratio. The spine of the sea urchin has remarkable strength that is achieved by the layered structure of an inorganic–organic composite, namely, single-crystalline calcium carbonate that grows on ordered layers of acidic macromolecules deposited on layers of protein (Figure 1.6).

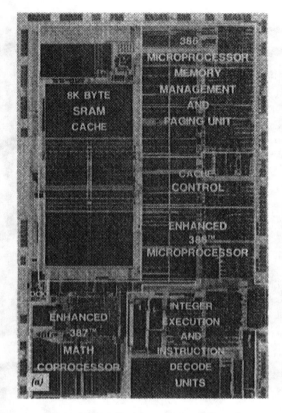

Figure 1.3. Integrated microelectronic circuits are the heart of computers and other electronic devices. Miniaturization increases their speed and permits the performance of more functions per unit area. (**a**) The Intel I486 microprocessor; (**b**) a close-up view of a circuit element; and (**c**) a cross-sectional view. As the integrated circuits are made increasingly more compact, their surface-to-volume ratio increases and essentially makes them surface devices. (Courtesy of Paul Davies, Intel Corporation.)

1.3 EXTERNAL SURFACES

1.3.1 Surface Concentration

The concentration of atoms or molecules at the surface of a solid or liquid can be estimated from the bulk density. For a bulk density of 1 g/cm^3 (such as ice or water), the molecular density ρ, in units of molecules per cm^3, is $\approx 5 \times 10^{22}$. The surface concentration of molecules σ (molecules/cm^2) is proportional to $\rho^{2/3}$, assuming a cube-like packing, and is thus on the order of 10^{15} molecules/cm^2. Because the densities of most solids or liquids are all within a factor of 10 or so of each other, 10^{15} molecules/cm^2 is a good order-of-magnitude estimate of the surface concentration of atoms or molecules for most solids or liquids. Of course, surface atom concentration of crystalline solids may vary by a factor of two or three, depending on the type of packing of atoms at a particular crystal face.

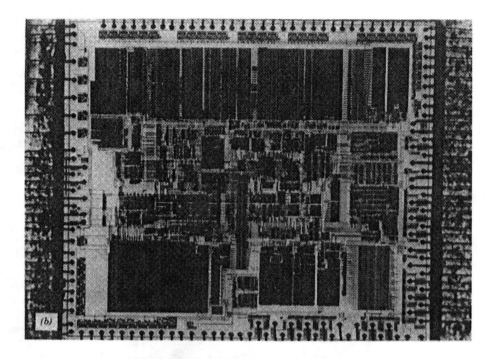

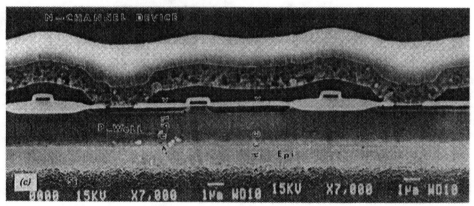

Figure 1.3. (*Continued*)

1.3.1.1 Clusters and Small Particles All of the atoms in a three- or four-atom cluster are by necessity "surface atoms." As a cluster grows in size, some atoms may become completely surrounded by neighboring atoms and are thus no longer on the "surface" (Figure 1.7). We frequently describe a particle of finite size by its dispersion $\mathfrak{D}$, where $\mathfrak{D}$ is the ratio of the number of surface atoms to the total number of atoms:

$$\mathfrak{D} = \frac{\text{number of surface atoms}}{\text{total number of atoms}} \tag{1.1}$$

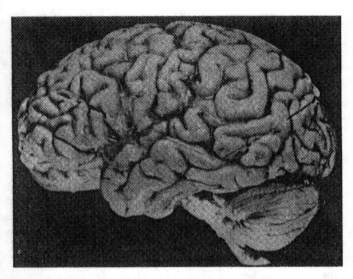

Figure 1.4. The intricate folds of the human brain expose the large interface area of this remarkable organ. The brain may be viewed as a device with enormous solid–liquid interface area.

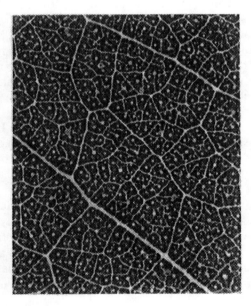

Figure 1.5. The green leaf. Photosynthesis involves the absorbtion of sunlight and the reactions of water and carbon dioxide to produce organic molecules and oxygen. High-surface-area systems, such as the green leaf, are most efficient to carry out photosynthesis.

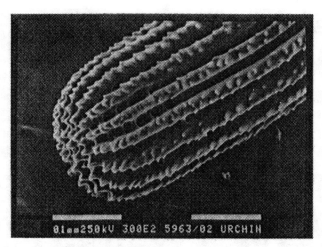

Figure 1.6. Spine of a sea urchin. Schematic diagram of the composite layer structure that makes up the spine of the sea urchin. Crystalline calcium carbonate grows on an acidic macromolecular layer that is bound to protein layers. The spine is a single crystal of $CaCO_3$ with its $00\bar{1}$ axis parallel to the growth axis. (Courtesy of S. Weiner, L. Addadi, and A. Berman, Weizmann Institute of Science, Rehovot, Israel.)

For very small particles, $\mathfrak{D}$ is unity. As the particle grows and some atoms become surrounded by their neighbors, the dispersion decreases. Of course, $\mathfrak{D}$ also depends somewhat on the shape of the particle and how the atoms are packed [4] (compare Figures 1.7 and 1.8). The dispersion is already as low as 10^{-3} for particles of 10-nm (100-Å) radius.

Many chemical reactions are facilitated by surface atoms of heterogeneous catalysts. These catalysts increase the rates of formation of product molecules and modify the relative distributions of the products. Most catalysts are in small particle form, including those used to produce fuels and chemicals ranging from high-octane gasoline to polyethylene.

1.3.1.2 Thin Films   Consider a monolayer of gold atoms (a layer of gold atoms one atom thick) deposited on iron (Figure 1.9). This film has a dispersion of unity, since all of the gold atoms are on the surface. About 50 layers of gold atoms ($\mathfrak{D} = 1/50$) are needed to obtain the optical properties that impart the familiar yellow color characteristic of bulk gold.

Thin films are of great importance to many real-world problems. Their material costs are very little as compared to the bulk material, and they perform the same function when it comes to surface processes. For example, a monolayer of rhodium, a very expensive metal, which contains only about 10^{15} metal atoms per cm^2, can catalyze the reduction of NO to N_2 by its reaction with CO in the catalytic converter of an automobile, or it can catalyze the conversion of methanol to acetic acid by the insertion of a CO molecule.

Thin ordered silicon layers optimize electron transport in integrated electronic circuits (Figure 1.3), and thin films of organic molecules lubricate our skin or the moving parts of internal combustion engines. A green leaf is a high-surface-area

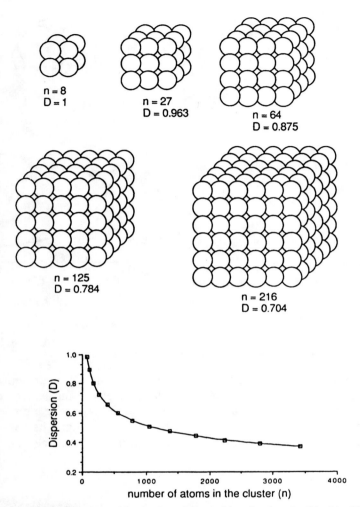

Figure 1.7. Clusters of atoms with single cubic packing having 8, 27, 64, 125, and 216 atoms. In an eight-atom cluster, all of the atoms are on the surface. However, the dispersion, $\mathcal{D}$, defined as the number of surface atoms divided by the total number of atoms in the cluster, declines rapidly with increasing cluster size. this is shown in the lower part of the figure.

system designed to maximize the absorption of sunlight in order to carry out chlorophyll-catalyzed photosynthesis at optimum rates (Figure 1.5).

Often the surface of a thin film is roughened deliberately. Automobile brake pads are designed to optimize the desired mechanical properties of surfaces in this way, as is the corrugated design of rubber soles of tennis shoes. The large number of folds of the human brain (Figure 1.4) helps to maximize the number of surface sites, which also facilitate charge transport and the transport of molecules. These are but some of the examples that show how external surfaces are frequently used in nature. External surfaces are a key element of technology, ranging from catalysts and passivating coatings, to computer-integrated circuitry and the storage and retrieval of information.

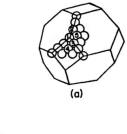

(a)

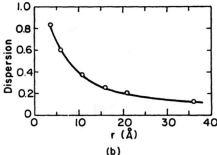

(b)

Figure 1.8. Dispersion as a function of size for particles with cubo-octahedral packing.

1.3.2 Internal Surfaces—Microporous Solids

Microporous solids are materials that are full of pores of molecular dimensions or larger. These materials have large internal surface areas. Many clays have layer structures that can accommodate molecules between the layers by a process called *intercalation*. Graphite will swell with water vapor to several times its original thickness (Figure 1.10) as water molecules become incorporated between the graphitic

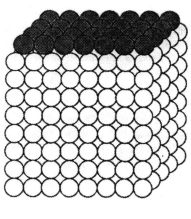

gold atom

iron atom

Figure 1.9 An iron particle with one surface covered with a monolayer of gold atoms. When it comes to surface properties such as adsorption or catalysis, one monolayer of atoms is all that is needed to carry out the necessary chemistry. Because of the very few atoms needed, such a device is inexpensive. For other applications a thin film of up to 50 layers is needed. For this reason, thin films are very often utilized to carry out surface reactions (adsorption, catalysis) and to impart mechanical properties (hardness, lubrication), optical properties (reflectivity, color), and electrical properties (conductivity, charge retention).

layers. Crystalline alumina silicates, often called *zeolites*, have ordered cages of molecular dimensions [5, 6] where molecules can adsorb or undergo chemical reactions (Figure 1.11). These materials are also called *molecular sieves*, because they may preferentially adsorb certain molecules according to their size or polarizability. This property is of great commercial importance and may be used to separate mixtures of gases (air) or liquids or to carry out selective chemical reactions. Bones of mammals are made out of calcium apatite, which has a highly porous structure, with pores on the order of 10 nm (100 Å) in diameter. Coal [7, 8] and char have porous structures, with pore diameters on the order of 10^2–10^3 nm (10^3–10^4 Å). These materials have very large internal surface areas, in the range of 100–400 m^2

Figure 1.10. Graphite swollen by intercalation of rubidium between layers of carbon (C$_8$Rb). Layer compounds like graphite, MoS$_2$, or TaS$_2$, and many clays have relatively weak bonding between the atomic sheets. This is an example of bonding at internal surfaces of microporous materials. (Courtesy of P. Gurth and M. Guntherrod, Institute of Physics, University of Basel.)

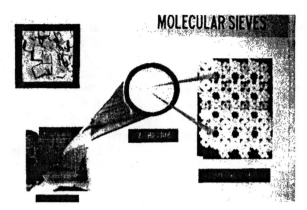

Figure 1.11. Microporous molecular sieve. There are many alumina silicates in nature that have pores with molecular dimensions. These are called *zeolites*. Synthetic zeolites are also produced in large numbers, mostly from silicates, phosphates, and borates. They are used as selective absorbers of gases or liquids, and they are the catalysts utilized in the largest volume in the chemical and petroleum technologies.

per gram of solid. As this short survey has shown, nature has provided us with many useful microporous materials; and many synthetic microporous substances are used in technology, both to separate gas and liquid mixtures by selective adsorption and to carry out surface reactions selectively in their pores, which are often of molecular dimensions. Because surface reaction rate (product molecules formed per second) is proportional to surface area, materials with high internal surface areas carry out surface reactions at very high rates.

1.4 CLEAN SURFACES

In order to study atomically clean surfaces, we must work under so-called ultrahigh vacuum (UHV) conditions [9–12], as the following rough calculation shows. We know that the concentration of atoms on the surface of a solid is on the order of 10^{15} cm^{-2}. To keep the surface clean for 1 sec or for 1 hr, then, the flux of molecules incident on the initially clean surface must therefore be less than $\approx 10^{15}$ molecules/cm^2/sec or $\approx 10^{12}$ molecules/cm^2/sec, respectively. From the kinetic theory of gases [13], the flux F of molecules striking the surface of unit area at a given ambient pressure P is

$$F = \frac{N_A P}{\sqrt{2\pi MRT}} \qquad (1.2)$$

or

$$F(\text{atoms}/\text{cm}^2 \cdot \text{sec}) = 2.63 \times 10^{20}\, \frac{P(\text{Pa})}{\sqrt{M(\text{g}/\text{mole})\,T}} \qquad (1.3)$$

or

$$F(\text{atoms}/\text{cm}^2 \cdot \text{sec}) = 3.51 \times 10^{22} \frac{P(\text{torr})}{\sqrt{M(\text{g}/\text{mole})\, T}} \qquad (1.4)$$

where M is the average molar weight of the gaseous species, T is the temperature, and N_A is Avogadro's number. Substituting $P = 4 \times 10^{-4}$ Pa (3×10^{-6} torr) and using the values $M = 28$ g/mole and $T = 300$ K, we obtain $F \approx 10^{15}$ molecules/cm^2/sec. Thus, at this pressure the surface is covered with a monolayer of gas within seconds, assuming that each incident gas molecule ''sticks.'' For this reason the unit of gas exposure is 1.33×10^{-4} Pa-sec (10^{-6} torr-sec), which is called the *langmuir* (L). Thus, a 1-L exposure will cover a surface with a monolayer amount of gas molecules, assuming a sticking coefficient of unity. At pressures on the order of 1.33×10^{-7} Pa (10^{-9} torr), it may take 10^3 sec before a surface is covered completely.

In practice, one usually wants to study a surface without worrying about contamination from ambient gases. Current surface science techniques can easily detect contamination on the order of 1% of a monolayer. This then will be our operational definition of ''clean.'' Thus, ultrahigh vacuum conditions ($< 1.33 \times 10^{-7}$ Pa $= 10^{-9}$ torr) are required to maintain a clean surface for about 1 hr, the time usually needed to perform experiments on clean surfaces.

1.5 INTERFACES

In most circumstances, however, and certainly in our earth's environment, surfaces are continually exposed to gases or liquids or placed in contact with other solids. As a result, we end up investigating the properties of interfaces—that is, between solids and gases, between solids and liquids, between solids and solids, and even between two immiscible liquids. Thus, unless specifically prepared otherwise, surfaces are always covered with a layer of atoms or molecules from the ambient (Figure 1.12).

1.5.1 Adsorption

On approaching the surface, each atom or molecule encounters an attractive potential that ultimately will bind it to the surface under proper circumstances. The process that involves trapping of atoms or molecules that are incident on the surface is called *adsorption*. It is always an exothermic process. For historical reasons, the heat of adsorption ΔH_{ads} is always denoted as having a positive sign—unlike the enthalpy ΔH, which for an exothermic process would be negative according to the usual thermodynamic convention.

The residence time τ of an adsorbed atom is given [14] by

$$\tau = \tau_0 \exp\left(\frac{\Delta H_{\text{ads}}}{RT}\right) \qquad (1.5)$$

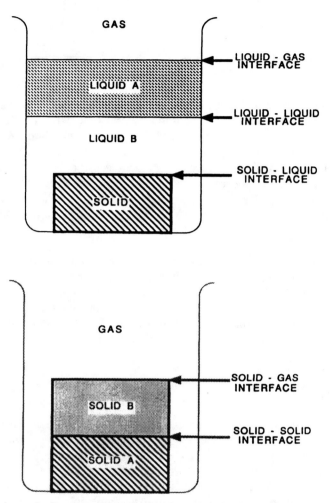

Figure 1.12. Schematic diagram of interfaces—for example, solid–liquid, liquid–liquid, liquid–gas, solid–solid, and solid–gas interfaces.

where τ_0 is correlated with the surface atom vibration times (it is frequently on the order of 10^{-12} sec), T is the temperature, and R is the gas constant. The value of τ can be 1 sec or longer at 300 K for $\Delta H_{ads} > 63$ kJ/mole (15 kcal/mole). The surface concentration σ (in molecules/cm^2) of adsorbed molecules on an initially clean surface is given by the product of the incident flux F and the residence time τ:

$$\sigma = F\tau \tag{1.6}$$

The surface of the material on which adsorption occurs is often called the *substrate*. Substrate-adsorbate bonds are usually stronger than the bonds between adsorbed molecules. As a result, the monolayer of adsorbate bonded to the substrate is held most tenaciously and is difficult to remove. Therefore, the properties of real surfaces are usually determined in the presence of an adsorbed monolayer. For this reason, in the chapters that follow we will discuss the various properties of surfaces when clean and also when covered with a monolayer of adsorbate.

1.5.2 Thickness of Surface Layers

A surface or interface may be defined as comprising either one atomic layer or several layers in the near-surface region. Usually the phenomena or the systems studied define the number of atomic layers that must be considered as part of the surface. For example, the chemical bond between an adsorbed molecule and atoms in the topmost atomic layer of a solid can be described rather well by considering the properties of one monolayer of adsorbate and one monolayer of substrate atoms. However, the build-up of charge at the surface of an electrical insulator may induce an electric field that extends over a hundred layers into the solid. When such a surface is in contact with a liquid, the electric field due to the surface charge accumulation also extends into the liquid several molecular layers deep. In this circumstance the interface must be defined as many atomic layers thick on both sides of the surface in order to properly describe the electrical properties of the interface.

1.6 THE TECHNIQUES OF SURFACE SCIENCE

Over the last three decades, a large number of techniques have been developed to study various surface properties, including structure, composition, oxidation states, and changes of chemical, electronic, and mechanical properties. The emphasis has been on surface probes that monitor properties on the molecular level and are sensitive enough to detect ever smaller numbers of surface atoms. The frontiers of surface instrumentation are constantly being pushed toward detection of finer detail: atomic spatial resolution, ever smaller energy resolution, and shorter time scales. Because no one technique provides all necessary information about surface atoms, the tendency is to use a combination of techniques. The most commonly used techniques involve the scattering, absorption, or emission of photons, electrons, atoms, and ions, although some important surface-analysis techniques cannot be classified this way.

Electrons, atoms, and ions are used primarily to investigate external surfaces and require low ambient pressures during their application. Photons can be used to study both internal and external surfaces because of their much-lower-scattering cross sections. They can also be employed at high gas pressures and for studies of solid–liquid and solid–solid interfaces.

Because many surface probes require high vacuum during their application, most surface science instruments are also equipped with high-pressure or environmental cells. The sample to be analyzed is first subjected to the usual high-pressure and/or high-temperature conditions encountered during reactions in the environmental cell. Then it is transferred into the evacuated chamber where the surface probe is located for surface analysis. One such apparatus is shown in Figure 1.13.

Sample preparation is always an important part of surface studies. Single crystals are oriented by X-ray back-diffraction, cut, and polished. They are then ion-bombarded or chemically treated to remove undesirable impurities from their surfaces.

Figure 1.13. Photograph of a stainless steel chamber used for surface studies. It is equipped with surface characterization instruments that are used in ultrahigh vacuum and with a high-pressure cell that is shown in both open (**top**) and in closed (**bottom**) positions. The cell is used to expose the samples to high pressures and high temperatures. The chamber can be evacuated to 10^{-10} torr and equipped with windows on steel flanges with glass-to-metal seals for easy viewing. The flanges are mounted using copper gaskets to avoid the use of lubricated seals so that the chamber can be "baked" at high temperatures ($\approx 200°C$) to clean its internal surfaces. A manipulator that is used to mount the sample provides motion in three dimensions and permits cooling and heating. Gas analysis is provided by a mass spectrometer mounted on the chamber. The pressure is measured by ionization and thermocouple gauges.

Thin films are deposited from vapor by sublimation, sputtering, or the use of plasma-assisted chemical vapor deposition. Materials of high internal surface area are prepared from a sol–gel or by calcination at high temperatures. The genesis and environmental history of the surface is primarily responsible for its structure and composition and must always be carefully monitored.

Table 1.1 lists many of the surface science techniques that have been used most frequently in recent years to learn about the interface on the atomic scale. The names of the techniques, their acronyms, and brief descriptions are provided, along with references, if a more detailed study of the capabilities and limitations of a particular technique is desired. We also indicate the primary surface information that can be obtained by the application of each technique. Detailed discussions of these techniques are outside the scope of this book. The reader is referred to review papers that describe the principles of operation for each, the instrumentation, and some of the findings of the experiments that used that technique. Many of the surface science techniques are used in combination to obtain a more complete characterization of the structure (atomic or molecular, electronic) and the composition (including oxidation states) of atoms and molecules at the interface with increasing spatial and time resolution.

1.7 SUMMARY AND CONCEPTS

- The surface concentration of atoms or molecules is on the order of 10^{15} cm^{-2} for most solids and liquids.
- Small particles used in surface studies are frequently described in terms of their dispersion.
- Thin films and microporous solids are systems with high surface-to-volume ratios.
- Many surface science studies focus on interfaces (solid–gas, solid–liquid, and solid–solid), since surfaces are covered with adsorbates under the practical conditions in which they are utilized.
- The definition of how many atomic layers constitute the ''surface region'' depends on the surface phenomena under investigation. For example, one atomic or molecular layer can be responsible for most surface chemical properties, whereas almost 10^3 layers are required to investigate surface effects in electron or photon transport.
- Most techniques provide information only about one side of the surface–adsorbate bond. Future instrumentation developments aim for molecular level studies at buried interfaces, of both sides of the surface chemical bonds, and on an ever shorter time scale (time resolved studies).
- Photon, electron, atom, and ion scattering are utilized most frequently to study surface atomic and electronic structures and composition.
- Vacuum or reduced pressures at the interface is needed during experiments using electrons, atoms, and ions. As a result, we know more about the properties of the solid–vacuum and solid–gas interfaces than about solid–liquid, solid–solid, and liquid–liquid interfaces.
- Clean, adsorbate-free surfaces must be prepared in ultrahigh vacuum.
- Selective adsorption of atoms and molecules are also important tools for studies of surface composition and bonding.

1.8 PROBLEMS

1.1 Calculate the concentration of surface atoms (atoms/cm^2) for mercury, copper, and benzene.

1.2 (a) What is the gas flux striking a surface in air at 1 atm and 300 K?
(b) Calculate the pressure necessary to keep a 1 cm^2 Cu surface clean for 1 hr at 300 K, assuming a sticking coefficient of 1 and no dissociation of the gas upon adsorption.

1.3   In most cases, of course, the sticking coefficients are much less than 1. Using the following sticking coefficients for O_2, calculate the pressure necessary to keep a 1-cm^2 Cu surface "clean" for 1 sec, 1 hr, and 8 hr at 300 K (the latter times correspond to times for "fast" or "relaxed" experiments, respectively). According to reference [15], σ = 0.01, 0.2, and 0.001 for the (100), (110), and (111) faces of Cu, respectively.

1.4   At 1 atm air pressure, compute the volume of nitrogen gas that is adsorbed on 10 g of zeolite with surface area of 400 m^2/g. Assume a surface area of 16.2 $Å^2$ per nitrogen molecule.

1.5   What is the residence time of molecular nitrogen in a zeolite at 77 K and at 300 K? Assume a heat of adsorption of 15 kJ/mole.

REFERENCES

[1]   J. Berzelius. *Jahres-Bericht über die Fortschritte der Physichen Wissenschaften*, Volume 15. Tübingen, 1836.

[2]   J.M. Thomas. *Michael Faraday and the Royal Institution*. IOP Publishing, Bristol, 1991.

[3]   W. Ostwald. *Electrochemistry: History and Theory*. Amerind Publishing Co. Pvt. Ltd., New Delhi, 1980, translated by N.P. Date. Originally: *Electrochemie: Ihre Geschichte und Lehre*. Verlag von Veit und Gesellschaft, Leipzig, 1896.

[4]   R. van Hardeveld and F. Hartog. The Statistics of Surface Atoms and Surface Sites on Metal Crystals. *Surf. Sci.* **15**:189 (1969).

[5]   W. Hölderich, M. Hesse, and F. Nämann. Zeolites: Catalysts for Organic Syntheses. *Angew. Chem.* (International Edition in English) 27:226 (1988).

[6]   S.M. Csicsery. Shape-Selective Catalysis. In: J.A. Rabo, editor, *Zeolite Chemistry and Catalysis. ACS Monographs*, Volume 171. American Chemical Society, Washington, D.C., 1976.

[7]   J.G. Speight. The Chemistry and Technology of Coal, Volume 12. Marcel Dekker, New York, 1983.

[8]   D.D. Whitehurst, T.O. Mitchell, and M. Farcasiu. *Coal Liquefaction: The Chemistry and Technology of Thermal Processes*. Academic Press, New York, 1980.

[9]   C. Klauber. UHV Basics. In: D.J. O'Conner, B.A. Sexton, and R. St. C. Smart, editors, *Surface Analysis Methods in Materials Science. Springer Series in Surface Sciences*, Volume 23. Springer-Verlag, Berlin, 1992.

[10]   J.F. O'Hanlon. *A User's Guide to Vacuum Technique*, 2nd edition. John Wiley & Sons, New York, 1989.

[11]   G.F. Weston. *Ultrahigh Vacuum Practice*. Butterworths, London, 1985.

[12]   S. Dushman. *Scientific Foundations of Vacuum Technique*, 2nd edition. John Wiley & Sons, New York, 1962. Revised by the General Electric Research Staff, J.M. Lafferty, editor.

[13]   J.H. de Boer. *The Dynamical Character of Adsorption*. Oxford University Press, New York, 1968.

[14]   F.C. Tompkins. *Chemisorption of Gases on Metals*. Academic Press, New York, 1978.

[15]   F.H.P.M. Habraken, C.M.A.M. Mesters, and G.A. Bootsma. The Adsorption and Incorporation of Oxygen on Cu(100) and Its Reaction with Carbon Monoxide; Comparison with Cu(111) and Cu(110). *Surf. Sci.* 97:264 (1980).

TABLE 1.1. Surface Science Techniques

Acronym	Name	Description	Primary Surface Information
	Adsorption or selective chemisorption (1)	Atoms or molecules are physisorbed into a porous structure—such as a zeolite or a sample of coal—or onto a surface, and the amount of gas adsorbed is a measure of the surface area available for adsorption. Chemisorption of atoms or molecules on surfaces yields surface concentration of selected elements or adsorption sites.	Surface area, adsorption site concentration
AD	Atom or helium diffraction (2–13)	Monoenergetic beams of atoms are scattered from ordered surfaces and detected as a function of scattering angle. This gives structural information on the outermost layer of the surface. Atom diffraction is extremely sensitive to surface ordering and defects.	Surface structure
AEAPS	Auger electron appearance potential spectroscopy (2–4, 14–17)	A monoenergetic beam of electrons is used to excite atoms in the near surface region. As the beam energy is swept, variations in the sample emission current occur as the beam energy sweeps over the energy of an Auger transition in the sample. Also known as APAES.	Chemical composition
AES	Auger electron spectroscopy (2–4, 14, 16, 18–29)	Core-hole excitations are created, usually by 1 to 10-keV incident electrons; Auger electrons of characteristic energies are emitted through a two-electron process as excited atoms decay to their ground state. AES gives information on the near-surface chemical composition.	Chemical composition
AFM	Atomic force microscopy (30–37)	Very similar to scanning tunneling microscopy (STM). In this technique, however, the attractive van der Waals forces between the surface and the probe cause a bending of the probe. This deflection is measurable by a variety of means. Because this technique does not require a current between the probe and the surface, nonconducting surfaces may be imaged.	Surface structure
APAES	Appearance potential auger electron spectroscopy	See AEAPS.	
APXPS	Appearance potential X-ray	The EAPFS excitation cross section is monitored by fluorescence from core-	Chemical composition

TABLE 1.1. (*Continued*)

Acronym	Name	Description	Primary Surface Information
	photoemission spectroscopy (2–4, 16)	hole decay (also known as SXAPS).	
ARAES	Angle-resolved auger electron spectroscopy (38)	Auger electrons are detected as a function of angle to provide information on the spatial distribution or environment of the excited atoms (see AES).	Surface structure
ARPEFS	Angle-resolved photoemission extended fine structure (38–40)	Electrons are detected at given angles after being photoemitted by polarized synchrotron radiation. The interference in the detected photoemission intensity as a function of electron energy ≈ 100–500 eV above the excitation threshold gives structural information.	Surface structure
ARPES	Angle-resolved photoemission spectroscopy (3, 24, 41–44)	A geneal term for structure-sensitive photoemission techniques, including AR-PEFS, ARXPS, ARUPS, and ARXPD.	Electronic structure, surface structure
ARUPS	Angle-resolved ultraviolet photoemission spectroscopy (3, 42, 45–48)	Electrons photoemitted from the valence and conduction bands of a surface are detected as a function of angle. This gives information on the dispersion of these bands (which is related to surface structure) and also gives structural information from the diffraction of the emitted electrons.	Valence band structure, bonding
ARXPD	Angle-resolved X-ray photoelectron diffraction (3, 38, 39, 49–51)	Similar to ARXPS and ARPEFS. The angular variation in the photoemission intensity is measured at a fixed energy above the excitation threshold to provide structural information.	Surface structure
ARXPS	Angle-resolved X-ray photoemission spectroscopy (3, 38, 39, 49, 50)	The diffraction of electrons photoemitted from core levels gives structural information on the surface.	Surface structure
CEM	Conversion electron Mössbauer spectroscopy (4, 52–55)	A surface-sensitive version of Mössbauer spectroscopy. Like Mössbauer spectroscopy, this technique is limited to some isotopes of certain metals. After a nucleus is excited by γ-ray absorption, it can undergo inverse β-decay, creating a core hole. The decay of core holes by Auger processes within an electron mean free path of the surface produces a signal. Detecting emitted electrons as a	Chemical environment, oxidation state

TABLE 1.1. (*Continued*)

Acronym	Name	Description	Primary Surface Information
		function of energy gives some depth-profile information because the changing electron mean free path.	
DAPS	Disappearance potential spectroscopy (2–4, 16)	The EAPFS cross section is monitored by variations in the intensity of electrons back-scattered from the surface.	Chemical composition
EAPFS	Electron appearance potential fine structure (3, 56)	A fine-structure technique (see EXAFS). Core holes are excited by monoenergetic electrons. The modulation in the excitation cross section as the beam energy is varied may be monitored through absorption, fluorescence, or Auger emission.	Surface structure
ELNES	Electron energy loss near edge structure	Similar to NEXAFS, except monoenergetic high-energy electrons ≈ 60–300 keV excite core holes.	Surface structure
ELS or EELS	Electron energy loss spectroscopy (3, 4, 20, 23, 41, 57–60)	Monoenergetic electrons are scattered off a surface, and the energy losses are determined. This gives information on the electronic excitations of the surface and the adsorbed molecules.	Electronic structure, surface structure
ESCA	Electron spectroscopy for chemical analysis (2–4, 16, 22, 61–63)	Now generally called XPS.	Composition, oxidation state
ESDIAD or PSD	Electron (photon)-stimulated ion angular distribution (2–4, 8, 64–69)	Electrons or photons break chemical bonds in absorbed atoms or molecules, causing ionized atoms or radicals to be ejected from the surface along the axis of the broken bond by Coulomb repulsion. The angular distribution of these ions gives information on the bonding geometry of adsorbed molecules.	Bonding geometry, molecular orientation
	Ellipsometry (70)	Used to determine thickness of an adsorbed film. A circular polarized beam of light is reflected from a surface, and the change in the polarization characteristics of the light gives information about the surface film.	Layer thickness
EXAFS	Extended X-ray absorbtion fine structure (3, 8, 71–77)	Monoenergetic photons excite a core hole. The modulation of the absorption cross section with energy at 100–500 eV above the excitation threshold yields information on the radial distances to the	Local surface structure and coordination numbers

TABLE 1.1. (*Continued*)

Acronym	Name	Description	Primary Surface Information
		neighboring atoms. The cross section can be measured by fluorescence as the core holes decay or by attenuation of the transmitted photon beam. EXAFS is one of the many "fine-structure" techniques.	
EXELFS	Extended X-ray energy loss fine structure (3)	Monoenergetic electrons excite a core hole. The modulation of the absorption cross section with energy 100–500 eV above the excitation threshold yields information on the radial distances to the neighboring atoms. The cross section can be measured by fluorescence as the core holes decay or by attenuation of the transmitted photon beam.	Local surface structure and coordination numbers
FEM	Field emission microscopy (2–4, 11, 20, 22, 78, 79)	A strong electric field (on the order of volts/angstrom) is applied to the tip of a sharp, single-crystal wire. The electrons tunnel into the vacuum and are accelerated along radial trajectories by Coulomb repulsion. When the electrons impinge on a fluorescent screen, variations of the electric field strength across the surface of the tip are displayed.	Surface structure
FIM	Field ionization microscopy (2–4, 11, 20, 22, 79, 80)	A strong electric field (on the order of volts/angstrom) is created at the tip of a sharp, single-crystal wire. Gas atoms, usually He, are polarized and attracted to the tip by the strong electric field, and are ionized by electrons tunneling from the gas atoms into the tip. These ions, accelerated along radial trajectories by Coulomb repulsion, map out the variations in the electric field strength across the surface, showing the surface topography with atomic resolution.	Surface structure and surface diffusion
FTIR	Fourier transform infrared spectroscopy (81–83)	Broad-band IRAS experiments are performed, and the IR adsorption spectrum is deconvoluted using a Doppler-shifted source and the Fourier analysis of the data. This technique is not restricted to surfaces.	Bonding geometry and strength
HEIS	High-energy ion scattering spectroscopy (3, 11, 84, 85)	High-energy ions, above ≈ 500 keV, are scattered off a single-crystal surface. The "channeling" and "blocking" of scattered ions within the crystal can be used to triangulate deviations from the	Surface structure

TABLE 1.1. (*Continued*)

Acronym	Name	Description	Primary Surface Information
		bulk structure. HEIS has been especially used to study surface reconstructions and the thermal vibrations of surface atoms. See also MEIS and ISS.	
HREELS	High-resolution electron energy loss spectroscopy (2, 3, 86–88)	A monoenergetic electron beam, ≈ 2–10 eV, is scattered off a surface; and the energy losses between ≈ 0.5 eV to bulk and surface phonons and vibrational excitations of adsorbates are measured as a function of angle and energy (also called EELS).	Bonding geometry, surface atom vibrations
INS	Ion-neutralization spectroscopy (2, 3, 89)	Slow ionized atoms, usually He^+, strike a surface, where they are neutralized in a two-electron process that can eject a surface electron—a process similar to Auger emission from the valence band. The ejected electrons are detected as a function of energy, and the surface density of states can be determined from the energy distribution. The interpretation is more complicated than for SPI or UPS.	Valence bands
IP	Inverse photo-emission (90–95)	The absorption of electrons by a surface is measured as a function of energy and angle. This technique gives information about conduction bands and unoccupied levels.	Electronic structure
IRAS	Infrared reflection adsorption spectroscopy (3, 59, 60, 83, 96, 97)	The vibrational modes of adsorbed molecules on a surface are studied by monitoring the absorbtion or emission of IR radiation from thermally excited modes as a function of energy.	Molecular structure
ISS	Ion scattering spectroscopy (2–4, 8, 98, 99)	Ions are scattered from a surface, and the chemical composition of the surface may be determined by the momentum transfer to surface atoms. The energy range is ≈ 1 keV to 10 MeV, and the lower energies are more surface-sensitive. At higher energies this technique is also known as *Rutherford back-scattering* (RBS). A compilation of surface structures determined with ion scattering summarizing the pre-1988 literature appears in reference [100].	Surface structure, composition
LEED	Low-energy electron diffraction (2–4, 8, 10–12,	Monoenergetic electrons below ≈ 500 eV are elastically back-scattered from a surface and detected as a function of en-	Surface structure

TABLE 1.1. (*Continued*)

Acronym	Name	Description	Primary Surface Information
	20, 22, 23, 101–104)	ergy and angle. This gives information on the structure of the near-surface region. A compilation of surface structures summarizing the pre-1986 literature appears in reference [105].	
LEIS	Low-energy ion scattering (3, 4, 11, 106, 107)	Low-energy ions below ≈ 5 eV are scattered from a surface, and the ion "shadowing" gives information on the surface structure. At these low energies the surface-atom ion-scattering cross section is very large, resulting in large surface sensitivity. Accuracy is limited because the low-energy ion-scattering cross sections are not well known.	Surface structure
LEPD	Low-energy positron diffraction (108, 109)	Similar to LEED with positrons as the incident particle. The interaction potential is somewhat different than for electrons, so the form of the structural information is modified.	Surface structure
MEED	Medium-energy electron diffraction (11)	Similar to LEED, except the energy range is higher, ≈ 300–1000 eV.	Surface structure
MEIS	Medium-energy ion scattering (4, 11)	Similar to HEIS, except that incident ion energies are ≈ 50–500 keV.	Surface structure
	Neutron diffraction (110–112)	Neutron diffraction is not an explicitly surface sensitive technique, but neutron diffraction experiments on large-surface-area samples have provided important structural information on adsorbed molecules and also on surface phase transitions.	Surface structure
NEXAFS	Near-edge X-ray absorption fine structure (71, 72, 113–115)	A core hole is excited as in fine-structure techniques (see EXAFS, SEXAFS, AR-PEFS, NPD, APD, EXELFS, SEELFS) except that the fine structure within ≈ 30 eV of the excitation threshold is measured. Multiple scattering is much stronger at low electron energies, so this technique is sensitive to the local three-dimensional geometry, not just the radial separation between the source atom and its neighbors. The excitation cross section may be monitored by detecting the photoemitted electrons or the Auger electrons emitted during the core-hole decay.	Surface structure

TABLE 1.1. (*Continued*)

Acronym	Name	Description	Primary Surface Information
NMR	Nuclear magnetic resonance (116, 117)	NMR is not an explicitly surface-sensitive technique, but NMR data on large surface area samples (≥ 1 m^2) have provided useful data on molecular adsorption geometries. The nucleus magnetic moment interacts with an externally applied magnetic field and provides spectra highly dependent on the nuclear environment of the sample. The signal intensity is directly proportional to the concentration of the active species. This method is limited to the analysis of magnetically active nuclei.	Chemical state
NPD	Normal photo-electron diffraction (38, 39)	Similar to ARPEFS, but with a somewhat lower energy range.	Surface structure
RBS	Rutherford back-scattering (2, 3, 118, 119)	Similar to ISS, except that the main focus is on depth-profiling and composition. The momentum transfer in back-scattering collisions between nuclei is used to identify the nuclear masses in the sample, and the smaller, gradual momentum loss of the incident nucleus through electron–nucleus interactions provides depth-profile information.	Composition
RHEED	Reflection high-energy electron diffraction (3, 4, 10, 11, 22, 120)	Monoenergetic electrons below ≈ 1–20 keV are elastically scattered from a surface at glancing incidence, and detected as a function of energy and angle for small forward-scattering angles. Back-scattering is less important at high energies, and glancing incidence is used to enhance surface sensitivity.	Surface structure, structure of thin films
SEELFS	Surface electron energy loss fine structure (77, 121, 122)	A fine-structure technique similar to EX-ELFS, except the incident electron is more surface-sensitive because of the lower excitation energy. A compilation of surface structures determined using SEELFS and SEXAFS summarizing the pre-1990 literature appears in reference (123).	Surface structure
SERS	Surface enhanced Raman spectroscopy (59, 124, 125)	Some surface geometries (rough surfaces) concentrate the electric fields of Raman scattering cross section so that it is surface-sensitive. This gives information on surface vibrational modes, and some	Surface structure

TABLE 1.1. (*Continued*)

Acronym	Name	Description	Primary Surface Information
		information on geometry via selection rules.	
SEXAFS	Surface extended X-ray absorption fine structure (3, 8, 72, 121, 126–128)	A more surface-sensitive version of EX-AFS where the excitation cross-section fine structure is monitored by detecting the photoemitted electrons (PE-SEX-AFS), Auger electrons emitted during core-hole decay (Auger-SEXAFS), or ions excited by photoelectrons and desorbed from the surface (PSD-SEX-AFS). A compilation of surface structures determined using SEELFS and SEXAFS summarizing the pre-1990 literature appears in reference 123.	Surface structure
SFA	Surface force apparatus (129–132)	Two bent mica sheets with atomically smooth surfaces are brought together with distance of separation in the nanometer range. The forces acting on molecular layers between the mica plates perpendicular and parallel to the plate surfaces can be measured.	Forces acting on molecules squeezed between mica plates are measured.
SFG	Sum frequency generation (133–135)	Similar to SHG. One of the lasers has a tunable frequency that permits variation of the second harmonic signal. In this way the vibrational excitation of the adsorbed molecules is achieved.	Surface structure
SHG	Second harmonic generation (133, 136, 137)	A surface is illuminated with a high-intensity laser, and photons are generated at the second harmonic frequency through nonlinear optical processes. For many materials, only the surface region has the appropriate symmetry to produce the SHG signal. The nonlinear polarizability tensor depends on the nature and geometry of adsorbed atoms and molecules.	Electronic structure, molecular orientation
SIMS	Secondary ion mass spectrometry (2–4, 99, 138–144)	Ions and ionized clusters ejected from a surface during ion bombardment are detected with a mass spectrometer. Surface chemical composition and some information on bonding can be extracted from SIMS ion fragment distributions.	Surface composition
SPI	Surface penning ionization (2, 23)	Neutral atoms, usually He, in electronically excited states collide with a surface at thermal energies. A surface electron may tunnel into an unoccupied	Electronic structure

TABLE 1.1. (*Continued*)

Acronym	Name	Description	Primary Surface Information
		electronic level of the incoming gas atom, causing the incident atom to ionize and eject an electron, which is then detected. This technique measures the density of states near the Fermi level of the substrate and is highly surface-sensitive.	
SPLEED	Spin-polarized low-energy electron diffraction (24, 145)	Similar to LEED, except the incident electron beam is spin-polarized. This is particularly useful for the study of surface magnetism and magnetic ordering.	Magnetic structure
STM	Scanning tunneling microscopy (2, 35, 146–152)	The topography of a surface is measured by mechanically scanning of a probe over a surface. The distance from the probe to the surface is measured by the probe–surface tunneling current. Angstrom resolution of surface features is routinely obtained.	Surface structure
SXAPS	Soft X-ray appearance potential spectroscopy	Another name for APXPS.	
TEM	Transmission electron microscopy (11, 12, 153, 154)	TEM can provide surface information for carefully prepared and oriented bulk samples. Real images have been formed of the edges of crystals where surface planes and surface diffusions have been observed. Diffraction patterns of reconstructed surfaces, superimposed on the bulk diffraction pattern, have also provided surface structural information.	Surface structure
TDS	Thermal desorption spectroscopy (3, 155–159)	An adsorbate-covered surface is heated, usually at a linear rate, and the desorbing atoms or molecules are detected with a mass spectrometer. This gives information on the nature of adsorbate species and some information on adsorption energies and the surface structure.	Composition, heat of adsorption, surface structure
TDP	Temperature programmed desorption (3, 157–159)	Similar to TDS, except the surface may be heated at a nonuniform rate to obtain more selective information on adsorption energies.	Composition, heat of adsorption, surface structure
UPS	Ultraviolet photoemission spectroscopy	Electrons photoemitted from the valence and conduction bands are detected as a function of energy to measure the elec-	Valence band structure

TABLE 1.1. (*Continued*)

Acronym	Name	Description	Primary Surface Information
	(2–4, 20, 22, 23, 42, 89, 160, 161)	tronic density of states near the surface. This gives information on the bonding of adsorbates to the surface (see ARUPS).	
	Work function measurements (3, 20, 22, 162, 163)	Changes in a substrate's work function during the adsorption of atoms and molecules provide information about charge transfer between the adsorbate and the substrate and also about chemical bonding.	Electronic structure
XANES	X-ray absorbtion near-edge structure	Another name for NEXAFS.	
XPS	X-ray photoemission spectroscopy (2–4, 16, 22, 61–63, 164, 165)	Electrons photoemitted from atomic core levels are detected as a function of energy. The shifts of core-level energies give information on the chemical environment of the atoms (see ARXPS, ARXPD).	Composition, oxidation state
XRD	X-ray diffraction (166–168)	X-ray diffraction has been carried out at extreme glancing angles of incidence where total reflection ensures surface sensitivity. This provides structural information that can be interpreted by well-known methods. An extremely high X-ray flux is required to obtain useful data from single-crystal surfaces. Bulk X-ray diffraction is used to determine the structure of organometallic clusters, which provide comparisons to molecules adsorbed on surfaces.	Surface structure

REFERENCES

1. S.J. Gregg and K.S.W. Sing. *Adsorption, Surface Area, and Porosity.* Academic Press, New York, 1967.
2. J.B. Hudson. *Surface Science: An Introduction.* Butterworth-Heinemann, Boston, 1992.
3. D.P. Woodruff and T.A. Delchar. *Modern Techniques of Surface Science.* Cambridge Solid State Science Series. Cambridge University Press, New York, 1986.
4. M.J. Higatsberger. Solid Surfaces Analysis. In: C. Marton, editor, *Advances in Electronics and Electron Physics*, Volume 56. Academic Press, New York, 1981.
5. E. Hulpke, editor, *Helium Atom Scattering from Surfaces. Springer Series in Surface Sciences*, Volume 27. Springer-Verlag, Berlin, 1992.
6. B. Poelsema and G. Comsa. Scattering of Thermal Energy Atoms from Disordered Surfaces. In: G. Höhler, editor, *Springer Tracts in Modern Physics*, Volume 115. Springer-Verlag, Berlin, 1989.
7. J.A. Barker and D.J. Auerbach. Gas–Surface Interactions and Dynamics; Thermal Energy Atomic and Molecular Beam Studies. *Surf. Sci. Rep.* **4**:1 (1984).
8. D.P. Woodruff, G.C. Wang, and T.M. Lu. Surface Structure and Order–Disorder Phenomena. In: D.A.

King and D.P. Woodruff, editors, *Adsorption at Solid Surfaces, The Chemical Physics of Solid Surfaces and Heterogeneous Catalysis*, Volume 2. Elsevier, New York, 1983.

9. D. Frankl. Atomic Beam Scattering from Single Crystal Surfaces. *Prog. Surf. Sci.* **13**:285 (1983).

10. D.P. Woodruff. Surface Periodicity, Crystallography and Structure. In: D.A. King and D.P. Woodruff, editors, *Clean Solid Surfaces. The Chemical Physics of Solid Surfaces and Heterogeneous Catalysis*, Volume 1. Elsevier, New York, 1981.

11. G.A. Somorjai and M.A. Van Hove. Adsorbed Monolayers on Solid Surfaces. In: J.D. Dunitz, J.B. Goodenough, P. Hemmerich, J.A. Albers, C.K. Jørgensen, J.B. Neilands, D. Reinen, and R.J.P. Williams, editors, *Structure and Bonding*, Volume 38. Springer-Verlag, Berlin, 1979.

12. H. Wagner. Physical and Chemical Properties of Stepped Surfaces. In: G. Höhler, editor, *Solid Surface Physics. Springer Tracts in Modern Physics*, Volume 85. Springer-Verlag, Berlin, 1979.

13. F. O. Goodman. Scattering of Atoms and Molecules by Solid Surfaces. In: R. Vanselow, editor, *Critical Reviews in Solid State and Material Sciences*, Volume 7. CRC Press, Boca Raton, FL, 1977.

14. R. Browning. Auger Spectroscopy and Scanning Auger Microscopy. In: D.J. O'Conner, B.A. Sexton, and R. St. C. Smart, editors, *Surface Analysis Methods in Materials Science, Springer Series in Surface Sciences*, Volume 23. Springer-Verlag, Berlin, 1992.

15. J.M. Slaughter, W. Weber, G. Güntherot, and C.M. Falco. Quantitative Auger and XPS Analysis of Thin Films. *MRS Bull.* **December:**39 (1992).

16. R.L. Park. Core-Level Spectroscopies. In: R.L. Park and M.G. Lagally, editors, *Solid State Physics: Surfaces, Methods of Experimental Physics*, Volume 22. Academic Press, New York, 1985.

17. R.L. Park. Surface Spacings from the Secondary Electron Yield. *Appl. Surf. Sci.* **4**:250 (1980).

18. L.E. Davis, N.C. MacDonald, P.W. Palmberg, G.E. Riach, and R.E. Weber. *Handbook of Auger Electron Spectroscopy*. Perkin Elmer, Eden Prairie, MN, 1978.

19. B.V. King. Sputter Depth Profiling. In: D.J. O'Conner, B.A. Sexton, and R. St. C. Smart, editors, *Surface Analysis Methods in Materials Science. Springer Series in Surface Sciences*, Volume 23. Springer-Verlag, Berlin, 1992.

20. R.P.H. Gasser. *An Introduction to Chemisorption and Catalysis by Metals*. Oxford University Press, New York, 1985.

21. R. Weissmann and K. Müller. Auger Electron Spectroscopy—A Local Probe for Solid Surfaces. *Surf. Sci. Rep.* **1**:251 (1981).

22. M. Prutton. *Surface Physics*. Oxford University Press, New York, 1975.

23. G. Ertl and J. Küpper. *Low Energy Electrons and Surface Chemistry*. VCH, Weinheim, Germany, 1985.

24. J. Kirschner. *Polarized Electrons at Surfaces*. In G. Höhler, editor, *Springer Tracts in Modern Physics*, Volume 106. Springer-Verlag, Berlin, 1985.

25. G.E. McGuire and P.H. Holloway. Applications of Auger Spectroscopy in Materials Analysis. In: C.R. Brundle and A.D. Baker, editors, *Electron Spectroscopy: Theory, Techniques and Applications*, Volume 4. Academic Press, New York, 1981.

26. J.C. Fuggle. High Resolution Auger Spectroscopy of Solids and Surfaces. In: C.R. Brundle and A.D. Baker, editors, *Electron Spectroscopy: Theory, Techniques and Applications*, Volume 4. Academic Press, New York, 1981.

27. C.C. Chang. Analytical Auger Electron Spectroscopy. In: P.F. Kane and G.B. Larrabee, editors, *Characterization of Solid Surfaces*. Plenum Press, New York, 1974.

28. P.M. Hall and J.M. Morabito. Compositional Depth Profiling by Auger Electron Spectroscopy. In: R. Vanselow, editor, *Chemistry and Physics of Solid Surfaces*, Volume 2. CRC Press, Boca Raton, FL, 1979.

29. A. Joshi, L.E. Davis, and P.W. Palmberg. Auger Electron Spectroscopy. In: A.W. Czanderna, editor, *Methods of Surface Analysis. Methods and Phenomena: Their Applications in Science and Technology*, Volume 1. Elsevier, New York, 1975.

30. E. Meyer. Atomic Force Microscopy. *Prog. Surf. Sci.* **41**:1 (1992).

31. D. Rugar and P. Hansma. Atomic Force Microscopy. *Phys. Today*, **October:**23 (1990).

32. S.A.C. Gould, B. Drake, C.B. Prater, A.L. Weisenhorn, S. Manne, H.G. Hansma, J. Massie, M. Longmire, V. Elings, B. Dixon-Northern, B. Mukergee, C.M. Peterson, W. Stoeckenius, T.R. Albrecht, and C.F. Quate. From Atoms to Integrated Circuit Chips, Blood Cells, and Bacteria with the Atomic Force Microscope. *J. Vacuum Sci. Technol.* A **8**:369 (1990).

33. E. Meyer, H. Heinzelmann, P. Grütter, T. Jung, H.R. Hidber, R. Rudin, and H.J. Güntherodt. Atomic Force Miscroscopy for the Study of Tribology and Adhesion. *Thin Solid Films* **181**:527 (1989).

34. N.A. Burnham and R.J. Colton. Measuring the Nanomechanical Properties and Surface Forces of Materials Using an Atomic Force Microscope. *J. Vacuum Sci. Technol. A* **7**:2906 (1989).

35. P.K. Hansma, V.B. Elings, O. Marti, and C.E. Bracker. Scanning Tunneling Microscopy and Atomic Force Microscopy: Application to Biology and Technology. *Science* **242**:157 (1988).

36. P.K. Hansma and J. Tersoff. Scanning Tunneling Microscopy. *J. Appl. Phys.* **61**(2):R1 (1987).

37. G. Binnig, C.F. Quate, and C. Gerber. Atomic Force Microscope. *Phys. Rev. Lett.* **56**:930 (1986).

38. C.S. Fadley. The Study of Surface Structures by Photoelectron Diffraction and Auger Electron Diffraction. In: R.Z. Bachrach, editor, *Synchrotron Radiation Research. Advances in Surface and Interface Science, Volume I: Techniques.* Plenum Press, New York, 1992.

39. D.P. Woodruff. Photoelectron Diffraction. In: S.D. Kevan, editor, *Angle-Resolved Photoemission: Theory and Current Applications, Studies in Surface Science and Catalysis*, Volume 74. Elsevier, New York, 1992.

40. J.J. Barton, S.W. Robey, and D.A. Shirley. Theory of Angle Resolved Photoemission Extended Fine Structure. *Phys. Rev. B* **34**:778 (1986).

41. N.V. Richardson and A.M. Bradshaw. Symmetry and the Electron Spectroscopy of Surfaces. In: C.R. Brundle and A.D. Baker, editors, *Electron Spectroscopy: Theory, Techniques and Applications*, Volume 4. Academic Press, New York, 1981.

42. G. Margaritondo and J.H. Weaver. Photoemission Spectroscopy of Valence States. In: R.L. Park and M.G. Lagally, editors, *Solid State Physics: Surfaces, Methods of Experimental Physics*, Volume 22. Academic Press, New York, 1985.

43. W. Eberhardt, Angle-Resolved Photoemission Spectroscopy. In: R.Z. Bachrach, editor, *Synchrotron Radiation Research. Advances in Surface and Interface Science, Volume I: Techniques.* Plenum Press, New York, 1992.

44. K.E. Smith and S.D. Kevan. The Electronic Structure of Solids Studied Using Angle Resolved Photoemission Spectroscopy. *Prog. Solid State Chem.* **21**:49 (1991).

45. N.V. Smith and S.D. Kevan. Introduction. In: S.D. Kevan, editor, *Angle-Resolved Photoemission: Theory and Current Applications, Studies in Surface Science and Catalysis*, Volume 74. Elsevier, New York, 1992.

46. F.J. Himpsel. Angle-Resolved Measurements of the Photoemission of Electrons in the Study of Solids. *Adv. Phys.* **32**:1 (1983).

47. M. Scheffler and A.M. Bradshaw. The Electronic Structure of Adsorbed Layers. In: D.A. King and D.P. Woodruff, editors, *Adsorption at Solid Surfaces, The Chemical Physics of Solid Surfaces and Heterogeneous Catalysis*, Volume 2. Elsevier, New York, 1983.

48. J.E. Inglesfield and B.W. Holland. Electrons at Surfaces. In: D.A. King and D.P. Woodruff, editors, *Clean Solid Surfaces, The Chemical Physics of Solid Surfaces and Heterogeneous Catalysis*, Volume 1. Elsevier, New York, 1981.

49. A.P. Kaduwela, D.J. Friedman, and C.S. Fadley. Application of a Novel Multiple Scattering Approach to Photoelectron Diffraction and Auger Electron Diffraction. *J. Electron Spectros. Relat. Phenom.* **57**:223 (1991).

50. C.S. Fadley. Angle Resolved X-Ray Photoelectron Spectroscopy. *Prog. Surf. Sci.* **16**:275 (1984).

51. S.Y. Tong and C.H. Li. Diffraction Effects in Angle-Resolved Photoemission Spectroscopy. In: R. Vanselow and W. England, editors, *Chemistry and Physics of Solid Surfaces*, Volume 3. CRC Press, Boca Raton, FL, 1982.

52. F.J. Berry. Mössbauer Spectroscopy. In: B.W. Rossiter and R.C. Baetzold, editors, *Determination of Structural Features of Crystalline and Amorphous Solids. Physical Methods of Chemistry*, Volume 5, 2nd edition. John Wiley & Sons, New York, 1990.

53. T.E. Cranshaw, B.W. Dale, G.O. Longworth, and C.E. Johnson. *Mössbauer Spectroscopy and Its Applications.* Cambridge University Press, New York, 1985.

54. W. Jones, J.M. Thomas, R.K. Thorpe, and M.J. Tricker. Conversion Electron Mössbauer Spectroscopy and the Study of Surface Properties and Reactions. *Appl. Surf. Sci.* **1**:388 (1978)

55. M.J. Tricker, J.M. Thomas, and A.P. Winterbottom. Conversion Electron Mössbauer Spectroscopy for the Study of Solid Surfaces. *Surf. Sci.* **45**:601 (1974).

56. R.L. Park. Extended Fine Structure Analysis of Materials. *Appl. Surf. Sci.* **13**:231 (1982).

57. N. Sheppard. Vibrational Spectroscopic Studies of the Structure of Species Derived from the Chemisorption of Hydrocarbons on Metal Single-Crystal Surfaces. *Ann. Rev. Phys. Chem.* **39**:589 (1988).

58. H. Ibach and D.L. Mills. *Electron Energy Loss Spectroscopy and Surface Vibrations.* Academic Press, New York, 1982.

59. R.F. Willis, A.A. Lucas, and G.D. Mahan. Vibrational Properties of Adsorbed Molecules. In: D.A. King and D.P. Woodruff, editors, *Adsorption at Solid State Surfaces. The Chemical Physics of Solid Surfaces and Heterogeneous Catalysis*, Volume 2. Elsevier, New York, 1983.

60. W.H. Weinberg. Vibrations in Overlayers. In: R.L. Park and M.G. Lagally, editors, *Solid State Physics: Surfaces. Methods of Experimental Physics*, Volume 22. Academic Press, New York, 1985.

61. D.T. Clark. Structure, Bonding, and Reactivity of Polymer Surfaces Studies by Means of ESCA. In: R. Vanselow, editor, *Chemistry and Physics of Solid Surfaces*, Volume 2. CRC Press, Boca Raton, FL, 1979.

62. D. Briggs. Analytical Applications of XPS. In: C.R. Brundle and A.D. Baker, editors, *Electron Spectroscopy: Theory, Techniques and Applications*, Volume 3. Academic Press, New York, 1979.

63. W.M. Riggs and M.J. Parker. Surface Analysis by X-Ray Photoelectron Spectroscopy. In: A.W. Czanderna, editor, *Methods of Surface Analysis. Methods and Phenomena: Their Applications in Science and Technology*, Volume 1. Elsevier, New York, 1975.

64. R.D. Ramsier and J.T. Yates, Jr. Electron Stimulated Desorption: Principles and Applications. *Surf. Sci. Rep.* **12**:243 (1991).

65. V. Rehn and R.A. Rosenberg. Photon-Stimulated Desorption. In: R.Z. Bachrach, editor, *Synchrotron Radiation Research. Advances in Surface and Interface Science, Volume I: Techniques.* Plenum Press, New York, 1992.

66. R.H. Stulen. Recent Developments in Angle-Resolved Electron-Stimulated Desorption of Ions from Surfaces. *Prog. Surf. Sci.* **32**:1 (1989).

67. M.L. Knotek. Stimulated Desorption. *Rep. Prog. Phys.* **47**:1499 (1984).

68. M.L. Knotek. Stimulated Desoprtion from Surfaces. *Phys. Today* **September**:24 (1984).

69. T.E. Madey and R. Stockbauer. Experimental Methods in Electron- and Photon-Stimulated Desorption. In: R.L. Park and M.G. Lagally, editors, *Solid State Physics: Surfaces. Methods of Experimental Physics*, Volume 22. Academic Press, New York, 1985.

70. F.K. Urban III. Ellipsometer Measurement of Thickness and Optical Properties of Thin Absorbing Films. *Appl. Surf. Sci.* **33/34**:934 (1988).

71. S.M. Heald and J.M. Tranquada. X-Ray Absorbtion Spectroscopy: EXAFS and XANES. In: B.W. Rossiter and R.C. Baetzold, editors, *Determination of Structural Features of Crystalline and Amorphous Solids. Physical Methods of Chemistry*, Volume 5, 2nd edition. John Wiley & Sons, New York, 1990.

72. D.P. Woodruff. Fine Structures in Ionisation Cross Sections and Applications to Surface Science. *Rep. Prog. Phys.* **49**:683 (1986).

73. P. Lagarde and H. Dexpert. EXAFS in Catalysis. *Adv. Phys.* **33**:567 (1984).

74. S.M. Heald. Design of an EXAFS Experiment. In: D.C. Koningsberger and R. Prins, editors, *X-Ray Absorption. Chemical Analysis*, Volume 92. John Wiley & Sons, New York, 1988.

75. B.K. Teo. *EXAFS: Basic Principles and Data Analysis.* Springer-Verlag, Berlin, 1986.

76. E.A. Stern. Theory of EXAFS. In: D.C. Koningsberger and R. Prins, editors, *X-Ray Absorption. Chemical Analysis*, Volume 92. John Wiley & Sons, New York, 1988.

77. J. Derrien, E. Chaînet, M. de Crescenzi, and C. Noguera. Fine Structure in Electron Energy Loss and Auger Spectra. *Surf. Sci.* **189/190**:590 (1987).

78. R. Gomer. Recent Applications of Field Emission Microscopy. In: R. Vanselow, editor, *Chemistry and Physics of Solid Surfaces*, Volume 2. CRC Press, Boca Raton, FL, 1979.

79. J.A. Panitz. High-Field Techniques. In: R.L. Park and M.G. Lagally, editors, *Solid State Physics: Surfaces. Methods of Experimental Physics*, Volume 22. Academic Press, New York, 1985.

80. G. Ehrlich. Wandering Surface Atoms and the Field Ion Microscope. *Phys. Today* **June**:44 (1981).

81. N.K. Roberts. Fourier Transform Infrared Spectroscopy of Surfaces. In: D.J. O'Conner, B.A. Sexton, and R. St. C. Smart, editors, *Surface Analysis Methods in Materials Science. Springer Series in Surface Sciences*, Volume 23. Springer-Verlag, Berlin, 1992.

82. D.M. Back. Fourier Transform Infrared Analysis of Thin Films. In: M.H. Francombe and J.L. Vossen,

editors, *Thin Films for Advanced Electronic Devices. Physics of Thin Films*, Volume 15. Academic Press, New York, 1991.

83. Y. J. Chabal. Surface Infrared Spectroscopy. *Surf. Sci. Rep.* **8**:211 (1988).

84. I. Stensgaard. Surface Studies with High Energy Ion Beams. *Rep. Prog. Phys.* **55**:989 (1992).

85. L.C. Feldman. MeV Ion Scattering for Surface Structure Determination. In: R. Vanselow and W. England, editors, *Critical Reviews in Solid State and Material Sciences*, Volume 10. CRC Press, Boca Raton, FL: 1981, page 411.

86. W. Ho, High Resolution Electron Energy Loss Spectroscopy. In: B.W. Rossiter and R.C. Baetzold, editors, *Investigations of Surfaces and Interfaces*, Part A. of *Physical Methods of Chemistry*, Volume 9A, 2nd edition. John Wiley & Sons, New York, 1993.

87. J.L. Erskine. High Resolution Electron Energy Loss Spectroscopy. In: J.E. Greene, editor, *Critical Reviews in Solid State and Material Sciences*, Volume 13. CRC Press, Boca Raton, FL, 1987.

88. J.J. Pireaux, P.A. Thiry, R. Sporken, and R. Caudano. Analysis of Semiconductors and Insulators by High Resolution Electron Energy Loss Spectroscopy—Prospects for Quantification. *Surf. Interface Anal.* **15**:189 (1990).

89. H.D. Hagstrum. Studies of Adsorbate Electronic Structure Using Ion Neutralization and Photoemission Spectroscopies. In: L. Fiermanns, J. Vennik, and V. Dekeyser, editors, *NATO Advanced Study Institutes Series: Series B, Physics*, Volume 32. Plenum Press, 1978.

90. P.D. Johnson. Inverse Photoemission. In: S.D. Kevan, editor, *Angle-Resolved Photoemission: Theory and Current Applications. Studies in Surface Science and Catalysis*, Volume 74. Elsevier, New York, 1992.

91. F.J. Himpsel. Inverse Photoemission from Semiconductors. *Surf. Sci. Rep.* **12**:1 (1990).

92. G. Borstel and G. Thörner. Inverse Photoemission from Solids: Theoretical Aspects and Applications. *Surf. Sci. Rep.* **8**:1 (1987).

93. N.V. Smith and D.P. Woodruff. Inverse Photoemission from Metal Surfaces. *Prog. Surf. Sci.* **21**:295 (1986).

94. V. Dose. Momentum-Resolved Inverse Photoemission. *Surf. Sci. Rep.* **5**:337 (1985).

95. V. Dose. Ultraviolet Bremsstrahlung Spectroscopy. *Prog. Surf. Sci.* **13**:225 (1983).

96. F.M. Hoffmann. Infrared Reflection Absorbtion Spectroscopy of Adsorbed Molecules. *Surf. Sci. Rep.* **3**:107 (1983).

97. H.G. Tompkins. Infrared Reflection–Absorption Spectroscopy. In: A.W. Czanderna, editor, Methods of Surface Analysis. *Methods and Phenomena: Their Applications in Science and Technology*, Volume 1. Elsevier, New York, 1975.

98. E. Taglauer. Ion Scattering Spectroscopy. In: A.W. Czanderna and D.M. Hercules, editors, *Ion Spectroscopies for Surface Analysis. Methods of Surface Characterization*, Volume 2. Plenum Press, New York, 1991.

99. W. Heiland and E. Taglauer. Ion Scattering and Secondary-Ion Mass Spectrometry. In: R.L. Park and M.G. Lagally, editors, *Solid State Physics: Surfaces. Methods of Experimental Physics*, Volume 22. Academic Press, New York, 1985.

100. P.R. Watson. Critical Compilation of Surface Structures Determined by Ion Scattering Methods. *J. Phys. Chem. Ref. Data* **19**:85 (1990).

101. M.A. Van Hove. Surface Crystallography with Low-Energy Electron Diffraction. *Proc. R. Soc. London Ser. A* (1993).

102. G.A. Somorjai and M.A. Van Hove. Surface Crystallography by Low-Energy Electron Diffraction. In: B.W. Rossiter, J.F. Hamilton, and R.C. Baetzold, editors, *Investigation of Interfaces and Surfaces*. Interscience Publishers, Rochester, NY, 1990.

103. M.A. Van Hove, W.H. Weinberg, and C.M. Chan. Low Energy Electron Diffraction. In G. Ertl, editor, *Springer Series in Surface Sciences*, Volume 6. Springer-Verlag, Berlin, 1986.

104. J.B. Pendry. *Low Energy Electron Diffraction*. Academic Press, New York, 1974.

105. P.R. Watson. Critical Compilation of Surface Structures Determined by Low Energy Electron Diffraction Crystallography. *J. Phys. Chem. Ref. Data* **16**:953 (1987).

106. D.J. O'Conner. Low Energy Ion Scattering. In: D.J. O'Conner, B.A. Sexton, and R. St. C. Smart, editors, *Surface Analysis Methods in Materials Science. Springer Series in Surface Sciences*, Volume 23. Springer-Verlag, Berlin, 1992.

107. T.M. Buck. Low-Energy Ion Scattering Spectrometry. In: A.W. Czanderna, editor, *Methods of Surface Analysis. Methods and Phenomena: Their Applications in Science and Technology*, Volume 1. Elsevier, New York, 1975.

108. K.F. Canter. Current Status of Low Energy Positron Diffraction. In: P.J. Schultz, G.R. Massoumi, and P.J. Simpson, editors, *Positron Beams for Solids and Surfaces. AIP Conference Proceedings*, Volume 218. American Institute of Physics, New York, 1990.

109. K.F. Canter, C.B. Duke, and A.P. Mills, Jr. Current Status of Low Energy Positron Diffraction. In: *Proceedings of the Ninth International Summer Institute in Surface Science*. Springer-Verlag, Berlin, 1989.

110. R.E. Lechner and C. Riekel. Applications of Neutron Scattering in Chemistry. In: G. Höhler, editor, *Neutron Scattering and Muon Spin Rotation. Springer Tracts in Modern Physics*, Volume 101. Springer-Verlag, Berlin, 1983.

111. J.P. McTague, M. Nielson, and L. Passell. Neutron Scattering by Adsorbed Monolayers. In: R. Vanselow, editor, *Chemistry and Physics of Solid Surfaces*, Volume 2. CRC Press, Boca Raton, FL, 1979.

112. C.J. Wright and C.M. Sayer. Inelastic Neutron Scattering from Adsorbates. *Rep. Prog. Phys.* **46**:773, 1983.

113. J. Stöhr. *NEXAFS Spectroscopy. Springer Series in Surface Sciences*, Volume 25. Springer-Verlag, Berlin, 1992.

114. A. Bianconi and A. Marcelli. X-Ray Absorbtion Near-Edge Structure: Surface XANES. In: R.Z. Bachrach, editor, *Synchrotron Radiation Research. Advances in Surface and Interface Science, Volume I: Techniques*. Plenum Press, New York, 1992.

115. A. Bianconi. XANES Spectrocopy. In: D.C. Koningsberger and R. Prins, editors, *X-Ray Absorption. Chemical Analysis*, Volume 92. John Wiley & Sons, 1988.

116. T.M. Duncan and C. Dybowski. Chemisorption and Surfaces Studied by Nuclear Magnetic Resonance Spectroscopy. *Surf. Sci. Rep.* **1**:157 (1981).

117. P.J. Barrie and J. Klinowski. ^{129}Xe NMR as a Probe for the Study of Microporous Solids: A Critical Review. *Prog. Nucl. Magn. Reson.* **24**:91 (1992).

118. L.C. Feldman. Rutherford Backscattering and Nuclear Reaction Analysis. In: A.W. Czanderna and D.M. Hercules, editors, *Ion Spectroscopies for Surface Analysis. Methods of Surface Characterization*, Volume 2. Plenum Press, New York, 1991.

119. J.F. van der Veen. Ion Beam Crystallography of Surfaces and Interfaces. *Surf. Sci. Rep.* **5**:199 (1985).

120. G.L. Price. Reflection High Energy Electron Diffraction. In: D.J. O'Conner, B.A. Sexton, and R. St. C. Smart, editors, *Surface Analysis Methods in Materials Science. Springer Series in Surface Sciences*, Volume 23. Springer-Verlag, Berlin, 1992.

121. D.P. Woodruff. From SEXAFS to SEELFS. *Surf. Interface Anal.* **11**:25 (1988).

122. M. De Crescenzi. Extended Energy Loss Fine Structure Analysis. In: J.E. Greene, editor, *Critical Reviews in Solid State and Material Sciences*, Volume 15. CRC Press, Boca Raton, FL, 1989.

123. P.R. Watson. Critical Compilation of Surface Structures Determined by Surface Extended X-Ray Absorption Fine Structure (SEXAFS) and Surface Extended Electron Energy Loss Spectroscopy (SEELFS). *J. Phys. Chem. Ref. Data* **21**:123 (1992).

124. I. Pockrand. Surface Enhanced Raman Vibrational Studies at Solid/Gas Interfaces. In: G. Höhler, editor, *Springer Tracts in Modern Physics*, Volume 104. Springer-Verlag, 1984.

125. M. Moskovits. Surface-Enhanced Spectroscopy. *Rev. Mod. Phys.* **57**:783 (1985).

126. J.E. Rowe. Surface EXAFS. In: R.Z. Bachrach, editor, *Synchrotron Radiation Research. Advances in Surface and Interface Science, Volume I: Techniques*. Plenum Press, New York, 1992.

127. J. Stöhr. SEXAFS: Everything You Always Wanted to Know about SEXAFS but Were Afraid to Ask. In: D.C. Koningsberger and R. Prins, editors, *X-Ray Absorption. Chemical Analysis*, Volume 92. John Wiley & Sons, New York, 1988.

128. P. Eisenberger, P. Citrin, R. Hewitt, and B. Kincaid. SEXAFS: New Horizons in Surface Structure Determinations. In: R. Vanselow and W. England, editors, *Chemistry and Physics of Solid Surfaces*, Volume 3. CRC Press, Boca Raton, FL, 1982.

129. J.N. Israelachvili. Adhesion Forces between Surfaces in Liquids and Condensable Vapours. *Surf. Sci. Rep.* **14**:109 (1992).

130. J.N. Israelachvili and P.M. McGuiggan. Adhesion and Short-Range Forces between Surfaces. Part I: New Apparatus for Surface Force Measurements. *J. Mater. Res.* **5**:2223 (1990).

131. J.L. Parker, H.K. Christenson, and B.W. Ninham. Device for Measuring the Force and Separation between Two Surfaces Down to Molecular Separations. *Rev. Sci. Instrum.* **60**:3135 (1989).

132. J.N. Israelachvili. *Intermolecular and Surface Forces.* Academic Press, London, 1985.

133. Y.R. Shen. Surface Properties by Second-Harmonic and Sum-Frequency Generation. *Nature* **337**:519 (1989).

134. P. Guyot-Sionnest, R. Superfine, J.H. Hunt, and Y.R. Shen. Sum-Frequency Generation for Surface Vibrational Spectroscopy. In: Z. Wang and Z. Zhang, editors, *Proceedings of the Topical Meeting on Laser Materials and Laser Spectroscopy.* World Scientific Press, Singapore, 1989.

135. Y.R. Shen. Optical Second Harmonic Generation at Interfaces. *Annu. Rev. Phys. Chem.* **40**:327 (1989).

136. W. Chen, M. Feller, P. Guyot-Sionnest, C.S. Mullin, H. Hsiung, and Y.R. Shen. Second Harmonic Generation Studies of Liquid Crystal Monolayers and Films. In: T.F. George, editor, *Photochemistry in Thin Films. Proceedings of the Society of Photo-Optical Instrumentation Engineers,* Volume 1056. Society of Photo-Optical Instrumentation Engineers, Bellingham, WA, 1989.

137. G.L. Richmond, J.M. Robinson, and V.L. Shannon. Second Harmonic Generation Studies of Interfacial Structure and Dynamics. *Prog. Surf. Sci.* **28**:1 (1988).

138. C.M. Greenlief and J.M. White. Secondary Ion Mass Spectroscopy. In: B.W. Rossiter and R.C. Baetzold, editors, *Investigations of Surfaces and Interfaces,* Part A. *Physical Methods of Chemistry,* Volume 9A, 2nd edition. John Wiley & Sons, New York, 1993.

139. R.J. MacDonald and B.V. King. SIMS—Secondary Ion Mass Spectroscopy. In: D.J. O'Conner, B.A. Sexton, and R. St. C. Smart, editors, *Surface Analysis Methods in Materials Science. Springer Series in Surface Sciences,* Volume 23. Springer-Verlag, Berlin, 1992.

140. J.C. Vickerman, A. Brown, and N.M. Reed, editors. *Secondary Ion Mass Spectrometry. International Series of Monographs in Chemistry,* Volume 17. Clarendon Press, Oxford, 1989.

141. R.G. Wilson, F.A. Stevie, and C.W. Magee. *Second Ion Mass Spectrometry: A Practical Handbook for Depth Profiling and Bulk Impurity Analysis.* John Wiley & Sons, New York, 1989.

142. A. Benninghoven, F.G. Rüdenauer, and H.W. Werner. *Secondary Ion Mass Spectrometry: Basic Concepts, Instrumental Aspects, Applications and Trends. Chemical Analysis,* Volume 86. John Wiley & Sons, New York, 1987.

143. H.W. Werner. Introduction to Secondary Ion Mass Spectrometry (SIMS). In: L. Fiermanns, J. Vennik, and V. Dekeyser, editors, *NATO Advanced Study Institutes Series: Series B, Physics,* Volume 32. Plenum Press, 1978.

144. J.A. McHugh. Secondary Ion Mass Spectrometry. In: A.W. Czanderna, editor, *Methods of Surface Analysis. Methods and Phenomena: Their Applications in Science and Technology,* Volume 1. Elsevier, New York, 1975.

145. R. Feder, H. Pleyer, P. Bauer, and N. Müller. Spin Polarization in Low Energy Electron Diffraction: Surface Analysis of Pt(111). *Surf. Sci.* **444**:419 (1981).

146. A.L. de Lozanne. Scanning Tunneling Microscopy. In: B.W. Rossiter and R.C. Baetzold, editors, *Investigations of Surfaces and Interfaces,* Part A. *Physical Methods of Chemistry,* Volume 9A, 2nd edition. John Wiley & Sons, New York, 1993.

147. B.A. Sexton. Scanning Tunneling Microscopy. In: D.J. O'Conner, B.A. Sexton, and R. St. C. Smart, editors, *Surface Analysis Methods in Materials Science. Springer Series in Surface Sciences,* Volume 23. Springer-Verlag, Berlin, 1992.

148. H.J. Günterodt and R. Wiesendanger, editors. *Scanning Tunneling Microscopy I: General Principles and Applications to Clean and Adsorbate Covered Surfaces. Springer Series in Surface Sciences,* Volume 20. Springer-Verlag, Berlin, 1992.

149. L.E.C. van de Leemput and H. van Kempen. Scanning Tunneling Microscopy. *Rep. Prog. Phys.* **55**:165 (1992).

150. M. Tsukada, K. Kobayashi, N. Isshiki, and H. Kagashima. First-Principles Theory of Scanning Tunneling Microscopy. *Surf. Sci. Rep.* **13**:265 (1991).

151. F. Ogletree and M. Salmerón. Scanning Tunneling Microscopy and the Atomic Structure of Solid Surfaces. *Prog. Solid State Chem.* **20**:235 (1990).

152. G. Binnig and H. Rohrer. Scanning Tunneling Microscopy—from Birth to Adolescence. *Rev. Mod. Phys.* **59**:615 (1987).

153. J.M. Cowley. Electron Microscopy of Surface Structure. *Prog. Surf. Sci.* **21**:209 (1986).

154. G. Thomas. Some Applications of Electron Microscopy in Materials Science. *Ultramicroscopy* **20**:239 (1986).

155. A.M. de Jong and J.W. Niemantverdriet. Thermal Desorption Analysis: Comparative Test of Ten Commonly Applied Procedures. *Surf. Sci.* **233**:355 (1990).

156. D.A. King. Thermal Desorption from Metal Surfaces: A Review. *Surf. Sci.* **47**:384 (1975).

157. J.T. Yates. The Thermal Desorption of Adsorbed Species. In: R.L. Park and M.G. Lagally, editors, *Solid State Physics: Surfaces. Methods of Experimental Physics*, Volume 22. Academic Press, New York, 1985.

158. R.J. Madix. The Application of Flash Desorption Spectroscopy to Chemical Reactions on Surfaces; Temperature Programmed Reaction Spectroscopy. In: R. Vanselow, editor, *Chemistry and Physics of Solid Surfaces*, Volume 2. CRC Press, Boca Raton, FL, 1979.

159. J.L. Beeby. The Theory of Desorption. In: R. Vanselow, editor, *Critical Reviews in Solid State and Material Sciences*, Volume 7. CRC Press, Boca Raton, FL, 1977.

160. R. Leckey. Ultraviolet Photoelectron Spectroscopy of Solids. In: D.J. O'Conner, B.A. Sexton, and R. St. C. Smart, editors, *Surface Analysis Methods in Materials Science. Springer Series of Surface Sciences*, Volume 23. Springer-Verlag, Berlin, 1992.

161. W.E. Spicer. The Use of Synchrotron Radiation in UPS: Theory and Results. In: L. Fiermanns, J. Vennik, and V. Dekeyser, editors, *NATO Advanced Study Institutes Series: Series B, Physics*, Volume 32. Plenum Press, New York, 1978.

162. J. Hölzl and F.K. Schulte. Work Function of Metals. In: G. Höhler, editor, *Solid Surface Physics. Springer Tracts in Modern Physics*, Volume 85. Springer-Verlag, Berlin, 1979.

163. L.W. Swanson and P.R. Davis. Work Function Measurements. In: R.L. Park and M.G. Lagally, editors, *Solid State Physics: Surfaces. Methods of Experimental Physics*, Volume 22. Academic Press, New York, 1985.

164. J.F. Moulder, W.F. Stickle, P.E. Sobol, and K.D. Bomben. *Handbook of X-Ray Photoelectron Spectroscopy*. Perkin Elmer, Eden Prairie, MN, 1992.

165. M.H. Kibel. X-Ray Photoelectron Spectroscopy. In: D.J. O'Conner, B.A. Sexton, and R. St. C. Smart, editors, *Surface Analysis Methods in Materials Science. Springer Series in Surface Sciences*, Volume 23. Springer-Verlag, Berlin, 1992.

166. I.K. Robinson and D.J. Tweet. Surface X-Ray Diffraction. *Rep. Prog. Phys.* **55**:599 (1992).

167. P.H. Fuoss, K.S. Liang, and P. Eisenberger. Grazing-Incidence X-Ray Scattering. In: R.Z. Bachrach, editor, *Synchrotron Radiation Research. Advances in Surface and Interface Science, Volume I: Techniques.* Plenum Press, New York, 1992.

168. R. Feidenhans'l. Surface Structure Determination by X-Ray Diffraction. *Surf. Sci. Rep.* **10**:105 (1989).

2

THE STRUCTURE OF SURFACES

2.1 INTRODUCTION

Throughout history, people have been intrigued and delighted by surfaces because of their smoothness, high reflectivity, and color. The face of a smiling baby, the surfaces of glittering diamond crystals and of gold jewelry, and the surfaces of polished leather or wood all look smooth and perfect to the naked eye. In fact, the first physical model of a surface was one of a smooth discontinuity [1]. A closer inspection of any surface by an optical microscope with fairly large magnification, however, reveals the presence of irregularities and a great deal of roughness. Inspection

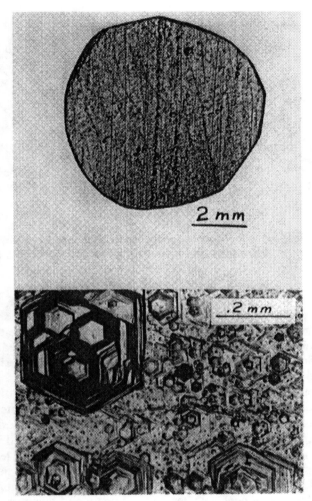

Figure 2.1. A cadmium sulfide crystal surface imaged by an optical microscope at two different magnifications.

of any crystal surface reveals large regions where atoms in parallel planes are separated by ledges 10^4 Å high (10^3 nm) (Figure 2.1). The growth of these terraces of parallel atomic planes is largely due to a small mismatch of atomic planes called a *dislocation*. One type, a screw dislocation, is shown in Figure 2.2. The growth of new atomic planes (*terraces*) can begin where these line defects appear. Dislocation densities on the order of 10^6–10^8 cm^{-2} are common at metal or ionic crystal surfaces, whereas smaller dislocation densities (on the order of 10^4–10^6 cm^{-2}) are common in most semiconductor or insulator crystals, because of their different chemical bonding properties. These concentrations may be compared with the surface concentration of atoms, which is on the order of 10^{15} cm^{-2}. Thus, each terrace may contain roughly $(10^{15}/10^6) = 10^9$ atoms in a low-dislocation-density single-crystal surface. By using an electron microscope in what is called the *back-reflection mode*,

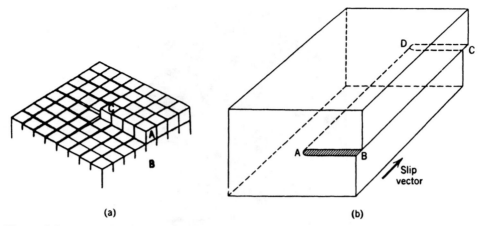

(a)   (b)

Figure 2.2. One type of screw dislocation giving rise to (a) atomic steps at the surface and (b) the slip plane that produces the dislocation and, ultimately, the defects at the surface (steps and kinks).

one can see these surface irregularities even more closely. Figure 2.3 shows a scanning electron microscope picture of a zinc crystal plane at a magnification of about 10^5. The surface is full of ledges, small stacks of terraces separated by steps of 5–10 Å (0.5–1 nm) high. Thus the surface is heterogeneous (not a single plane), even at this submicroscopic scale.

Let us continue to look at this zinc crystal. Typically in an area of 1 μm^2 (10^4 μm = 1 cm), one can distinguish several types of surface sites for atoms, each of which differs by the number of neighbors surrounding the atom. A surface atom is surrounded by the largest number of neighbor atoms when it is located in an atomic plane; this number is reduced substantially for surface atoms along a ledge or step.

Now let us consider an experimental technique that can view the same 1-μm^2 area on the atomic scale. The scanning tunneling microscope (STM) is capable of atomic-scale resolution. This instrument operates on the principle of quantum tunneling of electrons between a very sharp metallic tip that is brought within atomic distances [≈ 2 Å (0.2 nm)] to the surface under study. Using rapid-response electronic feedback circuits, the tip can be held steady at this close atomic range. It can also be moved along the surface when it is mounted on a piezoelectric ($BaTiO_3$) holder that expands in the 10^{-8}-cm (angstrom) range under an applied potential. Because the tunnel current varies exponentially with the distance from the surface, atomic-size bumps can be detected readily from fluctuations in the tunnel current. As the metal tip is moved along the surface while keeping the tunnel current constant, it tracks the atomic-scale fluctuations of the charge density about the surface atoms. Figure 2.4 shows the surface structure of a rhenium single crystal of (0001) orientation as identified by an STM when scanning over regions of various size. As the scanned area decreases, the surface roughness becomes a less dominant feature and the regularity of the atomic structure emerges. This is shown for the arrangement of carbon atoms for the (0001) face of graphite (Figure 2.5). With the experimental techniques that are presently available, the ordered atomic arrays in atomic planes, the periodicity of atomic steps, and the presence of kinks in the steps can all be identified.

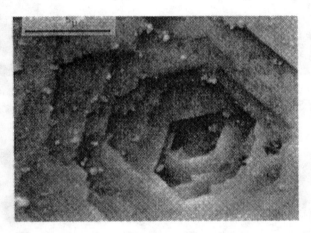

Figure 2.3. Scanning electron microscope picture of a zinc crystal surface at two different magnifications.

Using STM and other techniques—such as field-ion microscopy (FIM) and low-energy electron diffraction (LEED)—an atomic-scale model of the surface structure of solids can be constructed, as shown in Figure 2.6. On a heterogeneous solid surface, atoms in terraces are surrounded by the largest number of nearest neighbors. Atoms in steps have fewer neighbors, and atoms in kink sites have even fewer.

Kink, step, and terrace atoms have large equilibrium concentrations on any real surface. On a rough surface, 10–20% of the atoms are often in step sites, with about 5% in kink sites. Steps and kinks are also called *line defects*, to distinguish them from atomic vacancies, or adatoms, which are called *point defects*. These point defects are also present in most surfaces and are important participants of atom transport along the surface, although their equilibrium concentrations are much less than 1% of a monolayer even at the melting point. Thus the available data indicate

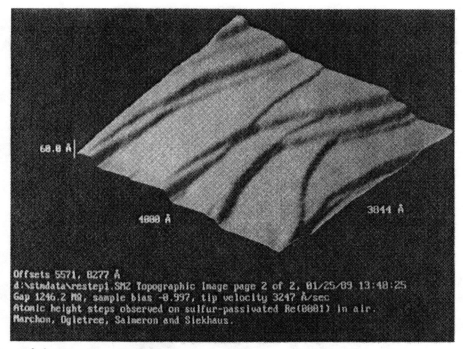

Figure 2.4. Scanning tunneling microscope (STM) picture of the (0001) face of rhenium over a 4000-Å^2 area.

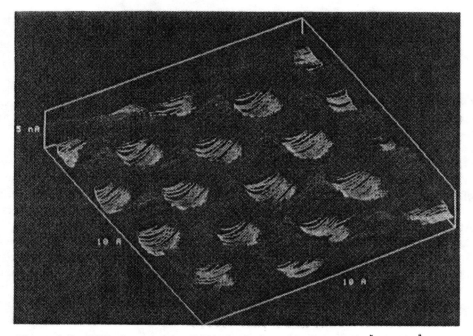

Figure 2.5. STM picture of the (0001) face of graphite over a 10-Å × 10-Å area.

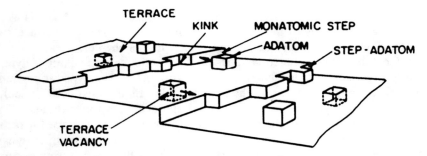

Figure 2.6. Model of a heterogeneous solid surface depicting different surface sites. These sites are distinguishable by their number of nearest neighbors.

that solid surfaces are heterogeneous on the atomic scale. The relative concentrations of atoms in terraces (representing ordered domains), in line defects, and in point defects can be altered, depending on the methods of surface preparation.

A major flaw of the terrace–step–kink model of the rough surface shown in Figure 2.6 is the assumption of a rigid lattice where every surface atom is located in its bulk-like equilibrium position and can be located by the projection of the bulk structure to that surface. Rather, surface atoms usually occupy sites that are shifted from the atomic positions in the bulk. There is a significant contraction of the interlayer distance at a clean surface between the first and second layer of atoms (Figure 2.7). As the surface structure exhibiting atomic roughness becomes more open (where roughness is defined as the inverse of the packing density), the contraction perpendicular to the surface becomes larger. Surface atoms often relocate along a surface as well. When atoms or molecules adsorb, forming chemical bonds with the substrate, the surface atoms again relocate to optimize the strength of the adsorbate–substrate bond. As a consequence, surface atoms may move outward, rotate, or be

Figure 2.7. Schematic representation of the contraction in interlayer spacing usually observed at clean solid surfaces.

displaced along the surface. Adsorbate-induced restructuring of surfaces is a common occurrence that leads to the formation of new and unexpected surface structures. These new structures will be discussed later in this chapter.

The surface, therefore, is heterogeneous on the atomic scale and exhibits dynamic restructuring responding to its changing local environment. The dynamic rearrangement of surface atoms can occur on the chemisorption time scale ($\approx 10^{-13}$ sec), on the time scale of catalytic reactions (seconds), and at longer times (hours). These longer times are needed when the atoms must diffuse along the surface for restructuring to occur. The most recent model of surfaces is one that permits dynamic restructuring of surface atoms at each surface site, with the restructuring occurring more easily at more open sites (open, low-packing-density surfaces, steps, and kinks). Many of the unique surface properties that we will discuss in this book are a consequence of the ability of surface atoms to readjust their local atomic structure and bonding according to the changing physical–chemical environment at the interface.

This structural heterogeneity, along with the varied composition of solid surfaces, introduces a great deal of complexity into surface studies of all types. In order to help us understand these complex surfaces, it is essential to study less complex surfaces. An appropriate starting point for this chapter on surface structure is to discuss the clean face of a single crystal of a monatomic solid, where most of the atoms are in identical equilibrium positions in a well-ordered surface. We can then introduce surface irregularities, steps, and kinks to this surface systematically and study their atomic structure. Our investigations may then extend to clean diatomic and polyatomic solid surfaces, where possible variations of surface composition that lead to nonstoichiometry may add to the structural complexity. With this knowledge in hand, we can then more easily understand the surface structure of polycrystalline foils, thin films, and small particles.

Then we will turn our attention to adsorbed monolayers. First we will study the surface structures of adsorbed atoms and small molecules. We will also explore how, by forming a strong chemical bond between adsorbate and substrate, we can alter the surface structure of the initially clean substrate. Then we will explore the surface structures of more complex organic molecules. We will study the formation of two-dimensional surface compounds whose existence either is restricted to two dimensions or is often a precursor to bulk compound formation.

2.2 SURFACE DIFFRACTION

Surfaces are usually ordered on the atomic scale if, during suitable preparation, the atoms are allowed to move to find their equilibrium positions. Thus LEED, X-ray diffraction, and atom diffraction are among the most useful techniques for studies of their structure. The deBroglie wavelength of a particle is given by

$$\lambda = \frac{h}{\sqrt{2mE}} \tag{2.1}$$

where h is Planck's constant, m is the mass of the particle, and E is the energy of the particle. For electrons and He atoms, Eq. 2.1 is more conveniently expressed:

$$\lambda_{e^-}(\text{Å}) = \sqrt{150/E(\text{eV})} \quad \text{and} \quad \lambda_{\text{He atom}}(\text{Å}) = \sqrt{0.02/E(\text{eV})} \qquad (2.2)$$

For X-rays, $\lambda(\text{Å}) \approx 1.24 \times 10^4/E(\text{eV})$. Thus, electrons with energies in the range of 10–200 eV and helium atoms with thermal energies have wavelengths that satisfy the atomic diffraction condition (λ smaller than or equal to interatomic distances). Because of their low energy and strong scattering by the atomic potentials, these particles are ideally suited for surface diffraction/surface structure determination studies. X-rays, with their high energies, are less so. However, at the high intensities available at synchrotron radiation facilities, the scattering at interfaces with differing electron densities makes X-rays suitable for surface and interface structure studies. Additionally, the X-ray-bombardment-induced emission of electrons also shows strong diffraction effects (photoelectron diffraction).

LEED surface crystallography produced most of the quantitative data on bond distance and bond angles, as well as on location of surface atoms and of adsorbed molecules. Much of the data that were obtained are presented in this book. Usually small (≈ 1 cm^2) single-crystal surfaces are used in LEED studies. After appropriate chemical or ion-bombardment cleaning in an ultrahigh vacuum chamber, the crystal is heated to permit the ordering of surface atoms by diffusion to their equilibrium positions. An electron beam in the energy range of 10–200 eV is back-scattered from the surface. The diffracted electrons that retain their incident kinetic energy in the scattering process are separated from the inelastically scattered electrons by retarding grids held at the appropriate potentials and then accelerated to strike a fluorescent screen (video-LEED) or some other type of electron detector that monitors their spatial distribution and interfaced with a computer (digital LEED). The intensity I of the diffracted beams is then monitored as a function of their kinetic energy (eV) (a so-called I–V curve). From these data using a suitable theory [2] that properly treats the strong (multiple) scattering of low-energy electrons, the atomic structure of the clean surface or the structure of atoms and molecules in the adsorbed monolayer can be obtained.

2.3 NOTATION OF SURFACE STRUCTURES

The structures of some of the most close-packed surfaces of face-centered cubic (fcc) crystals are shown in Figure 2.8. These surfaces are shown in their unreconstructed form—that is, with a unit cell that is predicted by projecting the bulk X-ray unit cell onto that surface. The surface unit-cell vectors permit the translational operation that can generate an infinite array of atoms in two dimensions: the surface structure. A surface may have periodic translations different from those predicted from the bulk projection—that is, a different unit cell with unit-cell vectors $\vec{a}\,'$ and $\vec{b}\,'$ that differ from those, $\vec{a}$ and $\vec{b}$, obtained from the bulk projection. The surface unit-cell vectors $\vec{a}\,'$ and $\vec{b}\,'$ can be expressed as

$$\vec{a}\,' = m_{11}\vec{a} + m_{12}\vec{b}$$

$$\vec{b}\,' = m_{21}\vec{a} + m_{22}\vec{b}$$

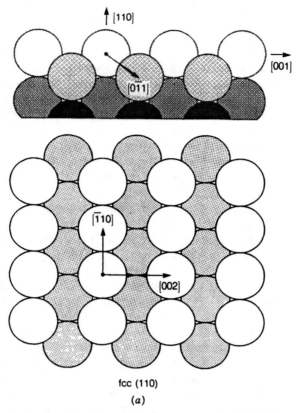

fcc (110)

(a)

Figure 2.8. Top views and side views of the face-centered cubic (fcc) crystal surfaces: (a) (110), (b) (100), and (c) (111)

where the coefficients m_{11}, m_{12}, m_{21}, and m_{22} define a matrix

$$M = \begin{pmatrix} m_{11} & m_{12} \\ m_{21} & m_{22} \end{pmatrix}$$

that defines any unit cell unambiguously. For example, for all of the unit cells in Figure 2.8,

$$M = \begin{pmatrix} 1 & 0 \\ 0 & 1 \end{pmatrix}$$

Let us consider the surface structure of an adsorbate that has a unit cell twice as long as the substrate unit cell and parallel to it. This is shown in part B of Figure 2.9a. The coefficients of its unit-cell vectors define the matrix

$$M = \begin{pmatrix} 2 & 0 \\ 0 & 2 \end{pmatrix}$$

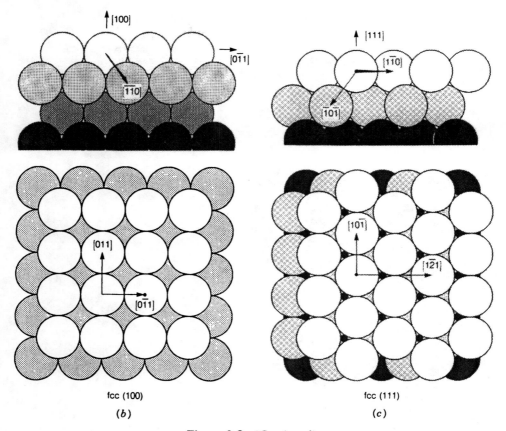

Figure 2.8. (*Continued*)

that identifies the unit cell. A surface structure that has a unit cell different from the bulk-projected substrate unit cell is often called a *superlattice*. Several superlattices are shown in Figure 2.9, and Table 2.1 (p. 87) lists the matrix notation that identifies these and other superlattices.

2.3.1 Abbreviated Notation of Simple Surface Structures

The simplest LEED patterns are most frequently characterized by a shorthand notation in which the unit cell of the surface structure is designated with respect to the bulk unit cell. An arrangement of surface atoms (the "surface net") identical to that in the bulk unit cell is called the *substrate structure* and is designated (1 × 1). For example, the substrate structure of platinum on the (111) surface is designated Pt(111)-(1 × 1). If the surface structure that forms in the presence of an adsorbed gas is characterized by a unit cell identical to the primitive unit cell of the substrate, the surface structure is denoted (1 × 1)-S, where S is the chemical symbol or formula for the adsorbate. For example, a monolayer of oxygen adsorbed on the (111) face of silicon is denoted Si(111)-(1 × 1)-O.

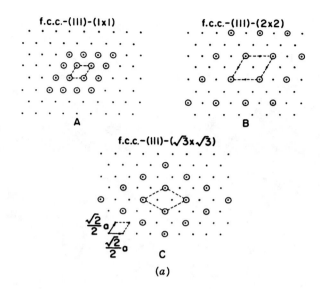

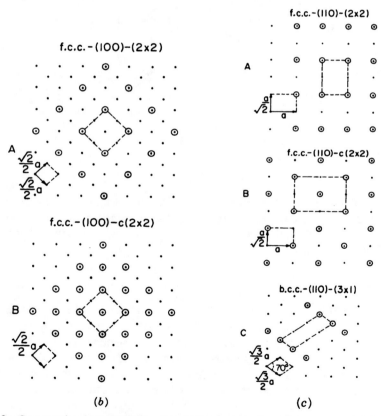

Figure 2.9. Commonly observed unit cells of adsorbate surface structures on (**a**) the fcc(111) crystal face, (**b**) the fcc(100) crystal face, and (**c**) the fcc(110) crystal face.

It is a common observation that the surface structures are frequently characterized by unit cells that are integral multiples of the substrate unit cell. If the unit cell of the surface structure is twice as large as the underlying bulk unit cell, it is designated (2×2), corresponding to the $\left(\begin{smallmatrix} 2 & 0 \\ 0 & 2 \end{smallmatrix}\right)$ matrix notation. A (2×2) surface structure formed by an adsorbed gas such as hydrogen on the (211) face of tungsten is designated W(211)-(2×2)-H. If the unit cell that characterizes the surface structure is twice as long as the bulk unit cell along one major crystallographic axis and has the same length along the other, the surface structure is designated (2×1). In Figure 2.9, examples are shown for the most frequently occurring different surface structures on substrates having sixfold, fourfold, and twofold rotational symmetry, respectively, where $|a|$ is the magnitude of the X-ray unit-cell vector.

This simple notation is adequate to give the size of the unit cell of the surface structure as long as it is in registry with the substrate unit cell. However, the notation is not easily applicable if the surface structure is rotated with respect to the bulk unit cell or if the unit-cell dimensions of the substrate and the surface net are not integer multiples of each other. For the simplest rotated surface structures, the abbreviated notation may still be applicable. For example, if every third lattice site on a hexagonal face is distinguished from the other sites, as shown in Figure 2.9(a), then a $(\sqrt{3} \times \sqrt{3})R30°$ surface structure may arise. The angle given after the notation and the letter R indicate the orientation of the surface structure, which is rotated with respect to the original substrate unit cell. If every other lattice site on a square face is unique, then a $(\sqrt{2} \times \sqrt{2})R45°$ surface structure could be formed, as shown in Figure 2.9(b).* In Table 2.1 several superlattices that are commonly detected on close-packed surfaces are listed by both shorthand and matrix notations.

2.3.2 Notation of High-Miller-Index, Stepped Surfaces

The atomic structures of high-Miller-index surfaces are composed of terraces separated by steps, which may also have kinks in them (Figure 2.10). For example, the unreconstructed (755) surface of an fcc crystal consists of (111) terraces, six atoms wide, separated by steps of (100) orientation and single-atom height. The surface structures of several high-Miller-index surfaces are displayed in Figure 2.11.

A notation for these surfaces, the compact-step notation, devised by Lang and Somorjai [3], gives the surface structure in the general form $w(h_t k_t l_t) \times (h_s k_s l_s)$, where $(h_t k_t l_t)$ and $(h_s k_s l_s)$ are the Miller indices of the terrace plane and the step plane, respectively, while w is the number of atoms that are counted in the width of the terrace, including the step-edge atom and the in-step atom. Thus, the fcc(755) surface is denoted by $7(111) \times 1(100)$, or also by $7(111) \times (100)$ for simplicity. A stepped surface with steps that are themselves high-Miller-index faces is termed a *kinked surface*. For example, the fcc(10, 8, 7) = $7(111) \times (310)$ surface is a kinked surface. The step notation is, of course, equally applicable to surfaces of bcc, hcp, and other crystals, in addition to surfaces of fcc crystals. Stepped surfaces of several orientations are listed in Table 2.2 (p. 88). Here the crystal faces are denoted both by their Miller indices and by their stepped-surface notation.

---

*This structure is sometimes denoted as $c(2 \times 2)$, where c stands for "centered," since this structure may be viewed as a (2×2) surface structure with an extra atom in its center.

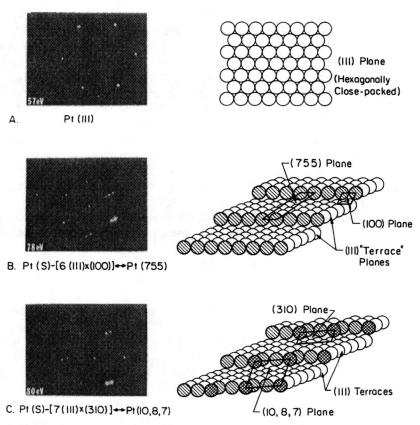

A.   Pt (III)

(III) Plane
(Hexagonally
Close-packed)

B. Pt (S)-[6 (III)x(IOO)]↔Pt (755)

┌(755) Plane

└(IOO) Plane

(III)"Terrace"
Planes

C. Pt (S)-[7 (III)x(3IO)]↔Pt (IO,8,7)

(3IO) Plane

(III) Terraces

(IO, 8, 7) Plane

Figure 2.10. Surface structures in real space and LEED diffraction patterns of the flat Pt(111), stepped Pt(755), and kinked Pt(10,8,7) crystal faces.

2.4 THE STRUCTURE OF CLEAN SURFACES

Two important structural changes that occur at solid surfaces are unique and are associated with the two-dimensional and anisotropic environment to which the surface atoms must adjust. These structural changes are bond-length contraction or relaxation and reconstruction. They are discussed below.

2.4.1 Bond-Length Contraction or Relaxation

Surface crystallography studies have shown that, in vacuum, virtually all surfaces relax (Figure 2.7). That is, the spacing between the first and second atomic layers is significantly reduced from the spacing characterizing the bulk. The lower the atomic packing and density of the surface, the larger the inward contraction. This is shown in Figure 2.12.

This trend fits long-established principles [4], if one relates coordination number to bond order [5, 6]. The clearest manifestation is provided by bond-length relaxations at clean metal surfaces. Compare close-packed with less close-packed surfaces

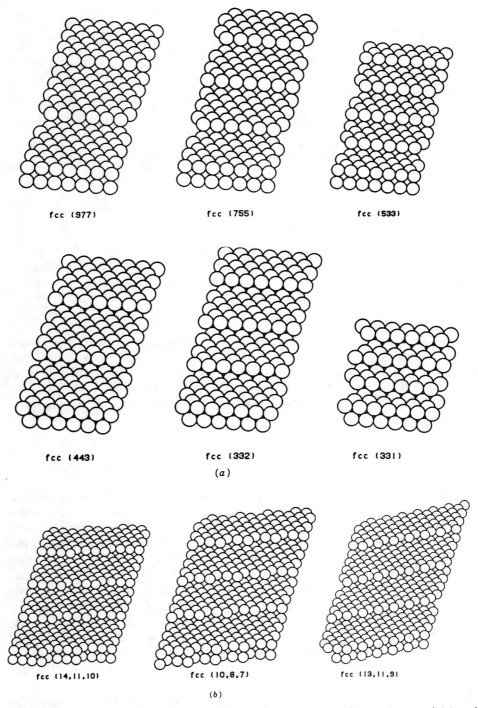

fcc (977)   fcc (755)   fcc (533)

fcc (443)   fcc (332)   fcc (331)

(a)

fcc (14,11,10)   fcc (10,8,7)   fcc (13,11,9)

(b)

Figure 2.11. Schematic representation of the surface structures of several stepped (**a**) and kinked (**b**) crystal faces deduced from the bulk unit cell. Contraction of interlayer spacing and other modes of restructuring that are commonly observed are not shown.

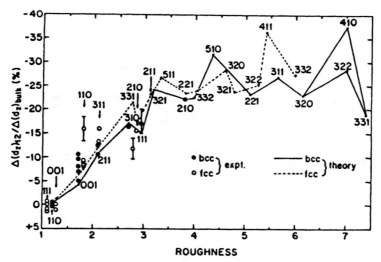

Figure 2.12. Contraction of interlayer spacing as a function surface roughness (defined as 1/packing density) for several fcc and bcc metal surfaces. The points indicate experimental data, and the lines are theoretical fits [165].

of metals: in less close-packed surfaces, the interlayer spacing between the topmost and the second atomic layers is smaller than the bulk spacing. Moreover, the perturbation caused by this surface relaxation propagates a few layers into the bulk. In fact, there is often a compensating expansion between the second and third metal layers (on the order of 1%), accompanied by a small but detectable change in the next layer.

2.5   RECONSTRUCTION

The forces that lead to surface relaxation and result in a change in the equilibrium position and bonding of surface atoms can give rise to more drastic reconstruction of the outermost layers; that is, the surface can assume an atomic structure that differs fundamentally from the structure one would expect if the bulk structure terminated abruptly at the surface. For semiconductor surfaces (Si, Ge, GaAs, InSb, etc.), which are covalently bonded, the surface atoms find it difficult to compensate for the loss of nearest neighbors. The dangling bonds created at the surface cannot easily be satisfied except through more drastic rearrangements of these atoms. Therefore, most semiconductor surfaces reconstruct, and major rebonding between surface atoms occurs in this process. The associated perturbation propagates several layers into the surface until the bulk lattice is recovered. Figure 2.13 illustrates the (2 × 1) reconstruction of the Si(100) face. The outermost plane consists of buckled dimers, and relaxation extends to the fourth layer.

Many metal surfaces also reconstruct. For example, at the (100) surfaces of Ir [7–11], Pt [10–13], and Au [10, 11, 14–16], the interatomic distance in the topmost layer shrinks by a few percent, parallel to the surface. It then becomes more favorable for this square unit cell to collapse into a hexagonally close-packed layer rather

Top View: Si(100)

Ideal   p(2x1) reconstructed

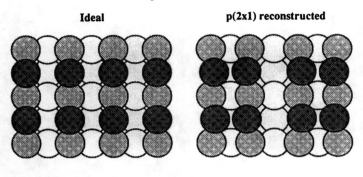

Side View: Si(100)

Ideal   p(2x1) reconstructed

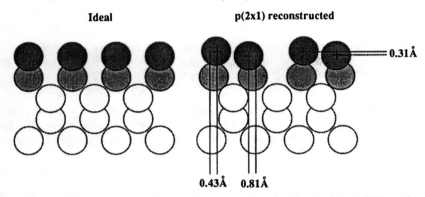

0.31Å

0.43Å   0.81Å

Figure 2.13. The reconstructed silicon (100) crystal face as obtained by LEED surface crystallography. Note that surface relaxation extends to three atomic layers into the bulk [166].

than to maintain the square lattice of the underlying layers. This is shown in Figure 2.14. The (110) crystal face of face centered cubic metals often exhibits the so-called "missing row" reconstruction. In this circumstance, a whole row of metal atoms is periodically missing giving rise to a (1 × 2) surface-structure (Figure 2.15).

Table 2.3 (p. 89) lists many reconstructed surface structures of metals and semiconductors, with brief descriptions of the nature of the restructuring that occurs. In

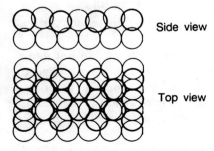

Side view

Top view

Ir(100) – (1 × 5)

Figure 2.14. The structure of the reconstructed iridium (100) crystal face obtained from LEED surface crystallography. Hexagonal packing in the surface layer induces buckling. The second layer retains its square unit cell [167, 168].

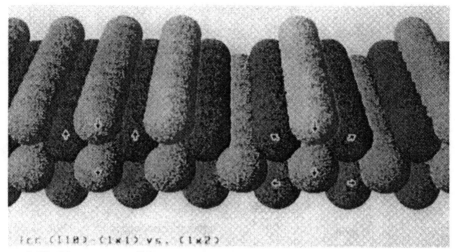

Ir. (110)-(1×1) vs. (1×2)

Figure 2.15. The reconstructed iridium (110) crystal face obtained by LEED surface crystallography. Every second row of atoms is missing. Note that relaxation extends to three atomic layers toward the bulk [169].

Table 2.4 (p. 104) we tabulate the sizes and symmetries of two-dimensional surface structures. Quantitative results of bond distances and bond angles are given in Table 2.5 (p. 115), for those systems for which such surface crystallography data are available.

2.5.1 Atomic Steps and Kinks

The presence of atomic steps and kinks even on nominally perfect low-index crystal faces has been revealed by several imaging techniques, but recent developments in scanning tunneling microscopy (STM) in particular have greatly increased our atomic-level understanding of their local surface structure. Figure 2.4 is an STM image of a rhenium (0001) surface that displays a large numbers of kinks and steps even on this close-packed surface.

Most high-Miller-index surfaces have close-packed terraces separated by steps one atom in height. This step–terrace arrangement is usually ordered and exhibits high thermal stability until the surface is heated to elevated temperatures, nearer to the melting point. The steps then become curved and often break up into small islands. When this occurs, it is called the *roughening transition* [17]. When a stepped surface below the roughening transition is annealed, the ordered step–terrace surface structure regenerates.

The ordering of steps at surfaces is due to an excess of electron charge at those sites. This charge excess may be viewed as a dipole. Because similarly aligned dipoles repel each other, the repulsive interaction between steps imposes ordering. More will be said in Chapter 5, where we shall discuss the electrical properties of surfaces. Adsorbates that modify the charge density at the step edges may also change the step structure and spacing.

Relaxation of the interlayer spacing at step edges can be large. The surface atoms relocate at the step edge to smooth out the structure at the surface irregularity. A schematic representation of two relaxed step structures is shown in Figure 2.16.

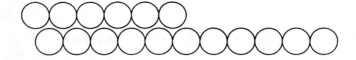

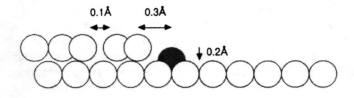

"Cracks" open close to the step edges

Each atom attempts to optimize its coordination

(a)

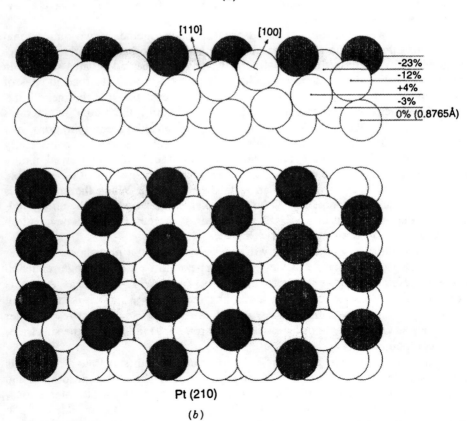

Pt (210)

(b)

Figure 2.16. Relaxation at two different stepped surfaces: (a) Cu(410) [170]; (b) Pt(210) [171].

2.6   THE STRUCTURE OF ADSORBED MONOLAYERS

2.6.1   Ordered Monolayers and the Reasons for Ordering

When atoms or molecules adsorb on ordered crystal surfaces, they usually form ordered surface structures over a wide range of temperatures and surface coverages. The driving force for ordering originates, just as with three-dimensional crystal formation, in mutual atomic interactions.

With adsorbates, an important distinction must be made between adsorbate–adsorbate and adsorbate–substrate interactions. In chemisorption, the adsorbate–adsorbate forces are usually small compared to the adsorbate–substrate binding forces, so that the adsorbate locations or sites are determined by the optimum adsorbate–substrate bonding. But the adsorbate–adsorbate interactions among adsorbates dominates the long-range ordering of the overlayer. These interactions can be studied by examining the changes in the overlayer structure as a function of temperature or coverage.

The surface coverage of an adsorbate is another important parameter in ordering. We shall use the common definition of coverage, in which one monolayer corresponds to one adsorbate atom or molecule for each unit cell of the clean, unreconstructed substrate surface. Thus, if undissociated carbon monoxide molecules bond to every other top-layer metal atom exposed at the Ni(100) surface, we have a coverage of one-half a monolayer.

At very low coverages, some of the adsorbates bunch together in two-dimensional islands. This effect results from short-range attractive adsorbate–adsorbate interactions combined with easy diffusion along the surface. Other adsorbates repel each other and form disordered overlayers with atoms or molecules adsorbed in sites of a given type (a lattice gas). When the coverage is increased so that the mean interadsorbate distance decreases to about 5–10 Å (0.5–1.0 nm), the mutual interactions often strongly influence the ordering, favoring certain adsorbate configurations over others. As a result, the structure can develop a unit cell that repeats periodically across the surface. For example, atomic oxygen on Ni(100) orders very well at one-quarter of a monolayer. The oxygen atoms then occupy one-quarter of the available hollow sites of Ni(100) in a square array labeled (2 × 2). When the coverage is doubled to one-half of a monolayer, the extra oxygen atoms occupy the empty hollow sites at the center of each of the (2 × 2) squares, thereby creating a new pattern labeled $c(2 \times 2)$, where c stands for centered.

Most nonmetallic adsorbed atoms will not compress into a monolayer overlayer on the closest-packed metal substrates. There appears to be a short-range repulsion that keeps these atoms apart by approximately a van der Waals distance. Attempts to compress the overlayer further by increasing the coverage (which is carried out by exposing the surface to higher pressures of the corresponding gas) result either in no further adsorption or in diffusion of the adatoms into the bulk of the substrate, forming compounds.

Some adsorbates do not form strong chemical bonds with substrate atoms. This situation is called *physical adsorption* or *physisorption*. For these adsorbates the adsorbate–adsorbate interactions can dominate the adsorbate–substrate interactions, and the optimum adsorbate–substrate bonding geometry can be overridden by the lateral adsorbate–adsorbate interactions, yielding, for example, incommensurate structures in which the overlayer and the substrate have independent lattices. When adsorbates that physisorb are used, the van der Waals distance determines the densest overlayer packing.

With metallic adsorbates, very close-packed overlayers can form, because metal adsorbate atoms attract each other relatively strongly and coalesce with covalent interatomic distances. When the atomic sizes of the overlayer and substrate metals are nearly the same, one may observe a one-monolayer (1 × 1) structure where adsorbate atoms occupy every unit cell of the substrate. This is called *epitaxial growth*. With less equal atomic radii, other structures are formed, dominated by the covalent closest-packing distance of the adsorbate. Beyond one close-packed overlayer or even before a monolayer is completed, metal adsorbates frequently form multilayers or even three-dimensional crystallites. Alloy formation by interdiffusion is also observed in a number of cases, even in the submonolayer regime.

2.6.2 Adsorbate-Induced Restructuring

Before we focus on the structures of monolayers of adsorbed atoms or molecules, we must consider the effects of forming a strong adsorbate–substrate bond on the surface structure of the substrate. The effects are dramatic indeed. The presence of a chemisorbed layer removes the relaxation and often the reconstruction observed for clean surfaces: The substrate surface atoms usually return to their bulk-like equilibrium position. However, the adsorbate can also induce a new surface restructuring. For example, low coverages of hydrogen induce a $c(2 \times 2)$ surface reconstruction of W(100) at 300 K; and a one-quarter monolayer of carbon on Ni(100) induces restructuring of the topmost nickel atoms both parallel and perpendicular to the surface in such a manner that the four nickel atoms surrounding each carbon atom are rotated with respect to the underlying layers (Figure 2.17). When sulfur is chemi-

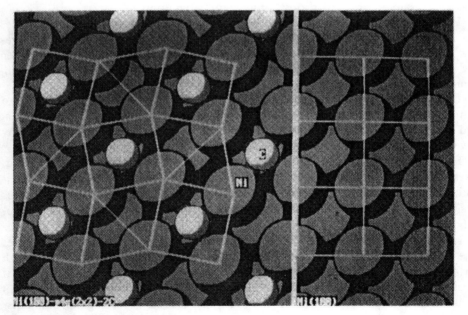

Figure 2.17. Carbon chemisorption induced restructuring of the Ni(100) surface [172, 173].

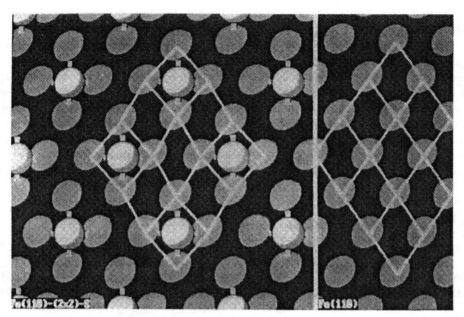

Figure 2.18. S–Fe(110). Sulfur-chemisorption-induced restructuring of the Fe(110) surface [174].

sorbed on the close-packed (110) iron surface, it creates its own fourfold site by restructuring the iron surface (Figure 2.18). The formation of the four strong Fe—S bonds readily compensates for the weakening of the nearest-neighbor iron–iron bonds that also occurs during the adsorbate-induced restructuring.

Table 2.6 (p. 256) lists surface structures on metal and semiconductor surfaces that form as a result of chemisorption-induced restructuring. Many of the surfaces that have been studied by surface crystallography so far restructure markedly. When the adsorbate is removed, the substrate surface usually returns to its clean-state configuration.

The ordered periodic terrace configuration with one-atom-height steps can undergo restructuring in the presence of adsorbates such as oxygen or carbon. The step height and the terrace width may double, or different step-and-terrace orientations may develop. This is shown schematically in Figure 2.19.

The clustering or merging of the one-atom-height steps into multiple-height structures upon chemisorption occurs because of the modification of the charge distribution of the step edges. The repulsive dipole–dipole interaction that keeps the steps apart can become attractive due to adsorption leading to step clustering or aggregation.

The more open the surface, the fewer nearest neighbors the surface atoms have. The rougher the surface is, the more steps and kinks the surface has. Open and rough surfaces restructure more readily upon chemisorption. For example, (110) crystal faces of fcc crystals (Ir, Pt, Ni) restructure more frequently upon chemisorption than do the closer-packed crystal faces. The (110) surfaces have rectangular unit cells, and restructuring leads to missing rows of atoms, such as shown in Figure 2.20. By

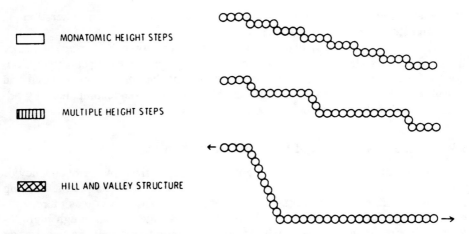

Figure 2.19. Adsorbate-induced restructuring of steps to multiple-height step–terrace configurations.

the relocation of a whole row of atoms, a trough two or three atomic layers deep forms. The adsorbate atoms are often located on the new crystal planes produced by the formation of the trough. The thermodynamic driving force for adsorbate-induced restructuring is the formation of strong adsorbate–substrate bonds that are comparable to or stronger than the bonds between substrate atoms in the clean substrate surface. The surface structures that form as a result of adsorbate-induced restructuring optimize the number of adsorbate–substrate bonds and their binding energy.

The time scale of adsorbate-induced restructuring may be short, on the order of adsorption times ($\approx 10^{-6}$ sec) of a chemisorbed monolayer. However, restructuring may occur slowly (hours) if massive diffusion-controlled atom transport along the surface is needed. One example of slow diffusion-controlled restructuring is the coalescence of surface steps into multiatomic-height steps (Figure 2.20) that ultimately lead to faceting (development of new crystal orientation). This type of behavior is

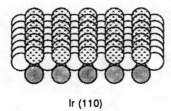

Ir (110)

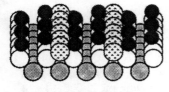

Ir (110) - (2x2) - 2S

Figure 2.20. Sulfur-chemisorption-induced restructuring of the Ir(110) surface obtained by LEED surface crystallography [175].

likely to be irreversible, and it leads to changes in crystal shape and alterations in the size and structure of small particles.

As the discussion above indicates, chemisorption can alter the surface structure of the substrate. We should not assume a rigid substrate lattice during chemisorption, as was usually assumed in the past. The formation of the chemisorption bond and the resulting adsorbate surface structure could very likely be accompanied by adsorbate-induced restructuring of the substrate as well.

2.6.3   Atomic Adsorption and Penetration into Substrates

The adsorption of atoms such as Na, S, and Cl is characterized by the occupancy of high-coordination surface sites on metal surfaces. These locations permit bonding to as many substrate atoms as possible. This bonding situation becomes more complicated with the smaller atomic adsorbates (H, C, N, and O). Although high coordination is still preferred, the small size of these atoms often allows penetration within and even below the first substrate layer.

For example, chemisorbed oxygen is located at the fourfold hollow site on Ni(100) above the metal atoms (Figure 2.21a). On the Cu(110) surface, however, oxygen is coplanar with the surface metal atoms. This surface structure may be viewed as a single-layer metal–adatom compound—that is, a surface compound (Figure 2.21b). The next stage of penetration is illustrated by N on Ti(0001), where the adatom occupies interstitial sites between the first and second metal layers, thereby forming a three-layer film of TiN (Figure 2.21c). Deeper penetration is often observed in the

Ni(100)-c(2x2)-O

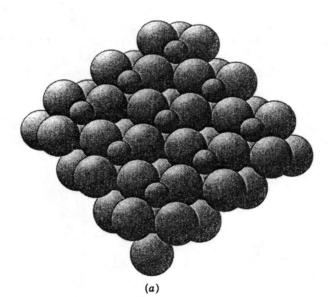

(a)

Figure 2.21. (a) Oxygen surface structure on Ni(100) [176]; (b) oxygen chemisorption induced surface structure on Cu(110) [177]; (c) nitrogen surface structure on Ti(0001) [178]. All of these structures were obtained by LEED surface crystallography.

Cu(110)-(2x1)-O

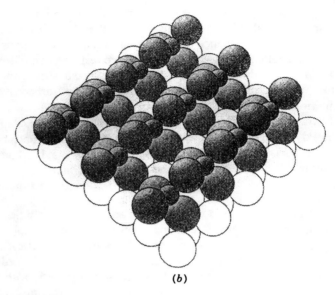

(b)

Ti(0001)-(1x1)-N

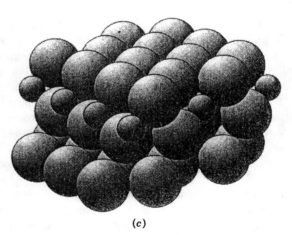

(c)

Figure 2.21. (*Continued*)

form of thicker compound films that show very different surface and near-surface structures than the parent substrate.

Semiconductors have more interstitial space available for diffusion of adatoms into the surface because of their more open substrate surface structures. Nevertheless, the very few cases of structure analysis indicate a preference for atom substitution; for example, adsorbed Al atoms on the GaAs(110) substrate surface tend to exchange places with Ga and occupy substrate atom positions.

2.6.4 Metals on Metals: Epitaxial Growth

At low coverages, most metallic adsorbates form ordered overlayers with a (1×1) surface structure on metal substrates. This implies that the substrate acts as a template and has a significant influence on the growth mode of the deposited material. This effect is usually called *epitaxial growth*. A more restricted definition of epitaxial growth would include only those examples where the substrate imposes its own crystal structure, orientation, and lattice parameter on the adsorbed overlayer. (This restricted definition is also called *pseudomorphic growth*.)

A good illustration of epitaxy is the behavior shown by the Pd/Ag(100) [18–21] system. In the Pd/Ag(100) system where the two metals have the same bulk structure (fcc), the Pd initially grows in perfect epitaxy, with a 5.1% lateral expansion of the interatomic spacing imposed by the substrate. This strained layer-by-layer growth persists to beyond three monolayers before relaxation to the bulk structure is seen.

The surface structures of single-crystal alloys are also being studied with increasing frequency and exhibit many interesting properties. For example, an α-Cu/Al alloy single crystal with a bulk concentration of 16 atomic % Al exhibits no long-range order in the bulk; the surface, by contrast, is completely ordered [22], as shown in Figure 2.22. Furthermore, due to aluminum segregation, the ordered surface phase contains equal numbers of copper and aluminum atoms. Thus the structure and composition of the surface may differ significantly from that of the bulk.

2.6.5 Growth Modes at Metal Surfaces

When metal atoms are condensed on surfaces of other metals, three type of growth modes can be distinguished by experiments.* The deposited metal may form a thin film in a *layer-by-layer* fashion, with the second layer beginning only after the first layer is completed (e.g., see references [23–44]). This is known as *Frank–van der Merwe growth* [45] and is shown schematically in Figure 2.23. In some systems, the subsequent thin film growth continues instead with the formation of three-dimensional islands—the so-called Stranski–Krastanov growth mechanism (Figure 2.23) (e.g., see references [24, 26, 34, 35, 37–40, 43, 44, 46–83]). In the third type of growth mode (Volmer–Weber), three-dimensional islands form from the very beginning of metal deposition (Figure 2.23) (e.g., see references [38, 44, 54, 63, 67, 79, 84–94]).

Auger electron spectroscopy is a valuable technique frequently employed to monitor [95–98] the growth of metal, oxide, and sulfide films. The Auger electron emission peaks from the substrate metal atoms decrease in intensity, whereas the Auger emission peaks from the deposited film increase in intensity as a function of deposition quantity. Depending on the growth mode, the coverage-dependent changes in the Auger peak intensities exhibit breaks (layer-by-layer growth) or vary smoothly (three-dimensional growth).

2.6.6 Molecular Adsorption

Carbon monoxide has proved a popular and convenient molecular adsorbate. It provides a rich variety of surface behaviors and is relatively easy to study by various

---

*Exceptions to these growth patterns occur when alloy or compound formation takes place between the condensed and substrate metals. This phenomenon is outside the scope of this discussion.

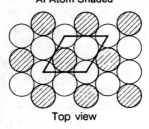

α - Cu Al (III) - (√3 x √3) R30°
Substitutional Al Model Parameters
Al Atom Shaded

Top view

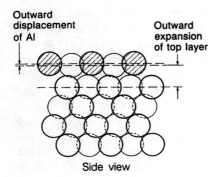

Outward displacement of Al

Outward expansion of top layer

Side view

Figure 2.22. Surface structure of the Cu–Al (16 atomic %) alloy (111) crystal face. Note that the surface composition is 50% [22].

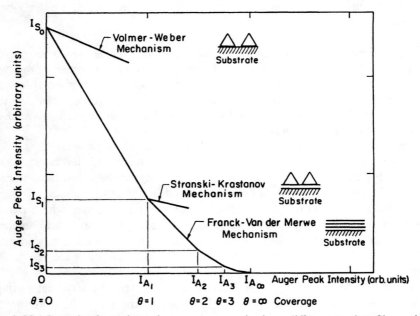

Figure 2.23. Growth of metal overlayers can occur in three different modes. Shown here is the behavior of the ratio of substrate and adsorbate Auger signals as a function of the deposition time for films that grow by the Volmer–Weber, Frank–van der Merwe, and Stranski–Krastanov types of mechanisms.

methods. As one goes toward the left of the periodic table, metal surfaces increasingly tend to dissociate CO. Then the separate C and O atoms bond directly to individual hollow sites, as in atomic adsorption. On other metal substrates (those not on the left side of the periodic table), CO remains intact and bonds through its carbon end to the surface, with the C–O axis perpendicular to the surface. However, the adsorption site varies considerably, depending on the surface structure. CO most commonly adsorbs in onefold coordinated top sites or in twofold coordinated bridge sites. Occasionally, threefold coordinated hollow adsorption sites are also found.

The preferred adsorption site of CO depends at least on three factors: the metal, the crystallographic face, and the CO coverage. For example, on the Ni(111) face, CO occupies the bridge sites first, while on Rh(111) [99] and Pt(111) [100] the top sites are preferred at low coverages. But the threefold site is occupied first on Pd(111). At higher coverages there are often two or more adsorption sites occupied simultaneously. The repulsive adsorbate–adsorbate interaction often forces the CO molecules onto unusual sites of lower symmetry to maximize the distance between the adsorbed molecules. Table 2.7 (p. 261) lists the two-dimensional surface structures of small chemisorbed molecules (CO and NO) that were observed.

Small organic molecules have been the subject of most surface-structure studies of molecular adsorbates. Most of these molecules are alkenes (such as ethylene) and aromatic molecules (such as benzene and substituted benzenes). Saturated hydrocarbons physisorb only at low temperatures; and although some ordered surface structures have been obtained using graphite as a substrate, these structures exhibit van der Waals packing without forming stronger chemical bonds.

Perhaps the best way to illustrate the diverse surface-structure chemistry of organic monolayer adsorbates is to review the adsorption behavior of ethylene and benzene on various transition metal surfaces. These two molecules are described in detail below.

2.6.6.1 Ethylene A general feature of unsaturated hydrocarbon adsorption on clean transition-metal surfaces is that it is largely irreversible. In other words, when one adsorbs an unsaturated hydrocarbon onto a transition-metal surface at low temperature and then heats the surface, the adsorbed molecule, rather than desorbing molecularly, will decompose to evolve hydrogen and leave the surface covered with the partially dehydrogenated fragments or carbon.

From measurement of desorption rates as a function of temperature, it is evident that these adsorbed alkenes dehydrogenate sequentially over a wide temperature range as shown in Figure 2.24 for different ethylene coverages; these typical desorption patterns of ethylene feature several peaks separated by valleys (plateaus). The areas between hydrogen-desorption peaks represent temperature regimes where partially dehydrogenated intermediates are stable on the surface. It is important to determine the structure and bonding of these various surface fragments in order to understand why C—H bond breaking occurs sequentially and over such a wide temperature range.

The adsorption of ethylene on the Rh(111) surface provides a typical example. The high-resolution electron-energy-loss (HREEL) spectrum at 77 K in Figure 2.25 has been attributed to ethylene adsorbed molecularly intact on the Rh(111) surface [101]. However, vibrational frequencies measured are markedly different from those for gas-phase ethylene, indicating a strong interaction between ethylene and the rhodium surface.

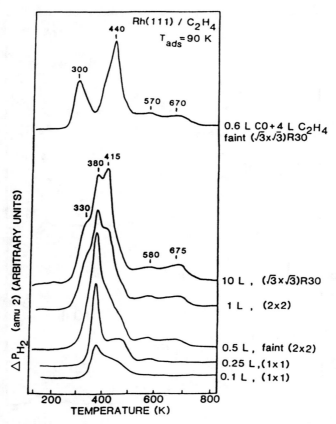

Figure 2.24. Thermal desorption of hydrogen from chemisorbed ethylene on Rh(111) due to thermal dehydrogenation starting from different coverages, ranging from 0.1 L to 10 L.

When one increases the temperature of a Rh(111) surface covered with molecularly adsorbed ethylene to above 220 K, ethylidyne (CCH_3) is formed. One hydrogen atom is eliminated in order to produce this surface fragment. Both LEED [102] and HREELS [103] confirm the bonding geometry for this species, as shown in Figure 2.26.

Ethylidyne chemisorption also restructures the metal surface. This is shown for Rh(111) in Figure 2.27. The metal–metal distances expand for those rhodium atoms that bind to the carbon of the ethylidyne molecule located in the threefold site. This expansion forces the next nearest-neighbor rhodium atom more into the surface, which becomes corrugated as a result. The rhodium atom in the second layer directly underneath the ethylidyne binding site moves upward, closer to the organic molecule which is now bound more strongly to four instead of only three metal atoms.

While the surface bonding of ethylidyne has been most extensively studied on the Pt(111) [104] and Rh(111) [105, 106] surfaces, this species has been isolated on the close-packed faces of several other metals, including Pd(111) [107–110] and Ru(0001) [111].

Returning to the HREEL spectrum of Figure 2.25, we see that ethylidyne decomposes on a Rh(111) surface when heated to 450 K. As indicated, this spectrum has

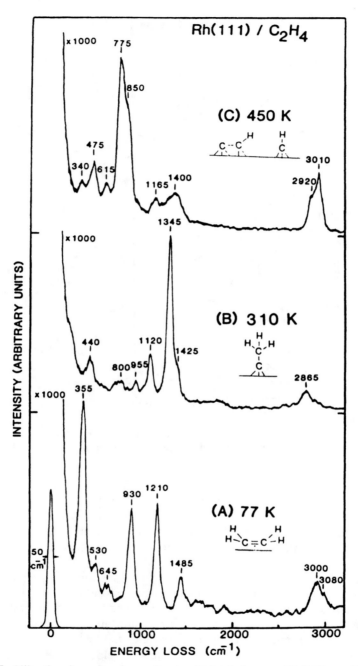

Figure 2.25. Vibrational spectra from chemisorbed ethylene on Rh(111) at different temperatures obtained by HREELS. Note the sequential dehydrogenation process.

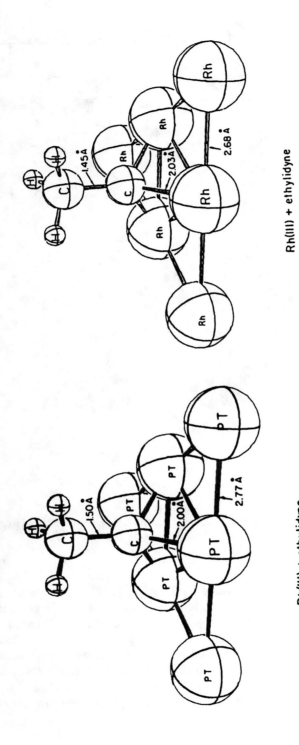

Pt (III) + ethylidyne

Rh(III) + ethylidyne

Figure 2.26. Bonding geometry of ethylidyne on the Rh(111) and Pt(111) crystal faces. Ethylidyne forms as a result of ethylene chemisorption.

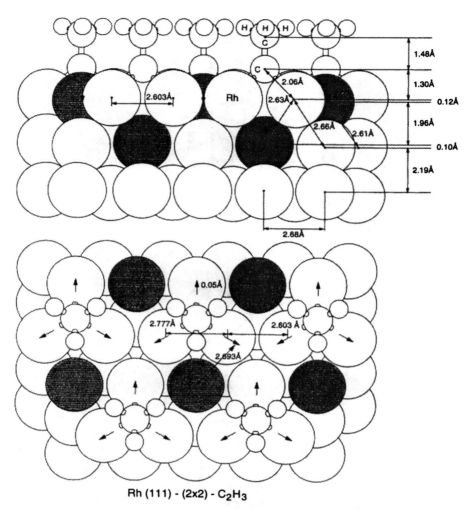

Figure 2.27. Ethylidyne-chemisorption-induced restructuring of the Rh(111) crystal face. Note the expansion of metal atoms around the adsorption site.

been attributed to a mixture of CH and C_2H species [101]. The general features of this spectrum remain unchanged throughout the 450–800 K temperature range, despite the continuous evolution of hydrogen from the surface (see Figure 2.24). This behavior is consistent with a mixture of surface fragments, all of whom have similar vibrational spectra, but whose relative concentrations change throughout this dehydrogenation process. The species that form are polymeric carbon chains terminated with hydrogen atoms and have the general formula C_xH. Increasing the surface temperature causes concurrent dehydrogenation and polymerization, eventually resulting in a graphitic monolayer.

2.6.6.2 Benzene Like ethylene, benzene readily adsorbs on most clean transition-metal surfaces and largely decomposes with heating, as opposed to desorbing molecularly. However, benzene decomposes at higher temperatures than does ethylene (generally above 350 K, compared with 200–300 K for ethylene). As a result, molecular benzene adsorption can be easily studied at room temperature.

The surface vibrational spectrum of benzene adsorption implies that the adsorbed benzene molecule bonds with its π ring parallel to the surface. This bonding orientation is supported by LEED crystallography (and other) studies of adsorbed benzene on the Pd(111) [112–115], Rh(111) [116–120], and Pt(111) [121–123] surfaces. Bonding within the benzene molecule, however, varies from one ordered structure to another, as shown in Table 2.8 (p. 269). In all cases, the benzene ring expands upon adsorption, as indicated by the increase in the ring radii. While the degree of expansion is not always greater than experimental uncertainty, the trend appears to be Pd(111) < Rh(111) < Pt(111) (Figure 2.28).

The symmetry of the benzene ring expansion varies with the adsorption site—twofold for the bridge site on Pt(111) and threefold for the hollow sites on Rh(111) and Pd(111). In the latter cases, the adsorbed benzene has alternating long and short C—C bonds in the carbon ring, resulting in a distorted cyclohexatriene structure. There is evidence that benzene buckles to assume a boat-like configuration when it is chemisorbed without long-range order on bridge sites of the Pt(111) surface. Two of the carbon atoms on opposite sides are closer to the metal atoms than the other four carbon atoms [124] (Figure 2.29).

It is intriguing to consider how this observed distortion of the adsorbed benzene molecule may correlate with its subsequent decomposition pathways. For benzene decomposition on Rh(111), the decomposition fragments determined by HREELS are compared in Figure 2.30 with those discussed above for ethylene. The significant result of this comparison is that CH and C_2H are the first stable decomposition intermediates for benzene that have been identified. These fragments are also the stable decomposition intermediates for acetylene at this temperature. It appears likely that, consistent with the threefold distortion determined by LEED, benzene decomposes via three acetylenes, which are unstable at the decomposition temperature and immediately dehydrogenate to CH and C_2H. In this regard, it is interesting that Pd(111), which induces the least distortion in adsorbed benzene (Table 2.8), is an active surface for the trimerization of acetylene to benzene [125–131].

2.6.7 Coadsorbed Monolayers

Large changes in the heat of adsorption are frequently found with increasing coverage of the adsorbed monolayer. These changes lead to a marked reduction in the average heat of adsorption per molecule. This effect is usually caused by a repulsive (predominantly dipolar) adsorbate–adsorbate interaction that becomes increasingly important as the interadsorbate separation decreases at higher coverages, and it results in a weakening of the bonding of the molecules to the surface. This effect is but one example of repulsion between ''like'' molecules, a behavior well-illustrated by CO–metal systems (Figure 2.31). These systems show a delicate interplay between the repulsive interadsorbate forces and structural changes within the adsorbed layer. This interplay results in modifications in the CO–substrate bonding strength and geometry. In Figure 2.31a the CO/Pt(111) structure at one-half monolayer coverage (in which the CO molecules occupy well-defined sites) is compared to that observed at higher coverages on a Rh(111) substrate, where, to minimize mutual

Substrate	(Gas Phase)	Pd(111)	Rh(111)		Pt(111)
Surface Structure		(3x3)-C$_6$H$_6$ + 2CO	(3x3)-C$_6$H$_6$ + 2CO	c(2$\sqrt{3}$x4)rect-C$_6$H$_6$ + CO	(2$\sqrt{3}$x4)rect-2C$_6$H$_6$ +4CO
The Structure of Benzene	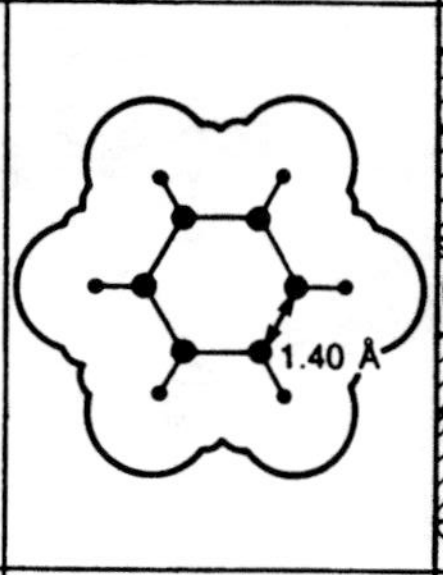	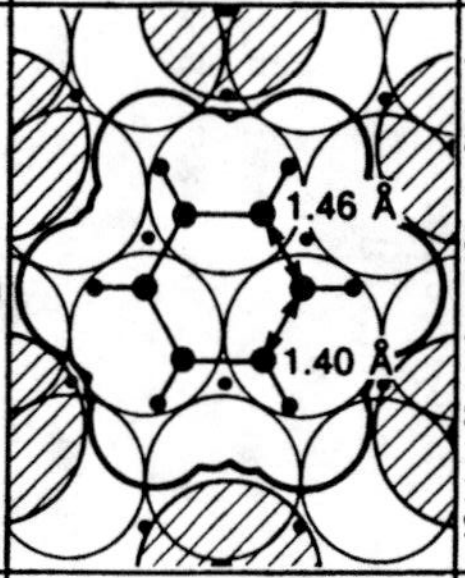	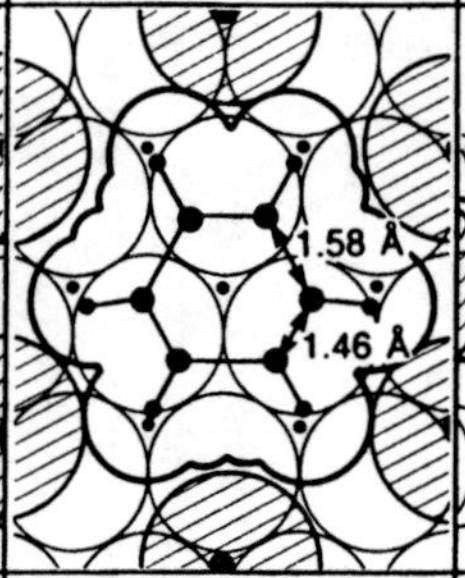	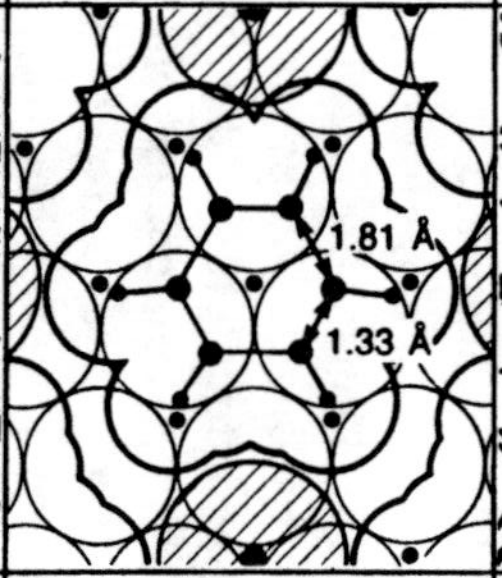	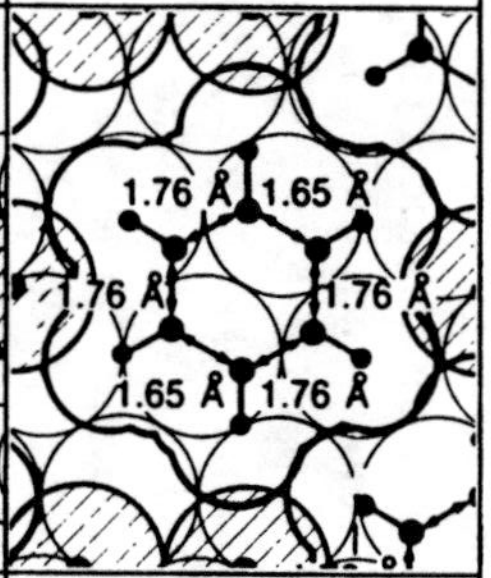
C$_6$ Ring Radius (Å)	1.40	1.43±0.10	1.51±0.15	1.65±0.15	1.72±0.15
d$_{M-C}$(Å)	-	2.39±0.05	2.30±0.05	2.35±0.05	2.25±0.05
γ_{CH}(cm^{-1})*	670	720-770	780-810		830-850

Figure 2.28. The surface structures of benzene of the Pd(111), Rh(111), and Pt(111) crystal faces when coadsorbed with CO that induces ordering. The gas-phase benzene molecular structure is shown for comparison. The center of the flat benzene molecule is located either at a threefold or at a bridge site. Note the lateral distortion of the benzene ring.

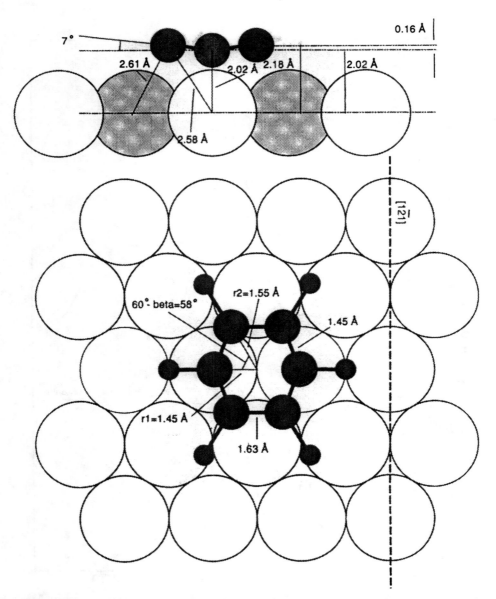

Figure 2.29. The surface structure of benzene in a disordered monolayer on Pt(111). The chemisorbed benzene layer remains disordered in the absence of coadsorbed CO. Note the bending of the benzene molecule into a boat-like surface structure.

repulsion, the adsorbed molecules adopt a pseudohexagonal structure [132] (Figure 2.31b).

Attractive adsorbate–adsorbate interactions, upon coadsorption of two different molecules, may also lead to pronounced structural effects. An example of the latter type is illustrated in Figure 2.32. LEED and HREELS studies show that benzene molecularly adsorbs at 300 K in a very poorly ordered manner on a clean Rh(111)

Rh(111)

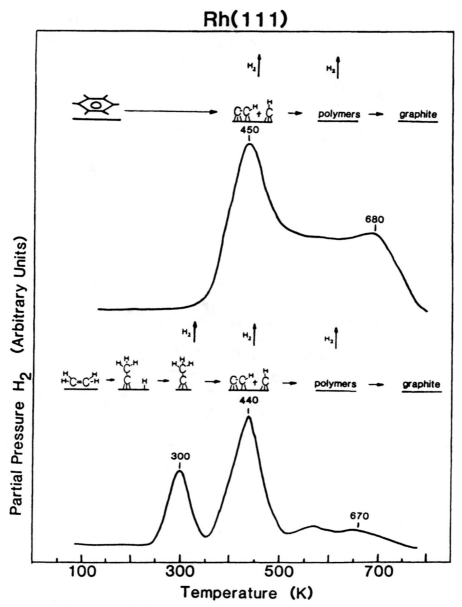

Figure 2.30. Comparison of benzene and ethylene thermal decomposition on Rh(111) as a function of temperature. The dehydrogenation sequences are different, but the chemisorbed species that form at high temperatures are the same.

surface [118, 119]. It can be readily ordered, however, by coadsorption with other molecules that are electron acceptors, such as CO [133–137] and NO [138], as is shown in Figure 2.32 for the Rh(111) surface. Like most organic molecules, benzene is a strong electron donor to metal surfaces. Apparently, therefore, the presence of electron acceptor–donor interactions induces both ordering and the formation of surface structures containing both benzene and CO molecules in the same unit cell.

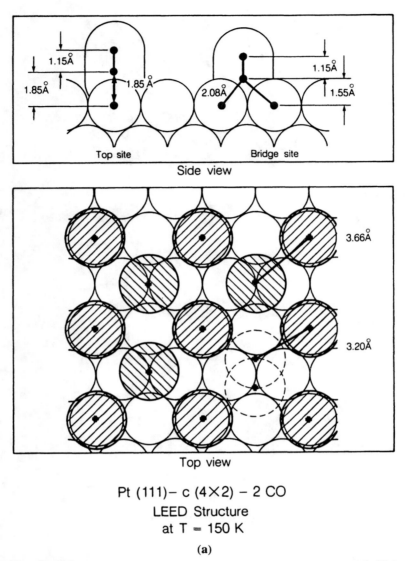

Figure 2.31. Ordered CO surface structures on Pt(111). The structure at one-half monolayer coverage (**a**) and at higher coverages (**b**). While the bridge and top site occupancy predominate at the lower coverages, repulsion forces the molecules to relocate at higher coverages.

The example given above is not an isolated phenomenon. Table 2.9 (p. 269) gives examples of several systems where the coadsorption of an electron donor and an acceptor leads to the formation of ordered structures, while the coadsorption of two electron donors or two electron acceptors yields disordered surface monolayers. Thus, in these systems at least, it is clear that the attractive forces arising from donor–acceptor interaction are crucially important in determining the stability and structure of the coadsorption system. With the coadsorption of benzene with CO on Rh(111), there is little change in the decomposition/desorption temperatures of either the CO or benzene. By contrast, the coadsorption of CO with alkali metals can have

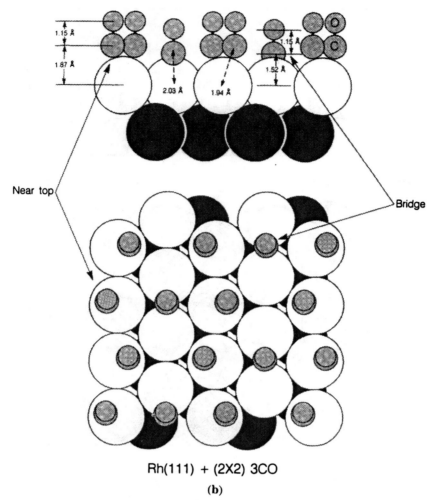

Rh(111) + (2X2) 3CO

(b)

Figure 2.31. (*Continued*)

a dramatic influence on the CO binding strength. For example, CO desorbs completely from a clean Cu(110) surface at temperatures below 200 K [139–141], whereas, in the presence of coadsorbed potassium, two new binding sites are populated, yielding CO desorption at 480 K and 550 K [142]. This corresponds to an increase in the heat of adsorption from around 11 kcal/mole to > 26 kcal/mole (45 kJ/mole to > 110 kJ/mole). Coadsorption and other aspects of the nature of surface chemical bonds will be discussed in more detail in Chapter 6.

2.6.8 Physisorbed Monolayers

At low enough temperatures, inert gas atoms and many diatomic or polyatomic molecules that cannot form strong chemical bonds with the surface will physisorb on surfaces. With inert gases and with saturated hydrocarbons, physisorption is com-

Figure 2.32. Schematic representation of CO–benzene coadsorbed surface structures on the Rh(111) and Pd(111) surfaces.

monplace and stable on many types of substrates. These substrates include metals, as well as inert surfaces such as the graphite basal plane. Over 30 such structures are listed in Table 2.10.

In physisorption the adsorbate–adsorbate interactions are usually comparable in strength to the adsorbate–substrate interactions, all of which are dominated by van der Waals forces. One can therefore examine large parts of the phase diagrams of these adsorption systems.

Many phases have been observed in physisorption, and new classes of phases continue to be discovered. There are commensurate and incommensurate phases, disordered lattice-gas phases, and fluid or liquid phases. There are out-of-phase domain structures, including striped-domain phases (e.g., see references [143–148]), pinwheel structures (e.g., see references [149–153]), herringbone structures (e.g., see references [149, 152, 154–161]), and hexatic phases (e.g., see references [162, 163]), among others (e.g., see references [164]).

The known van der Waals sizes of the adsorbed species lead to satisfactory models of structures that are more-or-less close-packed. With molecules, the best structural models usually involve flat-lying species, which are arranged in a close-packed superlattice. The flat geometry provides the greatest attractive van der Waals interaction with the substrate.

2.7   SUMMARY AND CONCEPTS

- Atomic scale studies of surface structure reveal both order and roughness, the presence of ordered terraces separated by line defects, steps, and kinks.
- Surface atoms move inward at a clean surface, and the magnitude of contraction or "relaxation" increases with lower packing density (increased surface roughness).
- Reconstruction, to form new surface structures that are very different from the bulk-like atomic arrangements, is commonly observed.
- The chemical bonds formed during adsorption between the adsorbate and the substrate lead to epitaxy or may induce surface restructuring.
- The surface atoms move outward, rotate, or are displaced along the surface in order to optimize the strength of the adsorbate–substrate bonds. Thus, the structure of the surface is not rigid but responds to the changing physical-chemical environment at the interface.
- Most atoms and molecules in the adsorption monolayer form ordered surface structures at appropriate temperature and coverage regimes.
- Adsorption of a molecule may take place at different surface sites and in diverse molecular orientations.
- Coadsorption of molecules produces new surface structures due to repulsive or attractive interactions between the adsorbed species.
- When the adsorbate–substrate bonds are weak (of the van der Waals type), the properties of the adsorbed monolayer are less dependent on the substrate's atomic structure. In this circumstance, two-dimensional adsorbate phases exhibit lattice-gas–liquid–solid transitions as a function of coverage and temperature changes.

2.8   PROBLEMS

2.1   Describe the operating principles of the scanning tunneling microscope [179–187] and the field-ion microscope [181, 188–194].

2.2   Describe the operating principles of low-energy electron diffraction [181, 188, 190–193, 195–202].

2.3   Calculate and sketch the diffraction patterns of ordered surface structures with (7 × 7) and (5 × 1) unit cells on the Si(111) and Pt(100) crystal faces, respectively.

2.4   Draw the geometric structures of the (332) and (430) face-centered-cubic metal surfaces.

*2.5   Jona and Marcus [165] plotted the relaxation detected at metal surfaces as a function of surface roughness (defined as $1/$packing density). Using this correlation, calculate the relaxation you would expect for the fcc(311) and (997) crystal faces.

*2.6   The reconstruction of the (100) crystal faces of platinum, iridium and gold were discussed in references [7–11]. Discuss the structures in light of the experimental findings, and describe how they change upon adsorption of various gases.

*2.7   The reconstruction of the (111) and (100) crystal faces of silicon were discussed in references [203–209]. Discuss the structures in light of the experimental findings, and describe how they change upon adsorption of various gases.

*2.8   The adsorption of hydrogen on the W(100) crystal face is described in references [210–213]. Discuss what happens with the H–W(100) system with changing H coverage and with changing temperature.

*2.9   The adsorption of carbon on Ni(100) is described in references [173, 214]. Discuss the nature of surface restructuring induced by carbon adsorption, and find values for the strength of the Ni–Ni and Ni–C bonds [215, 216].

*2.10   Carbon monoxide adsorbs in several binding sites on metal surfaces as described in references [132, 217–219]. Discuss the change of location of CO with coverage and its molecular orientation.

*2.11   Benzene forms a variety of coadsorbed surface structures with CO as described in Table 2.8 and references [135, 220, 221]. Discuss the reasons for ordering and for the location and orientation of the benzene molecule.

*2.12   The adsorption of xenon on graphite has been studied and reported in Table 2.10 and reference [222]. Discuss the changes of surface structure that were observed as a function of coverage and temperature.

*2.13   When pyridine adsorbs on various metal surfaces, it changes its orientation as a function of coverage and temperature and also is also dependent on the nature and structure of the substrate [223]. Describe what is known about the bonding and orientation of pyridine to metals, and give the reasons for the diverse bonding geometries.

**2.14   The coadsorption of CO and potassium on metals leads to large increases in the heat of adsorption of CO and to its dissociation in some cases [224–226]. Discuss the nature of bonding between the coadsorbates that leads to such a strong interaction.

REFERENCES

[1]   N.D. Lang and W. Kohn. Theory of Metal Surfaces: Charge Density and Surface Energy. *Phys. Rev. B* 1:4555 (1970).

[2]   M.A. Van Hove, W.H. Weinberg, and C.M. Chan. Low Energy Electron Diffraction. In: G. Ertl, editor, *Springer Series in Surface Sciences*, Volume 6. Springer-Verlag, Berlin, 1986.

[3]   B. Lang, R.W. Joyner, and G.A. Somorjai. Low Energy Electron Diffraction Studies of High Index Crystal Surfaces of Platinum. *Surf. Sci.* **30**:440 (1972).

[4]   L. Pauling. *The Nature of the Chemical Bond.* Cornell University Press, 1960.

[5]   K.A.R. Mitchell. Analysis of Surface Bond Lengths Reported for Chemisorption on Metal Surfaces. *Surf. Sci.* **149**:93 (1985).

[6]   K.A.R. Mitchell. On the Bond Lengths Reported for Chemisorption on Metal Surfaces. *Surf. Sci.* **100**:225 (1980).

[7]   S. Lehwald, J.G. Chen, G. Kisters, E. Preuss, and H. Ibach. Surface Phonon Dispersion Investigation of the $(1 \times 1) \rightarrow (5 \times 1)$ Reconstruction of the Ir(100) Surface. *Phys. Rev. B* **43**:3920 (1991).

[8]   K. Heinz, G. Schmidt, L. Hammer, and K. Mueller. Dynamics of the Reconstruction Process Ir(100) $1 \times 1 \rightarrow 1 \times 5$. *Phys. Rev. B* **32**:6214 (1985).

[9]   E. Lang, K. Mueller, K. Heinz, M.A. Van Hove, R.J. Koestner, and G.A. Somorjai. LEED Intensity Analysis of the (1×5) Reconstruction of Ir(100). *Surf. Sci.* **127**:347 (1983).

[10] M.A. Van Hove, R.J. Koestner, P.C. Stair, J.P. Biberian, L.L. Kesmodel, I. Bartos, and G.A. Somorjai. The Surface Reconstructions of the (100) Crystal Faces of Iridium, Platinum, and Gold. II. Structural Determination by LEED Intensity Analysis. *Surf. Sci.* **103**:218 (1981).

[11] M.A. Van Hove, R.J. Koestner, P.C. Stair, J.P. Biberian, L.L. Kesmodel, I. Bartos, and G.A. Somorjai. The Surface Reconstructions of the (100) Crystal Faces of Iridium, Platinum, and Gold. I. Experimental Observations and Possible Structural Models. *Surf. Sci.* **103**:189 (1981).

[12] R.J. Behm, W. Hoesler, E. Ritter, and G. Binnig. Correlation between Domain Boundaries and Surface Steps: A Scanning Tunneling Microscopy Study on Reconstructed Pt(100). *Phys. Rev. Lett.* **56**:228 (1986).

[13] K. Heinz, E. Lang, K. Strauss, and K. Mueller. Metastable (1×5) Structure of Pt(100). *Surf. Sci.* **120**:L401 (1982).

[14] W. Telieps, M. Mundschau, and E. Bauer. Surface Domain Structure of Reconstructed Au(100) Observed by Dark Field Low Energy Electron Microscopy. *Surf. Sci.* **225**:87 (1990).

[15] G.K. Binnig, H. Rohrer, C. Gerber, and E. Stoll. Real-Space Observation of the Reconstruction of Au(100). *Surf. Sci.* **144**:321 (1984).

[16] K.H. Rieder, T. Engel, R.H. Swendsen, and M. Manninen. A Helium Diffraction Study of the Reconstructed Au(100) Surface. *Surf. Sci.* **127**:223 (1983).

[17] J.D. Weeks and G.H. Gilmer. Dynamics of Crystal Growth. In: I. Prigogine and S.A. Rice, editors, *Advances in Chemical Physics*, Volume 40. John Wiley & Sons, New York, 1979.

[18] R.L. Fink, C.A. Ballentine, J.L. Erskine, and J.A. Araya-Pochet. Experimental Probe for Thin-Film Magnetism in p(1 $\times$ 1)Pd and V on Ag(100). *Phys. Rev. B* **41**:10175 (1990).

[19] C. Rau. Ferromagnetic Order and Critical Behavior at Surfaces of Ultrathin Epitaxial Films. *Appl. Phys. A* **49**:579 (1989).

[20] S.A. Chambers, T.R. Greenlee, C.P. Smith, and J.H. Weaver. Quantitative Characterization of Abrupt Interfaces by Angle-Resolved Auger Electron Emission. *Phys. Rev. B* **32**:4245 (1985).

[21] M. Pessa and M. Vulli. A Study of Thin Film Effects in Photoemission from Ag Overlayers on a Pd(100) Substrate. *J. Phys. C* **16**:L629 (1983).

[22] R.J. Baird, D.F. Ogletree, M.A. Van Hove, and G.A. Somorjai. The Structure of

the $(\sqrt{3} \times \sqrt{3})R30°$ Superlattice Phase on $(111)\alpha$Cu–16% Al; a LEED Intensity Analysis. *Surf. Sci.* **165**:345 (1986).

[23] G. Polanski and J.P. Toennies. The Growth Mode of Aluminum on Silver (111). *Surf. Sci.* **260**:250 (1992).

[24] J.M. Heitzinger, S.C. Gebhard, D.H. Parker, and B.E. Koel. Growth Mechanism and Structure of Ultrathin Palladium Films Formed by Deposition on Mo(110). *Surf. Sci.* **260**:151 (1992).

[25] L.Q. Jiang and M. Strongin. Structural and Electronic Properties of Rh Overlayers on Mo(110). *Phys. Rev. B* **42**:3282 (1990).

[26] M. Tikhov and E. Bauer. Growth, Structure, and Energetics of Ultrathin Ferromagnetic Single Crystal Films on Mo(110). *Surf. Sci.* **232**:73 (1990).

[27] C. Egawa, T. Aruga, and Y. Iwasawa. Epitaxial Growth of Fe Overlayers on the Ru(001) Surface. *Surf. Sci.* **188**:563 (1989).

[28] D.A. Steigerwald and W.F. Egelhoff, Jr. Observation of Intensity Oscillations in RHEED During the Epitaxial Growth of Cu and fcc Fe on Cu(100). *Surf. Sci.* **192**:L887 (1987).

[29] J.G. Tobin, S.W. Robey, L.E. Klebanoff, and D.A. Shirley. Development of a Three-Dimensional Valence-Band Structure in Ag Overlayers on Cu(001). *Phys. Rev. B* **35**:9056 (1987).

[30] P.A. Dowben, Y.J. Kime, S. Varma, M. Onellion, and J.L. Erskine. Interaction of Hg Overlayers with an Ag(100) Surface. *Phys. Rev. B* **36**:2519 (1987).

[31] R.M. Nix and R.M. Lambert. Surface Crystallography and Growth Modes of Rare Earth Metals and Alloys on Single Crystal Copper: Nd on Cu(001). *Surf. Sci.* **186**:163 (1987).

[32] Y. Borensztein. The Growth of Palladium on (111) Polycrystalline Silver Surface. *Surf. Sci.* **177**:353 (1986).

[33] U. Bardi, A. Santucci, and G. Rovida. Study of the Growth Mechanism of Platinum Layers on the $Na_{0.7}WO_3$ Single Crystal Surface. *Surf. Sci.* **162**:337 (1985).

[34] C. Park, E. Bauer, and H. Poppa. Growth and Alloying of Pd Films on Mo(110) Surfaces. *Surf. Sci.* **154**:371 (1985).

[35] E. Daugy, P. Mathiez, F. Salvan, and J.M. Layet. 7 × 7 Si(111)–Cu Interfaces. Combined LEED, AES, and EELS Measurements. *Surf. Sci.* **154**:267 (1985).

[36] H. Asonen, C.J. Barnes, A. Salokatve, and M. Pessa. The Effect of Interdiffusion on the Growth Mode of Copper on Al{111}. *Surf. Sci.* **152/153**:262 (1985).

[37] H.C. Peebles, D.D. Beck, and J.M. White. Structure of Ag on Rh and Its Effect on the Adsorption of D_2 and CO. *Surf. Sci.* **150**:120 (1985).

[38] J.C. Vickerman, K. Christmann, G. Ertl, P. Heimann, F.J. Himpsel, and D.E. Eastman. Geometric Structure and Electronic States of Copper Films on a Ruthenium (0001) Surface. *Surf. Sci.* **134**:367 (1983).

[39] B.C. de Cooman, V.D. Vankar, and R.W. Vook. AES Studies of Palladium Films Formed on Copper and Mica. *Surf. Sci.* **128**:128 (1983).

[40] K. Gürtler and K. Jacobi. Coverage and Adsorption-Site Dependence of Core-Level Binding Energies for Tin and Lead on Al(111) and Ni(111). *Surf. Sci.* **134**:309 (1983).

[41] G.C. Smith, C. Norris, C. Binns, and H.A. Padmore. A Photoemission Study of Ultra-Thin Palladium Overlayers on Low-Index Faces of Silver. *J. Phys. C* **15**:6481 (1982).

[42] U. Gradmann and G. Waller. Periodic Lattice Distortions in Epitaxial Films of Fe(110) on W(110). *Surf. Sci.* **116**:539 (1989).

[43] B. Gruzza and E. Gillet. Condensation and Early Stages of Gold Growth on an Mo(110) Surface. *Thin Solid Films* **68**:345 (1980).

[44] B. Goldstein and D.J. Szostak. Growth of Thin Platinum Films on Hydrogenated Amorphous Silicon and Its Oxide. *J. Vacuum Sci. Technol.* **17**:718 (1980).

[45] F.C. Frank and J.H. van der Merwe. One Dimensional Dislocations. II. Misfitting Monolayers and Oriented Overgrowth. *Proc. R. Soc. London* **A198**:216 (1949).

[46] J.A. Carlisle, T. Miller, and T.C. Chiang. Photoemission Study of the Growth, Desorption, Schottky–Barrier Formation, and Atomic Structure of Pb on Si(111). *Phys. Rev. B* **45**:3400 (1992).

[47] D. Tian, H. Li, and F. Jona. Study of the Growth of Fe on Ru(0001) by Low Energy Electron Diffraction. *Solid State Commun.* **80**:783 (1991).

[48] M. Prietsch, A. Samsavar, and R. Ludeke. Structural and Electronic Properties of the Bi/GaP(110) Interface. *Phys. Rev. B* **43**:11850 (1991).

[49] P.J. Schmitz, H.C. Kang, W.Y. Leung, and P.A. Thiel. Growth Mode and CO Adsorption Properties of Au Films on Pd(110). *Surf. Sci.* **248**:287 (1991).

[50] M. Tikhov, G. Boishin, and L. Surnev. Sodium Adsorption on a Si(001)-(2 × 2) Surface. *Surf. Sci.* **241**:103 (1991).

[51] W.C. Fan, A. Ignatiev, and N.J. Wu. Growth of Bismuth on the Si(100) Surface: AES and LEED Study. *Surf. Sci.* **235**:169 (1990).

[52] B. Vogt, B. Schmiedeskamp, and U. Heinzmann. Spin-Resolved Core and Valence Electron Photoemission from Non-epitaxially Grown Pb Layers on Pt(111). *Vacuum* **41**:1118 (1990).

[53] N. Boutaoui, H. Roux, and M. Tholomier. Ag Mass Transport on Si(111) in the 350–450°C Temperature Range. *Surf. Sci.* **239**:213 (1990).

[54] S. Andrieu, F. Arnaud d'Avitaya, and J.C. Pfister. Ga Adsorption on Si(111) Analysed by *in situ* Ellipsometry: 2D and 3D Growth. *Surf. Sci.* **238**:53 (1990).

[55] S. Tougaard and H.S. Hansen. Yb Growth on Ni(001) Studied by Inelastic Background Analysis. *Surf. Sci.* **236**:271 (1990).

[56] J.J. Joyce, M.M. Nelson, M. Tang, Y. Meng, J. Anderson, and G.J. Lapeyre. Local Order, Epitaxy, and Electronic Structure of the Bi/III-V Semiconductor Interfaces. *J. Vacuum Sci. Technol. A* **8**:3542 (1990).

[57] B. Oral and R.W. Vook. Growth Mode Determinations for the Epitaxial Cu/Pd(111)/Mica and Pd/Cu(111)/Mica Thin Film Systems. *J. Vacuum Sci. Technol. A* **8**:2682 (1990).

[58] L.Q. Jiang and M. Strongin. Characterization of Rh Films on Ta(110). *J. Vacuum Sci. Technol. A* **8**:2682 (1990).

[59] C.J. Spindt, R. Cao, K.E. Miyano, I. Lindau, W.E. Spicer, and Y.C. Pao. Morphological Study of Ag, In, Sb, and Bi Overlayers on GaAs(100). *J. Vacuum Sci. Technol. B* **8**:974 (1990).

[60] B. Frick and K. Jacobi. Growth and Electronic Structure of Ultrathin Palladium Films on Al(111) and Their Interaction with Oxygen and Carbon Monoxide. *Phys. Rev. B* **37**:4408 (1988).

[61] B. Bourgignon, K.L. Carleton, and S.R. Leone. Surface Structures and Growth Mechanism of Ga on Si(100) Determined by LEED and Auger Electron Spectroscopy. *Surf. Sci.* **204**:455 (1988).

[62] G. Le Lay, J. Peretti, and M. Hanbücken. Surface Spectroscopy of Pb Monolayers on Si(111). *Surf. Sci.* **204**:57 (1988).

[63] A.J. Melmed and N.D. Shinn. Nucleation and Epitaxial Growth of Cr Crystals on Stepped W Surfaces with Low-Index Facets. *Surf. Sci.* **193**:475 (1988).

[64] G. Quentel, M. Gauch, and A. Degiovanni. *In Situ* Ellipsometry Studies of the Growth of Pb on Si(111) Surfaces. *Surf. Sci.* **193**:212 (1988).

[65] T.N. Taylor and M.A. Hoffbauer. The Growth and Stability of Ag Layers on Cu(110). *J. Vacuum Sci. Technol. A* **5**:1625 (1987).

[66] C. Argile and G.E. Rhead. Adsorption of Lithium and Oxygen on Gallium Arsenide. *Thin Solid Films* **152**:545 (1987).

[67] P.M.J. Marée, K. Nakagawa, F.M. Mulders, and J.F. van der Veen. Thin Epitaxial Ge-Si(111) Films. Study and Control of Morphology. *Surf. Sci.* **191**:305 (1987).

[68] M.C. Muñoz, F. Soria, and J.L. Sacedón. The Interaction of Si with Al(111) Surfaces above Room Temperature. *Surf. Sci.* **189/190**:204 (1987).

[69] H.J. Gossmann. Growth and Electronic Structure of Ultra-thin Pd and Ag Films on Al(111). *Surf. Sci.* **179**:453 (1987).

[70] B. Frick and K. Jacobi. A Micro-Quantitative AES Analysis using SEM–SAM Apparatus. Applied to Ag/Si Interfaces on Si(111) and Si(100) Surfaces. *Surf. Sci.* **178**:907 (1986).

[71] Q.G. Zhu, A.D. Zhang, E.D. Williams, and R.L. Park. On the Detailed Growth Mode of Thin Silver Films on Si(111). *Surf. Sci.* **172**:433 (1986).

[72] J. Knall, J.E. Sundgren, G.V. Hansson, and J.E. Greene. Indium Overlayers on Clean Si(100)-(2 × 1). Surface Structure, Nucleation, and Growth. *Surf. Sci.* **166**:512 (1986).

[73] M. Saitoh, K. Oura, K. Asano, F. Shoji, and T. Hanawa. Low Energy Ion Scattering Study of Adsorption and Desorption Processes of Pb on Si(111) Surfaces. *Surf. Sci.* **154**:394 (1985).

[74] M. Hanbücken, M. Futamoto, and J.A. Venables. Nucleation, Growth, and the Intermediate Layer in Ag/Si(100) and Ag/Si(111). *Surf. Sci.* **147**:433 (1984).

[75] A. Rolland, J. Bernadini, and M.G. Barthes-Labrousse. Vapour Deposition of Lead on Ag(111) and Equilibrium Surface Segregation from Ag–Pb(111) Solid Solutions: A LEED–AES Comparative Study. *Surf. Sci.* **143**:579 (1984).

[76] E.J. van Loenen, M. Iwami, R.M. Tromp, and J.F. van der Veen. The Adsorption of Ag on the Si(111) 7 × 7 Surface at Room Temperature Studied by Medium Energy Ion Scattering, LEED, and AES. *Surf. Sci.* **137**:1 (1984).

[77] J.J. Métois and G. Le Lay. Complementary Data Obtained on the Metal–Semiconductor Interface by LEED, AES, and SEM: Pb/Ge(111). *Surf. Sci.* **133**:422 (1983).

[78] G.D.T. Spiller, P. Akhter, and J.A. Venables. UHV–SEM Study of the Nucleation and Growth of Ag/W(110). *Surf. Sci.* **131**:517 (1983).

[79] P.W. Davies, M.A. Quinlan, and G.A. Somorjai. The Growth and Chemisorptive Properties of Ag and Au Monolayers on Platinum Single Crystal Surfaces. An AES, TDS and LEED Study. *Surf. Sci.* **121**:290 (1982).

[80] C. Binns and C. Norris. The Epitaxial Growth of Thallium on Cu(100). A Study by LEED, AES, UPS and EELS. *Surf. Sci.* **115**:395 (1982).

[81] K. Takayanagi, D.M. Kolb, K. Kambe, and G. Lehmpfuhl. Deposition of Monolayer and Bulk Lead on Ag(111) Studied in Vacuum and in an Electrochemical Cell. *Surf. Sci.* **100**:407 (1980).

[82] E. Gillet and B. Gruzza. Stranski–Krastanov Growth Mode of Au on Mo(110) Surfaces: Investigation by Auger Electron Spectroscopy. *Surf. Sci.* **97**:553 (1980).

[83] K.J. Rawlings, M.J. Gibson, and P.J. Dobson. The Epitaxial Growth of Lead and Thallium on (111) Silver and Copper. *J. Phys. D* **11**:2059 (1978).

[84] C. Morant, L. Galán, J.M. Sanz, and F. Rueda. An XPS Study of the Adsorption of Re onto ZrO_2: Interface Formation and Film Growth. *Surf. Sci.* **251/252**:1023 (1991).

[85] P. Akhter and A. Baig. Comments on Stranski–Krastanov Growth of Bi and Ag on Si(100). *Jpn. J. Appl. Phys., Part 2* **30**:1203 (1991).

[86] T. Ichinokawa, T. Inoue, H. Izumi, and Y. Sakai. Epitaxial Growth in Cu/Si(001) 2 × 1 at High Temperatures. *Surf. Sci.* **241**:416 (1991).

[87] J.E.T. Andersen and P.J. Møller. Ultrathin Deposition of Copper on Room Temperature SrTiO₃(100). *Thin Solid Films* **186**:137 (1990).

[88] C. Binns and S.C. Bayliss. Growth of Silver Particles on Graphite Studied by Optical Scattering. *Surf. Sci.* **214**:165 (1989).

[89] B. Bourguignon, R.V. Smilgys, and S.R. Leone. AES and LEED Studies Correlating Desorption Energies with Surface Structures and Coverages for Ga on Si(100). *Surf. Sci.* **204**:473 (1988).

[90] W.C. Fan, J. Strozier, and A. Ignatiev. Island Formation of Aluminum on the Graphite (0001) Surface. LEED and AES Study. *Surf. Sci.* **195**:226 (1988).

[91] A. Taleb-Ibrahimi and C.A. Sébenne. Effect of Valency in Metal Adsorption on Si(111). The Case of Sb on the Cleaved Surface. *Surf. Sci.* **168**:114 (1986).

[92] J.S. Foord and P.D. Jones. Model Studies on Bimetallic Cu-Rh Catalysts. *Surf. Sci.* **152/153**:487 (1985).

[93] M. Hanbücken and H. Neddermeyer. A LEED-AES Study of the Growth of Ag Films on Si(100). *Surf. Sci.* **114**:563 (1982).

[94] J.W.A. Sachtler, M.A. Van Hove, J.P. Biberian, and G.A. Somorjai. The Structure of Epitaxially Grown Metal Films on Single Crystal Surfaces of Other Metals. Gold on Pt(100) and Platinum on Au(100). *Surf. Sci.* **110**:19 (1981).

[95] J.M. Slaughter, W. Weber, G. Güntherot, and C.M. Falco. Quantitative Auger and XPS Analysis of Thin Films. *MRS Bull.* **December**:39 (1992).

[96] C. Argile and G.E. Rhead. Adsorbed Layer and Thin Film Growth Modes Monitored by Auger Electron Spectroscopy. *Surf. Sci. Rep.* **10**:277 (1989).

[97] S. Ossicini, R. Memeo, and F. Ciccacci. AES Analysis of the Growth Mechanism of Metal Layers on Metal Surfaces. *J. Vacuum Sci. Technol. A* **3**:387 (1985).

[98] G.E. Rhead. Probing Surface Properties with Adsorbed Metal Overlayers. *J. Vacuum Sci. Technol.* **13**:603 (1976).

[99] R.J. Koestner, M.A. Van Hove, and G.A. Somorjai. A Surface Crystallography Study by Dynamical LEED of the $(\sqrt{3} \times \sqrt{3})R30°$ CO Structure on the Rh(111) Crystal Surface. *Surf. Sci.* **107**:439 (1981).

[100] D.F. Ogletree, M.A. Van Hove, and G.A. Somorjai. LEED Intensity Analysis of the Structures of Clean Pt(111) and of CO Adsorbed on Pt(111) in the c(4 × 2) Arrangement. *Surf. Sci.* **173**:351 (1986).

[101] B.E. Bent. *Bonding and Reactivity of Unsaturated Hydrocarbons on Transition Metal Surfaces: Spectroscopic and Kinetic Studies of Platinum and Rhodium Single Crystal Surfaces.* PhD thesis, University of California, Berkeley, 1986.

[102] R.J. Koestner, M.A. Van Hove, and G.A. Somorjai. A LEED Crystallography Study of the (2 × 2)-C₂H₃ Structure Obtained after Ethylene Adsorption on Rh(111). *Surf. Sci.* **121**:321 (1982).

[103] L.H. Dubois, D.G. Castner, and G.A. Somorjai. A Low Energy Electron Diffraction (LEED), High Resolution Electron Energy Loss (ELS), and Thermal Desorption Mass Spectrometry (TDS) Study. *J. Chem. Phys.* **72**:5234 (1980).

[104] L.L. Kesmodel, L.H. Dubois, and G.A. Somorjai. LEED Analysis of Acetylene and Ethylene Chemisorption on the Pt(111) Surface: Evidence for Ethylidyne Formation. *J. Chem. Phys.* **70**:2180 (1979).

[105]   D. Wang, K. Wu, Y. Cao, R. Zhai, and X. Guo. The Room Temperature Phase of Propene on Rh(111): A Caveat on Thermal Processing for the Production of Adsorbed Phases at Definite Temperatures. *Surf. Sci.* **223**:L927 (1989).

[106]   B.E. Bent, C.M. Mate, J.E. Crowell, B.E. Koch, and G.A. Somorjai. Bonding and Thermal Decomposition of Propylene, Propadiene, and Methyl Acetylene on the Rh(111) Single-Crystal Surface. *J. Phys. Chem.* **91**:1493 (1987).

[107]   P.A.P. Nascente, M.A. Van Hove, and G.A. Somorjai. Induced Ordering of Ethylidyne on the Pd(111) Surface by the Preadsorption of Oxygen: a LEED and HREELS Study. *Surf. Sci.* **253**:167 (1991).

[108]   L.L. Kesmodel, G.D. Waddill, and J.A. Gates. Vibrational Spectroscopy of Acetylene Decomposition on Palladium (111) and (100) Surfaces. *Surf. Sci.* **138**:464 (1984).

[109]   J.A. Gates and L.L. Kesmodel. Thermal Evolution of Acetylene and Ethylene on Pd(111). *Surf. Sci.* **124**:68 (1983).

[110]   D.R. Lloyd and F.P. Netzer. The Electronic Structure of the High Temperature Phase of Ethylene Adsorbed on Pd(111). *Surf. Sci.* **129**:L249 (1983).

[111]   C.M. Greenlief, P.L. Radloff, X.L. Zhou, and J.M. White. The Formation and Decomposition Kinetics of Ethylidyne on Ru(0001). *Surf. Sci.* **191**:93 (1987).

[112]   J.F.M. Aarts and N.R.M. Sassen. Vibrational Structure of C_6H_6 and C_6H_5D Adsorbed on Pd(111). *Surf. Sci.* **214**:257 (1989).

[113]   V.H. Grassian and E.L. Muetterties. Vibrational Electron Energy Loss Spectroscopic Study of Benzene, Toluene, and Pyridine Adsorbed on Pd(111) at 180 K. *J. Chem. Phys.* **91**:389 (1987).

[114]   G.D. Waddill and L.L. Kesmodel. Benzene Chemisorption on Palladium Surfaces. I. High Resolution Electron Energy Loss Vibrational Spectra and Structural Models. *Phys. Rev. B* **31**:4940 (1985).

[115]   F.P. Netzer and J.U. Mack. The Electronic Structure of Aromatic Molecules Adsorbed on Pd(111). *J. Chem. Phys.* **79**:1017 (1983).

[116]   F.P. Netzer, G. Rosina, E. Bertel, and H. Saalfeld. Azimuthal Orientation of Benzene on Rh(111) Determined by Angle Resolved Photoemission. *Surf. Sci.* **184**:L397 (1987).

[117]   M. Neumann, J.U. Mack, E. Bertel, and F.P. Netzer. The Molecular Structure of Benzene on Rh(111). *Surf. Sci.* **155**:629 (1985).

[118]   B.E. Koel, J.E. Crowell, C.M. Mate, and G.A. Somorjai. A High Resolution Electron Energy Loss Spectroscopy Study of the Surface Structure of Benzene Adsorbed on the Rhodium (111) Crystal Face. *J. Chem. Phys.* **88**:1988 (1984).

[119]   B.E. Koel and G.A. Somorjai. Vibrational Spectroscopy Using HREELS of Benzene Adsorbed on the Rh(111) Crystal Surface. *J. Electron Spectrosc. Relat. Phenom.* **29**:287 (1983).

[120]   R.F. Lin, R.J. Koestner, M.A. Van Hove, and G.A. Somorjai. The Adsorption of Benzene and Naphthalene on the Rh(111) Surface: A LEED, AES and TDS Study. *Surf. Sci.* **134**:161 (1983).

[121]   A. Wander, G. Held, R.Q. Hwang, G.S. Blackman, M.L. Xu, P. de Andres, M.A. Van Hove, and G.A. Somorjai. A Diffuse LEED Study of the Adsorption Structure of Disordered Benzene on Pt(111). *Surf. Sci.* **249**:21 (1991).

[122]   M. Abon, J.C. Bertolini, J. Billy, J. Massardier, and B. Tardy. Adsorption States of Benzene and Toluene on Pt(111): A Vibrational EELS, $\Delta\phi$, AES and TDS Study. *Surf. Sci.* **162**:395 (1985).

[123]   S. Lehwald, H. Ibach, and J.E. Demuth. Vibration Spectroscopy of Benzene Adsorbed on Pt(111) and Ni(111). *Surf. Sci.* **78**:577 (1978).

[124] G.A. Somorjai. Modern Concepts in Surface Science and Heterogeneous Catalysis. *J. Phys. Chem.* **94:**1013 (1990).

[125] C.H. Patterson, J.M. Mundenar, P.Y. Timbrell, A.J. Gellman, and R.M. Lambert. Molecular Pathways in the Cyclotrimerization of Acetylene on Pd(111): Vibrational Spectra of the C_4H_4 Intermediate and Its Thermal Decomposition Products. *Surf. Sci.* **208:**93 (1989).

[126] T.G. Rucker, M.A. Logan, T.M. Gentle, E.L. Muetterties, and G.A. Somorjai. Conversion of Acetylene to Benzene over Palladium Single-Crystal Surfaces. 1. The Low Pressure Stoichiometric and the High-Pressure Catalytic Reactions. *J. Phys. Chem.* **90:**2703 (1986).

[127] B. Marchon. HREELS Study of the Cyclomerization of Acetylene on Clean and Phosphorus-Covered Palladium (111) Surfaces. *Surf. Sci.* **162:**382 (1985).

[128] T.M. Gentle and E.L. Muetterties. Acetylene, Ethylene, and Arene Chemistry of Palladium Surfaces. *J. Phys. Chem.* **87:**2469 (1983).

[129] W. Sesselmann, B. Woratschak, G. Ertl, J. Küppers, and H. Haberland. Low Temperature Formation of Benzene from Acetylene on a Pd(111) Surface. *Surf. Sci.* **130:**245 (1983).

[130] W.T. Tysoe, G.L. Nyberg, and R.M. Lambert. Photoelectron Spectroscopy and Heterogeneous Catalysis: Benzene and Ethylene from Acetylene on Palladium 111. *Surf. Sci.* **135:**128 (1983).

[131] W.T. Tysoe, G.L. Nyberg, and R.M. Lambert. Low Temperature Catalytic Chemistry of the Pd(111) Surface: Benzene and Ethylene from Acetylene Photoelectron Spectroscopy and Heterogeneous Catalysis: Benzene and Ethylene from Acetylene on Palladium 111. *J. Chem. Soc. Chem. Commun.* **11:**623 (1983).

[132] M.A. Van Hove, R.J. Koestner, J.C. Frost, and G.A. Somorjai. The Structure of Rh(111)(2×2)-CO from LEED Intensities: Simultaneous Bridge and Near-Top Adsorption in a Distorted Compact Hexagonal Overlayer. *Surf. Sci.* **129:**482 (1983).

[133] H. Ohtani, R.J. Wilson, S. Chiang, and C.M. Mate. Scanning Tunneling Microscopy Observations of Benzene Molecules on the Rh(111)-(3 × 3)(C_6H_6 + 2CO) Surface. *Phys. Rev. Lett.* **60:**2398 (1988).

[134] C.M. Mate, C.T. Kao, and G.A. Somorjai. Carbon Monoxide Induced Ordering of Adsorbates on the Rh(111) Crystal Surface: Importance of Surface Dipole Moments. *Surf. Sci.* **206:**145 (1988).

[135] R.F. Lin, G.S. Blackman, M.A. Van Hove, and G.A. Somorjai. LEED Intensity Analysis of the Structure of Coadsorbed Benzene and CO on Rh(111). *Acta Crystallogr. B* **43:**368 (1987).

[136] E. Bertel, G. Rosina, and F.P. Netzer. The Structure of Benzene on Rh(111) Coadsorption with CO. *Surf. Sci.* **172:**L515 (1986).

[137] C.M. Mate and G.A. Somorjai. Carbon Monoxide Induced Ordering of Benzene on Pt(111) and Rh(111) Crystal Surfaces. *Surf. Sci.* **160:**542 (1985).

[138] H. Ohtani, M.A. Van Hove, and G.A. Somorjai. The Structures of CO, NO, and Benzene on Various Transition Metal Surfaces: Overview of LEED and HREELS Results. In: J.F. van der Veen and M.A. Van Hove, editors, *The Structure of Surfaces II. Springer Series in Surface Sciences*, Volume 11. Springer-Verlag, Berlin, 1988.

[139] C. Harendt, J. Goschnick, and W. Hirschwald. The Interaction of CO with Copper (110) Studied by TDS and LEED. *Surf. Sci.* **152:**453 (1985).

[140] A. Spitzer and H. Lüth. The Adsorption of CO on Cu Surfaces Studied by Electron Energy Loss Spectroscopy. *Surf. Sci.* **102:**29 (1981).

[141]   J.F. Wendelken and M.V.K. Úlehla. An ELS Vibrational Study of Carbon Monoxide Adsorbed on the Cu(110) Surface. *J. Vacuum Sci. Technol.* **16**:441 (1979).

[142]   D. Lackey, M. Surman, S. Jacobs, D. Grider, and D.A. King. Surface Reaction Between Coadsorbed K and CO on Cu(110). *Surf. Sci.* **152**:513 (1985).

[143]   J.C. Patrin, Y.Z. Li, M. Chander, and J.H. Weaver. Orientational Ordering and Domain-Wall Formation in Sb Overlayers on GaAs(110). *Phys. Rev. B* **45**:3918 (1992).

[144]   J. Cui, S.C. Fain, Jr., H. Freimuth, H. Wiechert, H.P. Schildberg, and H.J. Lauter. Modulated Structures for Incommensurate Monolayer Solid Phases of D_2 Physisorbed on Graphite. *Phys. Rev. Lett.* **60**:1848 (1988).

[145]   K. Kern, R. David, P. Zeppenfeld, R. Palmer, and G. Comsa. Symmetry Breaking Commensurate–Incommensurate Transition of Monolayer Xe Physisorbed on Pt(111). *Solid State Commun.* **62**:391 (1987).

[146]   H. Freimuth, H. Wiechert, and H.J. Lauter. The Commensurate–Incommensurate Transition of Hydrogen Monolayers Physisorbed on Graphite. *Surf. Sci.* **189/190**:548 (1987).

[147]   A.Q.D. Faisal, M. Hamichi, G. Raynerd, and J.A. Venables. Low Temperature Structures of Xenon Monolayers Adsorbed on Graphite. *Phys. Rev. B* **34**:7440 (1986).

[148]   H. Hong, R.J. Birgeneau, and M. Sutton. Low-Temperature Structures of Xe on Graphite in the One- to Two-Layer Regime. *Phys. Rev. B* **33**:3344 (1986).

[149]   S.K. Wang, J.C. Newton, R. Wang, H. Taub, and J.R. Dennison. Multilayer Structure of Nitrogen Adsorbed on Graphite. *Phys. Rev. B* **39**:10331 (1989).

[150]   H. You and S.C. Fain, Jr. Low Energy Electron Diffraction Study of Submonolayer Mixtures of Carbon Monoxide and Nitrogen with Argon Physisorbed on Graphite. *Phys. Rev. B* **34**:2840 (1986).

[151]   H. You, S.C. Fain, Jr., S. Satija, and L. Passell. Observation of Two-Dimensional Compositional Ordering of a Carbon Monoxide and Argon Monolayer Mixture Physisorbed on Graphite. *Phys. Rev. Lett.* **56**:244 (1986).

[152]   H. You and S.C. Fain, Jr. Structure of Carbon Monoxide Monolayers Physisorbed on Graphite. *Surf. Sci.* **151**:361 (1985).

[153]   P.R. Kubik, W.N. Hardy, and H. Glattli. Orientational Ordering of Hydrogen Molecules Adsorbed on Graphite. *Can. J. Phys.* **63**:605 (1985).

[154]   J. Schimmelpfennig, S. Fölsch, and M. Henzler. LEED Studies of the Adsorption of CO_2 on Thin Epitaxial NaCl(100) Films. *Surf. Sci.* **250**:198 (1991).

[155]   Y.P. Joshi, D.J. Tildesley, J.S. Ayres, and R.K. Thomas. The Structure of CS_2 Adsorbed on Graphite. *Mol. Phys.* **65**:991 (1988).

[156]   F.Y. Hansen and H. Taub. Theoretical and Experimental Study of the Vibrational Excitations in Ethane Monolayers Adsorbed on Graphite (0001) Surfaces. *J. Chem. Phys.* **87**:3232 (1987).

[157]   D. Schmeisser, F. Greuter, E.W. Plummer, and H.-J. Freund. Photoemission from Ordered Physisorbed Adsorbate Phases: N_2 on Graphite and CO on Ag(111). *Phys. Rev. Lett.* **54**:2095 (1985).

[158]   J. Suzanne, J.L. Seguin, H. Taub, and J.P. Biberian. A LEED Study of Ethane Films Adsorbed on Graphite in the Monolayer Range. *Surf. Sci.* **125**:153 (1983).

[159]   R.D. Diehl and S.C. Fain, Jr. Structure and Orientational Ordering of Nitrogen Molecules Physisorbed on Graphite. *Surf. Sci.* **125**:116 (1983).

[160]   R.D. Diehl and S.C. Fain, Jr. Observation of a Uniaxial Incommensurate Phase for Orientationally Ordered Nitrogen Molecules Physisorbed on Graphite and Proposal of a Phase Diagram. *Phys. Rev. B* **26**:4785 (1982).

[161] R.D. Diehl, M.F. Toney, and S.C. Fain, Jr. Orientational Ordering of Nitrogen Molecular Axes for a Commensurate Monolayer Physisorbed on Graphite. *Phys. Rev. Lett.* **48**:177 (1982).

[162] W.C. Fan and A. Ignatiev. Growth of an Orientationally Ordered Incommensurate Potassium Overlayer and Its Order–Disorder Transition on the Cu(111) Surface. *Phys. Rev. B* **37**:5274 (1988).

[163] W.C. Fan and A. Ignatiev. Ordered Phases and Phase Transitions of Cesium on the Cu(111) Surface. *J. Vacuum Sci. Technol. A* **6**:735 (1988).

[164] R. Duszak and R.H. Prince. Anti-Phase Domain Formation During Cesium Adsorption on Ru(0001). *Surf. Sci.* **226**:33 (1990).

[165] F. Jona and P.M. Marcus. Surface Structures from LEED: Metal Surfaces and Metastable Phases. In: J.F. van der Veen and M.A. Van Hove, editors, *The Structure of Surfaces II. Springer Series in Surface Sciences*, Volume 11. Springer-Verlag, Berlin, 1988.

[166] W.S. Yang, F. Jona, and P.M. Marcus. Atomic Structure of Si(001)2 × 1. *Phys. Rev. B* **28**:2049 (1983).

[167] M.A. Van Hove, R.J. Koestner, P.C. Stair, J.P. Biberian, L.L. Kesmodel, I. Bartos, and G.A. Somorjai. The Surface Reconstructions of the (100) Crystal Faces of Iridium, Platinum, and Gold. II. Structural Determination by LEED Intensity Analysis. *Surf. Sci.* **103**:218 (1981).

[168] M.A. Van Hove, R.J. Koestner, P.C. Stair, J.P. Biberian, L.L. Kesmodel, I. Bartos, and G.A. Somorjai. The Surface Reconstructions of the (100) Crystal Faces of Iridium, Platinum, and Gold. I. Experimental Observations and Possible Structural Models. *Surf. Sci.* **103**:189 (1981).

[169] C.M. Chan, M.A. Van Hove, W.H. Weinberg, and E.D. Williams. An R-Factor Analysis of Several Models of the Reconstructed Ir(110)-(1 × 2) Surface. *Surf. Sci.* **91**:440 (1980).

[170] P. Jiang, F. Jona, and P.M. Marcus. The Problem of Edge-Atom Depression on a Cu{410} Surface. *Surf. Sci.* **185**:L520 (1987).

[171] X.G. Zhang, M.A. Van Hove, G.A. Somorjai, P.J. Rous, D. Tobin, A. Gonis, J.M. MacLaren, K. Heinz, M. Michl, H. Lindner, K. Müller, M. Ehsasi, and J.H. Block. Efficient Determination of Multilayer Relaxation in the Pt(210) Stepped and Densely Kinked Surface. *Phys. Rev. Lett.* **67**:1298 (1991).

[172] Y. Gauthier, R. Baudoing-Savois, K. Heinz, and H. Landskron. Structure Determination of p4g Ni(100)-(2 × 2)C by LEED. *Surf. Sci.* **251**:493 (1991).

[173] J.H. Onuferko, D.P. Woodruff, and B.W. Holland. LEED Structure Analysis of the Ni{100}(2 × 2)C(p4g) Structure; A Case of Adsorbate-Induced Substrate Distortion. *Surf. Sci.* **87**:357 (1979).

[174] H.D. Shih, F. Jona, D.W. Jepsen, and P.M. Marcus. Metal–Surface Reconstruction Induced by Adsorbate: Fe(110)p(2 × 2)-S. *Phys. Rev. Lett.* **46**:731 (1981).

[175] C.M. Chan and M.A. Van Hove. The Structure of a (2 × 2) Sulfur Overlayer on the Ir(110)-(1 × 2) Reconstructed Surface. *Surf. Sci.* **183**:303 (1987).

[176] W. Oed, H. Lindner, U. Starke, K. Heinz, K. Müller, and J.B. Pendry. Adsorbate-Induced Reconstruction of c(2 × 2)O/Ni(100): A Reinvestigation by LEED Structure Analysis. *Surf. Sci.* **224**:179 (1989).

[177] S.R. Parkin, H.C. Zeng, M.Y. Zhou, and K.A.R. Mitchell. Low-Energy Electron-Diffraction Crystallographic Determination for the Cu(110) 2 × 1-O Surface Structure. *Phys. Rev. B* **41**:5432 (1990).

[178] H.D. Shih, F. Jona, D.W. Jepsen, and P.M. Marcus. Atomic Underlayer Formation during the Reaction of Ti{0001} with Nitrogen. *Surf. Sci.* **60**:445 (1976).

[179] A.L. de Lozanne. Scanning Tunneling Microscopy. In: B.W. Rossiter and R.C. Baetzold, editors, *Investigations of Surfaces and Interfaces, Part A. Physical Methods of Chemistry*, Volume 9A, 2nd edition. John Wiley & Sons, New York, 1993.

[180] B.A. Sexton. Scanning Tunneling Microscopy. In: D.J. O'Conner, B.A. Sexton, and R. St. C. Smart, editors, *Surface Analysis Methods in Materials Science. Springer Series in Surface Sciences*, Volume 23. Springer-Verlag, Berlin, 1992.

[181] J.B. Hudson. *Surface Science: An Introduction*. Butterworth-Heinemann, Boston, 1992.

[182] H.J. Günterodt and R. Wiesendanger, editors. *Scanning Tunneling Microscopy I: General Principles and Applications to Clean and Adsorbate Covered Surfaces. Springer Series in Surface Sciences*, Volume 20. Springer-Verlag, Berlin, 1992.

[183] P.K. Hansma, V.B. Elings, O. Marti, and C.E. Bracker. Scanning Tunneling Microscopy and Atomic Force Microscopy: Application to Biology and Technology. *Science* **242:**157 (1988).

[184] L.E.C. van de Leemput and H. van Kempen. Scanning Tunneling Microscopy. *Rep. Prog. Phys.* **55:**1165 (1992).

[185] M. Tsukada, K. Kobayashi, N. Isshiki, and H. Kagashima. First-Principles Theory of Scanning Tunneling Microscopy. *Surf. Sci. Rep.* **13:**265 (1991).

[186] F. Ogletree and M. Salmerón. Scanning Tunneling Microscopy and the Atomic Structure of Solid Surfaces. *Prog. Solid State Chem.* **20:**235 (1990).

[187] G. Binnig and H. Rohrer. Scanning Tunneling Microscopy—from Birth to Adolescence. *Rev. Mod. Phys.* **59:**615 (1987).

[188] D.P. Woodruff and T.A. Delchar. *Modern Techniques of Surface Science. Cambridge Solid State Science Series*. Cambridge University Press, New York, 1986.

[189] G. Ehrlich. Wandering Surface Atoms and the Field Ion Microscope. *Phys. Today* **June:**44 (1981).

[190] M.J. Higatsberger. Solid Surfaces Analysis. In C. Marton, editor, *Advances in Electronics and Electron Physics*, Volume 56. Academic Press, New York, 1981.

[191] G.A. Somorjai and M.A. Van Hove. Adsorbed Monolayers on Solid Surfaces. In: J.D. Dunitz, J.B. Goodenough, P. Hemmerich, J.A. Albers, C.K. Jørgensen, J.B. Neilands, D. Reinen, and R.J.P. Williams, editors, *Structure and Bonding*, Volume 38. Springer-Verlag, Berlin, 1979.

[192] R.P.H. Gasser. *An Introduction to Chemisorption and Catalysis by Metals*. Oxford University Press, New York, 1985.

[193] M. Prutton. *Surface Physics*. Oxford University Press, New York, 1975.

[194] J.A. Panitz. High-Field Techniques. In: R.L. Park and M.G. Lagally, editors, *Solid State Physics: Surfaces. Methods of Experimental Physics*, Volume 22. Academic Press, New York, 1985.

[195] M.A. Van Hove. Surface Crystallography with Low-Energy Electron Diffraction. *Proc. R. Soc. A* (1993).

[196] D.P. Woodruff, G.C. Wang, and T.M. Lu. Surface Structure and Order–Disorder Phenomena. In: D.A. King and D.P. Woodruff, editors, *Adsorption at Solid Surfaces. The Chemical Physics of Solid Surfaces and Heterogeneous Catalysis*, Volume 2. Elsevier, New York, 1983.

[197] H. Wagner. Physical and Chemical Properties of Stepped Surfaces. In: G. Höhler, editor, *Solid Surface Physics. Springer Tracts in Modern Physics*, Volume 85. Springer-Verlag, Berlin, 1979.

[198] D.P. Woodruff. Surface Periodicity, Crystallography and Structure. In: D.A. King and D.P. Woodruff, editors, *Clean Solid Surfaces. The Chemical Physics of Solid Surfaces and Heterogeneous Catalysis*, Volume 1. Elsevier, New York, 1981.

[199] G. Ertl and J. Küpper. *Low Energy Electrons and Surface Chemistry*. VCH, Weinheim, Germany, 1985.

[200] G.A. Somorjai and M.A. Van Hove. Surface Crystallography by Low-Energy Electron Diffraction. In: B.W. Rossiter, J.F. Hamilton, and R.C. Baetzold, editors, *Investigation of Interfaces and Surfaces*. Interscience, Rochester, NY, 1990.

[201] M.A. Van Hove, W.H. Weinberg, and C.M. Chan. Low Energy Electron Diffraction. In: G. Ertl, editor, *Springer Series in Surface Sciences*, Volume 6. Springer-Verlag, Berlin, 1986.

[202] J.B. Pendry. *Low Energy Electron Diffraction*. Academic Press, New York, 1974.

[203] I.P. Batra. Atomic Structure of the Si(001)-(2 × 1) Surface. *Phys. Rev. B* **41**:5048 (1990).

[204] R.S. Becker, J.A. Golovchenko, G.S. Higashi, and B.S. Swartzentruber. New Reconstructions on Silicon (111) Surfaces. *Phys. Rev. Lett.* **57**:1020 (1986).

[205] R.J. Hamers, R.M. Tromp, and J.E. Demuth. Scanning Tunneling Microscopy of Si(001). *Phys. Rev. B* **34**:5343 (1986).

[206] K. Takayanagi, Y. Tanishiro, S. Takahashi, and M. Takahashi. Structure Analysis of Si(111)-(7 × 7) Reconstructed Surface by Transmission Electron Diffraction. *Surf. Sci.* **164**:367 (1985).

[207] B.W. Holland, C.B. Duke, and A. Paton. The Atomic Geometry of Si(100)-(2 × 1) Revisited. *Surf. Sci.* **140**:L269 (1984).

[208] A. Kahn. Semiconductor Surface Structures. *Surf. Sci. Rep.* **3**:193 (1983).

[209] D.J. Chadi. Semiconductor Surface Reconstruction. *Vacuum* **33**:613 (1983).

[210] M.A. Passler, B.W. Lee, and A. Ignatiev. Surface Structure Determination of W(001)p(1 × 1)-2H. Chemisorption Induced Substrate Relaxation. *Surf. Sci.* **150**:263 (1985).

[211] R.A. Barker and P.J. Estrup. Hydrogen on Tungsten (100): Adsorbate Induced Surface Reconstruction. *Phys. Rev. Lett.* **41**:1307 (1978).

[212] M.K. Debe and D.A. King. New Evidence for a Clean Thermally Induced c(2 × 2) Surface Structure on W{100}. *J. Phys. C* **10**:L303 (1977).

[213] H. Froitzheim, H. Ibach, and S. Lehwald. Surface Sites of H on W(100). *Phys. Rev. Lett.* **36**:1549 (1976).

[214] A. Atrei, U. Bardi, M. Maglietta, G. Rovida, M. Torrini, and E. Zanazzi. SEELFS Study of Ni(001)(2 × 2)C p4g Structure. *Surf. Sci.* **211/212**:93 (1989).

[215] K.W. Frese, Jr. Calculation of Surface Binding Energies for Hydrogen, Oxygen, and Carbon Atoms on Metallic Surfaces. *Surf. Sci.* **182**:85 (1987).

[216] W.F. Egelhoff, Jr. Summary Abstract: Heats of Adsorption of Atomic C, N, and O on Ni(100) and Cu(100) from a (Z + 1) Core-Level Shift Analysis. *J. Vacuum Sci. Technol. A* **5**:700 (1987).

[217] J.C. Campuzano. The Adsorption of Carbon Monoxide by the Transition Metals. In: D.A. King and D.P. Woodruff, editors, *Chemisorption Systems, Part A. The Chemical Physics of Solid Surfaces and Heterogeneous Catalysis*, Volume 3. Elsevier, New York, 1990.

[218] S.D. Kevan, R.F. Davis, D.H. Rosenblatt, J.G. Tobin, M.G. Mason, and D.A. Shirley. Structural Determination of Molecular Overlayer Systems with Normal Photoelectron Diffraction: c(2 × 2)CO-Ni(001) and ($\sqrt{3} \times \sqrt{3}$)R30° CO-Ni(111). *Phys. Rev. Lett.* **46**:1629 (1981).

[219] J.L. Gland and G.A. Somorjai. Low Energy Electron Diffraction and Work Function Studies of Benzene, Naphthalene and Pyridine Adsorbed on Pt(111) and Pt(100) Single Crystal Surfaces. *Surf. Sci.* **38**:157 (1973).

[220] D.F. Ogletree, M.A. Van Hove, and G.A. Somorjai. Benzene and Carbon Monoxide Adsorbed on Pt(111). A LEED Intensity Analysis. *Surf. Sci.* **183**:1 (1987).

[221] M.A. Van Hove, R.F. Lin, and G.A. Somorjai. Surface Structure Determination of Coadsorbed Benzene and Carbon Monoxide on the Rhodium (111) Single Crystal Surface Analyzed with Low-Energy Electron Diffraction Intensities. *J. Am. Chem. Soc.* **108**:2532 (1986).

[222] H. Hong, C.J. Peters, A. Mak, R.J. Birgeneau, P.M. Horn, and H. Suematsu. Synchrotron X-Ray Study of the Structures and Phase Transitions of Monolayer Xenon on Single-Crystal Graphite. *Phys. Rev. B* **40**:4797 (1989).

[223] C.M. Mate, G.A. Somorjai, H.W.K. Tom, X.D. Zhu, and Y.R. Shen. Vibrational and Electronic Spectroscopy of Pyridine and Benzene Adsorbed on the Rh(111) Crystal Face. *J. Chem. Phys.* **88**:441 (1988).

[224] K.J. Uram, L. Ng, and J.T. Yates, Jr. Electrostatic Effects between Adsorbed Species—The K · · · CO Interaction on Ni(111) as Studied by Infrared Reflection-Absorbtion Spectroscopy. *Surf. Sci.* **177**:253 (1986).

[225] E.L. Garfunkel, M.H. Farias, and G.A. Somorjai. The Modification of Benzene and Carbon Monoxide Adsorption on Pt(111) by the Coadsorption of Potassium or Sulfur. *J. Am. Chem. Soc.* **107**:349 (1985).

[226] J.E. Crowell, W.T. Tysoe, and G.A. Somorjai. Potassium Coadsorption Induced Dissociation of CO on the Rh(111) Crystal Face: An Isotope Mixing Study. *J. Phys. Chem.* **89**:1598 (1985).

TABLE 2.1. (Abbreviated) and Matrix Notations for a Variety of Superlattices on Low-Miller-Index Crystal Surfaces

Substrate	Superlattice Unit Cell	
	Abbreviated Notation	Matrix Notation
fcc(100), bcc(100)	$p(1 \times 1)$	$\begin{vmatrix} 1 & 0 \\ 0 & 1 \end{vmatrix}$
	$c(2 \times 2) = (\sqrt{2} \times \sqrt{2})R45°$	$\begin{vmatrix} 1 & -1 \\ 1 & 1 \end{vmatrix}$
	$p(2 \times 1)$	$\begin{vmatrix} 2 & 0 \\ 0 & 1 \end{vmatrix}$
	$p(1 \times 2)$	$\begin{vmatrix} 1 & 0 \\ 0 & 2 \end{vmatrix}$
	$p(2 \times 2)$	$\begin{vmatrix} 2 & 0 \\ 0 & 2 \end{vmatrix}$
	$(2\sqrt{2} \times \sqrt{2})R45°$	$\begin{vmatrix} 2 & 2 \\ -1 & 1 \end{vmatrix}$
fcc(111)(60° between basis vectors)	$p(2 \times 1)$	$\begin{vmatrix} 2 & 0 \\ 0 & 1 \end{vmatrix}$

TABLE 2.1. (*Continued*)

| | Superlattice Unit Cell | |
Substrate	Abbreviated Notation	Matrix Notation
	p(2 × 2)	$\begin{vmatrix} 2 & 0 \\ 0 & 2 \end{vmatrix}$
	$(\sqrt{3} \times \sqrt{3})R30°$	$\begin{vmatrix} 1 & 1 \\ -1 & 2 \end{vmatrix}$
fcc(110)	p(2 × 1)	$\begin{vmatrix} 2 & 0 \\ 0 & 1 \end{vmatrix}$
	p(3 × 1)	$\begin{vmatrix} 3 & 0 \\ 0 & 1 \end{vmatrix}$
	c(2 × 2)	$\begin{vmatrix} 1 & -1 \\ 1 & 1 \end{vmatrix}$
bcc(110)	p(2 × 1)	$\begin{vmatrix} 2 & 0 \\ 0 & 1 \end{vmatrix}$

TABLE 2.2. Correspondence Between the Miller-Index Notation and Stepped-Surface Notation

Miller Index	Stepped-Surface Designation
(544)	(S)-[9(111) × (100)]
(755)	(S)-[6(111) × (100)]
(533)	(S)-[4(111) × (100)]
(211)	(S)-[3(111) × (100)]
(311)	(S)-[2(111) × (100)]
	(S)-[2(100) × (111)]
(511)	(S)-[3(100) × (111)]
(711)	(S)-[4(100) × (111)]
(665)	(S)-[12(111) × (111)]
(997)	(S)-[9(111) × (111)]
(332)	(S)-[6(111) × (111)]
(221)	(S)-[4(111) × (111)]
(331)	(S)-[3(111) × (111)]
	(S)-[2(110) × (111)]
(771)	(S)-[4(110) × (111)]
(610)	(S)-[6(100) × (110)]
(410)	(S)-[4(100) × (110)]
(310)	(S)-[3(100) × (110)]
(210)	(S)-[2(100) × (110)]
	(S)-[2(110) × (100)]
(430)	(S)-[4(110) × (100)]
(10,8,7)	(S)-[7(111) × (310)]

TABLE 2.3a. Clean Metal Surface Structures (Unreconstructed)[a]

Substrate Face	Bulk Spacing (Å)	Surface Spacing (Å)	Expansion (%)	Method
Ag(110) fcc	1.44	1.34	−7	LEED (1)
Ag(110) fcc		1.33 ± 0.04	−7.6	HEIS (2)
		1.50 ± 0.04	4.2	
Ag(111) fcc	2.35	2.35 ± 0.1	0	HEIS (3)
Al(100) fcc	2.02	2.025 ± 0.10	0.0	LEED (4)
Al(100) fcc		2.02	0.0	LEED (5)
Al(100) fcc		2.052	1.0	MEED (6)
Al(100) fcc	1.43	1.30	−9.1	LEED (5)
Al(110) fcc		1.304 ± 0.012	−8.8	LEED (7)
		1.499 ± 0.15	4.8	
		1.404 ± 0.017	−1.8	
		1.429 ± 0.018	0.0	
Al(110) fcc		1.310 ± 0.014	−8.4	LEED (8)
		1.510 ± 0.016	5.6	
		1.463 ± 0.019	2.3	
		1.455 ± 0.022	1.7	
Al(111) fcc	2.33	2.350 ± 0.012	0.9	LEED (9)
Al(111) fcc		2.41 ± 0.05	3.1	LEED (10)
Al(311) fcc	1.23	1.68 ± 0.01	−13.0	LEED (11)
		1.335 ± 0.02	8.8	
Al(331) fcc	0.93	0.82 ± 0.02	−11.7	LEED (12)
		0.89 ± 0.03	−4.1	
		1.05 ± 0.03	10.3	
		0.88 ± 0.03	−4.8	
First-layer registry shifts of ~0.06 ± 0.08 found				
Au(100) fcc (metastable)	2.04	2.04	0.0	LEED (13)
Cd(100) fcc	2.81	2.81	0	LEED (14)
Co(100) fcc	1.77	1.70	−4.0	LEED (15)
Co(111) fcc	2.05	2.05 ± 0.05	0.0	LEED (16)
Co(0001) hcp	2.05	2.05 ± 0.05	0.0	LEED (16)
Co(11$\bar{2}$0) hcp	1.25	1.14 ± 0.04	−8.8	LEED (17)
Cu(100) fcc	1.81	1.785	−1.1	LEED (18)
		1.836	1.7	
		1.832	1.5	
Cu(110) fcc	1.28	1.159	−9.2	LEED (18)
		1.305	2.3	
Cu(110) fcc		1.21 ± 0.02	−5.3	MEIS (19)
		1.32 ± 0.02	3.3	
Cu(110) fcc		1.170 ± 0.008	−8.5	LEED (20)
		1.307 ± 0.010	2.3	
Cu(111) fcc	2.09	2.076 ± 0.02	−0.7	LEED (21)
Cu(311) fcc	1.09	1.035 ± 0.02	−5.0	LEED (22)
Fe(100) bcc	1.43	1.41 ± 0.04	−1.6	LEED (23)
Fe(110) bcc	2.02	2.04 ± 0.04	0.5	LEED (24)
Fe(111) bcc	0.83	0.70 ± 0.03	−15.4	LEED (25)
Fe(111) bcc		0.69 ± 0.025	−16.6	LEED (26)
		0.75 ± 0.025	−9.3	
		0.86 ± 0.03	4.0	
		0.81 ± 0.025	−2.1	

TABLE 2.3a. (*Continued*)

Substrate Face	Bulk Spacing (Å)	Surface Spacing (Å)	Expansion (%)	Method
Fe(210)[b] bcc	0.64	0.50 ± 0.03	−21.9	LEED (27)
		0.57 ± 0.03	−10.9	
		0.61 ± 0.03	−4.7	
		0.64 ± 0.03	0.0	
Fe(211)[b] bcc	1.17	1.05 ± 0.03	−10.3	LEED (28)
		1.23 ± 0.03	5.1	
		1.15 ± 0.04	−1.7	
Fe(310) bcc	0.906	0.76 ± 0.03	−16.1	LEED (29)
		1.02 ± 0.03	12.6	
		0.87 ± 0.04	−4.0	
Ir(100) fcc (metastable)	1.92	1.85 ± 0.01	−3.6	LEED (30)
Ir(110) fcc	1.36	1.26 ± 0.10	−7.4	LEED (31)
Ir(111) fcc	2.22	2.16 ± 0.10	−2.6	LEED (32)
Na(110) bcc	3.03	3.03	0.0	LEED (33)
Na(110) bcc		3.0 ± 0.01	−1.0	LEED (34)
Na(0001) hcp	2.87	2.87	0.0	LEED (35)
Ni(100) fcc	2.22	1.78 ± 0.02	1.1	LEED (36)
Ni(100) fcc		1.604 ± 0.008	−8.9	MEIS (37)
Ni(110) fcc	1.25	1.195 ± 0.01	−4.0	MEIS (38)
Ni(110) fcc		1.18 ± 0.02	−4.8	HEIS (39)
		1.27 ± 0.02	2.4	
Ni(110) fcc		1.14 ± 0.01	−8.1	LEED (40)
		1.28 ± 0.01		
Ni(110) fcc		1.123 ± 0.02	−9.8	LEED (41)
		1.292 ± 0.02	3.8	
Ni(110) fcc		1.138 ± 0.006	−8.6	LEED (42)
		1.284 ± 0.007	3.1	
		1.240 ± 0.009	−0.4	
Ni(110) fcc		1.121 ± 0.012	−9.0	MEIS (43)
		1.290 ± 0.019	3.5	
Ni(111) fcc	2.03	2.005 ± 0.025	−1.2	LEED (36)
Ni(111) fcc		2.033 ± 0.020	0.0	HEIS (44)
Ni(311) fcc	1.06	0.894 ± 0.010	−15.9	LEED (45)
		1.106 ± 0.016	4.1	
		1.045 ± 0.017	−1.6	
Mo(100) bcc	1.25	1.424 ± 0.03	−9.5	LEED (46)
Mo(110) bcc	2.22	2.19 ± 0.04	−1.6	LEED (47)
Pd(100) fcc	1.94	1.95 ± 0.05	0.3	LEED (48)
Pd(110) fcc	1.37	1.29 ± 0.03	−5.8	LEED (49)
		1.38 ± 0.03	0.7	
Pd(111) fcc	2.246	2.25 ± 0.05	0.0	HEIS (50)
Pd(111) fcc		2.276	1.3	LEED (51)
		2.216	−1.3	
		2.296	2.2	
		2.296	2.2	
Pt(100) fcc (metastable)	1.96	1.96	0.2	SPLEED (52)

TABLE 2.3a. (*Continued*)

Substrate Face	Bulk Spacing (Å)	Surface Spacing (Å)	Expansion (%)	Method
Pt(100) fcc (metastable)		1.963	0.2	HEIS (53)
Pt(100) fcc (metastable)		1.96	0.0	LEED (13)
Pt(111) fcc	2.26	2.29 ± 0.10	1.1	LEED (54)
Pt(111) fcc		2.30 ± 0.02	1.4	HEIS (55)
Pt(111) fcc		2.276 ± 0.02	0.5	SPLEED (56)
Pt(111) fcc		2.265 ± 0.05	0.0	LEED (57)
Re(10$\bar{1}$0) hcp		0.67	−16.3	LEED (58)
Rh(100) fcc	1.90	1.91 ± 0.02	0.5	LEED (59)
Rh(110) fcc	1.34	1.33 ± 0.02	−0.7	LEED (59)
Rh(111) fcc	2.19	2.16 ± 0.02	−1.4	LEED (59)
Rh(111) fcc		2.192 ± 0.10	0.0	LEED (60)
Ru(0001) hcp	2.14	2.10 ± 0.02	−1.9	LEED (61)
Sc(0001) hcp	2.64	2.59 ± 0.02	−1.9	LEED (62)
Ta(100) bcc	1.65	1.47 ± 0.03	−10.9	LEED (63)
		1.67 ± 0.03	1.2	
Te(10$\bar{1}$0) hcp		0.34 ± 0.10		LEED (64)
Ti(0001) hcp	2.34	2.29 ± 0.05	−2.1	LEED (65)
V(100) bcc	1.51	1.41 ± 0.01	−6.6	LEED (66)
V(110) bcc	2.14	2.13 ± 0.10	−0.5	LEED (67, 68)
W(100) fcc	1.58	1.46 ± 0.03	−7.6	LEED (69)
W(110) fcc	2.23	2.23 ± 0.10	0.0	LEED (70)
Zn(0001) hcp	2.44	2.39 ± 0.05	−2.0	LEED (71)
Zr(0001) hcp	2.57	2.54 ± 0.05	−1.2	LEED (72)

[a] Where multiple-layer spacing changes have been investigated, these are listed in the table on successive lines.
[b] There are relaxations in the layer registries for the stepped iron (211) and (210) surfaces in addition to layer spacing relaxations.

REFERENCES

1. E. Zanazzi, F. Jona, D.W. Jepsen, and P.M. Marcus. *J. Phys. C* **10**:375 (1977).

2. Y. Kuk and L.C. Feldman. *Phys. Rev. B* **30**:5811 (1984).

3. R.J. Culbertson, L.C. Feldman, P.J. Silverman, and H. Boehm. *Phys. Rev. Lett.* **47**:657 (1981).

4. M.A. Van Hove, S.Y. Tong, and N. Stoner. *Surf. Sci.* **54**:259 (1976).

5. D. Aberdam, R. Baudoing, and C. Ganbert. *Surf. Sci.* **62**:567 (1977).

6. N. Masud, R. Baudoing, D. Aberdam, and C. Gaubert. *Surf. Sci.* **133**:580 (1983).

7. J.N. Andersen, H.B. Nielsen, L. Petersen, and D.L. Adams. *J. Phys. C* **17**:173 (1984).

8. J.R. Noonan and H.L. Davis. *Phys. Rev. B* **29**:4349 (1984).

9. H.B. Nielsen, and D.L. Adams. *J. Phys. C* **15**:615 (1982).

10. V. Martinez, F. Soria, M.C. Munoz, and J.L. Sacedon. *Surf. Sci.* **128**:424 (1983).

11. J.R. Noonan, H.L. Davis, and W. Erley. *Surf. Sci.* **152/153**:142 (1985).

12. D.L. Adams and C.S. Sorensen. *Surf. Sci.* **166**:495 (1986).

13. E. Lang, W. Grimm, and K. Heinz. *Surf. Sci.* **177**:169 (1982).

14. H.D. Shih, F. Jona, D.W. Jepsen, and P.M. Marcus. *Phys. Rev. B* **15**:5561 (1977).

15. M. Maglietta, E. Zanazzi, F. Jona, D.W. Jepsen, and P.M. Marcus. *Appl. Phys.* **15**:409 (1978).

16. B.W. Lee, R. Alsenz, A. Ignatiev, and M.A. Van Hove. *Phys. Rev. B* **17**:1510 (1978).

17. M. Weltz, W. Moritz, and D. Wolf. *Surf. Sci.* **125**:473 (1983).

18. H.L. Davis and J.R. Noonan. *Surf. Sci.* **126**:245 (1983).

19. I. Stensgaard, R. Feidenhans'l and J.E. Sørensen. *Surf. Sci.* **128**:281 (1983).

20. D.L. Adams, H.B. Nielsen, and J.N. Andersen. *Surf. Sci.* **128**:294 (1983).

21. S.A. Lindgren, L. Walldén, J. Rundgren, and P. Westrin. *Phys. Rev. B* **29**:576 (1984).

22. R.W. Streater, W.T. Moore, P.R. Watson, D.C. Frost, and K.A.R. Mitchell. *Surf. Sci.* **72**:744 (1978).

23. K.O. Legg, F. Jona, D.W. Jepsen, and P.M. Marcus. *J. Phys. C* **10**:937 (1977).

24. H.D. Shih, F. Jona, and P.M. Marcus. *J. Phys. C* **13**:3801 (1980).

25. F.E. Shih, F. Jona, D.W. Jepsen, and P.M. Marcus. *Phys. Rev. Lett.* **46**:731 (1981).

26. J. Sokolov, F. Jona, and P.M. Marcus. *Phys. Rev. B* **33**:1397 (1986).

27. J. Sokolov, F. Jona, and P.M. Marcus. *Phys. Rev. B* **31**:1929 (1985).

28. J. Sokolov, H.D. Shih, U. Bardi, and F. Jona. *J. Phys. C* **17**:371 (1984).

29. J. Sokolov, F. Jona, and P.M. Marcus. *Phys. Rev. B* **29**:5402 (1984).

30. K. Heinz and G. Besold. *Surf. Sci.* **125**:515 (1983).

31. C.M. Chan, S.M. Cunningham, K.L. Luke, W.H. Weinberg, and S.P. Withrow. *Surf. Sci.* **78**:15 (1978).

32. C.M. Chan, S.L. Cunningham, M.A. Van Hove, W.H. Weinberg, and S.P. Withrow. *Surf. Sci.* **66**:394 (1977).

33. P.M. Echenique. *J. Phys. C* **9**:3193 (1976).

34. S. Andersson, J.B. Pendry, and P.M. Echenique. *Surf. Sci.* **65**:539 (1977).

35. S.A. Lindgren, J. Paul, L. Walldén, and P. Westrin. *J. Phys. C* **15**:6285 (1982).

36. J.E. Demuth, P.M. Marcus, and D.W. Jepsen. *Phys. Rev. B* **11**:1460 (1975).

37. J.W.M. Frenken, J.F. van der Veen, and G. Allan. *Phys. Rev. Lett.* **51**:1876 (1983).

38. J.F. van der Veen, R.M. Tromp, R.G. Smeenk, and F.W. Saris. *Surf. Sci.* **82**:468 (1979).

39. R. Feidenhans'l, J.E. Sørensen, and I. Stensgaard. *Surf. Sci.* **134**:329 (1983).

40. Y. Gauthier, R. Baudoing, Y. Joly, and C. Gaubert. *J. Phys. C* **17**:4547 (1984).

41. M.L. Xu and S.Y. Tong. *Phys. Rev. B* **31**:6332 (1985).

42. D.L. Adams, L.E. Peterson, and C.S. Sorenson. *J. Phys. C* **18**:1753 (1985).

43. S.M. Yalisove, W.R. Graham, E.D. Adams, M. Copel, and T. Gustafsson. *Surf. Sci.* **171**:400 (1986).

44. T. Narusawa, W.M. Gibson, and E. Törnquist. *Phys. Rev. Lett.* **47**:417 (1981).

45. D.L. Adams, W.T. Moore, and K.A.R. Mitchell. *Surf. Sci.* **149**:407 (1985).

46. L.J. Clarke. *Surf. Sci.* **91**:131 (1980).

47. L. Morales, D.O. Garza, and L.J. Clarke. *J. Phys. C* **14**:5391 (1981).

48. R.J. Behm, K. Christmann, G. Ertl, and M.A. Van Hove. *J. Chem. Phys.* **73**:2984 (1980).

49. C.J. Barnes, M.Q. Ding, M. Lindroos, R.D. Diehl, and D.A. King. *Surf. Sci.* **162**:59 (1985).

50. Y. Kuk, L.C. Feldman, and P.J. Silverman. *Phys. Rev. Lett.* **50**:511 (1983).

51. H. Ohtani, M.A. Van Hove, and G.A. Somorjai. *Surf. Sci.* **187**:372 (1987). To be published.

52. R. Feder. *Surf. Sci.* **68**:229 (1977).

53. J.A. Davies, T.E. Jackman, D.P. Jackson, and P.R. Norton. *Surf. Sci.* **109**:20 (1981).

54. D.L. Adams, H.B. Nielsen, and M.A. Van Hove. *Phys. Rev. B* **20**:4789 (1979).

55. J.F. van der Veen, R.G. Smeenk, R.M. Tromp, and F.W. Saris. *Surf. Sci.* **79**:219 (1979).

56. R. Feder, H. Pleyer, P. Bauer, and N. Müller. *Surf. Sci.* **109**:419 (1981).

57. K. Hayek, H. Glassl, A. Gutmann, H. Leonhard, M. Prutton, S.P. Tear, and M.R. Welton-Cook. *Surf. Sci.* **152**:419 (1985).

58. H.L. Davis and D.M. Zehner. *J. Vacuum Sci. Technol.* **17**:190 (1980).

59. S. Hengrasmee, K.A.R. Mitchell, P.R. Watson, and S.J. White. *Can. J. Phys.* **58**:200 (1980).

60. M.A. Van Hove, and R.J. Koestner. In: P.M. Marcus and F. Jona, editors, *Surface Structure by LEED*. Plenum, New York, 1984.
61. G. Michalk, W. Moritz, H. Pfnür, and D. Menzel. *Surf. Sci.* **129**:92 (1983).
62. S. Tougaard, A. Ignatiev, and D.L. Adams. *Surf. Sci.* **115**:270 (1982).
63. A. Titov and W. Moritz. *Surf. Sci.* **123**:L709 (1982).
64. R.J. Meyer, W.R. Salaneck, C.B. Duke, A. Paton, C.H. Griffiths, L. Kovnat, and L.E. Meyer. *Phys. Rev. B* **21**: 4542 (1980).
65. H.D. Shih, F. Jona, D.W. Jepsen, and P.M. Marcus. *J. Phys. C* **9**:1405 (1976).
66. V. Jensen, J.N. Andersen, H.B. Nielsen, and D.L. Adams. *Surf. Sci.* **116**:66 (1982).
67. D.L. Adams and H.B. Nielsen. *Surf. Sci.* **107**:305 (1981).
68. D.L. Adams and H.B. Nielsen. *Surf. Sci.* **116**:598 (1982).
69. M.K. Debe and D.A. King. *J. Phys. C* **15**:2257 (1982).
70. M.A. Van Hove and S.Y. Tong. *Surf. Sci.* **54**:91 (1976).
71. W.N. Unertl and H.V. Thapliyal. *J. Vacuum Sci. Technology.* **12**:263 (1975).
72. W.T. Moore, P.R. Watson, D.C. Frost, and K.A.R. Mitchell. *J. Phys. C* **12**:L887 (1979).

TABLE 2.3b. Reconstructed Clean Metals and Alloy Surface Structures

Substrate and Unit Cell	Structure	Method
Au(110)(fcc) (2 × 1)	Missing-row reconstruction confirmed by X-ray diffraction. Lateral displacement of second-layer rows toward missing row position by 0.12 ± 0.02 Å. Antiphase domains due to monatomic steps observed. Evidence for top-layer expansion or contraction of ~0.25 Å.	X-ray diffraction (1)
Au(110)(fcc) (2 × 1)	Missing-row reconstruction. Every other top layer (1̄10) row is missing. The second-layer lateral displacements are ~0.07 Å, and the second row is buckled by −0.24 Å. The first-layer spacing is −20.1%, and the second- and third-layer spacing −6.3% relative to the 1.44-Å bulk-layer spacing.	LEED (2)
Au(110)(fcc) (2 × 1)	Missing-row reconstruction studied by channeling and blocking, top-layer spacing contracted by 18%, second-layer spacing expanded by 4% from bulk value of 1.44 Å. Third layer buckled by ±0.10 Å or ±7%, with atoms under missing row moving up, alternate rows down.	MEIS (3)
Au(110)(fcc) (2 × 1)	Missing-row reconstruction studied with top-layer spacing contracted by 0.15 ± 0.10 Å (−11%) from bulk value of 1.44 Å.	LEIS (4)

TABLE 2.3b. (*Continued*)

Substrate and Unit Cell	Structure	Method
α-CuAl(111) (fcc) ($\sqrt{3} \times \sqrt{3}$)$R30°$	Al substituted in one-third of top-layer Cu sites, second layer pure Cu, no buckling in top layer, layer spacing is 2.05 ± 0.05 Å, the same as bulk copper. Alloy composition 16% Al atoms in Cu.	LEED (5)
Ir(100)(fcc) (5 × 1)	The top layer of the surface reconstructs to form a compact hexagonal surface, with six-fifths the density of unreconstructed surface. The layer spacing expands by 14.6 ± 5.2% from the bulk value of 1.92 Å. Some top-layer atoms are buckled outward by up to an additional 0.2 ± 0.02 Å so the hexagonal layer can fit the square layer below. This is one-half to two-thirds of the buckling required to have top layer atoms in hard-sphere contact with all substrate atoms.	LEED (6)
Ir(100)(fcc) (5 × 1)	The top layer of the surface reconstructs to form a compact hexagonal surface, with six-fifths the density of unreconstructed surface. The layer spacing expands by 7.3 ± 2.6% from the bulk value of 1.92 Å. Some top-layer atoms are buckled outward by up to an additional 0.48 ± 0.02 Å so the hexagonal layer can fit the square layer below. This puts the top-layer atoms in hard-sphere contact with all substrate atoms.	LEED (7)
Ir(110)(fcc) (2 × 1)	Missing-row reconstruction similar to gold (110). One top-layer row is missing, the second-layer lateral displacements are ~0.04 Å, and the second row is buckled by −0.23 Å. The first-layer spacing is −12.4%, and the second- and third-layer spacing −5.8% relative to the 1.36-Å bulk-layer spacing.	LEED (8)
NiAl(110) (CsCl)(1 × 1)	Top-layer spacing 1.92 Å, contracted 6% from bulk value of 2.04 Å. Al atoms buckled out by 0.22 Å.	LEED (9)
Ni_3Al (Cu_3 Au)(100)	Top layer is 50–50 nickel and aluminum, second layer nickel, etc. Top-layer spacing is 1.73 ± 0.03 Å with Al atoms buckled outward by	LEED (10)

TABLE 2.3b. (*Continued*)

Substrate and Unit Cell	Structure	Method
	0.02 $\pm$ 0.03 Å, second-layer spacing is the bulk value of 1.78 Å within $\pm$0.03 Å.	
Pd(110)(fcc) (2 × 1)	Missing-row model with third layer assumed bulk-like. Second layer found to be bulk-like, first-layer spacing −1% relative to 1.37-Å bulk spacing. Saw-tooth model almost as good.	LEED (11)
Pt(100)(fcc)	Top-layer reconstruction, quasi-hexagonal surface given by $\left(\begin{smallmatrix}14&1\\-1&15\end{smallmatrix}\right)$ also called "1 × 5" or "4 × 20" reconstruction. Degree of top-layer buckling and top-layer registry proposed.	LEED (6)
Pt(110)(fcc) (2 × 1)	Missing-row model with the top layer expanded by 23% to 1.A Å from bulk value of 1.387 Å. Lateral shift of second-layer rows of 0.05 Å.	LEED (12)
Pt(110)(fcc) (2 × 1)	Missing-row model is better than buckled-row, paired-row, or saw-tooth models. Terminated-bulk positions assumed for remaining atoms.	LEIS (13)
Pt_xNi_{1-x}(110) (fcc)	LEED study of random substitutional alloy shows Pt enhancement in first and third layers, along with a Pt depletion in the second layer.	LEED (14)
W(100)(bcc) (2 × 1)	Below room temperature the "zig-zag" reconstruction occurs on the W(100) surface. Alternate atoms move along the (011) and $(0\bar{1}\bar{1})$ directions by $\pm$0.22 $\pm$ 0.07 Å, while the top layer spacing contracts by 0.05 $\pm$ 0.05 Å from the bulk value of 1.58 Å. There are two domains, since atoms may also move along the $(0\bar{1}1)$ and $(01\bar{1})$ directions.	LEED (15)
W(100)(bcc) ($\sqrt{2}$ × $\sqrt{2}$)R45°	Below room temperature the "zig-zag" reconstruction occurs on the W(100) surface. Alternate atoms move along the (011) and $(0\bar{1}\bar{1})$ directions by $\pm$0.16 Å, while the top-layer spacing contracts by 0.09 Å (6%) from the bulk value of 1.58 Å.	LEED (16)

REFERENCES

1. I.K. Robinson, Y. Kuk, and L.C. Feldman. *Phys. Rev.* **B29**:4762 (1984).
2. W. Moritz and D. Wolf. *Surf. Sci.* **163**:L655 (1985).
3. M. Copel and T. Gustafsson. *Phys. Rev. Lett.* **57**:723 (1986).
4. J. Möller, H. Niehaus, and W. Heiland. *Surf. Sci.* **166**:L111 (1986).
5. R.J. Baird, D.F. Ogletree, M.A. Van Hove, and G.A. Somorjai. *Surf. Sci.* **165**:345 (1986).
6. M.A. Van Hove, R.J. Koestner, P.C. Stair, J.P. Bibérian, L.L. Kesmodel, I. Bartoš, and G.A. Somorjai. *Surf. Sci.* **103**:218 (1981).
7. E. Lang, K. Müller, K. Heinz, M.A. Van Hove, R.J. Koestner, and G.A. Somorjai. *Surf. Sci.* **127**:347 (1983).
8. C.-M. Chan and M.A. Van Hove. *Surf. Sci.* **171**:226 (1986). (to be published).
9. H.L. Davis and J.R. Noonan. *Phys. Rev. Lett.* **54**:566 (1985).
10. D. Sondericker, F. Jona, and P.M. Marcus. *Phys. Rev. B* **33**:900 (1986).
11. C.J. Barnes, M.Q. Ding, M. Lindroos, R.D. Diehl, and D.A. King. *Surf. Sci.* **162**:59 (1985).
12. D.L. Adams, H.B. Nielsen, M.A. Van Hove, and A. Ignatiev. *Surf. Sci.* **104**:47 (1981).
13. H. Niehaus and G. Comsa. *Surf. Sci.* **140**:18 (1984).
14. Y. Gauthier, R. Baudoing, Y. Joly, J. Rundgren, J.C. Bertolini, and J. Massardier. *Surf. Sci.* **162**:348 (1985).
15. R.A. Barker, P.J. Estrup, F. Jona, and P.M. Marcus. *Solid State Commun.* **25**:375 (1978).
16. J.A. Walker, M.K. Debe, and D.A. King. *Surf. Sci.* **104**:405 (1981).

TABLE 2.3c. Semiconductor Surface Structures

Substrate	Unit Cell	Structure	Method
AlP(110) (zincblende)	(1 × 1)	Surface reconstructed with 25° bond angle rotation preserving bond lengths. Top layer buckled with P +0.06 Å and Al −0.57 Å, and layer spacing contracted to 1.33 Å (bulk 1.927 Å). Second layer buckled with P −0.035 Å and Al +0.035 Å, and layer expanded to 1.92 Å. First layer in-plane displacements of ~0.25 Å to preserve bond lengths.	LEED (1)
CdTe(110) (zincblende)	(1 × 1)	Te is buckled out from the top layer by 0.82 ± 0.05 Å. Top-layer Cd atoms contract by ~0.5 Å, and there are lateral motions of ~0.4 Å to conserve bond lengths.	LEED (2)
GaAs(110) (zincblende)	(1 × 1)	As is buckled out from the top layer by 0.70 Å. Top-layer Ga atoms contract by ~0.5 Å, and there are lateral motions of ~0.4 Å to conserve bond lengths. The lateral motions may be reduced	LEED (3)

TABLE 2.3c. (*Continued*)

Substrate	Unit Cell	Structure	Method
		by approximately three-quarters to give a better agreement with MEIS results without significantly worsening the LEED fit.	
GaAs(110) (zincblende)	(1 × 1)	As is buckled out from the top layer by 0.69 Å. Top-layer Ga atoms contract by ~0.5 Å, and there are lateral motions of ~0.3 Å to conserve bond lengths.	LEED (4)
GaAs(110) (zincblende)	(1 × 1)	As is buckled out from the top layer by 0.40 ± 0.30 Å. Top-layer Ga atoms contract by ~0.30 ± 0.30 Å, with no lateral motions.	HEIS (5)
GaAs(111) (zincblende) (Ga terminated)	(2 × 2)	One-quarter of the top-layer Ga atoms are missing, and the remaining atoms are almost coplanar with the first As layer, within 0.20 Å. Ga bonding is sp^2 rehybridized, instead of the normal sp^3 configuration. There are first bilayer lateral motions of ~0.2 Å, and some buckling in the third layer to maintain optimum bond lengths and angles.	LEED (6)
GaP(110) (zincblende)	(1 × 1)	Surface reconstructed with 27.5° bond angle rotation preserving bond lengths. Top layer buckled with Ga +0.09 Å and As −0.54 Å, and layer spacing contracted to 1.39 Å. second-layer spacing expanded to 1.93 Å.	LEED (7)
GaP(111) (zincblende) (Ga terminated)	(2 × 2)	One-quarter of the top-layer Ga atoms are missing, and the remaining atoms are almost coplanar with the first P layer. Ga bonding is sp^2 rehybridized, instead of the normal sp^3 configuration. There are lateral motions up to ~2 Å, and some buckling in the third layer to maintain optimum bond lengths and angles.	LEED (6)

TABLE 2.3c. (*Continued*)

Substrate	Unit Cell	Structure	Method
GaSb(110) (zincblende)	(1 × 1)	Sb is buckled out from the top layer by 0.77 ± 0.05 Å. Top-layer Ga atoms contract by ~0.5 Å, and there are lateral motions of ~0.4 Å to conserve bond lengths.	LEED (8, 9)
GaSb(110) (zincblende)	(1 × 1)	Consistent with LEED results (above) for layer displacements and lateral motions.	MEIS (10)
GaSb(110) (zincblende)	(1 × 1)	Bond-length-conserving rotations of 28.5 ± 2.6° bring Sb atoms up, Ga atoms down.	MEIS (11)
Ge(100) (diamond)	(2 × 1)	The top layer buckles, with one Ge atom moving out by 0.62 ± 0.04 Å and the other moving in by 0.66 ± 0.04 Å. There are lateral displacements of ~0.9 Å in the first layer and ~0.1 Å in the second layer.	XRD (12)
InAs(110) (zincblende)	(1 × 1)	The top layer has As buckled outward by ~0.8 Å with lateral motions of ~0.6 Å in the first three layers to conserve bond lengths.	LEED (13)
InAs(110) (zincblende)	(1 × 1)	Bond-length-conserving rotations of 30.0 ± 2.4° bring As atoms up, In atoms down.	MEIS (11)
InSb(111) (zincblende)	(2 × 2)	One-quarter of the top-layer In atoms are missing in the "vacancy buckling" model. Top layer is not buckled, in-plane Sb atoms expand radially by 0.45 ± 0.04 Å from their normal positions, and In atoms contract radially by 0.23 ± 0.05 Å from their normal positions. No data on layer spacing changes.	XRD (14)
InP(110) (zincblende)	(1 × 1)	Preliminary results show layer buckling with the P atom buckled out, and lateral and vertical shifts in the top layer ~0.4 Å. Second and deeper layer shifts were not investigated.	LEED (15)
InP(110) (zincblende)	(1 × 1)	Top layer buckling with the P atom buckled out by 0.69 ± 0.10 Å, and lateral and verti-	LEED (16)

TABLE 2.3c. (*Continued*)

Substrate	Unit Cell	Structure	Method
		cal shifts in the top layer ~0.4 Å. The second-layer spacing is contracted by 0.41 ± 0.10 Å with a slight buckling of 0.07 ± 0.10 Å. Only first-layer lateral displacements were investigated.	
Si(100)(diamond)	(2 × 1)	Buckled dimer model is best fit to data. Surface atoms dimerize to take up dangling bonds; surface buckles for best bond angles. There are lateral and vertical ion-core motions for at least four layers into the bulk crystal. Authors conclude that (2 × 1), (2 × 2), and c(4 × 2) buckled dimer domains may exist on the surface.	MEIS (17)
Si(100)(diamond)	(2 × 1)	LEED analysis considering vertical and lateral displacements in the top three atomic layers supports a buckled dimer model.	LEED (18)
Si(111)(diamond)	(1 × 1) laser annealed	The first two layers are almost coplanar, instead of the normal 0.78-Å separation. The first-layer spacing is 0.08 ± 0.02, a contraction of 90%; and the second-layer spacing is 2.95 ± 0.20 Å, a 25.5% expansion from 2.35 Å.	LEED (19)
Si(111)(diamond)	(2 × 1)	The top layer is buckled by 0.30 ± 0.05 Å, the second-layer spacing is 0.70 ± 0.05 (a change of +2.9%), and the third-layer spacing is contracted by 3.4% to 2.27 ± 0.02 from 2.35 Å. There are second-layer lateral shifts of ~0.12 Å. (More recent results favor the "π-bonded chain" model; see below.)	LEED (20)
Si(111)(diamond)	(2 × 1)	Analysis of LEED data supports a "π-bonded chain" model for the (2 × 1) reconstruction. Trial geometries are based on strain-minimization calculations for the model, involving vertical motions four	LEED (21)

TABLE 2.3c. (*Continued*)

Substrate	Unit Cell	Structure	Method
		layers deep with lateral motions along the long side of the unit cell.	
Si(111)(diamond)	(2 × 1)	Best fit to "π-bonded chain" model. This model involves buckling in layers 2 to 6 of up to 0.27 Å and small lateral shifts in the first six layers.	MEIS (22)
GaP(110) (zincblende)	(1 × 1)	Surface reconstructed with 28.0° bond angle rotation preserving bond lengths. Top layer buckled with Zn +0.08 Å and S −0.51 Å, and layer spacing is contracted to 1.40 Å. Second-layer spacing is expanded to 1.91 Å.	LEED (7)
ZnSe(110) (zincblende)	(1 × 1)	Two different models were consistent with LEED data. First, Se is buckled outward by ~0.70 Å and the second-layer spacing is contracted by ~0.60 Å, with lateral motions of ~0.7 Å. Second, Se buckled out by 0.10 Å and the second-layer spacing contracted by 0.09 Å, with lateral motions of ~0.07 Å.	LEED (23)
ZnTe(110) (zincblende)	(1 × 1)	Te is buckled out from the top layer by 0.71 ± 0.05 Å. Top-layer Zn atoms contract by ~0.5 Å, and there are lateral motions of ~0.4 Å to conserve bond lengths.	LEED (9)

REFERENCES

1. C.B. Duke, A. Paton, and C.R. Bonpace. *Phys. Rev.* **B28**:852 (1983).
2. C.B. Duke, A. Paton, W.K. Ford, A. Kahn, and G. Scott. *J. Vacuum Sci. Technol.* **20**:778 (1982).
3. C.B. Duke and A. Paton. *J. Vacuum Sci. Technol. B* **2**:327 (1984).
4. S.Y. Tong, W.M. Mei, and G. Xu. *J. Vacuum Sci. Technol. B* **2**:393 (1984).
5. H.J. Grossman and W.M. Gibson. *J. Vacuum Sci. Technol. B* **2**:343 (1984).
6. G. Xu, W.Y. Hu, M.W. Puga, S.Y. Tong, J.L. Yeh, S.R. Wang, and B.W. Lee. *Phys. Rev. B* **32**:8473 (1985).
7. C.B. Duke, A. Paton, and A. Kahn. *J. Vacuum Sci. Technol.* **A2**:515 (1984).
8. C.B. Duke, A. Paton, and A. Kahn. *Phys. Rev. B* **27**:3436 (1983).
9. C.B. Duke, A. Paton, and A. Kahn. *J. Vacuum Sci. Technol.* **A1**:672 (1983).
10. L. Smit, R.M. Tromp, and J.F. van der Veen. *Phys. Rev. B* **29**:4814 (1984).
11. L. Smit and J.F. van der Veen. *Surf. Sci.* **166**:183 (1986).
12. P. Eisenberger and W.C. Marra. *Phys. Rev. Lett.* **46**:1081 (1981).

13. C.B. Duke, A. Paton, A. Kahn, and C.B. Bonapace. *Phys. Rev. B* 27:6189 (1983).
14. J. Bohr, R. Feidenhans'l, M. Nielsen, M. Toney, R.L. Johnson, and I.K. Robinson. *Phys. Rev. Lett.* 54:1275 (1985).
15. S.P. Tear, M.R. Welton-Cook, M. Prutton, and J.A. Walker. *Surf. Sci.* 99:598 (1980).
16. R.J. Meyer, C.B. Duke, A. Paton, J.C. Tsang, J.L. Yeh, A. Kahn, and P. Mark. *Phys. Rev. B* 22:6171 (1980).
17. R.M. Tromp, R.G.Smeenk, F.W. Saris, and D.J. Chadi. *Surf. Sci.* 133:137 (1983).
18. B.W. Holland, C.D. Duke, and A. Paton. *Surf. Sci.* 140:L269 (1984).
19. G.J.R. Jones and B.W. Holland. *Solid State Commun.* 53:45 (1985).
20. R. Feder, W. Mönch, and P.P. Auer. *J. Phys. C* 12:L179 (1979).
21. FJ. Himpsel, P.M. Marcus, R. Tromp, I.P. Batra, M.R. Cook, F. Jona, and H. Liu. *Phys. Rev. B* 30:2257 (1984).
22. L. Smit, R.M. Tromp, and J.F. van der Veen. *Surf. Sci.* 163:315 (1985).
23. C.B. Duke, A. Paton, A. Kahn, and D.W. Tu. *J. Vacuum Sci. Technol. B* 2:366 (1984).

TABLE 2.3d. Insulator and Other Compound Surface Structures

Substrate	Unit Cell	Structure	Method
C(111) (diamond)	(1×1)	Terminated bulk diamond; no relaxation in layer spacing.	LEED (1)
C(111)–H (diamond)	(1×1)	Hydrogen-terminated diamond; surface layer spacing relaxed by -0.05 ± 0.05 Å.	MEIS (2)
C(111)–H (diamond)	(1×1)	Hydrogen in (1×1) arrangement. Determined to be in top site, assumed 1.09-Å H–C bond length.	Helium diffraction (3)
C(111) (diamond)	(2×2)	Evidence for "π-bonded chain reconstruction.	MEIS (2)
C(0001) (graphite)	(1×1)	Normal graphite layer stacking; first layer contracted 1.4% to 3.30 Å from bulk spacing of 3.35 Å.	LEED (4)
C(0001)–K (graphite)	Intercalated	When K is adsorbed on C(0001), it is intercalated between layers, changing the carbon stacking sequence from ABAB... to AAAA... and increasing the layer spacing to 5.35 Å from 3.35 Å.	LEED (5)
C(0001)–Ar (graphite)	Incomensurate	Ar forms an incomensurate hexagonal overlayer on graphite 3.2 ± 0.10 Å above the graphite surface.	LEED (6)

TABLE 2.3d. (*Continued*)

Substrate	Unit Cell	Structure	Method
C(0001)–Kr (graphite)	$(\sqrt{3} \times \sqrt{3})R30°$	Kr adsorbs in six-fold hollow sites on the graphite basal plane 3.35 ± 0.01 Å above the graphite surface. No nearest-neighbor sites are occupied. At higher coverages, next-nearest-neighbor sites are occupied.	LEED (7)
CaO(100) (rocksalt)	(1×1)	Top layer contracts by 1.2% to 2.38 from 2.41 Å. No top-layer buckling.	LEED (8)
CoO(111) (rocksalt)	(1×1)	Oxygen termination with fcc stacking; top-layer contraction of 17% to 1.06 Å from bulk 1.27 Å.	LEED (9)
CoO(100) (rocksalt)	(1×1)	Terminated bulk structure; top-layer spacing is 2.85 ± 0.08 Å.	LEED (10)
MgO(100) (rocksalt)	(1×1)	Top-layer oxygen is buckled out by 0.04 ± 0.05, and top layer is contracted by 0.02 ± 0.07 Å. Bulk-layer spacing is 2.10 Å.	LEED (11)
MgO(100) (rocksalt)	(1×1)	Terminated bulk structure; no change from bulk layer spacing of 2.10 Å.	LEED (12)
MoS_2(0001) (layer compound)	(1×1)	Normal stacking; S–Mo–S termination; top-layer contraction by 5% to 1.51 Å from bulk 1.59 Å. No second layer contraction.	LEED (13)
Na_2O(111) (fluorite)	(1×1)	Oxidation of epitaxial Na(110) on Ni(100) substrate. Determine fluorite lattice with Na–O–Na termination.	LEED (14)
$NbSe_2$(0001) (layer compound)	(1×1)	Normal layer stacking with Se–Nb–Se termination; no evidence for relaxation of first two layer spacings.	LEED (13)
NiI_2(0001)	(1×1)	Bulk NiI_2 is a layer compound with hexagonal metal layers and octo-	SEXAFS (15)

TABLE 2.3d. (*Continued*)

Substrate	Unit Cell	Structure	Method
		hedral coordination in a cubic unit cell. At the surface the bulk Ni–I bond length of 2.78 Å decreases by 0.036 ± 0.010 Å and the I–I separation of 3.89 Å decreases by −0.48 ± 0.010.	
NiO(100) (rocksalt)	(1 × 1)	Top layer is unbuckled, and it is contracted 2% to 2.04 Å from bulk value of 2.08 Å.	LEED (16)
Sb$_2$Te$_2$Se(0001) (layer compound)	(1 × 1)	This is an elemental layer compound with Te–Sb–Se–Sb–Te stacking. The Te–Te interfaces are cleavage planes. At the surface the bulk Te–Sb layer spacing of 1.733 Å is contracted by ~3% to 1.683 ± 0.05 Å.	ARUPS (17)
TiS$_2$(0001) (layer compound)	(1 × 1)	This is an elemental layer compound with Ti–S–Ti stacking. The Ti–Ti interfaces are cleavage planes. At the surface the bulk Ti–S layer spacing is contracted by ~5% and the first van der Waals spacing (Ti–S–Ti|Ti–S–Ti) is also contracted by 5%.	LEED (18)
TiSe$_2$(0001) (layer compound)	(1 × 1)	This is an elemental layer compound with Ti–Se–Ti stacking. The Ti–Ti interfaces are cleavage planes. At the surface the bulk Ti–Se layer spacing is expanded by ~5% and the first van der Waals spacing (Ti–Se–Ti|Ti–Se–Ti) is contracted by 5%.	LEED (18)
ZnO(0001) (wurtzite)	(1 × 1)	Zn termination; top-layer spacing is 0.60 ± 0.10 Å, a 25% contraction from the bulk value of 0.80 Å.	LEED (19)

TABLE 2.3d. (*Continued*)

Substrate	Unit Cell	Structure	Method
ZnO(10$\bar{1}$0) (wurtzite)	(1 × 1)	Oxygen is buckled outward by ~0.8 Å, and Zn is contracted by 1.2 Å in the first layer. Bulk layer spacing is 1.88 Å.	LEED (20)
ZnO(11$\bar{2}$0) (wurtzite)	(1 × 1)	Terminated bulk structure; no evidence for reconstruction or relaxation in layer spacings.	LEED (20)

REFERENCES

1. W.S. Yang, J. Sokolov, F. Jona, and P.M. Marcus. *Solid State Commun.* **41**:191 (1982).
2. T.E. Derry, L. Smit, and J.F. van der Veen. *Surf. Sci.* **167**:502 (1986).
3. G. Vidali, M.W. Cole, W.H. Weinberg, and W.A. Steele. *Phys. Rev. Lett.* **51**:118 (1983).
4. N.J. Wu and A. Ignatiev. *Phys. Rev. B* **25**:2983 (1982).
5. N.J. Wu and A. Ignatiev. *Phys. Rev. B* **28**:7288 (1983).
6. C.G. Shaw, S.C. Fain, Jr., M.D. Chinn, and M.F. Toney. *Surf. Sci.* **97**:128 (1980).
7. C. Bouldin and E.A. Stern. *Phys. Rev. B* **25**:3462 (1982).
8. M. Prutton, J.A. Ramsey, J.A. Walker, and M.R. Welton-Cook. *J. Phys. C* **12**:5271 (1979).
9. A. Ignatiev, B.W. Lee, and M.A. Van Hove. In: *Proceedings of the 7th International Vacuum Congress and 3rd International Conference on Solid Surfaces*, Vienna, 1977.
10. R.C. Felton, M.P. Prutton, S. Tear, and M.R. Welton-Cook. *Surf. Sci.* **88**:474 (1979).
11. M.R. Welton-Cook and W. Berndt. *J. Phys. C* **15**:569 (1982).
12. T. Urano, T. Kanaji, and M. Kaburagi. *Surf. Sci.* **134**:109 (1983).
13. B.J. Mrstik, R. Kaplan, T.L. Reinecke, M.A. Van Hove, and S.Y. Tong. *Phys. Rev. B* **15**:897 (1977).
14. S. Andersson, J.B. Pendry, and P.M. Echenique. *Surf. Sci.* **65**:539 (1977).
15. R.G. Jones, S. Ainsworth, M.D. Crapper, C. Somerton, D.P. Woodruff, R.S. Brooks, J.C. Campuzano, D.A. King, and G.M. Lambe. *Surf. Sci.* **152/153**:443 (1985).
16. M.R. Welton-Cook and M. Prutton. *J. Phys. C* **13**:3993 (1980).
17. R.L. Benbow, M.R. Thuler, Z. Hurych, K.H. Lau, and S.Y. Tong. *Phys. Rev. B* **28**:4161 (1983).
18. B. Lau, B.J. Mrstik, S.Y. Tong, and M.A. Van Hove. To be published.
19. A.R. Lubinsky, C.B. Duke, S.C. Chang, B.W. Lee, and P. Mark. *J. Vacuum Sci. Technol.* **13**:189 (1976).
20. C.B. Duke, A.R. Lubinsky, B.W. Lee, and P. Mark. *J. Vacuum Sci. Technol.* **13**:761 (1976).

TABLE 2.4a. Atomic Adsorbates on Metal Surfaces[a]

Absorption System	Overlayer Unit Cell	Absorption Site	Adsorbate Spacing	Method
Ag(100)–Cl	c(2 × 2)	4-fold	1.62	LEED (1)
Ag(111)–Au	(1 × 1)	3-fold fcc	2.35 ± 0.10	HEIS (2)
Ag(111)–I	($\sqrt{3}$ × $\sqrt{3}$)$R30°$	3-fold fcc and hcp	2.29 ± 0.06	LEED (3)
	Top layer	0.0%	2.36 ± 0.06	

TABLE 2.4a. (*Continued*)

Adsorption System	Overlayer Unit Cell	Adsorption Site	Adsorbate Spacing	Method
Ag(111)–I	$(\sqrt{3} \times \sqrt{3})R30°$	3-fold	2.34 ± 0.02	SEXAFS (4)
Ag(111)–Xe	Hexagonal (1×1)	Incommensurate	3.55 ± 0.10	LEED (5)
	Randomly oriented hexagonal Xe lattice, Xe–Xe spacing 1.51 Å			
Al(100)–Na	$c(2 \times 2)$	4-fold	2.08 ± 0.12	LEED (6)
Al(100)–Na	$c(2 \times 2)$	4-fold	2.05 ± 0.10	LEED (7)
Au(111)–Ag	(1×1)	3-fold fcc	2.35 ± 0.10	HEIS (2)
Cu(100)–Cl	$c(2 \times 2)$	4-fold	1.60 ± 0.03	LEED (8)
	Top layer	+3%	1.85 ± 0.03	
Cu(111)–Cs	(2×2)	Top	3.01 ± 0.05	LEED (9)
Cu(100)–I	(2×2)	4-fold	1.98 ± 0.02	SEXAFS (10)
Cu(111)–I	$(\sqrt{3} \times \sqrt{3})R30°$	3-fold hcp	2.21 ± 0.02	SEXAFS (10)
Cu(111)–Ni	(1×1)	3-fold fcc	2.04 ± 0.02	LEED (11)
Cu(100)–Pb	$c(2 \times 2)$	4-fold	2.05 ± 0.05	LEED (12)
Cu(100)–3Pb	$c(5\sqrt{2} \times \sqrt{2})R45°$	4-fold	2.4 ± 0.05	LEED (12)
	Overlayer relaxed with antiphase domain walls and off-center atoms			
Fe(100)–C + O	$c(2 \times 2)$	4-fold	0.48	LEED (13)
	CO decomposes on Fe(100), C and O occupy random sites in (2×2) lattice			
Fe(110)–H	(2×1)	Quasi 3-fold	0.90 ± 0.10	LEED (14)
Fe(110)–2H	(3×1)	Quasi 3-fold	1.00 ± 0.10	LEED (14)
Fe(100)–N	$c(2 \times 2)$	4-fold	0.25 ± 0.05	LEED (15)
	Top layer	+8%	1.54 ± 0.05	
Ir(110)–S	(2×2)			LEED (16)
	Sulfur is adsorbed in 3-fold sites on the (111) facets left by the clean-surface missing-row reconstruction. The first Ir–Ir spacing is contracted by 3.3%, and the second one is expanded by 1.3%. Ir–S is 2.39 Å for first-layer Ir atoms and 2.26 Å for second-layer Ir atoms.			
Mo(100)–N	$c(2 \times 2)$	4-fold	1.02	LEED (17)
Mo(100)–Si	(1×1)	4-fold	1.16 ± 0.10	LEED (18)
Ni(100)–Br	$c(2 \times 2)$	4-fold	1.51 ± 0.03	EXAFS (19)
Ni(100)–C + O	$c(2 \times 2)$	4-fold	0.93 ± 0.10	LEED (20)
	CO decomposes on Ni(100); C and O occupy random sites in (2×2) lattice			
Ni(100)–2C	(2×2)	4-fold	0.10 ± 0.10	LEED (21)
	Top layer	+22%	1.96 ± 0.05	
	Reconstructed	Lateral motions	0.25 ± 0.35 Å	
Ni(100)–Cu	(1×1)	4-fold	1.80 ± 0.03	LEED (22)
Ni(100)–Na	$c(2 \times 2)$	4-fold	2.2 ± 0.10	LEED (23)
Ni(100)–Na	$c(2 \times 2)$	4-fold	2.23 ± 0.10	LEED (24)
Ni(100)–Na	$c(2 \times 2)$	4-fold	1.90	XPD (25)
Ni(110)–C				SEELFS (26)
	Graphite layer on Ni(110), 3 carbon sites: A—off top, Ni–C = 1.95 Å B—4-fold, NiC = 2.49 Å, C—off bridge, Ni–C = 1.95, Ca–Cb = 1.49 Å			
Ni(111)–2C	(1×1)	3-fold hcp and fcc	2.80 ± 0.08	SEELFS (27)
Ni(111)–2H	(2×2)	3-fold hcp and fcc	1.15 ± 0.10	LEED (28)
Pd(111)–Au	(1×1)	3-fold fcc	2.25 ± 0.19	HEIS (29)
Ti(0001)–Cd	(1×1)	3-fold fcc	2.57	LEED (30)
Ti(0001)–Cd	(1×1)	3-fold fcc	2.63	LEED (31)

TABLE 2.4a. (*Continued*)

Adsorption System	Overlayer Unit Cell	Adsorption Site	Adsorbate Spacing	Method
	Two Cd layers: Cd–Cd distance 2.81 Å, Cd–Ti 2.63 Å			
Ti(0001)–N	(1 × 1)	Subsurface tetrahedral	−1.22 ± 0.05	LEED (32)
W(100)–H	c(2 × 2)	4-fold	1.32	LEED (33)
W(100)–2H	(1 × 1)	Short bridge		HREELS (34)
W(100)–2H	(1 × 1)	Short bridge	1.74	HREELS (35)
W(100)–2H	(1 × 1)	2-fold	1.17 ± 0.04	LEED (36)
	Top layer	−1.3%	1.56 ± 0.02	
W(100)–N	c(2 × 2)	4-fold	0.49 ± 0.06	LEED (37)
	Top layer	+1.3%	1.60 ± 0.06	

a Where relaxation of the metal layer spacing has investigated, this is listed in the table after the adsorbate information.

REFERENCES

1. F. Jona and P.M. Marcus. *Phys. Rev. Lett.* **50**:1823 (1983).
2. R.J. Culbertson, L.C. Feldman, P.J. Silverman, and H. Boehm. *Phys. Rev. Lett.* **47**:657 (1981).
3. M. Maglietta, E. Zanazzi, U. Bardi, D. Sondericker. F. Jona, and P.M. Marcus. *Surf. Sci.* **123**:141 (1982).
4. P.H. Citrin, P. Eisenberger, and R.C. Hewitt. *Surf. Sci.* **89**:28 (1979).
5. N. Stoner, M.A. Van Hove, S.Y. Tong, and M.B. Webb. *Phys. Rev. Lett.* **40**:243 (1978).
6. M.A. Van Hove, S.Y. Tong, and N. Stoner. *Surf. Sci.* **54**:259 (1976).
7. B.A. Hutchins, T.N. Rhodin, and J.E. Demuth. *Surf. Sci.* **54**:419 (1976).
8. F. Jona, D. Westphal, A. Goldman, and P.M. Marcus. *J. Phys. C* **16**:3001 (1983).
9. S.A. Lindgren, L. Walldén, J. Rundgren, P. Westrin, and J. Neve. *Phys. Rev. B* **28**:6707 (1983).
10. P.H. Citrin, P. Eisenberger, and R.C. Hewitt. *Phys. Rev. Lett.* **45**:1948 (1980).
11. S.P. Tear and K. Roell. *J. Phys. C* **15**:5521 (1982).
12. W. Hoesler and W. Moritz. *Surf. Sci.* **117**:196 (1982).
13. K.O. Legg, F. Jona, D.W. Jepsen, and P.M. Marcus. *Phys. Rev. B* **16**:5271 (1977).
14. W. Moritz, R. Imbihl, R.J. Behm, G. Ertl, and T. Matsushima. *J. Chem. Phys.* **83**:1959 (1985).
15. R. Imbihl, R.J. Behm, G. Ertl, and W. Moritz. *Surf. Sci.* **123**:129 (1982).
16. C.-M. Chan, and M.A. Van Hove. *Surf. Sci.* **183**:303 (1987).
17. A. Ignatiev, F. Jona, D.W. Jepsen, and P.M. Marcus. *Surf. Sci.* **49**:189 (1975).
18. A. Ignatiev, F. Jona, D.W. Jepsen, and P.M. Marcus. *Phys. Rev. B* **11**:4780 (1975).
19. B. Lairson, T.N. Rhodin, and W. Ho. *Solid State Commun.* **55**:925 (1985).
20. M.A. Passler, T.H. Lin, and A. Ignatiev. *J. Vacuum Sci. Technol.* **18**:481 (1981).
21. J.H. Onuferko, D.P. Woodruff, and B.W. Holland. *Surf. Sci.* **87**:357 (1979).
22. M. Abu-Joudeh, P.P. Vaishnava, and P.A. Montano. *J. Phys. C* **17**:6899 (1984).
23. S. Andersson and J.B. Pendry. *Solid State Commun.* **16**:563 (1975).
24. J.E. Demuth, D.W. Jepsen, and P.M. Marcus. *J. Phys. C* **8**:L25 (1975).
25. N.V. Smith, H.H. Farrell, M.M. Traum, D. P. Woodruff, D. Norman, M.S. Woolfson, and B.W. Holland. *Phys. Rev. B* **21**:3119 (1980).
26. L. Papagno and L.S. Caputi. *Phys. Rev. B* **29**:1483 (1984).
27. R. Rosei, M. de Crescenzi, F. Sette, C. Quaresima, A. Savoia, and P. Perfetti. *Phys. Rev. B* **28**:1161 (1983).
28. K. Christmann, R.J. Behm, G. Ertl, M.A. Van Hove, and W.H. Weinberg. *J. Chem. Phys.* **70**:4168 (1979).

29. Y. Kuk, L.C. Feldman, and P.J. Silverman. *Phys. Rev. Lett.* **50**:511 (1983).
30. H.D. Shih, F. Jona, D.W. Jepsen, and P.M. Marcus. *Phys. Rev. B* **15**:5550 (1977).
31. H.D. Shih, F. Jona, D.W. Jepsen, and P.M. Marcus. *Phys. Rev. B* **15**:5561 (1977).
32. H.D. Shih, F. Jona, D.W. Jepsen, and P.M. Marcus. *Surf. Sci.* **60**:445 (1976).
33. R.F. Willis. *Surf. Sci.* **89**:457 (1979).
34. W. Ho, R.F. Willis, and E.W. Plummer. *Phys. Rev. B* **21**:4202 (1980).
35. J.P. Woods and J.L. Erskine. *Phys. Rev. Lett.* **55**:2595 (1985).
36. M.A. Passler, B.W. Lee, and A. Ignatiev. *Surf. Sci.* **150**:263 (1985).
37. K. Griffiths, D.A. King, G.C. Aers, and J.B. Pendry. *J. Phys. C* **15**:4921 (1982).

TABLE 2.4b. Chalcogen Chemisorption on Metals[a]

Substrate Face	Overlayer Unit Cell	Adsorption Site	Adsorbate Spacing	Bond Length	Method
		Oxygen			
Ag(110)	(2 × 1)	Long bridge	0.20	2.05	SEXAFS (1)
Al(111)	(1 × 1)	3-fold fcc	0.70 ± 0.12	1.79 ± 0.05	LEED (2)
Al(111)	(1 × 1)	3-fold fcc	0.60 ± 0.10	1.75 ± 0.03	NEXAFS (3)
Al(111)	(1 × 1)	Subsurface tetrahedral	0.60 ± 0.10	1.75 ± 0.03	NEXAFS (3)
Al(111)	(1 × 1)	3-fold fcc	0.70	1.79	LEED (4)
Al(111)	Undetermined	3-fold fcc	0.98 ± 0.10	1.92 ± 0.05	EXAFS (5)
Co(100)	c(2 × 2)	4-fold	0.80	1.94	LEED (6)
Cu(100)	c(2 × 2)	4-fold	1.40	2.28	LEED (7)
Cu(100)	c(2 × 2)	4-fold	0.70 ± 0.01	1.94 ± 0.01	SEXAFS (8)
Cu(100)	c(2 × 2)	4-fold	0.80	1.97	NPD (9)
Cu(100)	(2 × 1)				ALICISS (10)
		Confirm surface missing-row model			
Cu-(100)	c(2 × 2)			1.97	ALICISS (11)
		Bulk layer spacing 1.28 Å. first-layer spacing expanded 25 ± 1.0%, second and third layers contracted by 10 ± 5.			
Cu(110)	Disordered	Long bridge	0.35		SEXAFS (12)
Cu(110)	(2 × 1)	Long bridge			HEIS (13)
		Top layer buckled, [001] rows alternate +0.27 ± 0.05 Å and −0.02 ± 0.03 Å, second-layer spacing expanded by 0.06 ± 0.03 Å			
Cu(410)	(1 × 1)	Quasi 4-fold at step edge	0.4 ± 0.2	1.85 ± 0.04	ARXPD (14)
Cu(410)	(1 × 1)	Quasi 4-fold at step edge and 4-fold	0.4 ± 0.2	1.85 ± 0.04	ARXPD (14)
Fe(100)	(1 × 1)	4-fold	0.48 ± 0.10	2.08 ± 0.02	LEED (15)
		Top layer	1.54 ± 0.10	+7.7%	
Ir(110)	c(2 × 2)	Short bridge	1.37 ± 0.05	1.93 ± 0.04	LEED (16)
		Top layer	1.33 ± 0.07	−2.2%	
Ir(111)	(2 × 2) or (2 × 1)	3-fold fcc	1.30 ± 0.05	2.04 ± 0.03	LEED (17)
Ni(100)	(2 × 2)	4-fold	0.86 ± 0.07	1.96 ± 0.03	EXAFS (18)
Ni(100)	c(2 × 2)	4-fold	0.86 ± 0.07	1.96 ± 0.03	EXAFS (18)
Ni(100)	(2 × 2)	4-fold	0.90 ± 0.10	1.98 ± 0.05	LEED (19)
Ni(100)	c(2 × 2)	4-fold	0.92 ± 0.05	1.99 ± 0.02	LEED (20)
Ni(100)	c(2 × 2)	4-fold	0.85 ± 0.05	1.96 ± 0.02	EXAFS (21)
Ni(100)	c(2 × 2)	4-fold	0.90 ± 0.04	1.98 ± 0.02	NPD (22)
Ni(100)	c(2 × 2)	4-fold	0.85 ± 0.04	1.96 ± 0.02	NPD (23)

TABLE 2.4b. *(Continued)*[a]

Substrate Face	Overlayer Unit Cell	Adsorption Site	Adsorbate Spacing	Bond Length	Method
		Oxygen (Continued)			
Ni(100)	c(2 × 2)	4-fold	0.86 ± 0.10	1.96 ± 0.04	HEIS (24)
		Top layer	1.85 ± 0.10	+5.1%	
Ni(100)	c(2 × 2)	4-fold	0.90 ± 0.10	1.98 ± 0.05	XANES (25)
Ni(100)	c(2 × 2)	4-fold	0.90	1.98	HREELS (26, 27)
Ni(100)	c(2 × 2)	4-fold	0.922		HREELS (28)
Ni(110)	(2 × 1)	Missing row, long bridge	0.25		ALICISS (29)
Ni(111)	$(\sqrt{3} \times \sqrt{3})R30°$	3-fold hcp	1.20	1.87	HEIS (30)
Ta(100)	(3 × 1)	Subsurface tetrahedral	−0.43	1.95	LEED (31)
		Top layer	1.55 ± 0.05	−6.1%	
W(100)	Disordered	4-fold	0.55 ± 0.10	2.30 ± 0.02	LEED (32)
W(100)	(2 × 1)	3-fold	1.25 ± 0.01	2.09 ± 0.01	LEED (33)
Zr(0001)	(2 × 2)	Subsurface octahedral	1.37	2.31	LEED (34)
		Sulfur			
Co(100)	c(2 × 2)	4-fold	1.30	2.20	LEED (35)
Cu(100)	(2 × 2)	4-fold	1.39	2.28	ARXPS (36)
Fe(100)	c(2 × 2)	4-fold	1.09 ± 0.05	2.30 ± 0.02	LEED (37)
Fe(110)	(2 × 2)	4-fold reconstructed	1.43	2.02	LEED (38)
Ir(111)	$(\sqrt{3} \times \sqrt{3})R30°$	3-fold fcc	1.65 ± 0.07	2.28 ± 0.05	LEED (39)
Ni(100)	(2 × 2)	4-fold	1.30 ± 0.10	2.19 ± 0.06	LEED (19)
Ni(100)	c(2 × 2)	4-fold	1.28 ± 0.05	2.18 ± 0.03	LEED (20)
Ni(100)	c(2 × 2)	4-fold	1.30 ± 0.05	2.19 ± 0.03	LEED (40)
Ni(100)	c(2 × 2)	4-fold	1.30 ± 0.04	2.19 ± 0.02	NPD (22)
Ni(100)	c(2 × 2)	4-fold	1.39 ± 0.04	2.24 ± 0.02	SEXAFS (41)
Ni(100)	c(2 × 2)	4-fold	1.37 ± 0.05	2.23 ± 0.03	ARPEFS (42)
Ni(100)	c(2 × 2)	4-fold	1.35 ± 0.10	2.22 ± 0.06	ARXPS (43)
Ni(100)	c(2 × 2)	4-fold	1.35	2.22	ARXPS (36)
Ni(100)	Disordered	4-fold	1.36 ± 0.03	2.22 ± 0.02	SEXAFS (44)
Ni(110)	c(2 × 2)	4-fold	0.9	2.34	ARXPS (45)
Ni(110)	c(2 × 2)	4-fold	0.84 ± 0.03	2.31 ± 0.01	LEED (46)
		Top layer	1.372 ± 0.02	+9.6%	
		Next layer	1.201 ± 0.02	−4.0%	
Ni(110)	c(2 × 2)	4-fold	0.87 ± 0.03	2.32 ± 0.01	MEIS (47)
		Top layer	1.31 ± 0.04	+4.8%	
Mo(100)	c(2 × 2)	4-fold	1.04		LEED (48)
		Top layer	1.42	−10%	
Pd(100)	c(2 × 2)	4-fold	1.30 ± 0.05	2.33 ± 0.03	LEED (49)
Pd(111)	$(\sqrt{3} \times \sqrt{3})R30°$	3-fold fcc	1.53 ± 0.05	2.20 ± 0.03	LEED (50)
Pt(111)	$(\sqrt{3} \times \sqrt{3})R30°$	3-fold fcc	1.62 ± 0.05	2.28 ± 0.04	LEED (51)
Rh(100)	(2 × 2)	4-fold	1.29	2.30	LEED (52)
Rh(110)	c(2 × 2)	4-fold	0.77	2.45	LEED (53)
Rh(111)	$(\sqrt{3} \times \sqrt{3})R30°$	3-fold fcc	1.53	2.18	LEED (54)
		Selenium			
Ag(100)	c(2 × 2)	4-fold	1.91 ± 0.04	2.80 ± 0.03	LEED (55)
Ni(100)	(2 × 2)	4-fold	1.55 ± 0.10	2.35 ± 0.07	LEED (19)
Ni(100)	c(2 × 2)	4-fold	1.45 ± 0.10	2.28 ± 0.06	NPD (56)

TABLE 2.4b. *(Continued)*[a]

Substrate Face	Overlayer Unit Cell	Adsorption Site	Adsorbate Spacing	Bond Length	Method
		Selenium (Continued)			
Ni(100)	c(2 × 2)	4-fold	1.55	2.35	NPD (57)
Ni(100)	c(2 × 2)	4-fold	1.10 ± 0.04	2.42 ± 0.02	NPD (58)
Ni(111)	(2 × 2)	3-fold fcc	1.80 ± 0.04	2.30 ± 0.03	NPD (58)
		Tellurium			
Cu(100)	(2 × 2)	4-fold	1.70 ± 0.15	2.48 ± 0.10	LEED (59)
Cu(100)	(2 × 2)	4-fold	1.90 ± 0.04	2.62 ± 0.03	SEXAFS (60)
Ni(100)	(2 × 2)	4-fold	1.80 ± 0.10	2.52 ± 0.07	LEED (19)
Ni(100)	c(2 × 2)	4-fold	1.90 ± 0.10	2.59 ± 0.07	LEED (61)
Ni(100)	c(2 × 2)	4-fold	1.9		XPD (62)
Ni(100)	c(2 × 2)	4-fold	1.9		SPLEED (63)

[a]Where relaxation of the metal layer spacing has been investigated, this is listed in the table after the adsorbate information.

REFERENCES

1. A. Puschmann and J. Haase. *Surf. Sci.* **144**:559 (1984).
2. V. Martinez, F. Soria, M.C. Munoz, and J.L. Sacedon. *Surf. Sci.* **128**:424 (1983).
3. D. Norman, S. Brennan, R. Jaeger, and J. Stöhr. *Surf. Sci.* **105**:L297 (1981).
4. J. Neve, J. Rundgren, and P. Westrin. *J. Phys. C* **15**:43911(1982).
5. R.Z. Bachrach, G.V. Hansson, and R.S. Bauer. *Surf. Sci.* **109**:L560 (1981).
6. M. Maglietta, E. Zanazzi, U. Bardi, and F. Jona. *Surf. Sci.* **77**:101 (1978).
7. J. Onuferko and D.P. Woodruff. *Surf. Sci.* **95**:555 (1980).
8. U. Döbler, K. Baberschke, J. Stöhr, and D.A. Outka. *Phys. Rev. B* **31**:2532 (1985).
9. J.G. Tobin, L.E. Klebanoff, D.H. Rosenblatt, R.F. Davis, E. Umbach, A.G. Baca, D.A. Shirley, Y. Huang, W.M. Kang, and S.Y. Tong. *Phys. Rev. B* **26**:7076 (1982).
10. H. Niehaus and G. Comsa. *Surf. Sci.* **140**:18 (1984).
11. J.A. Yarmoff and R.S. Williams. *J. Vacuum Sci. Technol.* **A4**:1274 (1985).
12. U. Döbler, K. Baberschke, J. Haase, and A. Puschmann. *Surf. Sci.* **152/153**:569 (1985).
13. I. Stensgaard, R. Feidenhans'l and J.E. Sørensen. *Surf. Sci.* **128**:281 (1983).
14. K.A. Thompson and C.S. Fadley. *Surf. Sci.* **146**:281 (1984).
15. K.O. Legg, F. Jona, D.W. Jepsen, and P.M. Marcus. *Phys. Rev. B* **16**:5271 (1977).
16. C.M. Chan, K.L. Luke, M.A. Van Hove, W.H. Weinberg, and S.P. Withrow. *Surf. Sci.* **78**:386 (1978).
17. C.-M. Chan and W.H. Weinberg. *J. Chem. Phys.* **71**:2788 (1979).
18. J. Stöhr, R. Jaeger, and T. Kendelewicz. *Phys. Rev. Lett.* **49**:142 (1982).
19. M.A. Van Hove and S.Y. Tong. *J. Vacuum Sci. Technol.* **12**:230 (1975).
20. P.C. Marcus, J.E. Demuth, and D.W. Jepsen. *Surf. Sci.* **53**:501 (1975).
21. M. de Crescenzi, F. Antonangeli, C. Bellini, and R. Rosei. *Phys. Rev. Lett.* **50**:1949 (1983).
22. D.H. Rosenblatt, J.G. Tobin, M.G. Mason, R.F. Davis, S.D. Kevan, D.A. Shirley, C.H. Li, and S.Y. Tong. *Phys. Rev. B* **23**:3828 (1981).
23. S.Y. Tong, W.M. Kang, D.H. Rosenblatt, J.G. Tobin, and D.A. Shirley. *Phys. Rev. B* **27**:4632 (1983).
24. J.W.M. Frenken, J.F. van der Veen, and G. Allan. *Phys. Rev. Lett.* **51**:1876 (1983).
25. D. Norman, J. Stöhr, R. Jaeger, P.J. Durham, and J.B. Pendry. *Phys. Rev. Lett.* **51**:2052 (1983).
26. T.S. Rahman, J.E. Black, and D.L. Mills. *Phys. Rev. Lett.* **46**:1469 (1981).

27. T.S. Rahman, D.L. Mills, J.E. Black, J.M. Szeftel, S. Lehwald, and H. Ibach. *Phys. Rev. B* **30**:589 (1984).

28. R.L. Strong and J.L. Erskine. *Phys. Rev. Lett.* **54**:346 (1985).

29. H. Hiehaus and G. Comsa. *Surf. Sci.* **151**:L171 (1985).

30. T. Narusawa and W.M. Gibson. *Surf. Sci.* **114**:331 (1981).

31. A.V. Titov and H. Jagodzinski. *Surf. Sci.* **152/153**:409 (1985).

32. K. Heinz, D.K. Saldin, and J.B. Pendry. *Phys. Rev. Lett.* **55**:2312 (1985).

33. M.A. Van Hove and S.Y. Tong. *Phys. Rev. Lett.* **35**:1092 (1975).

34. K.C. Hui, R.H. Milne, K.A.R. Mitchell, W.T. Moore, and M.Y. Zhou. *Solid State Commun.* **56**:83 (1985).

35. M. Maglietta. *Solid State Commun.* **43**:395 (1982).

36. E.L. Bullock, C.S. Fadley, and P.J. Orders. *Phys. Rev. B* **28**:4867 (1983).

37. K.O. Legg, F. Jona, D.W. Jepsen, and P.M. Marcus. *Surf. Sci.* **66**:25 (1977).

38. F.E. Shih, F. Jona, D.W. Jepsen, and P.M. Marcus. *Phys. Rev. Lett.* **46**:731 (1981).

39. C.-M. Chan and W.H. Weinberg. *J. Chem. Phys.* **71**:3988 (1979).

40. Y. Gauthier, D. Aberdam, and R. Baudoing. *Surf. Sci.* **78**:339 (1978).

41. J. Stöhr, R. Jaeger, and S. Brennan. *Surf. Sci.* **117**:503 (1982).

42. J.J. Barton, C.C. Bahr, Z. Hussain, S.W. Robey, L.E. Klebanoff, and D.A. Shirley. *J. Vacuum Sci. Technol. A* **2**:847 (1984).

43. P.J. Orders, B. Sinkovic, C.S. Fadley, R. Trehan, Z. Hussain, and J. Lecante. *Phys. Rev. B* **30**:1838 (1984).

44. J. Stöhr, E.B. Kollin, D.A. Fischer, J.B. Hastings, F. Zaera, and F. Sette. *Phys. Rev. Lett.* **55**:1468 (1985).

45. R. Baudoing, E. Blanc, C. Gaubert, Y. Gautheir, and N. Gnuchev. *Surf. Sci.* **128**:22 (1983).

46. Y. Gauthieer, R. Baudoing, Y. Joly, J. Rundgren, J.C. Bertolini, and J. Massardier. *Surf. Sci.* **162**:348 (1985).

47. J.F. van der Veen, R.M. Tromp, R.G. Smeenk, and F.W. Saris. *Surf. Sci.* **82**:468 (1979).

48. L.J. Clarke. *Surf. Sci.* **102**:331 (1981).

49. W. Berndt, R. Hora, and M. Scheffler. *Surf. Sci.* **117**:188 (1982).

50. F. Maca, M. Scheffler, and W. Berndt. *Surf. Sci.* **160**:467 (1985).

51. K. Hayek, H. Glassl, A. Gutmann, H. Leonhard, M. Prutton, S.P. Tear, and M.R. Welton-Cook. *Surf. Sci.* **152**: 419 (1985).

52. S. Hengrasmee, P.R. Watson, D.C. Frost, and K.A.Ṙ. Mitchell. *Surf. Sci.* **87**:L249 (1979).

53. S. Hengrasmee, P.R. Watson, D.C. Frost, and K.A.R. Mitchell. *Surf. Sci.* **92**:71 (1980).

54. P.C. Wong, M.Y. Zhou, K.C. Hui, and K.A.R. Mitchell. *Surf. Sci.* **163**:172 (1985).

55. A. Ignatiev, F. Jona, D.W. Jepsen, and P.M. Marcus. *Surf. Sci.* **40**:439 (1973).

56. J.E. Demuth, D.W. Jepsen, and P.M. Marcus. *Phys. Rev. Lett.* **31**:540 (1973).

57. D.H. Rosenblatt, S.D. Kevan, J.G. Tobin, R.F. Davis, M.G. Mason, D.A. Shirley, J.C. Tang, and S.Y. Tong. *Phys. Rev. B* **26**:3181 (1982).

58. D.H. Rosenblatt, S.D. Kevan, J.G. Tobin, R.F. Davis, M.G. Mason, D.R. Denley, D.A. Shirley, Y. Huang, and S.Y. Tong. *Phys. Rev. B* **26**:1812 (1982).

59. A. Salwén and J. Rundgren. *Surf. Sci.* **53**:523 (1975).

60. F. Comin, P.H. Citrin, P. Eisenberger, and J.E. Rowe. *Phys. Rev. B* **26**:7060 (1982).

61. J.E. Demuth, D.W. Jepsen, and P.M. Marcus. *J. Phys. C* **6**:L307 (1973).

62. N.V. Smith, H.H. Farrell, M.M. Traum, D.P. Woodruff, D. Norman, M.S. Woolfson, and B.W. Holland. *Phys. Rev. B* **21**:3119 (1980).

63. J.K. Lang, K.D. Jamison, F.B. Dunning, G.K. Walters, M.A. Passier, A. Ignatiev, E. Tamura, and R. Feder. *Surf. Sci.* **123**:247 (1982).

TABLE 2.4c. Atomic Adsorption on Semiconductor Surfaces[a]

Adsorption System	Overlayer Cell	Adsorbate Site	Layer Spacing	Method
GaAs(110)–Sb	(1 × 1)	2-fold	See note 1	LEED (1)
GaAs(110)–Al	(1 × 1)	Subsurface	See note 2	LEED (2)
Ge(111)–Cl	(2 × 8)	Top	2.07 ± 0.03	SEXAFS (3)
Ge(111)–I	(1 × 1)	Top	2.50 ± 0.04	SEXAFS (4)
Ge(111)–Br	Undetermined	Top	2.10 ± 0.04	X-ray resonance (5)
Ge(111)–Te	(2 × 2)	hcp hollow		SEXAFS (4)
Si(111)–Au	$(\sqrt{3} \times \sqrt{3})$	See note 3	0.30	ALICISS (6)
Si(111)–Br	Undetermined	Top	2.18 ± 0.06	X-ray resonance (7, 8)
Si(111)–Ag	(7 × 7)	Top	0.70 ± 0.04	SEXAFS (9)
Si(111)–Ag	$(2\sqrt{3} \times 2\sqrt{3})$	Subsurface	-0.68 ± 0.15	SEXAFS (9)
	Layer 1–2		1.36 ± 0.30	
	Layer 2–3		-0.30	
Si(111)–Cl	$(\sqrt{19} \times \sqrt{19})$	Top	1.98 ± 0.04	SEXAFS (3)
Si(111)–Cl	(7 × 7)	Top	2.03 ± 0.03	SEXAFS (3)
Si(111)–I	(7 × 7)	Top	2.44 ± 0.03	SEXAFS (4)
	Top layer	+15%	0.90 ± 0.05	
Si(111)–NiSi$_2$	(1 × 1)	Tetrahedral	See note 4	LEED (10)
Si(111)–NiSi$_2$	(1 × 1)	Tetrahedral	See note 5	MEIS (11)
Si(111)–Se	Undetermined	4-fold	1.68	X-ray resonance (12)
Si(111)–Te	(7 × 7)	2-fold	1.51	SEXAFS (4)

[a]Where substrate relaxations or reconstructions have been investigated, this information is listed in the table after the adsorbate information.

Note 1. Sb atoms fill As- and Ga-type sites in a slightly buckled (0.10 Å) first layer spaced 2.3 Å above the GaAs surface, with lateral distortions to form an sp^3 bonded chain.

Note 2. Aluminum substitutes for subsurface Ga atoms. for 0.5 monolayers, second-layer Ga atoms are replaced; for 1.0 monolayers, second- and third-layer Ga atoms are replaced. Above 1.5 monolayers, all near-surface Ga atoms are replaced by Al, forming an epitaxial AlAs(110) surface. In all cases the first interlayer spacing contracts by ~0.10 Å.

Note 3. Modified triplet cluster model. Au triplets are substituted for Si atoms with a 2.9-Å Au–Au bond length between the triplet Au atoms.

Note 4. Structure of NiSi$_2$, grown on Si(111) substrate. Forms fluorite structure layer compound Si–Ni–Si with nickel in tetrahedral sites. Silicon layer terminates crystal, with a first-layer contraction of ~25%.

Note 5. Ion scattering investigation of NiSi$_2$–bulk-Si interface. The Si–Ni–Si layer is determined to be 3.06 ± 0.08 Å above the next noncollinear Si atom (the bulk value is $0.77 + 2.35 = 3.12$). Of the two possible terminations, this most closely matches the bulk silicon structure.

REFERENCES

1. C.B. Duke, A. Paton, W.K. Ford, A. Kahn, and J. Carelli. *Phys. Rev. B* **26**:803 (1982).

2. A. Kahn, J. Carelli, D. Kanani, C.B. Duke, A. Paton, and L. Brillson. *J. Vacuum Sci. Technol.* **19**:331 (1981).

3. P.H. Citrin, J.E. Rowe, and P. Eisenberger. *Phys. Rev. B* **28**:2299 (1983).

4. P.H. Citrin, P. Eisenberger, and J.E. Rowe. *Phys. Rev. Lett.* **48**:802 (1982).

5. M. Bedzyk and G. Materlik. *Surf. Sci.* **152/153**:10 (1985).

6. K. Oura, M. Katayama, F. Shoji, and T. Hanawa. *Phys. Rev. Lett.* **55**:1486 (1985).

7. J.A. Golovchenko, J.R. Patel, D.R. Kaplan, P.L. Cowan, and M.J. Bedzyk. *Phys. Rev. Lett.* **49**:1560 (1982).

8. G. Materlik, A. Frohm, and M.J. Bedzyk. *Phys. Rev. Lett.* **52**:441 (1984).

9. J. Stöhr, R. Jaeger, G. Rossi, T. Kendelewicz, and I. Lindau. *Surf. Sci.* **134**:813 (1983).

10. W.S. Yang, F. Jona, and P.M. Marcus. *Phys. Rev. B* **28**:7377 (1983).

11. E.J. van Loenen, J.W.M. Frenken, J.F. van der Veen, and S. Valeri. *Phys. Rev. Lett.* **54**:827 (1985).

12. B.N. Dev, T. Thundat, and W.M. Gibson. *J. Vacuum Sci. Technol. A* **3**:946 (1985).

TABLE 2.4d. Carbon Monoxide, Di-Nitrogen, and Nitric Oxide Chemisorption on Metals[a]

Substrate Face	Overlayer Unit Cell	Adsorption Site	C-Metal $\perp$ Spacing	C-O Bond Length	Method
Cu(100)	c(2 × 2)–CO	Top	1.90 ± 0.10	1.13 ± 0.10	LEED (1)
Cu(100)	c(2 × 2)–CO	Top	1.92 ± 0.05		NEXAFS (2)
Ni(100)	c(2 × 2)–CO	Top	1.72	1.15	LEED (3)
Ni(100)	c(2 × 2)–CO	Top	1.80 ± 0.10	1.15 ± 0.05	ARXPS (4)
Ni(100)	c(2 × 2)–CO	Top	1.80 ± 0.10	1.15 ± 0.05	LEED (5)
Ni(100)	c(2 × 2)–CO	Top	1.70 ± 0.10	1.13 ± 0.10	LEED (6)
Ni(100)	c(2 × 2)–CO	Top	1.71 ± 0.10	1.15 ± 0.10	LEED (7)
Ni(100)	c(2 × 2)–CO	Top	1.80 ± 0.04	1.13	NPD (8)
Ni(111)	$(\sqrt{3} \times \sqrt{3})R30°$–CO	Bridge	1.27 ± 0.05	1.13 ± 0.05	NPD (8)
Ni(100)	Disordered–CO	Top			NEXAFS (9)
	Molecular axis $\perp$ to surface $\pm 10°$				
Ni(100)	Disordered–NO	Top			NEXAFS (9)
	Molecular axis $\perp$ to surface $\pm 10°$				
Ni(100)	Disordered–N_2	Top			NEXAFS (9)
	Molecular axis $\perp$ to surface $\pm 10°$				
Pd(100)	$(2\sqrt{2} \times \sqrt{2})R45°$–2CO	Bridge	1.36 ± 0.10	1.15 ± 0.10	LEED (10)
		Top layer	1.945 ± 0.10	$+0.4\%$	
Pd(111)	$(\sqrt{3} \times \sqrt{3})R30°$–2CO	3-fold fcc	1.29 ± 0.05	1.15 ± 0.05	LEED (11)
Pd(111)	(3 × 3)–2CO*	3-fold fcc	1.30 ± 0.05	1.17 ± 0.05	LEED (12)
Pt(111)	c(4 × 2)–2CO	Top	1.85 ± 0.05	1.15 ± 0.10	LEED (13)
		Bridge	1.55 ± 0.05	1.15 ± 0.10	
		Top layer	2.26 ± 0.025	0.0%	
Pt(111)	$(2\sqrt{3} \times 4)$rect–4CO[b]	Bridge	1.45	1.15	LEED (14)
Rh(111)	(3 × 3)–2CO[b]	3-fold hcp	1.30	1.17	LEED (15)
Rh(111)	$c(2\sqrt{3} \times 4)$rect–CO[b]	3-fold fcc	1.50 ± 0.05	1.21 ± 0.10	LEED (16)
Rh(111)	c(4 × 2)–CO[b]	3-fold hcp	1.30 ± 0.05	1.17 ± 0.05	LEED (17)
Rh(111)	c(4 × 2)–NO[b]	3-fold fcc	1.30 ± 0.05	1.17 ± 0.05	LEED (17)
Rh(111)	(2 × 2)–3CO	Quasi-top	1.87 ± 0.10	1.15 ± 0.10	LEED (18)
		Quasi-top	1.87 ± 0.10	1.15 ± 0.10	
		Bridge	1.52 ± 0.10	1.15 ± 0.10	
Rh(111)	$(\sqrt{3} \times \sqrt{3})R30°$–CO	Top	1.95 ± 0.10	1.07 ± 0.10	LEED (19)
		Top layer	2.19 ± 0.10	0.0%	
Ru(0001)	$(\sqrt{3} \times \sqrt{3})R30°$–CO	Top	2.00 ± 0.10	1.09 ± 0.10	LEED (20)

[a] In all the listed structures the CO or NO molecule is believed to adsorb perpendicular to the surface with the oxygen end away from the surface. For CO structures with multiple nonequivalent adsorption sites, these are listed on consecutive lines. If the metal layer spacing has been investigated, this is listed after the adsorbate information.

[b] These structures involve carbon monoxide co-adsorbed with other molecules. See Table 2.4e for more details.

REFERENCES

1. S. Andersson and J.B. Pendry. *J. Phys. C* 13:2547 (1980).
2. C.F. McConville, D.P. Woodruff, K.C. Prince, G. Paolucci, V. Chab, M. Surman, and A.M. Bradshaw. *Surf. Sci.* **166**:221 (1986).
3. M. Passler, A. Ignatiev, F. Jona, D.W. Jepsen, and P.M. Marcus. *Phys. Rev. Lett.* **43**:360 (1979).
4. L.G. Petersson, S. Kono, N.F.T. Hall, C.S. Fadley, and J.B. Pendry. *Phys. Rev. Lett.* **42**:1545 (1979).
5. K. Heinz, E. Lang, and K. Müller. *Surf. Sci.* **87**:595 (1979).
6. S.Y. Tong, A. Maldonado, C.H. Li, and M.A. Van Hove. *Surf. Sci.* **94**:73 (1980).
7. S. Andersson and J.B. Pendry. *J. Phys. C* 13:3547 (1980).
8. S.D. Kevan, R.F. Davis, D.H. Rosenblatt, J.G. Tobin, M.G. Mason, D.A. Shirley, C.H. Li, and S.Y. Tong. *Phys. Rev. Lett.* **46**:1629 (1981).
9. J. Stöhr and R. Jaeger. *Phys. Rev. B* **26**:4111 (1982).
10. R.J. Behm, K. Christmann, G. Ertl, and M.A. Van Hove. *J. Chem. Phys.* **73**:2984 (1980).
11. H. Ohtani, M.A. Van Hove, and G.A. Somorjai. *Surf. Sci.* **187**:372 (1987).
12. H. Ohtani, M.A. Van Hove, and G.A. Somorjai. *J. Phys. Chem.* **92**:3974 (1988).
13. D.F. Ogletree, M.A. Van Hove, and G.A. Somorjai. *Surf. Sci.* **173**:351 (1986).
14. D.F. Ogletree, M.A. Van Hove, and G.A. Somorjai. *Surf. Sci.* **187**:1 (1987).
15. R.F. Lin, G.A. Blackman, M.A. Van Hove, and G.A. Somorjai. *Acta Crystall.* **B43**:368 (1987).
16. M.A. Van Hove, R.F. Lin, and G.A. Somorjai. *J. Amer. Chem. Soc.* **108**:2532 (1980).
17. G.S. Blackman, C.T. Kao, B.E. Bent, C.M. Mate, M.A. Van Hove, G.A. Somorjai. *Surf. Sci.* **207**:66 (1988).
18. M.A. Van Hove, R.J. Koestner, J.C. Frost, and G.A. Somorjai. *Surf. Sci.* **129**:482 (1983).
19. R.J. Koestner, M.A. Van Hove, and G.A. Somorjai. *Surf. Sci.* **107**:439 (1981).
20. G. Michalk, W. Moritz, H. Pfnür, and D. Menzel. *Surf. Sci.* **129**:92 (1983).

TABLE 2.4e. Molecular Chemisorption Structures

System	Structure	Method
Cu(100)—HCO_2 disordered	The formate radical is in a plane perpendicular to the surface with the two oxygens closet to the surface and a formate O—C—O bond angle of 125° is assumed. The O atoms are slightly off-center above two adjacent 4-fold hollow sites. The O atoms are 1.54 Å above the Cu surface and separated by 2.17 Å. The C atom is 2.11 Å above the surface.	NEXAFS (1)
Cu(110)—HCO_2 disordered	The formate radical is in a plane perpendicular to the surface along the [001] direction. The oxygen atoms are closest to the surface, slightly off-center from two adjacent bridge sites, 1.51 Å above the surface and 2.29 Å apart. The O atom is 2.04 Å above the Cu atom.	NEXAFS (2)
Ni(111)—C_2H_2 (2 × 2)	The acetylene molecules are absorbed with the C—C bond parallel to the surface, and the center of the C—C bond is over a bridge site. The C—C bond is perpendicular to the Ni—Ni bridge. The C—C bond length is 1.50 Å, and the carbon atoms are 2.1 ± 0.10 Å above the surface.	LEED (3)

TABLE 2.4e. (*Continued*)

System	Structure	Method
Pd(111)$-C_6H_6$ + 2CO (3 × 3)	One benzene and two CO molecules per unit cell. CO is perpendicular to the surface, adsorbed 1.30 Å above 3-fold fcc sites. Benzene is parallel to the surface, centered 2.25 Å over a 3-fold hcp site.	LEED (4)
Pt(111)$-C_2H_2$ disordered	Acetylene bonded parallel to surface with C—C bond length 1.45 ± 0.03 Å.	NEXAFS (5)
Pt(111)$-C_2H_3$ disordered	Ethylidyne (CCH$_3$) bonded perpendicular to surface with C—C bond 1.47 ± 0.03 Å.	NEXAFS (6)
Pt(111)-C_2H_3 disordered	Ethylidyne (CCH$_3$) bonded perpendicular to surface with C—C bond 1.49 ± 0.02 Å.	NMR (7)
Pt(111)$-C_2H_3$ (2 × 2)	Ethylidyne (CCH$_3$) bonded perpendicular to surface in 3-fold fcc sites; C—C bond 1.50 ± 0.05 Å and C-surface perpendicular distance 1.20 ± 0.05 Å.	LEED (8)
Pt(111)$-C_2H_4$ disordered	Ethylene bonded parallel to surface with C—C bond length 1.49 ± 0.03 Å.	NEXAFS (5)
Pt(111)$-2C_6H_6$ disordered	Benzene ring parallel to surface, C—C bond length 1.40 ± 0.02 Å.	NEXAFS (9)
Pt(111)$-2C_6H_6$ + 4CO (2√3 × 4) rect	Benzene ring parallel to surface over bridge sites, 2.10 Å above surface, with two benzenes and four CO molecules per unit cell. CO also over bridge sites, 1.45 Å above the metal surface. Benzene ring is expanded with small in-plane distortions consistent with local symmetry.	LEED (10)
Rh(111)$-C_2H_3$ (2 × 2)	Ethylidyne (CCH$_3$) is adsorbed with the C—C axis perpendicular to the surface with a 1.45 ± 0.10-Å bond length. The terminal carbon atom is 1.31 ± 0.10 Å above a 3-fold hcp hollow site.	LEED (11)
Rh(111)$-C_2H_3$ + CO c(4 × 2)	Ethylidyne (CCH$_3$) is adsorbed in 3-fold fcc sites with a 1.30 ± 0.05-Å metal–terminal-carbon distance. The C—C axis is perpendicular to the surface with a 1.45 ± 0.05-Å bond length. CO molecules are adsorbed perpendicular to the surface in 3-fold hcp sites, carbon atom down, with a carbon–metal bond length of 1.30 ± 0.05 Å and a C—O bond length of 1.17 ± 0.05 Å.	LEED (12)
Rh(111)$-C_2H_3$ + NO c(4 × 2)	Ethylidyne (CCH$_3$) is absorbed in 3-fold hcp sites with a 1.30 ± 0.05-Å metal–terminal-carbon distance. The C—C axis is perpendicular to the surface with a 1.45 ± 0.05-Å bond length. NO molecules are absorbed perpendicular to the surface in 3-fold hcp sites, nitrogen atom down, with a nitrogen–metal bond length of 1.30 ± 0.05 Å and a N—O bond length of 1.17 ± 0.05 Å.	LEED (12)
Rh(111)$-C_6H_6$ + 2CO (3 × 3)	One benzene and two CO molecules per unit cell. CO is perpendicular to the surface, adsorbed	LEED (13)

TABLE 2.4e. (*Continued*)

System	Structure	Method
	130 Å above 3-fold hcp sites. Benzene is parallel to the surface, centered 2.20 Å over a 3-fold hcp site. Slight in-plane Kekulé distortion of the benzene molecule.	
Rh(111)–C_6H_6 + CO c($2\sqrt{3}$ × 4) rect	Benzene is coadsorbed with CO, each with one molecule per unit cell, both centered over 3-fold hcp sites; benzene is parallel to and 2.25 ± 0.05 Å above the surface. CO is perpendicular to the surface, and the metal–carbon spacing is 1.50 ± 0.05 Å. The benzene molecule has an in-plane Kekulé distortion, with alternating long and short bonds.	LEED (14)

REFERENCES

1. D.A. Outka, R.J. Madix, and J. Stöhr. *Surf. Sci.* **164**:235 (1985).
2. A. Puschman, J. Haase, M.D. Crapper, C.E. Riley, and D.P. Woodruff. *Phys. Rev. Lett.* **54**: 2250 (1985).
3. G. Casalone, M.G. Cattania, F. Merati, and M. Simonetta. *Surf. Sci.* **120**:171 (1982).
4. H. Ohtani, M.A. Hove, and G.A. Somorjai. *J. Phys. Chem.* **92**:3974 (1988).
5. J. Stöhr, F. Sette, and A.L. Johnson. *Phys. Rev. Lett.* **53**:1684 (1984).
6. J.A. Horsley, J. Stöhr, and R.J. Koestner. *J. Chem. Phys.* **83**:3146 (1985).
7. P.-K. Wang, C.P. Slichter, and J.J. Sinfelt. *J. Phys. Chem.* **89**:3606 (1985).
8. L.L. Kesmodel, L.H. Dubois, and G.A. Somorjai. *J. Chem. Phys.* **70**:2180. (1979).
9. J.A. Horsley, J. Stöhr, A.P. Hitchcock, D.C. Newbury, A.L. Johnson, and F. Sette. *J. Chem. Phys.* **83**:6099 (1985).
10. D.F. Ogletree, M.A. Van Hove, and G.A. Somorjai. *Surf. Sci.* **187**:1 (1987).
11. R.J. Koestner, M.A. Van Hove, and G.A. Somorjai. *Surf. Sci.* **121**:321 (1982).
12. G.S. Blackman, C.T. Kao, B.E. Bent, C.M. Mate, M.A. Van Hove, and G. A. Somorjai. *Surf. Sci.* **207**:66 (1988).
13. R.F. Lin, G.S. Blackman, M.A. Van Hove, and G.A. Somorjai. *Acta Crystallogr.* **B43**:368 (1987).
14. M.A. Van Hove, R.F. Lin, and G.A. Somorjai. *J. Amer. Chem. Soc.* **108**:2532 (1986).

TABLE 2.5a. Surface Structures on Substrates with Onefold Rotational Symmetry[a]

Substrate	Adsorbate	Surface Structure	References[b]
(AlGa)As(110)	[Clean]	(1 × 1)	1375
AlP(110)	[Clean]	(1 × 1)	1521, 1863
CdTe(110)	[Clean]	(1 × 1)	1367, 1382, 1495
CoSi(100)	[Clean]	(2 × 1)	1312
CoSi$_2$(100)	[Clean]	(1 × 1)	1312
GaAs(110)	[Clean]	(1 × 1)	896, 949, 1008, 1090, 1124, 1182 1449, 1480, 1519 1524, 1567, 1572

TABLE 2.5a. (*Continued*)[a]

Substrate	Adsorbate	Surface Structure	References[b]
			1575, 1576, 1702
			1764, 1765
		(2 × 2)	1567
	Ag	(1 × 1)	1575
		[001] streaks	1575
		c(4 × 4) [multilayer]	1344
	Al	(1 × 1)	1375, 1432, 1607
	Al (low coverage)	(1 × 1)–Al	1871
	Al (medium coverage)	(1 × 1)–Al	1871
	Al (high coverage)	(1 × 1)–Al	1871
		(1 × 4) [multilayer]	1344
	As	(1 × 1)	579, 1375
	Au	Disordered	1008
	Cu	Polycrystalline	1182
	Ga	Polycrystalline	1305
		(1 × 1) [multilayer]	1432
	Ge	(1 × 1)–Ge	1089, 1520, 1572
		(3 × 1)–Ge	588
		(2 × 1)–Ge	588
		(3 × 1) + (1 × 4) with streaks	1089
		(1 × 1) + blurred (8 × 10)	1089
	H_2O	(1 × 1)	1006
	In	(1 × 1)	1432
	$Fe(CO)_5$	Facet{100}	1377
	O_2	Disordered	744
	Pd	Disordered	1008
	Sb	(1 × 1)–Sb	1320, 1367, 1383
	ZnSe	(1 × 1)	1444
		(1 × 2)	1444
GaP(110)	[Clean]	(1 × 1)	819, 1445, 1495
	Al	AlP(110)	1426
		(1 × 1)	1521
GaSb(110)	[Clean]	(1 × 1)	1378, 1420, 1766
InAs(110)	[Clean]	(1 × 1)	1423
InP(110)	[Clean]	(1 × 1)	1495, 1521, 1568
			1570, 1573, 1618
			1784
	Al	(1 × 1)	1521
		(1 × 1) diffuse	1570
	Cl_2	Disordered	1660
	Cu	(1 × 1)	1568

TABLE 2.5a. *(Continued)*[a]

Substrate	Adsorbate	Surface Structure	References[b]
	H_2O	Ordered	1573
		Disordered	1573
	Ni	Disordered	1568
InSb(110)	[Clean]	(1×1)	1181, 1420, 1888
	Sn	Amorphous	1181
Te(10$\bar{1}$0)	[Clean]	(1×1)	1801
ZnO(10$\bar{1}$0)	[Clean]	(1×1)	1104, 1239, 1612
	H_2O	Disordered	1104
	O_2	(1×1)-O	392
	Xe	Hexagonal	1026
		Disordered	1026
ZnO(11$\bar{2}$0)	[Clean]	(1×1)	1772
ZnS(110)	[Clean]	(1×1)	1445
ZnSe(110)	[Clean]	(1×1)	1478
	O_2	ZnO(000$\bar{1}$)	642
ZnTe(110)	[Clean]	(1×1)	1378, 1495, 1504

[a] Organic overlayer structures and high-Miller-index surface structures are not included. See Table 2.5g and Table 2.5j, respectively, for these structures.
[b] References for Table 2.5a to 2.5j are listed at the end of Table 2.5j.

TABLE 2.5b. Surface Structures on Substrates with Twofold Rotational Symmetry[a]

Substrate	Adsorbate	Surface Structure	References[b]
Ag(110)	[Clean]	(1×1)	707, 888, 1522, 1534
	Br_2	(2×1)-Br	888
		c(4×2)-Br	888
		AgBr	888
	C_2N_2	Disordered	407
	Cl_2	Adsorbed	732
	Cs	(1×2)	859, 1534
		(1×3)	859, 1534
	HCN	Disordered	407
	H_2O	Disordered	878
	$H_2O + Li^+$	Complex	1557
	H_2S	(3×2)-S	627
		c(10×2)-S	627
		(3×4)-S	627
	I_2	Pseudohexagonal-I	1145
		c(2×2)-I	1145
	K	(1×2)	1534
		(1×3)	1534
	Li	(1×2)	1534
	Na	(1×1)	489
	NO	Disordered	345

TABLE 2.5b. (*Continued*)[a]

Substrate	Adsorbate	Surface Structure	References[b]
	O_2	(2×1)–O	146, 341–343
			344, 695, 878, 943
			974, 1027, 1047
			1140, 1143, 1160
			1300, 1690, 1751
		(1×2)–O	1376
		(3×1)–O	146, 341–343
			695, 878, 974, 1143
			1160, 1300, 1690
		(4×1)–O	146, 341, 342, 878
			974, 1143, 1300 1690
		(5×1)–O	146, 341, 695, 1300
		(6×1)–O	146, 341, 1300
		(7×1)–O	146
		$c(6 \times 2)$–O	974
	$O(a) + CO_2$	(2×2)	1143
	$O(a) + SO_2$	$c(6 \times 2)$–SO_3	1027, 1371
		(1×2)–SO_4 etc.	1371
	$O_2 + H_2O$	(1×2)–OH	878
		(1×3)–OH	878
	SO_2	(1×2)–SO_2	1027, 1371
		$c(4 \times 2)$–SO_2	1027, 1371
		(1×1)	1371
		$\begin{bmatrix} \frac{2}{3} & 1 \\ \frac{4}{3} & 0 \end{bmatrix}$–$SO_2$	1027
	Xe	Hexagonal overlayer	159
Al(110)	[Clean]	$(1 \times 1) + (1 \times 2)$	965
		(1×1)	1354, 1409, 1464
			1498, 1566, 1721
	CO	Not adsorbed	1273
	O_2	(331) facets	123
		(111) facets	122
		Disordered	709
Au(110)	[Clean]	(1×2)	754, 1009, 1098, 1166
		$(1 \times 2) + (1 \times 1)$	965
		(1×3)	754, 1098
	Bi	$\begin{bmatrix} 1 & 1 \\ 1 & -1 \end{bmatrix}$	498
		$\begin{bmatrix} 2 & 1 \\ -1 & 1 \end{bmatrix}$	498
		(2×1)	498
	H_2S	(1×2)–S	251
		$c(4 \times 2)$–S	251

TABLE 2.5b. *(Continued)*[a]

Substrate	Adsorbate	Surface Structure	References[b]
	Pb	(1 × 3)	444, 495, 683
		(1 × 1)	444, 495, 683
		(7 × 1)	444, 495, 683
		(7 × 3)	444, 495, 683
		(4 × 4)	444, 495, 683
C(10$\bar{1}$0)	[Clean]	c(2 × 2/3)	1033
C(110), diamond	O_2	Not adsorbed	164
	N_2	Not adsorbed	164
	NH_3	Not adsorbed	164
	H_2S	Not adsorbed	164
CdTe(100)	[Clean]	(3 × 1), (1 × 1), {110} f	1393
Co(10$\bar{1}$0)	O_2	(2 × 1)–O	1070
Co(11$\bar{2}$0)	[Clean]	(1 × 1)	1197, 1768, 1848
	CO	(3 × 1)–CO	902
		Disordered	902
	H_2O	Disordered	1310
		(4 × 1)–O	1310
		Complicated	1310
Cr(110)	[Clean]	(1 × 1)	1343
	CO	Disordered	1343
	Br_2	$\begin{bmatrix} 1 & 1 \\ -1 & 2 \end{bmatrix}$	1016
		$\begin{bmatrix} 1 & \dfrac{1}{1+X} \\ -1 & \dfrac{2}{1+X} \end{bmatrix}$–Br	1016
		$(0 < X \leq \frac{1}{3})$	
		$\begin{bmatrix} 1 & 1 \\ -2 & 2 \end{bmatrix}$–Br	1016
	O_2	(3 × 1)–O	140
		(100) facets	140, 256
		Cr_2O_3(0001)	140, 256
		Cr_2O_3	1261
		Disordered	1261
Cu(110)	[Clean]	(1 × 1)	725, 925, 995, 1023, 1131 1135, 1136, 1436 1498, 1723
	Au	$\begin{bmatrix} 1 & 0 \\ -\frac{1}{2} & \frac{3}{2} \end{bmatrix}$	479
		(1 × 2)	479
		(2 × 2)	479
		Complex structures	479
	Br_2	c(2 × 2)	1557
	Br(a) + H_2O	(3 × 2)	1557
	C	Disordered	1695
	C + O_2	(2 × 1)	1695

TABLE 2.5b. (*Continued*)[a]

Substrate	Adsorbate	Surface Structure	References[b]
	CO	One-dimensional order	26
		(2 × 3)–CO	26
		(2 × 1)–CO	255, 876, 1234
		c(5/4 × 2)	876
		c(1.3 × 2)–CO	1234
		Hexagonal overlayer	255
	H_2	Not adsorbed	7
	H_2O	Disordered	26
		Not specified	1131
		c(2 × 2)–H_2O	1023, 1178, 1270
		c(2 × 2)	1557
		(2 × 1)–OH	1270
		(1 × 1)–H_2O	1178
	H_2S	c(2 × 3)–S	35
		Adsorbed	35
	I_2	c(2 × 2)–I	572, 1915
		c(2 × 2) compressed	1915
	Kr	c(2 × 8)–Kr	1304, 1331
		Quasiperfect hexagonal	1331
	N_2O	(2 × 1)–O	879
	$Ni(CO_4)$ + CO	(1 × 1) + disordered	1048
	O_2	(2 × 1)–O	7, 8, 9, 45, 46, 246
			656, 750, 879, 885
			920, 953, 982, 1053
			1066, 1076, 1095
			1257, 1285, 1695
			1916, 1917, 1918
		c(2 × 1)–O	1270
		Streaks along ⟨110⟩	1023
		(1 × 1)–O	656, 885
		(3 × 1)–O_2	885
		c(6 × 2)–O	8, 9, 45, 246, 656
			885, 920, 1066, 1076
			1095, 1257, 750, 953
			1285, 1695
		(5 × 3)–O	8, 115
		c(14 × 7)–O	1332
	O(a) + CO	Disordered	1066
	O(a) + H_2O	(2 × 1)–O, H_2O	1023
		c(2 × 2)–O, H_2O	1023
		(2 × 1)–H_2O	1270
		(1 × 1)–H_2O, OH	1270
		(2 × 1)–OH, O	1270
	Pb	$\begin{bmatrix} 1 & 1 \\ -1 & 1 \end{bmatrix}$	481, 482

TABLE 2.5b. *(Continued)^a*

Substrate	Adsorbate	Surface Structure	References^b
		(5×1)	481, 482
		(4×1)	482
	Pd	(2×1)–Pd	726
		(1×1)–Pd	726
	Xe	$c(2 \times 2)$–Xe	159, 1331, 1611
		Hexagonal overlayer	159, 1611
		Pseudohexagonal	1331
Cu/Au(110)	[Clean]	Streak	737
		(1×2)	737
		Complex pattern	737
		$c(3 \times 1)$	737
		(2×2)	737
Cu/Ni(110)	CO	(2×1)–CO	134
		(2×2)–CO	134
		(1×2)–CO	787
	H_2	(1×3)–H	787
	H_2S	$c(2 \times 2)$–S	134
	O_2	(2×1)–O	134, 872
		(2×2)–O	872
Cu(110)–Ni(1°)	O_2	(2×1)–O	1311
		$c(6 \times 2)$–O	1311
Cu/Pd(110)	[Clean]	(2×1)	737
Cu–25% Zn(110)	[Clean]	(1×1)	1152
	O_2	Disordered	1152
Fe(110)	[Clean]	(1×1)	1015, 1639
	CO	$\begin{bmatrix} 3 & -2 \\ 0 & 4 \end{bmatrix}$–CO	346
		$c(2 \times 4)$	810
		(1×2)	810
		$c(2 \times 2)$	687
		(1×4)	687
		$\begin{bmatrix} 4 & 0 \\ -1 & 3 \end{bmatrix}$	687
	CO_2	$c(2 \times 2)$	687
		(1×4)	687
		$\begin{bmatrix} 4 & 0 \\ -1 & 3 \end{bmatrix}$	687
	Fe_3O_4	Not well-ordered	1189
	H_2	(2×1)–H	177, 1753
		(3×1)–2H	177, 1753
		(1×1)–H	177
		$c(2 \times 2)$–H	687, 1298
		(3×3)–6H	1298
		$\begin{bmatrix} 1 & -1 \\ 1 & 2 \end{bmatrix}$	687
	H_2S or S	(2×4)–S	114
		(1×2)–S	114
		(2×2)–S	836, 1000, 1015, 1608

TABLE 2.5b. (*Continued*)[a]

Substrate	Adsorbate	Surface Structure	References[b]
		c(3 × 1)–S	836
		c(18 × 3)–S	836
	K	Hexagonal array	728
	K + O$_2$	c(4 × 2)	786
	N$_2$	$\begin{bmatrix} 3 & -2 \\ 0 & 4 \end{bmatrix}$–N$_2$	346
	NH$_3$	(2 × 2)	687
		Disordered	1686
		$\begin{bmatrix} 4 & 1 \\ -3 & 3 \end{bmatrix}$	687
		(2 × 2)–NH	1686
	O$_2$	c(2 × 2)–O	87, 88, 99
		(2 × 2)–O	1015
		c(3 × 1)–O	87, 88, 99
		(2 × 8)–O	98
		FeO(111)	87, 88, 99, 269
		(2 × 1)–O	141
		(5 × 12)–O	1664
Fe/Cr(110)	O$_2$	Cr$_2$O$_2$(0001)	280
		Amorphous oxide	279
Ge(110)	[Clean]	c(8 × 10)	804, 1683
		Ge(17, 15, 1)–(2 × 1)	804, 1683
	H$_2$S	(10 × 5)–S	178
	O$_2$	Disordered	17, 18
		(1 × 1)–O	17, 18
GaAs(100)	[Clean]	(2 × 4)	1085, 1090, 1274, 1387
		(4 × 2)	1274
		c(4 × 4)	697, 1085, 1240 1387, 1541
		(4 × 6)	697, 1090, 1213, 1240 1377, 1448, 1541
		c(6 × 4)	697
		(1 × 1)	1213, 1519
		(2 × 8)	1449
		c(2 × 8)	697, 1090, 1213, 1240 1541
		c(8 × 2)	697, 1090, 1213, 1448 1541
		(8 × 2)–Ga	1519
		(1 × 2)	1448
		(1 × 6)	697, 1541
	[Laser process]	(1 × 1) + steps	1446
	Ag	c(8 × 2)	710
		c(2 × 8)	710

TABLE 2.5b. *(Continued)[a]*

Substrate	Adsorbate	Surface Structure	References[b]
		c(4 × 4)	710
		c(6 × 6) [multilayer]	1344
	Al	c(8 × 2)	710
		c(2 × 8)	710
		c(2 × 2) [multilayer]	1344
	As$_4$	c(4 × 4)	1422
	As$_4$, Ga	(2 × 4)	1365
		(4 × 6)	1365
		c(8 × 2)	1365
		(4 × 1)	1365
		(3 × 1)	1365
	Fe(CO)$_5$	$(1/\sqrt{2} \times 1/\sqrt{2})R45°$ [multilayer]	1377
	Bi	(1 × 1)–Bi	1491
		(1 × 2)–Bi	1491
		(3 × 1)–Bi	1491
		(8 × 2)–Bi	1491
	Ge	(1 × 2)–Ge	1213
		(1 × 2) + (2 × 1)	1213
	H$_2$	(1 × 1)	1541
	H$_2$S	c(2 × 8)–H$_2$S	589
		(2 × 1)	589
	HCl, H$_2$O	(1 × 1)	1518
	Pb	(1 × 4)–Pb	1387
	Pb, As$_4$	(1 × 2)–Pb	1387
	Sn	(1 × 3)–Sn	1387
	Sn	(1 × 2)–Sn	1387
GaAs(100)–As rich	[Clean]	c(4 × 4)	1214, 1524
GaAs(100)–Ga rich	[Clean]	(4 × 6)	1214, 1524
GaP(100)	[Clean]	(4 × 2)	694
	Cs	(1 × 4)–Cs	694
		(7 × 1)–Cs	694
		(1 × 4)–Cs	694
	PH$_3$	(1 × 2)	694
	Si	(2 × 1)	1554
Ge(110)	H$_2$S	(10 × 5)–S	178
	O$_2$	Disordered	17, 18
		(1 × 1)–O	17, 18
InP(100)	[Clean]	(4 × 2)	1170, 1384
	[Laser annealed]	(1 × 1)	1170, 1384
	[Laser annealed]	(1 × 1) + steps	1446
	Sb	(1 × 1)	1919, 1920
InSb(001)	[Clean]	c(2 × 8)	1159, 1421
	Sn	α-Sn(001)–(2 × 1)	1159
Ir(110)	[Clean]	(1 × 2)	701, 1321, 1665 1787, 1875
		(1 × 1)	1786
	CO	(2 × 2)–CO	347, 348
		(4 × 2)–CO	348

TABLE 2.5b. (*Continued*)[a]

Substrate	Adsorbate	Surface Structure	References[b]
	H_2	(1×2)	830
		Adsorbed	347
	H_2O	Adsorbed	615
	H_2S	(2×2)–2S	715, 1875
		(1×2)–S	715
		$c(2 \times 4)$–S	715
	N_2	(1×2)	678
		(2×2)–N_2	678
		Not adsorbed	347
	NO	Disordered	677
		Streaks	677
	O_2	(1×2)–O	347, 1687
		(1×4)–oxide	571, 706
		Disordered	571
		(2×2)–O	571, 706
		$c(2 \times 2)$–O	571, 706, 1788
		(3×2)	706
		(1×1)–O	1676, 1687
$LaB_6(110)$	[Clean]	$c(2 \times 2)$	775, 1328
	O_2	(1×1)–O	349, 1328
Mo(110)	[Clean]	(1×1)	1634
	Al	Hexagonal	515
	Au	Disordered	1681
	C	(4×6)–C	1250
	Cl_2	(2×1)–Cl	1250
		(1×1)–Cl	1250
		(1×2)–Cl	1250
		(1×3)–Cl	1250
	CO	(1×1)–CO	62, 100
		$c(2 \times 2)$–CO	94
		Disordered	406
	CO_2	Disordered	94
	Cs	Hexagonal	512
	H_2	Adsorbed	100
	H_2S	(2×2)–S	351
		$c(2 \times 2)$–S	351
		(1×1)–S	351
		$c(1 \times 3)$–S	351
		$c(1 \times 5)$–S	351
		(1×3)–S	351
		$c(1 \times 7)$–S	351
		(1×4)–S	351
		(1×5)–S	351
		$c(1 \times 11)$–S	351
		$\begin{bmatrix} 2 & 2 \\ -1 & 1 \end{bmatrix}$	351
	K	Hexagonal	512
	KCl	Disordered	781
	N_2	(1×1)–N	62

TABLE 2.5b. *(Continued)*[a]

Substrate	Adsorbate	Surface Structure	References[b]
	Na	No ordered structure	512
	O_2	(2×2)-O	62, 63, 100, 1154
		(2×1)-O	62, 63, 100
		(1×1)-O	62, 63
		Disordered	350
		Complex	1154
	Rb	Hexagonal	512
$MoO_3(010)$	[Clean]	(1×1)	922, 1128
	H_2	3D-MoO_2	922
Na(110)	[Clean]	(1×1)	1754, 1755
	O_2	$Na_2O(111)$	352
$Na_{2/3}WO_3(110)$	[Clean]	(3×1)	906
Nb(110)	CO	Disordered	101
		(3×1)-O	101
	H_2	(1×1)-H	111
	O_2	(3×1)-O	101
		NbO(111)	192
		NbO(110)	192
		NbO(220)	192
		Oxide	101
		Complex pattern	1688
	Sn	Disordered	505
		(3×1)	505
Ni(110)	[Clean]	(2×1)	882
		(1×1)	890, 1061, 1459 1468, 1469, 1470 1756, 1757, 1758, 1853
	C	c(4×5)-C	1198
	Cl_2	c(2×2)-Cl	1341
		(10×2)-Cl diffuse	1341
	CO	(1×1)-CO	2, 94
		Adsorbed	198
		c(2×1)-CO	353, 356, 359, 645
		(2×1)-CO	356, 357, 358
		c(2×2)-CO	359, 645
		(4×2)-CO	359, 645
	CO + O_2	(3×1)-CO + O_2	91
	Cs	Disordered	455
	D_2	(2×1)-D	869, 944, 1097
		(1×2)-D	869, 944, 1097
	H_2	(1×2)-H	59, 81, 94, 110, 198
			203, 353, 360, 867
			941, 927, 1031, 1074, 1527, 1673

TABLE 2.5b. (*Continued*)[a]

Substrate	Adsorbate	Surface Structure	References[b]
		(2×1)–2H	867
		(2×1)–H	941, 890, 1031,
			1074
			1527
		$c(2 \times 6)$–H	890, 1031
		(2×6)–H	1074
		(2×3) with streaks	1031
		$c(2 \times 4)$–H	1031
	H_2O	(2×1)–H_2O	110
	H_2S (or S)	$c(2 \times 2)$–S	36, 198, 205, 294
			1079, 1142, 1370
			1759, 1853
		(2×2)–S	1079, 1867
		(3×2)–S	36
		(1×1)–S	1142
	H_2Se	$c(2 \times 2)$–Se	137, 1708
	K	Disordered	455
	N_2	(2×1)–N_2	1074, 1309
		$(\frac{2}{3} \times \frac{1}{3})$–$N_2$	1074
		(2×3)–N	759
		$c(2 \times 4)$–N_2	759
	N_2^+	Disordered	1290
		(2×3)–N	1290
	Na	Disordered	455, 458, 460
		Hexagonal	455, 458, 460
	NH_3	(1×1)–NH_3	840, 880, 1560
		(4×2)–NH_3	840
		$c(6 \times 2)$–NH_3	840
		$c(4 \times 2)$–NH_3	840, 880
		(3×2)–NH	1107
	$NH_3 + e^-$	$c(2 \times 2)$–NH_2	840
		(2×3)–N	840
	NO	(2×3)–N	361
		(2×1)–O	361
	O_2	(2×1)–O	2, 3, 51, 57, 83,
			89
			91, 92, 99, 198,
			353
			354, 355, 729,
			912, 968
			1069, 1074, 1140
			1164, 1168, 1290
			1292, 1370, 1437
		(3×1)–O	2, 51, 83, 89, 91,
			92
			94, 198, 353, 354
			355, 912, 1011,
			1041
			1437

TABLE 2.5b. *(Continued)^a*

Substrate	Adsorbate	Surface Structure	References^b
		$(2 \times 1) + (3 \times 1)$–O	968
		(5×1)–O	2, 89
		(9×4)–O	51, 354, 355, 1437
		Disordered	1437
		NiO(100)	6, 51, 83, 91, 198 354, 355
		Disordered oxide	1437
	$O_2 + H_2O$	(2×1)–OH	1011
	Pb	$\begin{bmatrix} 1 & 1 \\ 1 & -1 \end{bmatrix}$	471
		(3×1)	471
		(4×1)	471
		(5×1)	471
	Se	$c(2 \times 2)$–Se	1370
	Te	$c(2 \times 2)$–Te	1370
	Yb	(2×1)–Yb	844
NiAl(110)	[Clean]	(1×1)	1771
Ni–25% Fe(110)	H_2S, H_2	(2×3)–S	1121
Ni$_4$Mo(101)	[Clean]	Ordered	1115
Pd(110)	[Clean]	(1×1)	1760
	Cl_2	$c(16 \times 2)$–Cl	1341
	CO	(5×2)–CO	95
		(2×1)–CO	95, 209
		(4×2)–CO	209
		$c(2 \times 2)$–CO	209
		$c(4 \times 2)$–CO	1359
		(4×1)–CO	1359
		$c(4 \times 2)$–CO imperfect	1251
	Cs	(1×2) + disordered Cs	1760
	H_2	(1×2)–H	212, 1173
		(2×1)–H	1173
	H_2S	(2×3)	625
		$c(2 \times 2)$	625
		$c(8 \times 2)$	625
		(3×2)	625
	Na	(1×2) + disordered Na	1760
	O_2	(1×3)–O	95
		(1×2)–O	95
		$c(2 \times 4)$–O	95
	Xe	Hexagonal	743
Pt(110)	[Clean]	(1×2)	960, 1062, 1080, 1166, 1187 1271, 1279, 1297 1761, 1890
		(1×1)	1279
	C_2N_2	(1×1)	407, 435
	C_3O_2	(1×1)–C$_3$O$_2$	365
	Cl_2	(1×2)–Cl	1341

TABLE 2.5b. (*Continued*)[a]

Substrate	Adsorbate	Surface Structure	References[b]
		(1 × 1)–Cl	1341
		(2 × 1)–Cl diffuse	1341
	CO	(2 × 1)–CO	366, 981, 1271
			1279, 1297, 1360
		(1 × 1)–CO	139, 364, 763,
			1271
			1279, 1297, 1360
		(1 × 2)–CO	1360
		(1 × 1) + (1 × 2)	1297, 1360
		c(8 × 4)–CO	1271, 1279, 1360
	CO + NO	(1 × 1)–CO + NO	364
	H$_2$	(1 × 1)	1279
		(1 × 2)	1279
	H$_2$S	c(2 × 6)–S	247, 367, 368,
			1114
		(2 × 3)–S	247, 367, 368,
			1114
		(4 × 3)–S	247, 367, 368,
			1114
			1116
		c(2 × 4)–S	247, 367, 368,
			1114
		(4 × 4)–S	247, 367, 1114
	HCN	$\begin{bmatrix} 1 & \frac{2}{3} \\ -1 & \frac{2}{3} \end{bmatrix}$	434
		c(2 × 4)	434
		(1 × 1)	434
	HNCO	(2 × 2)–NCO	657
		(1 × 2)–NCO	657
	NO	(1 × 1)–NO	222, 364
		(2 × 1)–NO	614
		c(4 × 8)–NO	614
		Disordered	614
	O$_2$	(2 × 1)–O	11,363
		(4 × 2)–O	11
		Adsorbed	362
		c(2 × 2)–O	363
		PtO(100)	363
		(1 × 1)–CO	139, 364
		(1 × 3)	763
		(1 × 5)	763
		(1 × 7)	763
		Satellite spots	1279
Pt–2% Cu(110)	[Clean]	(1 × 3)	1062
	CO	(1 × 1)–CO	1063
Re(10$\bar{1}$0)	[Clean]	(1 × 1)	584
	Ba	c(2 × 2)	1591
	Mg	(1 × 3)	1675

TABLE 2.5b. *(Continued)*[a]

Substrate	Adsorbate	Surface Structure	References[b]
Re(11$\bar{2}$0)	After NH$_3$ synthesis	(1 × 1)	977
Rh(110)	[Clean]	(1 × 1)	1800
	CO	(2 × 1)–CO	369
		c(2 × 2)–C	369
		Disordered	569
	H$_2$S	c(2 × 2)–S	769
	NO	Disordered	569, 791
		(2 × 2)–N, O	569, 791
		(2 × 1)–N, O	569, 791
	O$_2$	Disordered	96, 97
		c(2 × 4)–O	96, 97
		c(2 × 8)–O	96, 97
		(2 × 2)–O	96, 97
		(2 × 3)–O	96, 97
		(1 × 2)–O	96, 97
		(1 × 3)–O	96, 97
	S or H$_2$S	c(2 × 2)–S	769, 1473
Ru(101)	CO	$\begin{bmatrix} 1 & 1 \\ 3 & 0 \end{bmatrix}$–CO	372
		$\begin{bmatrix} 0 & 1 \\ 2 & 0 \end{bmatrix}$–C	372
	NO	Disordered	373
	O$_2$	$\begin{bmatrix} 1 & 1 \\ 3 & 0 \end{bmatrix}$–O	374
		$\begin{bmatrix} 2 & 1 \\ 5 & 0 \end{bmatrix}$–O	374
		$\begin{bmatrix} 4 & 1 \\ 9 & 0 \end{bmatrix}$–O	374
Ru(10$\bar{1}$0)	Cl$_2$	(1 × 1)–Cl	1052
		(2 × 3)–Cl	1052
		$\begin{bmatrix} 2 & 0 \\ -1 & 3 \end{bmatrix}$–Cl	1052
		$\begin{bmatrix} 2 & 0 \\ -1 & 4 \end{bmatrix}$–Cl	1052
		(2 × 1)–Cl	1052
	CO	Disordered	371
	H$_2$	Not adsorbed	371
	N$_2$	Not adsorbed	371
	NO	c(4 × 2)–N + O	370, 371
		(2 × 1)–N + O	370, 371
		(2 × 1)–O	371
		c(4 × 2)–O	371
		c(2 × 6)–O	370
		(7 × 1)–O	370
		c(4 × 8)–O	370

TABLE 2.5b. (*Continued*)[a]

Substrate	Adsorbate	Surface Structure	References[b]
		(2×1)–N	371
		$c(4 \times 2)$–N	371
	O_2	$c(4 \times 2)$–O	370, 371
		(2×1)–O	370, 371
		$c(2 \times 6)$–O	370
		(7×1)–O	370
		$c(4 \times 8)$–O	370
Si(110)	[Clean]	(4×5)	803, 1685
		(2×1)	803, 1685
		(5×1)	803, 1685
	Bi	(2×3)–Bi	659
		Disordered	659
	H_2	(1×1)–H	375
	H_2O	Adsorbed	903
	Si, laser	(1×2)	1392
$SnO_2(101)$	[Clean]	(1×1)	1183
$SrTiO_3(110)$	[Clean]	(3×2)	1490
Ta(110)	Al	Hexagonal	508, 509
		Square	508, 509
	Cl_2	(1×1)–Cl	1180
		(1×2)–Cl	1180
		$c(1 \times 5)$–Cl	1180
		$c(1 \times 7)$–Cl	1180
		Streak $\langle 001 \rangle$	1180
		Complicated	1180
	CO	Disordered	101, 102
		(3×1)–O	101, 102
	H_2	(1×1)–H	102
	I_2	$\begin{bmatrix} 3 & 0 \\ -1 & 1 \end{bmatrix}$–I	1180
		$(1 \times 1) + c(1 \times 3)$–I	1180
		(1×1) with ring	1180
		$c(4 \times 4)$–I	1180
	N_2	Not adsorbed	101
	O_2	(3×1)–O	101, 102
		Oxide	101, 102
$TiO_2(100)$	[Clean]	(011)–(2×1) facet	1318, 1615
		(114) facet	1318
	H_2O	Disordered	376
	O_2	Disordered	376
$TiO_2(110)$	[Clean]	(1×1)	1615
V(110)	[Clean]	(1×1)	649, 1498, 1762
	CO	Disordered	101
		(3×1)–O	101
	O_2	(3×1)–O	101
$V_6O_{13}(001)$	K	No superstructure	1186
W(110)	[Clean]	(1×1)	1123, 1247, 1763
	Ag	Hexagonal structures	546, 547
		Ag(111)	1151

TABLE 2.5b. *(Continued)*[a]

Substrate	Adsorbate	Surface Structure	References[b]
	Au	Hexagonal structures	546, 548
	Ba	Disordered hexagonal	533–535
		Hexagonal	533–535
		$\begin{bmatrix} 2 & 2 \\ 0 & 6 \end{bmatrix}$	533–535
		$\begin{bmatrix} 2 & 2 \\ 0 & 5 \end{bmatrix}$	533–535
		$\begin{bmatrix} 3 & 3 \\ 1 & 5 \end{bmatrix}$	533–535
		Hexagonal compact	533–535
	Be	(1×9)	529
		(1×1)	529
		$\begin{bmatrix} 9 & 0 \\ -1 & 1 \end{bmatrix}$	529
	Cl_2	(5×2)–Cl	796
	CO	Disordered	109
		$c(9 \times 5)$–CO	109
		(1×1)–CO	379
		$c(2 \times 2)$–CO	379
		(2×7)–CO	389
		$c(4 \times 1)$–CO	389
		(3×1)–CO	389
		(4×1)–CO	389
		(5×1)–CO	389, 390
		(2×1)–C + O	389, 390
		$c(9 \times 5)$–C + O	389
	CO + O_2	$c(11 \times 5)$–CO + O_2	93
	Cs	Disordered hexagonal	523, 527, 528
		Hexagonal	523, 527, 528, 1677
	Cu	Hexagonal	543–545
		Cu(111)	1151
	Fe	Three-dimensional crystals	451
		Fe(110)	1325
		(1×1)	1325
	H_2 or D_2	(2×1)–H	136, 1516, 1674
		(1×2)–H	672
		(2×2)–H	1516, 1674
		(1×1)–H	1516
		Ordered	1438
		(2×2)–H_2	1516
		(2×1)–H_2	1516
	I_2	(2×2)–I	391
		(2×1)–I	391
	Li	$\begin{bmatrix} 1 & 5 \\ -2 & 2 \end{bmatrix}$	517–519

TABLE 2.5b. (*Continued*)[a]

Substrate	Adsorbate	Surface Structure	References[b]
		(2×2)	517–519
		$\begin{bmatrix} 1 & 1 \\ -1 & 2 \end{bmatrix}$	517–519
		(2×3)-Li	1269
		$c(2 \times 2)$-Li	1269
		$c(3 \times 1)$-Li	1269
		$c(1 \times 1)$-Li	1269
	N_2	(2×2)-N	758
	Na	$\begin{bmatrix} 1 & 5 \\ -2 & 2 \end{bmatrix}$	445, 446
		(2×2)	445, 446
		$\begin{bmatrix} 1 & 1 \\ -1 & 2 \end{bmatrix}$	445, 446
		$\begin{bmatrix} 1 & 1 \\ 0 & 8 \end{bmatrix}$	445, 446
		$\begin{bmatrix} 1 & 1 \\ 0 & 5 \end{bmatrix}$	445, 446
		Hexagonal	445, 446
	Ni	(1×1)-Ni	970
		(8×2)-Ni	970
		(7×2)-Ni	970
	NO	(1×1) streaked	661
		$c(11 \times 5)$	661, 799
		(2×2)	661, 799
	O_2	(2×1)-O	57, 103, 377–385, 386, 387 599, 699, 1277, 1418 1587, 1651
		$c(2 \times 2)$-O	104
		(2×2)-O	104, 387, 599, 699
		(1×1)-O	104
		$c(14 \times 7)$-O	57, 103, 104, 628
		$c(21 \times 7)$-O	104
		$c(48 \times 16)$-O	104
		$WO_3(100)$	388
	Pd	(1×3)	542
		Hexagonal	542
	Pd(1ML) + CO	Not adsorbed	1218
	Pd(2.2ML) + O_2	(2×2)-O	1218
	Pb	Split $\begin{bmatrix} 3 & 0 \\ -1 & 1 \end{bmatrix}$	551, 552
		$\begin{bmatrix} 3 & 0 \\ -1 & 1 \end{bmatrix}$	551, 552

TABLE 2.5b. *(Continued)*[a]

Substrate	Adsorbate	Surface Structure	References[b]
	S_2	(2 × 2)–S	1246
		(7 × 2)–S	1246
		Rotated structure	1246
		(1 × N)–S (N ≥ 3)	1246
	Sb	$\begin{bmatrix} 1 & 1 \\ 0 & 4 \end{bmatrix}$	553, 555
		$\begin{bmatrix} 2 & 0 \\ -1 & 1 \end{bmatrix}$	553, 555
		$\begin{bmatrix} 3 & 0 \\ -1 & 1 \end{bmatrix}$	553, 555
	Sc	$\begin{bmatrix} 1 & 1 \\ 0 & 3 \end{bmatrix}$	536–538
		$\begin{bmatrix} 2 & 2 \\ 0 & 8 \end{bmatrix}$	536–538
	Se	(5 × 2)–Se	1228
		(1 × 3)–Se	1228
		Complex	1228
	Sr	$\begin{bmatrix} 3 & 3 \\ -2 & 5 \end{bmatrix}$	530
		$\begin{bmatrix} 2 & 2 \\ 0 & 6 \end{bmatrix}$	530
		$\begin{bmatrix} 2 & 2 \\ 1 & 6 \end{bmatrix}$	530
		$\begin{bmatrix} 1 & 0 \\ 0 & 3 \end{bmatrix}$	530
		Hexagonal	530
	Te	(4 × 2)–Te	1280
		(20 × 2)–Te	1280
		(17 × 2)–Te	1280
		(5 × 2)–Te	1280
		(22 × 2)–Te	1280
	W	Ring pattern	1623
	WO_2	(2 × 2)	1501
	Xe	(2 × 2)–Xe	713
		Disordered	713
	Y	Hexagonal	539, 540
ZnSe(100)	[Clean]	($\sqrt{2} \times \sqrt{2}$)$R45°$	1393
		(5 × 1)	1393
ZnTe(100)	[Clean]	(1 × 3), (1 × 1) + {110} f	1393
	Au	(1 × 1)–Au	1188

[a]Organic overlayer structure are not included. See Table 2.5g for these structures.
[b]References for Table 2.5a to 2.5j are listed at the end of Table 2.5j.

TABLE 2.5c. Surface Structures on Substrates with Threefold Rotational Symmetry[a]

Substrate	Adsorbate	Surface Structure	References[b]
Ag(111)	[Clean]	(1×1)	975, 1894, 1895, 1896
	Al	Disordered	491
	Au	(1×1)	491, 1355, 1825 1826
	Bi	Disordered	491
	Br$_2$	$(\sqrt{3} \times \sqrt{3})R30°$–Br	155
		(3×3)–Br	155
	Cd	No condensation	491
	Cl$_2$	(1×1) + disordered	1050
		$(\sqrt{3} \times \sqrt{3})R30°$–Cl	151, 1050
		(10×10)–Cl	151
		AgCl(111)	152, 732, 1050
	Co	Disordered	491
	CO + O$_2$	$(2 \times \sqrt{3})$–(CO + O$_2$)	27
	Cr	Disordered	491
	Cu	Hexagonal overlayer	1822–1824
	H$_2$O	Disordered	1034
	H$_2$S	(4×4)–S	627
		$\begin{bmatrix} 3 & 2 \\ -2 & 1 \end{bmatrix}$–S	627
	I$_2$	$(\sqrt{3} \times \sqrt{3})R30°$–I	145, 149, 150, 1145, 1225, 1259, 1440
		Hexagonal overlayer	1145
	K	Hexagonal overlayer	1345
	Kr	Hexagonal overlayer	156
	Mg	Disordered	491
	Na	(1×1)	488
	Ni	Hexagonal overlayer	491, 1821
	NO	Disordered	163
	O$_2$	(2×2)–O	1
		$(\sqrt{3} \times \sqrt{3})R30°$–O	1
		Not adsorbed	146
		(4×4)–O	147, 148
	Pb	$(\sqrt{3} \times \sqrt{3})R30°$–Pb	975
		Pb(111)	975
		Hexagonal overlayer	491, 1827
	Pd	(1×1)	1463
		Disordered	491
	Rb	(1×1)–Rb	490, 705
		(9×9)	705
	S$_2$	$(\sqrt{39} \times \sqrt{39})R16.1°$–S	714
		$(\sqrt{7} \times \sqrt{7})R10.9°$ of γ–Ag$_2$S(111)	714
	Sb	Disordered	491
	Sn	Disordered	491
	Tl	Hexagonal overlayer	491
	Xe	Hexagonal overlayer	156–159 160

TABLE 2.5c. *(Continued)*[a]

Substrate	Adsorbate	Surface Structure	References[b]
		Incommensurate	1599, 1845
	Zn	No condensation	491
Ag(111)–Rb dosed	O$_2$	$(2\sqrt{3} \times 2\sqrt{3})R30°-$ Rb/O	653
		(4 × 4)–Rb/O	653
		Complex structures	653
		(9 × 9)–Rb/O	653
Al(111)	[Clean]	(1 × 1)	863, 951, 1141, 1354, 1467, 1472, 1498, 1640
	Ag	Ag–Al(0001)	1161
	CO	Disordered	1175
		Not adsorbed	1273
	Cu	(1 × 1)	863
		Disordered [multilayer]	863
		Cu(111) [multilayer]	863
	H$_2$O	Disordered	1157
	Mn	$\begin{bmatrix} 6 & 0 \\ -1 & 2 \end{bmatrix}$	502
		Hexagonal rotated ± 9°	502
	Na	$\begin{bmatrix} 1 & -1 \\ 1 & 2 \end{bmatrix}$	500
		(2 × 1)	500
	Ni	(1 × 1)	1680
		$\begin{bmatrix} 1 & 1 \\ -1 & 2 \end{bmatrix}$	1839
	O$_2$	(4 × 4)–O	123
		(1 × 1)–O	638, 709, 756, 951, 1141, 1397, 1467, 1621, 1637, 1774, 1775
		Oxide-like	1141
	Pb	Hexagonal rotated ± 9°	504
		Hexagonal overlayer	1060
	Pd	Hexagonal overlayer	1682
		$(\sqrt{3} \times \sqrt{3})R30°-$Pd	1661
	Sn	Hexagonal rotated ± 9°	504
		Hexagonal overlayer	1060, 1682
	Tl	$(\sqrt{3} \times \sqrt{3})R30°-$Tl	1661
Au(111)	[Clean]	(23 × 1)	861, 1146, 1558, 1889
		(5 × 1)	1146
	Ag	(1 × 1)	491, 1825
		fcc(111)	1689
	Ag, air	Ag$_2$O(110)–(2 × 1)	997
	Bi	$\begin{bmatrix} 10 & 10 \\ -10 & 20 \end{bmatrix}$	498
		(2 × 2)	924
	Cl$_2$	(1 × 1)–Cl	647

TABLE 2.5c. *(Continued)*[a]

Substrate	Adsorbate	Surface Structure	References[b]
	Cr	Hexagonal	493
	Cu	$(\sqrt{3} \times \sqrt{3})R30°$–Cu	861, 1558, 1582
		(1×1)	1558
		Extra lines (RHEED)	1836, 1837
	Fe	(1×1)	1828, 1830–1832
	O_2	Oxide	161
		Not adsorbed	162
		Adsorbed	162
	Pb	Hexagonal rotated $\pm 5°$	444, 495
		$(\sqrt{3} \times \sqrt{3})R30°$	924
		(1×1)	683
	Pd	(1×1)	1807, 1834
	Pt	(1×1)	1835
	Si	(2×2)–AuSi	1622
		(3×3)–AuSi	1622
		Hexagonal silicide	1622
Be(0001)	[Clean]	(1×1)	911, 1900, 1901
	CO	Disordered	22
	H_2	Not adsorbed	22
	N_2	Not adsorbed	22
	O_2	Disordered	22
		BeO(0001)–(1×1)	911
		BeO(0001)–(2×2)	911
Bi(0001)	O_2	$(\sqrt{3} \times \sqrt{3})R30°$	576, 1065, 1288
		(1×1)	576
		Coincidence lattice	576, 1065
		BiO	1288
	O_2 + K	$(\sqrt{3} \times \sqrt{3})R30°$ + BiO(0001) layer	1288
	Cl_2	(1×1)–Cl	1242
		$(2\sqrt{3} \times 2\sqrt{3})R30°$– $BiCl_3$	1242
		(4×4)	1242
C(111), diamond	[Clean]	(2×2)	820
		(2×1)	820
		(1×1)	1551
	H_2 (or D_2)	(1×1)–H(or D)	30, 1386, 1697
	H_2S	Not adsorbed	164
	N_2	Not adsorbed	164
	NH_3	Not adsorbed	164
	O_2	Adsorbed	16
		Not adsorbed	164
	P	$(\sqrt{3} \times \sqrt{3})R30°$–P	30
C(0001), graphite	[Clean]	"(2×2)"	1190
		(1×1)	1373, 1439, 1846
	Ar	$(\sqrt{3} \times \sqrt{3})R30°$–Ar	720, 960
		Incommensurate	1882, 1903
	Ar + Xe	$(\sqrt{3} \times \sqrt{3})R30°$–Ar, Xe	1193

TABLE 2.5c. *(Continued)*[a]

Substrate	Adsorbate	Surface Structure	References[b]
	CF_4	$(2 \times 2)-CF_4$	1192, 1194
		Close to (2×2)	1404
	CO	$(\sqrt{3} \times \sqrt{3})R30°-CO$	884
		$(2\sqrt{3} \times 2\sqrt{3})R30°-CO$	889
		$(2\sqrt{3} \times \sqrt{3})R30°$ [herringbone]	884
		Triangular incommensurate (2×2)	884
	H_2	$(\sqrt{3} \times \sqrt{3})R30°-H_2$	1283
		(2×2)	1373
		$(\sqrt{3} \times \sqrt{3})R30°$	1373
	K (Intercalated)	Disordered	1847
	KOH	Disordered	1083
	Kr	$(\sqrt{3} \times \sqrt{3})R30°-Kr$	166, 167, 174, 721, 828, 960, 1616, 1904
		Incommensurate	1413, 1616
		Disordered	1413
	N_2	$(2\sqrt{3} \times 2\sqrt{3})R30°-N_2$	889
		$(\sqrt{3} \times \sqrt{3})R30°$	1064, 1190, 1512
		$(\sqrt{3} \times \sqrt{3})R30° + (2 \times 1)$	1435
		Commensurate	1190, 1883
		Incommensurate	1443, 1883
	NaOH	$\frac{1}{2}$ Order ring	1083
	Ne	Incommensurate	629
	Ne	$(\sqrt{3} \times \sqrt{3})R30°$ rotated by $\pm 17°$	629, 960
		Ordered	1338
	NO	Incommensurate	1602
	O_2	Triangular	1883
		Centered-parallelogram-O_2	1425, 1883
		$(\sqrt{3} \times \sqrt{3})R30°-O_2$	1200
		Physisorbed	1411
	Xe	$(\sqrt{3} \times \sqrt{3})R30°-Xe$	165, 618, 960, 1038, 1201
Cd(0001)	[Clean]	(1×1)	1902
CdS(0001)	O_2	Disordered	25
CdTe(111)	[Clean]	(2×2)	1393
CdTe(111)	[Clean]	$(1 \times 1), (1 \times 1) + \{110\}f$	1393
Co(0001)	[Clean]	(1×1)	1130, 1580, 1613
	CO	$(\sqrt{3} \times \sqrt{3})R30°-CO$	168, 1130, 1362
		$(2\sqrt{3} \times 2\sqrt{3})R30°-CO$	1130, 1362
		$c(4 \times 2)-CO$	1362, 1581
		$(\sqrt{7}/2 \times \sqrt{7}/2)R19.10°-CO$	1362
		Hexagonal overlayer	168

TABLE 2.5c. *(Continued)*[a]

Substrate	Adsorbate	Surface Structure	References[b]
		$(\sqrt{7/3} \times \sqrt{7/3})$ $R10.9°$–CO	1130
	H_2O	Disordered	1310
	NO	$(\sqrt{39} \times \sqrt{39})R16.1°$– N, O	788
	O_2	No superstructure	1235
Co(111)	[Clean]	(1×1)	1613
CoO(111)	[Clean]	(1×1)	1769
Cr(111)	Ag	(8×8)	46
	Au	$\begin{bmatrix} \frac{2}{3} & 3 \\ -\frac{2}{3} & \frac{4}{3} \end{bmatrix}$	52
	Bi	$\begin{bmatrix} 1 & 1 \\ -1 & 2 \end{bmatrix}$	61
		$\begin{bmatrix} 2 & -1 \\ 0 & 2 \end{bmatrix}$	61
		$\begin{bmatrix} 2 & 3 \\ 1 & 2 \end{bmatrix}$	61
	Fe	(1×1)	39
	Ni	(1×1)	43, 44
	O_2	$(\sqrt{3} \times \sqrt{3})R30°$–O	169
	Pb	(4×4)	55, 58
	Sn	$\begin{bmatrix} 1 & 1 \\ -1 & 2 \end{bmatrix}$	54
Cu(111)	[Clean]	(1×1)	1101, 1408, 1419, 1462, 1510, 1538, 1635
	Ag	(8×8)	477
		Three-dimensional crystals	1813, 1815–1818
		(1×1)	1526
	Au	$\begin{bmatrix} \frac{2}{3} & \frac{2}{3} \\ -\frac{2}{3} & \frac{4}{3} \end{bmatrix}$	479
		(2×2)	479
		Three-dimensional crystals	1815, 1819
	Bi	$\begin{bmatrix} 1 & 1 \\ -1 & 2 \end{bmatrix}$	487
		$\begin{bmatrix} 2 & -1 \\ 0 & 2 \end{bmatrix}$	487
		$\begin{bmatrix} 2 & 3 \\ 1 & 2 \end{bmatrix}$	487
	C_2N_2	Disordered	666
	C	Disordered	840, 1695

TABLE 2.5c. (*Continued*)[a]

Substrate	Adsorbate	Surface Structure	References[b]
	Cl_2	$(\sqrt{3} \times \sqrt{3})R30°$–Cl	151, 1102
		$(6\sqrt{3} \times 6\sqrt{3})R30°$–Cl	151
		$(12\sqrt{3} \times 12\sqrt{3})R30°$–Cl	151
		$(4\sqrt{7} \times 4\sqrt{7})R19.2°$–Cl	151
	CO	Not adsorbed	26
		$(\sqrt{3} \times \sqrt{3})R30°$	172, 173, 590, 1306
		$(1.5 \times 1.5)R18°$	590
		(1.39×1.39)	591–593
		$(\sqrt{7}/3 \times \sqrt{7}/3)R49.1°$	172, 173
		$(\frac{3}{2} \times \frac{3}{2})$	173
	Co	(1×1)–Co	1299
	Cs	(2×2)–Cs	711, 1430
	Fe	(1×1)	474
	H_2	Not adsorbed	7
	H_2S	$(\sqrt{3} \times \sqrt{3})R30°$–S	35
		Adsorbed	35
	HNCO	Disordered	624
	I_2	$(\sqrt{3} \times \sqrt{3})R30°$–I	574, 1259, 1779
	Na	(2×2)–Na	1571
	Ni	(1×1)–Ni	475, 476, 1466, 1813
	$Ni(CO)_4$ + CO	(1×1) + disordered	1048
	O_2	Disordered	7, 170, 171, 1095, 1244
		(7×7)–O	7, 8
		$(\sqrt{3} \times \sqrt{3})R30°$–O	7, 8, 1286
		(2×2)–O	7, 8, 115
		(3×3)–O	8
		$(11 \times 5)R5°$–O	9
		$(2 \times 2)R30°$–O	115, 119
		$(2 \times 2)R30°$–O	115, 119
		$\begin{bmatrix} 3 & 2 \\ -1 & 2 \end{bmatrix}$–O	1066
		Hexagonal	246
	O_2 + HCN	Disordered	1244
	O(a) + CO	Disordered	1066
	Pb	(4×4)	481, 484
	Pd	(1×1)	726, 1144, 1538
	Sn	$\begin{bmatrix} 1 & 1 \\ -1 & 2 \end{bmatrix}$	480
	Te	$(2\sqrt{3} \times \sqrt{3})R30°$	1905
	Xe	$(\sqrt{3} \times \sqrt{3})R30°$–Xe	159
Cu/Al(111)	[Clean]	(1×1)	813
		$(\sqrt{3} \times \sqrt{3})R30°$	813
Cu–5.7% Al(111)	[Clean]	(1×1)	813
Cu–10% Al(111)	[Clean]	(1×1)	1303
		$(\sqrt{3} \times \sqrt{3})R30°$–Al	1303

TABLE 2.5c. *(Continued)*[a]

Substrate	Adsorbate	Surface Structure	References[b]
Cu–11% Al(111)	[Clean]	(1×1)	1506
Cu–12.5% Al(111)	[Clean]	$(\sqrt{3} \times \sqrt{3})R30°$	813
Cu–16% Al(111)	[Clean]	$(\sqrt{3} \times \sqrt{3})R30°$	1699
Cu/Au(111)	[Clean]	$(\frac{2}{3}\sqrt{3} \times \frac{2}{3}\sqrt{3})R30°$	737
		(2×2)	737
Cu/Ni(111)	CO	Disordered	173, 734
Cu/Pd(1$\bar{1}$1)	[Clean]	(1×1)	737
Cu–25% Zn(111)	[Clean]	(1×1)	1152
	O_2	Disordered	1152
Fe(111)	[Clean]	(1×1)	1700, 1701
	CO	Disordered	1004
		(1×1)	687
		(5×5)	687
		(3×3)	687
	CO_2	(1×1)	687
		(5×5)	687
		(3×3)	687
	H_2	Adsorbed	177
		(1×1)	687
	H_2O	(1×1)–H_2O	1588
	K	(3×3)–K	665, 1350
	N_2	$c(2 \times 2)$–N	1350
		(3×3)–N	1350
	N(a) + K	(3×3)–K, N	1350
	NH_3	Disordered	176, 687
		(3×3)–N	176, 687
		(5×5)	687
		$(\sqrt{19} \times \sqrt{19})R23.4°$–N	176
		$(\sqrt{21} \times \sqrt{21})R10.9°$–N	176, 687
	O_2	(6×6)–O	175
		(5×5)–O	175
		(4×4)–O	175
		$(2\sqrt{7} \times 2\sqrt{7})R19.1°$–O	175
		$(2\sqrt{3} \times 2\sqrt{3})R30°$–O	175
	S	(1×1)–S	1577, 1655
Fe–18% Cr–12% Ni(111)	[Clean]	(1×1)	1249
	I_2	$(\sqrt{3} \times \sqrt{3})R30°$–I	1249
	H_2O	Ordered	1249
	O_2	Ordered	1249
	I(a) + H_2O	Oxide not formed	1249
	H_2O(a) + I_2	Adsorbed	1249
α-Fe_2O_3(001)	[Clean]	(2×2)	1118
		Incommensurate	1118
		$(\sqrt{3} \times \sqrt{3})R30°$	1118
FeTi(111)	[Clean]	(1×1)	1241
GaAs(111)	[Clean]	$c(8 \times 2)$	1170
	Laser-annealed	(1×1)	1170
		(2×2)	1090, 1702
GaAs($\overline{111}$)	[Clean]	(1×1)	1090
GaAs($\overline{111}$)	$Fe(CO)_5$	Facet{100}	1377

TABLE 2.5c. *(Continued)*[a]

Substrate	Adsorbate	Surface Structure	References[b]
GaAs($\overline{111}$)–As rich	[Clean]	(2×2)	1524, 1541
	H_2	(1×1)	1541
GaAs($\overline{111}$)–Ga rich	[Clean]	$(\sqrt{19} \times \sqrt{19})R23, 4°$	1541
		(1×1)	1541
GaP(111)	[Clean]	(2×2)	819, 1703
Ge(111)	[Clean]	(2×8)	804, 996, 1046, 1075
		(2×1)	1046, 1075, 1086, 1296, 1374, 1683
		(1×1)	1075, 1296, 1374
		$c(2 \times 8)$	1550
	Laser process	(1×1)	1492
	Al	(2×1)	1550
	Au	$(\sqrt{3} \times \sqrt{3})R30°$–Au	1223
	Cl_2 or Cl	(7×7)–Cl	1088
		(1×1)–Cl	1704
	H_2O	(1×1)–H_2O	121, 179, 1662
	H_2S	(2×2)–S	37
		(2×1)–S	178
	H_2Se	(2×2)–Se	37
	I_2 or I	(1×1)–I	19, 1088
	In	$(n \times 2\sqrt{3}$–In, $n = 10$– 13	802
		$(4\sqrt{3} \times 4\sqrt{3})R30°$–In	802
		$(\sqrt{31} \times \sqrt{31})R(\pm9°)$– In	802
		$(\sqrt{61} \times \sqrt{61})R(30 \pm 4°)$–In	802
		(4.3×4.3)–In	802
		(4×4)–In	802
	O_2	Disordered	17, 18
		(1×1)	19, 21
	P	(1×1)–P	19
	Pb	$(\sqrt{3} \times \sqrt{3})R30°$–Pb	1075, 1400, 1474
		(1×1)–Pb	1075, 1474
	S	$(\sqrt{3} \times \sqrt{3})R30°$	1589
		(3×3)	1589
		(2×8)–Ge_2S	1589
	Si	(1×1) with streaks	1029
	Sn	(2×8)–Sn	639
		$(\sqrt{3} \times \sqrt{3})R30°$–Sn	639, 1049
		(7×7)–Sn	639, 1049
		(5×5)–Sn	639, 1049
		$(3 \times 2\sqrt{3})$–Sn	996
		$(\sqrt{91} \times \sqrt{3})$–Sn	996
		(1×1)–Sn	996
	Te	(2×2)–Te	1088
InSb(111)	[Clean]	(2×2)	849, 852, 1906
	a–Sn	a–Sn(111) (1×1) [multilayer]	849

TABLE 2.5c. *(Continued)*[a]

Substrate	Adsorbate	Surface Structure	References[b]
InSb($\overline{111}$)	[Clean]	(3×3)	849, 852
	a–Sn	a–Sn(111) (1×1) [multilayer]	849
Ir(111)	[Clean]	(1×1)	1705
	Au	(1×1)	453
	CO	$(\sqrt{3} \times \sqrt{3})R30°$–CO	124, 180, 182, 183 185, 186
		$(2\sqrt{3} \times 2\sqrt{3})R30°$–CO	180, 182, 183, 185, 186
	Cr	Hexagonal	453
	H_2	Adsorbed	187
	H_2O	Not adsorbed	182
	H_2S	$(\sqrt{3} \times \sqrt{3})R30°$–S	822
	NO	(2×2)–NO	188
	O_2	(2×2)–O or (2×1)– O	124, 180–182, 183, 184, 827
		Ir oxide	181
LaB_6(111)	[Clean]	(1×1)	775, 1328
	O_2	(1×1)	1328
Mg(0001)	[Clean]	(1×1)	1289
	O_2	$(1 \times 1)R30°$– MgO(111)	655
		$(\sqrt{7} \times \sqrt{7})R19°$	655
		Disordered	1289
		(1×1)	797, 1289
		MgO(111)	1671
Mo(111)	[Clean]	(1×1)	1203
	H_2S	$c(4 \times 2)$–H_2S	191
		MoS_2(0001)	191
	KCl	Disordered	781
	$N_2 + NH_3$	Disordered	1203
	$N_2 + NH_3$	(433) facet	1203
		$c(3 \times 2)$–N/Mo(433)	1203
	O_2	(211) facets	14, 189
		(110) facets	189
		(4×2)–O	190
		(4×4)	898
		(1×3)	898
		(112)–(1×2) facets	898
		(112)–(1×3) facets	898
		MoO_2(100)	898
MoS_2(0001)	[Clean]	(1×1)	1706
	Cs	Amorphous layer	686, 855
MoSe(0001)	[Clean]	(1×1)	1035
	H_2O	Not adsorbed	1035
	$HClO_4$	Not adsorbed	1035
	I_2	Slightly adsorbed	1035
	NaI_3	Slightly adsorbed	1035
Na(0001)	[Clean]	(1×1)	1731

TABLE 2.5c. (*Continued*)

Substrate	Adsorbate	Surface Structure	References[b]
$Na_2O(111)$	[Clean]	(1×1)	1755
Nb(111)	O_2	(2×2)–O	192
		(1×1)–O	192
$NbSe_2(0001)$	[Clean]	(1×1)	1706
Ni(111)	[Clean]	(1×1)	1707, 1794
	Ag	(6×6)	465, 466
	Au	(6×6)	467, 468, 469, 470
		(13×13)	467, 468, 469
	Bi	$(\sqrt{3} \times \sqrt{3})R30°$–Bi	864
		(7×7)–Bi	864
		$(\sqrt{7/4} \times \sqrt{7/4})R19°$– Bi	864
	Cl_2	$\sqrt{3} \times \sqrt{3})R30°$–Cl	206
		$\begin{bmatrix} 2 & 1 \\ 4 & 7 \end{bmatrix}$–Cl	206
	CO	$(\sqrt{3} \times \sqrt{3})R30°$–CO	195, 196, 199, 200, 1314, 1795
		Hexagonal overlayer	200
		(2×2)–CO	3
		$(\sqrt{3} \times \sqrt{3})R30°$–O	5
		$(2 \times \sqrt{3})$–CO	5
		$(\sqrt{39} \times \sqrt{39})$–C	5, 27
		(1×1)–C (graphite)	1907
		Disordered	198, 1402
		$(\sqrt{7} \times \sqrt{7})R19.1°$	195, 196
		$(\sqrt{7}/2 \times \sqrt{7}/2)R19°$– CO	957, 1402
		$c(4 \times 2)$–CO	195, 196, 957, 1402
		$c(2 \times 2)$–CO	1314
		Complex pattern	1402
	CO_2	(2×2)–CO_2	5
		$(\sqrt{3} \times \sqrt{3})R30°$–O	5
		$(2 \times \sqrt{3})$–CO_2	5
		$(\sqrt{39} \times \sqrt{39})$–C	5, 27
	GeH_4	$(\sqrt{3} \times 2\sqrt{3})R30°$	907
		$(\sqrt{3} \times \sqrt{3})R30°$–Ge	907
		(1×1)–Ge	907
	H_2	(1×1)–H	3
		(2×2)–(2*)H	29, 201, 202, 204, 823, 1585, 1666
		(2×1)	1667
		Disordered	203
	H_2S	(2×2)–S	36, 118, 197, 198, 205, 294, 577, 990, 992, 1264, 1493
		$(\sqrt{3} \times \sqrt{3})R30°$–S	36, 118, 577, 1264
		(5×5)–S	36
		Adsorbed	36

TABLE 2.5c. *(Continued)*[a]

Substrate	Adsorbate	Surface Structure	References[b]
		$(5\sqrt{3} \times 2)$	606, 992
		$(8\sqrt{3} \times 2)$-S	607–609
		Complex	1493
	H_2Se	(2×2)-Se	137, 577, 1708
		(4×4)-Se	577
		$(\sqrt{3} \times \sqrt{3})R30°$-Se	137, 577
	H_2O	$(\sqrt{3} \times \sqrt{3})R30°$	1308
	Mo	(5×5)	447, 448
		(4×4)	447, 448
		$\begin{bmatrix} 2 & 0 \\ 5 & 10 \end{bmatrix}$	447, 448
		$\begin{bmatrix} 1 & 0 \\ 5 & 10 \end{bmatrix}$	447, 448
	N_2	Not adsorbed	131
	Na	Hexagonal	455, 458, 460
	NH_3	(2×2)-NH	778
		(6×2)-N	778
		$(\sqrt{7} \times \sqrt{7})R19°$	811, 818
		Disordered	811
		$(\sqrt{7}/2 \times \sqrt{7}/2)R19°$	1282
	$Ni(CO)_4$, CO	$(\sqrt{7}/2 \times \sqrt{7}/)R19°$-CO	1150
		$c(4 \times 2)$-CO	1150
	NO	$c(4 \times 2)$-NO	193
		Hexagonal overlayer	193
		(2×2)-O	193
		(6×2)-N	193
		(1×1)-NO	676
		$c(4 \times 2)$-NO	676
		Complex	676
		(2×2)-O	676
		$\begin{bmatrix} 2 & 1 \\ 4 & 7 \end{bmatrix}$-Cl	206
	O_2	(2×2)-O	2–4, 116, 193, 194, 195, 196, 197, 198, 577, 883, 990, 1282, 1308, 1346, 1351, 1652
		$(\sqrt{3} \times \sqrt{3})R30°$-O	2.5, 195, 577, 1346, 1351, 1652, 1796
		$(\sqrt{3} \times \sqrt{21})$-O	116
		NiO(111)	4, 6, 116, 193, 194
		NiO	1351
	$O(a) + H_2O$	No new features	1308
	$O(a) + NO$	(2×2)	676
	Pb	$\begin{bmatrix} 1 & 1 \\ -1 & 2 \end{bmatrix}$	471, 472
		(7×7)	471, 864

TABLE 2.5c. (*Continued*)

Substrate	Adsorbate	Surface Structure	References[b]
		(13×13)	471, 472
		(3×3)	471, 864, 1060
		(4×4)–Pb	864, 1060
		Hexagonal rotated $\pm 3°$	472
		$(\sqrt{3} \times \sqrt{3})R30°$–Pb	864, 1060
	PF_3	(2×2)	833
Ni–17% Cu(111)	SiH_4	(2×2)	907
		$(\sqrt{3} \times \sqrt{3})R30°$–Si	907
Ni–25% Fe(111)		(2×2)–Si	907
$NiI_2(0001)$	Sn	(2×2)–Sn	1060
NiO(111)		$(\sqrt{3} \times \sqrt{3})R30°$–Sn	1060
$NiSi_2(111)$	Te	$(2\sqrt{3} \times 2\sqrt{3})R30°$–Te	577
Os(0001)		$(\sqrt{3} \times \sqrt{3})R30°$–Te	577
	[Clean]	(1×1)	868
	H_2	(2×2)–H	868
Pd(111)	H_2S, H_2	(3×3)–S	1121
	[Clean]	(1×1)	1908
	Si	$(\sqrt{3} \times \sqrt{3})R30°$–Si	1185
	[Clean]	(1×1)	1770
	CO	$(\sqrt{3} \times \sqrt{3})R30°$–CO	1169
		$(2\sqrt{3} \times 2\sqrt{3})R30°$–CO	1169
		$(3\sqrt{3} \times 3\sqrt{3})R30°$–CO	1169
	[Clean]	(1×1)	1208, 1509, 1709, 1710, 1861
	Au	(1×1)–Au	1709, 1807
	Br_2	$(\sqrt{3} \times \sqrt{3})R30°$–Br	1103, 1208, 1327
		Ring pattern	1327
	C	Ring pattern	762
	Cl_2	$(\sqrt{3} \times \sqrt{3})R30°$–Cl	785
		(3×3)–Cl	785
	CO	$(\sqrt{3} \times \sqrt{3})R30°$–CO	209, 210, 691, 1042, 1861
		Hexagonal overlayer	209
		$c(4 \times 2)$–CO	210
		$c(4 \times 2)$	691
		Disordered	1208
		(1×1)	1509
	CO_2	Not adsorbed	691
	Fe	(1×1)	1546
	H_2	(1×1)–H	211, 212
	H_2S	$(\sqrt{3} \times \sqrt{3})R30°$–S	1710
	NO	$c(4 \times 2)$–NO	208
		(2×2)–NO	208
	O_2	(2×2)–O	207, 691, 1670
		$(\sqrt{3} \times \sqrt{3})R30°$–O	207
		(2×2)–PdO	207
		(1×1)	1509, 1670
	$O_2 + CO$	$(\sqrt{3} \times \sqrt{3})R30°$	691
		(2×1)	691
	PF_3	(2×2)	833

TABLE 2.5c. (*Continued*)

Substrate	Adsorbate	Surface Structure	References[b]
Pd–33% Ag(111)	[Clean]	(1 × 1)	877
	CO	(1 × 1)	877
Pd–25% Cu(111)	[Clean]	(1 × 1)	877
	CO	(1 × 1)	877
Pd$_2$Si(0001)	[Clean]	(3 × 3)	1555
		(1 × 1)	1555
Pt(111)	[Clean]	(1 × 1)	1226, 1556, 1614, 1711, 1712, 1799, 1874
	Ag	Disordered	1254
	Au	Disordered	1254
	Br$_2$	(3 × 3)-Br	724
	Cu	(12 × 12)-Cu	1054
		(2 × 2)-Cu	1054
	C$_2$N$_2$	Disordered	1002
	Cl$_2$ + Br$_2$	c(2 × 4)-Cl, Br	610
		($\sqrt{3} \times \sqrt{3}$)$R30°$-Cl, Br	610
		(3 × 3)-Cl, Br	610
		($\sqrt{7} \times \sqrt{7}$)$R19.1°$	610
	CO	($\sqrt{3} \times \sqrt{3}$)$R30°$-CO	218, 696, 1205, 1452
		c(4 × 2)-2CO	28, 107, 120, 218, 219, 696, 981, 1196, 1205, 1232, 1237, 1452, 1711
		Hexagonal overlayer	218
		(2 × 2)-CO	120
		($\sqrt{2/3} \times \sqrt{2/3}$)$R15°$-CO	1232
		Ordered	1232
	CO + O$_2$	($\sqrt{3} \times \sqrt{3}$)$R30°$ (misfit)	909
	Cu	(1 × 1)-Cu	842
		Cu(111) multilayers	842
		Alloy formation	842
	F	Streak pattern	1694
	H$_2$	Not absorbed	120
		Adsorbed	220, 221
		(1 × 1)-H	1110
	H$_2$ + C$_2$N$_2$	Disordered	1002
	H$_2$ + O$_2$	($\sqrt{3} \times \sqrt{3}$)$R30°$	11
	H$_2$O	($\sqrt{3} \times \sqrt{3}$)$R30°$-H$_2$O	223, 224, 929
		H$_2$O(111)	224
		Not adsorbed	580
	H$_2$S or S$_2$	(2 × 2)-S	225–227, 247, 933, 1114, 1248
		($\sqrt{3} \times \sqrt{3}$)$R30°$-S	225–227, 247, 874, 933, 1114, 1712, 1248
		Complex structure	1114

TABLE 2.5c. (*Continued*)

Substrate	Adsorbate	Surface Structure	References[b]
		$\begin{bmatrix} 4 & -1 \\ -1 & 2 \end{bmatrix}$-S	225, 226
		Hexagonal	227
	HBr	c(3 × 3)–3Br, HBr	806, 1258
		(3 × 3)	806
	HCl	Disordered	806
	HI	$(\sqrt{3} \times \sqrt{3})R30°$–I	774
		$(\sqrt{7} \times \sqrt{7})R19.1°$–I	580, 774
	I_2	$(\sqrt{7} \times \sqrt{7})R19.1°$–I	580, 937, 1106, 1258, 1391, 1556
		$(3\sqrt{3} \times 9\sqrt{3})R30°$–I	937
		$(\sqrt{3} \times \sqrt{3})R30°$–I	937, 1258, 1391
		(3 × 3)–I	937
	$I_2(a)$ + HBr	HBr not adsorbed	1258
	I(a) + Cu	(3 × 3)–I, Cu	1556
		(10 × 10)–I, Cu	1556
	I(a) + Ag	(3 × 3)–Ag + I	937, 1106, 1391
		(5 × 5)–Ag + I	937
		(12 × 12)–Ag + I	1106
		(17 × 17)–Ag + I	937
		(18 × 18)–Ag + I	1106
		(18 × 18) + (10 × 10)–Ag + I	1106
		$(\sqrt{7} \times \sqrt{7})$ + (3 × 3)	1391
		$(\sqrt{3} \times \sqrt{3})R30°$–Ag, I	1391
	K	$(\sqrt{3} \times \sqrt{3})R30°$–K	1071, 1238, 1337
		$\begin{bmatrix} 1.66 & 0 \\ 0 & 1.66 \end{bmatrix}$-K	1337
		Ring pattern	1337
	K + CO	Disordered	1255
	K + O_2	(4 × 4)–K, O	1238, 1337
		(8 × 2)	1337
		(10 × 2)	1337
		K_2O	1337
	N	Disordered	228
	NH_3	Disordered	599
		Adsorbed	626
		Not adsorbed	580
	NO	Disordered	222
		(2 × 2)–NO	690, 1030, 1096
	NO_2	Disordered	1227
		(2 × 2)–O, NO, (NO_2)	1227
	O_2	(2 × 2)–O	10, 11, 213—215, 216, 217, 581, 592, 1221, 1237, 1248, 1301
		(2 × 2)–O, (O_2)	1221
		$(\sqrt{3} \times \sqrt{3})R30°$–O	214, 215, 217, 708

TABLE 2.5c. (*Continued*)

Substrate	Adsorbate	Surface Structure	References[b]
		(1×1)-O	708, 1171
		Not adsorbed	120
		$(4\sqrt{3} \times 4\sqrt{3})R30°$-O	214, 215
		$PtO_2(0001)$	214, 215
		(3×15)-O	217
		disordered	581
		$(\frac{3}{2} \times \frac{3}{2})R15°$-O	1221
	SO_2	Disordered	1179
	$S + O_2$	(2×2)-O	1040
	O_2	Not adsorbed	1248
	Xe	$(\sqrt{3} \times \sqrt{3})R30°$-Xe	846
		Hexagonal	846
PtNi(111)	[Clean]	(1×1)	1909
Pt-22% Ni(111)	[Clean]	(1×1)	1162
Pt-50% Ni(111)	[Clean]	(1×1)	1162
Pt$_3$Ti(111)	[Clean]	(2×2)	935
Re(0001)	Ba	(2×2)	565, 566
		Hexagonal	565, 566
	CO	Not adsorbed	24
		(2×2)-CO	23
		Disordered	230, 1132
		$(\sqrt{3} \times 4)$	1176
		$(2 \times \sqrt{3})$	230
	H_2	Not adsorbed	24
		Disordered	664
	H_2O	$(\sqrt{3} \times \sqrt{3})R30°$-H$_2$O	1003
		(2×2)-H$_2$O	1003
	N_2	Not adsorbed	24
	O_2	(2×2)-O	23, 24, 229, 723
		Ordered	1515
		(2×1)-O	972, 1654
Re(0001) on Pt(111)	CO	$(\sqrt{3} \times 4)$rect	843
		(2×2)	843
	O_2	(2×1)-O	843
Rh(111)	[Clean]	(1×1)	1648, 1713, 1800
	C	$(2\sqrt{3} \times 2\sqrt{3})R30°$-C	1012
	C_2N_2	Adsorbed	926
	Cl_2	$(\sqrt{3} \times \sqrt{3})R30°$-Cl	654
		(4×4)-Cl	654
	CO	$(\sqrt{3} \times \sqrt{3})R30°$-CO	231, 652, 727, 931
		(2×2)-3CO	12, 231, 727, 931, 1122
		$(\sqrt{3} \times 7)$rect	1844
	CO + Na	c(4×2)-CO + Na	1844
	CO + NO	Disordered	1876
	CO_2	$(\sqrt{3} \times \sqrt{3})R30°$-CO	231
		(2×2)-CO	231
	H_2	Adsorbed	231

TABLE 2.5c. (*Continued*)

Substrate	Adsorbate	Surface Structure	References[b]
	H_2 + CO	(2×2)	829
		$(\sqrt{3} \times \sqrt{3})R30°$	829
		(2×2)	829
	H_2O	$(\sqrt{3} \times \sqrt{3})R30°–H_2O$	583
	H_2S or S_2	$c(2 \times 4)–S$	875
		$(\sqrt{3} \times \sqrt{3})R30°–S$	1768
	NO	$c(4 \times 2)–NO$	231
		$(2 \times 2)–NO$	231
	O_2	$(2 \times 2)–O$	12, 231
		Disordered	570, 583
		(2×2)	570
		$(2 \times 1)–O$	1692
		$(8 \times 8)–Rh_2O_3(0001)$	1692
Ru(0001)	[Clean]	(1×1)	914, 1127, 1233, 1380
	C_2N_2	$c(2 \times 2)–CN$	1217
		$(3 \times 3)–CN$	1217
		$c(4 \times 8)–CN$	1217
		(1×2)	1217
		(1×3)	1217
		(1×1)	1217
		Graphite	1217
	CO	$(\sqrt{3} \times \sqrt{3})R30°–CO$	12, 233, 248, 716, 825, 914, 1127, 1357
		$(2 \times 2)–CO$	12, 248
		$(2\sqrt{3} \times 2\sqrt{3})R30°–CO$	716, 825
		$(5\sqrt{3} \times 5\sqrt{3})R30°–CO$	825
	CO + O_2	(2×2)	768
	CO_2	$(\sqrt{3} \times \sqrt{3})R30°–CO_2$	12
		$(2 \times 2)–CO_2$	12
	Cu	Disordered	1679
	H_2	$(1 \times 1)–H$	234, 870
	H_2O	$(\sqrt{3} \times \sqrt{3})R30°$ + "halo"	1233, 1380
		$(\sqrt{3} \times \sqrt{3})R30°$	835, 1380
		$(\sqrt{3} \times \sqrt{3})R30°–H_2O$	1082
		"Hexagon"	1233, 1380
		$(2 \times 2)–O_2$	1233
		Complex	1233
	H_2S	(2×2)	740
		$(\sqrt{3} \times \sqrt{3})R30°$	740
		$c(4 \times 2)$	740
	Na	$(\frac{3}{2} \times \frac{3}{2})–Na$	1129, 1406
		Ring pattern	1129
		$(2 \times 2)–Na$	1082, 1129
		$(\sqrt{3} \times \sqrt{3})R30°–Na$	1082, 1129, 1406
		Hexagonal overlayer	1406

TABLE 2.5c. (*Continued*)

Substrate	Adsorbate	Surface Structure	References[b]
	Na + CO	$(2\sqrt{3} \times 2\sqrt{3})R30°$	976
	Na + H$_2$O	$(2\sqrt{3} \times 2\sqrt{3})R30°$	1082
		Complex	1082
	N$_2$	Adsorbed	234
		$(\sqrt{3} \times \sqrt{3})R30°$-N$_2$	1455
	N$_2$O	(2×2)	832
	NH$_3$	(2×2)-NH$_3$	234, 235, 1045
		$(\sqrt{3} \times \sqrt{3})R30°$-NH$_3$	235
		$(2\sqrt{3} \times 2\sqrt{3})R30°$-NH$_3$	1045
	NO	(2×2)-NO	598
		$(2\sqrt{3} \times 2\sqrt{3})R30°$-NH$_3$	1045
		(2×1)-NO	963
	O$_2$	(2×2)-O	12, 232, 248, 832, 1233
		(2×1)-O	963, 1233
		(1×2)-O	619, 631
	O(a) + NO	Disordered	963
Sb(0001)	Fe	(1×1)	567
	Th	$\begin{bmatrix} 1 & 1 \\ 1 & -1 \end{bmatrix}$	510, 511
Sc(0001)	[Clean]	(1×1)	1333
Si(111)	[Clean]	(2×1)	847, 856, 947, 954, 956, 1019, 1028, 1086, 1087, 1361, 1374, 1412, 1427, 1428, 1457, 1477, 1528, 1533, 1542, 1543, 1563, 1646, 1714
		(7×7)	851, 857, 921, 934, 954, 996, 1019, 1021, 1022, 1056, 1073, 1158, 1170, 1206, 1207, 1210, 1342, 1457, 1475, 1477, 1483, 1486, 1499, 1507, 1517, 1525, 1529, 1533, 1536, 1537, 1543, 1685
		(1×1)	568, 954, 1366, 1427, 1457, 1492, 1543, 1544
		$(\sqrt{19} \times \sqrt{19})$	1653
	Laser-annealed	(1×1)	1170, 1492, 1517, 1716
		(1×1) + Steps	1446
	Ag	(6×1)-Ag	795, 807, 948, 1158, 1342, 1696

TABLE 2.5c. (*Continued*)

Substrate	Adsorbate	Surface Structure	References[b]
		$(\sqrt{7} \times \sqrt{7})R19.1°$–Ag	1696
		$(\sqrt{3} \times \sqrt{3})R30°$–Ag	807, 923, 1037, 1073, 1108, 1158, 1206, 1322, 1342, 1536, 1650, 1696, 1910, 1911
		(3×1)–Ag	807, 948, 1037, 1158, 1342, 1536, 1696
		$\sqrt{3}(3 \times 1)$	1536
		(1×1)–Ag	1022, 1026
		Ag island	1911
	Ag(a) + H	$(\sqrt{3} \times \sqrt{3})R30°$	1536
	Al	$(\sqrt{7} \times \sqrt{7})R19.1°$–Al	816
		(1×1)	1563
		$(\sqrt{3} \times \sqrt{3})R30°$	816, 1563
	Au	(5×1)–Au	792, 1091, 1628, 1699
		$(\sqrt{3} \times \sqrt{3})R30°$–Au	792, 1091, 1215, 1322, 1499, 1628, 1669
		(6×6)–Au	792, 1091, 1215
		(1×1)–Au	956, 1091
	Bi	$(\sqrt{3} \times \sqrt{3})R30°$–Bi	736
		Bi(0001)–(1×1)	736
	Br	Not specified	1858, 1859
	Cl	(7×7)	1088
		$(\sqrt{19} \times \sqrt{19})$–Cl	1088, 1385
	Cl$_2$	Disordered	138
		(7×7)–Cl	138, 236, 1715
		(1×1)–Cl	138, 236, 1715
	Co	(1×1)	851
		(1×1)–CoSi$_2$	851, 1414
		$(\sqrt{7} \times \sqrt{7})$–2d silicide	851
		(2×2) or (2×1)–2d silicide	851
	Cr	(1×1)	1500
		$(\sqrt{3} \times \sqrt{3})R30°$	1500
		(7×7)	1500
	Cu	(1×1)–Cu	841
		Cu(111) or Cu–Si(111) [multilayers]	841
		5×5–Cu	841, 857, 858
		(4×1)	856
		(4×2) [Multilayer]	856
		Cu(111)–$\sqrt{3} \times \sqrt{3}$ [multilayer]	856
		Cu(111)–(1×1) [multilayer]	856

TABLE 2.5c. (*Continued*)

Substrate	Adsorbate	Surface Structure	References[b]
	Cu(a) + O_2	"5 × 5"	856
	D_2	(7 × 7)	1369
	Ga	$(\sqrt{3} \times \sqrt{3})R30°$-Ga	1028
		(1 × 1)-Ga	1028
		(7 × 7)-Ge	1029, 1537
		(1 × 1)-Ge	1029, 1486, 1544
		(5 × 5)	1475, 1483, 1486
		(5 × 5)-Ge	1029
		(5 × 5)-SiGe(111)	934, 1010
		Ordered	1507
		$(\sqrt{3} \times \sqrt{3})R30°$	1544
	H_2	(1 × 1)-H	237, 1216, 1477
		(7 × 7)-H	237, 904, 1610
		(2 × 1)	1477
	H_2O	(1 × 1)	1477
		Adsorbed	903
	I	(7 × 7)-I	1088
	I_2	(1 × 1)-I	133
		(7 × 7)-I	1857
	In	$(\sqrt{3} \times \sqrt{3})R30°$-In	1028
		(2 × 2)-In	1028
		Complicated	1028
	Kr	(7 × 7)	1125
	N_2	(8 × 8)-N	34, 586, 1229
		Doublet	586
		Diffuse	586
		Si(1 × 1)	1229
		"Quadruplet"	1229
	N	(1 × 1)	798
	NH_3	(8 × 8)-N	238
		(7 × 7) + "quadruplet"	1479
		(7 × 7) + (8 × 8)	1479
	Ni	(1 × 1)-$NiSi_2$	838, 1434, 1659, 1860
		(1 × 1)-Ni	850, 1366, 1434
		Disordered	964
		(1 × 1)-Ni with streaks	964
		$(\sqrt{3} \times \sqrt{3})R30°$	964, 969
		$(\sqrt{19} \times \sqrt{19})R \pm 23.5°$-Ni	964, 1366
		Si(111)-(7 × 7)	964
	NO	Disordered	1021
		(8 × 8)-N	1021
		Complex	1021
	O_2	Disordered	17, 20, 21
		(1 × 1)	847
	P	$(6\sqrt{3} \times 6\sqrt{3})$-P	132, 133
		(1 × 1)-P	132

TABLE 2.5c. (*Continued*)

Substrate	Adsorbate	Surface Structure	References[b]
		$(2\sqrt{3} \times 2\sqrt{3})$–P	132
		(4×4)–P	133
	Pd	Disordered	964, 1207
		(5×1)	964
		$(\sqrt{3} \times \sqrt{3})R30°$–Pd	964, 1456, 1484, 1487, 1496
		(3×1)–Pd	1456
		(1×1) + Streaks	964, 1456
		$(2\sqrt{3} \times 2\sqrt{3})R30°$–Pd	964, 1456
		Pd_2Si (epitaxial)	1134
	PH₃	(7×7)–P	239
		(1×1)–P	239
		$(6\sqrt{3} \times 6\sqrt{3})$–P	239
		$(2\sqrt{3} \times 2\sqrt{3})$–P	239
	Si	(1×1)	1517
	Si + laser	(1×1)	1392
	Sb	(1×1)–Sb	1019
	Sn	(1×1)–Sn	996
		$(\sqrt{3} \times \sqrt{3})R30°$–Sn	996
		$(2\sqrt{3} \times 2\sqrt{3})R30°$–Sn	996
		$(\sqrt{133} \times 4\sqrt{3})$–Sn	996
		$(3\sqrt{7} \times 3\sqrt{7})R(30 \pm 10.9)°$–Sn	996
		$(2\sqrt{91} \times 2\sqrt{91})R(30 \pm 3.0)°$–Sn	996
	Te	(7×7)–Te	1088, 1857
		(1×1)–Te	1617
	Yb	(2×1)–Yb	844, 1525
		(3×1)–Yb	844, 1525
		(2×1) + (7×7)	844
		(5×1)–Yb	1525
	Xe	(7×7)	1125
Ti(0001)	[Clean]	(1×1)	1007, 1717
	Cd	(1×1)	562, 563, 564
	CO	(1×1)–CO	18, 240
		(2×2)–CO	240
		$(\sqrt{3} \times \sqrt{3})R30°$–N	241, 242
	Cu	Extra spots	561
	N₂	(1×1)–N	241, 242
		$(\sqrt{3} \times \sqrt{3})R30°$–N	241, 242
	O₂	(1×1)–O	18
TiC(111)	[Clean]	(1×1)	1631
Th(111)	CO	Disordered	243
		$ThO_2(111)$	243
	O₂	Disordered	243
		$ThO_2(111)$	243
UO₂(111)	O₂	(3×3)–O	13
		$(2\sqrt{3} \times 2\sqrt{3})R30°$–O	13
W(111)	Cl₂	Facet surface	796

TABLE 2.5c. (*Continued*)

Substrate	Adsorbate	Surface Structure	References[b]
	CO	Disordered	746
		{211} facets	746
	O_2	Disordered	244, 746
		{211} facets	15, 746
		(4 × 4)–O	746
Y(0001)	[Clean]	(1 × 1)	1120
Xe(111)	[Clean]	(1 × 1)	1912
Zn(0001)	[Clean]	(1 × 1)	1267, 1870
	Cu	(1 × 1)	449, 450
	O_2	(1 × 1)–O	122
		ZnO(0001)	245
	SO_2	No LEED pattern	1569
		Oxide	1569
Zn(000$\bar{1}$)	O_2	($\sqrt{3} \times \sqrt{3}$)$R30°$–O	122
ZnO(0001)	[Clean]	(1 × 1)	1104, 1239, 1773
	H_2O	Disordered	1104
ZnO(000$\bar{1}$)	[Clean]	(1 × 1)	1104
	H_2O	Disordered	1104
	K	($2\sqrt{3} \times 2\sqrt{3}$)$R30°$–K	1629
	Xe	Disordered	1026
ZnSe(111)	[Clean]	(2 × 2)	1393
		(1 × 1) + {110}facet + (2 × 2)	1393
ZnSe($\overline{111}$)	[Clean]	(1 × 1)	1393
		(1 × 1) + {331}f + {110}f, {110}f	1393
ZnTe(111)	[Clean]	(2 × 2)	1393
		(1 × 1)	1393
ZnTe($\overline{111}$)	[Clean]	(1 × 1) + {331}f + {110}f	1393
Ar(0001)	[Clean]	(1 × 1)	1473, 1642
	O_2	(2 × 2)–O	1718

[a] Organic overlayer structures are not included. See Table 2.5g for these structures.
[b] References for Tables 2.5a to 2.5j are listed at the end of Table 2.5j.

TABLE 2.5d. Surface Structures on Substrates with Fourfold Rotational Symmetry[a]

Substrate	Adsorbate	Surface Structure	References[b]
Ag(100)	[Clean]	(1 × 1)	1272, 1562, 1897, 1898, 1899
	Au	(1 × 1)	1806
	Cl_2	c(2 × 2)–Cl	572, 605, 732, 962, 1719
	Cl_2 + K	c(2 × 2)–K/Cl	673
	Cu	Epitaxial	1167
		Cu(100)	1476
		(1 × 1)	1820
	Fe	(1 × 1)	1272
	H_2O	Disordered	1034

TABLE 2.5d. (*Continued*)^a

Substrate	Adsorbate	Surface Structure	References^b
	H$_2$S	(2 × 2)–S	627
		$\begin{bmatrix} 4 & -1 \\ 1 & 4 \end{bmatrix}$-S	627
		$\begin{bmatrix} 4 & 4 \\ -4 & 4 \end{bmatrix}$-S	627
		Partially disordered	1117
	I$_2$	c(2 × 2)–I	1145
	K + O$_2$	$\begin{bmatrix} 1 & 1 \\ -5 & 4 \end{bmatrix}$	658
		Hexagonal overlayer	658
	Ni	(1 × 1)	1820
	O$_2$	Disordered	146
	O(ad.) + H$_2$O	c(2 × 2)–OH	1034
	Pd	Epitaxial	1167
		(1 × 1)	1463
	Se	c(2 × 2)–Se	250
Al(100)	[Clean]	(1 × 1)	1077, 1354, 1532, 1720, 1721
	Ag	(5 × 1)–Ag	1363
		(1 × 1) [multilayer]	1363
	Au	Disordered	1363
	CO	Not adsorbed	1273
		Disordered	1368
	Cu	Disordered	1363
	Fe	Poor epitaxy	452
	H$_2$	(1 × 1)	1693
	Na	c(2 × 2)	499, 500, 501, 1720
		Hexagonal overlayer	500
	O$_2$	Disordered	42–44, 709
	Pb	$\begin{bmatrix} 2 & 0 \\ -1 & 2 \end{bmatrix}$	503
		c(2 × 2)–Pb	704
		$\begin{bmatrix} 2 & 0 \\ 1 & n \end{bmatrix}$-Pb $2 < n < 3$	704
	Sm	(1 × 1) disorder	1532
		Complicated	1532
	Sn	$\begin{bmatrix} 2 & 0 \\ -1 & 3 \end{bmatrix}$	503
		c(2 × 6)–Sn	704
		$\begin{bmatrix} 2 & 0 \\ 1 & n \end{bmatrix}$-Sn, $2 < n < 3$	704
Au(100)	[Clean]	(1 × 5)	1153
		c(26 × 68)	1153
		(5 × 20)	1170, 1361
		$\begin{bmatrix} X & 0 \\ Z & Y \end{bmatrix}$ $X = 24 \pm 3,$ $Y = 43$ or $48,$ $-5 \leq Z \leq 0$	967

TABLE 2.5d. *(Continued)[a]*

Substrate	Adsorbate	Surface Structure	References[b]
	Laser-annealed	(1×1)	1170, 1293, 1361, 1722
	Ag	(1×1)	473, 474, 494, 1838
	Bi	$\begin{bmatrix} 2 & 0 \\ -1 & 2 \end{bmatrix}$	498
	Br_2	(1×1)–Br	793
		$c(2 \times 2)$–Br	793
		$(\sqrt{2} \times 4\sqrt{2})R45°$	793
		$c(4 \times 2)$	793
	CO	Disordered	252
	Cu	(1×1)	473
	Fe	(1×1)	1828–1830
	H_2S	(2×2)–S	251
		$c(2 \times 2)$–S	251
		(6×6)–S	251
		$c(4 \times 4)$–S	251
	Na	Hexagonal	492
		Ordered	492
	Pb	$\begin{bmatrix} 1 & 1 \\ 1 & -1 \end{bmatrix}$	441–444
		$\begin{bmatrix} 1 & 1 \\ -3 & 4 \end{bmatrix}$	441–444
		$\begin{bmatrix} 1 & 1 \\ -1 & 2 \end{bmatrix}$	441–444
		$\begin{bmatrix} 2 & 0 \\ -1 & 3 \end{bmatrix}$	441–444
		(1×1)	683
	Pd	(1×1)	473, 1833
		$c(2 \times 2)$	683
		$c(7\sqrt{2} \times \sqrt{2})R45°$	683
		$c(3\sqrt{2} \times \sqrt{2})R45°$	683
		$c(6 \times 2)$	683
	Pt	(1×1)	438, 439, 440
	Xe	Disordered	252
BaTiO$_3$(100)	[Clean]	(2×2)	853, 1490
		(3×3)	1490
		(1×1)	853, 1490
	O_2	Disordered	853
C(100), diamond	H_2S	Not adsorbed	164
	N_2	Not adsorbed	164
	NH_3	Not adsorbed	164
	O_2	Disordered	16
		Not adsorbed	164
CaO(100)	[Clean]	(1×1)	755, 1641
Ce(100)	[Clean]	$\begin{bmatrix} \frac{3}{5} & \pm\frac{1}{5} \\ \frac{1}{5} & \pm\frac{2}{5} \end{bmatrix}$	845
Co(100)	[Clean]	(1×1)	1539, 1776
	C	(2×2)–C	1539
	CO	$c(2 \times 2)$–CO	253, 1553
		(2×2)–C	253
	H_2S or S	(2×2)–S	1539
		$c(2 \times 2)$–S	1539, 1548, 1698

TABLE 2.5d. *(Continued)*[a]

Substrate	Adsorbate	Surface Structure	References[b]
	H_2S + C	(2×2)-S, C	1539
	O_2	(2×2)-O	254
		$c(2 \times 2)$-O	254, 1777
	S	$c(2 \times 2)$-S	1698
CoO(100)	[Clean]	(1×1)	1778
Cr(100)	[Clean]	(1×1)	1126, 1330, 1606
	C, O, N	$c(2 \times 2)$	1126
	Br_2	$c(2 \times 2)$-Br	1051
		$c(2 \times 4)$-Br	1051
		Pseudohexagonal $CrBr_2$	1051
	Cl_2	$c(2 \times 2)$-Cl	1330
		(2×5)-Cl	1330
		$c(2 \times 4)$-Cl	1330
	H_2S	$c(2 \times 2)$-S	1245
	N	(1×1)-N	1330
		$(\sqrt{2}R45° \times \sqrt{5}R27°)$-N	1330
		$c(2 \times 2)$-N	1330
	N_2	$c(2 \times 2)$-N	1245
		$c(\sqrt{2} \times 3\sqrt{2})R \pm 45°$-N	1245
		(1×1)-N	1245
	O_2	$c(2 \times 2)$-O	255, 634, 1245
		$Cr_2O_3(310)$	256
		(1×1)-O	1245
		$c(2 \times 4)$-O	1245
	Br_2	$c(2 \times 2)$-Br	1051
		$c(2 \times 4)$-Br	1051
		CrB_2	1051
Cu(100)	[Clean]	(1×1)	585, 1101, 1419, 1723
	Ag	$\begin{bmatrix} 2 & 0 \\ -1 & 5 \end{bmatrix}$	473, 477
	Au	$\begin{bmatrix} 1 & 1 \\ 1 & -1 \end{bmatrix}$	473, 478
		$\begin{bmatrix} 2 & 0 \\ -1 & 7 \end{bmatrix}$	473, 478
	Bi	(2×2)	483, 486
		$\begin{bmatrix} 1 & 1 \\ 1 & -1 \end{bmatrix}$	483, 486
		$\begin{bmatrix} 1 & 1 \\ -4 & 5 \end{bmatrix}$	483, 486
		$\begin{bmatrix} 5 & 4 \\ -4 & 5 \end{bmatrix}$	483, 486
		(1×1)	703
		$c(2 \times 2)$	703
	Cl_2	$c(2 \times 2)$-Cl	1081, 1102, 1461, 1545, 1724, 1869
	Co	(1×1)-Co	983, 1105, 1810, 1811
	Co (multilayer) + CO	$c(2 \times 2)$-CO	983
	CO	$c(2 \times 2)$-CO	125, 126, 265, 1005, 1353, 1407, 1605, 1663, 1780

TABLE 2.5d. *(Continued)*[a]

Substrate	Adsorbate	Surface Structure	References[b]
		$(7\sqrt{2} \times \sqrt{2})R45°$–CO	1407
		$(\sqrt{2} \times \sqrt{2})R45°$	1626
		Hexagonal overlayer	126, 127, 265
		(2×2)–C	26, 125
	Cs	Disordered	865
	Cs	Hexagonal overlayer	865
		Quasi-hexagonal	865
	Fe	(1×1)	452, 1808, 1809
	H$_2$S	Adsorbed	35
		(2×2)–S	35, 260, 262, 1211, 1725
		(2×1)–S	128
		Partially disordered	1117
	I$_2$	(2×2)–I	1779
	K	$\begin{bmatrix} 2 & 3 \\ 0 & 5 \end{bmatrix}$	1405
		$\begin{bmatrix} 2 & 2 \\ 0 & 3 \end{bmatrix}$	1405
		Incommensurate	1405
	Mn	c(2×2)–Mn	1319
	N$_2$	(1×1)–N	49
		c(2×2)–N	47, 132, 258, 261, 266
	Nb	Incommensurate	667
	Ni	(1×1)	1812
	O$_2$	(1×1)–O	9, 45
		(2×1)–O	9, 45, 46
		$(2 \times 4)R45°$–O	7, 47, 246, 261
		(2×3)–O	119
		c(4×4)–O	119
		c(2×2)–O	171, 246, 257, 258, 259, 260, 263, 264, 1417, 1726, 1727, 1781
		(2×2)	171
		$(2 \times 2\sqrt{2})R45°$	259, 262–264
		Hexagonal	259
		(410) facets	259
		$(\sqrt{2} \times \sqrt{2})R45°$–O	641, 1095, 1285, 1598, 1633
		$(\sqrt{2} \times \sqrt{2})R45°$–O	1095, 1781
		$(2\sqrt{2} \times 2\sqrt{2})R45°$–O	1633
		$(\sqrt{2} \times 0.46$ nm$)R45°$–O [coincidence]	1691
	Pb	$\begin{bmatrix} 2 & 2 \\ -2 & 2 \end{bmatrix}$	481–485
		$\begin{bmatrix} 1 & 1 \\ -2 & 3 \end{bmatrix}$	481–485
		c$(5\sqrt{2} \times \sqrt{2})R45°$–Pb	703, 1295
		c(2×2)–Pb	1295
		$(2\sqrt{2} \times 2\sqrt{2})R45°$–Pb	1295
	Sn$(\theta > 1)$ + Pb	Disordered	1041
	Pd	c(2×1)–Pd	726

TABLE 2.5d. *(Continued)*[a]

Substrate	Adsorbate	Surface Structure	References[b]
		(1 × 1)–Pd	726, 1649
		c(2 × 2)–Cu$_3$Pd	1649
	Sn	(2 × 2)	480
	Te	(2 × 2)–Te	267, 1119, 1728
	Tl	$\begin{bmatrix} 2 & 2 \\ 2 & -2 \end{bmatrix}$	1167, 1564
		$\begin{bmatrix} 4 & 0 \\ 2 & 7 \end{bmatrix}$–Tl	1167, 1336, 1564
		$\begin{bmatrix} 4 & 0 \\ 2 & 6 \end{bmatrix}$–Tl	1336, 1564
		$\begin{bmatrix} 6 & 6 \\ 2 & -2 \end{bmatrix}$	1564
		c(4 × 4)–Tl	1336
	Xe	Hexagonal overlayer	159
		Disordered	741
Cu–3% Al(100)	O$_2$	c(2 × 2)–O	1632
		Disordered	1632
Cu–5.7% Al(100)	[Clean]	(1 × 1)	813
Cu–12.5% Al(100)	[Clean]	(1 × 1)	813
Cu$_3$Au(100)	[Clean]	c(2 × 2)	916
Cu/Au(100)	[Clean]	c(2 × 2)	737
Cu/Pd(100)	[Clean]	Streak	737
		c(2 × 2)	737
CuSn(100)	[Clean]	c(2 × 2)	1481
		$(3\sqrt{2} \times \sqrt{2})R45°$	1481
		c(3 × 2) + (2 × 2)	1481
Cu–25% Zn(100)	[Clean]	(1 × 1)	1152
	O$_2$	Disordered	1152
EuO(100)	[Clean]	(1 × 1)	1502
Fe(100)	[Clean]	(1 × 1)	989, 1078, 1729
	Br$_2$	c(2 × 2)	752
		(2 sin α × 2 sin α') $R\alpha'$ $\alpha' = 26.57°,$ 37.49°, 40.5°	752
		$\begin{bmatrix} 1 & \dfrac{1}{\tan \alpha} \\ -1 & \dfrac{1}{\tan \alpha} \end{bmatrix}$ $\alpha = 53.13°, 53.47°, 56.31°$	752
		$(\sqrt{41}/5 \times \sqrt{41}/5)$ $R38.7°$	752
		c(2 × 4)	752
	CBr$_4$	c(2 × 2)	757
		(2 sin α' × 2 sin α') $R\alpha'$	757
	CCl$_4$	(2 sin α' × 2 sin α') $R\alpha'$	753
		$(\sqrt{13} \times \sqrt{13})R \tan^{-1}(\tfrac{2}{3})$	753
		(6 × 6)–Cl	753

TABLE 2.5d. (*Continued*)[a]

Substrate	Adsorbate	Surface Structure	References[b]
		$\begin{bmatrix} 1 & \dfrac{1}{\tan\alpha} \\ -1 & \dfrac{1}{\tan\alpha} \end{bmatrix}$	753
		c(2 × 2)	
	CO	c(2 × 2)–CO	275, 1596
		c(2 × 2)–C, O–disordered	783, 893, 1601
		Disordered	783
	Fe_3O_4	(1 × 1)–like	1189
	H_2	Adsorbed	177
	H_2O	c(2 × 2)	278
	H_2S or S	c(2 × 2)–S	276, 277, 1552, 1630
		Complex	1552
		c(6 × 2)	1552
	I_2	c(2 × 2)–I	751
		(2 sin 40.5 × 2 sin 40.5) R40.5°–I	751
		($\sqrt{85} \times \sqrt{85}$) R40.6°–I	751
	K	Disordered	665
		(2 × 2)–K	784
		Hexagonal close pack	784
	N	c(2 × 2)–N	893
	NH_3	Disordered	176
		c(2 × 2)–N	176, 1224
	O_2	c(2 × 2)–O	60, 269–271, 274, 635, 893, 1596
		(1 × 1)–O	144, 268, 271, 272, 1596
		FeO(100)	60, 269, 270, 272, 273, 635
		FeO(111)	270, 635
		FeO(110)	272
		Disordered	273, 276
	Se	c(2 × 2)–Se	1078
	Si	c(2 × 2)–Si	985, 986
	Te	(2 × 2)–Te	1204
		c(2 × 2)–Te	1204
Fe/Cr(100)	O_2	c(2 × 2)–O	279, 636
		c(4 × 4)–O	279
		Oxide	280
FeTi(100)	[Clean]	(1 × 1)	1241
	S	c(2 × 2)–S	1241
Ge(100)	[Clean]	(2 × 1)	1094, 1213, 1522, 1636, 1645, 1783
		(2 × 2)	1449
		(4 × 2)	804
	Ag	(1 × 1)–Ag	1522
	Bi	(1 × 1)–Bi	1449
	I_2	(3 × 3)–I	19
	O_2	Disordered	17, 18
		(1 × 1)	1094

TABLE 2.5d. (*Continued*)[a]

Substrate	Adsorbate	Surface Structure	References[b]
Ir(100)	[Clean]	(1×1)	866, 1199, 1293, 1361, 1381
		(5×1)	866, 1156, 1361, 1381, 1785
	Ba	(2×1)-Ba	1399
	CO	$c(2 \times 2)$-CO	48
		(2×2)-CO	48
		(1×1)-CO	282
	CO_2	$c(2 \times 2)$-CO_2	48
		(2×2)-CO_2	48
		(7×20)-CO_2	48
	Cs	$c(4 \times 2)$-Cs with streak	1395
		Close-packed layer	1395
		Compressed layer	1395
		(5×5)-Cs	1395
	H_2	Adsorbed	281
	K	$c(2\sqrt{2} \times 4\sqrt{2})R45°$	866
		$\begin{bmatrix} 2 & 1 \\ -1 & 2 \end{bmatrix}$	866
		$c(4 \times 2)$	866
		(3×2)	866
		$c(2 \times 2)$	866
		$\begin{bmatrix} \frac{5}{2} & 0 \\ -\frac{5}{4} & \frac{5}{3} \end{bmatrix}$	866
		$\begin{bmatrix} 2 & 0 \\ -1 & \frac{5}{3} \end{bmatrix}$	866
		$\begin{bmatrix} \frac{10}{7} & 0 \\ -\frac{5}{7} & \frac{5}{3} \end{bmatrix}$	866
	Kr	(3×5)-Kr	283
		Kr(111)	283
	NO	(1×1)-NO	188
	O_2	(2×1)-O	48, 281
		(5×1)-O	48
	O_2	(1×1)-O	797
KBr(100)	[Clean]	(1×1)	1592
KCl(100)	[Clean]	(1×1)	1592
LaB₆(001)	[Clean]	(1×1)	738, 1625
	O_2	(1×1)	770
MgO(100)	[Clean]	(1×1)	755, 908, 918, 1067, 1139, 1284, 1574
	Ag	(1×1)	1559
Mo(100)	[Clean]	(1×1)	761, 1379
	Ag	Ag(100)	513, 514
		Ag(110)	513, 514
	CO	Disordered	62, 693
		(1×1)-CO	62, 64, 285, 286, 1379
		$c(2 \times 2)$-CO	64, 285, 286
		(4×1)-CO	64
	Cs	$(\sqrt{2} \times \sqrt{2})R45°$	932
		(2×2)	932

TABLE 2.5d. *(Continued)*[a]

Substrate	Adsorbate	Surface Structure	References[b]
		c(2 × 2)	932
		Rectangular centered mesh	932
		Quasi-hexagonal	932
		Hexagonal overlayer	932
	Cs(a) + O$_2$	c(2 × 2) + (4 × 1)	932
		(4 × 1)	932
		Disordered	932
	O(a) + Cs	c(2 × 2)–Cs + O	932
		(110) microfacets	932
		Disordered	932
	Ga	(1 × 1)–Ga	789
	H$_2$	c(4 × 2)–H	77
		(3 × 2)–H	780
		($\sqrt{2}$ × $\sqrt{2}$)–H	780, 814
		(1 × 1)–H	77
	H$_2$S, S, or S$_2$	(1 × 1)–S	130, 1149
		(1 × 1)–S diffuse	1174
		($\sqrt{5}$ × $\sqrt{5}$)–S	130, 288
		c(2 × 2)–S	130, 578, 917, 998, 1039, 1149, 1174, 1913
		MoS$_2$(100)	288
		(2 × 1)–S	578, 612, 917, 998, 1039, 1149, 1174
		($\sqrt{5}$ × $\sqrt{5}$)$R26.6°$	578, 613
		c(4 × 4)–S	578, 613, 998
		c(4 × 2)–S	578, 917, 998, 1039, 1149, 1174
		$\begin{bmatrix} 1 & 1 \\ 2 & -1 \end{bmatrix}$–S	578, 917, 998, 1039, 1174, 1149
	H$_2$S + O$_2$	($\sqrt{5}$ × $\sqrt{5}$)$R26.6°$–S, O	917
	N$_2$	(1 × 1)–N	62
		c(2 × 2)–N	287
	O$_2$	Disordered	61, 62
		c(2 × 2)–O	61, 62, 63, 64, 284, 660, 898
		($\sqrt{5}$ × $\sqrt{5}$)$R26°$–O	61, 62, 189, 190, 284, 660, 898, 1155, 1379
		(2 × 2)–O	61, 189, 190, 660, 1379
		c(4 × 4)–O	62, 189, 284, 660, 898
		(2 × 1)–O	189, 190, 660, 748, 898
		(5 × 5)–O	748
		(4 × 1)	898
		(6 × 1)–O	748
		(6 × 2)–O	284, 660, 7484
		(3 × 1)–O	284, 748
		(1 × 1)–O	284, 660, 898, 1155
		c(4 × 4) + (2 × 1)–O	748
		(2 × 1) + c(2 × 2)	748
		Microfacet	660
		Streak (1 × 1)–O	660
		Diffuse (1 × 1)–O	660
		Facet	748

TABLE 2.5d. (*Continued*)[a]

Substrate	Adsorbate	Surface Structure	References[b]
		(110), (112) facets	898
		$MoO_2(110)$	898
	O(a) + CO	(2 × 1)–O	817
	O(a) + CO_2	(2 × 1)–O	817
	Si	(1 × 1)–Si	1730
	Sn	$\begin{bmatrix} 1 & 1 \\ 1 & -1 \end{bmatrix}$	516
		(1 × 2)	516
		(1 × 1)–Sn	789
		c(2 × 2)–Sn	789
NaCl(100)	Ag	Ag(111)	1678
	Xe	Hexagonal overlayer	289
$Na_{0.47} WO_3(100)$	[Clean]	(3 × 1)	808
$Na_{0.72} WO_3(100)$	[Clean]	(2 × 1)	808
		c(2 × 2)	808
$Na_{0.79} WO_3(100)$	[Clean]	(2 × 1)	1485
Nb(100)	N_2	(5 × 1)–N	290
	O_2	c(2 × 2)–O	192, 290
		(1 × 1)–O	192, 290
		(3 × 10)–NbO_2	290
Ni(100)	[Clean]	(1 × 1)	973, 1092, 1231, 1458, 1561, 1565, 1739, 1864
	Ba	Disordered	454
	C	(2 × 2)–C	640, 745, 1231, 1673
	C(a) + O_2	c(2 × 2)–O	1177
	Cl_2	(2 × 2)–Cl	662
		c(2 × 2)–Cl	662
	Co	(1 × 1)	1803
	CO	c(2 × 2)–CO	54, 55, 68, 129, 198, 300–302, 747, 782, 950, 981, 993, 1202, 1604, 1605, 1789, 1790, 1791, 1795
		c(2 × 2)	1281
		(2 × 2)–CO	69
		c($\sqrt{2} \times 3\sqrt{2}$)$R45°$–CO	1202
		Hexagonal overlayer	129, 301, 302
		(2 × 2)–C	198
		Disordered	1291
	CO + H_2	c(3 × 3)	301
	CO_2	(2 × 2)–O + c(2 × 2)–CO	76
	Cr	(1 × 1)	463, 464
	Cs	(2 × 2)	454
		Hexagonal	463, 464
	Cu	(1 × 1)–Cu	1113, 1458, 1804
	e beam	(2 × 2)	1401
	Fe	c(2 × 2)–Fe	973
		(2 × 2) [multilayer]	973
		c(2 × 2) [multilayer]	973
		Fe(110) [multilayer]	452, 973
	H_2	Disordered	198, 203, 211
		c(2 × 2)–H	301

TABLE 2.5d. *(Continued)[a]*

Substrate	Adsorbate	Surface Structure	References[b]
		(1×1)–H	1092, 1202, 1658
		(1×1)–H streaked	663
	H_2 + CO	$c(2 \times 2)$–CO, H	1202
		$c(\sqrt{2} \times \sqrt{2})R45°$–CO, H	1202
	H_2S, S, or S_2	(2×2)–S	36, 118, 197, 198, 621, 622, 623, 637, 979, 1329, 1508, 1732
		$c(2 \times 2)$–S	36, 118, 197, 198, 293, 294, 303, 304, 340, 621–623, 681, 979, 1121, 1329, 1482, 1725, 1734, 1735, 1736, 1792, 1793, 1852
		(2×1)–S	198
		$c(2 \times 2)$–H_2S	304
		Ni_3S_2 island	681
	H_2S, H_2	$c(2 \times 2)$–S	1121
	H_2S + Na	$c(2 \times 2)Na + c(2 \times 2)S$	1887
		$(2 \times 2)Na + c(2 \times 2)S$	1887
		$(2 \times 2)Na + (2 \times 2)S$	1887
	H_2Se	(2×2)–Se	197, 198, 1732, 1866
		$c(2 \times 2)$–Se	142, 197, 198, 293, 294, 305, 340, 1733
		(2×1)–Se	198
		$c(2 \times 2)$–Se	305
	H_3P	Disordered	662
	HCl	$c(2 \times 2)$–HCl	1644
		$c(2 \times 2)$–Cl	1644
	I_2	$c(2 \times 2)$–I	1036, 1148
		NiI_2	1148
		$\begin{bmatrix} 1 - \dfrac{1}{\tan \theta} & - 1 - \dfrac{1}{\tan \theta} \\ \dfrac{2}{\tan \theta} & \dfrac{2}{\tan \theta} \end{bmatrix}$, $\theta \sim 61°$	1148
		$\begin{bmatrix} 5 & -3 \\ 3 & -5 \end{bmatrix}$	1148
		$\begin{bmatrix} 7 & -5 \\ 3 & \frac{3}{5} \end{bmatrix}$	1148
		(2×4)–I_2	1148
	K	(4×2)	454
		Hexagonal	457, 461, 462
	N_2	Not adsorbed	80, 81
		(2×2)–N	772, 1578
		$c(2 \times 2)$–N_2	984, 987
	N_2H_2	(2×2)–N	690
	Na	$c(2 \times 2)$–Na	452, 454–459, 1737, 1865
	NH_3	$c(2 \times 2)$–N	1137
		$c(2 \times 2)$–Na	

TABLE 2.5d. (*Continued*)

Substrate	Adsorbate	Surface Structure	References[b]
	NO	(1×1)	767, 663, 812
		$c(2 \times 2)$-N + O	767, 812
		$c(2 \times 2)$ streaked	663
		(2×2)	767, 812
		Disordered	779
	O_2	(2×2)-O	2, 49, 50, 51, 198, 296–299, 310, 766, 978, 1095, 1138, 1168, 1195, 1356, 1358, 1364, 1417, 1732, 1738, 1743
		$c(2 \times 2)$-O	2, 6, 52–57, 197, 198, 290–299, 310, 340, 640, 766, 978, 1044, 1095, 1138, 1168, 1195, 1220, 1356, 1358, 1417, 1441, 1565, 1738, 1739, 1740, 1741, 1742, 1743, 1793
		(2×1)-O	198
		NiO(100)	6, 297–299, 310
		NiO(111)	298, 299
		Disordered	1291
	O(a) + CO	$c(2 \times 2)$-C, O	1356
	P	$(\sqrt{5} \times \sqrt{5})R26.7°$-P	773, 1644
		$\begin{bmatrix} 1 & -1 \\ 2 & 3 \end{bmatrix}$-P	773
		$\begin{bmatrix} 1 & -1 \\ 2 & 1 \end{bmatrix}$-P	773
	Pb	$\begin{bmatrix} 1 & 1 \\ 1 & -1 \end{bmatrix}$	471
		$\begin{bmatrix} 1 & 1 \\ -2 & 3 \end{bmatrix}$	471
		$(5\sqrt{2} \times \sqrt{2})$ $R45°$-Pb	773
	Si	$c(2 \times 2)$-Si	1644
		$(2\sqrt{2} \times \sqrt{2})$ $R45°$-Si	1644
	Sn	$c(2 \times 2)$-Sn	773
	SO_2	$c(2 \times 2)$-SO_2	86
		(2×2)-SO_2	86
	S, C	(1×1)	1561
	Te	(2×2)-Te	197, 198, 306, 1119, 1732
		$c(2 \times 2)$-Te	197, 198, 294, 305, 340, 1119, 1231, 1597, 1744
		(2×1)-Te	198
		$c(4 \times 2)$-Te	305, 306
	Xe	Partially ordered	1268
Ni$_3$Al(001)	[Clean]	(1×1)	1868

TABLE 2.5d. *(Continued)*[a]

Substrate	Adsorbate	Surface Structure	References[b]
NiCu(100) (Ni < 50%)	S	c(2 × 2)–S	905
Ni–24% Fe(100)	O_2	c(2 × 2)–O	573
Ni–25% Fe(100)	H_2S, H_2	c(2 × 2)–S	1121
Ni–41% Fe(100)	[Clean]	(1 × 1)	1263
	O_2	c(2 × 2)–O	1263
		Oxide	1263
NiO(100)	[Clean]	(1 × 1)	755, 894, 1638
	Cl_2	Disordered	309
	H_2	Adsorbed	307
		Ni(100)	307
		(1 × 1)	894
		Coincidence	894
		(2 × 2)	894
	H_2S	Ni(100)–c(2 × 2)–S	308
	S	c(2 × 2)–S	1185
	SO_2	Disordered	1583
Pb(100)	O_2	PbO(100)	1691
Pd(100)	[Clean]	(1 × 1)	1797
	Ag	(1 × 1)	473, 1797
	Au	(1 × 1)	473, 1806, 1807
	C	c(4 × 2)–C	762
	CO	Disordered	70
		c(4 × 2)–CO	70
		c(2 × 2)–CO	210
		(2 × 4)$R45°$–CO	71, 209, 210
		c($2\sqrt{2} \times \sqrt{2}$)$R45°$–CO	1276, 1294
		c($\sqrt{2} \times \sqrt{2}$)$R45°$–2CO	910, 1797
		Incommensurate	1276
		Hexagonal overlayer	209, 210
	Cu	(1 × 1)	862
	Fe	Fe(100) and Fe(110)	452
	H_2	c(2 × 2)–H	595, 919, 1163, 1454
		(1 × 1)–H	919
	H(a) + O_2	Adsorbed	1163, 1454
	H_2 + O_2	Disordered	1163
		(2 × 2)–O, H	1163
	H_2S	(2 × 2)–S	1294
		c(2 × 2)–S	1294
	Kr	Liquid-like	913
	Ni	(1 × 1)	1805
	NO	(2 × 2)	910
	O_2	(2 × 2)–O	596, 597, 891, 939, 1163, 1334, 1454
		c(2 × 2)–O	596, 682, 939, 1334
		(2 × 2) + (7 × 7)–O	682
		(5 × 5)–O	596, 1334
		($\sqrt{5} \times \sqrt{5}$)$R27°$	596, 1334
		Oxide	1334
		Hexagonal	1334
		Disordered O_2	1454
	O(a) + CO	Disordered	939
	O(a) + H_2O	(2 × 1)–OH	940
	O(a) + H_2	Not adsorbed	1163, 1454
		(2 × 2)	1454

TABLE 2.5d. *(Continued)*[a]

Substrate	Adsorbate	Surface Structure	References[b]
	Xe	Hexagonal overlayer	311
		Liquid-like	913
		Island	913
Pt(100)	[Clean]	(5 × 20) or "hex"	862, 946, 1265, 1394
		(1 × 1)	1265, 1293, 1394, 1745, 1798
		(1 × 5)	1265, 1394, 1171
	Au	$\begin{bmatrix} 14 & 1 \\ -1 & 5 \end{bmatrix}$	671
		(1 × 1)–Au	671
		(1 × 5)–Au	671
		(1 × 7)–Au	671
	C	Ring pattern	762
	C_2N_2	(1 × 1)	433
	CO	c(4 × 2)–CO	28, 72, 73, 120, 314, 316, 663, 952, 1307
		$(3\sqrt{2} \times \sqrt{2})R45°$–CO	28, 72, 73, 316
		$(\sqrt{2} \times \sqrt{5})R45°$–CO	72, 73
		(2 × 4)–CO	10
		(1 × 3)–CO	10
		(1 × 1)–CO	120, 312, 314, 316, 663
		c(2 × 2)–CO	312, 316, 928, 1059, 1252, 663
		Reconstructed hexagonal (or 5 × 20)	1059
	$CO + H_2$	c(2 × 2)–CO + H_2	72, 74
	$CO + O_2$	(1 × 1) diffuse	909
		c(2 × 2)–CO + (3 × 1)–O	928
	Cu	(1 × 1)	862
	F_2	(1 × 1)–F	886
	H_2	Adsorbed	312, 317
		(2 × 2)–H	72, 74
		Not adsorbed	312
		(1 × 1)–H	582
	H_2O	Not adsorbed	1258
	$H_2O + HBr$	$c(2\sqrt{2} \times \sqrt{2})R45°$–Br, HBr	1258
	H_2S or S_2	(2 × 2)–S	225, 226, 247, 320
		c(2 × 2)–S	225, 226, 247, 320, 321
	HBr	$c(2\sqrt{2} \times \sqrt{2})R45°$–(Br + HBr)	806, 1258
	Br, HBr(a) + H_2O	Not adsorbed	1258
	Br, HBr(a) + NH_3	No affinity	1258
	HCl	(2 × 2)–(Cl + HCl)	806
	HI	$c(\sqrt{2} \times \sqrt{2})R45°$–I	580
		c(2 × 4)–I	774
		Ring pattern	774
		$c(2\sqrt{2} \times dn\sqrt{2})R45°, n \geq 7$	774
		$c(2\sqrt{2} \times \sqrt{2})R45°$–I	774
	I_2	$c(\sqrt{2} \times \sqrt{2})R45°$–I	580
		Incommensurate–I	1390
		$c(\sqrt{2} \times 5\sqrt{2})R45°$–I	1390
		$c(\sqrt{2} \times 2\sqrt{2})R45°$–I	1390
		Hexagonal overlayer	1390

TABLE 2.5d. *(Continued)*[a]

Substrate	Adsorbate	Surface Structure	References[b]
		c(2 × 4)	1390
		$(\sqrt{7} \times \sqrt{7})R19.1°$-I	1390
	I(a) + Ag	$(\sqrt{2} \times \sqrt{2})R45°$-I, Ag	1390
		$(10\sqrt{2} \times 10\sqrt{2})R45°$-I, Ag	1390
		$(\sqrt{34} \times \sqrt{34})R31°$-I, Ag	1390
	N	Disordered	228
	NH_3	Poorly ordered	1258
	NO	(1 × 1)-NO	318
		c(4 × 2)-NO	319
		(5 × 1)-NO	826
		c(2 × 4)-NO	826
		(1 × 1) + c(2 × 4)	946
	NO_2	(1 × 1)-N, NO	881
		(5 × 20)-NO_2	881
	O_2	Not adsorbed	120, 312
		Adsorbed	312, 315
		$(2\sqrt{2} \times 2\sqrt{2})R45°$-O	215, 313
		PtO_2(0001)	215
		(5 × 1)-O	315
		(5 × 1)-(1 × 1)-O	708, 1171
		$(2\sqrt{2} \times \sqrt{2})R45°$-O	708
		(2 × 1)-O	315
		(3 × 1)-O	928, 1014
		Complex	1014
	SO_2	(1 × 1) diffuse	1258
	SO_2(a) + NH_3	(1 × 1) diffuse	1258
$Pt_3Ti(100)$	[Clean]	c(2 × 2)	935
Rh(100)	[Clean]	(1 × 1)	895, 1024, 1348, 1800
		(2 × 2)	1147
	Ag	(1 × 1)-Ag	895
		Complex [multilayer]	895
	CO	Hexagonal overlayer	231
		(4 × 1)-CO	58
		c(2 × 2)-CO	231, 1348
	CO(a) + D_2	Compressed (CO)	1348
	CO_2	c(2 × 2)-CO	231
		Hexagonal overlayer	231
	D_2	(1 × 1)-D	1348
	D(a) + CO	c(2 × 2)	1348
	Fe	Fe(100) and Fe(110)	452
	H_2	Adsorbed	231
	H_2S or S	c(2 × 2)-S	1403
		(2 × 2)-S	742, 1403, 1473
	N_2O	(2 × 2)	801
	NO	c(2 × 2)-NO	231
		Disordered	1024
	NO + D_2	Disordered	1025
	O_2	(2 × 2)-O	231, 1403
		c(2 × 2)-O	231
		c(2 × 2)	801
		c(2 × 8)-O	58
		(3 × 1)	1403
Si(100)	[Clean]	(2 × 1)	848, 980, 1017, 1019, 1084, 1207, 1222,

TABLE 2.5d. *(Continued)*[a]

Substrate	Adsorbate	Surface Structure	References[b]
			1428, 1451, 1477, 1494, 1483, 1505, 1514, 1517, 1523, 1535, 1547, 1549, 1645, 1746
		(2×2)	1579
		$c(4 \times 2)$	1600
	Ag	(2×1)	923
		Ag(111)	1352
	Au	Au(111)	Si(100)
	Ge	(2×1)	1483
	H	(1×1)–H	325
		(2×1)–H	325
	H_2	(1×1)–H	322, 323, 324, 633, 680, 1494
		(1×1)–2H	848, 999, 1222, 1477, 1488, 1535
		(2×1)–H	237, 848, 999, 1222, 1477, 1488, 1489, 1494, 1535
	H_2O	(2×1)	1477
		Adsorbed	903
	I_2	(3×3)–I	326
	In	(2×1)–In	971
		(4×3)–In	971
		(1×1)–In	971
	K	(2×1)–K	1184
	N	Not ordered	1230
	NH_3	(111) facets	238
	Ni	$(NiSi_2(100)$	854, 1659
	O_2	(1×1)–O	17, 18, 20
	Pd	Pd_2Si (not epitaxial)	1134
		(2×1)	1207
		(111) facets	17, 18, 20
	PH_3	(2×1)	1451
	Sb	(1×1)–Sb	1019
	Si	(1×1)	1517
	Si + laser	(2×1)	1392
	Sn	$c(4 \times 4)$–Sn	959
		(6×2)–S	959
		$c(8 \times 4)$–Sn	959, 971
		(5×1)–Sn	959, 971
		(2×1)–Sn	959
		(1×1)–Sn	959
SiC(100)	[Clean]	(1×1)	1450
$SmB_6(001)$	[Clean]	(2×2)	738
		(3×3)	738
Sn(100)	[Clean]	(2×1)	1421, 1497, 1513
	H_2	(2×1)	1497, 1513
Sr(100)	O_2	SrO(100)	327
$SrTiO_3(100)$	O_2	(1×1)	1672
		(2×2)	1672
		(2×1)	1672
	Ga	(3×2)–Ga	800

TABLE 2.5d. (*Continued*)[a]

Substrate	Adsorbate	Surface Structure	References[b]
		(5 × 2)–Ga	800
		(2 × 2)–Ga	800
		(8 × 1)–Ga	800, 809
Ta(100)	[Clean]	(1 × 1)	966, 1219
	Au	Split $\begin{bmatrix} 1 & 1 \\ -1 & 2 \end{bmatrix}$	506, 507
	CO	c(3 × 1)–O	328
	CO$_2$	c(3 × 1)–O	328
	H$_2$	(1 × 1)	1219
	I$_2$	Amorphous	1180
		c(2 × 10)	1180
		c(2 × 2)	1180
	N$_2$	Adsorbed	328
	NO	c(3 × 1)–O	328
	O$_2$	$(2 \times \frac{8}{9})$–O	328
		c(3 × 1)–O	328
		(4 × 1)–O	328
		(3 × 3)–O	873
		(1 × 2)–O	873
		(1 × 3)–O	873
	Th	$\begin{bmatrix} 1 & 1 \\ 1 & -1 \end{bmatrix}$	510, 511
		(1 × 1)	510, 511
Th(100)	CO	Disordered	329
	O$_2$	Disordered	329
		ThO$_2$	329
TiC(001)	[Clean]	(1 × 1)	1631
	O$_2$	Disordered	611
UO$_2$(100)	[Clean]	c(2 × 2)	648
V(100)	[Clean]	(1 × 1)	1126, 1315, 1498
	Br$_2$	(1 × 1)–Br	651
		(5 × 1)–Br	1126
		c(2 × 2)–Br	1126
		(6 × 4)–Br	1126
		Ring pattern	1126
	CO	(5 × 1)–O	1315
	H$_2$	Disordered	65
	O$_2$	(1 × 1)–O	65, 651
		(2 × 2)–O	65
		(5 × 1)–O	1315
	O	(5 × 1)	1126
	S	(2 × 2)–S	650, 1315
		(1 × 1)–S	650
		(5 × 1)	650
		$\sqrt{2} \times \sqrt{2})R27°$–S	1315
W(100)	[Clean]	c(2 × 2)	749, 1147, 1340, 1347, 1388, 1396, 1503, 1668
		(2 × 2)	1340
		(1 × 1)	1340, 1347, 1396, 1471, 1656, 1763, 1802
	Ag	(2 × 1)	546

TABLE 2.5d. *(Continued)*[a]

Substrate	Adsorbate	Surface Structure	References[b]
		$\begin{bmatrix} 1 & 1 \\ 1 & -1 \end{bmatrix}$	546
		(1×1)	546
	Au	(2×1)	546
		$\begin{bmatrix} 2 & 0 \\ -1 & 2 \end{bmatrix}$	546
		(1×1)	546
	Ba	$\begin{bmatrix} 2 & 0 \\ -8 & 2 \end{bmatrix}$	531, 532
		Split $\begin{bmatrix} 1 & 1 \\ -2 & 2 \end{bmatrix}$	531, 532
		$\begin{bmatrix} 1 & 1 \\ -2 & 2 \end{bmatrix}$	531, 532
		$\begin{bmatrix} 1 & 1 \\ 1 & -1 \end{bmatrix}$	531, 532
		c(4 × 2)–Ba	735
		c(2√2 × √2)–Ba	735
		c(√6 × √2)R45°–Ba	735
		c(2√2 × $\frac{2}{3}$√6)R45°–Ba	735
		(2 × 12)–Ba	994
		(2 × 10)–Ba	994
		(3 × 2)–Ba	994
		(3 × 2) + c(2 × 2)	994
		c(2 × 1.86)	994
		(10 × 21)–Ba	994
		c(2 × 2)–Ba	1399
		c(2 × k) (1.86 ≤ k ≤ 2√$\frac{7}{3}$)	994
		Hexagonal	994
	Bi	c(2 × 2)	1260
		(2 × 2)	1260
		(1 × 1)	1260
	Br₂	c(2 × 2)	604
		(0.75√2 × √2)R45°	604
		c(4 × 2)	604
		(5 × 2)	604
		c(6 × 2)	604
		(7 × 2)	604
		c(8 × 2)	604
	C	(5 × 1)–C	821
	Cl₂	c(2 × 2)	643
		c(4 × 1)	644
		(1 × 1)	644
		$\begin{bmatrix} -7 & 1 \\ 1 & 1 \end{bmatrix}$–Cl	733
		$\begin{bmatrix} -5 & 1 \\ 1 & 1 \end{bmatrix}$–Cl	733
		(3√2 × √2)R45°–Cl	733
	CO + N₂	(4 × 1)–CO + N₂	82

TABLE 2.5d. *(Continued)*[a]

Substrate	Adsorbate	Surface Structure	References[b]
	CO	Disordered	75
		$c(2 \times 2)$-CO	66, 75, 1465
		$c(2 \times 2)$-C + O	777
	CO_2	Disordered	338
		(2×1)-O	338
		$c(2 \times 2)$-CO	338
	Cs	$\begin{bmatrix} 1 & 1 \\ 1 & -1 \end{bmatrix}$	523–526
		(2×2)	523–526
		Split (2×2)	523, 524
		Hexagonal	525, 526
	Cu	(2×2)	543
		$\begin{bmatrix} 1 & 1 \\ 1 & -1 \end{bmatrix}$	543
	Ga	(1×1)-Ga	789
	H_2	$c(2 \times 2)$-H	66, 78, 79, 337, 411, 712, 771, 1275, 1347, 1361, 1388, 1603, 1668
		$(\sqrt{2} \times \sqrt{2})$-H	834, 1032, 1511
		Incommensurate $(\sqrt{2} \times \sqrt{2})$-H	1032
		One-dimensional order	1032
		(2×5)-H	79
		(4×1)-H	79
		(1×1)-(2*)H	411, 771, 897, 1032, 1165, 1347, 1388, 1747
		(2×2)-H	1361
		Incommensurate	834, 1511
		Disordered	771, 834
	H_2S	(2×1)	821
		$c(2 \times 2)$-S	887
	Hg	(1×1)	549
	K	$\begin{bmatrix} 1 & 1 \\ 1 & 1 \end{bmatrix}$	1841
	N_2	$c(2 \times 2)$-N	68, 82, 131, 776, 1099, 1465, 1748
		Contracted domain	1609
	N_2O	(1×1)-N_2O	143
		(4×1)-N_2O	143
	Na	$\begin{bmatrix} 1 & 1 \\ 1 & -1 \end{bmatrix}$	1840
	NH_3	Disordered	84
		$c(2 \times 2)$-NH_2	84
		$c(2 \times 2)$-N	1099
		(1×1)-NH_2	84
	NO	(2×2)-NO	339
		(4×1)-NO	339
		(2×2)-O	339
		(4×1)-O	339
		(2×1)-O	339
	O_2	Disordered	330, 1749

TABLE 2.5d. *(Continued)*[a]

Substrate	Adsorbate	Surface Structure	References[b]
		(4×1)-O	66, 330–333, 336, 821, 1058
		(2×2)-O	330–334
		(2×1)-O	66, 67, 330–336, 821, 1058
		(3×3)-O	331, 333, 335
		$c(2 \times 2)$-O	333
		$c(8 \times 2)$-O	333
		(3×1)-O	333
		(1×1)-O	333
		(8×1)-O	333
		(4×4)-O	333, 335
		(110) facets	333
	$O_2 + H_2$	$(\sqrt{2} \times \sqrt{2}) + (4 \times 1)$-O, H	815
	Pb	Disordered (2×2)	550
		Split $\begin{bmatrix} 1 & 1 \\ 1 & -1 \end{bmatrix}$	550
		$\begin{bmatrix} 1 & 1 \\ 1 & -1 \end{bmatrix}$	550, 551
		$\begin{bmatrix} 1 & 1 \\ -2 & 2 \end{bmatrix}$	550
		Hexagonal	550
		(2×2)	551
		$\begin{bmatrix} 2 & 0 \\ -1 & 2 \end{bmatrix}$	551
		(1×1)	551, 702
		$c(4 \times 2)$-Pb	702
		$c(2 \times 2)$-Pb	702
	Pd	$(2\sqrt{2} \times \sqrt{2})R45°$-Pd	646
		(2×1)-Pd	646
		$c(2 \times 2)$-Pd	646
	Rb	$\begin{bmatrix} 1 & 1 \\ 1 & -1 \end{bmatrix}$	522
		(2×2)	522
		Hexagonal	522
	S	(2×2)-S	1324
		(3×3)-S	1324
		(4×2)-S	1324
		(2×1)-S	1324
		(5×5)-S	1324
	Sb	(2×2)	553, 554
		$\begin{bmatrix} 1 & 1 \\ 1 & -1 \end{bmatrix}$	553, 554
		(1×1)	553, 554
	Se	(2×1)	1326
		(8×1)	1326
		(6×1)	1326
		(3×1)	1326
		$c(2 \times 2)$	1326
	Sn	(1×1)-Sn	789
		$c(2 \times 2)$-Sn	789

TABLE 2.5d. *(Continued)*[a]

Substrate	Adsorbate	Surface Structure	References[b]
	Te	(3 × 3)–Te	1323
		(2 × 2)–Te	1323
		Complex	1323
		(2 × 1)–Te	1323
	Th	$\begin{bmatrix} 1 & 1 \\ 1 & -1 \end{bmatrix}$	556–560
		(1 × 1)	556–560
		$\begin{bmatrix} 2 & 0 \\ -1 & 3 \end{bmatrix}$	559, 560
		Hexagonal	559, 560
	Rb	$\begin{bmatrix} 1 & 1 \\ 1 & -1 \end{bmatrix}$	522
		(2 × 2)	522
		Hexagonal	522
	Zr	(1 × 1)	541, 764, 789
		$\begin{bmatrix} 2 & 0 \\ -1 & 2 \end{bmatrix}$	541
		c(2 × 2)	764, 789
WO₃(100)	[Clean]	Split (1 × 1)	991
		(5 × 1)	1393
		$(\sqrt{2} \times \sqrt{2})R45°$, (5 × 1)	1393
Xe(100)	[Clean]	(1 × 1)	1914

[a]Organic overlayer structures are not included. See Table 2.5g for these structures.
[b]References for Tables 2.5a to 2.5j are listed at the end of Table 2.5j.

TABLE 2.5e. Surface Structures of Metallic Monolayers on Metal Crystal Surfaces

Substrate	Adsorbate	Surface Structure	References[a]
Ag(100)	Au	(1 × 1)	1806
	Cu	Epitaxial	1167
		Cu (100)	1476
		(1 × 1)	1820
	Fe	(1 × 1)	1272
	Ni	(1 × 1)	1820
	Pd	Epitaxial	1167
		(1 × 1)	1463
Ag(110)	Cs	(1 × 2)	859, 1534
		(1 × 3)	859, 1543
	K	(1 × 2)	1534
		(1 × 3)	1534
	Li	(1 × 2)	1534
	Na	(1 × 1)	489
Ag(111)	Al	Disordered	491
	Au	(1 × 1)	491, 1355, 1825, 1826
	Bi	Disordered	491
	Cd	No condensation	491

TABLE 2.5e. (*Continued*)

Substrate	Adsorbate	Surface Structure	References
	Co	Disordered	491
	Cr	Disordered	491
	Cu	Hexagonal overlayer	1822–1824
	K	Hexagonal overlayer	1345
	Mg	Disordered	491
	Na	(1×1)	488
	Ni	Hexagonal overlayer	491, 1821
	Pb	$(\sqrt{3} \times \sqrt{3})R30°$–Pb	975
		Pb(111)	975
		Hexagonal overlayer	491, 1827
	Pd	(1×1)	1463
		Disordered	491
	Rb	(1×1)–Rb	490, 705
		(9×9)	705
	Sb	Disordered	491
	Sn	Disordered	491
	Tl	Hexagonal overlayer	491
	Zn	No condensation	491
Al(100)	Ag	(5×1)–Ag	1363
		(1×1) [multilayer]	1363
	Au	Disordered	1363
	Cu	Disordered	1363
	Fe	Poor epitaxy	452
	Na	$c(2 \times 2)$	499, 500, 501, 1926
		Hexagonal overlayer	500
	Pb	$\begin{bmatrix} 2 & 0 \\ -1 & 2 \end{bmatrix}$	503
		$c(2 \times 2)$–Pb	704
		$\begin{bmatrix} 2 & 0 \\ 1 & n \end{bmatrix} - \text{Pb}, \ 2 < n$ < 3	704
	Sm	(1×1) disorder	1532
		Complicated	1532
	Sn	$\begin{bmatrix} 2 & 0 \\ -1 & 3 \end{bmatrix}$	503
		$c(2 \times 6)$–Sn	704
		$\begin{bmatrix} 2 & 0 \\ 1 & n \end{bmatrix} 3 < n < 3$	704
Au(100)	Ag	(1×1)	473, 474, 494, 1838
	Bi	$\begin{bmatrix} 2 & 0 \\ -1 & 2 \end{bmatrix}$	498
	Cu	(1×1)	473
	Fe	(1×1)	1828–1830
	Na	Hexagonal	492
		Ordered	492
	Pb	$\begin{bmatrix} 1 & 1 \\ 1 & -1 \end{bmatrix}$	441–444

TABLE 2.5e. (*Continued*)

Substrate	Adsorbate	Surface Structure	References
		$\begin{bmatrix} 1 & 1 \\ -3 & 4 \end{bmatrix}$	441–444
		$\begin{bmatrix} 1 & 1 \\ -1 & 2 \end{bmatrix}$	441–444
		$\begin{bmatrix} 2 & 0 \\ -1 & 3 \end{bmatrix}$	441–444
		(1 × 1)	683
	Pd	(1 × 1)	473, 1833
		c(2 × 2)	683
		c(7$\sqrt{2}$ × $\sqrt{2}$)$R45°$	683
		c(3$\sqrt{2}$ × $\sqrt{2}$)$R45°$	683
		c(6 × 2)	683
	Pt	(1 × 1)	438–440
Au(110)	Bi	$\begin{bmatrix} 1 & 1 \\ 1 & -1 \end{bmatrix}$	498
		$\begin{bmatrix} 2 & 1 \\ -1 & 1 \end{bmatrix}$	498
		(2 × 1)	498
	Pb	(1 × 3)	444, 495, 683
		(1 × 1)	444, 495, 683
		(7 × 1)	444, 495, 683
		(7 × 3)	444, 495, 683
		(4 × 4)	444, 495, 683
Au(111)	Ag	(1 × 1)	491, 1825
		fcc(111)	1689
	Ag, Air	Ag$_2$O(110)–(2 × 1)	997
	Bi	$\begin{bmatrix} 10 & 10 \\ -10 & 20 \end{bmatrix}$	498
		(2 × 2)	924
	Cr	Hexagonal	493
	Cu	($\sqrt{3}$ × $\sqrt{3}$)R–Cu	861, 1558, 1582
		(1 × 1)	1558
		Extra Lines (RHEED)	1836, 1837
	Fe	(1 × 1)	1828, 1830–1832
	Pb	Hexagonal rotated ± 5°	444, 495
		($\sqrt{3}$ × $\sqrt{3}$)$R30°$	924
		(1 × 1)	683
	Pd	(1 × 1)	1807, 1834
	Pt	(1 × 1)	1835
Au(311)	Pb	(5 × 3)	496
		(3 × 3)–Pb	730
		(3 × 4)–Pb	730
Au(511)	Pb	$\begin{bmatrix} 1 & 1 \\ 1 & 1 \end{bmatrix}$	444
		$\begin{bmatrix} 2 & 0 \\ 1 & 3 \end{bmatrix}$	444

TABLE 2.5e. (*Continued*)

Substrate	Adsorbate	Surface Structure	References
	Pd	$c(2 \times 2)$	683
		$c(7\sqrt{2} \times \sqrt{2})R45°$	683
		$c(3\sqrt{2} \times \sqrt{2})R45°$	683
		$c(6 \times 2)$	683
Au(711)	Pb	$\begin{bmatrix} 1 & 1 \\ 1 & -1 \end{bmatrix}$	444
		$\begin{bmatrix} 2 & 0 \\ -1 & 3 \end{bmatrix}$	444
	Pd	$c(2 \times 2)$	683
		$c(7\sqrt{2} \times \sqrt{2})R45°$	683
		$c(3\sqrt{2} \times \sqrt{2})R45°$	683
		$c(6 \times 2)$	683
Au(911)	Pb	$\begin{bmatrix} 1 & 1 \\ 1 & -1 \end{bmatrix}$	444
		$\begin{bmatrix} 2 & 0 \\ -1 & 3 \end{bmatrix}$	444
	Pd	$c(2 \times 2)$	683
		$c(7\sqrt{2} \times \sqrt{2})R45°$	683
		$c(3\sqrt{2} \times \sqrt{2})R45°$	683
		$c(6 \times 2)$	683
Au(11, 1, 1)	Pb	$\begin{bmatrix} 1 & 1 \\ 1 & -1 \end{bmatrix}$	444
		$\begin{bmatrix} 2 & 0 \\ -1 & 3 \end{bmatrix}$	444
Au(210)	Pb	(1×1)	497
Au(320)	Pb	(3×3)	496
		$(1 \times 1)-Pb$	730
Cr(111)	Ag	(8×8)	46
	Au	$\begin{bmatrix} \frac{2}{3} & 3 \\ -\frac{2}{3} & \frac{4}{3} \end{bmatrix}$	52
	Bi	$\begin{bmatrix} 1 & 1 \\ -1 & 2 \end{bmatrix}$	61
		$\begin{bmatrix} 2 & -1 \\ 0 & 2 \end{bmatrix}$	61
		$\begin{bmatrix} 2 & 3 \\ 1 & 2 \end{bmatrix}$	61
	Fe	(1×1)	39
	Ni	(1×1)	43, 44
	Pb	(4×4)	55, 58
	Sn	$\begin{bmatrix} 1 & 1 \\ -1 & 2 \end{bmatrix}$	54
Cu(100)	Ag	$\begin{bmatrix} 2 & 0 \\ -1 & 5 \end{bmatrix}$	473, 477

TABLE 2.5e. (*Continued*)

Substrate	Adsorbate	Surface Structure	References
	Au	$\begin{bmatrix} 1 & 1 \\ 1 & -1 \end{bmatrix}$	473, 478
		$\begin{bmatrix} 2 & 0 \\ -1 & 7 \end{bmatrix}$	473, 478
	Bi	(2×2)	483, 486
		$\begin{bmatrix} 1 & 1 \\ 1 & -1 \end{bmatrix}$	483, 486
		$\begin{bmatrix} 1 & 1 \\ -4 & 5 \end{bmatrix}$	483, 486
		$\begin{bmatrix} 5 & 4 \\ -4 & 5 \end{bmatrix}$	483, 486
		(1×1)	703
		$c(2 \times 2)$	703
	Co	(1×1)–Co	983, 1105, 1810 1811
	Cs	Hexagonal overlayer	865
		Quasi-hexagonal	865
		Disordered	865
	Fe	(1×1)	452, 1808, 1809
	K	$\begin{bmatrix} 2 & 3 \\ 0 & 5 \end{bmatrix}$	1405
		$\begin{bmatrix} 2 & 2 \\ 0 & 3 \end{bmatrix}$	1405
		Incommensurate	1405
	Mn	$c(2 \times 2)$–Mn	1319
	Nb	Incommensurate	667
	Ni	(1×1)	1812
	Pb	$\begin{bmatrix} 2 & 2 \\ -2 & 2 \end{bmatrix}$	481–485
		$\begin{bmatrix} 1 & 1 \\ -2 & 3 \end{bmatrix}$	481–485
		$c(5\sqrt{2} \times \sqrt{2})R45°$–Pb	703, 1295
		$c(2 \times 2)$–Pb	1295
		$(2\sqrt{2} \times 2\sqrt{2})R$–Pb	1295
	Pd	$c(2 \times 1)$–Pd	726
		(1×1)–Pd	726, 1649
		$c(2 \times 2)$–Cu_3Pd	1649
	Sn	(2×2)	480
	Te	(2×2)–Te	267, 1119, 1728
	Tl	$\begin{bmatrix} 2 & 2 \\ 2 & -2 \end{bmatrix}$	1167, 1564
		$\begin{bmatrix} 4 & 0 \\ 2 & 7 \end{bmatrix}$–Tl	1167, 1336, 1564

TABLE 2.5e. (*Continued*)

Substrate	Adsorbate	Surface Structure	References
		$\begin{bmatrix} 4 & 0 \\ 2 & 6 \end{bmatrix}$–Tl	1336, 1564
		$\begin{bmatrix} 6 & 6 \\ 2 & -2 \end{bmatrix}$	1564
		c(4 × 4)–Tl	1336
Cu(110)	Au	$\begin{bmatrix} 1 & 0 \\ -\frac{1}{2} & \frac{3}{2} \end{bmatrix}$	479
		(1 × 2)	479
		(2 × 2)	479
	Pb	$\begin{bmatrix} 1 & 1 \\ -1 & 1 \end{bmatrix}$	481, 482
		(5 × 1)	481, 482
		(4 × 1)	482
	Pd	(2 × 1)–Pd	726
		(1 × 1)–Pd	726
Cu(111)	Ag	(8 × 8)	477
		Three-dimensional crystals	1813, 1815–1818
		(1 × 1)	1526
	Au	$\begin{bmatrix} \frac{2}{3} & \frac{2}{3} \\ -\frac{2}{3} & 4\frac{1}{3} \end{bmatrix}$	479
		(2 × 2)	479
		Three-dimensional crystals	1815, 1819
	Bi	$\begin{bmatrix} 1 & 1 \\ -1 & 2 \end{bmatrix}$	487
		$\begin{bmatrix} 2 & -1 \\ 0 & 2 \end{bmatrix}$	487
		$\begin{bmatrix} 2 & 3 \\ 1 & 2 \end{bmatrix}$	487
	Co	(1 × 1)–Co	1299
	Cs	(2 × 2)–Cs	711, 1430
	Fe	(1 × 1)	474
	Na	(2 × 2)–Na	1571
	Ni	(1 × 1)–Ni	475, 476, 1466, 1813
	Pb	(4 × 4)	481, 484
	Pd	(1 × 1)	726, 1538
	Sn	$\begin{bmatrix} 1 & 1 \\ -1 & 2 \end{bmatrix}$	480
Cu(211)	Pb	(4 × 1)	484
Cu(311)	Pb	$\begin{bmatrix} 3 & 1 \\ -2 & 1 \end{bmatrix}$	484
		(4 × 2)	484

TABLE 2.5e. (*Continued*)

Substrate	Adsorbate	Surface Structure	References
Cu(511)	Pb	(4 × 1)	482
Cu(711)	Pb	(4 × 1)	482, 484
Fe(100)	K	Disordered	665
		(2 × 2)–K	784
		Hexagonal close pack	784
Fe(110)	K	Heaxgonal array	728
Fe(111)	K	(3 × 3)–K	665, 1350
Ge(111)	Al	(2 × 1)	1550
	Au	$(\sqrt{3} \times \sqrt{3})R30°$–Au	1223
Ir(100)	Ba	(2 × 1)–Ba	1399
	Cs	c(4 × 2)–Cs with streak	1395
		Close-packed layer	1395
		Compressed layer	1395
		(5 × 5)–Cs	1395
	K	$c(2\sqrt{2} \times 4\sqrt{2})R45°$	866
		$\begin{bmatrix} 2 & 1 \\ -1 & 2 \end{bmatrix}$	866
		c(4 × 2)	866
		(3 × 2)	866
		c(2 × 2)	866
		$\begin{bmatrix} \frac{5}{2} & 0 \\ -\frac{5}{4} & \frac{5}{3} \end{bmatrix}$	866
		$\begin{bmatrix} 2 & 0 \\ -1 & \frac{5}{3} \end{bmatrix}$	866
		$\begin{bmatrix} \frac{10}{7} & 0 \\ -\frac{5}{7} & \frac{5}{3} \end{bmatrix}$	866
Ir(111)	Au	(1 × 1)	453
	Cr	Hexagonal	453
Mo(100)	Ag	Ag(100)	513, 514
		Ag(110)	513, 514
	Cs	$(\sqrt{2} \times \sqrt{2})R45°$	932
		(2 × 2)	932
		c(2 × 2)	932
		Rectangular centered mesh	932
		Quasi-hexagonal	932
		Hexagonal overlayer	932
	Ga	(1 × 1)–Ga	789
	Sn	$\begin{bmatrix} 1 & 1 \\ 1 & -1 \end{bmatrix}$	516
		(1 × 2)	516
		(1 × 1)–Sn	789
		c(2 × 2)–Sn	789
Mo(110)	Al	Hexagonal	515
	Au	Disordered	1681
	Cs	Hexagonal	512

TABLE 2.5e. (*Continued*)

Substrate	Adsorbate	Surface Structure	References
	K	Hexagonal	512
	Na	No disordered structure	512
	Rb	Hexagonal	512
Mo(211)	Ba	(1×5)	1591, 1675
		(4×2)	1591
	Cs	$c(2 \times 1/J)\ 0.15$ $< J < 0.64$	1590
		$c(2 \times 2)$	1590
	La	Linear chains	1447
		$c(2 \times 2)$	1447
		$c(2 \times \frac{4}{3})$	1447
	Li	(1×4)–Li	1593
		(1×2)–Li	1593
		(1×1)–Li	1593
	Na	(1×4)–Na	1684
		(1×3)–Na	1684
		(1×2)–Na	1684
		$(1 \times \frac{3}{2})$–Na	1684
		(1×9)–Sr	1594
		(1×5)–Sr	1594
		(4×2)–Sr	1594
Nb(110)	Sn	Disordered	505
		(3×1)	505
Ni(100)	Ba	Disordered	454
	Co	(1×1)	1803
	Cr	(1×1)	463, 464
	Cs	(2×2)	454
		Hexagonal	463, 464
	Cu	(1×1)–Cu	1113, 1458, 1804
	Fe	$c(2 \times 2)$–Fe	973
		Fe(110)	452, 973
	K	(4×2)	454
	Na	$\begin{bmatrix} 1 & 1 \\ 1 & -1 \end{bmatrix}$	452, 454–459
		$c(2 \times 2)$–Na	1737, 1865
	Pb	$\begin{bmatrix} 1 & 1 \\ 1 & -1 \end{bmatrix}$	471
		$\begin{bmatrix} 1 & 1 \\ -2 & 3 \end{bmatrix}$	471
		$(5\sqrt{2} \times \sqrt{2})R45°$–Pb	773
	Sn	$c(2 \times 2)$–Sn	773
Ni(110)	Cs	Disordered	455
	K	Disordered	455
	Na	Disordered	455, 458, 460
		Hexagonal	455, 458, 460
	Pb	$\begin{bmatrix} 1 & 1 \\ 1 & -1 \end{bmatrix}$	471

TABLE 2.5e. (*Continued*)

Substrate	Adsorbate	Surface Structure	References
		(3×1)	471
		(4×1)	471
		(5×1)	471
	Yb	(2×1)–Yb	844
Ni(111)	Ag	(6×6)	465, 466
	Au	(6×6)	467–470
		(13×13)	467–469
	Bi	$(\sqrt{3} \times \sqrt{3})R30°$–Bi	864
		(7×7)–Bi	864
		$(\sqrt{\frac{7}{4}} \times \sqrt{\frac{7}{4}})R19°$–Bi	864
	Mo	(5×5)	447, 448
		(4×4)	447, 448
		$\begin{bmatrix} 2 & 0 \\ 5 & 10 \end{bmatrix}$	447, 448
		$\begin{bmatrix} 1 & 0 \\ 5 & 10 \end{bmatrix}$	447, 448
	Na	Hexagonal	455, 458, 460
	Pb	$\begin{bmatrix} 1 & 1 \\ -1 & 2 \end{bmatrix}$	471, 472
		(7×7)	471, 864
		(13×13)	471, 472
		(3×3)	471, 864, 1060
		(4×4)–Pb	864, 1060
		Hexagonal rotated $\pm$ 3°	472
		$(\sqrt{3} \times \sqrt{3})R30°$–Pb	864, 1060
	Sn	(2×2)–Sn	1060
		$(\sqrt{3} \times \sqrt{3})R30°$–Sn	1060
	Te	$(2\sqrt{3} \times 2\sqrt{3})R30°$– Te	577
Pd(100)	Ag	(1×1)	473, 1797
	Au	(1×1)	473, 1806, 1807
	Cs	(1×2) + disordered C_s	1760
	Cu	(1×1)	862
	Fe	Fe(100) and Fe(110)	452
	Na	(1×2) + disordered Na	1760
	Ni	(1×1)	1805
Pd(111)	Au	(1×1)–Au	1709
	Fe	(1×1)	1546
Pt(100)	Au	$\begin{bmatrix} 14 & 1 \\ -1 & 5 \end{bmatrix}$	671
		(1×1)–Au	671
		(1×5)–Au	671
		(1×7)–Au	671
Pt(111)	Ag	Disordered	1254

TABLE 2.5e. (*Continued*)

Substrate	Adsorbate	Surface Structure	References
	Au	Disordered	1254
	Cu	(1×1)–Cu	842
		Cu(111) multilayers	842
		Alloy formation	842
	K	$(\sqrt{3} \times \sqrt{3})R30°$–K	1071, 1238, 1337
		$\begin{bmatrix} 1.66 & 0 \\ 0 & 1.66 \end{bmatrix}$	1337
		Ring pattern	1337
Re(0001)	Ba	(2×2)	565, 566
		Hexagonal	565, 566
Re(10$\bar{1}$0)	Ba	$c(2 \times 2)$	1591
	Mg	(1×3)	1675
Rh(100)	Ag	(1×1)	895
		Complex	895
	Fe	Fe(100) and Fe(110)	452
Ru(0001)	Cu	Disordered	1679
	Na	$(\frac{3}{2} \times \frac{3}{2})$–Na	1129, 1406
		Ring pattern	1129
		(2×2)–Na	1082, 1129
		$(\sqrt{3} \times \sqrt{3})R30°$–Na	1082, 1129, 1406
		Hexagonal overlayer	1406
Sb(0001)	Fe	(1×1)	567
	The	$\begin{bmatrix} 1 & 1 \\ 1 & -1 \end{bmatrix}$	510, 511
		(1×1)	510, 511
Ta(110)	Al	Hexagonal	508, 509
		Square	508, 509
Ti(0001)	Cd	(1×1)	562, 563, 564
	Cu	Extra spots	561
W(100)	Ag	(2×1)	546
		$\begin{bmatrix} 1 & 1 \\ 1 & -1 \end{bmatrix}$	546
		(1×1)	546
	Au	(2×1)	546
		$\begin{bmatrix} 2 & 0 \\ -1 & 2 \end{bmatrix}$	546
		(1×1)	546
	Ba	$\begin{bmatrix} 2 & 0 \\ -8 & 2 \end{bmatrix}$	531, 532
		Split $\begin{bmatrix} 1 & 1 \\ -2 & 2 \end{bmatrix}$	531, 532
		$\begin{bmatrix} 1 & 1 \\ -2 & 2 \end{bmatrix}$	531, 532
		$\begin{bmatrix} 1 & 1 \\ 1 & -1 \end{bmatrix}$	531, 532
		$c(4 \times 2)$–Ba	735

TABLE 2.5e. (*Continued*)

Substrate	Adsorbate	Surface Structure	References
		$c(2\sqrt{2} \times \sqrt{2})$–Ba	735
		$c(\sqrt{6} \times \sqrt{2})R45°$–Ba	735
		$c(2\sqrt{2} \times \frac{2}{3}\sqrt{6})R45°$–Ba	735
		(2×12)–Ba	994
		(2×10)–Ba	994
		(3×2)–Ba	994
		$(3 \times 2) + c(2 \times 2)$	994
		$c(2 \times 1.86)$	994
		(10×21)–Ba	994
		$c(2 \times 2)$–Ba	1399
		$c(2 \times K), (1.86 \leq K \leq 2\sqrt{\frac{2}{3}})$	994
		Hexagonal	994
	Bi	$c(2 \times 2)$	1260
		(2×2)	1260
		(1×1)	1260
	Cs	$\begin{bmatrix} 1 & 1 \\ 1 & -1 \end{bmatrix}$	523–526
		(2×2)	523–526
		Split (2×2)	523, 524
		Hexagonal	525, 526
	Cu	(2×2)	543
		$\begin{bmatrix} 1 & 1 \\ 1 & -1 \end{bmatrix}$	543
	Ga	(1×1)–Ga	789
	Hg	(1×1)	549
	K	$\begin{bmatrix} 1 & 1 \\ 1 & 1 \end{bmatrix}$	1841
	Na	$\begin{bmatrix} 1 & 1 \\ 1 & -1 \end{bmatrix}$	1840
	Pb	Disordered (2×2)	550
		Spilt $\begin{bmatrix} 1 & 1 \\ 1 & -1 \end{bmatrix}$	550
		$\begin{bmatrix} 1 & 1 \\ 1 & -1 \end{bmatrix}$	550, 551
		$\begin{bmatrix} 1 & 1 \\ -2 & 2 \end{bmatrix}$	550
		Hexagonal	550
		(2×2)	551
		$\begin{bmatrix} 2 & 0 \\ -1 & 2 \end{bmatrix}$	551
		(1×1)	551, 702
		$c(4 \times 2)$–Pb	702
		$c(2 \times 2)$–Pb	702
	Pd	$(2\sqrt{2} \times \sqrt{2})R45°$–Pd	646
		(2×1)–Pd	646
		$c(2 \times 2)$–Pd	646

TABLE 2.5e. (*Continued*)

Substrate	Adsorbate	Surface Structure	References
	Rb	$\begin{bmatrix} 1 & 1 \\ 1 & -1 \end{bmatrix}$	522
		(2×2)	522
		Hexagonal	522
	Sb	(2×2)	553, 554
		$\begin{bmatrix} 1 & 1 \\ 1 & -1 \end{bmatrix}$	553, 554
		(1×1)	553, 554
	Sn	(1×1)–Sn	789
		$c(2 \times 2)$–Sn	789
	Th	$\begin{bmatrix} 1 & 1 \\ 1 & -1 \end{bmatrix}$	556, 560
		(1×1)	556–560
		$\begin{bmatrix} 2 & 0 \\ -1 & 3 \end{bmatrix}$	559, 560
		Hexagonal	559, 560
	Rb	$\begin{bmatrix} 1 & 1 \\ 1 & -1 \end{bmatrix}$	522
		(2×2)	522
		Hexagonal	522
	Zr	(1×1)	541, 764, 789
		$\begin{bmatrix} 2 & 0 \\ -1 & 2 \end{bmatrix}$	541
		$c(2 \times 2)$	764, 789
W(110)	Ag	Hexagonal structures	546, 547
		Ag(111)	1151
	Au	Hexagonal structures	546, 548
	Ba	Disordered Hexagonal	533–535
		Hexagonal	533–535
		$\begin{bmatrix} 2 & 2 \\ 0 & 6 \end{bmatrix}$	533–535
		$\begin{bmatrix} 2 & 2 \\ 0 & 5 \end{bmatrix}$	533–535
		$\begin{bmatrix} 3 & 3 \\ 1 & 5 \end{bmatrix}$	533–535
		Hexagonal compact	533–535
	Be	(1×9)	529
		(1×1)	529
		$\begin{bmatrix} 9 & 0 \\ -1 & 1 \end{bmatrix}$	529
	Cs	Disordered hexagonal	523, 527, 528
		Hexagonal	523, 527, 528, 1677
	Cu	Hexagonal	543–545
		Cu(111)	1151
	Fe	Three-dimensional crystals	451
		Fe(110)	1325

TABLE 2.5e. (*Continued*)

Substrate	Adsorbate	Surface Structure	References
		(1×1)	1325
	Li	$\begin{bmatrix} 1 & 5 \\ -2 & 2 \end{bmatrix}$	517–519
		(2×2)	517–519
		$\begin{bmatrix} 1 & 1 \\ -1 & 2 \end{bmatrix}$	517–519
		(2×3)–Li	1269
		$c(2 \times 2)$–Li	1269
		$c(3 \times 2)$–Li	1269
		$c(1 \times 1)$–Li	1269
	Na	$\begin{bmatrix} 1 & 5 \\ -2 & 2 \end{bmatrix}$	445, 446
		(2×2)	445, 446
		$\begin{bmatrix} 1 & 1 \\ -1 & 2 \end{bmatrix}$	445, 446
		$\begin{bmatrix} 1 & 1 \\ 0 & 8 \end{bmatrix}$	445, 446
		$\begin{bmatrix} 1 & 1 \\ 0 & 5 \end{bmatrix}$	445, 446
		Hexagonal	445, 446
	Ni	(1×1)–Ni	970
		(8×2)–Ni	970
		(7×2)–Ni	970
	Pd	(1×3)	542
		Hexagonal	542
	Pb	Split $\begin{bmatrix} 3 & 0 \\ -1 & 1 \end{bmatrix}$	551, 552
		$\begin{bmatrix} 3 & 0 \\ -1 & 1 \end{bmatrix}$	551, 552
	Sb	$\begin{bmatrix} 1 & 1 \\ 0 & 4 \end{bmatrix}$	553, 555
		$\begin{bmatrix} 2 & 0 \\ -1 & 1 \end{bmatrix}$	553, 555
		$\begin{bmatrix} 3 & 0 \\ -1 & 1 \end{bmatrix}$	553, 555
		$\begin{bmatrix} 4 & 0 \\ -1 & 1 \end{bmatrix}$	553, 555
	Sc	$\begin{bmatrix} 1 & 1 \\ 0 & 3 \end{bmatrix}$	536–538
		$\begin{bmatrix} 2 & 2 \\ 0 & 8 \end{bmatrix}$	536–538
	Sr	$\begin{bmatrix} 3 & 3 \\ -2 & 5 \end{bmatrix}$	530

TABLE 2.5e. (*Continued*)

Substrate	Adsorbate	Surface Structure	References
		$\begin{bmatrix} 2 & 2 \\ 0 & 6 \end{bmatrix}$	530
		$\begin{bmatrix} 2 & 2 \\ 1 & 6 \end{bmatrix}$	530
		$\begin{bmatrix} 1 & 0 \\ 0 & 3 \end{bmatrix}$	530
	W	Ring pattern	1623
	Y	Hexagonal	539, 540
W(211)	Ag	(1 × 1)–Ag	969
	Au	(1 × 1)–Au	969
		(1 × 2)–Au	969
		(1 × 3)–Au	969
		(1 × 4)–Au	969
	Li	(4 × 1)	518, 520
		(3 × 1)	518, 520
		(2 × 1)	518, 520
		Incoherent	518, 520
		(1 × 1)	518, 520
	[Clean]	(2 × 2)	518, 520
	Mg	(1 × 7)–Mg	1657
		(3 × 3)–Mg	1657
	Na	(2 × 1)	521
		compressed (2 × 1)	521
	Sb	(2 × 1)	553
		(1 × 1)	553
W(221)	Na	Compressed (2 × 1)	521
		(2 × 1)	521
	Ni	(1 × 1)–Ni	970
		(6 × 1)–Ni	970
Zn(0001)	Cu	(1 × 1)	449, 450

[a] References for Tables 2.5a to 2.5 j are listed at the end of Table 2.5j.

TABLE 2.5f. Surface Structures of Alloys

Substrate	Adsorbate	Surface Structure	Reference[a]
Ag(111)–Rb dosed	O_2	$(2\sqrt{3} \times 2\sqrt{3})R30°$–Rb/O	653
		(4 × 4)–Rb/O	653
		Complex structures	653
		(9 × 9)Rb/O	653
cu—3% Al(100)	O_2	c(2 × 2)-O	1632
		Disordered	1632
Cu–5.7% Al(100)	[Clean]	(1 × 1)	813
Cu–12.5% Al(100)	[Clean]	(1 × 1)	813
Cu_3Al(100)	[Clean]	c(2 × 2)	916
Cu/Al(111)	[Clean]	(1 × 1)	835
		$(\sqrt{3} \times \sqrt{3})R30°$	835

TABLE 2.5f. (*Continued*)

Substrate	Adsorbate	Surface Structure	Reference[a]
Cu–5.7% Al(111)	[Clean]	(1×1)	813
Cu–10% Al(111)	[Clean]	(1×1)	1303
		$(\sqrt{3} \times \sqrt{3})R30°$–Al	1303
Cu–11% Al(111)	[Clean]	(1×1)	1506
Cu–12.5% Al(111)	[Clean]	$(\sqrt{3} \times \sqrt{3})R30°$	813
Cu–16% Al(111)	[Clean]	$(\sqrt{3} \times \sqrt{3})R30°$	1699
Cu/Au(100)	[Clean]	$c(2 \times 2)$	737
Cu/Au(110)	[Clean]	Streak	737
		(1×2)	737
		Complex pattern	737
		$c(3 \times 1)$	737
		$c(2 \times 2)$	737
Cu/Au(111)	[Clean]	$(\frac{2}{3}\sqrt{3} \times \frac{2}{3}\sqrt{3})R30°$	737
		(2×2)	737
Cu/Au(1$\bar{1}$1)	[Clean]	$(\frac{2}{3}\sqrt{3} \times \frac{2}{3}\sqrt{3})R30°$	737
		(2×2)	737
Cu(110)–Ni(1°)	O_2	(2×1)–O	1311
		$c(6 \times 2)$–O	1311
Cu/Ni(110)	CO	(2×1)–CO	134
		(2×2)–CO	134
		(1×2)–CO	787
	H_2	(1×3)–H	787
	H_2S	$c(2 \times 2)$–S	134
	O_2	(2×1)–O	134, 872
		(2×2)–O	872
Cu/Ni(111)	CO	Distorted	173, 734
Cu/Pd(100)	[Clean]	Streak	737
		$c(2 \times 2)$	737
Cu/Pd(110)	[Clean]	(2×1)	737
Cu/Pd(111)	[Clean]	(1×1)	737
Cu/Pd(1$\bar{1}$1)	[Clean]	(1×1)	737
CuSn(100)	[Clean]	$c(2 \times 2)$	1481
		$(3\sqrt{2} \times \sqrt{2})R45°$	1481
		$c(3 \times 2) + (2 \times 2)$	1481
Cu–25% Zn(100)	[Clean]	(1×1)	1152
	O_2	Disordered	1152
Cu–25% Zn(110)	[Clean]	(1×1)	1152
	O_2	Disordered	1152
Cu–25% Zn(111)	[Clean]	(1×1)	1152
	O_2	Disordered	1152
Fe/Cr(100)	O_2	$c(2 \times 2)$–O	279, 636
		$c(4 \times 4)$–O	279
		Oxide	280
Fe/Cr(110)	O_2	$Cr_2O_3(0001)$	280
		Amorphous oxide	279
Fe–18% Cr–12% Ni(111)	[Clean]	(1×1)	1249
	I_2	$(\sqrt{3} \times \sqrt{3})R30°$–I	1249
	H_2O	Ordered	1249
	O_2	Ordered	1249

TABLE 2.5f. (*Continued*)

Substrate	Adsorbate	Surface Structure	Reference[a]
	I(a) + H_2O	Oxide nor formed	1249
	H_2O(a) + I_2	Adsorbed	1249
FeTi(100)	[Clean]	(1 × 1)	1241
	S	c(2 × 2)–S	1241
FeTi(111)	[Clean]	(1 × 1)	1241
Ni_3Al(001)	[Clean]	(1 × 1)	1868
NiAl(110)	[Clean]	(1 × 1)	1771
NiCu(100)(Ni < 50%)	S	C(2 × 2)–S	905
Ni–17% Cu(111)	[Clean]	(1 × 1)	868
	H_2	(2 × 2)–H	868
Ni–24% Fe(100)	O_2	C(2 × 2)–O	573
Ni–25% Fe(100)	H_2S, H_2	c(2 × 2)–S	1121
Ni–25% Fe(110)	H_2S, H_2	(2 × 3)–S	1121
Ni–25% Fe(111)	H_2S, H_2	(3 × 3)–S	1121
Ni–41% Fe(100)	[Clean]	(1 × 1)	1263
	O_2	c(2 × 2)–O	1263
		Oxide	1263
Ni_4Mo(211)	[Clean]	Ordered	1115
Pd–33% Ag(111)	[Clean]	(1 × 1)	877
	Co	(1 × 1)	877
Pd–25% Cu(111)	[Clean]	(1 × 1)	877
	CO	(1 × 1)	877
Pt–2% Cu(110)	[Clean]	(1 × 3)	1062
	CO	(1 × 1)–CO	1063
Pt–22% Ni(111)	[Clean]	(1 × 1)	1162
Pt–50% Ni(111)	[Clean]	(1 × 1)	1162
Pt_3 Ti(100)	[Clean]	c(2 × 2)	935
Pt_3 Ti(111)	[Clean]	(2 × 2)	935

[a]References for Tables 2.5a to 2.5j are listed at the end of Table 2.5j.

TABLE 2.5g. Surface Structure Formed by Adsorption of Organic Molecules

Substrate	Adsorbate	Surface Structure	Reference[a]
Ag(100)	$C_2H_4Cl_2$	c(2 × 2)–Cl	154, 249, 1873
Ag(110)	C_2H_2	Not adsorbed	812
	C_2H_4	Not adsorbed	1690
	C_2H_4 + O_2	(2 × 1)–O	1001
	$C_2H_4Cl_2$	(2 × 1)–Cl	154
		c(4 × 2)–Cl	154
	O(a) + C_2H_2	c(2 × 6)–acetylide	1335, 1398
		(2 × 2)–acetylide	1335, 1398
		(2 × 3)–acetylide	1335, 1398
		(1 × 1)–C	1335
Ag(111)	CH_2Br_2	(1 × 1)	594
	CH_3I	$(\sqrt{3} \times \sqrt{3})R30°$–I	594
	$CHCl_3$	(1 × 1)	594
	$C_2H_4CL_2$	$(\sqrt{3} \times \sqrt{3})R30°$–Cl	153, 154
		(3 × 3)–Cl	153, 154

TABLE 2.5g. *(Continued)*

Substrate	Adsorbate	Surface Structure	Reference[a]
	Acetic acid	$\begin{bmatrix} 2 & -0.7 \\ 2 & 2.7 \end{bmatrix} + \begin{bmatrix} 2.8 & 1.4 \\ 0 & 2.5 \end{bmatrix}$	587
		Ring pattern	587
	Propanoic acid	$\begin{bmatrix} 4 & 2 \\ 0 & 4.3 \end{bmatrix} + \begin{bmatrix} 3.9 & 1.3 \\ 1 & 4.6 \end{bmatrix}$	587
		$\begin{bmatrix} 4 & 2 \\ 0 & 4.3 \end{bmatrix}$	587
Ag[3(111) × (100)]	CH$_2$Br$_2$	(1 × 1)	594
	CH$_3$I	($\sqrt{3} \times \sqrt{3}$)$R30°$-I	594
	CHCl$_3$	(1 × 1)	594
Al(100)	C$_2$H$_1$	(1 × 1)	1693
Au(111)	C$_2$H$_4$	Not adsorbed	161
	Benzene	Not adsorbed	161
	Cyclohexene	No adsorbed	161
	Naphthalene	Disordered	161
	N-Heptane	Not adsorbed	161
Au(S)-[6(111) × (100)]	C$_2$H$_4$	Not adsorbed	161
	Benzene	Not adsorbed	161
	Cyclohexene	No adsorbed	161
	Naphthalene	Disordered	161
	N-Heptane	Not adsorbed	161
		(4 × $\sqrt{3}$)-C$_2$H$_6$	1191
		(2 × 2)-C$_2$H$_6$	1191
		(10 × 2$\sqrt{3}$)-C$_2$H$_6$	1191
		($\sqrt{3} \times \sqrt{3}$)-C$_2$H$_6$	1191
Cr(100)	C$_2$H$_4$	c(2 × 2)-C	1245
		($\sqrt{2} \times 3\sqrt{2}$)$R \pm 45°$-C	1245
Cu(100)	C$_2$H$_4$	(2 × 2)	26
	O(a) + HCOOH	Disordered-HCO$_2$	1921, 1922
	O(a) + HC$_3$OH	Disordered-CH$_3$O$_2$	1922
	Cu-phthalocyanine	$\begin{bmatrix} 5 & -2 \\ 2 & 5 \end{bmatrix}$	408
	D-Tryptophan	(4 × 4)	409
	Fe-phthalocyanine	$\begin{bmatrix} 5 & -2 \\ 2 & 5 \end{bmatrix}$	408
	Glycine	(4 × 2)	409
		$\begin{bmatrix} 8 & -4 \\ 0.8 & 1.6 \end{bmatrix}$	409
	H-phthalocyanine	$\begin{bmatrix} 5 & -2 \\ 2 & 5 \end{bmatrix}$	408
	L-Alanine	$\begin{bmatrix} 2 & 1 \\ 2 & -1 \end{bmatrix}$	409
	L-Tryptophan	(4 × 4)	409
	D-Tryptophan	(4 × 4)	409
Cu(110)	C$_2$H$_4$	One-dimensional order	26
	HCOOH	HCO$_2$-disordered	1849
Cu(111)	C$_2$H$_4$	Not adsorbed	26
	Cu-phthalocyanine	Adsorbed	408
	D-Tryptophan	$\begin{bmatrix} -8 & 1 \\ -2 & 4 \end{bmatrix}$	409
	Fe-phthalocyanine	Adsorbed	408

TABLE 2.5g. (*Continued*)

Substrate	Adsorbate	Surface Structure	Reference[a]
	Glycine	(8×8)	409
	H-phtalocyanine	Adsorbed	408
	L-Alanine	$(2\sqrt{13} \times \sqrt{13})R13°40'$	409
	L-Tryptophan	$\begin{bmatrix} 7 & 1 \\ -2 & 4 \end{bmatrix}$	409
Cu(S)–[3(100) × (100)]	CH_4	Not adsorbed	132
	C_2H_4	Not adsorbed	132
Cu(S)–[4(100) × (100)]	CH_4	Not adsorbed	132
	C_2H_4	Not adsorbed	132
Fe(100)	C_2H_4	$c(2 \times 2)$–C	274, 893
Fe(110)	C_2H_2	(2×2)	687
		(2×3)	687
		Coincidence	687
		$\begin{bmatrix} 4 & 0 \\ -1 & 3 \end{bmatrix}$	687
Fe(111)	C_2H_2	(1×1)	687
		(5×5)	687
		(3×3)	687
	C_2H_4	(1×1)	687
		(5×5)	687
		(3×3)	687
GaAs(110)	HCOOH	$c(2 \times 2)$–H + HCOO	1124, 1302
Ir(100)	C_2H_2	Disordered	281, 410
		$c(2 \times 2)$–C	281, 410
	C_2H_4	Disordered	410
		$c(2 \times 2)$–C	410
	Benzene	Disordered	410
Ir(110)	C_2H_4	Disordered	347
		(1×1)–C	347
	Benzene	Disordered	347
		(1×1)–C	347
Ir(111)	C_2H_2	$(\sqrt{3} \times \sqrt{3})R30°$	187
		(9×9)–C	187
	C_2H_4	$(\sqrt{3} \times \sqrt{3})R30°$	187
		(9×9)–C	187
	Benzene	(3×3)	187
		(9×9)–C	187
	Cyclohexane	Disordered	187
		(9×9)–C	187
Ir(S)–[6(111) × (100)]	C_2H_2	(2×2)	187
	C_2H_4	(2×2)	187
	Benzene	Disordered	187
	Cyclohexane	(2×2)	187
Mo(100)	CH_4	$c(4 \times 4)$–C	286
		$c(2 \times 2)$–C	286
		$c(6\sqrt{2} \times 2\sqrt{2})R45°$–C	286
		(1×1)–C	286
	C_2H_4	$c(2 \times 2)$–carbide	602, 603, 659
		$\begin{bmatrix} 3 & 0 \\ 1 & -1 \end{bmatrix}$	602, 603, 660
		(1×1)	602, 603, 660
		(1×1) with streaks	1379
		$c(2 \times 2)$–C	660
	HCOOH	Disordered	1155

TABLE 2.5g. (*Continued*)

Substrate	Adsorbate	Surface Structure	Reference[a]
	$O(a) + C_2H_4$	(2×1)-O	817
	$O(a) + HCOOH$	(2×1)-O, C	1155
Ni(100)	CH_4	$c(2 \times 2)$	117
		(2×2)	117
	C_2H_2	$c(2 \times 2)$	416, 1923
		(2×2)	416
		$c(4 \times 2)$	417
		(2×2)-C	417
	C_2H_4	$c(2 \times 2)$	88, 416
		Quasi-$c(2 \times 2)$	1561
		(2×2)	416
		(2×2)-C (p4g)	417, 670, 745, 1092, 1177
		$c(4 \times 2)$	417
		$(\sqrt{7} \times \sqrt{7})R19°$-C	88
	C_2H_6	$c(2 \times 2)$	117
		(2×2)	117
	CH_3OH	Disordered	601
		$c(2 \times 2)$	601
	Benzene	$c(4 \times 4)$	415
Ni(110)	CH_4	(2×2)	117
		(4×3)	117
		(4×5)-C	117, 418
		(2×3)-C	418, 679
	C_2H_2	$c(2 \times 2)$-C_2H_2	915
	C_2H_4	(2×1)-C	419, 420, 421
		(4×5)-C	419, 420, 915
		"$c(2 \times 4)$"-C_2H_4	915
		$c(2 \times 2)$-CCH	915
		Graphite overlayer	420
	C_2H_6	(2×2)	117
	CH_3OH	$c(2 \times 4)$-CH_3O	890
		$c(2 \times 6)$-CH_3O	890
		$c(2 \times 2)$-CO	890
	C_5H_{12}	(4×3)	422
		(4×5)	422
Ni(111)	CH_4	(2×2)	117
		(2×2)-C	990
		Graphite	990
		$(2 \times \sqrt{3})$	117
		$(16\sqrt{3} \times 16\sqrt{3})$-$R30°$-C	739
		(4×5)	739
	C_2H_2	(2×2)-C_2H_2	412, 413, 719, 1262, 1925
		$(\sqrt{3} \times \sqrt{3})R30° + (2 \times 2)$	719
		Disordered	1627
	C_2H_4	(2×2)	29, 39, 412
	C_2H_6	(2×2)	39, 117
		$(2 \times \sqrt{3})$	117
		$(\sqrt{7} \times \sqrt{7})R19°$-C	29
		(2×2)-C	990
		Disordered graphite	990
	Benzene	$(2\sqrt{3} \times 2\sqrt{3})R30°$	414, 415
	Cyclohexane	$(2\sqrt{3} \times 2\sqrt{3})R30°$	414
Pb(100)	HCOOH	Not adsorbed	1691

TABLE 2.5g. (*Continued*)

Substrate	Adsorbate	Surface Structure	Reference[a]
Pd(100)	Benzene	c(4 × 4)	616, 617
		(2 × 2)$R45°$–C_6H_6	616, 617, 630
Pd(111)	C_2H_2	($\sqrt{3}$ × $\sqrt{3}$)$R30°$–Diffuse	1043, 1209
	C_2H_2	($\sqrt{3}$ × $\sqrt{3}$)$R30°$–C_2H_2	1209
		Disordered	1266
		($\sqrt{3}$ × $\sqrt{3}$)$R30°$–C_2H_3	1266
	CH_3OH	($\sqrt{3}$ × $\sqrt{3}$)$R30°$–CO_2H_2	1042
		Complex	1042
	Benzene	Disordered	1961
	Benzene + CO	(2$\sqrt{3}$ × 2$\sqrt{3}$)$R30°$	961
		Complex	961
		(3 × 3)–C_6H_6 + 2CO	961, 1862
Pt(100)	C_2H_2	c(2 × 2)	28, 72, 321, 431, 432
	C_2H_4	c(2 × 2)	28, 72, 313, 321, 431
		Graphite overlayer	313, 426
		(511), (311) Facets	426
	Acrylic acid	(1 × 1) diffuse	1258
	Acrylic acid(a) + NH_3	(1 × 1) diffuse	1258
	Aniline	Disordered	430
	Benzene	Disordered	432
		2(One-dimensional order)	429
	Cyanobenzene	Disordered	430
	Mesitylene	3(One-dimensional order)	430
	m-Xylene	3(One-dimensional order)	430
	Naphthalene	(1 × 1)	429
	Nitrobenzene	Disordered	430
	N-Butylbenzene	Disordered	430
	Pyridine	(1 × 1)	429
		c(2 × 2)	429
	Toluene	3(One-dimensional order)	430
	t-Butylbenzene	Disordered	430
Pt(110)	HCOOH	One-dimensional disordered	1080
	CH_3NCO	(1 × 2)	938
	Benzene	Disordered	1172
Pt(111)	C_2H_2	(2 × 1)	28
		(2 × 2)	423–425, 1316, 1196
	C_2H_2 + H_2	(2 × 2)–C_2H_3	824, 1316, 1587
	C_2H_4	(2 × 2)	40, 424, 425, 1196, 1313
		(2 × 2)–C_2H_3	824, 1316, 1586
		(2 × 1)	28
		2(One-dimensional order)–C	221
		Disordered	1316
		Complex	1313
		Graphite overlayer	221, 426, 1093
	(O) + C_2H_4	(2 × 2)	1313
		Ordered	1313
	C_3H_4	(2 × 2)	1316
	C_3H_4 + H_2	(2 × 2)	1316
	C_3H_6	Disordered	1316
		(2 × 2)	1316

TABLE 2.5g. (*Continued*)

Substrate	Adsorbate	Surface Structure	Reference[a]
	cis-2-C$_4$H$_8$	$(2\sqrt{3} \times 2\sqrt{3})R30°$	1316
	trans-2-C$_4$H$_8$	(8×8)	1316
	C$_{10}$H$_8$	(6×3)-C$_{10}$H$_8$	668
	Acetic acid	(2×2)	580, 1287
		Disordered	587, 1287
	Acetonitrile	(1×1) diffuse	1258
		(2×2)	580, 1287
		Disordered	1287
	Acetonitrile + I$_2$	I$_2$ adsorbed	1258
	Aniline	3 (One-dimensional order)	430
	Azulene	Disordered	1349
		(3×3)	1349
		$(3 \times 3) + (3 \times 3)R30°$	1349
		(10×10)	1349
	Benzene	Disordered	429, 1854, 1891
		Graphite	1924
	Benzene + CO	$\begin{bmatrix} -2 & 2 \\ 5 & 5 \end{bmatrix} = \begin{bmatrix} 4 & -2 \\ 0 & 4 \end{bmatrix}$	221, 428, 429
		$\begin{bmatrix} 4 & -2 \\ 0 & 5 \end{bmatrix} + \begin{bmatrix} -2 & 2 \\ 4 & 4 \end{bmatrix}$ $= (2\sqrt{3} \times 4)\,\text{rect-}$ $2C_6H_6 + 4CO$	428, 429, 1854
	Cyanobenzene	3 (One-dimensional order)	430
	Cyclohexane	$\begin{bmatrix} 4 & -1 \\ 1 & 5 \end{bmatrix}$	427
		Disordered	221
		(2×2)	221
		Graphite overlayer	221
	Dichloromethane	Not adsorbed	580
	Dimethylsulfoxide (DMSO)	(2×2)	580, 1287
		$(\sqrt{3} \times \sqrt{3})R30°$	1287
		(1×1)	1287
	Dimethylformamide	(2×2) diffuse	580, 1258
		Disordered	1287
	DMSO	(1×1)-DMSO	1258
	DMSO(a) + I$_2$	Ordered	1258
	DMSO(a) + pyridine	Not adsorbed	1258
	I(a) + pyridine	Not adsorbed	1258
	I(a) + acetonitrile	Not adsorbed	1258
	I(a) + DMSO	DMSO not adsorbed	1258
	I$_2$ + DMSO	$c(2 \times 4\sqrt{3}/3)$-I, DMSO	1258
	Mesitylene	3.4 (dimensional order)	430
	m-Xylene	2.6 (One-dimensional order)	430
	Naphthalene	(6×6)	224, 429
	Naphthalene (001)		224
		(6×3)	1349
		Disordered	1349
	Nitrobenzene	3 (One-dimensional order)	430
	n-Butane	$\begin{bmatrix} 2 & 1 \\ -1 & 2 \end{bmatrix}$	427

TABLE 2.5g. (*Continued*)

Substrate	Adsorbate	Surface Structure	Reference[a]
		$\begin{bmatrix} 2 & 2 \\ -5 & 5 \end{bmatrix}$	427
		$\begin{bmatrix} 3 & -2 \\ 2 & 5 \end{bmatrix}$	427
	n-Butylbenzene	Disordered	430
	n-Heptane	$\begin{bmatrix} 2 & 1 \\ 0 & 8 \end{bmatrix}$	427
		(2 × 2)	221
	n-Hexane	$\begin{bmatrix} 2 & 1 \\ -1 & 3 \end{bmatrix}$	427
	n-Octane	$\begin{bmatrix} 2 & 1 \\ -1 & 4 \end{bmatrix}$	427
	n-Pentane	$\begin{bmatrix} 2 & 1 \\ 0 & 6 \end{bmatrix}$	427
	Propylene Carbonate	(2 × 2)	580, 1287
		Disordered	1287
	Propanoic acid	Disordered	587
	Pyridine	(2 × 2)	429, 580, 1287
		(1 × 1) diffuse	1258
		Disordered	1287
	Pyridine (a) + DMSO	Not adsorbed	1258
	Pyridine (a) + H_2O	Not adsorbed	1258
	Pyridine (a) + I_2	I_2 adsorbed	1258
	p-Dioxane	(2 × 2)	580, 1287
		Disordered	1287
	Sulfolane	(2 × 2)	580, 1287
		$(\sqrt{3} \times \sqrt{3})R30°$	580, 1287
		(1 × 1)	580, 1287
	Toluene	3 (One-dimensional order)	221, 430
		(4 × 2)	430
		Graphite overlayer	430
	t-Butylbenzene	Disordered	430
Pt (S)–[7 (111) × (100)]	Azulene	1/3 order ring	1057
Pt (S)–[4 (111) × (100)]	Naphthalene	1/3 order spots	1057
	C_2H_4	Disordered	221
		Graphite overlayer	221
		Facets	221
	Benzene	Disordered	221
		Graphite overlayer	221
		Facets	221
	Cyclohexane	Disordered	221
		(4 × 2)–C	221
	n-Heptane	(4 × 2)	221
		(4 × 2)–C	221
	Toluene	Disordered	221
		2 (One-dimensional order)–C	221
Pt (S)–[6 (111) × (100)]	C_2H_4	(2 × 2)	120, 221
		$\begin{bmatrix} 3 & 2 \\ -2 & 5 \end{bmatrix}$–C	221

TABLE 2.5g. (*Continued*)

Substrate	Adsorbate	Surface Structure	Reference[a]
		$\begin{bmatrix} 6 & 1 \\ -1 & 7 \end{bmatrix}$-C	221
		$(\sqrt{19} \times \sqrt{19})\,R\,22.4°$-C	426
		Graphite overlayer	426
	Benzene	3 (One-dimensional order)	221
		(9 × 9)-C	221
	Cyclohexane	2 (One-dimensional order)	221
	n-Heptane	(2 × 2)	221
		$\begin{bmatrix} 1 & 1 \\ -1 & 2 \end{bmatrix}$	221
		(9 × 9)-C	221
	Toluene	Disordered	221
		(9 × 9)-C	221
Pt(S)-[7(111) × (310)]	C_2H_4	Disordered	221
		Graphite overlayer	221
	Benzene	Disordered	221
	Cyclohexane	Disordered	221
	n-Heptane	Disordered	221
	Toluene	Disordered	221
		Graphite overlayer	221
Pt(S)-[9(111) × (100)]	C_2H_4	Adsorbed	221
	Benzene	Disordered	221
		$\begin{bmatrix} 1 & 1 \\ -1 & 2 \end{bmatrix}$-C	221
		Graphite overlayer	221
	Cyclohexane	Disordered	221
	n-Heptane	(2 × 2)	221
		$\begin{bmatrix} 1 & 1 \\ -1 & 2 \end{bmatrix}$	221
		(5 × 5)-C	221
		(2 × 2)-C	221
		$\begin{bmatrix} 1 & 1 \\ -1 & 2 \end{bmatrix}$-C	221
		2 (One-dimensional order)-C	221
	Toluene	3 (One-dimensional order)	221
		Graphite overlayer	221
Pt(S)-[9(111) × (111)]	C_2H_4	Disordered	120
		Graphite overlayer	398, 399
		(2 × 2)	685
Pt(S)-[9(111) × (100)]	C_2H_4	Graphite overlayer	426
		(511), (311), and (731) facets	426
Re(0001)	C_2H_2	Disordered	436, 664
		$(2 \times \sqrt{3})\,R\,30°$-C	436
	C_2H_4	Disordered	436, 664
		$(2 \times \sqrt{3})\,R\,30°$-C	436
Re(S)-[14(0001) × (10$\bar{1}$1)]	C_2H_2	Disordered	664
	C_2H_4	$(2 \times \sqrt{3})\,R\,30°$	664
Re(S)-[6(0001) × (16$\bar{7}$1)]	C_2H_2	Disordered	664
	C_2H_4	Disordered	664
Rh(100)	C_2H_2	c(2 × 2)	231
		c(2 × 2)-C_2H + C_2H_3	1880
	C_2H_4	c(2 × 2)	231

TABLE 2.5g. (*Continued*)

Substrate	Adsorbate	Surface Structure	Reference[a]
		$c(2 \times 2)$-C_2H + C_2H_3	1878
		(2×2)-C_2H	1878
		$c(2 \times 2)$-C	231, 1403
		Graphite overlayer	231, 1403
	$CO + C_2H_4$	$c(4 \times 2)$-CO + C_2H_3	1878
		Split $c(2 \times 2)$-CO + C_2H_3	1878
	C_6H_6	$c(4 \times 4)$	1879
	$C_6H_6 + CO$	$c(2\sqrt{2} \times 4\sqrt{2})R45°$-CO $+ C_6H_6$	1879
		$c(2 \times 2)$	1879
Rh(111)	C_2H_2	$c(4 \times 2)$	231, 831
		(2×2)	831
	$C_2H_2 + CO$	$c(4 \times 2)$-CO + C_2H_2	1844, 1881
	$C_2H_2 + Na$	Disordered	1876
	C_2H_4	$c(4 \times 2)$	231, 831, 1256, 1372
		(2×2)-C_2H_3 (ethylidyne)	831, 1256, 1372
		Partially ordered	955
		(8×8)-C	231
		$(2 \times 2)R30°$-C	231
		$(\sqrt{19} \times \sqrt{19})R23.4°$-C	231
		$(2\sqrt{3} \times 2\sqrt{3})R30°$-C	231
		(12×12)-C	231
	$C_2H_4 + CO$	$c(4 \times 2)$-CO + C_2H_3	1881
	$C_2H_4 + NO$	$c(4 \times 2)$-NO + C_2H_3	1877
	$C_2H_4 + H_2$	$c(4 \times 2)$-CCH_3	955
		$c(4 \times 2)$	1372
		$(2 \times 2) + c(4 \times 2)$	1256
	$C_3H_6 + CO$	$(2\sqrt{3} \times 2\sqrt{3})R30°$-CO $+ C_3H_5$	1884
	CH_3OH	Disordered	988
	Benzene	$(2\sqrt{3} \times 3)$ rect	1842, 1892
		$(\sqrt{7} \times \sqrt{7})R19.1°$	1892
	Benezene + CO	$c(2\sqrt{3} \times 4)$ rect	1068, 1416, 1453, 1842, 1856, 1892
		$= \begin{bmatrix} 3 & 1 \\ 1 & 3 \end{bmatrix}$-$C_6H_6$ + CO	
		(3×3)-C_6H_6 + CO	1068, 1842, 1855, 1892
	Benzene + Na	$(\sqrt{3} \times \sqrt{3})R30° + (2\sqrt{3} \times 3)$ rect	1876
	$C_6H_5F + CO$	(3×3)	1844
	Methylacetylene	$c(4 \times 2)$	1372
	Naphthalene	$(3\sqrt{3} \times 3\sqrt{3})R30°$	1068
		(3×3)	1068
	Propylene	$(2 \times 2) + (2\sqrt{3} \times 2\sqrt{3})R30°$	1372
		$(2\sqrt{3} \times 2\sqrt{3})R30°$	1372
Rh(331)	C_2H_2	$\begin{bmatrix} -1 & 1 \\ 3 & 0 \end{bmatrix}$	402
	C_2H_2	$\begin{bmatrix} -1 & 1 \\ 3 & 0 \end{bmatrix}$	402, 722
		Graphite overlayer	402, 722
Rh(S)-[6(111) × (100)]	C_2H_2	Disordered	402

TABLE 2.5g. (*Continued*)

Substrate	Adsorbate	Surface Structure	Reference[a]
	C_2H_4	Disordered	402, 722
		(111), (100) facets	402, 722
Ru (0001)	C_2H_6	Disordered	692
	Cyclopropane	Disordered	692, 1111
	Cyclohexane	Disordered	692, 1112
		(1 × 1)	692
	Cyclooctane	(1 × 1)	692
Si (111)	C_2H_2	Disordered	437
	CH_3OH	(7 × 7)–CH_3O + H + CH_3OH	837
		Disordered	988
		Disordered CH_3O + H	942
Si (311)	C_2H_2	c(1 × 1)	135
		(2 × 1)	135
		(3 × 1)	135
	C_2H_4	c(1 × 1)	135
		(2 × 1)	135
		(3 × 1)	135
Ta (100)	C_2H_4	Adsorbed	328
W (100)	CH_4	(5 × 1)–C	41
	C_2H_2	Disordered	700
		(5 × 1)–C	700, 901
		c(3 × 2)–C	700
		c(2 × 2)–C	700
	Propylene	$\begin{bmatrix} 3 & 0 \\ 1 & -1 \end{bmatrix}$–C	887
		(5 × 1)–C	887
W (110)	C_2H_2	(2 × 2)–C_2H_2	1072
		c(2 × 2)–C_2H_2	1072
		(15 × 3)R14°–C	1072
	C_2H_4	(15 × 3)$R\alpha$–C	41
		(15 × 12)$R\alpha$–C	41
W (111)	CH_4	(6 × 6)–C	41
	C_2H_4	(1 × 1)	1595
	C_2H_6	(1 × 1)	1595
W (211)	Propylene	c(6 × 4)–C	887
ZnO (10$\bar{1}$0)	C_6H_6	c(2 × 2)–C_6H_6	632, 620
		c(4 × 3)–C_6H_6	632, 620

[a]References for Tables 2.5a to 2.5j are listed at the end of Table 2.5j.

TABLE 2.5h. Coadsorbed Overlayer Structures[a]

Substrate	Adsorbate	Surface Structure	References[a]
Ag (100)	Cl_2 + K	c(2 × 2)–K/Cl	673
	K + O_2	$\begin{bmatrix} 1 & 1 \\ -5 & 4 \end{bmatrix}$	658
		Hexagonal overlayer	658
	O(ad.) + H_2O	c(2 × 2)–OH	1034
Ag (110)	C_2H_4 + O_2	(2 × 1)–O	1001

TABLE 2.5h. (*Continued*)[a]

Substrate	Adsorbate	Surface Structure	References[a]
	$H_2O + Li^+$	Complex	1557
	$O(a) + C_2H_2$	$c(2 \times 6)$–acetylide	1335, 1398
		(2×2)–acetylide	1335, 1398
		(2×3)–acetylide	1335, 1398
		(1×1)–C	1335
	$O(a) + SO_2$	$c(6 \times 2)$–SO_3	1027, 1371
		(1×2)–SO_4	1371
	$O_2 + H_2O$	(1×2)–OH	878
		(1×3)–OH	878
Ag(111)	$CO + O_2$	$(2 \times \sqrt{3})$–$(CO + O_2)$	27
Bi(0001)	$O_2 + K$	$\sqrt{3}$ + BiO(0001)layer	1288
C(0001), graphite	Ar + Xe	$(\sqrt{3} \times \sqrt{3})R30°$–Ar, Xe	1193
Co(100)	$H_2S + C$	(2×2)–S, C	1529
Cr(100)	C, O, N	$c(2 \times 2)$	1126
Cu(110)	$Br(a) + H_2O$	(3×2)	1557
	$C + O_2$	(2×1)	1695
	$Ni(CO_4) + CO$	(1×1)	1048
	$O(a) + CO$	Disordered	1066
	$O(a) + H_2O$	(2×1)–O, H_2O	1023
		$C(2 \times 2)$–O, H_2O	1023
		(2×1)–H_2O	1270
		(1×1)–H_2O, OH	1270
		(2×1)–OH, O	1270
Cu(111)	$Ni(CO)_4 + CO$	(1×1)	1048
	$O_2 + HCN$	Disordered	1244
	$O(a) + CO$	Disordered	1066
	$O(a) + CO_2$	(2×2)	1142
Fe(110)	$K + O_2$	$c(4 \times 2)$	786
Fe(111)	$N(a) + K$	(3×3)–K, N	1350
Fe–18% Cr–12% Ni(111)	$I(a) + H_2O$	Oxide not formed	1249
	$H_2O(a) + I_2$	Adsorbed	1249
GaAs(100)	As_4, Ga	(2×4)	1365
		(4×6)	1365
		$c(8 \times 2)$	1365
		(4×1)	1365
		(3×1)	1365
	HCl, H_2O	(1×1)	1518
	Pb, As_4	(1×2)–Pb	1387
Hg(110)	$O(ad) + SO_2$	$c(6 \times 2)$–SO_3	1027, 1371
		(1×2)–SO_3 etc.	1371
	$O_2 + H_2O$	(1×2)–OH	878
		(1×3)–OH	878
Mo(100)	$Cs + O_2$	$c(2 \times 2)$ + (4×1)	932
		(4×1)	932
		$c(2 \times 2)$	932
	$H_2S + O_2$	$(\sqrt{5} \times \sqrt{5})R26.6°$–S, O	917
	$O(a) + CO$	(2×1)–O	817
	$O(a) + CO_2$	(2×1)–O	817
	$O(a) + C_2H_4$	(2×1)–O	817
	$O(a) + HCOOH$	(2×1)–O, C	1155

TABLE 2.5h. (*Continued*)[a]

Substrate	Adsorbate	Surface Structure	References[a]
Mo(111)	$N_2 + NH_3$	Disordered	1203
	$N_2 + NH_3$	(433) facet	1203
		$c(3 \times 2)$–N/Mo(433)	1203
Ni(100)	$C(a) + O_2$	$c(2 \times 2)$–O	1177
	$CO + H_2$	$c(3 \times 3)$	301
	$H_2 + CO$	$c(2 \times 2)$–CO, H	1202
		$c(\sqrt{2} \times 2\sqrt{2})R45°$–CO, H	1202
	H_2S, H_2	$c(2 \times 2)$–S	1121
	$H_2S + Na$	$c(2 \times 2)$Na + $c(2 \times 2)$S	1887
		(2×2)Na + $c(2 \times 2)$S	1887
		(2×2)Na + (2×2)S	1887
	$O(a) + CO$	$c(2 \times 2)$–C, O	1356
	S, C	(1×1)	1561
Ni(110)	$CO + O_2$	(3×1)–$(CO + O_2)$	91
	$O_2 + H_2O$	(2×1)–OH	1011
Ni(111)	$NI(CO)_4$, CO	$(\sqrt{7}/2 \times \sqrt{7}/2)R19°$–CO	1150
		$c(4 \times 2)$–CO	1150
Ni(111)	$O(a) + H_2O$	No new features	1308
	$O(a) + NO$	(2×2)	676
Ni(331)	O, CO	(2×3)	1318
Pd(100)	$H(a) + O_2$	Adsorbed	1163, 1454
	H_2, O_2	Disordered	1163
		(2×2)–O, H	1163
	$O_2 + CO$	Disordered	939
	$O_2 + H_2O$	(2×1)–OH	940
	$O(a) + H_2$	Not adsorbed	1163, 1454
Pd(111)	$O_2 + CO$	$(\sqrt{3} \times \sqrt{3})R30°$	691
		(2×1)	691
Pt(100)	$CO + H_2$	$c(2 \times 2)$–$(CO + H_2)$	72, 74
	$CO + O_2$	(1×1) diffuse	909
		$c(2 \times 2)$–CO + (3×1)–O	928
	$H_2O + HBr$	$c(2\sqrt{2} \times \sqrt{2})R45°$–Br, HBr	1258
	Br, HBr(a) + H_2O	Not adsorbed	1258
	Br, HBr(a) + NH^3	No affinity	1258
	$I(a) + Ag$	$(\sqrt{2} \times \sqrt{2})R45°$–I, Ag	1390
		$(10\sqrt{2} \times 10\sqrt{2})R45°$–I, Ag	1390
		$(\sqrt{34} \times \sqrt{34})R31°$–I, Ag	1390
	$SO_2(a) + NH_3$	(1×1) diffuse	1258
Pt(110)	$CO + NO$	(1×1)–CO + NO	364
Pt(111)	$C_2H_2 + H_2$	(2×2)–C_2H_3	824, 1316, 1587
	(O) + C_2H_4	(2×2)	1313
		Ordered	1313
	Benzene + CO	$\begin{bmatrix} -2 & 2 \\ 5 & 5 \end{bmatrix} = \begin{bmatrix} 4 & -2 \\ 0 & 4 \end{bmatrix}$	221, 428, 429

TABLE 2.5h. (*Continued*)[a]

Substrate	Adsorbate	Surface Structure	References[a]
		$\begin{bmatrix} 4 & -2 \\ 0 & 5 \end{bmatrix} = \begin{bmatrix} -2 & 2 \\ 4 & 4 \end{bmatrix}$ $= (2\sqrt{3} \times 4)\text{rect-}2C_6H_6$ $+ 4CO$	428, 429, 1854
	DMSO(a) + I$_2$	Ordered	1258
	DMSO(a) + pyridine	Not adsorbed	1258
	I(a) + pyridine	Not adsorbed	1258
		Pyridine adsorbed	1258
	I(a) + acetonitrile	Acetonitrile not adsorbed	1258
	I(a) + DMSO	DMSO not adsorbed	1258
	I$_2$ + DMSO	c(2 × 4 $\sqrt{3}$/3)—I, DMSO	1258
	Pyridine(a) + DMSO	Adsorbed	1258
	Pyridine(a) + H$_2$O	Adsorbed	1258
	Pyridine(a) + I$_2$	I$_2$ adsorbed	1258
	H$_2$ + C$_2$N$_2$	Disordered	1002
	Cl$_2$ + Br$_2$	c(2 × 4)–Cl, Br	610
		($\sqrt{3} \times \sqrt{3}$)R30°-Cl, Br	610
		(3 × 3)-Cl, Br	610
		($\sqrt{7} \times \sqrt{7}$)R19.1°	610
	CO + O$_2$	($\sqrt{3} \times \sqrt{3}$)R30°(misfit)	909
	H$_2$ + O$_2$	($\sqrt{3} \times \sqrt{3}$)R30°	11
	I$_2$(a) + HBr	HBr not adsorbed	1258
	I(a) + Cu	(3 × 3)–I, Cu	1556
		(10 × 10)–I, Cu	1556
	I$_2$ + Ag	(3 × 3)–Ag, I	937, 1106, 1391
		(5 × 5)–Ag, I	937
		(17 × 17)–Ag, I	937
		($\sqrt{7} \times \sqrt{7}$) + (3 × 3)	1391
		($\sqrt{3} \times \sqrt{3}$)R30°–Ag, I	1391
	K + CO	Disordered	1255
	K + O$_2$	(4 × 4)–K, O	1238, 1337
		(8 × 2)	1337
		(10 × 2)	1337
		K$_2$O	1337
	S + O$_2$	(2 × 2)–O	1040
Pt(S)-[6(111) × (100)]	K(a) + O$_2$	(4 × 4) potassium oxide	1337
		(8 × 2) potassium oxide	1337
		(10 × 2) potassium oxide	1337
Rh(100)	CO(a) + D$_2$	Compressed (CO)	1348
	CO + CH$_4$	c(4 × 2)-CO + C$_2$H$_3$	1878
		split c(2 × 2)–CO + C$_2$H$_3$	1878
	C$_6$H$_6$ + CO	c(2$\sqrt{2}$ × 4 $\sqrt{2}$)R45°–CO $+ C_6H_6$	1879
	D(a) + CO	c(2 × 2)	1348
	NO + D$_2$	Disordered	1025
	Na + H$_2$O	(2 $\sqrt{3}$ × $\sqrt{3}$)R30°	1082
	Na + H$_2$O	Complex	1082
Rh(111)		($\sqrt{3}$ × 7)rect	1844

TABLE 2.5h. (*Continued*)[a]

Substrate	Adsorbate	Surface Structure	References[a]
	CO + Na	c(4 × 2)–CO + Na	1844
	H$_2$ + CO	(2 × 2)	829
		($\sqrt{3}$ × $\sqrt{3}$)R30°	829
	C$_2$H$_2$ + H$_2$	c(4 × 2)	231, 831
	C$_2$H$_2$ + CO	c(4 × 2)–CO + C$_2$H$_2$	1844, 1881
	C$_2$H$_2$ + Na	Disordered	1876
	C$_2$H$_4$ + H$_2$	c(4 × 2)–CCH$_3$	955
		c(4 × 2)	1372
		(2 × 2) + c(4 × 2)	1256
	C$_3$H$_6$ + CO	(2 $\sqrt{3}$ × 2 $\sqrt{3}$)R30°–CO + C$_3$H$_5$	1884
	Benzene + CO	c(2 $\sqrt{3}$ × 4)rect	1068, 1416, 1453, 1842, 1856
		$= \begin{bmatrix} 3 & 1 \\ 1 & 2 \end{bmatrix}$–C$_6H_6$ + CO	
		(3 × 3)–C$_6$H$_6$ + 2CO	1068, 1842, 1855
	Benzene + Na	($\sqrt{3}$ × $\sqrt{3}$)R30° + (2 $\sqrt{3}$ × 3)rect	1876
	C$_6$H$_5$F + CO	(3 × 3)	1844
	NO + CO	Disordered	1876
Ru(0001)	CO + O$_2$	(2 × 2)	768
	Na + CO	(2 $\sqrt{3}$ × 2 $\sqrt{3}$)R30°	976
Si(111)	Ag(a) + H	($\sqrt{3}$ × $\sqrt{3}$)R30°	1536
W(100)	CO + N$_2$	(4 × 1)–(CO + N$_2$)	82
	O$_2$ + H$_2$	($\sqrt{2}$ × $\sqrt{2}$) + (4 × 1)–O, H	815
W(110)	CO + O$_2$	c(11 × 5)–(CO + O$_2$)	93
	Pd(1 ML) + CO	Not adsorbed	1218
	Pd(2.2 ML) + O$_2$	(2 × 2)–O	1218
W(221)	CO + O$_2$	(1 × 1)–(CO + O$_2$)	108
		(1 × 2)–(CO + O$_2$)	108

[a]References for Tables 2.5a to 2.5j are listed at the end of Table 2.5j.

TABLE 2.5i. Physisorbed Overlayer Structures

Substrate	Adsorbate	Surface Structure	References[a]
Ag(110)	Xe	Hexagonal overlayer	159
Ag(111)	Kr	Hexagonal overlayer	156
	Xe	Hexagonal overlayer	156–159, 160
Ag(211)	Xe	Hexagonal overlayer	159
Au(100)	Xe	Disordered	252
C(0001), graphite	Ar	($\sqrt{3}$ × $\sqrt{3}$)R30°–Ar	720, 960
	Ar	Incommensurate	1882
	Ar + Xe	($\sqrt{3}$ × $\sqrt{3}$)R30°–Ar, Xe	1193
	CF$_4$	(2 × 2)–CF$_4$	1194
		Close to (2 × 2)	1404

TABLE 2.5i. (*Continued*)

Substrate	Adsorbate	Surface Structure	References[a]
	CH_4	$(\sqrt{3} \times \sqrt{3})R30°$	1018
	C_2H_6	$(4 \times \sqrt{3})C_2H_6$	1191
		(2×2)–C_2H_6	1191
		$(10 \times 2\sqrt{3})$–C_2H_6	1191
		$(\sqrt{3} \times \sqrt{3})$–$C_2H_6$	1191
	CO	$(\sqrt{3} \times \sqrt{3})R30°$–CO	884
		$(2\sqrt{3} \times 2\sqrt{3})R30°$–CO	889
		$(2\sqrt{3} \times \sqrt{3})R30°$	884
		Incommensurate(2×2)	884
	H_2	$(\sqrt{3} \times \sqrt{3})R30°$–$H_2$	1283
	Kr	$(\sqrt{3} \times \sqrt{3})R30°$–Kr	166, 167, 174, 721
			828, 960, 1616
		Incommensurate	1616
	N_2	$(2\sqrt{3} \times 2\sqrt{3})R30°$–$N_2$	889
		$(\sqrt{3} \times \sqrt{3})R30°$	1064
		$(\sqrt{3} \times \sqrt{3})R30°$ + (2×1)	1435
		Commensurate	1190, 1512, 1883
		Incommensurate	1443, 1190, 1512, 1883
	Ne	Incommensurate	629
		$(\sqrt{3} \times \sqrt{3})R30°$ rotated by 17°	629
		Layer + Island	960
		Ordered	1338
	NO	Incommensurate	1602
	O_2	Triangular	1883
		Centered-parallelogram–O_2	1200, 1425, 1883
		Physisorbed	1411
	Xe	$(\sqrt{3} \times \sqrt{3})R30°$–Xe	165, 618, 960, 1038, 1201
Cu(100)	Xe	Hexagonal overlayer	159
		Disordered	741
Cu(110)	Kr	c(2×8)–Kr	1304, 1331
	Xe	c(2×2)–Xe	159, 1331, 1611
		Hexagonal overlayer	159, 1611
Cu(111)	Xe	$(\sqrt{3} \times \sqrt{3})R30°$–Xe	159
Cu(211)	Kr	Hexagonal overlayer	156
	Xe	Hexagonal overlayer	156
Cu(311)	Xe	Hexagonal overlayer	394
Cu(610)	Xe	(2×6)–Xe	790
Ir(100)	Kr	(3×5)–Kr	283
		Kr(111)	283
NaCl(100)	Xe	Hexagonal overlayer	289
Ni(100)	Xe	Partially ordered	1268
Pd(100)	Kr	Liquid-like	913
	Xe	Hexagonal overlayer	311
		Liquid-like	913

TABLE 2.5i. (*Continued*)

Substrate	Adsorbate	Surface Structure	References[a]
Pd(110)	Xe	Hexagonal	743
Pd(S)–[8(100) × (110)]	Xe	One-dimensional periodicity	1100
Pt(111)	Xe	($\sqrt{3} \times \sqrt{3}$)$R30°$–Xe	846
		Hexagonal overlayer	846
Si(111)	Kr	(1 × 1)	1125
	Xe	(1 × 1)	1125
W(110)	Xe	(2 × 2)–Xe	713
		Disordered	713
ZnO(000$\bar{1}$)	Xe	Disordered	1026
ZnO(10$\bar{1}$0)	Xe	Hexagonal	1026

[a]References for Tables 2.5a to 2.5j are listed at the end of Table 2.5j.

TABLE 2.5j. Surface Structures on High-Miller-Index (Stepped) Crystal Faces

Substrate	Adsorbate	Surface Structure	References
Ag(211)	Xe	Hexagonal overlayer	159
Ag(331)	Cl$_2$	(6 × 1)–Cl	393
	O$_2$	Disordered	393
		Ag(110)–(2 × 1)–O	393
Al(311)	[Clean]	(1 × 1)	860
Au(210)	Pb	(1 × 1)	497
Au(311)	Pb	(5 × 3)	496
		(3 × 3)–Pb	730
		(3 × 4)–Pb	730
Au(320)	Pb	(3 × 3)	496
		(1 × 1)	730
Au(511)	Pb	$\begin{bmatrix} 1 & 1 \\ 1 & 1 \end{bmatrix}$	444
		$\begin{bmatrix} 2 & 0 \\ 1 & 3 \end{bmatrix}$	444
	Pd	c(2 × 2)	683
		c(7 $\sqrt{2} \times \sqrt{2}$)$R45°$	683
		c(3 $\sqrt{2} \times \sqrt{2}$)$R45°$	683
		c(6 × 2)	683
Au(711)	Pb	$\begin{bmatrix} 1 & 1 \\ 1 & 1 \end{bmatrix}$	444
		$\begin{bmatrix} 2 & 0 \\ -1 & 3 \end{bmatrix}$	444
	Pd	c(2 × 2)	683
		c(7 $\sqrt{2} \times \sqrt{2}$)$R45°$	683
		c(3 $\sqrt{2} \times \sqrt{2}$)$R45°$	683
		c(6 × 2)	683
Au(911)	Pb	$\begin{bmatrix} 1 & 1 \\ 1 & -1 \end{bmatrix}$	444

TABLE 2.5j. *(Continued)*[a]

Substrate	Adsorbate	Surface Structure	References
		$\begin{bmatrix} 2 & 0 \\ -1 & 3 \end{bmatrix}$	444
	Pd	c(2 × 2)	683
		c(7 $\sqrt{2}$ × $\sqrt{2}$)R45°	683
		c(3 $\sqrt{2}$ × $\sqrt{2}$)R45°	683
		c(6 × 2)	683
Au(11, 1, 1)	Pb	$\begin{bmatrix} 1 & 1 \\ 1 & -1 \end{bmatrix}$	444
		$\begin{bmatrix} 2 & 0 \\ -1 & 3 \end{bmatrix}$	444
Au(S)–[6(111) × (100)]	O_2	Oxide	161
Bi(1, 0, 1, 16)	[Clean]	(1 × 1)	1109
C(0001) [stepped]	K	(1 × 1)-K	1433
	K	(2 × 2)-K	1433
	K	No LEED superstructure	1540
CO(10$\bar{1}$2)	[Clean]	(1 × 1)	698, 1584
	CO	Co_3C(001)–(2 × 3)	698
		(3 × 1)-CO	698
Cu(210)	O_2	(410), (530) facets	259
		Streak pattern	688
		(2 × 1)-O	688
		(3 × 1)-O	688
	N	c(11 $\sqrt{2}$ × $\sqrt{2}$)R45°-N	794
		(2 × 3)-N	794
Cu(211)	Kr	Hexagonal overlayer	156
	O_2	Cu(S)[5(111) × 2(100)]	958
	Facet		958
	Pb	(4 × 1)	484
	Xe	Hexagonal overlayer	156
Cu(311)	[Clean]	(1 × 1)	925, 1473, 1782
	CO	Adsorbed	394
	Pb	$\begin{bmatrix} 3 & 1 \\ -2 & 1 \end{bmatrix}$	484
		(4 × 2)	484
	Xe	Hexagonal overlayer	394
Cu(322)	O_2	(1 × 1)-O	1257
Cu(410)	O_2	(1 × 1) streaked	958
		(1 × 1)-O["c(2 × 2)-O" on a terrace]	958, 1133, 1257
		(1 × 1)-20	958
Cu(511)	[Clean]	(1 × 1)	925
	Pb	(4 × 1)	482
Cu(530)	O_2	(1 × 1)-O	1257
Cu(610)	Xe	(2 × 6)-Xe	790
Cu(711)	[Clean]	(1 × 1)	925
	Pb	(4 × 1)	482, 484
Cu(841)	O_2	(410), (100) facets	259

TABLE 2.5j. (*Continued*)*a*

Substrate	Adsorbate	Surface Structure	References
Cu(S)–[3(100) × (100)]	CO	Not adsorbed	132
	N_2	(1 × 2)–N	132
Cu(S)–[4(100) × (100)]	CO	Not adsorbed	132
	N_2	(1 × 3)–N	132
	O_2	(1 × 1)–O	132
Cu(S)–[4(100) × (111)]	H_2S	8(1d)–S	35
Fe(210)	[Clean]	(1 × 1)	1530, 1767
Fe(211)	[Clean]	(1 × 1)	1460, 1531
Fe(310)	[Clean]	(1 × 1)	1410, 1530
Fe(12, 1, 0)	N_2	Reconstruction by nitride formation	669
GaAs(211)	[Clean]	(110) facets	936
Ge(210)	[Clean]	(2 × 2)	804, 1683
Ge(211)	[Clean]	(3 × 1)(311) facets	936
		(1 × 2)	804, 1683
Ge(311)	[Clean]	(3 × 1)	804, 1683
Ge(331)	[Clean]	(5 × 1)	804, 1683
Ge(510)	[Clean]	(1 × 2)	804, 1683
Ge(511)	[Clean]	(3 × 1)	804, 1683
Ge(551)	[Clean]	(5 × 2)	804, 1683
Ir(S)–[6(111) × (100)]	CO	Disordered	182
	H_2	Adsorbed	187
	H_2O	Not adsorbed	182
	O_2	(2 × 1)–O	182
LaB_6(210)	O_2	Disordered	1624
Mo(100)[stepped]	Cs	(2 × 2)	932
		c(2 × 2)	932
	Cs(a) + O_2	c(2 × 2)	932
		Disordered	932
Mo(211)	Ba	(1 × 5)	1591, 1675
		(4 × 2)	1591
	Cs	c(2 × 1/J), 0.15 < J < 0.64	1590
		c(2 × 2)	1590
	CO	Disordered	105
	H_2	(1 × 2)–H	105
	La	Linear chains	1447
		c(2 × 2)	1447
		c(2 × 4/3)	1447
	Li	(1 × 4)–Li	1593
		(1 × 2)–Li	1593
		(1 × 1)–Li	1593
	N_2	Not adsorbed	105
	Na	(1 × 4)–Na	1684
		(1 × 3)–Na	1684
		(1 × 2)–Na	1684
		(1 × 3/2)–Na	1684
	O_2	(2 × 1)–O	105
		(1 × 2)–O	105

TABLE 2.5j. *(Continued)*[a]

Substrate	Adsorbate	Surface Structure	References
		(1×3)–O	105
		$c(4 \times 2)$–O	105
	Sr	(1×9)–Sr	1594
		(1×5)–Sr	1594
		(4×2)–Sr	1594
Nb(750)	O_2	(110) terrace + (310) step	1688
Ni(210)	N_2	Ni(100)– $(6\sqrt{2} \times \sqrt{2})R45°$–N	395
		$c(11\sqrt{2} \times \sqrt{2})R45°$–N	794
		(2×3)–N	794
		Ni(110)–(2×3)–N	395
	O_2	Facets	395, 794
Ni(211)	O_2	NiO	1351
Ni(311)	[Clean]	(1×1)	900, 1473
			1885, 1886
Ni(331)	[Clean]	(1×1)	1247, 1893
		(1×2)–S	1893
		(2×5)–S	1893
		(2×1)–S	1893
	O, CO	(2×3)	1893
Ni(hkO)[(210) to (410)]	[Clean]	Ordered	871
	O_2	Facets	871
Ni(S)–[3(100) × (111)]	H_2S	(2×2)	1055
Ni(S)–[5(100) × (111)]	[Clean]	Streaks	1055
	CO	Streaks disappear	1055
	H_2S	Streaks disappear	1055
Ni₄Mo(211)	[Clean]	Ordered	1115
Pd(111) [stepped]	NO	$c(4 \times 2)$–NO	1236
		(2×2)–NO	1236
Pd(210)	CO	(1×1)–CO	209, 210
		(1×2)–CO	209, 210
Pd(311)	CO	(2×1)–CO	209
		3(one-dimensional order)–CO	209
Pd(331)	O_2	Disordered	675
		2(one-dimensional order)	675
		$\begin{bmatrix} 1 & 2 \\ 2 & 0 \end{bmatrix}$–O	675
	NO	Disordered	675
Pd(S)–[8(100) × (110)]	Xe	One-dimensional periodicity	1100
Pd(S)–[9(111) × (111)]	CO	$(\sqrt{3} \times \sqrt{3})R30°$–CO	209
		Hexagonal overlayer	209
Pt(321)	[Clean]	Ordered	1212
	O_2	Disordered	760
Pt(654)	O_2	$(\sqrt{3} \times \sqrt{3})R30°$–O	760
Pt(997)	[Clean]	(1×1)	1226, 1278
	O_2	Pt(S)–$(17(111) \times 2(11\bar{1})$–O)	1226, 1278

TABLE 2.5j. *(Continued)*[a]

Substrate	Adsorbate	Surface Structure	References
Pt(12, 9, 8)	O_2	$(\sqrt{3} \times \sqrt{3})R30°$–O	760
Pt(12, 11, 9)	[Clean]	(1×1)	1226
Pt(62, 62, 60)	[Clean]	(1×1)	1226
		(2×2)–O	760
Pt(S)–[4(111) × (100)]	CO	Disordered	899
	H_2	Facets	221
Pt(S)–[5(100) × (111)]	O_2	(1×1)–O	708, 1171
		$(2\sqrt{2} \times \sqrt{2})R45°$–O	708, 1171
		Terrace broadening and diffused background	1171
Pt(S)–[6(111) × (111)]	I_2	(3×3) or $(\sqrt{3} \times \sqrt{3})R30°$ domains	930
	NH_3	Adsorbed	626
	O_2	$(\sqrt{3} \times \sqrt{3})R30°$–$PtO_2(0001)$	805
		$(4 \times \sqrt{3} \times 2\sqrt{3})R30°$–$PtO_3(0001)$	805
Pt(S)–[6(111) × (100)]	CO	Disordered	120
	H_2	2(one-dimensional order)–H	120, 221
		Adsorbed	396
		Pt(S)–[11(111) × 2(100)]	396
	K(a) + O_2	(4×4) Potassium oxide	1337
		(8×2) Potassium oxide	1337
		(10×2) Potassium oxide	1337
	O_2	2(one-dimensional order)–O	120
		Pt(111)–(2×2)–O	215, 708, 1171
		Pt(111)–$(\sqrt{3} \times \sqrt{3})R30°$–O	215
		Pt(111)–$(\sqrt{79} \times \sqrt{79})$–$R18°7'$–O	215
		Pt(111)–$(4 \times 2\sqrt{3})$–$R30°$–O	215
		Pt(111)–3(one-dimensional order)–O	215
		Reconstructed (2×2)	1171
		Terrace broadening	1171
Pt(S)–[7(111) × (310)]	O_2	$(\sqrt{3} \times \sqrt{3})R30°$–O	760
		(2×2)–O	760
Pt(S)–[9(111) × (100)]	H_2	2(one-dimensional order)–H	221
	H_2S	(2×2)–S	1339
		$(\sqrt{3} \times \sqrt{3})R30°$–S	1339
Pt(S)–[9(111) × (111)]	C_2N_2	Disordered	685
	CO	Disordered	120, 685
		$(\sqrt{3} \times \sqrt{3})R30°$–CO	685
		c(4×2)–CO	685

TABLE 2.5j. (*Continued*)[a]

Substrate	Adsorbate	Surface Structure	References
	H_2	(2 × 2)–H	120
		Adsorbed	400
	N	Disordered	228
	O_2	(2 × 2)–O	397–299, 685
		Not adsorbed	120
Pt(S)–[12(111) × (111)]	NH_3	Disordered	401
	NO	(2 × 2)–NO	401
	O_2	(2 × 2)–O	689
Pt(S)–[13(111) × (310)]	O_2	(2 × 2)–O	708, 1171
		($\sqrt{3} \times \sqrt{3}$)$R30°$–O	708
		Reconstructed (2 × 2)	1171
		Terrace broadening	1171
Pt(S)–[20(111) × (111)]	O_2	(2 × 2)–O	1013
Re(S)–[6(0001) × (16$\bar{7}$6)]	O_2	ReO_3 reconstruction	1654
Re(S)–[14(0001) × (10$\bar{1}$1)]	CO	(2 × 2)–CO	230
		(2 × 1)–C	230
	H_2	Disordered	663
Re(S)–[14(0001) × (16$\bar{7}$1)]	H_2	Disordered	664
Re(S)–[16(0001) × 2(10$\bar{1}$1)]	O_2	(2 × 2)	1515
Rh(331)	[Clean]	(1 × 1)	839
	CO	$\begin{bmatrix} 1 & 2 \\ 5 & -1 \end{bmatrix}$–CO	402, 722
		$\begin{bmatrix} 1 & 2 \\ 2 & 0 \end{bmatrix}$–CO	402
		Hexagonal overlayer	402
	CO_2	$\begin{bmatrix} 1 & 2 \\ 5 & -1 \end{bmatrix}$–CO	402
		$\begin{bmatrix} 1 & 2 \\ 2 & 0 \end{bmatrix}$–CO	402, 722
		Hexagonal overlayer	402, 722
	H_2	Adsorbed	402
		(1 × 1)	722
	NO	Disordered	402
		$\begin{bmatrix} -1 & 1 \\ 3 & 0 \end{bmatrix}$	402
	O_2	2(one-dimensional order)– O	402, 722
		$\begin{bmatrix} 1 & 2 \\ 2 & 0 \end{bmatrix}$–O	402, 722
		$\begin{bmatrix} 1 & 2 \\ 7 & -1 \end{bmatrix}$–O	402, 722
		Facets	402
Rh(S)–[6(111) × (100)]	CO	($\sqrt{3} \times \sqrt{3}$)–$R30°$–CO	402, 722
		(2 × 2)–CO	402
	CO_2	($\sqrt{3} \times \sqrt{3}$)–$R30°$–CO	402
		(2 × 2)–CO	402, 722

TABLE 2.5j. *(Continued)*[a]

Substrate	Adsorbate	Surface Structure	References
	H_2	Adsorbed	402
		(1×1)	722
	NO	(2×2)–NO	402, 722
	O_2	(2×2)–O	402, 722
		Rh(S)–[12(111) × 2(100)]–(2×2)–O	402
		Rh(111)–(2×2)–O	402
Si(111) [stepped]	Ni	Si(221)–(2×2)	964
	Pd	Si(221)–(2×2)	964
	Si + laser	Unchanged	1392
Si(210)	[Clean]	(2×2)	803, 1685
Si(211)	[Clean]	Complex	1317
		(4×2)	803, 936, 1685
	Ga	Ordered	1317
	H_2	Ordered (facet)	1317
Si(311)	[Clean]	(3×2)	803, 1685
	NH_3	Adsorbed	238
Si(320)	[Clean]	(1×2)	803, 1685
		(1×1)	803
		Facet	803
Si(331)	[Clean]	(13×1)	803
		(13×2)	1685
Si(510)	[Clean]	(1×2)	803, 1685
Si(511)	[Clean]	(3×1)	803, 1685
Si(S)–[14(111) × $(\overline{1}\overline{1}2)$]	Si	(1×1)	1517
Si(hkl)([001]zone)	[Clean]	Facets	892
	Au	Diffuse	892
		Facets	892
		3d–Au clusters	892
Ta(211)	CO	Disordered	101, 102
		(3×1)–O	102
	H_2	(1×1)–H	102
	N_2	Disordered	102
		(311) facets	102
	O_2	(3×1)–O	101, 102
Ti_2O_3 (047)	[Clean]	(1×1)	1020
		Oxide	101, 102
W(100)[stepped]	[Clean]	$(\sqrt{2} \times \sqrt{2})R45°$	1243
W(210)	CO	(2×1)–CO	138, 575, 1647
		(1×1)–CO	138
	N_2	(2×1)–N	131, 575, 887, 1647
W(211)	[Clean]	(1×1)	887
		(1×2)	1147
	Ag	(1×1)–Ag	969
	Au	(1×1)	969
		(1×2)	969
		(1×3)	969
		(1×4)	969

TABLE 2.5j. *(Continued)*[a]

Substrate	Adsorbate	Surface Structure	References
	C	c(10 × 4)–C	887
		c(6 × 4)–C	887
	CO	c(2 × 4)–C, O	887
	H_2S	c(2 × 6)–S	887
		c(2 × 2)–S	887
	Li	(4 × 1)	518, 520
		(3 × 1)	518, 520
		(2 × 1)	518, 520
		Incoherent	518, 520
		(1 × 1)	518, 520
	[Clean]	(2 × 2)	518, 520
	Mg	(1 × 7)–Mg	1657
		(3 × 3)–Mg	1657
	Na	(2 × 1)	521
		Compressed (2 × 1)	521
	O_2	(2 × 1)–O	15, 106–108, 403 404, 887, 1415, 1431
		(1 × 1)–O	106, 107, 403, 404, 887
		(1 × 2)–O	15, 106, 404, 887
		(1 × 3)–O	106
		(1 × 4)–O	106, 404
		(1 × n)–O (n = 3–7)	887
	Sb	(2 × 1)	553
		(1 × 1)	553
W(221)	CO	Disordered	108
		c(6 × 4)–CO	108
		(2 × 1)–CO	108
		c(2 × 4)–CO	108
	CO + O_2	(1 × 1)–CO + O_2	108
		(1 × 2)–CO + O_2	108
	H_2	(1 × 1)–H	112
	Na	Compressed (2 × 1)	521
		(2 × 1)	521
	Ni	(1 × 1)–Ni	970
		(6 × 1)–Ni	970
	NH_3	c(4 × 2)–NH_2	113
	O_2	(2 × 1)–O	15, 106–108 403, 404
		(1 × 2)–O	15, 106, 404
		(1 × 1)–O	106, 107, 403, 404
		(1 × 3)–k	106
		(1 × 4)–O	106, 404
W(310)	N_2	(2 × 1)–N	131
		c(2 × 2)–N	131
	O_2	(2 × 1)–O	1389
W(S)–[6(110) × (1$\bar{1}$0)]	O_2	(2 × 1)–O	382

TABLE 2.5j. (*Continued*)^a

Substrate	Adsorbate	Surface Structure	References
W(S)–[8(110) × (112)]	O_2	(2 × 1)–O	382
W(S)–[10(110) × (011)]	O_2	(2 × 1)–O	405
W(S)–[12(110) × (1$\bar{1}$0)]	O_2	(2 × 1)–O	382
W(S)–[13(001) × (1$\bar{1}$0)]	H_2	($\sqrt{2}$ × $\sqrt{2}$)R45°–H	945
		Incommensurate	945
		(1 × 1)–H	945
W(S)–[16(110) × (112)]	O_2	(2 × 1)–O	382
W(S)–[24(110) × (011)]	O_2	(2 × 1)–O	405
ZnO(40$\bar{4}$1)	[Clean]	Similar to 1 × 1	1239
ZnO(50$\bar{5}$1)	[Clean]	Similar to 1 × 1	1239

^{a}Organic overlayer structures are not included. See Table 2.5g for these structures.

REFERENCES FOR TABLES 2.5a–2.5j

1. K. Muller. *Z. Naturforschung.* **20A**:153 (1965).
2. A.U. MacRae. *Surf. Sci.* **1**:319 (1964).
3. L.H. Germer, E.J. Schneiber, and C.D. Hartman. *Philos. Mag.* **5**:222 (1960).
4. R.L. Park and H.E. Farnsworth. *Appl. Phys. Lett.* **3**:167 (1963).
5. T. Edmonds and R.C. Pitkethly. *Surf. Sci.* **15**:137 (1969).
6. A.U. MacRae. *Science* **139**:379 (1963).
7. G. Ertl. *Surf. Sci.* **6**:208 (1967).
8. N. Takahasni et al. *C. R. Acad. Sci.* **269B**:618 (1969).
9. G.W. Simmons, D.F. Mitchell, and K.R. Lawless. *Surf. Sci.* **8**:130 (1967).
10. C.W. Tucker, Jr. *Surf. Sci.* **2**:516 (1964).
11. C.W. Tucker, Jr. *J. Appl. Phys.* **35**:1897 (1964).
12. J.T. Grant and T. W. Haas. *Surf. Sci.* **21**:76 (1970).
13. W.P. Ellis. *J. Chem. Phys.* **48**:5695 (1968).
14. J. Ferrante and G.C. Barton. NASA Technical Note D-4735 (1968).
15. N.J. Taylor. *Surf. Sci.* **2**:544 (1964).
16. J.B. Marsh and H.E. Farnsworth. *Surf. Sci.* **1**:3 (1964).
17. R.E. Schlier and H.E. Farnsworth. *J. Chem. Phys.* **30**:917 (1959).
18. H.E. Farnsworth, R.E. Schlier, T.H. George and R.M. Buerger. *J. Appl. Phys.* **29**:1150 (1958).
19. J.J. Lander and J. Morrison. *J. Appl. Phys.* **34**:1411 (1963).
20. J.J. Lander and J. Morrison. *J. Appl. Phys.* **33**:2089 (1962).
21. G. Rovida et al. *Surf. Sci.* **14**:93 (1969).
22. R.O. Adams. In: G.A. Somorjai, editor, *The Structure and Chemistry of Solid Surfaces*, John Wiley & Sons, New York, 1969.
23. H.E. Farnsworth and D.M. Zehner. *Surf. Sci.* **17**:7 (1969).
24. G.J. Dooley and T.W. Haas. *Surf. Sci.* **19**:1 (1970).
25. B.D. Campbell, C.A. Haque and H.E. Farnsworth. In: G.A. Somorjai, editor, *The Structure and Chemistry of Solid Surfaces*, John Wiley & Sons, New York, 1969.
26. G. Ertl. *Surf. Sci.* **7**:309 (1977).
27. T. Edmonds and R.C. Pitkethly. *Surf. Sci.* **17**:450 (1969).
28. A.E. Morgan and G.A. Somorjai. *J. Chem. Phys.* **51**:3309 (1969).
29. J.C. Bertolini and G. Dalmai-Imelik. Coll. Intern. CNRS, Paris, 7–11 July 1969.
30. J.J. Lander and J. Morrison. *Surf. Sci.* **4**:241 (1966).
31. L.H. Germer and A.U. MacRae. *J. Chem. Phys.* **36**:1555 (1962).

32. A.J. van Bommel and F. Meyer. *Surf. Sci.* **8**:381 (1967).
33. J.J. Lander and J. Morrison. *J. Chem. Phys.* **37**:729 (1962).
34. R. Heckingbottom. In: G.A. Somorjai, editor, *The Structure and Chemistry of Solid Surfaces*, John Wiley & Sons, New York, 1969.
35. J.L. Domange and J. Oudar. *Surf. Sci.* **11**:124 (1968).
36. M. Perdereau and J. Oudar. *Surf. Sci.* **20**:80 (1970).
37. A.J. van Bommel and F. Meyer. *Surf. Sci.* **6**:391 (1967).
38. J.V. Florio and W.D. Robertson. *Surf. Sci.* **18**:398 (1969).
39. J.C. Bertolini and G. Dalmai-Imelik. Rapport Institute de Recherche sur la Catalyse, Villeurbanne, 1969.
40. D.L. Smith and R.F. Merrill. *J. Chem. Phys.* **52**:5861 (1970).
41. M. Boudart and D.F. Ollis. In: G.A. Somorjai, editor, *The Structure and Chemistry of Solid Surfaces*, John Wiley & Sons, New York, 1969.
42. F. Jona. *J. Phys. Chem. Solids* **28**:2155 (1967).
43. S.M. Bedair, F. Hoffmann, and H.P. Smith, Jr. *J. Appl. Phys.* **39**:4026 (1968).
44. H.H. Farrell. Ph.D. Dissertation, University of California, Berkeley, 1969.
45. L.K. Jordan and E.J. Scheibner. *Surf. Sci.* **10**:373 (1968).
46. L. Trepte, C. Menzel-Kopp, and E. Mensel. *Surf. Sci.* **8**:223 (1967).
47. R.N. Lee and H.E. Farnsworth. *Surf. Sci.* **3**:461 (1965).
48. J.T. Grant. *Surf. Sci.* **18**:228 (1969).
49. R.E. Schlier and H.E. Farnsworth. *J. Appl. Phys.* **25**:1333 (1954).
50. H.E. Farnsworth and J. Tuul. *J. Phys. Chem. Solids* **9**:48 (1958).
51. J.W. May and L.H. Germer. *Surf. Sci.* **11**:443 (1968).
52. R.E. Schlier and H.E. Farnsworth. *Adv. Catal.* **9**:434 (1957).
53. L.H. Germer and C.D. Hartman. *J. Appl. Phys.* **31**:2085 (1960).
54. H.E. Farnsworth and H.H. Madden, Jr. *J. Appl. Phys.* **32**:1933 (1961).
55. R.L. Park and H.E. Farnsworth. *J. Chem. Phys.* **43**:2351 (1965).
56. L.H. Germer. *Adv. Catal.* **13**:191 (1962).
57. L.H. Germer, R. Stern, and A.U. MacRae. *Metal Surfaces ASM*, Metals Park, OH, 1963, p. 287.
58. C.W. Tucker, Jr. *J. Appl. Phys.* **37**:3013 (1966).
59. C.A. Haque and H.E. Farnsworth. *Surf. Sci.* **1**:378 (1964).
60. A.J. Pignocco and G.E. Pellissier. *J. Electrochem. Soc.* **112**:1188 (1965).
61. H.K.A. Kann and S. Feuerstein. *J. Chem. Phys.* **50**:3618 (1969).
62. K. Hayek and H.E. Farnsworth. *Surf. Sci.* **10**:429 (1968).
63. H.E. Farnsworth and K. Hayek. *Nuovo Cimento Suppl.* **5**:2 (1967).
64. G.J. Dooley and T.W. Haas. *J. Chem. Phys.* **52**:461 (1970).
65. K.K. Vijai and P.F. Packman. *J. Chem. Phys.* **50**:1343 (1969).
66. P.J. Estrup. In: G.A. Somorjai, editor, *The Structure and Chemistry of Solid Surfaces*, John Wiley & Sons, New York, 1969.
67. J. Anderson and W.E. Danforth. *J. Franklin Inst.* **279**:160 (1965).
68. M. Onchi and H.E. Farnsworth. *Surf. Sci.* **11**:203 (1968).
69. R.A. Armstrong. In: G.A. Somorjai, editor, *The Structure and Chemistry of Solid Surfaces*, John Wiley & Sons, New York, 1969.
70. J.C. Tracy and P.W. Palmberg. *J. Chem. Phys.* **51**:4852 (1969).
71. R.L. Park and H.H. Madden. *Surf. Sci.* **11**:188 (1968).
72. A.E. Morgan and G.A. Somorjai. *Surf. Sci.* **12**:405 (1968).
73. C. Burggraf and A. Mosser. *C. R. Acad. Sci.* **268B**:1167 (1969).
74. A.E. Morgan and G.A. Somorjai. *Trans. Am. Cryst. Assoc.* **4**:59 (1968).

75. J. Anderson and P.J. Estrup. *J. Chem. Phys.* **46**:563 (1968).

76. M. Onchi and H.E. Farnsworth. *Surf. Sci.* **13**:425 (1969).

77. G.J. Dooley and T.W. Haas. *J. Chem. Phys.* **53**:993 (1970).

78. P.W. Tamm and L.D. Schmidt. *J. Chem. Phys.* **51**:5352 (1969).

79. P.J. Estrup and J. Anderson. *J. Chem. Phys.* **45**:2254 (1966).

80. H.H. Madden and H.E. Farnsworth. *J. Chem. Phys.* **34**:1186 (1961).

81. J.W. May and L.H. Germer. In: G.A. Somorjai, editor, *The Structure and Chemistry of Surface,* John Wiley & Sons, New York, 1969.

82. P.J. Estrup and J. Anderson. *J. Chem. Phys.* **46**:567 (1967).

83. T.L. Park and H.E. Farnsworth. *J. Appl. Phys.* **35**:2220 (1964).

84. P.J. Estrup and J. Anderson. *J. Chem. Phys.* **49**:523 (1968).

85. E. Margot et al. *C. R. Acad. Sci.* **270C**:1261 (1970).

86. N.W. Wideswell and J.M. Ballingal. *J. Vacuum Sci. Technol.* **7**:496 (1970).

87. F. Portele. *Z. Naturforschung* **24A**:1268 (1969).

88. G. Dalmai-Imelik and J.C. Bertolini. *C. R. Acad. Sci.* **270**:1079 (1970).

89. L.H. Germer and A.U. MacRae. *J. Appl. Phys.* **33**:2923 (1962).

90. L.H. Germer, A.U. MacRae, and A. Robert. *Welch Foundation Research Bull.* No. 11, 1961, p. 5.

91. R.L. Park and H.E. Farnsworth. *J. Chem. Phys.* **40**:2354 (1964).

92. L.H. Germer, J.W. May, and R.J. Szostak. *Surf. Sci.* **7**:430 (1967).

93. J.W. May, L.H. Germer, and C.C. Chang. *J. Chem. Phys.* **45**:2383 (1966).

94. A.G. Jackson and M.P. Hooker. *Surf. Sci.* **6**:297 (1967).

95. G. Ertl and P. Rau. *Surf. Sci.* **15**:443 (1969).

96. C.W. Tucker, Jr. *J. Appl. Phys.* **38**:2696 (1967).

97. C.W. Tucker, Jr. *J. Appl. Phys.* **37**:4147 (1966).

98. A.J. Pignocco and G.E. Pellisier. *Surf. Sci.* **7**:261 (1967).

99. K. Moliere and F. Portele. In: G.A. Somorjai, editor, *The Structure and Chemistry of Solid Surfaces,* John Wiley & Sons, New York, 1969.

100. T.W. Haas and A.G. Jackson. *J. Chem. Phys.* **44**:2921 (1966).

101. T.W. Haas, A.G. Jackson, and M.P. Hooker. *J. Chem. Phys.* **46**:3025 (1967).

102. T.W. Haas. In: G.A. Somorjai, editor, *The Structure and Chemistry of Solid Surfaces,* John Wiley & Sons, New York, 1969.

103. L.H. Germer. *Phys. Today* **July**:19 (1964).

104. L.H. Germer and J.W. May. *Surf. Sci.* **4**:452 (1966).

105. G.J. Dooley and T.W. Haas. *J. Vacuum Sci. Technol.* **7**:49 (1970).

106. C.C. Chang and L.H. Germer. *Surf. Sci.* **8**:115 (1967).

107. T.C. Tracy and J.M. Blakeley. In: G.A. Somorjai, editor, *The Structure and Chemistry of Solid Surfaces,* John Wiley & Sons, New York (1969).

108. C.C. Chang. *J. Electrochem. Soc.* **115**:354 (1968).

109. J.W. May and L.H. Germer. *J. Chem. Phys.* **44**:2895 (1966).

110. L.H. Germer and A.U. MacRae. *Proc. Natl. Acad. Sci. U.S.A.* **48**:997 (1962).

111. T.W. Haas. *J. Appl. Phys.* **39**:5854 (1968).

112. D.L. Adams et al. *Surf. Sci.* **22**:45 (1970).

113. J.W. May, R.J. Szostak, and L.H. Germer. *Surf. Sci.* **15**:37 (1969).

114. D.H. Buckley. NASA Technical Note D-5689, 1970.

115. I. Marklund, S. Andersson, and J. Martinsson. *Ark. Fys.* **37**:127 (1968).

116. P. Legare and G. Marie. *J. Chim. Phys. Phys. Chim. Biol.* **68**(7–8):120 (1971).

117. G. Marie, J.R. Anderson, and B.B. Johnson. *Proc. R. Soc. London A* **320**:227 (1970).

118. T. Edmonds, J.J. McCarrol, and R.C. Pitkethly. *J. Vacuum Sci. Technol.* **8**(1):68 (1971).

119. K. Okado, T. Halsushika, H. Tomita, S. Motov, and N. Takalashi. *Shinku* **13**(11):371 (1970).

120. B. Lang, R.W. Joyner, and G.A. Somorjai. *Surf. Sci.* **30**:454 (1972).

121. M. Henzler and J. Topler. *Surf. Sci.* **40**:388 (1973).

122. H. Van Hove and R. Leysen. *Phys. Status Solidi A* **9**(1):361 (1972).

123. S.M. Bedair and H.P. Smith, Jr. *J. Appl. Phys.* **42**:3616 (1971).

124. J.T. Grant. *Surf. Sci.* **25**:451 (1971).

125. R.W. Joyner, C.S. McKee, and M.W. Roberts. *Surf. Sci.* **26**:303 (1971).

126. J.C. Tracy. *J. Chem. Phys.* **56**(6):2748 (1971).

127. M.A. Chester and J. Pritchard. *Surf. Sci.* **28**:460 (1971).

128. R.W. Joyner, C.S. McKee, and M.W. Roberts. *Surf. Sci.* **27**:279 (1971).

129. J.C. Tracy. *J. Chem. Phys.* **56**(6):2736 (1971).

130. D. Tabor and J.M. Wilson. *J. Cryst. Growth* **9**:60 (1971).

131. D.L. Adams and L.H. Germer. *Surf. Sci.* **27**:21 (1971).

132. J. Perdereau and G.E. Rhead. *Surf. Sci.* **24**:555 (1971).

133. P.W. Palmberg. *Surf. Sci.* **25**:104 (1971).

134. G. Ertl and J. Küppers. *Surf. Sci.* **24**:104 (1971).

135. R. Heckingbottom and P.R. Wood. *Surf. Sci.* **23**:437 (1970).

136. K.J. Matysik. *Surf. Sci.* **29**:324 (1972).

137. G.E. Becker and H.D. Hagstrum. *Surf. Sci.* **30**:505 (1972).

138. D.L. Adams and L.H. Germer. *Surf. Sci.* **32**:205 (1972).

139. H.P. Bonzel and R. Ku. *Surf. Sci.* **33**:91 (1972).

140. P. Michel and Ch. Jardin. *Surf. Sci.* **36**:478 (1973).

141. A. Melmed and J.J. Carroll. *J. Vacuum Sci. Technol.* **10**:164 (1973).

142. H.D. Hagstrum and G.E. Becker. *Phys. Rev. Lett.* **22**:1054 (1969); *J. Chem. Phys.* **54**:1015 (1971).

143. W.H. Weinberg and R.P. Merrill. *Surf. Sci.* **32**:317 (1972).

144. P.B. Sewell, D.F. Mitchell, and M. Cohen. *Surf. Sci.* **33**:535 (1972).

145. F. Forstmann, W. Berndt, and P. Buttner. *Phys. Rev. Lett.* **30**:17 (1973).

146. H.A. Engelhardt and D. Menzel. *Surf. Sci.* **57**:591 (1976).

147. H. Albers, W.J.J. VanderWal, and G.A. Bootsma. *Surf. Sci.* **68**:47 (1977).

148. G. Rovida, F. Pratesi, M. Maglietta, and E. Ferroni. *Surf. Sci.* **43**:230 (1974).

149. W. Berndt. *Proceedings of the 2nd International Conference on Solid Surfaces*, 1974, p. 653.

150. F. Forstmann. *Proceedings of the 2nd International Conference on Solid Surfaces*, 1974, p. 657.

151. P.J. Goddard and R.M. Lambert. *Surf. Sci.* **67**:180 (1977).

152. Y. Tu and J.M. Blakely. *J. Vacuum Sci. Technol.* **15**:563 (1978).

153. G. Rovida, F. Pratesi, M. Maglietta, and E. Ferroni. *Proceedings of the 2nd International Conference on Solid Surfaces*, 1974, p. 117.

154. G. Rovida and F. Pratesi. *Surf. Sci.* **51**:270 (1975).

155. P.J. Goddard, K. Schwaha, and R.M. Lambart. *Surf. Sci.* **71**:351 (1978).

156. R.H. Roberts and J. Pritchard. *Surf. Sci.* **54**:687 (1976).

157. N. Stone, M.A. VanHove, S.Y. Tong, and M.B. Webb. *Phys. Rev. Lett.* **40**:273 (1978).

158. G. McElhiney, H. Papp, and J. Pritchard. *Surf. Sci.* **54**:617 (1976).

159. M.A. Chesters, M. Hussain, and J. Pritchard. *Surf. Sci.* **35**:161 (1973).

160. P.I. Cohen, J. Unguris and M.B. Webb. *Surf. Sci.* **58**:429 (1976).

161. M.A. Chesters and G.A. Somorjai. *Surf. Sci.* **52**:21 (1975).

162. D.M. Zehner and J.F. Wendelken. *Proceedings of the 7th International Vacuum Congress and 3rd International Conference on Solid Surfaces*, 1977, p. 517.

163. P.J. Goddard, J. West, and R.M. Lambert. *Surf. Sci.* **71**:447 (1978).

164. P.G. Lurie and J.M. Wilson. *Surf. Sci.* **65**:453 (1977).

165. J. Suzanne, J.P. Coulomb, and M. Bienfait. *Surf. Sci.* **40**:414 (1973).

166. M.D. Chinn and S.C. Fain, Jr. *J. Vacuum Sci. Technol.* **14**:314 (1977).

167. H.M. Kramer and J. Suzanne. *Surf. Sci.* **54**:659 (1976).

168. M.E. Bridge, C.M. Comrie, and R.M. Lambert. *Surf. Sci.* **67**:393 (1977).

169. C. Jardin and P. Michel. *Surf. Sci.* **71**:575 (1978).

170. F.H.P.M. Habraken, E.P. Kieffer, and G.A. Bootsma. *Proceedings of the 7th International Vacuum Congress and 3rd International Conference on Solid Surfaces*, 1977, p. 877.

171. L. McDonnel and D.P. Woodruff. *Surf. Sci.* **46**:505 (1974).

172. J. Kessler and F. Thieme. *Surf. Sci.* **67**:405 (1977).

173. C. Benndorf, K.H. Gressman and F. Thieme. *Surf. Sci.* **61**:646 (1976).

174. M.D. Chinn and S.C. Fain, Jr. *Phys. Rev. Lett.* **39**:146 (1977).

175. S. Nakanishi and T. Horiguchi. *Proceedings of the 7th International Vacuum Congress and 3rd International Conference on Solid Surfaces*, 1977, p. A2727.

176. M. Grunze, F. Bozso, G. Ertl, and M. Weiss. *Appl. Surf. Sci.* **1**:241 (1978).

177. F. Bozso, G. Ertl, M. Grunze, and M. Weiss. *Appl. Surf. Sci.* **1**:103 (1978)

178. B.Z. Olshanetsky, S.M. Repinsky, and A.A. Shklyaev. *Surf. Sci.* **64**:224 (1977).

179. S. Sinharoy and M. Henzler. *Surf. Sci.* **51**:75 (1975).

180. V.P. Ivanov, G.K. Boreshov, V.I. Savchenko, W.F. Egelhoff, Jr., and W.H. Weinberg. *J. Catalysis* **48**:269 (1977).

181. H. Conrad, J. Küppers, F. Nitschke, and A. Plagge. *Surf. Sci.* **69**:668 (1977).

182. D.I. Hagen, B.E. Nieuwenhuys, G. Rovida, and G.A. Somorjai. *Surf. Sci.* **57**:632 (1976).

183. J. Küppers and A. Plagge. *J. Vacuum Sci. Technol.* **13**:259 (1976).

184. V.P. Ivanov, G.K. Boreskov, V.I. Savchenko, W.F. Egelhoff, Jr., and W.H. Weinberg. *Surf. Sci.* **61**:207 (1976).

185. C.M. Comrie and W.H. Weinberg. *J. Vacuum Sci. Technol.* **13**:264 (1976).

186. C.M. Comrie and W.H. Weinberg. *J. Chem. Phys.* **64**:250 (1976).

187. B.E. Nieuwenhuys, D.I. Hagen, G. Rovida, and G.A. Somorjai. *Surf. Sci.* **59**:155 (1976).

188. J. Kanski and T.N. Rhodin. *Surf. Sci.* **65**:63 (1977).

189. L.J. Clark, *Proceedings of the 7th International Vacuum Congress and 3rd International Conference on Solid Surfaces*, 1977, p. A2725.

190. H.M. Kennett and A.E. Lee. *Surf. Sci.* **48**:606 (1975).

191. J.M. Wilson, *Surf. Sci.* **59**:315 (1976).

192. R. Pantel, M. Bujor, and J. Bardolle. *Surf. Sci.* **62**:739 (1977).

193. H. Conrad, G. Ertl, J. Küppers, and E.E. Latta. *Surf. Sci.* **50**:296 (1975).

194. P.H. Holloway and J.B. Hudson. *Surf. Sci.* **43**:141 (1974)

195. H. Conrad, G. Ertl, J. Küppers, and E.E. Latta. *Surf. Sci.* **57**:475 (1976).

196. W. Erley, K. Besoche, and H. Wagner, *J. Chem. Phys.* **66**:5269 (1977).

197. P.M. Marcus, J.E. Demuth, and D.W. Jepsen. *Surf. Sci.* **53**:501 (1975).

198. J.E. Demuth and T.N. Rhodin. *Surf. Sci.* **45**:249 (1974).

199. G. Ertl. *Surf. Sci.* **47**:86 (1975).

200. K. Christmann, O. Schober, and G. Ertl. *J. Chem. Phys.* **60**:4719 (1974).

201. M.A. Van Hove, G. Ertl, W.H. Weinberg, K. Christmann, and H.J. Behm. *Proceedings of the 7th International Vacuum Congress and 3rd International Conference on Solid Surfaces*, 1977, p. 2415.

202. H. Conrad, G. Ertl, J. Küppers, and E.E. Latta. *Surf. Sci.* **58**:578 (1976).

203. K. Christmann, O. Schober, G. Ertl, and M. Neumann. *J. Chem. Phys.* **60**:4528 (1974).

204. G. Casalone, M.G. Cattania, M. Simonetta, and M. Tescari. *Surf. Sci.* **72**:739 (1978).

205. J.E. Demuth, D.W. Jepsen, and P.M. Marcus. *Phys. Rev. Lett.* **32**:1182 (1974).

206. W. Erley and H. Wagner. *Surf. Sci.* **66**:371 (1977).

207. H. Conrad, G. Ertl, J. Küppers, and E.E. Latta. *Surf. Sci.* **65**:245 (1977).
208. H. Conrad, G. Ertl, J. Küppers, and E.E. Latta. *Surf. Sci.* **65**:235 (1977).
209. H. Conrad, G. Ertl, J. Koch, and E.E. Latta. *Surf. Sci.* **43**:462 (1974).
210. A.M. Bradshaw and F.M. Hoffman. *Surf. Sci.* **72**:513 (1978).
211. K. Christmann, G. Ertl, and O. Schober. *Surf. Sci.* **40**:61 (1973).
212. H. Conrad, G. Ertl, and E.E. Latta. *Surf. Sci.* **41**:435 (1974).
213. H.P. Bonzel and R. Ku. *Surf. Sci.* **40**:85 (1973).
214. B. Carrière, J.P. Deville, G. Maire, and P. Légaré. *Science* **58**:578 (1976).
215. P. Légaré, G. Maire, B. Carière, and J.P. Deville. *Surf. Sci.* **68**:348 (1977).
216. J.A. Joebstl. *J. Vacuum Sci. Technol.* **12**:347 (1975).
217. W.H. Weinberg, D.R. Monroe, V. Lampton, and R.P. Merrill. *J. Vacuum Sci. Technol.* **14**:444 (1977).
218. G. Ertl, M. Neumann, and K.M. Streit. *Surf. Sci.* **64**:393 (1977).
219. S.L. Bernasek and G.A. Somorjai. *J. Chem. Phys.* **60**:4552 (1974).
220. K. Christmann, G. Ertl, and T. Pignet. *Surf. Sci.* **54**:365 (1976).
221. K. Baron, D.W. Blakely, and G.A. Somorjai. *Surf. Sci.* **41**:45 (1974).
222. C.M. Comrie, W.H. Weinberg, and R.M. Lambert. *Surf. Sci.* **57**:619 (1976).
223. L.E. Firment and G.A. Somorjai. *J. Chem. Phys.* **63**:1037 (1975).
224. L.E. Firment and G.A. Somorjai. *Surf. Sci.* **55**:413 (1976).
225. W. Heegemann, E. Bechtold, and K. Hayek. *Proceedings of the 2nd International Conference on Solid Surfaces*, 1974, p. 185.
226. W. Heegemann, K.H. Meister, E. Bechtold, and K. Hayek. *Surf. Sci.* **49**:161 (1975).
227. Y. Berthier, M. Perdereau, and J. Oudar. *Surf. Sci.* **44**:281 (1974).
228. K. Schwaka and E. Bechtold. *Surf. Sci.* **66**:383 (1977).
229. D.A. Gorodetsky and A.N. Knysh. *Surf. Sci.* **40**:651 (1973).
230. M. Housley, R. Ducros, G. Piquard, and A. Cassuto. *Surf. Sci.* **68**:277 (1977).
231. D.G. Castner, B.A. Sexton, and G.A. Somorjai. *Surf. Sci.* **71**:519 (1978).
232. T.E. Madey, H.A. Engelhardt, and D. Menzel. *Surf. Sci.* **48**:304 (1975).
233. T.E. Madey and D. Menzel. *Proceedings of the 2nd International Conference on Solid Surfaces*, 1974, p. 228.
234. L.R. Danielson, M.J. Dresser, E.E. Donaldson, and J.T. Dickinson. *Surf. Sci.* **71**:599 (1978).
235. L.R. Danielson, M.J. Dresser, E.E. Donaldson, and D.R. Sandstrom. *Surf. Sci.* **71**:615 (1978).
236. K.C. Pandey, T. Sakurai, and H.D. Hagstrum. *Phys. Rev. B.* **16**:3648 (1977).
237. H. Ibach and J.E. Rowe. *Surf. Sci.* **43**:481 (1974).
238. R. Heckingbottom and P.R. Wood. *Surf. Sci.* **36**:594 (1973).
239. A.J. van Bommel and J.E. Crombeen. *Surf. Sci.* **36**:773 (1973).
240. H.D. Shih, F. Jona, D.W. Jepsen, and P.M. Marcus. *J. Vacuum Sci. Technol.* **15**:596 (1978).
241. H.D. Shih, F. Jona, D.W. Jepsen, and P.M. Marcus. *Phys. Rev. Lett.* **36**:798 (1976).
242. H.D. Shih, F. Jona, D.W. Jepsen, and P.M. Marcus. *Surf. Sci.* **60**:445 (1976).
243. R. Bastasz, C.A. Colmenares, R.L. Smith, and G.A. Somorjai. *Surf. Sci.* **67**:45 (1977).
244. T.E. Madey, J.J. Czyzewski, and J.T. Yates, Jr. *Surf. Sci.* **57**:580 (1976).
245. W.N. Unertl and J.M. Blakely. *Surf. Sci.* **69**:23 (1977).
246. A. Oustry, L. Lafourcade, and A. Escaut. *Surf. Sci.* **40**:545 (1973).
247. Y. Berthier, M. Perdereau, and J. Oudar. *Surf. Sci.* **36**:225 (1973).
248. J.C. Fuggle, E. Umbach, P. Feulner, and D. Menzel. *Surf. Sci.* **64**:69 (1977).
249. E. Zanazzi, F. Jona, D.W. Jepsen, and P.M. Marcus. *Phys. Rev. B* **14**:432 (1976).
250. A. Ignatiev, F. Jona, D.W. Jepsen, and P.M. Marcus. *Surf. Sci.* **40**:439 (1973).
251. M. Kostelitz, J.L. Domange, and J. Oudar. *Surf. Sci.* **34**:431 (1973).
252. G. McElhiney and J. Pritchard. *Surf. Sci.* **60**:397 (1976).

253. M. Maglietta and G. Rovida. *Surf. Sci.* **71**:495 (1978).

254. G. Rovida and M. Maglietta. *Proceedings of the 7th International Vacuum Congress and 3rd International Conference on Solid Surfaces*, 1977, p. 963.

255. K. Horn, M. Hussain, and J. Pritchard. *Surf. Sci.* **63**:244 (1977).

256. S. Ekelund and C. Leygraf. *Surf. Sci.* **40**:179 (1973).

257. L. McDonnell, D.P. Woodruff, and K.A.R. Mitchell. *Surf. Sci.* **45**:1 (1974).

258. G.G. Tibbetts, J.M. Burkstrand, and J.C. Tracy. *Phys. Rev. B.* **15**:3652 (1977).

259. E. Legrand-Bonnyns and A. Ponslet. *Surf. Sci.* **53**:675 (1975).

260. G.G. Tibbetts, J.M. Burkstrand, and J.C. Tracy. *J. Vacuum Sci. Technol.* **13**:362 (1976).

261. E.G. McRae and C.W. Caldwell. *Surf. Sci.* **57**:77 (1976).

262. J.R. Noonan, D.M. Zehner, and L.H. Jenkins. *Surf. Sci.* **69**:731 (1977).

263. P. Hoffmann, R. Unwin, W. Wyrobisch, and A.M. Bradshaw. *Surf. Sci.* **72**:635 (1978).

264. U. Gerhardt and G. Franz-Moller. *Proceedings of the 7th International Vacuum Congress and 3rd International Conference on Solid Surfaces*, 1977, p. 897.

265. C.R. Brundle and K. Wandelt. *Proceedings of the 7th International Vacuum Congress and 3rd International Conference on Solid Surfaces*, 1977, p. 1176.

266. J.M. Burkstrand, G.G. Kleiman, G.G. Tibbetts, and J.C. Tracy. *J. Vacuum Sci. Technol.* **13**:291 (1976).

267. A. Salwen and J. Rundgren. *Surf. Sci.* **53**:523 (1975).

268. K.O. Legg, F. Jona, D.W. Jepsen, and P.M. Marcus. *Phys. Rev. B.* **16**:5271 (1977).

269. C. Leygraf and S. Ekelund. *Surf. Sci.* **40**:609 (1973).

270. G.W. Simmons and D.J. Dwyer. *Surf. Sci.* **48**:373 (1975).

271. C.F. Brucker and T.N. Rhodin. *Surf. Sci.* **57**:523 (1976).

272. T. Horiguchi and S. Nakanishi. *Proceedings of the 2nd International Conference on Solid Surfaces*, 1974, p. 89.

273. M. Watanabe, M. Miyamura, T. Matsudaira, and M. Onchi. *Proceedings of the 2nd International Conference on Solid Surfaces*, 1974, p. 501.

274. C. Brucker and T. Rhodin. *J. Catal.* **47**:214 (1977).

275. F. Jona, K.O. Legg, H.D. Shih, D.W. Jepsen, and P.M. Marcus. *Phys. Rev. Lett.* **40**:1466 (1978).

276. T. Matsudaira, M. Watanabe, and M. Onchi. *Proceedings of the 2nd International Conference on Solid Surfaces*, 181 (1974).

277. K.O. Legg, F. Jona, D.W. Jepsen, and P.M. Marcus. *Surf. Sci.* **66**:25 (1977).

278. D.J. Dwyer and G.W. Simmons. *Surf. Sci.* **64**:617 (1977).

279. C. Leygraf, G. Hultquist, and S. Ekelund. *Surf. Sci.* **51**:409 (1975).

280. C. Leygraf and G. Hultquist. *Surf. Sci.* **61**:69 (1976).

281. T.N. Rhodin and G. Brodén. *Surf. Sci.* **60**:466 (1976).

282. G. Brodén and T.N. Rhodin. *Solid State Commun.* **18**:105 (1976).

283. A. Ignatiev, T.N. Rhodin, and S.Y. Tong. *Surf. Sci.* **42**:37 (1974).

284. R. Riwan, C. Guillot, and J. Paigne. *Surf. Sci.* **47**:183 (1975).

285. J. Lecante, R. Riwan, and G. Guillot. *Surf. Sci.* **35**:271 (1973).

286. C. Guillot, R. Riwan, and J. Lecante. *Surf. Sci.* **59**:581 (1976).

287. A. Ignatiev, F. Jona, D.W. Jepsen, and P.M. Marcus. *Surf. Sci.* **49**:189 (1975).

288. J.M. Wilson. *Surf. Sci.* **53**:330 (1975).

289. A. Glachant, J.P. Coulomb, and J.P. Biberian. *Surf. Sci.* **59**:619 (1976).

290. H.H. Farrell and M. Strongin. *Surf. Sci.* **38**:18 (1973).

291. H.H. Brongersma and J.B. Theeten. *Surf. Sci.* **54**:519 (1976).

292. Y. Murata, S. Ohtani, and K. Terada. *Proceedings of the 2nd International Conference on Solid Surfaces*, 1974, p. 837.

293. J.E. Demuth, D.W. Jepsen, and P.M. Marcus. *J. Vacuum Sci. Technol.* **11**:190 (1974).

294. T.N. Rhodin and J.E. Demuth. *Proceedings of the 2nd International Conference on Solid Surfaces*, 1974, p. 167.

295. S. Andersson, B. Kasemo, J.B. Pendry, and M.A. Van Hove. *Phys. Rev. Lett.* **31**:595 (1973).

296. E.G. McRae and C.W. Caldwell. *Surf. Sci.* **57**:63 (1976).

297. P.H. Holloway and J.B. Hudson. *Surf. Sci.* **43**:123 (1974).

298. G. Dalmai-Imelik, J.C. Bertolini, and J. Rousseau. *Surf. Sci.* **63**:67 (1977).

299. D.F. Mitchell, P.B. Sewell, and M. Cohen. *Surf. Sci.* **61**:355 (1976).

300. S. Andersson and J.B. Pendry. *Surf. Sci.* **71**:75 (1978).

301. S. Andersson. *Proceedings of the 3rd International Vacuum Congress and 7th International Conference on Solid Surfaces*, 1977, p. 1019.

302. K. Horn, A.M. Bradshaw, and K. Jacobi. *Surf. Sci.* **72**:719 (1978).

303. J.E. Demuth, D.W. Jepsen, and P.M. Marcus. *Surf. Sci.* **45**:733 (1974).

304. T. Matsudaira, M. Nishijima, and M. Onchi. *Surf. Sci.* **61**:651 (1976).

305. H. Froitzheim and H.D. Hagstrum. *J. Vacuum Sci. Technol.* **15**:485 (1978).

306. G.E. Becker and H.D. Hagstrum. *J. Vacuum Sci. Technol.* **11**:234 (1974).

307. J.M. Rickard, M. Perdereau, and L.G. Dufour. *Proceedings of the 7th International Vacuum Congress and 3rd International Conference on Solid Surfaces*, 1977, p. 847.

308. A. Steinbrunn, P. Dumas, and J.C. Colson. *Surf. Sci.* **74**:201 (1978).

309. F.P. Netzer and M. Prutton. *Surf. Sci.* **52**:505 (1975).

310. C.A. Pagageorgopoulos and J.M. Chen. *Surf. Sci.* **52**:40 (1975).

311. P.W. Palmberg. *Surf. Sci.* **25**:104 (1971).

312. C.R. Helms, H.P. Bonzel, and S. Kelemen. *J. Chem. Phys.* **65**:1773 (1976).

313. B. Lang, P. Légaré, and G. Maire. *Surf. Sci.* **47**:89 (1975).

314. G. Kneringer and F.P. Netzer. *Surf. Sci.* **49**:125 (1975).

315. G. Pirug, G. Brodén, and H.P. Bonzel. *Proceedings of the 7th International Vacuum Congress and 3rd International Conference on Solid Surfaces*, 1977, p. 907.

316. G. Brodén, G. Pirug, and H.P. Bonzel. *Surf. Sci.* **72**:45 (1978).

317. F.P. Netzer and G. Kneringer. *Surf. Sci.* **51**:526 (1975).

318. H.P. Bonzel and G. Pirug. *Surf. Sci.* **62**:45 (1977).

319. H.P. Bonzel, G. Brodén, and G. Pirug. *J. Catal.* **53**:96 (1978).

320. T.E. Fischer and S.R. Kelemen. *Surf. Sci.* **69**:1 (1977).

321. T.E. Fischer and S.R. Kelemen. *J. Vacuum Sci. Technol.* **15**:607 (1978).

322. S.J. White and D.P. Woodruff. *Surf. Sci.* **63**:254 (1977).

323. S.J. White, D.P. Woodruff, B.W. Holland, and R.S. Zimmer. *Surf. Sci.* **74**:34 (1978).

324. S.J. White, D.P. Woodruff, B.W. Holland, and R.S. Zimmer. *Surf. Sci.* **68**:457 (1977).

325. T. Sakurai and H.D. Hagstrum. *Phys. Rev. B* **14**:1593 (1976).

326. J.J. Lander and J. Morrison. *J. Chem. Phys.* **37**:729 (1962).

327. A.P. Janssen and R.C. Schoonmaker. *Surf. Sci.* **55**:109 (1976).

328. M.A. Chesters, B.J. Hopkins, and M.R. Leggett. *Surf. Sci.* **43**:1 (1974).

329. T.N. Taylor, C.A. Colmenares, R.L. Smith, and G.A. Somorjai. *Surf. Sci.* **54**:317 (1976).

330. B.J. Hopkins, G.D. Watts, and A.R. Jones. *Surf. Sci.* **52**:715 (1975).

331. C.A. Papageorgopoulous and J.M. Chen. *Surf. Sci.* **39**:313 (1973).

332. A.M. Bradshaw, D. Menzel, and M. Steinkilberg. *Proceedings of the 2nd International Conference on Solid Surfaces*, 1974, p. 841.

333. E. Bauer, H. Poppa, and Y. Viswanath. *Surf. Sci.* **58**:578 (1976).

334. S. Prigge, H. Niehus, and E. Bauer. *Surf. Sci.* **65**:141 (1977).

335. J.L. Desplat. *Proceedings of the 2nd International Conference on Solid Surfaces*, 1974, p. 177.

336. P.E. Luscher and F.M. Propst. *J. Vacuum Sci. Technol.* **14**:400 (1977).

337. R. Jaeger and D. Menzel. *Surf. Sci.* **63**:232 (1977).

338. B.J. Hopkins, A.R. Jones, and R.I. Winton. *Surf. Sci.* **57**:266 (1976).

339. S. Usami and T. Nakagima. *Proceedings of the 2nd International Conference on Solid Surfaces*, 1974, p. 237.

340. J.E. Demuth, D.W. Jepsen, and P.M. Marcus. *Phys. Rev. Lett.* **31**:540 (1973).

341. H.A. Engelhardt, A.M. Bradshaw, and D. Menzel. *Surf. Sci.* **40**:410 (1973).

342. G. Rovida and F. Pratesi. *Surf. Sci.* **52**:542 (1975).

343. W. Heiland, F. Iberi, E. Taglauer, and D. Menzel. *Surf. Sci.* **53**:383 (1975).

344. E. Zanazzi, M. Maglietta, U. Bardi, F. Jona, D.W. Jepsen, and P.M. Marcus. *Proceedings of the 7th International Vacuum Congress and 3rd International Conference on Solid Surfaces*, 1977, p. 2447.

345. R.A. Marbrow and R.M. Lambert. *Surf. Sci.* **61**:317 (1976).

346. G. Gafner and R. Feder. *Surf. Sci.* **57**:37 (1976).

347. B.E. Nieuwenhuys and G.A. Somorjai. *Surf. Sci.* **72**:8 (1978).

348. J.L. Taylor and W.H. Weinberg. *J. Vacuum Sci. Technol.* **15**:590 (1978).

349. E.B. Bas, P. Hafner, and S. Klauser. *Proceedings of the 7th International Vacuum Congress and 3rd International Conference on Solid Surfaces*, 1977, p. 881.

350. T. Miura and Y. Tuzi. *Proceedings of the 2nd International Conference on Solid Surfaces*, 1974, p. 85.

351. L. Peralta, Y. Berthier, and J. Oudar. *Surf. Sci.* **55**:199 (1976).

352. S. Andersson, J.B. Pendry, and P.M. Echenique. *Surf. Sci.* **65**:539 (1977).

353. J. Küppers. *Surf. Sci.* **36**:53 (1973).

354. D.F. Mitchell and P.B. Sewell. *Proceedings of the 7th International Vacuum Congress and 3rd International Conference on Solid Surfaces*, 1977, p. 963.

355. D.F. Mitchell, P.B. Sewell, and M. Cohen. *Surf. Sci.* **69**:310 (1977).

356. H.H. Madden, J. Küppers, and G. Ertl. *J. Chem. Phys.* **58**:3401 (1973).

357. H.H. Madden and G. Ertl. *Surf. Sci.* **35**:211 (1973).

358. H.H. Madden, J. Küppers, and G. Ertl. *J. Vacuum Sci. Technol.* **11**:190 (1974).

359. T.N. Taylor and P.J. Estrup. *J. Vacuum Sci. Technol.* **10**:26 (1973).

360. T.N. Taylor and P.J. Estrup. *J. Vacuum Sci. Technol.* **11**:244 (1974).

361. G.L. Price, B.A. Sexton, and B.G. Baker. *Surf. Sci.* **60**:506 (1976).

362. M. Wilf and P.T. Dawson. *Surf. Sci.* **65**:399 (1977).

363. R. Ducros and R.P. Merrill. *Surf. Sci.* **55**:227 (1976).

364. R.M. Lambert and C.M. Comrie. *Surf. Sci.* **46**:61 (1974).

365. P.D. Reed and R.M. Lambert. *Surf. Sci.* **57**:485 (1976).

366. R.M. Lambert. *Surf. Sci.* **49**:325 (1975).

367. Y. Berthier, J. Oudar, and M. Huber. *Surf. Sci.* **65**:361 (1977).

368. H.P. Bonzel and R. Ku. *J. Chem. Phys.* **58**:4617 (1973).

369. R.A. Marbrow and R.M. Lambert. *Surf. Sci.* **67**:489 (1977).

370. T.W. Orent and R.S. Hansen. *Surf. Sci.* **67**:325 (1977).

371. R. Ku, N.A. Gjostein, and H.P. Bonzel. *Surf. Sci.* **64**:465 (1977).

372. P.D. Reed, C.M. Comrie, and R.M. Lambert. *Surf. Sci.* **59**:33 (1976).

373. P.D. Reed, C.M. Comrie, and R.M. Lambert. *Surf. Sci.* **72**:423 (1978).

374. P.D. Reed, C.M. Comrie, and R.M. Lambert. *Surf. Sci.* **64**:603 (1977).

375. T. Sakurai and H.D. Hagstrum. *J. Vacuum Sci. Technol.* **13**:807 (1976).

376. W.J. Lo, Y.W. Chung, and G.A. Somorjai. *Surf. Sci.* **71**:199 (1978).

377. M.A. Van Hove, S. Y. Tong, and M.H. Elconin. *Surf. Sci.* **64**:85 (1977).

378. G.C. Wang, T.M. Lu, and M.G. Lagally. *Proceedings of the 7th International Vacuum Congress and 3rd International Conference on Solid Surfaces*, 1974, p. A2726.

379. J.M. Baker and D.E. Eastman. *J. Vacuum Sci. Technol.* **10**:223 (1973).

380. J.C. Buchholz and M.G. Lagally. *J. Vacuum Sci. Technol.* **11**:194 (1974).

381. J.C. Buchholz and M.G. Lagally. *Phys. Rev. Lett.* **35**:442 (1975).

382. K. Besocke and S. Berger. *Proceedings of the 7th International Vacuum Congress and 3rd International Conference on Solid Surfaces*, 1977, p. 893.

383. T.E. Madey and J.T. Yates. *Surf. Sci.* **63**:203 (1977).

384. T. Engel, H. Niehus, and E. Bauer. *Surf. Sci.* **52**:237 (1975).

385. J.C. Buchholz, G.C. Wang, and M.G. Lagally. *Surf. Sci.* **49**:508 (1975).

386. M.A. Van Hove and S.Y. Tong. *Phys. Rev. Lett.* **35**:1092 (1975).

387. E. Bauer and T. Engel. *Surf. Sci.* **71**:695 (1978).

388. N.R. Avery. *Surf. Sci.* **41**:533 (1974).

389. Ch. Steinbruchel and R. Gomer. *Surf. Sci.* **67**:21 (1977).

390. Ch. Steinbruchel and R. Gomer. *J. Vacuum Sci. Technol.* **14**:484 (1977).

391. N.R. Avery. *Surf. Sci.* **43**:101 (1974).

392. W. Göpel. *Surf. Sci.* **62**:165 (1977).

393. R.A. Marbrow and R.M. Lambert. *Surf. Sci.* **71**:107 (1978).

394. H. Papp and J. Pritchard. *Surf. Sci.* **53**:371 (1975).

395. R.E. Kirby, C.S. McKee, and M.W. Roberts. *Surf. Sci.* **55**:725 (1976).

396. G. Maire, P. Bernhardt, P. Légaré, and G. Lindauer. *Proceedings of the 7th International Vacuum Congress and 3rd International Conference on Solid Surfaces*, 1977, p. 861.

397. K. Schwaha and E. Bechtold. *Surf. Sci.* **65**:277 (1977).

398. F.P. Netzer and R.A. Willie. *J. Catal.* **51**:18 (1978).

399. F.P. Netzer and R.A. Willie. *Proceedings of the 7th International Vacuum Congress and 3rd International Conference on Solid Surfaces*, 1977, p. 927.

400. K. Christmann and G. Ertl. *Surf. Sci.* **60**:365 (1976).

401. J. Gland. *Surf. Sci.* **71**:327 (1978).

402. D.G. Castner and G.A. Somorjai. *Surf. Sci.* **83**:60 (1979).

403. G. Ertl and M. Plancher. *Surf. Sci.* **48**:364 (1975).

404. B.J. Hopkins and G.D. Watts. *Surf. Sci.* **44**:237 (1974).

405. T. Engel, T. von dem Hagen, and E. Bauer. *Surf. Sci.* **62**:361 (1977).

406. E. Gillet, J.C. Chiarena, and M. Gillet. *Surf. Sci.* **67**:393 (1977).

407. M.E. Bridge, R.A. Marbrow, and R.M. Lambert. *Surf. Sci.* **57**:415 (1976).

408. J.C. Buchholz and G.A. Somorjai. *J. Chem. Phys.* **66**:573 (1977).

409. L.L. Atanasoska, J.C. Buchholz, and G.A. Somorjai. *Surf. Sci.* **72**:189 (1978).

410. G. Brodén, T. Rhodin, and W. Capehart. *Surf. Sci.* **61**:143 (1976).

411. C.A. Papageorgopoulos and J.M. Chen. *Surf. Sci.* **39**:283 (1973).

412. D.E. Eastman and J.E. Denuth. *Proceedings of the 2nd International Conference on Solid Surfaces*, 1974, p. 827.

413. J.E. Demuth. *Surf. Sci.* **69**:365 (1977).

414. G. Dalmai-Imelik, J.C. Bertolini, J. Massardier, J. Rousseau, and B. Imelik. *Proceedings of the 7th International Vacuum Congress and 3rd International Conference on Solid Surfaces*, 1977, p. 1179.

415. J.C. Bertolini, G. Dalmai-Imelik, and J. Rousseau. *Surf. Sci.* **67**:478 (1977).

416. C. Casalone, M.G. Cattania, M. Simonetta, and M. Tescari. *Surf. Sci.* **62**:321 (1977).

417. K. Horn, A.M. Bradshaw, and K. Jacobi. *J. Vacuum Sci. Technol.* **15**:575 (1978).

418. F.C. Schouter, E.W. Kaleveld, and G.A. Bootsma. *Surf. Sci.* **63**:460 (1977).

419. J. McCarty and R.J. Madix. *J. Catal.* **38**:402 (1975).

420. J.G. McCarty and R.J. Madix. *J. Catal.* **48**:422 (1977).

421. N.M. Abbas and R.J. Madix. *Surf. Sci.* **62**:739 (1977).

422. G. Maire, J.R. Anderson, and B.B. Johnson. *Proc. R. Soc. London A* **320**:227 (1970).

423. L.L. Kesmodel, R.C. Baetzold, and G.A. Somorjai. *Surf. Sci.* **66**:299 (1977).

424. P.C. Stair and G.A. Somorjai. *J. Chem. Phys.* **66**:573 (1977).

425. W.H. Weinberg, H.A. Deans, and R.P. Merrill. *Surf. Sci.* **41**:312 (1974).

426. B. Lang. *Surf. Sci.* **53**:317 (1975).

427. L.E. Firmet and G.A. Somorjai. *J. Chem. Phys.* **66**:2901 (1977).

428. P.C. Stair and G.A. Somorjai. *J. Chem. Phys.* **67**:4361 (1977).

429. J.L. Gland and G.A. Somorjai. *Surf. Sci.* **38**:157 (1973).

430. J.L. Gland and G.A. Somorjai. *Surf. Sci.* **41**:387 (1974).

431. T.E. Fischer and S.R. Kelemen. *Surf. Scie.* **69**:485 (1977).

432. T.E. Fischer, S.R. Kelemen, and H.P. Bonzel. *Surf. Sci.* **64**:85 (1977).

433. F.P. Netzer. *Surf. Sci.* **52**:709 (1975).

434. M.E. Bridge and R.M. Lambert. *J. Catal.* **46**:143 (1977).

435. M.E. Bridge and R.M. Lambert. *Surf. Sci.* **63**:315 (1977).

436. R. Ducros, M. Housley, M. Alnot, and A. Cassuot. *Surf. Sci.* **71**:433 (1978).

437. Y.W. Chung, W. Siekhaus, and G.A. Somorjai. *Surf. Sci.* **53**:341 (1976).

438. J.P. Biberian and G.A. Somorjai. *J. Vacuum Sci. Technol.* **16**:2073 (1979).

439. J.W. Matthews and W.A. Jesser. *Acta Metal.* **15**:595 (1967).

440. J.W. Matthews. *Philos. Mag.* **13**:1207 (1966).

441. J.P. Biberian and G. E. Rhead. *J. Phys.* **F3**:675 (1973).

442. J.P. Biberian and M. Huber. *Surf. Sci.* **55**:259 (1976).

443. A.K. Green, S. Prigge, and E. Bauer. *Thin Solid Films* **52**:163 (1978).

444. J.P. Biberian. *Surf. Sci.* **74**:437 (1978).

445. V.K. Medvedev, A.G. Nauvomets, and A.G. Fedorus. *Sov. Phys. Solid State* **12**:301 (1970).

446. A.G. Naumovets and A.G. Fedorus. *JETP Lett.* **10**:6 (1969).

447. L.G. Feinstein and E. Blanc. *Surf. Sci.* **18**:350 (1969).

448. T. Edmonds and J.J. McCarroll. *Surf. Sci.* **24**:353 (1971).

449. I. Abbati, L. Braicovich, C.M. Bertoni, C. Calandra, and F. Manghi. *Phys. Rev. Lett.* **40**:469 (1978).

450. J. Abbati and L. Braicovich. *Proceedings of the 7th International Vacuum Congress and 3rd International Conference on Solid Surfaces*, Vienna, 1977, p. 1117.

451. A.J. Melmed and J.J. McCarroll. *Surf. Sci.* **19**:243 (1970).

452. D.C. Hothersall. *Philos. Mag.* **15**:1023 (1967).

453. R.E. Thomas and G.A. Haas. *J. Appl. Phys.* **43**:4900 (1972).

454. S. Anderson and B. Kasemo. *Surf. Sci.* **32**:78 (1972).

455. R.L. Gerlach and T.N. Rhodin. *Surf. Sci.* **17**:32 (1969).

456. S. Anderson and J.B. Pendry. *J. Phys. C.* **6**:601 (1973).

457. S. Anderson and U. Jostell. *Surf. Sci.* **46**:625 (1974).

458. R.L. Gerlach and T.N. Rhodin. In: G.A. Somorjai, editor, *The Structure and Chemistry of Solid Surfaces*, John Wiley & Sons, New York, 1969, p. 55.

459. S. Anderson and J.B. Pendry. *J. Phys. C.* **5**:L41 (1972).

460. R.L. Gerlach and T.N. Rhodin. *Surf. Sci.* **10**:446 (1968).

461. S. Anderson and U. Jostell. *Solid State Commun.* **13**:829 (1973).

462. S. Anderson and U. Jostell. *Solid State Commun.* **13**:833 (1973).

463. C.A. Papageorgopoulos and J.M. Chen. *Surf. Sci.* **52**:40 (1975).

464. C.A. Papageorgopoulos and J.M. Chen. *Surf. Sci.* **52**:53 (1975).

465. L.G. Beinstein, E. Blanc, and D. Dufayard. *Surf. Sci.* **19**:269 (1970).

466. D.C. Jackson, T.E. Gallon, and A. Chambers. *Surf. Sci.* **36**:381 (1973).

467. J.J. Burton, C.R. Helms, and R.S. Polizzotti. *Surf. Sci.* **57**:425 (1976).

468. J.J. Burton, C.R. Helms, and R.S. Polizzotti. *J. Chem. Phys.* **65**:1089 (1976).

469. J.J. Burton, C.R. Helms, and R.S. Polizzotti. *J. Vacuum Sci. Technol.* **13**:204 (1976).

470. J.R. Wolfe and H.W. Weart. In: G.A. Somorjai, editor, *The Structure and Chemistry of Solid Surfaces*, John Wiley & Sons, New York, 1969, p. 32.

471. J. Perdereau and I. Szymerska. *Surf. Sci.* **32**:247 (1972).

472. E. Alkhoury Nemen, R.C. Cinti, and T.T.A. Nguyen. *Surf. Sci.* **30**:697 (1972).

473. P.W. Palmberg and T.N. Rhodin. *J. Chem. Phys.* **49**:134 (1968).

474. U. Gradmann, W. Kümmerle, and P. Tillmanns. *Thin Solid Films* **34**:249 (1976).

475. C.A. Haque and H.E. Farnsworth. *Surf. Sci.* **4**:195 (1966).

476. U. Gradmann. *Surf. Sci.* **13**:498 (1969).

477. E. Bauer. *Surf. Sci.* **7**:351 (1967).

478. P.W. Palmberg and T.N. Rhodin. *J. Appl. Phys.* **39**:2425 (1968).

479. Y. Fujinaga. *Surf. Sci.* **64**:751 (1977).

480. J. Erlewein and S. Hofmann. *Surf. Sci.* **68**:71 (1977).

481. J. Henrion and G.E. Rhead. *Surf. Sci.* **29**:20 (1972).

482. A. Sepulveda and G.E. Rhead. *Surf. Sci.* **66**:436 (1977).

483. C. Argile and G.E. Rhead. *Surf. Sci.* **78**:115 (1978).

484. M.G. Barthes and G.E. Rhead. *Surf. Sci.* **80**:421 (1979).

485. K. Reichelt and F. Müller. *J. Cryst. Growth* **21**:323 (1974).

486. F. Delamare and G.E. Rhead. *Surf. Sci.* **35**:172 (1973).

487. F. Delamare and G.E. Rhead. *Surf. Sci.* **35**:185 (1973).

488. P.J. Goddard, J. West, and R.M. Lambert. *Surf. Sci.* **71**:447 (1978).

489. R.A. Marbrow and R.M. Lambert. *Surf. Sci.* **61**:329 (1976).

490. P.J. Goddard and R.M. Lambert. *Surf. Sci.* **79**:93 (1979).

491. R.C. Newman. *Philos. Mag.* **2**:750 (1957).

492. E. Bauer. *Structure et Proprietés des Solides*, CNRS, Paris, 1969.

493. R.E. Thomas and G.A. Haas. *J. Appl. Phys.* **43**:4900 (1972).

494. H.E. Farnsworth. *Phys. Rev.* **40**:684 (1932).

495. J. Perdereau, J.P. Bibérian, and G.E. Rhead. *J. Phys. F.* **4**:798 (1974).

496. M.G. Barthes and G.E. Rhead. *Surf. Sci.* **85**:L211 (1979).

497. M.G. Barthes. Thesis, University of Paris, 1978.

498. A. Sepulveda and G.E. Rhead. *Surf. Sci.* **49**:669 (1975).

499. B.A. Hutchins, T.N. Rhodin, and J.E. Demuth. *Surf. Sci.* **54**:419 (1976).

500. J.O. Porteus. *Surf. Sci.* **41**:515 (1974).

501. M.A. Van Hove, S.Y. Tong, and N. Stoner. *Surf. Sci.* **54**:259 (1976).

502. I.A.S. Edwards and H.R. Thirsk. *Surf. Sci.* **39**:245 (1973).

503. C. Argile and G.E. Rhead. *Surf. Sci.* **78**:125 (1978).

504. C. Argile. Thesis, University of Paris, 1978.

505. A.G. Jackson and M.P. Hooker. In: G.A. Somorjai, editor, *The Structure and Chemistry of Solid Surfaces*, John Wiley & Sons, New York, 1969, p. 73.

506. A.G. Elliot. *Surf. Sci.* **51**:489 (1975).

507. J.P. Bibérian. *Surf. Sci.* **59**:307 (1976).

508. T.W. Haas, A.G. Jackson, and M.P. Hooker. *J. Appl. Phys.* **38**:4998 (1967).

509. A.G. Jackson, M.P. Hooker, and T.W. Haas. *Surf. Sci.* **10**:308 (1968).

510. J.H. Pollard and W.E. Danforth. In: G.A. Somorjai, editor, *The Structure and Chemistry of Solid Surfaces*, John Wiley & Sons, New York, 1969, p. 71.

511. J.H. Pollard and W.E. Danforth. *J. Appl. Phys.* **39**:4019 (1968).

512. S. Thomas and T.W. Haas. *J. Vacuum Sci. Technol.* **9**:840 (1972).

513. K. Hartig, A.P. Janssen, and J.A. Venables. *Surf. Sci.* **74**:69 (1978).

514. K. Hartig. Thesis, Ruhr-Universität, Bochum.

515. A.G. Jackson and M.P. Hooker. *Surf. Sci.* **28**:373 (1971).

516. A.G. Jackson and M.P. Hooker. *Surf. Sci.* **27**:197 (1971).

517. D.A. Gorodetsky, Yu.P. Melnik, and A.A. Yasko. *Ukr. Fiz. Zhurn.* **12**:649 (1967).

518. V.K. Medvedev and T.P. Smereka. *Sov. Phys. Solid State* **16**:1046 (1974).

519. A.G. Naumovets and A.G. Fedorus. *Sov. Phys. JETP* **41**:587 (1975).

520. V.K. Medvedev, A.G. Naumovets, and T.P. Smereka. *Surf. Sci.* **34**:368 (1973).

521. J.M. Chen and C.A. Papageorgopoulos. *Surf. Sci.* **21**:377 (1970).

522. S. Thomas and T.W. Haas. *J. Vacuum Sci. Technol.* **10**:218 (1973).

523. A.U. MacRae, K. Müller, J.J. Lander, and J. Morrison. *Surf. Sci.* **15**:483 (1969).

524. C.A. Papageorgopoulos and J.M. Chen. *Surf. Sci.* **39**:283 (1973).

525. V.B. Voronin, A.G. Nauvomets, and A.G. Fedorus. *JETP Lett.* **15**:370 (1972).

526. C.S. Wang. *J. Appl. Phys.* **48**:1477 (1977).

527. A.G. Fedorus and A.G. Naumovets. *Surf. Sci.* **21**:426 (1970).

528. A.G. Fedorus and A.G. Naumovets. *Sov. Phys. Solid State* **12**:301 (1970).

529. H. Niehus. Thesis, Clausthal, 1975.

530. O.V. Kanash, A.G. Neumovets, and A.G. Fedorus. *Sov. Phys. JETP* **40**:903 (1974).

531. D.A. Gorodetskii and Yu.P. Mel'nik. *Akad. Nauk SSSR* **33**:430 (1969)

532. D.A. Gorodetskii, Yu.P. Mel'nik, V.K. Skylar, and V.A. Usenko. *Surf. Sci.* **85**:L503 (1979).

533. D.A. Gorodetskii and Yu.P. Mel'nik. *Surf. Sci.* **52**:647 (1977).

534. D.A. Gorodetskii, A.D. Gorchinskii, V.I. Maksimenko, and Yu.P. Mel'nik. *Sov. Phys. Solid State* **18**:691 (1976).

535. D.A. Gorodetskii, A.M. Kornev, and Yu.P. Mel'nik. *Izv. Akad. Nauk SSSR Ser. Fiz.* **28**:1337 (1964).

536. V.B. Voronin and A.G. Naumovets. *Ukr. Fiz. Zhurn.* **13**:1389 (1968).

537. V.B. Voronin. *Soviet Phys. Solid State* **9**:1758 (1968).

538. D.A. Gorodetskii and A.A. Yas'ko. *Sov. Phys. Solid State* **10**:1812 (1969).

539. D.A. Gorodetskii, A.A. Yas'ko, and S.A. Shevlyakov. *Izv. Akad. Nauk SSSR, Ser. Fiz.* **35**:436 (1971).

540. V.B. Voronin and A.G. Naumovets. *Izv. Akad. Nauk SSSR Ser. Fiz.* **35**:325 (1971).

541. G.E. Hill, J. Marklund, and J. Martinson. *Surf. Sci.* **24**:435 (1971).

542. D. Paraschkevov, W. Schlenk, R.P. Bajpai, and E. Bauer. *Proceedings of the 7th International Vacuum Congress and 3rd International Conference on Solid Surfaces, Vienna, 1977*, p. 1737.

543. E. Bauer, H. Poppa, G. Todd, and F. Bonczek. *J. Appl. Phys.* **45**:5164 (1974).

544. N.J. Taylor. *Surf. Sci.* **4**:161 (1966).

545. A.R. Moss and B.H. Blott. *Surf. Sci.* **17**:240 (1969).

546. E. Bauer, H. Poppa, G. Todd, and P.R. Davis. *J. Appl. Phys.* **48**:3773 (1977).

547. J.B. Hudson and C.M. Lo. *Surf. Sci.* **36**:141 (1973).

548. P.D. Augustus and J.P. Jones. *Surf. Sci.* **64**:713 (1977).

549. R.G. Jones and D.L. Perry. *Surf. Sci.* **71**:59 (1978).

550. D.A. Gorodetskii and A.A. Yas'ko. *Sov. Phys. Solid State* **14**:636 (1972).

551. E. Bauer, H. Poppa, and G. Todd. *Thin Solid Films* **28**:19 (1975).

552. D.A. Gorodetskii and A.A. Yas'ko. *Sov. Phys. Solid State* **11**:640 (1969).

553. B.J. Hopkins and G.D. Watts. *Surf. Sci.* **47**:195 (1975).

554. B.J. Hopkins and G.D. Watts. *Surf. Sci.* **45**:77 (1974).

555. D.A. Gorodetskii and A.A. Yas'ko. *Sov. Phys. Solid State* **13**:1085 (1971).

556. P.J. Estrup, J. Anderson, and W.E. Danforth. *Surf. Sci.* **4**:286 (1966).

557. P.J. Estrup and J. Anderson. *Surf. Sci.* **7**:255 (1967).

558. P.J. Estrup and J. Anderson. *Surf. Sci.* **8**:101 (1967).

559. J.H. Pollard. *Surf. Sci.* **20**:269 (1970).

560. J. Anderson, P.J. Estrup, and W.E. Danforth. *Appl. Phys. Lett.* **7**:122 (1965).

561. R.E. Schlier and H.E. Farnsworth. *J. Phys. Chem. Solids* **6**:271 (1958).

562. H.D. Shih, F. Jona, D.W. Jepsen, and P.M. Marcus. *Phys. Rev. B.* **15**:5550 (1977).

563. H.D. Shih, F. Jona, D.W. Jepsen, and P.M. Marcus. *Phys. Rev. B* **15**:5561 (1971).

564. H.D. Shih, F. Jona, D.W. Jepsen, and P.M. Marcus. *Comm. on Physics* **1**:25 (1976).

565. D.A. Gorodetskii and A.N. Knysh. *Surf. Sci.* **40**:636 (1973).

566. D.A. Gorodetskii and A.N. Knysh. *Surf. Sci.* **40**:651 (1973).

567. T. Shigematsu, S. Hine, and T. Takada. *J. Cryst. Growth* **43**:531 (1978).

568. T. Narusawa, S. Shimizu, and S. Komiya. *J. Vacuum Sci. Technol.* **16**:366 (1979).

569. R.J. Baird, R.C. Ku, and P. Wynblatt. *J. Vacuum Sci. Technol.* **16**:435 (1979).

570. P.A. Thiel, J.T. Yates, and W.H. Weinberg. *J. Vacuum Sci. Technol.* **16**:438 (1979).

571. D.E. Ibbotson, J.C. Taylor, and W.H. Weinberg. *J. Vacuum Sci. Technol.* **16**:439 (1979).

572. S.P. Weeks and J.E. Rowe. *J. Vacuum Sci. Technol.* **16**:470 (1979).

573. C.R. Brundle, E. Silverman, and R.J. Madix. *J. Vacuum Sci. Technol.* **16**:474 (1979).

574. P.H. Citrin, P. Eisenberger, R.C. Hewitt, and H.H. Farrell. *J. Vacuum Sci. Technol.* **16**:537 (1979).

575. A. Ignatiev, H.B. Nielsen, and D.L. Adams. *J. Vacuum Sci. Technol.* **16**:552 (1979).

576. T.N. Taylor, J.W. Rogers, and W.P. Ellis. *J. Vacuum Sci. Technol.* **16**:581 (1979).

577. T.W. Capehart and T.N. Rhodin. *J. Vacuum Sci. Technol.* **16**:594 (1979).

578. L.J. Clarke. *J. Vacuum Sci. Technol.* **16**:651 (1979).

579. B.J. Mrstik, S.Y. Tong, and M.A. Van Hove. *J. Vacuum Sci. Technol.* **16**:1258 (1979).

580. A.T. Hubbard. *J. Vacuum Sci. Technol.* **17**:49 (1980).

581. G.B. Fischer, B.A. Sexton, and J.L. Gland. *J. Vacuum Sci. Technol.* **17**:144 (1980).

582. P.R. Norton, D.K. Creber, and J.A. Davies. *J. Vacuum Sci. Technol.* **17**:149 (1980).

583. J.J. Zinck and W.H. Weinberg. *J. Vacuum Sci. Technol.* **17**:188 (1980).

584. H.L. Davis and D.M. Zehner. *J. Vacuum Sci. Technol.* **17**:190 (1980).

585. J.R. Noonan and H.L. Davis. *J. Vacuum Sci. Technol.* **17**:194 (1980).

586. J.F. Delord, A.G. Schrott, and S.C. Fain, Jr. *J. Vacuum Sci. Technol.* **17**:517 (1980).

587. L.E. Firment and G.A. Somorjai. *J. Vacuum Sci. Technol.* **17**:574 (1980).

588. W. Mönch and H. Gant. *J. Vacuum Sci. Technol.* **17**:1094 (1980).

589. J. Massies, F. Dezaly, and N.T. Linh. *J. Vacuum Sci. Technol.* **17**:1134 (1980).

590. P. Hollins and T. Pritchard. *Surf. Sci.* **99**:L389-L394 (1980).

591. M.A. Chesters. Ph.D. Thesis, London University, 1972.

592. J. Pritchard. *J. Vacuum Sci. Technol.* **89**:486 (1979).

593. P. Hollins and J. Pritchard. *Surf. Sci.* **89**:486 (1979).

594. L.J. Gerenser and R.C. Baetzold. *Surf. Sci.* **99**:259 (1980).

595. R.J. Behm, K. Christmann, and G. Ertl. *Surf. Sci.* **99**:320 (1980).

596. S.D. Bader, T.W. Orent, and L. Richter. *Bull. Am. Phys. Soc.* **24**:468 (1979).

597. S.D. Bader. *Surf. Sci.* **99**:392 (1980).

598. P. Feulner, S. Kulkami, E. Umbach, and D. Menzel. *Surf. Sci.* **99**:489 (1980).

599. K.J. Rawlings. *Surf. Sci.* **99**:507 (1980).

600. B.A. Sexton and G.E. Mitchell. *Surf. Sci.* **99**:523 (1980).

601. F.L. Baudais, H.J. Borschke, J.D. Fedyk, and M.J. Digna. *Surf. Sci.* **100**:210 (1980).

602. G. Guillot, R. Riwan, and J. Lecante. *Surf. Sci.* **59**:581 (1976).

603. E.J. Ko and R.J. Madix. **100**:L449 (1980).

604. K.J. Rawlings, G.G. Price, and B.J. Hopkins. *Surf. Sci.* **100**:289 (1980).

605. M. Kitson and R.M. Lambert. *Surf. Sci.* **100**:368 (1980).

606. P. Delescluse and A. Masson. *Surf. Sci.* **100**:423 (1980).

607. M. Perdereau and J. Oudar. *Surf. Sci.* **20**:80 (1970).

608. T. Edmonds, J.J. McCarroll, and R.C. Pitkethly. *J. Vacuum Sci. Technol.* **8**:68 (1971).

609. P.H. Holloway and J.B. Hudson. *Surf. Sci.* **33**:56 (1972).

610. H.H. Farrell. *Surf. Sci.* **100**:613 (1980).

611. C. Oshima, M. Aono, T. Tanaka, S. Kawai, S. Zaima, and Y. Shimbata. *Surf. Sci.* **102**:312 (1981).

612. L.J. Clarke. *Surf. Sci.* **102**:331 (1981).

613. L.J. Clarke. Ph.D. Thesis, Cambridge, U.K., 1978.

614. R.J. Gorte and J.L. Gland. *Surf. Sci.* **102**:348 (1981).

615. T.S. Wittrig, D.E. Ibbotson, and W.H. Weinberg. *Surf. Sci.* **102**:506 (1981).

616. P. Hofmann, K. Horn, and A.M. Bradshaw. *Surf. Sci.* **105**:L260 (1980).

617. G.L. Nyberg and N.V. Richardson. *Surf. Sci.* **85**:335 (1979).

618. S. Calisti and J. Suzanne. *Surf. Sci.* **105**:L255 (1981).

619. S.K. Shi, J.A. Schreifels, and J.M. White. *Surf. Sci.* **105**:1 (1981).

620. D. Pöso, W. Ranke, and K. Jacobi. *Surf. Sci.* **105**:77 (1981).

621. D.W. Goodman, and M. Kiskinova. *Surf. Sci.* **105**:L265 (1981).

622. H.D. Hagstrum and G.E. Becker. *Proc. R. Soc. London* **A331**:395 (1971).

623. G.B. Fisher. *Surf. Sci.* **62**:31 (1977).

624. F. Solymosi and J. Kiss. *Surf. Sci.* **104**:181 (1981).

625. L. Peralta, Y. Berthier, and M. Huber. *Surf. Sci.* **104**:435 (1981).

626. J.L. Gland and E.B. Kollin. *Surf. Sci.* **104**:478 (1981).

627. G. Rouida and F. Pratesi. *Surf. Sci.* **104**:609 (1981).

628. E. Bauer and T. Engel. *Surf. Sci.* **71**:695 (1978).

629. S. Calisti and J. Suzanne. *Surf. Sci.* **105**:L255 (1981).

630. P. Hofmann, K. Horn, and A.M. Bradshaw. *Surf. Sci.* **105**:L260 (1981).

631. S.K. Shi, J.A. Schreifels, and J.M. White. *Surf. Sci.* **105**:1 (1981).

632. D. Pöss, W. Ranke and K. Jacobi. *Surf. Sci.* **105**:77 (1981).

633. H.H. Madden. *Surf. Sci.* **106**:129 (1981).

634. P. Michel and C. Jardin. *Surf. Sci.* **36**:478 (1973).

635. G.W. Simmons and D.J. Dwyer. *Surf. Sci.* **48**:373 (1975).

636. C. Leygraf and G. Hultquist. *Surf. Sci.* **61**:61 (1976).

637. D.W. Goodman and M. Kiskinova. *Surf. Sci.* **105**:L265 (1981).

638. S.A. Flodström, C.W.B. Martinson, R.Z. Bockrach, S.B.M. Hagstïm, and R.S. Bauer. *Phys. Rev. Lett.* **40**:907 (1978).

639. T. Ichikawa and S. Ino. *Surf. Sci.* **105**:395 (1981).

640. D.J. Godfrey and D.P. Woodruff. *Surf. Sci.* **105**:438 (1981).

641. D.J. Godfrey and D.P. Woodruff. *Surf. Sci.* **105**:459 (1981).

642. T. Takahashi, H. Takiguchi, and A. Ebina. *Surf. Sci.* **105**:475 (1981).

643. M.K. Debe and D.A. King. *Surf. Sci.* **81**:193 (1978).

644. H.M. Kramer and E. Bauer. *Surf. Sci.* **107**:1 (1981).

645. T.N. Taylor and P.J. Estrup. *J. Vacuum Sci. Technol.* **10**:26 (1973).

646. S. Prigge, H. Roux, and E. Bauer. *Surf. Sci.* **107**:101 (1981).

647. N.D. Spencer and R.M. Lambert. *Surf. Sci.* **107**:237 (1981).

648. T.N. Taylor and W.P. Ellis. *Surf. Sci.* **107**:249 (1981).

649. D.L. Adams and H.B. Nielsen. *Surf. Sci.* **107**:305 (1981).

650. P.W. Davies and R.M. Lambert. *Surf. Sci.* **107**:391 (1981).

651. P.W. Davies and R.M. Lambert. *Surf. Sci.* **95**:571 (1980).

652. R.J. Koestner, M.A. Van Hove, and G.A. Somorjai. *Surf. Sci.* **107**:439 (1981).

653. P.J. Goddard and R.M. Lambert. *Surf. Sci.* **107**:519 (1981).

654. M.P. Cox and R.M. Lambert. *Surf. Sci.* **107**:547 (1981).

655. H. Namba, J. Dennille, and J.M. Gilles. *Surf. Sci.* **108**:446 (1981).

656. J.F. Wendelken. *Surf. Sci.* **108**:605 (1981).

657. F. Solymosi and J. Kiss. *Surf. Sci.* **108**:641 (1981).

658. M. Kitson and K.M. Lambert. *Surf. Sci.* **109**:60 (1981).

659. T. Oyama, S. Ohi, A. Kawazu, and G. Tominaga, *Surf. Sci.* **109**:82 (1981).

660. E.I. Ko and R.J. Madix. *Surf. Sci.* **109**:221 (1981).

661. K.J. Rawlings, S.D. Foulias, and B.J. Hopkins. *Surf. Sci.* **108**:49 (1981).

662. M. Kiskinova and D.W. Goodman. *Surf. Sci.* **108**:64 (1981).

663. P.R. Norton, J.A. Davies, D.K. Creber, C.W. Sitter, and T.E. Jackman. *Surf. Sci.* **108**:205 (1981).

664. R. Ducros, M. Housley, G. Piquard, and M. Alnot. *Surf. Sci.* **108**:235 (1981).

665. S.B. Lee, M. Weiss, and G. Ertl. *Surf. Sci.* **108**:357 (1981).

666. F. Solymosi and J. Kiss. *Surf. Sci.* **108**:368 (1981).

667. W.S. Yang and F. Jona. *Surf. Sci.* **109**:L505 (1981).

668. D. Dahlgren and J.C. Hemminger. *Surf. Sci.* **109**:L513 (1981).

669. P.A. Dowben, M. Grunze, and R.G. Jones. *Surf. Sci.* **109**:L519 (1981).

670. M. Kiskinova and D.W. Goodman. *Surf. Sci.* **109**:L555 (1981).

671. J.W.A. Sachtler, M.A. Van Hove, J.P. Biberian, and G.A. Somorjai. *Surf. Sci.* **110**:19 (1981).

672. M.W. Holmes and D.A. King. *Surf. Sci.* **110**:120 (1981).

673. M. Kitson and R.M. Lambert. *Surf. Sci.* **110**:205 (1981).

674. C.J. Schramn, Jr., M.A. Langell, and S.L. Bernasek. *Surf. Sci.* **110**:217 (1981).

675. P.W. Davies and R.M. Lambert. *Surf. Sci.* **110**:227 (1981).

676. F.P. Netzer and T.E. Madey. *Surf. Sci.* **110**:251 (1981).

677. D.E. Ibbotson, T.S. Wittrig, and W.H. Weinberg. *Surf. Sci.* **110**:294 (1981).

678. D.E. Ibbotson, T.S. Wittrig, and W.H. Weinberg. *Surf. Sci.* **110**:313 (1981).

679. F.C. Schouten, E.T. Brake, O.L.J. Gijzeman, and G.A. Bootsma. *Surf. Sci.* **74**:1 (1978).

680. S.J. White, D.P. Woodruff, B.W. Holland, and R.S. Zimmer. *Surf. Sci.* **74**:34 (1978).

681. A. Steinbrunn, P. Dumas, and J.C. Colson. *Surf. Sci.* **74**:201 (1978).

682. S.D. Bader, J.M. Blakely, M.B. Brodsky, R.J. Friddle, and R.L. Panosh. *Surf. Sci.* **74**:405 (1978).

683. J.P. Biberian. *Surf. Sci.* **74**:437 (1978).

684. B. Goldstein and D.J. Szostak. *Surf. Sci.* **74**:461 (1978).

685. F.P. Netzer and R.A. Wille. *Surf. Sci.* **74**:547 (1978).

686. C.A. Papageargopoulos. *Surf. Sci.* **75**:17 (1978).

687. K. Yoshida and G.A. Somorjai. *Surf. Sci.* **75**:46 (1978).

688. C.S. McKee, L.V. Remny, and M.W. Roberts. *Surf. Sci.* **75**:92 (1978).

689. J.L. Gland and V.L. Korehak. *Surf. Sci.* **75**:733 (1978).

690. H. Ibach and S. Lehwald. *Surf. Sci.* **76**:1 (1978).

691. H. Conrad, G. Ertl, and J. Küppers. *Surf. Sci.* **76**:323 (1978).

692. T.E. Madey and J.T. Yates, Jr. *Surf. Sci.* **76**:397 (1978).

693. T.E. Felter and P.J. Estrup. *Surf. Sci.* **76**:464 (1978).

694. A.J. Van Boommel and J.E. Crombeen. *Surf. Sci.* **76**:499 (1978).

695. H. Albers, W.J.J. Van Der Wal, O.L.J. Gijzemann, and G.A. Bootsma. *Surf. Sci.* **77**:1 (1978).

696. H. Hopster and H. Ibach. *Surf. Sci.* **77**:109 (1978).

697. P. Drathen, W. Ranke, and K. Jacobi. *Surf. Sci.* **77**:L162 (1978).

698. K.A. Prior, K. Schwaha, and R.M. Lambert. *Surf. Sci.* **77**:193 (1978).

699. W.Y. Ching, D.L. Huber, M.G. Lagally, and G.C. Wang. *Surf. Sci.* **77**:550 (1978).

700. K.J. Rawlings, B.J. Hopkins, and S.D. Foulias. *Surf. Sci.* **77**:561 (1978).

701. C.M. Chan, S.L. Cunningham, K.L. Luke, W.H. Weinberg, and S.P. Withrow. *Surf. Sci.* **78**:15 (1978).

702. J.P. Jones and E.W. Roberts. *Surf. Sci.* **78**:37 (1978).

703. C. Argile and G.E. Rhead. *Surf. Sci.* **78**:115 (1978).

704. C. Argile and G.E. Rhead. *Surf. Sci.* **78**:125 (1978).

705. P.J. Goddard and R.M. Lambert. *Surf. Sci.* **79**:93 (1979).

706. J.L. Taylor, D.E. Ibbotson, and W.H. Weinberg. *Surf. Sci.* **79**:349 (1979).

707. M. Alff and W. Moritz. *Surf. Sci.* **80**:24 (1979).

708. G. Maire, P. Légaré, and Lindauer. *Surf. Sci.* **80**:238 (1979).

709. C.W.B. Martinson and S.A. Flodström. *Surf. Sci.* **80**:306 (1979).

710. J. Massies, P. Elienne, and N.T. Linh. *Surf. Sci.* **80**:550 (1979).

711. S.A. Lindgren and and L. Wallden. *Surf. Sci.* **80**:620 (1979).

712. M.K. Debe and D.A. King. *Surf. Sci.* **81**:193 (1979).

713. T. Engel, P. Bornemann, and E. Bauer. *Surf. Sci.* **81**:252 (1979).

714. K. Schwaha, N.D. Spencer, and R.M. Lambert. *Surf. Sci.* **81**:273 (1979).

715. E.D. Williams, C.M. Chan, and W.H. Weinberg. *Surf. Sci.* **81**:L309 (1979).

716. E.D. Williams and W.H. Weinberg. *Surf. Sci.* **82**:93 (1979).

717. K. Oura and T. Hamawa. *Surf. Sci.* **82**:202 (1979).

718. J.F. Van Der Veen, R.M. Tromp, R.G. Smeenk, and F.W. Saris. *Surf. Sci.* **82**:468 (1979).

719. M.G. Cattania, M. Simonetta, and M. Tescari. *Surf. Sci.* **82**:L615 (1979).

720. C.G. Shaw and S.C. Fain, Jr. *Surf. Sci.* **83**:1 (1979).

721. M.D. Chim and S.C. Fain, Jr. *Phys. Rev. Lett.* **39**:146 (1977).

722. D.G. Castner and G.A. Somorjai. *Surf. Sci.* **83**:60 (1979).

723. R. Pantel, M. Bujor, and J. Bardolle. *Surf. Sci.* **83**:228 (1979).

724. E. Bertel, K. Schwaha, and F.P. Netzer. *Surf. Sci.* **83**:439 (1979).

725. H.L. Danis, J.R. Noonan, and L.H. Jenkins. *Surf. Sci.* **83**:559 (1979).

726. Y. Fujunaga. *Surf. Sci.* **84**:1 (1979).

727. P.A. Thiel, E.D. Williams, J.T. Yates, Jr., and W.H. Weiberg. *Surf. Sci.* **84**:54 (1979).

728. G. Broden and H.P. Bonzel. *Surf. Sci.* **84**:106 (1979).

729. L.K. Verheij, J.A. Van Den Berg, and D.G. Armour. *Surf. Sci.* **84**:408 (1979).

730. M.G. Barthes and G.E. Rhead. *Surf. Sci.* **85**:L211 (1979).

731. P.R. Novtom, J.A. Davies, D.P. Jackson, and N. Matsunami. *Surf. Sci.* **85**:269 (1979).

732. Y.Y. Tu and J.M. Blakely. *Surf. Sci.* **85**:276 (1979).

733. G.G. Price, K.J. Rawlings, and B.J. Hopkins. *Surf. Sci.* **85**:379 (1979).

734. C. Benndorf, K.H. Gressmann, J. Kessler, W. Kirstein, and F. Thieme. *Surf. Sci.* **85**:389 (1979).

735. D.A. Gorodetsky, Y.P. Melnik, V.K. Sklyar, and V.A. Usenko. *Surf. Sci.* **85**:L503 (1979).

736. A. Kawazu, Y. Saito, N. Ogiwara, T. Otsuki, and G. Tominaga. *Surf. Sci.* **86**:108 (1979).

737. Y. Fujinaga. *Surf. Sci.* **86**:581 (1979).

738. M. Aono, R. Nishitani, C. Oshima, T. Tanaka, E. Bannai, and S. Kawai. *Surf. Sci.* **86**:631 (1979).

739. F.C. Schouten, O.L.J. Gijzeman, and G.A. Bootsma. *Surf. Sci.* **87**:1 (1979).

740. S.R. Kelemen and T.E. Fischer. *Surf. Sci.* **87**:53 (1979).

741. A. Glachant and U. Bardi. *Surf. Sci.* **87**:187 (1979).

742. S. Hengrasmee, P.R. Watson, D.C. Frost, and K.A.R. Mitchell. *Surf. Sci.* **87**:L249 (1979).

743. J. Küppers, F. Nitschke, K. Wandelt, and G. Ertl. *Surf. Sci.* **87**:295 (1979).

744. A. Kahn, D. Kanani, P. Mark, P.W. Chye, C.Y. Su, I. Lindau, and W.E. Spicer. *Surf. Sci.* **87**:325 (1979).

745. J.H. Onuferko, D.P. Woodruff, and B.W. Holland. *Surf. Sci.* **87**:357 (1979).

746. H. Nilhus. *Surf. Sci.* **87**:561 (1979).

747. K. Heinz, E. Lang, and K. Müller. *Surf. Sci.* **87**:595 (1979).

748. E. Bauer and H. Poppa. *Surf. Sci.* **88**:31 (1979).

749. M.N. Read and G.J. Russell. *Surf. Sci.* **88**:95 (1979).

750. F.H.P.M. Habraken, G.A. Bootsma, P. Hofmann, S. Hachicha, and A.M. Bradshaw. *Surf. Sci.* **88**:285 (1979).

751. R.G. Jones and D.L. Perry. *Surf. Sci.* **88**:331 (1979).

752. P.A. Dowben and R.G. Jones. *Surf. Sci.* **88**:348 (1979).

753. R.G. Jones. *Surf. Sci.* **88**:367 (1979).

754. W. Mortiz and D. Wolf. *Surf. Sci.* **88**:L29 (1979).

755. M. Prutton, J.A. Walker, M.R. Welton-Cook, R.C. Felton, and J.A. Ramsey. *Surf. Sci.* **89**:95 (1979).

756. C.W.B. Madinson, S.A. Flodström, J. Rundgren, and P. Westrin. *Surf. Sci.* **89**:102 (1979).

757. P.A. Dowben and R.G. Jones. *Surf. Sci.* **89**:114 (1979).

758. C. Somerton and D.A. King. *Surf. Sci.* **89**:391 (1979).

759. M. Grunze, R.K. Driscoll, G.N. Burland, J.C.L. Cornish, and J. Pritchard. *Surf. Sci.* **89**:381 (1979).

760. S.M. Davis and G.A. Somorjai. *Surf. Sci.* **91**:73 (1980).

761. L.J. Clarke. *Surf. Sci.* **91**:131 (1980).

762. J.C. Hamilton and J.M. Blakely. *Surf. Sci.* **91**:199 (1980).

763. M. Salmeron and G.A. Somorjai. *Surf. Sci.* **91**:373 (1980).

764. P.R. Davis. *Surf. Sci.* **91**:385 (1980).

765. C.M. Chan, M.A. Van Hove, W.H. Weiberg, and E.D. Williams. *Surf. Sci.* **91**:440 (1980).

766. G. Hanke, E. Lang, K. Heinz, and K. Müller. *Surf. Sci.* **91**:551 (1980).

767. G.L. Price and B.G. Baker. *Surf. Sci.* **91**:571 (1980).

768. H.I. Lee, G. Praline, and J.M. White. *Surf. Sci.* **91**:581 (1980).

769. S. Hengrasmee, P.R. Watson, D.C. Frost, and K.A.R. Mitchell. *Surf. Sci.* **92**:71 (1980).

770. P. Nishitani, S. Kawai, H. Iwasaki, S. Nakamura, M. Aono, and T. Tanaka. *Surf. Sci.* **92**:191 (1980).

771. D.A. King and G. Thomas. *Surf. Sci.* **92**:201 (1980).

772. R.S. Li and L.X. Tu. *Surf. Sci.* **92**:L71 (1980).

773. O. Oda and G.E. Rhead. *Surf. Sci.* **92**:617 (1980).

774. G.A. Garwood, Jr. and A.T. Hubbard. *Surf. Sci.* **92**:467 (1980).

775. R. Nishitani, M. Aono, T. Tanaka, C. Oshima, S. Kawai, H. Iwasaki, and S. Nakamura. *Surf. Sci.* **93**:535 (1980).

776. A.H. Mahan, T.W. Riddle, F.B. Duming, and G.K. Walters. *Surf. Sci.* **93**:550 (1980).

777. J. Anderson and P.J. Estrup. *J. Chem. Phys.* **46**:563 (1967).

778. C.W. Seabury, T.N. Rhodin, R.J. Purtell, and R.P. Merrill. *Surf. Sci.* **93**:117 (1980).

779. Y. Sakisaka, M. Miyamura, J. Tamaki, M. Nishijima, and M. Onchi. *Surf. Sci.* **93**:327 (1980).

780. R.A. Barker, S. Semancik, and P.J. Estrup. *Surf. Sci.* **94**:L162 (1980).

781. F. Bonczek, T. Engel, and E. Bauer. *Surf. Sci.* **94**:57 (1980).

782. S.Y. Tong, A. Maldonado, C.H. Li, and M.A. Van Hove. *Surf. Sci.* **94**:73 (1980).

783. J. Benziger and R.J. Madix. *Surf. Sci.* **94**:119 (1980).

784. M. Textor, I.D. Gay, and R. Mason. *Proc. R. Soc. London* **A356**:37 (1977).

785. W. Erley. *Surf. Sci.* **94**:281 (1980).

786. G. Pirug, G. Brodén, and H.P. Bonzel. *Surf. Sci.* **94**:323 (1980).

787. D.T. Ling and W.E. Spicer. *Surf. Sci.* **94**:403 (1980).

788. M.E. Bridge and R.M. Lambert. *Surf. Sci.* **94**:469 (1980).

789. O. Nishikawa, M. Wada, and M. Konishi. *Surf. Sci.* **97**:16 (1980).

790. U. Bardi, A. Glachant, and M. Bienfait. *Surf. Sci.* **97**:137 (1980).

791. R.J. Baird, R.C. Ku, and P. Wynblatt. *Surf. Sci.* **97**:346 (1980).

792. N. Osakahe, Y. Tamishiro, K.K. Yagi, and G. Honjo. *Surf. Sci.* **97**:393 (1980).

793. E. Bertel and F.P. Netzer. *Surf. Sci.* **97**:409 (1980).

794. R.G. Kirby, C.S. McKee, and L.V. Renny. *Surf. Sci.* **97**:457 (1980).

795. T. Ichikawa and S. Ino. *Surf. Sci.* **97**:489 (1980).

796. F. Bönczek, T. Engel, and E. Bauer. *Surf. Sci.* **97**:595 (1980).

797. B.E. Hayden, E. Schweizer, R. Kötz, and A.M. Bradshaw. *Surf. Sci.* **111**:26 (1981).

798. A.G. Schrott and S.C. Fain, Jr. *Surf. Sci.* **111**:39 (1981).

799. K.J. Rawlings, S.D. Foulias, and B.J. Hopkins. *Surf. Sci.* **111**:L690 (1981).

800. T. Sakamoto and H. Kawanami. *Surf. Sci.* **111**:177 (1981).

801. W.M. Daniel, Y. Kim, H.C. Peebles, and J.M. White. *Surf. Sci.* **111**:189 (1981).

802. T. Ichikawa. *Surf. Sci.* **111**:227 (1981).

803. B.Z. Olshanetsky and V.I. Mashanov. *Surf. Sci.* **111**:414 (1981).

804. B.Z. Olshanetsky, V.I. Mashanov, and A.I. Nikiforov. *Surf. Sci.* **111**:429 (1981).

805. M. Salmeron, L. Brewer, and G.A. Somorjai. *Surf. Sci.* **112**:207 (1981).

806. G.A. Garwood, Jr. and A.T. Hubbard, *Surf. Sci.* **112**:281 (1981).

807. M. Saitoh, F. Shoji, K. Oura, and T. Hanawa. *Surf. Sci.* **112**:306 (1981).

808. M.A. Langell and S.L. Bernasek. *J. Vacuum Sci. Technol.* **17**:1287 (1980).

809. S. Shimizu and S. Komiya. *J. Vacuum Sci. Technol.* **18**:765 (1981).

810. W. Erley. *J. Vacuum Sci. Technol.* **18**:472 (1981).

811. T.E. Madey, J.G. Houston, C.W. Seabury, and T.N. Rhodin. *J. Vacuum Sci. Technol.* **18**:476 (1981).

812. M.A. Passler, T.H. Lin, and A. Ignatiev. *J. Vacuum Sci. Technol.* **18**:481 (1981).

813. R.J. Baird and W. Eberhardt. *J. Vacuum Sci. Technol.* **18**:538 (1981).

814. S. Semancik and P.J. Estrup. *J. Vacuum Sci. Technol.* **18**:541 (1981).

815. R.A. Barker and P.J. Estrup. *J. Vacuum Sci. Technol.* **18**:546 (1981).

816. G.V. Hansson, R.Z. Bachrach, R.S. Bauer, and P. Chiaradia. *J. Vacuum Sci. Technol.* **18**:550 (1981).

817. B.W. Walker and P.C. Stair. *J. Vacuum Sci. Technol.* **18**:591 (1981).

818. C.W. Seabury, T.N. Rhodin, R.J. Purtell, and R.P. Merrill. *J. Vacuum Sci. Technol.* **18**:602 (1981).

819. B.W. Lee, R.K. Ni, N. Masud, X.R. Wang, D.C. Wang, and M. Rowe. *J. Vacuum Sci. Technol.* **19**:294 (1981).

820. B.B. Pate, P.M. Stefan, C. Binns, P.J. Jupiter, M.L. Shek, I. Lindau, and W.E. Spicer. *J. Vacuum Sci. Technol.* **19**:349 (1981).

821. E.I. Ko and R.J. Madix. *J. Phys. Chem.* **85**:4019 (1981).

822. C.M. Chan and W.H. Weinberg. *J. Chem. Phys.* **71**:3988 (1979).

823. K. Christmann, R.J. Behm, G. Ertl, M.A. Van Hove, and W.H. Weinberg. *J. Chem. Phys.* **70**:4168 (1979).

824. L.L. Kesmodel, L.H. Dubois, and G.A. Somorjai. *J. Chem. Phys.* **70**:2180 (1979).

825. G.E. Thomas and W.H. Weinberg. *J. Chem. Phys.* **70**:1437 (1979).

826. G. Pirug, H.P. Bonzel, H. Hopster, and H. Ibach. *J. Chem. Phys.* **71**:593 (1979).

827. C.M. Chan and W.H. Weinberg. *J. Chem. Phys.* **71**:2788 (1979).

828. Y. Larher and A. Terlain. *J. Chem. Phys.* **72**:1052 (1980).

829. E.D. Williams, P.A. Thiel, W.H. Weinberg, and J.T. Yates, Jr. *J. Chem. Phys.* **72**:3496 (1980).

830. D.E. Ibbotson, T.S. Wittrig, and W.H. Weinberg. *J. Chem. Phys.* **72**:4885 (1980).

831. L.H. Dubois, D.G. Castner, and G.A. Somorjai. *J. Chem. Phys.* **72**:5234 (1980).

832. S.K. Shi and J.M. White. *J. Chem. Phys.* **73**:5889 (1980).

833. F. Nitschke, G. Ertl, and J. Küppers. *J. Chem. Phys.* **74**:5911 (1981).

834. R.A. Barker and P.J. Estrup. *J. Chem. Phys.* **74**:1442 (1981).

835. P.A. Thiel, F.M. Hoffmann, and W.H. Weinberg. *J. Chem. Phys.* **75**:5556 (1981).

836. S.R. Kelemen and A. Kaldor. *J. Chem. Phys.* **75**:1530 (1981).

837. J.A. Stroscio, S.R. Bare, and W. Ho. *Surf. Sci.* **154**:35 (1985).

838. E.J. Van Loenen, A.E.M.J. Fischer, J.F. van der Veen, and F. Legoues. *Surf. Sci.* **154**:52 (1985).

839. L.A. DeLouise and N. Winograd. *Surf. Sci.* **154**:79 (1985).

840. C. Klauber, M.D. Alvey, and J.T. Yates, Jr. *Surf. Sci.* **154**:139 (1985).

841. E. Daugy, P. Mathiez, F. Salvan, and J. M. Layet. *Surf. Sci.* **154**:267 (1985).

842. M.T. Paffeit, C.T. Campbell, T.N. Taylor, and S. Srinivasan. *Surf. Sci.* **154**:284 (1985).

843. F. Zaera and G.A. Somorjai. *Surf. Sci.* **154**:303 (1985).

844. I. Chorkendorff, J. Kofoed, and J. Onsgaard. *Surf. Sci.* **152/153**:749 (1985).

845. G. Strasser, G. Rosina, E. Bertel, and F.P. Netzer. *Surf. Sci.* **152/153**:765 (1985).

846. B. Poelsema, L.K. Verheij and G. Comsa. *Surf. Sci.* **152/153**:851 (1985).

847. Y. Canivez, M. Wautelet, L.D. Laude, and R. Andrew. *Surf. Sci.* **152/153**:995 (1985).

848. P. Koke, A. Goldmann, W. Mönch, G. Wolfgarten, and J. Pollmann. *Surf. Sci.* **152/153**:1001 (1985).

849. I. Hernandez-Calderón and H. Höchst. *Surf. Sci.* **152/153**:1035 (1985).

850. P. Morgen, W. Wurth, and E. Umbach. *Surf. Sci.* **152/153**:1086 (1985).

851. C. Pirri, J.C. Peruchetti, G. Gewinner, and J. Derrien. *Surf. Sci.* **152/153**:1106 (1985).

852. I. Hernandez-Calderon. *Surf. Sci.* **152/153**:1130 (1985).

853. B. Cord and R. Courths. *Surf. Sci.* **152/153**:1141 (1985).

854. P.C. Pond and D. Cherns. *Surf. Sci.* **152/153**:1197 (1985).

855. S. Kennov, S. Ladas, and C. Papageorgopoulos. *Surf. Sci.* **152/153**:1213 (1985).

856. A. Taleb-Ibrahimi, V. Mercier, C.A. Sébenne, D. Bolmont, and P. Chen. *Surf. Sci.* **152/153**:1228 (1985).

857. E. Daugy, P. Mathiez, F. Salvan, J.M. Layet, and J. Derrien. *Surf. Sci.* **152/153**:1239 (1985).

858. J.T. Grant and T.W. Haas. *Surf. Sci.* **19**:347 (1970).

859. S.M. Francis and N.V. Richardson. *Surf. Sci.* **152/153**:63 (1985).

860. J.R. Noonan, H.L. Davis, and W. Erley. *Surf. Sci.* **152/153**:142 (1985).

861. M.S. Zei, Y. Nakai, D. Weick, and G. Lehmpfuhl. *Surf. Sci.* **152/153**:254 (1985).

862. C.J. Barnes, M. Lindroos, and M. Pessa. *Surf. Sci.* **152/153**:260 (1985).

863. H. Asonen, C.J. Barnes, A. Salokatve, and M. Pessa. *Surf. Sci.* **152/153**:262 (1985).

864. K. Gürtler and K. Jacobi. *Surf. Sci.* **152/153**:272 (1985).

865. J. Cousty, R. Riwan, and P. Soukiassian, *Surf. Sci.* *152/153*:297 (1985).

866. K. Heinz, H. Hertrich, L. Hammar, and K. Müller. *Surf. Sci.* **152/153**:303 (1985).

867. K. Christmann, F. Chehab, V. Penka, and G. Ertl. *Surface. Sci.* **152/153**:356 (1985).

868. F. Chehab, W. Krstein, and F. Thieme. *Surf. Sci.* **152/153**:367 (1985).

869. K. Griffiths, P.R. Norton, J.A. Davies, W.N. Unertl, and T.E. Jackman. *Surf. Sci.* **152/153**:374 (1985).

870. P. Hofmann and D. Menzel. *Surf. Sci.* **152/153**:382 (1985).

871. J.C. Boulliard and M. Sotto. *Surf. Sci.* **152/153**:392 (1985).

872. C. Benndorf, G. Klatte, and F. Thieme. *Surf. Sci.* **152/153**:399 (1985).

873. A.V. Titov and H. Jagodzinski. *Surf. Sci.* **152/153**:409 (1985).

874. K. Hayek, H. Glassl, A. Gutmann, H. Leonhard, M. Prutton, S.P. Tear, and M.R. Welton-Cook. *Surf. Sci.* **152/153**:419 (1985).

875. J.S. Foord and A.E. Reynolds. *Surf. Sci.* **152/153**:426 (1985).

876. C. Harendt, J. Goschnick, and W. Hirschwald. *Surf. Sci.* **152/153**:453 (1985).

877. G.A. Kok, A. Noordermeer, and B.F. Nieuwenhuys. *Surf. Sci.* **152/153**:505 (1985).

878. K. Bange, T.E. Madey, and J.K. Sass. *Surf. Sci.* **152/153**:550 (1985).

879. U. Dobler, K. Baberschke, J. Haase, and A. Puschmann. *Surf. Sci.* **152/153**:569 (1985).

880. T.E. Madey and C. Benndorf. *Surf. Sci.* **152/153**:587 (1985).

881. U. Schwalke, H. Niehus, and G. Comsa. *Surf. Sci.* **152/153**:596 (1985).

882. H. Hiehus and G. Comsa. *Surf. Sci.* **151**:L171 (1985).

883. W. Altmann, K. Desinger, M. Donath, V. Dose, A. Goldmann, and H. Scheidt. *Surf. Sci.* **151**:L185 (1985).

884. H. You and S.C. Fain, Jr. *Surf. Sci.* **151**:361 (1985).

885. G.R. Gruzalski, D.M. Zehner, J.F. Wendelken, and R.S. Hathcock. *Surf. Sci.* **151**:430 (1985).

886. E. Bechtold and H. Leonhard. *Surf. Sci.* **151**:521 (1985).

887. J.B. Benziger and R.E. Preston. *Surf. Sci.* **151**:183 (1985).

888. C. Benndorf and B. Krüger. *Surf. Sci.* **151**:271 (1985).

889. K. Morishige, C. Mowforth, and R.K. Thomas. *Surf. Sci.* **151**:289 (1985).

890. S.R. Bare, J.A. Stroscio, and W. Ho. *Surf. Sci.* **150**:399 (1985).

891. K.H. Rieder and W. Stocker. *Surf. Sci.* **150**:L66 (1985).

892. A.G. Schrott and J.M. Blakely. *Surf. Sci.* **150**:L77 (1985).

893. T.J. Vink, O.L.J. Gijzeman, and J.W. Geus. *Surf. Sci.* **150**:14 (1985).

894. R.P. Furstenau, G. McDougall, and M.A. Langell. *Surf. Sci.* **150**:55 (1985).

895. H.C. Peebles, D.D. Beck, J.M. White, and C.T. Campbell. *Surf. Sci.* **150**:120 (1985).

896. L. Smit, T.E. Derry, and J.F. van der Veen. *Surf. Sci.* **150**:245 (1985).

897. M.A. Passler, B.W. Lee, and A. Ignatiev. *Surf. Sci.* **150**:263 (1985).

898. C. Zhang. M.A. Van Hove, and G.A. Somorjai. *Surf. Sci.* **149**:326 (1985).

899. B.E. Hayden, K. Kretzschmar, and A.M. Bradshaw. *Surf. Sci.* **149**:394 (1985).

900. D.L. Adams, W.T. Moore, and K.A.R. Mitchell. *Surf. Sci.* **149**:407 (1985).

901. P.M. Stefan, M.L. Shek, and W.E. Spicer. *Surf. Sci.* **149**:423 (1985).

902. H. Papp. *Surf. Sci.* **149**:460 (1985).

903. W. Ranke and D. Schmeisser. *Surf. Sci.* **149**:485 (1985).

904. H. Froitzheim, U. Köhler, and H. Lammering. *Surf. Sci.* **149**:537 (1985).

905. M.L. Shek. *Surf. Sci.* **149**:L39 (1985).

906. R.G. Egdell, H. Innes, and M.D. Hill. *Surf. Sci.* **149**:33 (1985).

907. L.H. Dubois and R.G. Nuzzo. *Surf. Sci.* **149**:133 (1985).

908. P.A. Maksym. *Surf. Sci.* **149**:157 (1985).

909. R.C. Yeates, J.E. Turner, A.J. Gellman, and G.A. Somorjai. *Surf. Sci.* **149**:175 (1985).

910. J. Rogozik, J. Küppers, and V. Dose. *Surf. Sci.* **148**:L653 (1984).

911. D.E. Fowler and J.M. Blakely. *Surf. Sci.* **148**:265 and 283 (1984).

912. T. Engel, K.H. Rieder, and I.P. Batra. *Surf. Sci.* **148**:321 (1984).

913. E.R. Moog and M.B. Webb. *Surf. Sci.* **148**:338 (1984).

914. H. Pfnür and D. Menzel. *Surf. Sci.* **148**:411 (1984).

915. J.A. Stroscio, S.R. Bare, and W. Ho. *Surf. Sci.* **148**:499 (1984).

916. E.G. McRae and R.A. Malic, *Surf. Sci.* **148**:551 (1984).

917. V. Maurice, L. Peralta, Y. Berthier, and J. Oudar. *Surf. Sci.* **148**:623 (1984).

918. P. Cantini and E. Cevasco. *Surf. Sci.* **148**:37 (1984).

919. K.H. Rieder and W. Stocker. *Surf. Sci.* **148**:139 (1984).

920. G.R. Gruzalski, D.M. Zehner, and J.F. Wendelken. *Surf. Sci.* **147**:L623 (1984).

921. E.G. McRae and P.M. Petroff. *Surf. Sci.* **147**:385 (1984).

922. L.C. Dufour, O. Bertrand, and N. Floquet. *Surf. Sci.* **147**:396 (1984).

923. M. Hanbücken, M. Fukamoto, and J.A. Venables. *Surf. Sci.* **147**:433 (1984).

924. J.P. Ganon and J. Clavilier. *Surf. Sci.* **147**:583 (1984).

925. D. Gorse, B. Salanon, F. Fabre, A. Kara, J. Perreau, G. Armand, and J. Lapujoulade. *Surf. Sci.* **148**:611 (1984).

926. F. Solymosi and L. Bugyi. *Surf. Sci.* **147**:685 (1984).

927. G.J.R. Jones, J.H. Onuferko, D.P. Woodruff, and B.W. Holland. *Surf. Sci.* **147**:1 (1984).

928. R.J. Behm, P.A. Thiel, P.R. Norton, and P.E. Bindner. *Surf. Sci.* **147**:143 (1984).

929. E. Langenbach, A. Spitzer, and H. Lüth. *Surf. Sci.* **147**:179 (1984).

930. T. Solomun, A. Wieckowski, S.D. Rosasco, and A.T. Hubbard. *Surf. Sci.* **147**:241 (1984).

931. L.A. DeLouise, E.J. White, and N. Winograd. *Surf. Sci.* **147**:252 (1984).

932. R. Riwan, P. Soukiassian, S. Zuber, and J. Cousty. *Surf. Sci.* **146**:382 (1984).

933. J. Billy and M. Abon. *Surf. Sci.* **146**:L525 (1984).

934. E.G. McRae, H.-J. Gossmann, and L.C. Feldman. *Surf. Sci.* **146**:L540 (1984).

935. U. Bardi and P.N. Ross. *Surf. Sci.* **146**:L555 (1984).

936. P. Hren, D.W. Tu, and A. Kahn. *Surf. Sci.* **146**:69 (1984).

937. A. Wieckowski, B.C. Schardt, S.D. Rosasco, J.L. Stickney, and A.T. Hubbard. *Surf. Sci.* **146**:115 (1984).

938. M. Surman, F. Solymosi, R.D. Diehl, P. Hofmann, and D.A. King. *Surf. Sci.* **146**:135 (1984).

939. E.M. Stuve, R.J. Madix, and C.R. Brundle. *Surf. Sci.* **146**:155 (1984).

940. E.M. Stuve, S.W. Jorgensen, and R.J. Madix. *Surf. Sci.* **146**:179 (1984).

941. M. Nishijima, M. Jo, and M. Onchi. *Surf. Sci.* **151**:L179 (1985).

942. K. Edamoto, Y. Kubota, M. Onchi, and M. Nishijima. *Surf. Sci.* **146**:L533 (1984).

943. A. Puschmann and J. Haase. *Surf. Sci.* **144**:559 (1984).

944. T.E. Jackman, J.A. Davies, P.R. Norton, W.N. Unertl, and K. Griffiths. *Surf. Sci.* **141**:L313 (1984).

945. J.F. Wendelken and G.-C. Wang. *Surf. Sci.* **140**:425 (1984).

946. U. Schwalke, H. Niehus, and G. Comsa. *Surf. Sci.* **137**:23 (1984).

947. V.Y. Aristov, I.E. Batov, and V.A. Grazhulis. *Surf. Sci.* **132**:73 (1983).

948. S. Kono, H. Sakurai, K. Higashiyama, and T. Sagawa. *Surf. Sci.* **130**:L299 (1983).

949. C.B. Duke, S.L. Richardson, A. Paton, and A. Kahn. *Surf. Sci.* **L135**:127 (1983).

950. K.E. Foley and N. Winograd. *Surf. Sci.* **122**:541 (1982).

951. C.B. Bargeron, B.H. Nall, and A.N. Jette. *Surf. Sci.* **120**:L483 (1982).

952. P.R. Norton, J.W. Goodale, and D.K. Creber. *Surf. Sci.* **119**:411 (1982).

953. J. Lapujoulade, Y.L. Cruër, M. Lefort, Y. Lejay, and E. Maurel. *Surf. Sci.* **118**:103 (1982).

954. R.I.G. Uhrberg, G.V. Hansson, J.M. Nicholls, and S.A. Flodström. *Surf. Sci.* **117**:394 (1982).

955. B.E. Koel, B.E. Bent, and G.A. Somorjai. *Surf. Sci.* **146**:211 (1984).

956. A. Taleb-Ibrahimi, C.A. Sébenne, D. Bolmont, and P. Chen. *Surf. Sci.* **146**:229 (1984).

957. M. Trenary, K.J. Uram, F. Bozso, and J.T. Yates, Jr. *Surf. Sci.* **146**:269 (1984).

958. K.A. Thompson and C.S. Fadley. *Surf. Sci.* **146**:281 (1984).

959. K. Ueda, K. Kinoshita, and M. Mannami. *Surf. Sci.* **145**:261 (1984).

960. J.A. Venables, J.L. Seguin, J. Suzanne, and M. Bienfait. *Surf. Sci.* **145**:345 (1984).

961. H. Ohtani, B.E. Bent, C.M. Mate, and G.A. Somorjai. Unpublished results.

962. W.R. Lambert, M.J. Cardillo, P.L. Trevor, and R.D. Doak. *Surf. Sci.* **145**:519 (1984).

963. H. Conrad, R. Scala, W. Stenzel, and R. Unwin. *Surf. Sci.* **145**:1 (1984).

964. J.G. Clabes. *Surf. Sci.* **145**:87 (1984).

965. M.J. Yacamán and P. Schabes-Retchkiman. *Surf. Sci.* **144**:L439 (1984).

966. S.T. Ceyer, A.J. Melmed, J.J. Carroll, and W.R. Graham. *Surf. Sci.* **144**:L444 (1984).

967. G.K. Binnig, H. Rohrer, C. Gerber, and E. Stoll. *Surf. Sci.* **144**:321 (1984).

968. J.S. Villarrubia and W. Ho. *Surf. Sci.* **144**:370 (1984).

969. J. Kolaczkiewicz and E. Bauer. *Surf. Sci.* **144**:477 (1984).

970. J. Kolaczkiewicz and E. Bauer. *Surf. Sci.* **144**:495 (1984).

971. N. Kuwata, T. Asai, K. Kimura, and M. Mannami. *Surf. Sci.* **144**:L393 (1984).

972. J. Jupille, J. Fusy, and P. Pareja. *Surf. Sci.* **144**:L433 (1984).

973. Y.C. Lee, M. Abu-Joudeh, and P.A. Montano. *Surf. Sci.* **144**:469 (1984).

974. C.T. Campbell and M.T. Paffett. *Surf. Sci.* **144**:517 (1984).

975. A. Rolland, J. Bernardini, and M.G. Barthes-Labrousse. *Surf. Sci.* **143**:579 (1984).

976. F.P. Netzer, D.L. Doering, and T.E. Madey. *Surf. Sci.* **143**:L363 (1984).

977. M. Asscher and G.A. Somorjai. *Surf. Sci.* **143**:L389 (1984).

978. J. Szeftel and S. Lehwald. *Surf. Sci.* **143**:11 (1984).

979. J.L. Gland, R.J. Madix, R.W. McCabe, and C. DeMaggio. *Surf. Sci.* **143**:46 (1984).

980. F. Stucki, J. Anderson, G.J. Lapeyre, and H.H. Farrell. *Surf. Sci.* **143**:84 (1984).

981. D. Rieger, R.D. Schnell, and W. Steinmann. *Surf. Sci.* **143**:157 (1984).

982. T.M. Hupkens and J.M. Fluit. *Surf. Sci.* **143**:267 (1984).

983. F. Falo, I. Cano, and M. Salmerón. *Surf. Sci.* **143**:303 (1984).

984. W.F. Egelhoff. *Surf. Sci.* **141**:L324 (1984).

985. H. Viefhaus and W. Rossow. *Surf. Sci.* **141**:341 (1984).

986. B. Egert, H.J. Grabke, Y. Sakisaka, and T.N. Rhodin. *Surf. Sci.* **141**:397 (1984).

987. M. Grunze, P.A. Dowben, and R.G. Jones. *Surf. Sci.* **141**:455 (1984).

988. F. Solymosi, A. Berkó, and T.I. Tarnóczi. *Surf. Sci.* **141**:533 (1984).

989. L. Marchut, T.M. Buck, G.H. Wheatley, and C.J. McMahon, Jr. *Surf. Sci.* **141**:549 (1984).

990. J.B. Benziger and R.E. Preston. *Surf. Sci.* **141**:567 (1984).

991. P.A. Cox, M.D. Hill, F. Peplinskii, and R.G. Egdell. *Surf. Sci.* **141**:13 (1984).

992. D.A. Andrews and D.P. Woodruff. *Surf. Sci.* **141**:31 (1984).

993. R.G. Tobin, S. Chiang, P.A. Thiel, and P.L. Richards. *Surf. Sci.* **140**:393 (1984).

994. A.G. Fedorus and V.V. Gonchar. *Surf. Sci.* **140**:499 (1984).

995. H. Niehus and G. Comsa. *Surf. Sci.* **140**:18 (1984).

996. T. Ichikawa. *Surf. Sci.* **140**:37 (1984).

997. W. Krakow. *Surf. Sci.* **140**:137 (1984).

998. M.H. Farias, A.J. Gellman, G.A. Somorjai, R.R. Chianelli, and K.S. Liang. *Surf. Sci.* **140**:181 (1984).

999. J.A. Schaefer, F. Stucki, J.A. Anderson, G.J. Lapeyre, and W. Göpel. *Surf. Sci.* **140**:207 (1984).

1000. E. Tamura and R. Feder. *Surf. Sci.* **139**:L191 (1984).

1001. C.T. Campbell and M.T. Paffett. *Surf. Sci.* **139**:396 (1984).

1002. J.R. Kingsley, D. Dahlgren, and J.C. Hemminger. *Surf. Sci.* **139**:417 (1984).

1003. J. Jupille, P. Pareja, and J. Fusy. *Surf. Sci.* **139**:505 (1984).

1004. U. Seip, M.-C. Tsai Christmann, J. Küppers, and G. Ertl. *Surf. Sci.* **139**:29 (1984).

1005. C.F. McConville, C. Somerton, and D.P. Woodruff. *Surf. Sci.* **139**:75 (1984).

1006. W. Mokwa, P. Kohl, and G. Heiland. *Surf. Sci.* **139**:98 (1984).

1007. C.B. Bargeron, B.H. Nall, and A.N. Jette. *Surf. Sci.* **139**:219 (1984).

1008. H.-J. Gossmann and W.M. Gibson. *Surf. Sci.* **139**:239 (1984).

1009. Y. Kuk, L.C. Feldman, and I.K. Robinson. *Surf. Sci.* **138**:L168 (1984).

1010. H.-J. Gossmann, J.C. Bean, L.C. Feldman, and W.M. Gibson. *Surf. Sci.* **138**:L175 (1984).

1011. C. Benndorf, C. Nöbl, and T.E. Madey. *Surf. Sci.* **138**:292 (1984).

1012. L.A. DeLouise and N. Winograd. *Surf. Sci.* **138**:417 (1984).

1013. J. Segner, C.T. Campbell, G. Doyen, and G. Ertl. *Surf. Sci.* **138**:505 (1984).

1014. K. Griffiths, T.E. Jackman, J.A. Davies, and P.R. Norton. *Surf. Sci.* **138**:113 (1984).

1015. J. Kirschner. *Surf. Sci.* **138**:191 (1984).

1016. J.S. Foord and R.M. Lambert. *Surf. Sci.* **138**:258 (1984).

1017. M. Hanbücken, H. Neddermeyer, and J.A. Venables. *Surf. Sci.* **137**:L92 (1984).

1018. R. Beaume, J. Suzanne, J.P. Coulomb, A. Glachant, and G. Bomchil. *Surf. Sci.* **137**:L117 (1984).

1019. R.A. Metzger and F.G. Allen. *Surf. Sci.* **137**:397 (1984).

1020. J.M. McKay and V.E. Henrich. *Surf. Sci.* **137**:463 (1984).

1021. M. Nishijima, H. Kobayashi, K. Edamoto, and M. Onchi. *Surf. Sci.* **137**:473 (1984).

1022. E.J. van Loenen, M. Iwami, R.M. Tromp, and J.F. van der Veen. *Surf. Sci.* **137**:1 (1984).

1023. K. Bange, D.E. Grider, T.E. Madey, and J.K. Sass. *Surf. Sci.* **137**:38 (1984).

1024. P. Ho and J.M. White. *Surf. Sci.* **137**:103 (1984).

1025. P. Ho and J.M. White. *Surf. Sci.* **137**:117 (1984).

1026. A. Gutmann, G. Zwicker, D. Schmeisser, and K. Jacobi. *Surf. Sci.* **137**:211 (1984).

1027. D.A. Outka and R.J. Madix. *Surf. Sci.* **137**:242 (1984).

1028. D. Bolmont, P. Chen, C.A. Sébenne, and F. Proix. *Surf. Sci.* **137**:280 (1984).

1029. T. Ichikawa and S. Ino. *Surf. Sci.* **136**:267 (1984).

1030. M. Kiskinova, G. Pirug, and H.P. Bonzel. *Surf. Sci.* **136**:285 (1984).

1031. V. Penka, K. Christmann, and G. Ertl. *Surf. Sci.* **136**:307 (1984).

1032. A.H. Smith, R.A. Barker, and P.J. Estrup. *Surf. Sci.* **136**:327 (1984).

1033. S.R. Kelemen and C.A. Mims. *Surf. Sci.* **136**:L35 (1984).

1034. M. Klaua and T.E. Madey. *Surf. Sci.* **136**:L42 (1984).

1035. J.L. Stickney, S.D. Rosasco, B.C. Schardt, T. Solomun, A.T. Hubbard, and B.A. Parkinson. *Surf. Sci.* **136**:15 (1984).

1036. C. Somerton, C.F. McConville, D.P. Woodruff, and R.G. Jones. *Surf. Sci.* **136**:23 (1984).

1037. K. Horioka, H. Iwasaki, S. Maruno, S.T. Li, and S. Nakamura. *Surf. Sci.* **136**:121 (1984).

1038. G. Bracco, P. Cantini, E. Cavanna, R. Tatarek, and A. Glachant. *Surf. Sci.* **136**:169 (1984).

1039. A.J. Gellman, M.H. Farias, M. Salmeron and G.A. Somorjai. *Surf. Sci.* **136**:217 (1984).

1040. U. Köhler, M. Alavi, and H.-W. Wassmuth. *Surf. Sci.* **136**:243 (1984).

1041. C. Argile and G.E. Rhead. *Surf. Sci.* **135**:18 (1983).

1042. G.A. Kok, A. Noordermeer and B.E. Nieuwenhuys. *Surf. Sci.* **135**:65 (1983).

1043. W.T. Tysoe, G.L. Nyberg, and R.M. Lambert. *Surf. Sci.* **135**:128 (1983).

1044. J.W.M. Frenken, R.G. Smeenk, and J.F. van der Veen. *Surf. Sci.* **135**:147 (1983).

1045. C. Benndorf and T.E. Madey. *Surf. Sci.* **135**:164 (1983).

1046. G. Quentel and R. Kern. *Surf. Sci.* **135**:325 (1983).

1047. R. Kötz and B.E. Hayden. *Surf. Sci.* **135**:374 (1983).

1048. C.M.A.M. Mesters, G. Wermer, O.L.J. Gijzeman, and J.W. Geus. *Surf. Sci.* **135**:396 (1983).

1049. H. Sakurai, K. Higashiyama, S. Kono, and T. Sagawa. *Surf. Sci.* **134**:L550 (1983).

1050. M. Bowker and K.C. Waugh. *Surf. Sci.* **134**:639 (1983).

1051. A.P.C. Reed, R.M. Lambert, and J.S. Foord. *Surf. Sci.* **134**:689 (1983).

1052. N.J. Gudde and R.M. Lambert. *Surf. Sci.* **134**:703 (1983).

1053. P.S. Uy, J. Bardolle, and M. Bujor. *Surf. Sci.* **134**:713 (1983).

1054. R.C. Yeates and G.A. Somorjai. *Surf. Sci.* **134**:729 (1983).

1055. R.C. Cinti, T.T.A. Nguyen, Y. Capiomont, and S. Kennou. *Surf. Sci.* **134**:755 (1983).

1056. J. StLör, R. Jaeger, G. Rossi, T. Kendelewiczx, and I. Lindau. *Surf. Sci.* **134**:813 (1983).

1057. D. Dahlgren and J.C. Hemminger. *Surf. Sci.* **134**:836 (1983).

1058. J.-M. Baribeau and J.-D. Carette. *Surf. Sci.* **134**:886 (1983).

1059. M.P. Cox, G. Ertl, R. Imbihl, and J. Rhstig. *Surf. Sci.* **134**:L517 (1983).

1060. K. Gürtler and K. Jacobi. *Surface. Sci.* **134**:309 (1983).

1061. R. Feidenhans'l, J.E. Sorensen, and I. Stensgaard. *Surf. Sci.* **134**:329 (1983).

1062. M.L. Shek, P.M. Stefan, I. Lindau, and W.E. Spicer. *Surf. Sci.* **134**:399 (1983).

1063. M.L. Shek, P.M. Stefan, I. Lindau, and W.E. Spicer. *Surf. Sci.* **134**:427 (1983).

1064. Y. Larher. *Surf. Sci.* **134**:469 (1983).

1065. T.N. Tayler, C.T. Campbell, J.W. Rogers, Jr., W.P. Ellis, and J.M. White. *Surf. Sci.* **134**:529 (1983).

1066. P. Hollins and J. Pritchard. *Surf. Sci.* **134**:91 (1983).

1067. T. Urao, T. Kanaji, and M. Kaburagi. *Surf. Sci.* **134**:109 (1983).

1068. R.F. Lin, R.J. Koestner, M.A. Van Hove, and G.A. Somorjai. *Surf. Sci.* **134**:161 (1983).

1069. M. Schuster and C. Varelas. *Surf. Sci.* **134**:195 (1983).

1070. A. Bogen and J. Küppers. *Surf. Sci.* **134**:223 (1983).

1071. M. Kiskinova, G. Pirug, and H.P. Bonzel. *Surf. Sci.* **133**:321 (1983).

1072. S.D. Foulias, K.J. Rawlings, and B.J. Hopkins. *Surf. Sci.* **133**:377 (1983).

1073. Y. Horio and A. Ichimiya. *Surf. Sci.* **133**:393 (1983).

1074. K. Jacobi and H.H. Rotermund. *Surf. Sci.* **133**:401 (1983).

1075. J.J. Métois and G.L. Lay. *Surf. Sci.* **133**:422 (1983).

1076. R. Feidenhans'l and I. Stensgaard. *Surf. Sci.* **133**:453 (1983).

1077. N. Masud, R. Baudoing, D. Aberdam, and C. Gaubert. *Surf. Sci.* **133**:580 (1983).

1078. S. Nakanishi and T. Horiguchi. *Surf. Sci.* **133**:605 (1983).

1079. R.J. Madix, M. Thornburg, and S.-B. Lee. *Surf. Sci.* **133**:L477 (1983).

1080. P. Hofmann, S.R. Bare, N.V. Richardson, and D.A. King. *Surf. Sci.* **133**:L459 (1983).

1081. K.K. Kleinherbers and A. Goldmann. *Surf. Sci.* **133**:38 (1983).

1082. D.L. Doering, S. Semancik, and T.E. Madey. *Surf. Sci.* **133**:49 (1983).

1083. S.R. Kelemen and C.A. Mims. *Surf. Sci.* **133**:71 (1983).

1084. R.M. Tromp, R.G. Smeenk, F.W. Saris, and D.J. Chadi. *Surf. Sci.* **133**:137 (1983).

1085. J.H. Neave, P.K. Larsen, J.F. van der Veen, P.J. Dobson, and B.A. Joyce. *Surf. Sci.* **133**:267 (1983).

1086. G.V. Hansson, R.I.G. Uhrberg, and J.M. Nicholls. *Surf. Sci.* **132**:31 (1983).

1087. F. Houzay, G. Guichar, R. Pinchaux, G. Jezequel, F. Solal, A. Barsky, P. Steiner, and Y. Petroff. *Surf. Sci.* **132**:40 (1983).

1088. P.H. Citrin and J.F. Rowe. *Surf. Sci.* **132**:205 (1983).

1089. P. Chen, D. Bolmont, and C.A. Sébenne. *Surf. Sci.* **132**:505 (1983).

1090. J.R. Waldrop, E.A. Kraut, S.P. Kowalczyk, and R.W. Grant. *Surf. Sci.* **132**:513 (1983).

1091. Y. Yabuuchi, F. Shoji, K. Oura, and T. Hanawa. *Surf. Sci.* **131**:L412 (1983).

1092. K.H. Rieder and H. Wilsch. *Surf. Sci.* **131**:245 (1983).

1093. J. Segner, H. Robota, W. Vielhaber, G. Ertl, F. Frenkel, J. Häger, W. Krieger, and H. Walther. *Surf. Sci.* **131**:273 (1983).

1094. J.G. Nelson, W.J. Gignac, R.S. Williams, S.W. Robey, J.G. Tobin, and D.A. Shirley. *Surf. Sci.* **131**:290 (1983).

1095. S.M. Thurgate and P.J. Jennings. *Surf. Sci.* **131**:309 (1983).

1096. B.E. Hayden. *Surf. Sci.* **131**:419 (1983).

1097. I. Stensgaard and R. Feidenhans'l. *Surf. Sci.* **131**:L373 (1983).

1098. G. Binning, H. Rohrer, C. Gerber, and E. Weibel. *Surf. Sci.* **131**:L379 (1983).

1099. C. Egawa, S. Naito, and K. Tamaru. *Surf. Sci.* **131**:49 (1983).

1100. R. Miranda, S. Daiser, K. Wandelt, and G. Ertl. *Surf. Sci.* **131**:61 (1983).

1101. D. Westphal and A. Goldmann. *Surf. Sci.* **131**:92 (1983).

1102. D. Westphal and A. Goldmann. *Surf. Sci.* **131**:113 (1983).

1103. D.R. Lloyd and F.P. Netzer. *Surf. Sci.* **131**:139 (1983).

1104. G. Zwicker and K. Jacobi. *Surf. Sci.* **131**:179 (1983).

1105. R. Miranda, D. Chandesris, and J. Lecante. *Surf. Sci.* **130**:269 (1983).

1106. J.L. Stickney, S.D. Rosasco, D. Song, M.P. Soriaga, and A.T. Hubbard. *Surf. Sci.* **130**:326 (1983).

1107. M. Hüttinger and J. Küppers. *Surf. Sci.* **130**:L277 (1983).

1108. S. Kono, H. Sakurai, K. Higashiyama, and T. Sagawa. *Surf. Sci.* **130**:L299 (1983).

1109. W.P. Ellis, K.A. Thompson, and N.S. Nogar. *Surf. Sci.* **130**:L317 (1983).

1110. J. Lee, J.P. Cowin, and L. Wharton. *Surf. Sci.* **130**:1 (1983).

1111. T.E. Felter, F.N. Hoffmann, P.A. Thiel, and W.H. Weinberg. *Surf. Sci.* **130**:163 (1983).

1112. F.M. Hoffmann, T.E. Felter, P.A. Thiel, and W.H. Weinberg. *Surf. Sci.* **130**:173 (1983).

1113. P.A. Montano, P.P. Vaishnava, and E. Boling. *Surf. Sci.* **130**:191 (1983).

1114. C.M. Pradier, Y. Berthier, and J. Oudar. *Surf. Sci.* **130**:229 (1983).

1115. M. Yamamoto and D.N. Seidman. *Surf. Sci.* **129**:281 (1983).

1116. V. Maurice, J.J. Legendre, and M. Huber. *Surf. Sci.* **129**:301 (1983).

1117. V. Maurice, J.J. Legendre, and M. Huber. *Surf. Sci.* **129**:312 (1983).

1118. R.L. Kurtz and V.E. Henrich. *Surf. Sci.* **129**:345 (1983).

1119. P.D. Johnson, D.P. Woodruff, H.H. Farrell, N.V. Smith, and M.M. Traum. *Surf. Sci.* **129**:366 (1983).

1120. M.P. Cox, J.S. Foord, R.M. Lambert, and R.H. Prince. *Surf. Sci.* **129**:375 (1983).

1121. P. Marcus, A. Teissier, and J. Oudar. *Surf. Sci.* **129**:432 (1983).

1122. M.A. Van Hove, R.J. Koestner, J.C. Frost, and G.A. Somorjai. *Surf. Sci.* **129**:482 (1983).

1123. R. Opila and R. Gomer. *Surf. Sci.* **129**:563 (1983).

1124. M. Mattern-Klosson, X.M. Ding, H. Lüth, and A. Spitzer. *Surf. Sci.* **129**:1 (1983).

1125. E. Conrad and M.B. Webb. *Surf. Sci.* **129**:37 (1983).

1126. J.S. Foord, A.P.C. Reed, and R.M. Lambert. *Surf. Sci.* **129**:79 (1983).

1127. G. Michalk, W. Moritz, H. Pfnür, and D. Menzel. *Surf. Sci.* **129**:92 (1983).

1128. L.E. Firment and A. Ferretti. *Surf. Sci.* **129**:155 (1983).

1129. D.L. Doering and S. Semancik. *Surf. Sci.* **129**:177 (1983).

1130. H. Papp. *Surf. Sci.* **129**:205 (1983).

1131. P.S. Uy, J. Bardolle and M. Bujor. *Surf. Sci.* **129**:219 (1983).

1132. R. Ducros, B. Tardy, and J.C. Bertolini. *Surf. Sci.* **128**:L219 (1983).

1133. A.J. Algra, E.P.T.M. Suurmeijer, and A.L. Boers. *Surf. Sci.* **128**:207 (1983).

1134. R.M. Tromp, E.J. Van Loenen, M. Iwami, R.G. Smeenk, F.W. Saris, F. Nava, and G. Ottaviani. *Surf. Sci.* **128**:224 (1983).

1135. I. Stensgaard, R. Feidenhans'l, and J.E. Sørensen. *Surf. Sci.* **128**:281 (1983).

1136. D.L. Adams, H. E. Nielsen, and J.N. Andersen. *Surf. Sci.* **128**:294 (1983).

1137. M. Grunze, P.A. Dowben, and C.R. Brundle. *Surf. Sci.* **128**:311 (1983).

1138. K.H. Rieder. *Surf. Sci.* **128**:325 (1983).

1139. A. Ichimiya and Y. Takeuchi. *Surf. Sci.* **128**:343 (1983).

1140. C. Benndorf, M. Frank, and F. Thieme. *Surf. Sci.* **128**:417 (1983).

1141. V. Martines, F. Soria, M.C. Munoz, and J.L. Sacedon. *Surf. Sci.* **128**:424 (1983).

1142. R. Baudoing, E. Blanc, C. Gaubert, Y. Gauthier, and N. Gnuchev. *Surf. Sci.* **128**:22 (1983).

1143. C. Backx, C.P.M. de Groot, P. Biloen, and W.M.H. Sachtler. *Surf. Sci.* **128**:81 (1983).

1144. B.C. De Cooman, V.D. Vankar, and R.W. Vook. *Surf. Sci.* **128**:128 (1983).

1145. U. Bardi and G. Rovida. *Surf. Sci.* **128**:145 (1983).

1146. K. Truszkowska and M.J. Yacaman. *Surf. Sci.* **127**:L159 (1983).

1147. C.-F. Ai and T.T. Tsong. *Surf. Sci.* **127**:L165 (1983).

1148. R.G. Jones, C.F. McConville, and D.P. Woodruff. *Surf. Sci.* **127**:424 (1983).

1149. M. Salmeron, G.A. Somorjai, and R.R. Chianelli. *Surf. Sci.* **127**:526 (1983).

1150. J.L. Gland, R.W. McCabe, and G.E. Mitchell. *Surf. Sci.* **127**:L123 (1983).

1151. B.T. Jonker, N.C. Bartelt, and R.L. Park. *Surf. Sci.* **127**:183 (1983).

1152. S. Maroie, P.A. Thiry, R. Caudano, and J.J. Verbist. *Surf. Sci.* **127**:200 (1983).

1153. K.H. Rieder, T. Engel, R.H. Swendsen, and M. Manninen. *Surf. Sci.* **127**:223 (1983).

1154. E. Bauer and H. Poppa. *Surf. Sci.* **127**:243 (1983).

1155. S.L. Miles, S.L. Bernasek, and J.L. Gland. *Surf. Sci.* **127**:271 (1983).

1156. E. Lang, K. Müller, K. Heinz, M.A. Van Hove, R.J. Koestner, and G.A. Somorjai. *Surf. Sci.* **127**:347 (1983).

1157. F.P. Netzer and T.E. Madey. *Surf. Sci.* **127**:L102 (1983).

1158. T. Yokotsuka, S. Kono, S. Suzuki, and T. Sagawa. *Surf. Sci.* **127**:35 (1983).

1159. H. Höchst and I. Hernández-Calderón. *Surf. Sci.* **126**:25 (1983).

1160. K.C. Prince and A.M. Bradshaw. *Surf. Sci.* **126**:49 (1983).

1161. U.O. Karlsson, G.V. Hansson, and S.A. Flodström. *Surf. Sci.* **126**:58 (1983).

1162. J. Massardier, B. Tardy, M. Abon, and J.C. Bertolini. *Surf. Sci.* **126**:154 (1983).

1163. C. Nyberg and C.G. Tengstal. *Surf. Sci.* **126**:163 (1983).

1164. A.M. Baró and L. Ollé. *Surf. Sci.* **126**:170 (1983).

1165. E.F.J. Didham, W. Allison, and R.F. Willis. *Surf. Sci.* **126**:219 (1983).

1166. D.P. Jackson, T.E. Jackman, J.A. Davies, W.N. Unertl, and P.R. Norton. *Surf. Sci.* **126**:226 (1983).

1167. C. Binns, C. Norris, G.C. Smith, H.A. Padmore, and M.G. Barthès-Labrousse. *Surf. Sci.* **126**:258 (1983).

1168. C. Benndorf, C. Nöbl, and F. Thieme. *Surf. Sci.* **126**:265 (1983).

1169. N. Vennemann, E.W. Schwarz, and M. Neumann. *Surf. Sci.* **126**:273 (1983).

1170. J.M. Moison and M. Bensoussan. *Surf. Sci.* **126**:294 (1983).

1171. G. Lindauer, P. Légaré, and G. Maire. *Surf. Sci.* **126**:301 (1983).

1172. M. Surfman, S.R. Bare, P. Hofmann, and D.A. King. *Surf. Sci.* **126**:349 (1983).

1173. M.G. Cattania, V. Penka, R.J. Behm, K. Christmann, and G. Ertl. *Surf. Sci.* **126**:382 (1983).

1174. M. Salmerón and G.A. Somorjai. *Surf. Sci.* **126**:410 (1983).

1175. K. Khonde, J. Darville, and J.M. Gilles. *Surf. Sci.* **126**:414 (1983).

1176. S. Tatarenko, R. Ducros, and M. Alnot. *Surf. Sci.* **126**:422 (1983).

1177. F. Labohm, C.W.R. Engelen, O.L.J. Gijzeman, J.W. Geus, and G.A. Bootsma. *Surf. Sci.* **126**:429 (1983).

1178. K. Bange, D. Grider, and J.K. Sass. *Surf. Sci.* **126**:437 (1983).

1179. U. Köhler and H.-W. Wassmuth. *Surf. Sci.* **126**:448 (1983).

1180. Z.T. Stott and H.P. Hughes. *Surf. Sci.* **126**:455 (1983).

1181. M. Mattern and H. Lüth. *Surf. Sci.* **126**:502 (1983).

1182. D. Bolmont, V. Mercier, P. Chen, H. Lhth, and C.A. Sébenne. *Surf. Sci.* **126**:509 (1983).

1183. E. De Frésart, J. Darville, and J.M. Gilles. *Surf. Sci.* **126**:518 (1983).

1184. H. Tochihara. *Surf. Sci.* **126**:523 (1983).

1185. N. Floquet and L.-C. Dufour. *Surf. Sci.* **126**:543 (1983).

1186. J.P. Landuyt, L. Vandenbroucke, R.D. Gryse, and J. Vennik. *Surf. Sci.* **126**:598 (1983).

1187. A.M. Lahee, W. Allison, R.F. Willis, and K.H. Rieder. *Surf. Sci.* **126**:654 (1983).

1188. M. Shiojiri, N. Nakamura, C. Kaito, and T. Miyano. *Surf. Sci.* **126**:719 (1983).

1189. M. Domke, B. Kyvelos, and G. Kaindl. *Surf. Sci.* **126**:727 (1983).

1190. R.D. Diehl and S.C. Fain, Jr. *Surf. Sci.* **125**:116 (1983).

1191. J. Suzanne, J.L. Seguin, H. Taub, and J.P. Biberian. *Surf. Sci.* **125**:153 (1983).

1192. K. Kjaer, M. Nielsen, J. Bohr, H.J. Lauter, and J.P. McTague. *Surf. Sci.* **125**:171 (1983).

1193. J. Bohr, M. Nielsen, J. Als-Nielsen, K. Kjaer, and J.P. McTague. *Surf. Sci.* **125**:181 (1983).

1194. P. Bak and T. Bohr. *Surf. Sci.* **125**:279 (1983).

1195. D.E. Taylor and R.L. Park. *Surf. Sci.* **125**:L73 (1983).

1196. N. Freyer, G. Pirug, and H.P. Bonzel. *Surf. Sci.* **125**:327 (1983).

1197. M. Welz, W. Moritz, and D. Wolf. *Surf. Sci.* **125**:473 (1983).

1198. R.J. Madix, J.L. Gland, G.E. Mitchell, and B.A. Sexton. *Surf. Sci.* **125**:481 (1983).

1199. K. Heinz and G. Besold. *Surf. Sci.* **125**:515 (1983).

1200. P.A. Heiney, P.W. Stephens, S.G.J. Mochrie, J. Akimitsu, R.J. Birgeneau, and P.M. Horn. *Surf. Sci.* **125**:539 (1983).

1201. G. Bracco, P. Cantini, A. Glachant, and R. Tatarek. *Surf. Sci.* **125**:L81 (1983).

1202. H.C. Peebles, D.E. Peebles, and J.M. White. *Surf. Sci.* **125**:L87 (1983).

1203. C. Egawa, S. Naito, and K. Tamaru. *Surf. Sci.* **125**:605 (1983).

1204. S. Nakanishi and T. Horiguchi. *Surf. Sci.* **125**:635 (1983).

1205. B.E. Hayden and A.M. Bradshaw. *Surf. Sci.* **125**:787 (1983).

1206. F. Houzay, G.M. Guichar, A. Cros, F. Salvan, P. Pinchaux, and J. Derrien. *Surf. Sci.* **124**:L1 (1983).

1207. R.M. Tromp, E.J. van Loenen, M. Iwami, R.G. Smeenk, F.W. Saris, F. Nava, and G. Ottaviani. *Surf. Sci.* **124**:1 (1983).

1208. F.P. Netzer and M.M. El Gomati. *Surf. Sci.* **124**:26 (1983).

1209. J.A. Gates and L.L. Kesmodel. *Surf. Sci.* **124**:68 (1983).

1210. E.G. McRae. *Surf. Sci.* **124**:106 (1983).

1211. D.T. Ling, J.N. Miller, D.L. Weissman, P. Pianetta, P.M. Stefan, I. Lindau, and W.E. Spicer. *Surf. Sci.* **124**:175 (1983).

1212. M.R. McClellan, F.R. McFeely, and J.L. Gland. *Surf. Sci.* **124**:188 (1983).

1213. B.J. Mrstik. *Surf. Sci.* **124**:253 (1983).

1214. S.P. Svensson, J. Kanski, T.G. Andersson, and P.O. Nilsson. *Surf. Sci.* **124**:L31 (1983).

1215. J. Derrien and F. Ringeisen. *Surf. Sci.* **124**:L35 (1983).

1216. G. Schultze and M. Henzler. *Surf. Sci.* **124**:336 (1983).

1217. N.J. Gudde and R.M. Lambert. *Surf. Sci.* **124**:372 (1983).

1218. D. Prigge, W. Schlenk, and E. Bauer. *Surf. Sci.* **123**:L698 (1982).

1219. A. Titov and W. Moritz. *Surf. Sci.* **123**:L709 (1982).

1220. H. Scheidt, M. Glöbl, and V. Dose. *Surf. Sci.* **123**:L728 (1982).

1221. H. Steininger, S. Lehwald, and H. Ibach. *Surf. Sci.* **123**:1 (1982).

1222. S. Maruno, H. Iwasaki, K. Horioka, S.-T. Li, and S. Nakamura. *Surf. Sci.* **123**:18 (1982).

1223. G. Le Lay, M. Manneville, and J.J. Métois. *Surf. Sci.* **123**:117 (1982).

1224. R. Imbihl, R.J. Behm, G. Ertl, and W. Moritz. *Surf. Sci.* **123**:129 (1982).

1225. M. Maglietta, E. Zanazzi, U. Bardi, D. Sondericker, F. Jona, and P.M. Marcus. *Surf. Sci.* **123**:141 (1982).

1226. B. Poelsema, R.L. Palmer, and G. Comsa. *Surf. Sci.* **123**:152 (1982).

1227. D. Dahlgren and J.C. Hemminger. *Surf. Sci.* **123**:L739 (1982).

1228. G. Popov and E. Bauer. *Surf. Sci.* **123**:165 (1982).

1229. A.G. Schrott and S.C. Fain, Jr. *Surf. Sci.* **123**:204 (1982).

1230. A.G. Schrott. Q.X. Su, and S.C. Fain, Jr. *Surf. Sci.* **123**:223 (1982).

1231. J.K. Lang, K.D. Jamison, F.B. Dunning, G.K. Walters, M.A. Passler, A. Ignatiev, E. Tamura, and R. Feder. *Surf. Sci.* **123**:247 (1982).

1232. H. Steininger, S. Lehwald, and H. Ibach. *Surf. Sci.* **123**:264 (1982).

1233. D.L. Doering and T.E. Madey. *Surf. Sci.* **123**:305 (1982).

1234. D.P. Woodruff, B.E. Hayden, K. Prince, and A.M. Bradshaw. *Surf. Sci.* **123**:397 (1982).

1235. G.R. Castro and J. Küppers. *Surf. Sci.* **123**:456 (1982).

1236. H.-D. Schmick and H.-W. Wassmuth. *Surf. Sci.* **123**:471 (1982).

1237. P.R. Norton, J.A. Davies, and T.E. Jackmon. *Surf. Sci.* **122**:L593 (1982).

1238. G. Pirug, H.P. Bonzel, and G. Brodén. *Surf. Sci.* **122**:1 (1982).

1239. W.H. Cheng and H.H. Kung. *Surf. Sci.* **122**:21 (1982).

1240. G. Landgren, S.P. Svensson, and T.G. Andersson. *Surf. Sci.* **122**:55 (1982).

1241. T.E. Felter, S.A. Steward, and F.S. Uribe. *Surf. Sci.* **122**:69 (1982).

1242. C.T. Campbell and T.N. Taylor. *Surf. Sci.* **122**:119 (1982).

1243. G.-C. Wang and T.-M. Lu. *Surf. Sci.* **122**:L635 (1982).

1244. F. Solymosi and A. Berko. *Surf. Sci.* **122**:275 (1982).

1245. G. Gewinner, J.C. Peruchetti, and A. Jaéglé. *Surf. Sci.* **122**:383 (1982).

1246. G. Popov and E. Bauer. *Surf. Sci.* **122**:433 (1982).

1247. W.T. Moore, S.J. White, D.C. Frost, and K.A.R. Mitchell. *Surf. Sci.* **116**:261 (1982).

1248. S. Astegger and E. Bechtold. *Surf. Sci.* **122**:491 (1982).

1249. G.A. Garwood, Jr., A.T. Hubbard, and J.B. Lumsden. *Surf. Sci.* **121**:L524 (1982).

1250. L.J. Clarke and L. Morales de la Garza. *Surf. Sci.* **121**:32 (1982).

1251. H. Conrad, G. Ertl, J. Küppers, W. Sesselmann, and H. Haberland. *Surf. Sci.* **121**:161 (1982).

1252. P.A. Thiel, R.J. Behm, P.R. Norton, and G. Ertl. *Surf. Sci.* **121**:L553 (1982).

1253. A. Fujimori, F. Minami, and N. Tsuda. *Surf. Sci.* **121**:199 (1982).

1254. P.W. Davies, M.A. Quinlan, and G.A. Somorjai. *Surf. Sci.* **121**:290 (1982).

1255. J.E. Crowell, E.L. Garfunkel, and G.A. Somorjai. *Surf. Sci.* **121**:303 (1982).

1256. R.J. Koestner, M.A. Van Hove, and G.A. Somorjai. *Surf. Sci.* **121**:321 (1982).

1257. R.H. Milne. *Surf. Sci.* **121**:347 (1982).

1258. J.Y. Katekaru, G.A. Garwood, Jr., J.F. Hershberger, and A.T. Hubbard. *Surf. Sci.* **121**:396 (1982).

1259. S.B. DiCenzo, G.K. Wertheim, and D.N.E. Buchanan. *Surf. Sci.* **121**:411 (1982).

1260. J.P. Jones. *Surf. Sci.* **121**:487 (1982).

1261. Y. Sakisaka, H. Kato, and M. Onchi. *Surf. Sci.* **120**:150 (1982).

1262. G. Casalone, M.G. Cattania, F. Merati, and M. Simonetta. *Surf. Sci.* **120**:171 (1982).

1263. S.E. Greco, J.P. Roux, and J.M. Blakely. *Surf. Sci.* **120**:203 (1982).

1264. T.W. Capehart, C.W. Seabury, G.W. Graham, and T.N. Rhodin. *Surf. Sci.* **120**:L441 (1982).

1265. K. Heinz, E. Lang, K. Strauss, and K. Müller. *Surf. Sci.* **120**:L401 (1982).

1266. J.A. Gates and L.L. Kesmodel. *Surf. Sci.* **120**:L461 (1982).

1267. A. Fasana and L. Braicovich. *Surf. Sci.* **120**:239 (1982).

1268. K. Christmann and J.E. Demuth. *Surf. Sci.* **120**:291 (1982).

1269. A.G. Naumovets and A.G. Fedorus. *Zh. Eksper. Teor. Fiz.* **68**:1183 (1975) [*Soviet Phys.-JETP* **41**:587 (1976)].

1270. A. Spitzer and H. Lüth. *Surf. Sci.* **120**:376 (1982).

1271. T.E. Jackman, J.A. Davies, D.P. Jackson, W.N. Unertl, and P.R. Norton. *Surf. Sci.* **120**:389 1982.

1272. G.C. Smith, H.A. Padmore, and C. Norris. *Surf. Sci.* **119**:L287 (1982).

1273. C.B. Bargeron and B.H. Nall. *Surf. Sci.* **119**:L319 (1982).

1274. P.J. Dobson, J.H. Neave, and B.A. Joyce. *Surf. Sci.* **119**:L339 (1982).

1275. T.N. Gardiner and E. Bauer. *Surf. Sci.* **119**:L353 (1982).

1276. A. Ortega, F.M. Hoffman, and A.M. Bradshaw. *Surf. Sci.* **119**:79 (1982).

1277. M. Grunze, C.R. Brundle, and D. Tomanek. *Surf. Sci.* **119**:133 (1982).

1278. G. Comsa, G. Mechtersheimer, and B. Poelsema. *Surf. Sci.* **119**:159 (1982).

1279. S. Ferrer and H.R. Bonzel. *Surf. Sci.* **119**:234 (1982).

1280. C. Park, E. Bauer, and H.M. Kramer. *Surf. Sci.* **119**:251 (1982).

1281. P.J. Orders, S. Kono, C.S. Fadley, R. Trehan, and J.T. Lloyd. *Surf. Sci.* **119**:371 (1982).

1282. F.P. Netzer and T.E. Madey. *Surf. Sci.* **119**:422 (1982).

1283. J.L. Seguin and J. Suzanne. *Surf. Sci.* **118**:L241 (1982).

1284. K.H. Rieder. *Surf. Sci.* **118**:57 (1982).

1285. A. Spitzer and H. Lüth. *Surf. Sci.* **118**:121 (1982).

1286. A. Spitzer and H. Lüth. *Surf. Sci.* **118**:136 (1982).

1287. G.A. Garwood, Jr. and A.T. Hubbard. *Surf. Sci.* **118**:223 (1982).

1288. C.T. Campbell and T.N. Taylor. *Surf. Sci.* **118**:401 (1982).

1289. S.A. Flodström and C.W.B. Martinsson. *Surf. Sci.* **118**:513 (1982).

1290. E. Roman and R. Riwan. *Surf. Sci.* **118**:682 (1982).

1291. J. Ibañez, N. Garcia, J.M. Rojo, and N. Cabrera. *Surf. Sci.* **117**:23 (1982).

1292. W. Englert, E. Taglauer, and W. Heiland. *Surf. Sci.* **117**:124 (1982).

1293. E. Lang, W. Grimm, and K. Heinz. *Surf. Sci.* **117**:169 (1982).

1294. W. Berndt, R. Hora, and M. Scheffler. *Surf. Sci.* **117**:188 (1982).

1295. W. Hoesler and W. Mortiz. *Surf. Sci.* **117**:196 (1982).

1296. Y.Y. Aristov, N.I. Golovko, V.A. Grazhulis, Y.A. Ossipyan, and V.I. Talyanskii. *Surf. Sci.* **117**:204 (1982).

1297. P. Hofmann, S.R. Bare, and D.A. King. *Surf. Sci.* **117**:245 (1982).

1298. R. Imbihl, R.J. Behm, K. Christmann, G. Ertl, and T. Matsushima. *Surf. Sci.* **117**:257 (1982).

1299. R. Miranda, F. Yndurain, D. Chandesris, J. Lecante, and Y. Petroff. *Surf. Sci.* **117**:319 (1982).

1300. R. Kötz, B.E. Hayden, E. Schweizer, and A.M. Bradshaw. *Surf. Sci.* **117**:331 (1982).

1301. S. Lehwald, H. Ibach, and H. Steininger. *Surf. Sci.* **117**:342 (1982).

1302. R. Matz and H. Lüth. *Surf. Sci.* **117**:362 (1982).

1303. M. Pessa, H. Asonen, R.S. Rao, R. Prasad, and A. Bansil. *Surf. Sci* **117**:371 (1982).

1304. K. Horn, C. Mariani, and L. Cramer. *Surf. Sci.* **117**:376 (1982).

1305. D. Bolmont, P. Chen, and C.A. Sébenne. *Surf. Sci.* **117**:417 (1982).

1306. S.Å. Lindgren, J. Paul, and L. Walldén. *Surf. Sci.* **117**:426 (1982).

1307. R. Brooks, N.V. Richardson, and D.A. King. *Surf. Sci.* **117**:434 (1982).

1308. T.E. Madey and F.P. Netzer. *Surf. Sci.* **117**:549 (1982).

1309. Ya-Po Hsu, K. Jacobi, and H.H. Rotermund. *Surf. Sci.* **117**:581 (1982).

1310. J.M. Heras, H. Papp, and W. Spiess. *Surf. Sci.* **117**:590 (1982).

1311. C.M.A.M. Mesters, A.F.H. Wielers, O.L.J. Gijzeman, G.A. Bootsma, and J.W. Geus. *Surf. Sci.* **117**:605 (1982).

1312. G. Castro, J.E. Hulse, J. Küppers, and A. Rodriguez Gonzalez-Elipe. *Surf. Sci.* **117**:621 (1982).

1313. H. Steininger, H. Ibach, and S. Lehwald. *Surf. Sci.* **117**:685 (1982).

1314. K.E. Foley and N. Winograd. *Surf. Sci.* **116**:1 (1982).

1315. V. Jensen, J.N. Andersen, H.B. Nielsen, and D.L. Adams. *Surf. Sci.* **116**:66 (1982).

1316. R.J. Koestner, J.C. Frost, P.C. Stair, M.A. Van Hove, and G.A. Somorjai. *Surf. Sci.* **116**:85 (1982).

1317. P. Kaplan. *Surf. Sci.* **116**:104 (1982).

1318. L.E. Firment. *Surf. Sci.* **116**:205 (1982).

1319. C. Binns and C. Norris. *Surf. Sci.* **116**:338 (1982).

1320. J. Carelli and A. Kahn. *Surf. Sci.* **116**:380 (1982).

1321. T.S. Wittrig, P.D. Szuromi, and W.H. Weinberg. *Surf. Sci.* **116**:414 (1982).

1322. A. Cros, F. Houzay, G.M. Guichar, and R. Pinchaux. *Surf. Sci.* **116**:L232 (1982).

1323. C. Park, H.M. Kramer, and E. Bauer. *Surf. Sci.* **116**:456 (1982).

1324. C. Park, H.M. Kramer, and E. Bauer. *Surf. Sci.* **116**:467 (1982).

1325. U. Gradmann and G. Waller. *Surf. Sci.* **116**:539 (1982).

1326. C. Park, H.M. Kramer, and E. Bauer. *Surf. Sci.* **115**:1 (1982).

1327. W.T. Tysoe and P.M. Lambert. *Surf. Sci.* **115**:37 (1982).

1328. R. Nishitani, C. Oshima, M. Aono, T. Tanaka, S. Kawai, H. Iwasaki, and S. Nakamura. *Surf. Sci.* **115**:48 (1982).

1329. S.W. Johnson and R.J. Madix. *Surf. Sci.* **115**:61 (1982).

1330. J.S. Foord and R.M. Lambert. *Surf. Sci.* **115**:141 (1982).

1331. A. Glachant, M. Jaubert, M. Bienfait, and G. Boato. *Surf. Sci.* **115**:219 (1982).

1332. H. Poppa and F. Soria. *Surf. Sci.* **115**:L105 (1982).

1333. S. Tougaard and A. Ignatiev. *Surf. Sci.* **115**:270 (1982).

1334. T.W. Orent and S.D. Bader. *Surf. Sci.* **115**:323 (1982).

1335. M.A. Barteau and R.J. Madix. *Surf. Sci.* **115**:355 (1982).

1336. C. Binns and C. Norris. *Surf. Sci.* **115**:395 (1982).

1337. E.L. Garfunkel and G.A. Somorjai. *Surf. Sci.* **115**:441 (1982).

1338. S. Calisti, J. Suzanne, and J.A. Venables. *Surf. Sci.* **115**:455 (1982).

1339. G.E. Gdowski and R.J. Madix. *Surf. Sci.* **115**:524 (1982).

1340. R.T. Tung, W.R. Graham, and A.J. Melmed. *Surf. Sci.* **115**:576 (1982).

1341. W. Erley. *Surf. Sci.* **114**:47 (1982).

1342. Y. Terada, T. Yoshizuka, K. Oura, and T. Hanawa. *Surf. Sci.* **114**:65 (1982).

1343. H. Kato, Y. Sakisaka, T. Miyano, K. Kamei, M. Nishijima, and M. Onchi. *Surf. Sci.* **114**:96 (1982).

1344. J. Massies and N.T. Linh. *Surf. Sci.* **114**:147 (1982).

1345. Å. Fäldt. *Surf. Sci.* **114**:311 (1982).

1346. T. Narusawa, W.M. Gibson, and E. Törnqvist. *Surf. Sci.* **114**:331 (1982).

1347. G.-C. Wang, J. Unguris, D.T. Pierce, and R.J. Celotta. *Surf. Sci.* **114**:L35 (1982).

1348. Y. Kim, H.C. Peebles, and J.M. White. *Surf. Sci.* **114**:363 (1982).

1349. D. Dahlgren and J.C. Hemminger. *Surf. Sci.* **114**:459 (1982).

1350. G. Ertl, S.B. Lee, and M. Weiss. *Surf. Sci.* **114**:527 (1982).

1351. D.F. Mitchell and M.J. Graham. *Surf. Sci.* **114**:546 (1982).

1352. M. Hanbücken and H. Neddermeyer. *Surf. Sci.* **114**:563 (1982).

1353. R. Ryberg. *Surf. Sci.* **114**:627 (1982).

1354. P.E. Viljoen, B.J. Wessels, G.L.P. Berning, and J.P. Roux. *J. Vacuum Sci. Technol.* **20**:204 (1982).

1355. L.C. Feldman, R.J. Culbertson, and P.J. Silverman. *J. Vacuum Sci. Technol.* **20**:368 (1982).

1356. M. Oku and C.R. Brundle. *J. Vacuum Sci. Technol.* **20**:532 (1982).

1357. E.D. Williams and W.H. Weinberg. *J. Vacuum Sci. Technol.* **20**:534 (1982).

1358. J.E. Black, T.S. Rahman, and D.L. Mills. *J. Vacuum Sci. Technol.* **20**:567 (1982).

1359. S.-W. Wang. *J. Vacuum Sci. Technol.* **20**:600 (1982).

1360. W.N. Unertl, T.E. Jackman, P.R. Norton, D.P. Jackson, and J.A. Davies. *J. Vacuum Sci. Technol.* **20**:607 (1982).

1361. D.E. Eastman, F.J. Himpsel, and J.F. van der Veen. *J. Vacuum Sci. Technol.* **20**:609 (1982).

1362. D. Heskett, F. Greuter, H.-J. Freund, and E.W. Plummer. *J. Vacuum Sci. Technol.* **20**:623 (1982).

1363. W.F. Egelhoff, Jr. *J. Vacuum Sci. Technol.* **20**:668 (1982).

1364. P.H. Holloway and R.A. Outlaw. *J. Vacuum Sci. Technol.* **20**:671 (1982).

1365. J.M. Van Hove and P.I. Cohen. *J. Vacuum Sci. Technol.* **20**:726 (1982).

1366. Y.J. Chabal, J.E. Rowe, and S.B. Christman. *J. Vacuum Sci. Technol.* **20**:763 (1982).

1367. A. Kahn, J. Carelli, C.B. Duke, A. Paton, and W.K. Ford. *J. Vacuum Sci. Technol.* **20**:775 (1982).

1368. K. Khonde, J. Darville, and J.M. Gilles. *J. Vacuum Sci. Technol.* **20**:834 (1982).

1369. R.J. Culbertson, L.C. Feldman, P.J. Silverman, and R. Haight. *J. Vacuum Sci. Technol.* **20**:868 (1982).

1370. K. Jacobi, G.W. Graham, and T.N. Rhodin. *J. Vacuum Sci. Technol.* **20**:878 (1982).

1371. D.A. Outka and R.J. Madix. *J. Vacuum Sci. Technol.* **20**:882 (1982).

1372. M.A. Van Hove, R.J. Koestner, and G.A. Somorjai. *J. Vacuum Sci. Technol.* **20**:886 (1982).

1373. N.J. Wu and A. Ignatiev. *J. Vacuum Sci. Technol.* **20**:896 (1982).

1374. D. Haneman and R.Z. Bachrach. *J. Vacuum Sci. Technol.* **21**:337 (1982).

1375. A. Kahn, J. Carelli, D.L. Miller, and S.P. Kowalczyk. *J. Vacuum Sci. Technol.* **21**:380 (1982).

1376. E. Zanazzi, M. Maglietta, U. Bardi, F. Jona, and P.M. Marcus. *J. Vacuum Sci. Technol. A* **1**:7 (1983).

1377. R. Kaplan. *J. Vacuum Sci. Technol. A* **1**:551 (1983).

1378. C.B. Duke, A. Paton, and A. Kahn. *J. Vacuum Sci. Technol. A* **1**:672 (1983).

1379. S.H. Overbury and P.C. Stair. *J. Vacuum Sci. Technol. A* **1**:1055 (1982).

1380. E.D. Williams and D.L. Doering. *J. Vacuum Sci. Technol. A* **1**:1188 (1983).

1381. G. Besold, K. Heinz, E. Lang, and K. Müller. *J. Vacuum Sci. Technol. A* **1**:1473 (1983).

1382. C.B. Duke, A. Paton, W.K. Ford, A. Kahn, and G. Scott. *J. Vacuum Sci. Technol.* **20**:778 (1982).

1383. P. Skeath, C.Y. Su, I. Lindau, and W.E. Spicer. *J. Vacuum Sci. Technol.* **20**:779 (1982).

1384. J.M. Moison and M. Bensoussan. *J. Vacuum Sci. Technol.* **21**:315 (1982).

1385. J.E. Rowe and P.H. Citrin. *J. Vacuum Sci. Technol.* **21**:338 (1982).

1386. B.J. Waclawski, D.T. Pierce, N. Swanson, and R.J. Celotta. *J. Vacuum Sci. Technol.* **21**:368 (1982).

1387. J.F. van der Veen, L. Smit, P.K. Larsen, J.H. Neave, and B.A. Joyce. *J. Vacuum Sci. Technol.* **21**:375 (1982).

1388. J.F. Wendelken and G.-C. Wang. *J. Vacuum Sci. Technol. A* **2**:888 (1984).

1389. G.-C. Wang and T.-M. Lu. *J. Vacuum Sci. Technol. A* **2**:1048 (1984).

1390. J.L. Stickney, S.D. Rosasco, B.C. Schardt, and A.T. Hubbard. *J. Phys. Chem.* **88**:251 (1984).

1391. A.T. Hubbard, J.L. Stickney, S.D. Rosasco, M.P. Soriaga, and D. Song. *J. Electroanal. Chem. Interfac. Electrochem.* **150**:165 (1983).

1392. T. De Jong, L. Smit, V.V. Korablev, R.M. Tromp, and F.W. Saris. *Appl. Surf. Sci.* **10**:10 (1982).

1393. T. Takahashi and A. Ebina. *Appl. Surf. Sci.* **11/12**:268 (1982).

1394. K. Heinz, E. Lang, K. Strauss, and K. Müller. *Appl. Surf. Sci.* **11/12**:611 (1982).

1395. K. Müller, E. Lang, H. Endriss, and K. Heinz. *Appl. Surf. Sci.* **11/12**:625 (1982).

1396. M.A. Stevens-Kalceff and C.J. Russel. *Appl. Surf. Sci.* **13**:94 (1982).

1397. J.A. Ramsey. *Appl. Surf. Sci.* **13**:159 (1982).

1398. R. Madix. *Appl. Surf. Sci.* **14**:41 (1982).

1399. A. Shih, G.A. Haas, and C.R.K. Marrian. *Appl. Surf. Sci.* **16**:93 (1982).

1400. G. Le Lay and J.J. Metois. *Appl. Surf. Sci.* **17**:131 (1982).

1401. W. Wen-Hao and J. Verhoeven. *Appl. Surf. Sci.* **17**:331 (1982).

1402. F.P. Netzer and T.E. Madey. *J. Chem. Phys.* **76**:710 (1982).

1403. L.H. Dubois. *J. Chem. Phys.* **77**:5228 (1982).

1404. J. Suzanne, J.L. Seguin, M. Bienfait, and E. Lerner. *Phys. Rev. Lett.* **52**:632 (1984).

1405. T. Aruga, H. Tochihara, and Y. Murata. *Phys. Rev. Lett.* **52**:1794 (1984).

1406. D.L. Doering and S. Semancik. *Phys. Rev. Lett.* **53**:66 (1984).

1407. R. Ryberg. *Phys. Rev. Lett.* **53**:945 (1984).

1408. S.Å. Lindgren, L. Walldén, J. Rundgren, and P. Westrin. *Phys. Rev. B* **29**:576 (1984).

1409. J.R. Noonan and H.L. Davis. *Phys. Rev. B* **29**:4349 (1984).

1410. J. Sokolov, F. Jona, and P.M. Marcus. *Phys. Rev. B* **29**:5402 (1984).

1411. M.F. Toney and S.C. Fain, Jr. *Phys. Rev. B.* **30**:1115 (1984).

1412. F.J. Himpsel, P.M. Marcus, R. Tromp, I.P. Batra, M.R. Cook, F. Jona, and H. Liu. *Phys. Rev. B* **30**:2257 (1984).

1413. Z.P. Hu and A. Ignatiev. *Phys. Rev. B* **30**:4856 (1984).

1414. C. Pirri, J.C. Peruchetti, G. Gewinner, and J. Derrien. *Phys. Rev. B* **30**:6227 (1984).

1415. G.-C. Wang and T.-M. Lu. *Phys. Rev. Lett.* **50**:2014 (1983).

1416. M.A. Van Hove, R. Lin, and G.A. Somorjai. *Phys. Rev. Lett.* **51**:778 (1983).

1417. H. Richter and U. Gerhardt. *Phys. Rev. Lett.* **51**:1570 (1983).

1418. P.K. Wu, J.H. Perepezko, J.T. McKinney, and M.G. Lagally. *Phys. Rev. Lett.* **51**:1577 (1983).

1419. A.H. Weiss, I.J. Rosenberg, K.F. Canter, C.B. Duke, and A. Paton. *Phys. Rev. B* **27**:867 (1983).

1420. C.B. Duke, A. Paton, and A. Kahn. *Phys. Rev. B* **27**:3436 (1983).

1421. I. Hernández-Calderón and H. Höchst. *Phys. Rev. B* **27**:4961 (1983).

1422. P.K. Larsen, J.H. Neave, J.F. van der Veen, P.J. Dobson, and B.A. Joyce. *Phys. Rev. B* **27**:4966 (1983).

1423. C.B. Duke, A. Paton, A. Kahn, and C.R. Bonapace. *Phys. Rev. B* **27**:6189 (1983).

1424. P. Skeath, C.Y. Su, W.A. Harrison, I. Lindau, and W.E. Spicer. *Phys. Rev. B* **27**:6246 (1983).

1425. M.F. Toney, R.D. Diehl, and S.C. Fain, Jr. *Phys. Rev. B* **27**:6413 (1983).

1426. C.B. Duke, A. Paton, A. Kahn, and C.R. Bonapace. *Phys. Rev. B* **28**:852 (1983).

1427. W.S. Yang and F. Jona. *Phys. Rev. B* **28**:1178 (1983).

1428. W.S. Yang, F. Jona, and P.M. Marcus. *Phys. Rev. B* **28**:2049 (1983).

1429. H. Liu, M.R. Cook, F. Jona, and P.M. Marcus. *Phys. Rev. B* **28**:6137 (1983).

1430. S.Å. Lindgren, L. Walldén, J. Rundgren, P. Westrin, and J. Neve. *Phys. Rev. B* **28**:6707 (1983).

1431. G.-C. Wang and T.-M. Lu. *Phys. Rev. B* **28**:6795 (1983).

1432. P. Skeath, I. Lindau, C.Y. Su, and W.E. Spicer. *Phys. Rev. B* **28**:7051 (1983).

1433. N.J. Wu and A. Ignatiev. *Phys. Rev. B* **28**:7288 (1983).

1434. W.S. Yang, F. Jona, and P.M. Marcus. *Phys. Rev. B* **28**:7377 (1983).

1435. R.D. Diehl, M.F. Toney, and S.C. Fain, Jr. *Phys. Rev. Lett.* **48**:177 (1982).

1436. D.L. Adams, H.B. Nielsen, J.N. Andersen, I. Stensgaard, R. Feidenhans'l, and J.E. Sorensen. *Phys. Rev. Lett.* **49**:669 (1982).

1437. S. Masuda, M. Nishijima, Y. Sakisaka, and M. Onchi. *Phys. Rev. B* **25**:863 (1982).

1438. J.M. Baribeau and J.D. Carette. *Phys. Rev. B* **25**:2962 (1982).

1439. N.J. Wu and A. Ignatiev. *Phys. Rev. B* **25**:2983 (1982).

1440. G.K. Wertheim, S.B. DiCenzo, and D.N.E. Buchanan. *Phys. Rev. B* **25**:3020 (1982).

1441. S.Y. Tong and K.H. Lau. *Phys. Rev. B* **25**:7382 (1982).

1442. C.B. Duke, A. Paton, W.K. Ford, A. Kahn, and J. Carelli. *Phys. Rev. B* **26**:803 (1982).

1443. R.D. Diehl and S.C. Fain, Jr. *Phys. Rev. B* **26**:4785 (1982).

1444. D.-W. Tu and A. Kahn. *J. Vacuum Sci. Technol. A* **2**:511 (1984).

1445. C.B. Duke, A. Paton, and A. Kahn. *J. Vacuum Sci. Technol. A* **2**:515 (1984).

1446. J.M. Moison and M. Bensoussan. *Appl. Surf. Sci.* **20**:84 (1984).

1447. Y.B. Lozovyi, V.K. Medvedev, T.P. Smereka, B.M. Palyukh, and G.V. Babkin. *Fiz. Tverdogo Tela.* **24**:2130 (1982).

1448. L.G. Salmon and T.N. Rhodin. *Tenth International Symposium on Gallium Arsenide and Related Compounds*, 1983, p. 566.

1449. V.F. Dvoryankin, A.A. Komarov, and V.V. Panteleev. *Izv. Akad. Nauk SSSR Neorg. Mater.* **19**:186 (1983).

1450. R. Kaplan. *J. Appl. Phys.* **56**:1636 (1984).

1451. B.S. Meyerson and M.L. Yu. *J. Electrochem. Soc.* **131**:2366 (1984).

1452. B.E. Hayden and A.M. Bradshaw. *J. Electron Spectrosc. Relat. Phenom.* **30**:51 (1983).

1453. B.E. Koel and G.A. Somorjai. *J. Electron Spectrosc. and Relat. Phenom.* **29**:287 (1983).

1454. C. Nyberg and C.G. Tengstål. *J. Electron Spectrosc. and Relat. Phenom.* **29**:191 (1983).

1455. A.B. Anton, N.R. Avery, B.H. Toby and W.H. Weinberg. *J. Electron Spectrosc. and Relat. Phenom.* **29**:181 (1983).

1456. A. Oustry, J. Berty, M. Caumont, and M.J. David. *J. Microsc. Spectrosc. Electron.* **9**:49 (1984).

1457. N.P. Lieske. *J. Phys. Chem. Solids* **45**:821 (1984).

1458. M. Abu-Joudeh, P.P. Vaishnava, and P.A. Montano. *J. Phys. C* **17**:6899 (1984).

1459. Y. Gauthier, R. Baudoing, Y. Joly, C. Gaubert, and J. Rundgren. *J. Phys. C* **17**:4547 (1984).

1460. J. Sokolov, H.D. Shih, U. Bardi, F. Jona, and P.M. Marcus. *J. Phys. C* **17**:371 (1984).

1461. F. Jona, D. Westphal, A. Goldmann, and P.M. Marcus. *J. Phys. C* **16**:3001 (1983).

1462. J. Neve, P. Westrin, and J. Rundgren. *J. Phys. C* **16**:1291 (1983).

1463. G.C. Smith, C. Nooris, C. Binns, and H.A. Padmore. *J. Phys. C* **15**:6481 (1982).

1464. H.B. Nielsen, J.N. Andersen, L. Petersen, and D.L. Adams. *J. Phys. C* **15**:L1113 (1982).

1465. K. Griffiths, D.A. King, G.C. Aers, and J.B. Pendry. *J. Phys. C* **15**:4921 (1982).

1466. S.P. Tear and K. Roll. *J. Phys. C* **15**:5521 (1982).

1467. J. Neve, J. Rundgren, and P. Westrin. *J. Phys. C* **15**:4391 (1982).

1468. L.J. Clarke, R. Baudoing, and Y. Gauthier. *J. Phys. C* **15**:3249 (1982).

1469. Y. Gauthier, R. Baudoing, and L. Clarke. *J. Phys. C* **15**:3231 (1982).

1470. Y. Gauthier, R. Baudoing, C. Gaubert, and L. Clarke. *J. Phys. C* **15**:3223 (1982).

1471. M.K. Debe and D.A. King. *J. Phys. C* **15**:2257 (1982).

1472. H.B. Nielsen and D.L. Adams. *J. Phys. C* **15**:615 (1982).

1473. W.T. Moore, D.C. Frost, and K.A.R. Mitchell. *J. Phys. C* **15**:L5 (1982).

1474. G. Le Lay and J.J. Metois. *J. Phys. Colloq.* **45**(C5):427–433 (1984).

1475. K. Shoji, H. Ueba, and C. Tatsuyama. *J. Vacuum Soc. Jpn.* **26**:778 (1983).

1476. H.-J. Gossmann, L.C. Feldman, and W.M. Gibson. *J. Vacuum Sci. Technol. B* **2**:407 (1984).

1477. J.A. Schaefer, F. Stucki, D.J. Frankel, W. Gopel, and G.J. Lapeyre. *J. Vacuum Sci. Technol. B* **2**:359 (1984).

1478. C.B. Duke, A. Paton, A. Kahn, and D.W. Tu. *J. Vacuum Sci. Technol. B* **2**:366 (1984).

1479. C. Maillot, H. Roulet, and G. Dufour. *J. Vacuum Sci Technol. B* **2**:316 (1984).

1480. C.B. Duke and A. Paton. *J. Vacuum Sci. Technol. B* **2**:327 (1984).

1481. Y. Fujinaga. *J. Vacuum Soc. Jpn.* **25**:468 (1982).

1482. J. Tang. *J. Zhejiang Univ.* **18**:81 (1984).

1483. K. Shoji, M. Hyodo, H. Ueba, and C. Tatsuyama. *Jpn. J. Appl. Phys. Part 1* **22**:1482 (1983).

1484. Y. Yabuuchi, F. Shoji, K. Oura, T. Hanawa, Y. Kishikawa, and S. Okada. *Jpn. J. Appl. Phys. Part 1* **21**:L752 (1982).

1485. H. Nakamatsu, Y. Yamamoto, S. Kawai, K. Oura, and T. Hanawa. *Jpn. J. Appl. Phys. Part 2* **22**:L461 (1983).

1486. K. Shoji, M. Hyodo, H. Ueba, and C. Tatsuyama. *Jpn. J. Appl. Phys. Part 2* **22**:L200 (1983).

1487. Y. Yabuuchi, F. Shoji, K. Oura, T. Hanawa, Y. Kishikawa, and S. Okada. *Jpn. J. Appl. Phys. Part 2* **21**:L752 (1982).

1488. S. Maruno, H. Iwasaki, K. Horioka, Sung-te Li, and S. Nakamura. *Jpn. J. Appl. Phys. Part 2* **21**:L263 (1982).

1489. K. Horioka, H. Iwasaki, A. Ichimiya, S. Maruno, S. Te Li, and S. Nakamura. *Jpn. J. Appl. Phys. Part 2* **21**:L189 (1982).

1490. Y.Y. Tomashpol'skii, E.N. Lubnin, M.A. Sevost'yanov, and V.I. Kukuev. *Kristallografiya* **27**:1152 (1982).

1491. V.F. Dvoryankin, A.Y. Mityagin, and V.V. Panteleev. *Kristallografiya* **27**:349 (1982).

1492. D.M. Zehner and C.W. White. In J.M. Poate and James W. Mayer, editors, *Laser Annealing of Semiconductors*, Academic Press, New York, 1982, p. 281.

1493. R. Ramanathan and J.M. Blakeley. *Mater. Lett.* **2**:12 (1983).

1494. H. Iwasaki, S. Maruno, K. Horioka, S.-T. Li, and S. Nakamura. In: *Molecular Beam Epitaxy and Clean Surface Techniques. Collected Papers of 2nd International Symposium*, 1982, p. 305.

1495. H. Sato, A. Ebina, and T. Takahashi. In: *Molecular Beam Epitaxy and Clean Surface Techniques. Collected Papers of 2nd International Symposium*, 1982, p. 309.

1496. K. Oura, Y. Yabuuchi, F. Shoji, T. Hanawa, and S. Okada. In: *Proceedings of the 6th International Conference on Ion Beam Analysis*, 1983, p. 23.

1497. D.P. Woodruff and K. Horn. *Philos. Mag. A* **47**:L5 (1983).

1498. D.L. Adams, H.B. Nielsen, and J.N. Andersen. *Phys. Scr.* **T4**:22 (1983).

1499. F. Houzay, G.M. Guichar, A. Cros, F. Salvan, R. Pinchaux, and J. Derrien. *Surf. Sci.* **124**:L1 (1984).

1500. V.G. Lifshits, V.G. Zavodinskii, and N.I. Plyusnin. *Phys. Chem. Mech. Surf.* **2**:784 (1984).

1501. P.P. Lutsishin and T.N. Nakhodkin. *Phys. Chem. Mech. Surf.* **1**:3596 (1984).

1502. V.A. Grazhulis, A.M. Ionov, and V.F. Kuleshov. *Phys. Chem. Mech. Surf.* **2**:540 (1984).

1503. T. Matsubara. *Prog. Theor. Phys.* **71**:399 (1984).

1504. H. Sato, A. Ebina, and T. Takahashi. *Rec. Electr. Commun. Eng., Conversazione Tohoku Univ.* **51**:9 (1982).

1505. J.C. Dupuy, B. Vilotitch, and A. Sibai. *Rev. Phys. Appl.* **19**:965 (1984).

1506. G.L.P. Berning and W.J. Coleman. *S. Afr. J. Phys.* **7**:79 (1984).

1507. P. Chen, D. Bolmont, and C.A. Sébenne. *Thin Solid Films* **111**:367 (1984).

1508. P. Godowski and S. Mroz. *Thin Solid Films* **111**:129 (1984).

1509. V.D. Vankar, R.W. Vook, and B.C. De Cooman. *Thin Solid Films* **102**:313 (1983).

1510. V.E. de Carvalho, M.W. Cook, P.G. Cowell, O.S. Heavens, M. Prutton, and S.P. Tear. *Vacuum* **34**:893 (1984).

1511. B.J. Hinch, M.S. Foster, G. Jennings, and R.F. Willis. *Vacuum* **33**:864 (1983).

1512. R.D. Diehl, S.C. Fain, Jr., J. Talbot, D.J. Tildesley, and W.A. Steele. *Vacuum* **33**:857 (1983).

1513. D.P. Woodruff and K. Horn. *Vacuum* **33**:633 (1983).

1514. G.J.R. Jones and B.W. Holland. *Vacuum* **33**:627 (1983).

1515. S. Tatarenko and R. Ducros. *Vide Les Couches Minces* **38**:121 (1983).

1516. V.V. Gonchar, Y.M. Kagan, O.V. Kanash, A.G. Naumovets, and A.G. Fedorus. *Zh. Eksp. Teor. Fiz.* **84**:249 (1983).

1517. T. De Yong, W.A.S. Douma, L. Smit, V.V. Korablev, and F.W. Saris. *J. Vacuum Sci. Technol. B* **1**:888 (1983).

1518. J.M. Woodall, P. Oelhafen, T.N. Jackson, J.L. Freeouf, and G.D. Pettit. *J. Vacuum Sci. Technol. B* **1**:795 (1983).

1519. P. Oelhafen, J.L. Freeouf, G.D. Pettit, and J.M. Woodall. *J. Vacuum Sci. Technol. B* **1**:787 (1983).

1520. P. Zurcher, J. Anderson, D. Frankel, and G.J. Lapeyre. *J. Vacuum Sci. Technol. B* **1**:682 (1983).

1521. A. Kahn, C.R. Bonapace, C.B. Duke, and A. Paton. *J. Vacuum Sci. Technol. B* **1**:613 (1983).

1522. J.R. Lince, J.G. Nelson, and R.S. Williams. *J. Vacuum Sci. Technol. B* **1**:553 (1983).

1523. W.S. Yang, F. Jona, and P.M. Marcus. *J. Vacuum Sci. Technol. B* **1**:718 (1983).

1524. R.S. Bauer. *J. Vacuum Sci. Technol. B* **1**:314 (1983).

1525. J. Kofoed, I. Chorkendorff, and J. Onsgaard. *Solid State Commun.* **52**:283 (1984).

1526. B.T. Jonker and R.L. Park. *Solid State Commun.* **51**:871 (1984).

1527. K. Christemann, V. Penka, R.J. Behm, F. Chehab, and G. Ertl. *Solid State Commun.* **51**:487 (1984).

1528. R. Feder and W. Monch. *Solid State Commun.* **50**:311 (1984).

1529. J. Derrien and F. Ringeisen. *Solid State Commun.* **50**:627 (1984).

1530. J. Sokolov, F. Jona, and P.M. Marcus. *Solid State Commun.* **49**:307 (1984).

1531. J. Sokolov, H.D. Shih, U. Bardi, F. Jona, and P.M. Marcus. *Solid State Commun.* **48**:739 (1983).

1532. A. Faldt and H.P. Myers. *Solid State Commun.* **48**:253 (1983).

1533. W.S. Yang and F. Jona. *Solid State Commun.* **48**:377 (1983).

1534. B.E. Hayden, K.C. Prince, P.J. Davie, G. Paolucci, and A.M. Bradshaw. *Solid State Commun.* **48**:325 (1983).

1535. F. Stucki, J.A. Schaefer, J.R. Anderson, G.P. Lapeyre, and W. Gopel. *Solid State Commun.* **47**:795 (1983).

1536. K. Horioka, H. Iwasaki, S. Maruno, S. Te Li, and S. Nakamura. *Solid State Commun.* **47**:55 (1983).

1537. P. Chen, D. Bolmond, and C.A. Sébenne. *Solid State Commun.* **46**:689 (1983).

1538. M. Pessa and O. Jylha. *Solid State Commun.* **46**:419 (1983).

1539. M. Maglietta, A. Fallavollita, and G. Rovida. *Solid State Commun.* **46**:273 (1983).

1540. N.J. Wu and A. Ignatiev. *Solid State Commun.* **46**:59 (1983).

1541. R.D. Bringans, and B.Z. Bachrach. *Solid Sate Commun.* **45**:83 (1983).

1542. R. Feder. *Solid State Commun.* **45**:51 (1983).

1543. H. Kobayashi, K. Edamoto, M. Onchi, and M. Nishijima. *Solid State Commun.* **44**:1449 (1982).

1544. P. Chen, D. Bolmont, and C.A. Sébenne. *Solid State Commun.* **44**:1191 (1982).

1545. D. Westphal, A. Goldmann, F. Jona, and P.M. Marcus. *Solid State Commun.* **44**:685 (1982).

1546. C. Binns, C. Norris, I. Lindau, M.L. Shek, B. Pate, P.M. Stefan, and W.E. Spicer. *Solid State Commun.* **43**:853 (1982).

1547. W.S. Yang, F. Jona, and P.M. Marcus. *Solid State Commun.* **43**:847 (1982).

1548. M. Maglietta. *Solid State Commun.* **43**:395 (1982).

1549. S.J. White, D.C. Frost, and K.A.R. Mitchell. *Solid State Commun.* **42**:763 (1982).

1550. W.S. Yang and F. Jona. *Solid State Commun.* **42**:49 (1982).

1551. W.S. Yang, J. Sokolov, F. Jona, and P.M. Marcus. *Solid State Commun.* **41**:191 (1982).

1552. T. Weir and G.W. Simmons. *AIP Conf. Proc.* **84**:113 (1982).

1553. M. Maglietta. *Appl. Phys. A* **31**:165 (1983).

1554. T.D. Jong, W.A.S. Douma, J.F. Van der Veen, F.W. Saris, and J. Haisma. *Appl. Phys. Lett.* **42**:1037 (1983).

1555. K. Oura, S. Okada, Y. Kishikawa, and T. Hanawa. *Appl. Phys. Lett.* **40**:138 (1982).

1556. J.L. Stickney, S.D. Rosasco, and A.T. Hubbard. *J. Electrochem. Soc.* **131**:260 (1984).

1557. J.K. Sass, K. Bange, R. Dohl, E. Piltz, and R. Unwin. *Ber Bunsenges Phys. Chem.* **88**:354 (1984).

1558. Y. Nakai, M.S. Zei, D.M. Kolb, and G. Lehmpfuhl. *Ber. Bunsenges Phys. Chem.* **88**:340 (1984).

1559. M. Ohno, Y. Nakanishi, and G. Shimaoka. *Bull. Res. Inst. Electron Shizuoka Univ.* **19**:19 (1984).

1560. C. Klauber, M.D. Alvey, and J.T. Yates. *Chem. Phys. Lett.* **106**:477 (1984).

1561. R. Ramanathan, M. Quinlan, and H. Wise. *Chem. Phys. Lett.* **106**:87 (1984).

1562. G.C. Smith, C. Norris, and C. Binns. *J. Phys. C* **17**:4389 (1984).

1563. C. Ping, D. Bolmont, and C.A. Sébenne. *J. Phys. C* **17**:4897 (1984).

1564. C. Binns, M.G. Barthes-Labrousse, and C. Norris. *J. Phys. C* **17**:1465 (1984).

1565. M. Maglietta. *J. Phys. C* **17**:363 (1984).

1566. J.N. Andersen, H.B. Nielsen, L. Petersen, and D.L. Adams. *J. Phys. C* **17**:173 (1984).

1567. F. Proix, A. Akremi, and Z.T. Zhong. *J. Phys. C* **16**:5449 (1983).

1568. G.J. Hughes, A. Mckinley, and R.H. Williams. *J. Phys. C* **16**:2391 (1983).

1569. T.C. Gainey and B.J. Hopkins. *J. Phys. C* **16**:975 (1983).

1570. A. Mckinley, G.J. Hughes, and R.H. Williams. *J. Phys. C* **15**:7049 (1982).

1571. S.A. Lindgren, J. Paul, L. Wallden, and P. Westrin. *J. Phys. C* **15**:6285 (1982).

1572. P. Chen, D. Bolmont, and C.A. Sébenne. *J. Phys. C* **15**:6101 (1982).

1573. V. Montgomery and R.H. Williams. *J. Phys. C* **15**:5887 (1982).

1574. M.R. Welton-Cook and W. Berndt. *J. Phys. C* **15**:5691 (1982).

1575. D. Bolmont, P. Chen, F. Proix, and C.A. Sébenne. *J. Phys. C* **15**:3639 (1982).

1576. N. Masud. *J. Phys. C* **15**:3209 (1982).

1577. F. Storbeck. *Acta. Phys. Acad. Sci. Hung.* **49**:75 (1980).

1578. L.R. Sheng and L.-X. Tu. *Acta. Phys. Sin. (China)* **29**:524 (1980).

1579. D.J. Chadi. *Appl. Opt.* **19**:3971 (1980).

1580. M. Maglietta, E. Zanazzi, F. Jona, D.W. Jepsen, and P.M. Marcus. *Appl. Phys.* **15**:409 (1978).

1581. M. Maglietta. *Appl. Phys.* **A31**:165 (1983).

1582. Y. Nakai, M.S. Zei, D.M. Kolb, and G. Lehmpfuhl. *Ber Bunsenges PHys. Chem.* **88**:340 (1984).

1583. A. Steinbrum, P. Dumas, and J.C. Colson. *C.R. Acad. Sci. Ser. C* **290**:329 (1980).

1584. K.A. Prior, K. Schwaha, M.E. Bridge, and R.M. Lambert. *Chem. Phys. Lett.* **65**:472 (1979).

1585. G. Casalone, M.G. Cattania, M. Simonetta, and M. Tescari. *Chem. Phys. Lett.* **61**:36 (1979).

1586. I.L. Kesmodel, L.H. Dubois, and G.A. Somorjai. *Chem. Phys. Lett.* **56**:267 (1978).

1587. M.G. Lagally, Wang Gwo-Ching, and Lu Toh-Ming. *CRC Crit. Rev. Solid State and Mater. Sci.* **7**:233 (1979).

1588. H.J. Mussig, and W. Arabczyk. *Cryst. Res. Technol.* **16**:827 (1981).

1589. M. Klaua, K. Meinel, O.P. Pchelyakov, V.A. Ivanchenko, and S.I. Stenin. *Sov. Phys. Solid State* **23**:1501 (1981).

1590. M.S. Gupalo, V.K. Medvedev, B.M. Palyukh, and T.P. Smereka. *Sov. Phys. Solid State* **23**:1211 (1981).

1591. V.K. Medvedev and I.N. Yakovkin. *Sov. Phys. Solid State* **23**:379 (1981).

1592. S.A. Knyazev and G.K. Zyryanov. *Sov. Phys. Solid State* **22**:1554 (1980).

1593. M.S. Gupalo, V.K. Medvedev, B.M. Palyukh, and T.P. Smereka. *Sov. Phys. Solid State* **21**:568 (1979).

1594. V.K. Medvedev and I.N. Yakovkin. *Sov. Phys. Solid State* **21**:187 (1979).

1595. H.F. Winters. *IBM J. Res. Dev.* **22**:260 (1978).

1596. C.R. Brundle. *IBM J. Res. Dev.* **22**:235 (1978).

1597. A. Ignatiev. *IEEE Trans. Nuc. Sci.* **NS-26**:1824 (1979).

1598. S.A. Isa. *Iraqi J. Sci.* **20**:225 (1979).

1599. N. Stoner, M.A. Van Hove, S.Y. Tong, and M.B. Webb. *Phys. Rev. Lett.* **40**:243 (1978).

1600. M.J. Cardillo and G.E. Becker. *Phys. Rev. Lett.* **40**:1148 (1978).

1601. F. Jona, K.O. Legg, H.D. Shih, D.W. Jepsen, and P.M. Marcus. *Phys. Rev. Lett.* **40**:1466 (1978).

1602. J. Suzanne, J.P. Coulomb, M. Bienfait, M. Matecki, A. Thomy, B. Croset, and C. Marti. *Phys. Rev. Lett.* **41**:760 (1978).

1603. R.A. Barker and P.J. Estrup. *Phys. Rev. Lett.* **41**:1307 (1978).

1604. M. Passler, A. Ignatiev, F. Jona, D.W. Jepsen, and P.M. Marcus. *Phys. Rev. Lett.* **43**:360 (1979).

1605. S. Andersson and J.B. Pendry. *Phys. Rev. Lett.* **43**:363 (1979).

1606. G. Gewinner, J.C. Peruchetti, A. Jaegle, and R. Riedinger. *Phys. Rev. Lett.* **43**:935 (1979).

1607. R.J. Meyer, L.J. Brillson, A. Kahn, D. Kanani, J. Carelli, J.L. Yeh, G. Margaritondo, and A.D. Katnani. *Phys. Rev. Lett.* **46**:440 (1979).

1608. H.D. Shih, F. Jona, D.W. Jepsen, and P.M. Marcus. *Phys. Rev. Lett.* **46**:731 (1979).

1609. K. Griffiths, C. Kendon, D.A. King, and J.B. Pendry. *Phys. Rev. Lett.* **46**:1584 (1981).

1610. E.G. McRae and C.W. Caldwell. *Phys. Rev. Lett.* **46**:1632 (1981).

1611. M. Jaubert, A. Glachant, M. Bienfait, and G. Boato. *Phys. Rev. Lett.* **46**:1679 (1981).

1612. C.B. Duke, R.J. Meyer, A. Paton, and P. Marck. *Phy. Rev. B* **18**:4225 (1978).

1613. B.W. Lee, R. Alsenz, A. Ignatiev, and M.A. Van Hove. *Phys. Rev. B* **17**:1510 (1978).

1614. D.L. Adams, H.B. Nielsen, and M.A. Van Hove. *Phys. Rev. B* **20**:4789 (1979).

1615. R.H. Tait and R.V. Kasowski. *Phys. Rev. B* **20**:5178 (1979).

1616. S.C. Fain, Jr., M.D. Chinn, and R.D. Diehl. *Phys. Rev. B* **21**:4170 (1980).

1617. D.W. Jepsen, H.D. Shih, F. Jona, and P.M. Marcus. *Phys. Rev. B* **22**:814 (1980).

1618. R.J. Meyer, C.B. Duke, A. Paton, J.C. Tsant, J.L. Yeh, A. Kahn, and P. Mark. *Phys. Rev. B* **22**:6171 (1980).

1619. C.B. Duke, A. Paton, W.K. Ford, A. Kahn, and J. Carelli. *Phys. Rev. B* **24**:562 (1981).

1620. C.B. Duke, A. Paton, W.K. Ford, A. Kahn, and G. Scott. *Phys. Rev. B* **24**:3310 (1981).

1621. F. Soria, V. Martinez, M.C. Munoz, and J.L. Sacedon. *Phys. Rev. B* **24**:6926 (1981).

1622. A.K. Green and E. Bauer. *J. Appl. Phys.* **52**:5098 (1981).

1623. P. Hahn, J. Clabes, and M. Henzler. *J. Appl. Phys.* **51**:2079 (1980).

1624. C. Oshima, M. Aono, T. Tanaka, R. Nishitani, and S. Kawai. *J. Appl. Phys.* **51**:997 (1980).

1625. M. Aono, C. Oshima, T. Tanoka, E. Bannai, and S. Kawai. *J. Appl. Phys.* **49**:2761 (1978).

1626. S.A. Isa, R.W. Joyner, and M.W. Roberts. *J. Chem. Soc. Faraday Trans. I* **74**:546 (1978).

1627. J.C. Bertolini, J. Massardier, and G. Dalmai-Imelik. *J. Chem. Soc. Faraday Trans. I* **74**:1720 (1978).

1628. G. Le Lay. *J. Cryst. Growth.* **54**:501 (1981).

1629. M. Grunze, W. Hirschwald, and D. Hofmann. *J. Cryst. Growth* **52**:241 (1981).

1630. K. Nii and K. Yoshihara. *J. Jpn. Inst. Met.* **44**:100 (1980).

1631. C. Oshima, M. Oano, S. Zaima, Y. Shibata, and S. Kawai. *J. Less-Common Met.* **82**:69 (1981).

1632. A. Landet, M. Jardinier-Offergeld, and F. Bouillon. *J. Microsc. and Spectrosc. Electron* **30**:101 (1978).

1633. C. Benndorf, B. Egert, G. Keller, H. Seidel, and F. Thieme. *J. Phys. Chem. Solids* **40**:877 (1979).

1634. L. Morales, D.O. Garza, and L.J. Clarke. *J. Phys. C.* **14**:5391 (1981).

1635. S.P. Tear, K. Roll, and M. Prutton. *J. Phys. C* **14**:3297 (1981).

1636. J.C. Fernandez, W.S. Yang, H.D. Shih, F. Jona, P.W. Jepsen, and P.M. Marcus. *J. Phys. C* **14**:L55 (1981).

1637. F. Jona and P.M. Marcus. *J. Phys. C* **13**:L477 (1980).

1638. M.R. Welton-Cook and M. Prutton. *J. Phys. C* **13**:3993 (1980).

1639. H.D. Shih, F. Jona, U. Bardi, and P.M. Marcus. *J. Phys. C* **13**:3801 (1980).

1640. F. Jona, D. Sondericker, and P.M. Marcus. *J. Phys. C* **13**:L155 (1980).

1641. M. Prutton, J.A. Ramsey, J.A. Walker, and M.R. Welton-Cook. *J. Phys. C* **12**:5271 (1979).

1642. W.T. Moore, P.R. Watson, D.C. Frost, and K.A.R. Mitchell. *J. Phys. C* **12**:L887 (1979).

1643. K. Griffiths and D.A. King. *J. Phys. C* **12**:L755 (1979).

1644. T. Matsudaira and M. Onchi. *J. Phys. C* **12**:3381 (1979).

1645. F. Jona, H.D. Shih, D.W. Jepsen, and P.M. Marcus. *J. Phys. C* **12**:L455 (1979).

1646. R. Feder, W. Monch, and P.P. Auer. *J. Phys. C* **12**:L179 (1979).

1647. A. Ignatiev, H.B. Nielsen, and D.L. Adams. *J. Phys. C.* **11**:L837 (1978).

1648. F.R. Shepherd, P.R. Watson, D.C. Frost, and K.A.R. Mitchell. *J. Phys. C* **11**:4591 (1978).

1649. Y. Fujinaga. *J. Vacuum Soc. Jpn* **23**:253 (1980).

1650. Y. Terada, T. Yoshizuka, K. Oura, and T. Hanawa. *Jpn. J. Appl. Phys.* **20**:L333 (1981).

1651. M.G. Lagally, T.M. Lu, and G.C. Wang. In: *Ordering in Two Dimensions. Proceedings of an International Conference on Ordering in Two Dimensions*, 1981, p. 113.

1652. R.L. Park, T.L. Einstein, A.R. Kortan, and L.D. Roelofs. *In: Ordering in Two Dimensions. Proceedings of an International Conference on Ordering in Two Dimensions.* 1981, p. 17.

1653. Y.J. Chabal and J.E. Rowe. In: *Ordering in Two Dimensions: Proceedings of an International Conference on Ordering in Two Dimensions*, 1981, p. 256.

1654. R. Ducros, M. Housley, and G. Piquard. *Phys. Status Solidi A* **56**:187 (1979).

1655. W. Arabczyk, H.J. Mussig, and F. Storbeck. *Phys. Status Solidi A* **55**:437 (1979).

1656. R. Feder and J. Kirschner. *Phys. Status Solidi A* **45**:K117 (1978).

1657. V.K. Medvedev and V.N. Pogoreli. *UKR Fiz. Zh.* **25**:1524 (1980).

1658. K. Chritmann. *Z. Naturforsch A* **34A**:22 (1979).

1659. K.C.R. Chiu, J.M. Paate, J.E. Rowe, T.T. Sheng, and A.G. Cullis. *Appl. Phys. Lett.* **38**:988 (1981).

1660. V. Montgomery, R.H. Williams, and R.R. Varma. *J. Phys. C* **11**:1989 (1978).

1661. K.J. Rawlings, M.J. Gibson, and P.J. Dobson. *J. Phys. D* **11**:2059 (1978).

1662. A.A. Galaev, L.V. Gamosov, Yu.N. Parkhomenko, and A.V. Shirkov. *Sov. Phys. Crystallogr.* **24**:72 (1979).

1663. S. Ferrer, L. Gonzalez, M. Salmeron, J.A. Verges, and F. Ndurain. *Solid State Commun.* **38**:317 (1981).

1664. W. Erley and H. Ibach. *Solid State Commun.* **37**:937 (1981).

1665. C.M. Chan. M.A. Van Hove, W.H. Weinberg, and E.D. Williams. *Solid State Commun.* **30**:47 (1979).

1666. M.A. Van Hove, G. Ertl, K. Christmann, R.J. Behm, and W.H. Weinberg. *Solid State Commun.* **28**:373 (1978).

1667. J. Behm, K. Christmann, and G. Ertl. *Solid State Commun.* **25**:763 (1978).

1668. R.A. Barker, P.J. Estrup, F. Jona, and P.M. Marcus. *Solid State Commun.* **25**:375 (1978).

1669. V.G. Lifshits, V.B. Akilov, and Y.L. Gavriljuk. *Solid State Commun.* **40**:429 (1981).

1670. P. Légaré, Y. Holl, and G. Maire. *Solid State Commun.* **31**:307 (1979).

1671. H. Namba, J. Darville, and J.M. Gilles. *Solid State Commun.* **34**:287 (1980).

1672. R. Courths. *Phys. Status Solidi B* **100**:135 (1980).

1673. J.H. Onuferko. Ph.D. Thesis. University of Warwick, Coventry, England.

1674. I.F. Lyuksyutov and A.G. Fedorus. *Zh. Eksp. Teor. Fiz.* **80**:2511 (1981).

1675. F. Lyuksyutov, V.K. Medvedev, and I.N. Yakovkin. *Zh. Eksp. Teor. Fiz.* **80**:2452 (1981).

1676. V.P. Ivanov, V.I. Savchenko, and V.L. Tataurov. *Sov. Phys. Tech. Phys.* **26**:237 (1981).

1677. B.M. Zykov, V.K. Tskhakaya. *Sov. Phys. Tech. Phys.* **24**:948 (1979).

1678. A. Gutmann and K. Hayek. *Thin Solid Films* **58**:145 (1979).

1679. K. Christmann, G. Ertl, and H. Shimizu. *Thin Solid Films.* **57**:247 (1979).

1680. M.G. Barthes and A. Rolland. *Thin Solid Films* **76**:45 (1981).

1681. B. Gruzza and E. Gillet. *Thin Solid Films.* **68**:345 (1980).

1682. C. Argile and G.E. Rhead. *Thin Solid Films* **67**:299 (1980).

1683. B.Z. Olshanetskii, V.I. Mashanov, and A.I. Nikiforov. *Sov. Phys. Solid State.* **23**:1505 (1981).

1684. M.S. Gupalo, V.K. Medvedev, B.M. Palyukh, and T.P. Smereka. *Sov. Phys. Solid State* **22**:1873 (1980).

1685. B.Z. Olshanetskii and V.I. Mashanov. *Sov. Phys. Solid State* **22**:1705 (1980).

1686. M. Weiss, G. Ertl, and F. Nitschke. *Appl. Surf. Sci.* **2**:614 (1979).

1687. J. Küppers and H. Mitchel. *Appl. Surf. Sci.* **3**:179 (1979).

1688. K.H. Rieder. *Appl. Surf. Sci.* **4**:183 (1980).

1689. H. Geiger and P. Wissmann. *Appl. Surf. Sci.* **5**:153 (1980).

1690. G. Ronida, F. Pratesi, and E. Ferroni. *Appl. Surf. Sci.* **5**:121 (1980).

1691. S.A. Isa, R.W. Joyner, M.H. Matloob, and M. Wyn Roberts. *Appl. Surf. Sci.* **5**:345 (1980).

1692. D.G. Castner and G.A. Somorjai. *Appl. Surf. Sci.* **6**:29 (1980).

1693. K. Khonde, J. Darville, S.E. Donnelly, and J.M. Gilles. *Appl. Surf. Sci.* **6**:297 (1980).

1694. E. Bechtold. *Appl. Surf. Sci.* **7**:231 (1981).

1695. O. Oda, L.J. Harrekamp, and G.A. Bootsma. *Appl. Surf. Sci.* **7**:206 (1981).

1696. G. Le Lay, A. Chauvet, M. Manneville, and R. Kern. *Appl. Surf. Sci.* **9**:190 (1981).

1697. G. Vidali, M.W. Cole, W.H. Weinberg, and W.A. Steele. *Phys. Rev. Lett.* **51**:118 (1983).

1698. M. Maglietta. *Solid State Commun.* **43**:395 (1982).

1699. R.J. Baird, D.F. Ogletree, M.A. Van Hove, and G.A. Somorjai. *Surf. Sci.* **165**:345 (1986).

1700. J. Sokolov, F. Jona, and P.M. Marcus. *Phys. Rev. B* 1397 (1986).

1701. H.D. Shih, F. Jona, and P.M. Marcus. *Surf. Sci.* **104**:39 (1981).

1702. S.Y. Tong, W.M. Mei, and G. Xu. *J. Vacuum Sci. Technol. B* **2**:393 (1984).

1703. G. Xu, W.Y. Hu, M.W. Puga, S.Y. Tong, J.L. Yeh, S.R. Wang, and B.W. Lee. *Phys. Rev. B* **32**:8473 (1985).

1704. P.H. Citrin, J.E. Rowe, and P. Eisenberger. *Phys. Rev. B* **28**:2299 (1983).

1705. C.M. Chan, S.L. Cunningham, M.A. Van Hove, W.H. Weinberg, and S.P. Withrow. *Surf. Sci.* **66**:394 (1977).

1706. B.J. Mrstik, R. Kaplan, T.L. Reinecke, M.A. Van Hove, and S.Y. Tong. *Phys. Rev. B* **15**:897 (1977).

1707. J.E. Demuth, P.M. Marcus, and D.W. Jepsen. *Phys. Rev. B* **11**:1460 (1975).

1708. D.H. Rosenblatt, S.D. Kevan, J.G. Tobin, R.F. Davis, M.G. Mason, D.R. Denley, D.A. Shirley, Y. Huang, and S.Y. Tong. *Phys. Rev.* **B26**:1812 (1982).

1709. Y. Kuk, L.C. Feldman, and P.J. Silverman. *Phys. Rev. Lett.* **50**:511 (1983).

1710. F. Maca, M. Scheffler, and W. Berndt. *Surf. Sci.* **160**:467 (1985).

1711. D.F. Ogletree, M.A. Van Hove, and G.A. Somorjai. *Surf. Sci.* **173**:351 (1986).

1712. K. Hayek, H. Glassl, A. Gutmann, H. Leonhard, P. Prutton, S.P. Tear, and M.R. Welton-Cook. *Surf. Sci.* **152**:419 (1985).

1713. M.A. Van Hove and R.J. Koestner. In: P.M. Marcus and F. Jona, editors, *Determination of Surface Structure by LEED.* Plenum, New York, 1984.

1714. L. Smit, R.M. Tromp, and J.F. Van der Veen. *Surf. Sci.* **163**:315 (1985).

1715. P.H. Citrin, J.E. Rowe, and P. Eisenberger. *Phys. Rev. B* **28**:2299 (1983).

1716. G.J. Jones and B.W. Holland. *Solid State Commun.* **53**:45 (1985).

1717. H.D. Shih, F. Jona, D.W. Jepsen, and P.M. Marcus. *J. Phys. C* **9**:1405 (1976).

1718. K.C. Hui, R.H. Milne, K.A.R. Mitchell, W.T. Moore, and M.Y. Zhou. *Solid State Commun.* **56**:83 (1985).

1719. M.J. Cardillo, G.E. Becker, D.R. Hamann, J.A. Serri, L. Whitman, and L.F. Mattheiss. *Phys. Rev. B* **28**:494 (1983).

1720. M.A. Van Hove, S.Y. Tong, and N. Stoner. *Surf. Sci.* **54**:259 (1976).

1721. Groupe d'Etude des Surfaces. *Surf. Sci.* **62**:567 (1977).

1722. E. Lang, W. Grimm, and K. Heinz. *Surf. Sci.* **117**:169 (1982).

1723. H.L. Davis and R.J. Noonan. *Surf. Sci.* **126**:245 (1983).

1724. F. Jona, D. Westphal, A. Goldman, and P.M. Marcus. *J. Phys. C* **16**:3001 (1983).

1725. E.L. Bullock, C.S. Fadley, and P.J. Orders. *Phys. Rev. B* **28**:4867 (1983).

1726. J.G. Tobin, L.E. Klebanoff, D.H. Rosenblatt, R.F. Davis, E. Umbach, A.G. Baca, D.A. Shirley, Y. Huang, W.M. Kang, and S.Y. Tong. *Phys. Rev. B* **26**:7076 (1982).

1727. U. Döbler, K. Baberschke, J. Stör, and D.A. Outka. *Phys. Rev. B* **31**:2532 (1985).

1728. F. Comin, P.H. Citrin, P. Eisenberger, and J.E. Rowe. *Phys. Rev. B* **26**:7060 (1982).

1729. K.O. Legg, F. Jona, D.W. Jepsen, and P.M. Marcus. *J. Phys. C* **10**:937 (1977).

1730. A. Ignatiev, F. Jona, D.W. Jepsen, and P.M. Marcus. *Phys. Rev. B* **11**:4780 (1975).

1731. S.A. Lindgren, J. Paul, L. Wallden, and P. Westrin. *J. Phys. C* **15**:6285 (1982).

1732. M.A. Van Hove and S.Y. Tong. *J. Vacuum Sci. Technol.* **12**:230 (1975).

1733. D.H. Rosenblatt, S.D. Kevan, J.G. Tobin, R.F. Davis, M.G. Mason, D.A. Shirley, J.C. Tang, and S.Y. Tong. *Phys. Rev. B* **26**:3181 (1982).

1734. J.J. Barton, C.C. Bahr, Z. Hussain, S.W. Robey, L.E. Klebanoff, and D.A. Shirley. *J. Vacuum Sci. Technol. A* **2**:847 (1984).

1735. P.J. Orders, B. Sinkovic, C.S. Fadley, R. Trehan, Z. Hussain, and J. Lecante. *Phys. Rev. B* **30**:1838 (1984).

1736. J. Stöhr, R. Jaeger, and S. Brennan. *Surf. Sci.* **117**:503 (1982).

1737. J.E. Demuth, D.W. Jepsen, and P.M. Marcus. *J. Phys. C* **8**:L25 (1975).

1738. J. Stöhr, R. Jaeger, and T. Kendelewicz. *Phys. Rev. Lett.* **49**:142 (1982).

1739. J.W.M. Frenken, J.F. Van der Veen, and G. Allan. *Phys. Rev. Lett.* **51**:1876 (1983).

1740. M. de Crescenzi, F. Antonangeli, C. Bellini, and R. Rosei. *Phys. Rev. Lett.* **50**:1949 (1983).

1741. S.Y. Tong, W.M. Kang, D.H. Rosenblatt, J.G. Tobin, and D.A. Shirley. *Phys. Rev. B* **27**:4632 (1983).

1742. T.S. Rahman, D.L. Mills, J.E. Black, J.M. Szeftel, S. Lehwald, and H. Ibach. *Phys. Rev. B* **30**:589 (1984).

1743. D. Norman, J. Stöhr, R. Jaeger, P.J. Durham, and J.B. Pendry. *Phys. Rev. Lett.* **51**:2052 (1983).

1744. J.E. Demuth, D.W. Jepsen, and P.M. Marcus. *J. Phys. C* **6**:L307 (1973).

1745. R. Feder. *Surf. Sci.* **68**:229 (1977).

1746. B.W. Holland, C.D. Duke, and A. Paton. *Surf. Sci.* **140**:L269 (1984).

1747. M.A. Passler, A. Ignatiev, B.W. Lee, D. Adams, and M.A. Van Hove. In: P.M. Marcus and F. Jona, editors, *Determination of Surface Structure by LEED*. Plenum, New York, 1984.

1748. K. Griffiths, D.A. King, G.C. Aers, and J.B. Pendry. *J. Phys. C* **15**:4921 (1982).

1749. K. Heinz, D.K. Saldin, and J.B. Pendry. *Phys. Rev. Lett.* **55**:2312 (1985).

1750. Y. Kuk and L.C. Feldman. *Phys. Rev. B* **30**:5811 (1984).

1751. A. Puschmann and J. Haasse. *Surf. Sci.* **144**:559 (1984).

1752. W. Moritz and D. Wolf. *Surf. Sci.* **163**:L655 (1985).

1753. W. Moritz, R. Imbihl, R.J. Behm, G. Ertl, and T. Matsushima. *J. Chem. Phys.* **83**:1959 (1985).

1754. P.M. Echenique. *J. Phys. C* **9**:3193 (1976).

1755. S. Andersson, J.B. Pendry, and P.M. Echenique. *Surf. Sci.* **65**:539 (1977).

1756. D.L. Adams, L.E. Peterson, and C.S. Sorenson. *J. Phys. C* **18**:1753 (1985).

1757. M.L. Xu and S.Y. Tong. *Phys. Rev. B* **31**:6332 (1985).

1758. E. Tornquist, E.D. Adams, M. Copel, T. Gustafsson, and W.R. Graham. *J. Vacuum Sci. Technol. A* **2**:939 (1984).

1759. R. Baudoing, Y. Gauthier, and Y. Joly. *J. Phys. C* **18**:4061 (1985).

1760. C.J. Barnes, M.Q. Ding, M. Lindroos, R.D. Diehl, and D.A. King. *Surf. Sci.* **162**:59 (1985).

1761. H. Niehus. *Surf. Sci.* **145**:407 (1984).

1762. D.L. Adams and H.B. Nielsen. *Surf. Sci.* **116**:598 (1982).

1763. M.A. Van Hove and S.Y. Tong. *Surf. Sci.* **54**:91 (1976).

1764. C.B. Duke and A. Paton. *J. Vacuum Sci. Technol. B* **2**:327 (1984).

1765. H.J. Grossman and M.W. Gibson. *J. Vacuum Sci. Technol. B* **2**:343 (1984).

1766. L. Smit, R.M. Tromp, and J.F. Van der Veen. *Phys. Rev. B* **29**:4814 (1984).

1767. J. Sokolov, F. Jona, and P.M. Marcus. *Phys. Rev. B* **31**:1929 (1985).

1768. P.C. Wong, M.Y. Zhou, K.C. Hui, and K.A.R. Mitchell. *Surf. Sci.* **163**:172 (1985).

1769. A. Ignatiev, B.W. Lee, and M.A. Van Hove. In: *Proceedings of the 7th International Vacuum Congress and 3rd International Conference on Solid Surfaces, Vienna*, 1977.

1770. W.S. Yang, F. Jona, and P.M. Marcus. *Phys. Rev. B* **28**:7377 (1983).

1771. H.L. Davis and J.R. Noonan. *Phys. Rev. Lett.* **54**:566 (1985).

1772. C.B. Duke, A.R. Lubinsky, B.W. Lee, and P. Mark. *J. Vacuum Sci. Technol.* **13**:761 (1976).

1773. A.R. Lubinsky, C.B. Duke, S.C. Chang, B.W. Lee, and P. Mark. *J. Vacuum Sci. Technol.* **13**:89 (1976).

1774. D. Norman, S. Brennan, R. Jaeger, and J. Stöhr. *Surf. Sci.* **105**:L297 (1981).

1775. R.Z. Bachrach, G.V. Hansson, and R.S. Bauer. *Surf. Sci.* **109**:L560 (1981).

1776. M. Maglietta, E. Zanazzi, F. Jona, D.W. Jepsen, and P.M. Marcus. *Appl. Phys.* **15**:409 (1978).

1777. M. Maglietta, E. Zanazzi, U. Bardi, and F. Jona. *Surf. Sci.* **77**:101 (1978).

1778. R.C. Felton, M. Prutton, S.P. Tear, and M.R. Welton-Cook. *Surf. Sci.* **88**:474 (1979).

1779. P.H. Citrin, P. Eisenberger, and R.C. Hewitt. *Phys. Rev. Lett.* **45**:1948 (1980).

1780. S. Andersson and J.B. Pendry. *J. Phys. C* **13**:2547 (1980).

1781. J. Onuferko and D.P. Woodruff. *Surf. Sci.* **95**:555 (1980).

1782. R.W. Streater, W.T. Moore, P.R. Watson, D.C. Frost, and K.A.R. Mitchell. *Surf. Sci.* **72**:744 (1978).

1783. P. Eisenberger and W.C. Marra. *Phys. Rev. Lett.* **46**:1081 (1981).

1784. S.P. Tear, M.R. Welton-Cook, M. Prutton, and J.A. Walker. *Surf. Sci.* **99**:598 (1980).

1785. M.A. Van Hove, R.J. Koestner, P.C. Stair, J.P. Bibérian, L.L. Kesmodel, I. Bartos, and G.A. Somorjai. *Surf. Sci.* **103**:218 (1981).

1786. C.M. Chan, S.M. Cunningham, K.L. Luke, W.H. Weinberg, and S.P. Withrow. *Surf. Sci.* **78**:15 (1978).

1787. C.M. Chan, M.A. Van Hove, W.H. Weinberg, and E.D. Williams. *J. Vacuum Sci. Technol.* **16**:642 (1979).

1788. C.M. Chan, K.L. Luke, M.A. Van Hove, W.H. Weinberg, and S.P. Withrow. *Surf. Sci.* **78**:386 (1978).

1789. S.Y. Tong, A. Maldonado, C.H. Li, and M.A. Van Hove. *Surf. Sci.* **94**:73 (1980).

1790. S. Andersson and J.B. Pendry. *J. Phys.* C **13**:3547 (1980).

1791. L.G. Petersson, S. Kono, N.F.T. Hall, C.S. Fadley, and J.B. Pendry. *Phys. Rev. Lett.* **42**:1545 (1979).

1792. Y. Gauthier, D. Aberdam, and R. Baudoing. *Surf. Sci.* **78**:339 (1978).

1793. D.H. Rosenblatt, J.G. Tobin, M.G. Mason, R.F. Davis, S.D. Kevan, D.A. Shirley, C.H. Li, and S.Y. Tong. *Phys. Rev. B* **23**:3828 (1981).

1794. T. Narasawa, W.M. Gibson, and E. Tornquist. *Phys. Rev. Lett.* **47**:417 (1981).

1795. S.D. Kevan, R.F. Davis, D.H. Rosenblatt, J.G. Tobin, M.G. Mason, D.A. Shirley, C.H. Li, and S.Y. Tong. *Phys. Rev. Lett.* **46**:1629 (1981).

1796. T. Narasawa and W.M. Gibson. *Surf. Sci.* **114**:331 (1981).

1797. R.J. Behm, K. Christmann, G. Ertl, and M.A. Van Hove. *J. Chem. Phys.* **73**:2984 (1980).

1798. J.A. Davies, T.E. Jackman, D.P. Jackson, and P.R. Norton. *Surf. Sci.* **109**:20 (1981).

1799. R. Feder, H. Pleyer, P. Bauer, and N. Müller. *Surf. Sci.* **109**:419 (1981).

1800. S. Hengrasmee, A.R. Mitchell, P.R. Watson, and S.J. White. *Can. J. Phys.* **58**:200 (1980).

1801. R.J. Meyer, W.R. Salaneck, C.B. Duke, A. Paton, C.H. Griffiths, L. Kovnat, and L.E. Meyer. *Phys. Rev. B* **21**:4542 (1980).

1802. F.S. Marsh, M.K. Debe, and D.A. King. *J. Phys.* C **13**:2799 (1980).

1803. W.A. Jesser and J.W. Matthews. *Acta Metall.* **16**:1307 (1968).

1804. A. Chambers and D.C. Jackson. *Philos. Mag.* **31**:1357 (1975).

1805. J.W. Matthews. *Thin Solid Films* **12**:243 (1972).

1806. J.W. Matthews. *Philos. Mag.* **13**:1207 (1966).

1807. K. Yagi, K. Takanayagi, K. Kobayashi, and G. Honjo. *J. Cryst. Growth.* **9**:84 (1971).

1808. W.A. Jesser and J.W. Matthews. *Philos. Mag.* **17**:595 (1968).

1809. W.A. Jesser and J.W. Matthews. *Philos. Mag.* **15**:1097 (1967).

1810. W.A. Jesser and J.W. Matthews. *Philos. Mag.* **17**:461 (1968).

1811. A.I. Fedorenko and R. Vincent. *Philos. Mag.* **24**:55 (1971).

1812. R. Kuntze, A. Chambers, and M. Prutton. *Thin Solid Films* **4**:47 (1969).

1813. U. Gradmann. *Ann. Physik.* **13**:213 (1964).

1814. U. Gradmann. *Ann. Physik.* **17**:91 (1966).

1815. R.W. Vook, C.T. Horng, and J.E. Macur. *J. Cryst. Growth.* **31**:353 (1979).

1816. R.W. Vook and C.T. Horng. *Philos. Mag.* **33**:843 (1976).

1817. C.T. Horng and R.W. Vook. *Surf. Sci.* **54**:309 (1976).

1818. U. Gradmann. *Phys. Kondens. Mater.* **3**:91 (1964).

1819. R.W. Vook and J.E. Macur. *Thin Solid Films.* **32**:199 (1976).

1820. L.A. Bruce and H. Jaeger. *Philos. Mag.* **36**:1331 (1977).

1821. C. Gonzalez. *Acta Metall.* **15**:1373 (1967).

1822. C.T. Horng and R.W. Vook. *J. Vacuum Sci. Technol.* **11**:140 (1974).

1823. E. Grhnbaum, G. Kremer, and C. Reymond. *J. Vacuum Sci. Technol.* **6**:475 (1969).

1824. R.C. Newman and D.W. Pashley. *Philos. Mag.* **46**:927 (1955).

1825. F. Soria, J.L. Sacedon, P.M. Echenique, and D. Titterington. *Surf. Sci.* **68**:448 (1977).

1826. M. Klaua and H. Bethce. *J. Cryst. Growth* **3,4**:188 (1968).

1827. E. Grunbaum. *Proc. Phys. Soc. London* **72**:459 (1958).

1828. E.F. Wassermann and H.P. Jablonski. *Surf. Sci.* **22**:69 (1970).

1829. D.C. Hothersall. *Philos. Mag.* **15**:1023 (1967).

1830. P. Gueguen, C. Camoin, and M. Gillet. *Thin Solid Films* **26**:107 (1975).

1831. G. Honjo, K. Takayanagi, K. Kobayashi, and K. Yagi. *J. Cryst. Growth.* **42**:98 (1977).

1832. P. Gueguen, M. Cahareau, and M. Gillet. *Thin Solid Films* **16**:27 (1973).

1833. D. Cherns and M.J. Stowell. *Thin Solid Films* **29**:107 (1975).

1834. D. Cherns and M.J. Stowell. *Thin Solid Films* **29**:127 (1975).

1835. W.A. Jesser, J.W. Matthews, and D. Kuhlmann-Wilsdorf. *Appl. Phys. Lett.* **9**:176 (1966).

1836. J.E. Macur and R.W. Vook. In: C.J. Arceneaux, editor, *32nd Annual Proceedings of the Electron Microscopy Society of America*, St. Louis, Missouri, 1974.

1837. J.E. Macur. In: B.W. Bailey, editor, *33rd Annual Proceedings of the Electron Microscopy Society of America*, Las Vegas, Nevada, 1975, p. 98.

1838. J.W. Matthews. *Phys. Thin Films.* **4**:137 (1967).

1839. G. Dorey. *Thin Solid Films* **5**:69 (1970).

1840. A. Mlynczak and R. Niedermayer. *Thin Solid Films* **28**:37 (1975).

1841. P.W. Steinhage and H. Mayer. *Thin Solid Films* **28**:131 (1975).

1842. C.M. Mate and G.A. Somorjai. *Surf. Sci.* **160**:542 (1985).

1843. S. Thomas and T.W. Haas. *Surf. Sci.* **28**:632 (1971).

1844. C.M. Mate, B.E. Bent, and G.A. Somorjai. *J. Electron Spectrosc. Related Phenom.* **39**:205 (1986).

1845. P.I. Cohen, J. Unguris, and M.B. Webb. *Surf. Sci.* **58**:429 (1976).

1846. N.J. Wu and A. Ignatiev. *Phys. Rev. B* **25**:2983 (1982).

1847. N.J. Wu and A. Ignatiev. *Phys. Rev. B* **28**:7288 (1983).

1848. M. Weltz, W. Moritz, and D. Wolf. *Surf. Sci.* **125**:473 (1983).

1849. D.A. Outka, R.J. Madix, and J. Stöhr. *Surf. Sci.* **164**:235 (1985).

1850. A. Puschman, J. Hasse, M.D. Crapper, C.E. Riley, and D.P. Woodruff. *Phys. Rev. Lett.* **54**:2250 (1985).

1851. D. Sondericker, F. Jona, and P.M. Marcus. *Phys. Rev.* **B33**:900 (1986).

1852. J. Stöhr, E.B. Kollin, D.A. Fischer, J.B. Hasting, F. Zaera, and F. Sette. *Phys. Rev. Lett.* **55**:1468 (1985).

1853. J.F. Van der Veen, R.M. Tromp, R.G. Smeenk, and F.W. Saris. *Surf. Sci.* **82**:468 (1979).

1854. D.F. Ogletree, M.A. Van Hove, and G.A. Somorjai. *Surf. Sci.* **183**:1 (1987).

1855. R.F. Lin, G.S. Blackman, A.M. Van Hove, and G.A. Somorjai. *Acta Crystallogr. B* (1987).

1856. M.A. Van Hove, R.F. Lin, and G.A. Somorjai. *J. Am. Chem. Soc.* **108**:2532 (1986).

1857. P.H. Citrin, P. Eisenberger, and J.E. Rowe. *Phys. Rev. Lett.* **48**:802 (1982).

1858. G. Materlik, A. Frohm, and M.J. Bedzyk. *Phys. Rev. Lett.* **52**:441 (1984).

1859. J.A. Golovchenko, J.R. Patel, D.R. Kaplan, P.L. Cowan, and M.J. Bedzyk. *Phys. Rev. Lett.* **49**:1560 (1982).

1860. E.J. Van Loenen, J.W.M. Frenken, J.F. Van der Veen, and S. Valeri. *Phys. Rev. Lett.* **54**:827 (1985).

1861. H. Ohtani, M.A. Van Hove, and G.A. Somorjai. *Surf. Sci.* **187**:372 (1987).

1862. H. Ohtani, M.A. Van Hove, and G.A. Somorjai. *J. Phys. Chem.* **92**:3974 (1988).

1863. C.B. Duke, A. Paton, A. Kahn, and C.R. Bonapace. *Phys. Rev. B* **28**:852 (1983).

1864. J.E. Demuth, P.M. Marcus, and D.W. Jepsen. *Phys. Rev. B* **11**:1460 (1975).

1865. S. Anderson and J.B. Pendry. *Solid State Commun.* **16**:563 (1975).

1866. D.H. Rosenblatt, S.D. Kevan, J.G. Tobin, R.F. Davis, M.G. Mason, D.R. Denley, D.A. Shirley, Y. Huang, and S.Y. Tong. *Phys. Rev.* **B26**:1812 (1982).

1867. M.A. Van Hove and S.Y. Tong. *J. Vacuum Sci. Technol.* **12**:230 (1975).

1868. D. Sondericker, F. Jona, and P.M. Marcus. *Phys. Rev.* **B33**:900 (1986).

1869. P.H. Citrin, D.R. Hamann, L.F. Mattheiss, and J.E. Rowe. *Phys. Rev. Lett.* **49**:1712 (1982).

1870. W.N. Unertl and H.V. Thapliyal. *J. Vacuum Sci. Technol.* **12**:263 (1975).

1871. A. Kahn, J. Carelli, D. Kanani, C.B. Duke, A. Paton, and L. Brillson. *J. Vacuum Sci. Technol.* **19**:331 (1981).

1872. J.R. Noonan and H.R. Davis. *Vacuum* **31**:107 (1982).

1873. E. Zanazzi and F. Jana. *Surf. Sci.* **62**:61 (1977).

1874. J.F. Van der Veen, R.G. Smeenk, R.M. Tromp, and F.M. Saris. *Surf. Sci.* **79**:219 (1979).

1875. C.M. Chan and M.A. Van Hove. Unpublished results.

1876. C.M. Mate, C.-T. Kao, and G.A. Somorjai. Unpublished results.

1877. G.S. Blackman, C.-T. Kao, R.J. Koestner, B.E. Bent, C.M. Mate, M.A. Van Hove, and G.A. Somorjai. Unpublished results.

1878. A.L. Slavin, B.E. Bent, C.-T. Kao, and G.A. Somorjai. Unpublished results.

1879. B.E. Bent, C.-T. Kao, and G.A. Somorjai. Unpublished results.

1880. C.-T. Kao, B.E. Bent, and G.A. Somorjai. Unpublished results.

1881. C.-T. Kao, C.M. Mate, B.E. Bent, and G.A. Somorjai. Unpublished results.

1882. C.G. Shaw, S.C. Fain, Jr., and M.D. Chinn. *Phys. Rev. Lett.* **41**:955 (1978).

1883. S.C. Fain, Jr., M.F. Torey, and R.D. Diehl. In: J.L. de Segovia, editor, *Proceedings of the 9th International Vacuum Congress and 5th International Conference on Solid Surfaces*, Madrid, 1983, p. 129.

1884. B.E. Koel, J.E. Crowell, C.M. Mate, and G.A. Somorjai. *J. Phys. Chem.* **88**:1988 (1984).

1885. W.T. Moore, S.J. White, D.C. Frost, and K.A.R. Mitchell. *Surf. Sci.* **116**:253 (1982).

1886. W.T. Moore, S.J. White, D.C. Frost, and K.A.R. Mitchell. *Surf. Sci.* **116**:261 (1982).

1887. S. Andersson and J.B. Pendry. *J. Phys. C* **9**:2721 (1976).

1888. R.J. Meyer, C.B. Duke, A. Paton, J.L. Yeh, J.C. Tsung, A. Kahn, and P. Mark. *Phys. Rev. B* **21**:4740 (1980).

1889. H. Melle and E. Menzel. *Z. Naturforsch* **33a**:282 (1978).

1890. H. Niehus. *Surf. Sci.* **145**:407 (1984).

1891. S. Lehwald, H. Ibach, and J.E. Demuth. *Surf. Sci.* **78**:577 (1978).

1892. B.E. Koel, J.E. Crowell, C.M. Mate, and G.A. Somorjai. *J. Phys. Chem.* **88**:1988 (1984).

1893. W.T. Moore, S.J. White, D.C. Frost, and K.A.R. Mitchell. *Surf. Sci.* **116**:253 (1982).

1894. F. Frostmann. *Jpn. J. Appl. Phys.* [*Suppl.*] **2**(P2):657 (1974).

1895. N. Stoner. Ph.D. Thesis, University of Wisconsin, Milwaukee (1976).

1896. F. Soria, J.L. Sacedon, P.M. Echenique, and D. Titterington. *Surf. Sci.* **68**:448 (1977).

1897. D.W. Jepsen, P.M. Marcus, and F. Jona. *Phys. Rev. B* **5**:3933 (1972).

1898. D.W. Jepsen, P.M. Marcus, and F. Jona. *Phys. Rev. B* **8**:5523 (1973).

1899. W. Moritz. Ph.D. Thesis, University of Munich (1976).

1900. J.A. Strozier and R.O. Jones. *Phys. Rev. B* **3**:3228 (1971).

1901. J.A. Strozier and R.O. Jones. *Phys. Rev. Lett.* **25**:516 (1970).

1902. H.D. Shih, F. Jona, D.W. Jepsen, and P.M. Marcus. *Commun. Phys.* **1**:25 (1976).

1903. C.G. Shaw, S.C. Fain, M.D. Chinn, and M.F. Toney. *Surf. Sci.* **97**:128 (1980).

1904. C. Bouldin and E.A. Stern. *Phys. Rev. B* **25**:3462 (1982).

1905. F. Comin, P.H. Citrin, P. Eisenberger, and J.E. Rowe. *Phys. Rev. B* **26**:7060 (1982).

1906. J. Bohr, R. Feidenhansl, M. Nielsen, M. Toney, R.L. Johnson, and I.K. Robinson. *Phys. Rev. Lett.* **54**:1275 (1985).

1907. R. Rosei, M. de Crescenzi, F. Sette, C. Quaresima, A. Savoia, and P. Perfetti. *Phys. Rev. B* **28**:1161 (1983).

1908. R.G. Jones, S. Ainsworth, M.D. Crapper, C. Somerton, D.P. Woodruff, R.S. Brooks, J.C. Campuzano, D.A. King, and G.M. Lamble. *Surf. Sci.* **152/153**:443 (1985).

1909. Y. Gauthier, R. Baudoing, Y. Joly, J. Rundgren, J.C. Bertolini, and J. Massardier. *Surf. Sci.* **162**:342 (1985).

1910. M. Saitoh, F. Shoji, K. Oura, and T. Hanawa. *Jpn. J. Appl. Phys.* **19**:L421 (1980).

1911. J. Stöhr, R. Jaeger, G. Rossi, T. Kendelewicz, and I. Lindau. *Surf. Sci.* **134**:813 (1983).

1912. A. Ignatiev, T.N. Rhodin, S.Y. Tong, B.I. Lundqvist, and J.B. Pendry. *Solid State Commun.* **9**:1851 (1971).

1913. L.J. Clarke. *Surf. Sci.* **102**:331 (1981).

1914. A. Ignatiev, J.B. Pendry, and T.N. Rhodin. *Phys. Rev. Lett.* **26**:189 (1971).

1915. P.H. Citrin. *Bull. Am. Phys. Soc.* **25**:383 (1980).

1916. A.G.J. de Wit, R.P.N. Bronckers, Th. M. Hupkens, and J.M. Fluit. *Surf. Sci.* **90**:676 (1979).

1917. R. Feidenhansl and I. Stensgaard. *Surf. Sci.* **133**:453 (1983).

1918. U. Döbler, K. Baberschke, J. Haase, and A. Puschmann. *Phys. Rev. Lett.* **52**:1437 (1984).

1919. C.B. Duke, A. Paton, A. Kahn, and K. Li. *Bull. Am. Phys. Soc.* **30**:313 (1985).

1920. C.B. Duke, C. Mailhiot, A. Paton, K. Li, C. Bonapace, and A. Kahn. *Surf. Sci.* **163**:391 (1985).

1921. J. Stöhr, D.A. Outka, R.J. Madix, and U. Döbler. *Phys. Ref. Lett.* **54**:1256 (1985).

1922. D.A. Outka, R.J. Madix, and J. Stöhr. *Surf. Sci.* **164**:235 (1985).

1923. G. Casalone, M.G. Cattania, and M. Simonetta. *Surf. Sci.* **103**:L121 (1981).

1924. Z.P. Hu, D.F. Ogletree, M.A. Van Hove, and G.A. Somorjai. *Surf. Sci.* **180**:433 (1987).

1925. H. Kobayashi, H. Teramae, T. Yamabe, and M. Yamaguchi. *Surf. Sci.* **141**:580 (1984).

1926. B.A. Hutchins, T.N. Rhodin, and J.E. Demuth. *Surf. Sci.* **54**:419 (1976).

TABLE 2.6. Surface Structures Formed by Chemisorption-Induced Restructuring

Surface–Adsorbate System and Periodicity	Method and Reference of Investigation	Type of Restructuring
Cu(100)/O-$(2 * \sqrt{2} \times \sqrt{2})$	LEED (1) STM/LEED (2) Theory (effective medium) (3)	Missing-row reconstruction Pairing of (Cu) atoms next to row "Cu–O–Cu chains" similar to Cu(110)
Ni(100)/C-p4g(2 × 2) Ni(100)/N-p4g(2 × 2)	LEED (4) SEXAFS (5, 6)	4-fold site Clockwise rotation of first-layer atoms Buckling of second-layer atoms
Ni(100)/Cl-c(2 × 2) Cu(100)/Cl-c(2 × 2) Mo(100)/C-c(2 × 2)	SEXAFS/XSW (7) SEXAFS (8) LEED (45)	4-fold site First- to second-layer expansion
Cu(100)/N-c(2 × 2) Cu(100)/S-p(2 × 2)	LEED (9) XRD (10) MEIS (11) LEED (12)	4-fold site Second-layer buckling (atom underneath adsorbate moves down)
Ni(100)/O-c(2 × 2) Ni(100)/O-p(2 × 2) Ni(100)/O-disordered Ni(100)/S-c(2 × 2) Ni(100)/S-p(2 × 2) Ni(100)/S-disordered	LEED (13, 18) LEED (14, 18) DLEED (15, 18) LEED (16, 18) LEED (17, 18) DLEED (15, 18)	
Cu(110)/K-(1 × 2)	LEED (19)	Low coverage induced

TABLE 2.6. (*Continued*)

Surface–Adsorbate System and Periodicity	Method and Reference of Investigation	Type of Restructuring
Cu(110)/Cs–(1 × 2)	LEED (19)	missing-row reconstruction
Cu(110)/N–(2 × 3)	PED (20) LEIS (21) XPD/AES (22) STM (23)	Favoring "pseudo-square" model with square-like first-layer arrangement Favoring missing-row reconstruction with every third ⟨110⟩ row missing
Cu(110)/O–(2 × 1)	SEXAFS (24) XRD (25) LEED (26) LEED (27) Theory (effective medium) (3) ICISS (28)	Missing-row reconstruction Long bridge site "Cu–O–Cu chains" similar to Cu(100)
Ni(110)/O–(2 × 1)	LEED (29)	Missing-row reconstruction asymmetric long bridge site
Cu(110)/O–c(6 × 2)	STM/XRD (30)	Cu-adatom c(6 × 2) superstructure on (3 × 1) 2 per 3 missing-row reconstruction Two oxygen sites
Ni(110)/H–(1 × 2)	LEED (31)	Row-pairing, second-layer buckling
Ni(111)/O–(2 × 2)	LEED (32)	fcc-hollow site Clockwise rotation and buckling in first-layer
Ni(111)/O–p($\sqrt{3} \times \sqrt{3}$)	LEED (33)	fcc-hollow site
Ni(111)/S–(2 × 2)	LEED (34)	fcc-hollow site First- to second-, to third-layer expansion
Rh(100)/O–(2 × 2)	LEED (35)	4-fold hollow site, layer contraction
Rh(110)/H	LEED (36–41)	Five superstructure phases Local outwards movement of substrate atoms with H-bond
Ru(001)/O–p(2 × 1)	LEED (42)	3-fold hcp-hollow site Buckling and row pairing in first and second layer
Ru(001)/O–p(2 × 2)	LEED (43)	3-fold hcp-hollow site Buckling and lateral out-

TABLE 2.6. (*Continued*)

Surface–Adsorbate System and Periodicity	Method and Reference of Investigation	Type of Restructuring
		wards movement in first layer
Cr(110)/N–(1 × 1)	LEED (44)	4-fold hollow site First-layer expansion (24.8%)
Mo(100)/S–c(2 × 2)	LEED (45)	4-fold hollow site Second-layer buckling
Rh(111)/C$_2$H$_3$–(2 × 2)	LEED (46)	Buckling in first- and second-layer molecule pulls nearest metal neighbors outward; radial expansion parallel to surface of 3-fold site in top metal layer
Si(100)/H–(2 × 1)	STM (47) LEIS (48) Theory (SLAB-MINDO) (49)	Dangling bond adsorption Dimer relaxation (lengthening)
Si(100)/F–(2 × 1) Si(100)/Cl–(2 × 1)	ESDIAD (50) SEXAFS (51)	
Si(100)/O–(2 × 1)	SEXAFS (52) Theory (HF-Cluster) (53) Theory (DF-LDA) (54)	Dimer insertion (adsorption into dimer and dimer lengthening)
Si(100)/K–(2 × 1) Si(100)/Na–(2 × 1) Si(100)/Li–(2 × 1)	LEEDS (55) SEXAFS (56) Theory (DF-LDA) (57) LEED (58) Theory (DF-LDA) (59)	One-dimensional alkali chains Dimerization nearly removed Relaxation of substrate atom distances
Si(111)/Al– ($\sqrt{3} \times \sqrt{3}$)$R30°$ Si(111)/Ga– ($\sqrt{3} \times \sqrt{3}$)$R30°$ Si(111)/Sn–($\sqrt{3} \times \sqrt{3}$) Ge(111)/Pb–($\sqrt{3} \times \sqrt{3}$)	LEED (60) Theory (total energy) (61) LEED (62) STM (63) XSW, STM (64) XRD (65) LEED (66)	Removal of (7 × 7) T4 adsorption site (triangular site with 4-fold coordination)
Si(111)/B–($\sqrt{3} \times \sqrt{3}$)	XRD (67) LEED (68)	Removal of (7 × 7) T4 "upside down" site (triangular site with 4-fold coordination, B–Si substitution)
Si(111)/Fe–(1 × 1) Si(111)/As–(1 × 1)	LEED (69) XSW (70) MEIS (71)	Removal of (7 × 7) Missing top layer in Fe structure

TABLE 2.6. (*Continued*)

Surface–Adsorbate System and Periodicity	Method and Reference of Investigation	Type of Restructuring
Si(111)/GaAs-(1 × 1)	STM, MEIS (72) Theory (GVB) (73) XSW (74) Theory (DF-LDA) (75)	
Si(111)/Sb-($\sqrt{3} \times \sqrt{3}$) Si(111)/Bi-($\sqrt{3} \times \sqrt{3}$)	XPD (76) XRD (77)	Removal of (7 × 7) Milk-stool structure Adsorbate trimers Si honeycomb layer on top in Bi structure

REFERENCES

1. H.C. Zeng, R.A. McFarlane, and K.A.R. Mitchell. *Surf. Sci.* **208**:L7 (1989).

2. Ch. Wöll, R.J. Wilson, S. Chiang, H.C. Zeng and K.A.R. Mitchell. *Phys. Rev. B* **42**:11926 (1990).

3. K.W. Jacobsen and J.K. Norskov. *Phys. Rev. Lett.* **65**:1788 (1990).

4. Y. Gauthier, R. Baudoing-Savois, K. Heinz, and H. Landskron. *Surf. Sci.* **276**:1 (1992); *Proceedings of ECOSS*, Salamanca, Spain, 1990.

5. L. Wenzel, D. Arvanitis, W. Daum, H.H. Rotermund, J. Stöhr, K. Baberschke, and H. Ibach. *Phys. Rev. B* **36**:7689 (1987).

6. D. Arvanitis, K. Baberschke, and L. Wenzel. *Phys. Rev. B* **37**:7143 (1988).

7. T. Yokoyama, Y. Takata, T. Ohta, M. Funagashi, Y. Kitajima, and H. Kuroda. *Phys. Rev. B* **42**:7000 (1990).

8. J.R. Patel, D.W. Berreman, F. Sette, P.H. Citrin, J.E. Rowe, P.L. Cowan, T. Jack, and B. Karlin. *Phys. Rev. B* **40**:1330 (1989).

9. H.C. Zeng and K.A.R. Mitchell. *Langmuir* **5**:829 (1989).

10. E. Vlieg, I.K. Robinson, and R. McGrath. *Phys. Rev. B* **41**:7896 (1990).

11. Q.T. Jiang, P. Fenter, and T. Gustafsson. *Phys. Rev. B* **42**:9291 (1990).

12. H.C. Zeng, R.A. McFarlane and K.A.R. Mitchell. *Can J. Phys.* **68**:353 (1990).

13. W. Oed, H. Lindner, U. Starke, K. Heinz, and K. Müller. *Surf. Sci.* **224**:179 (1989).

14. W. Oed, H. Lindner, U. Starke, K. Heinz, K. Müller, D.K. Saldin, P. de Andres, and J.B. Pendry. *Surf. Sci.* **225**:242 (1990).

15. U. Starke, W. Oed, P. Bayer, F. Bothe, G. Fürst, P.L. de Andres, K. Heinz, and J.B. Pendry. In: S.Y. Tong, M.A. Van Hove, K. Takayanagi, and X.D. Xie, editors, *The Structure of Surfaces III*. Springer-Verlag, Berlin, 1991.

16. U. Starke, F. Bothe, W. Oed, and K. Heinz. *Surf. Sci.* **232**:56 (1990).

17. W. Oed, U. Starke, F. Bothe, and K. Heinz. *Surf. Sci.* **234**:72 (1990).

18. W. Oed, U. Starke, K. Heinz, K. Müller, and J.B. Pendry. *Surf. Sci.* **251-2**:488 (1991); *Proceedings of ECOSS*, Salamanca, Spain, 1990.

19. Z.P. Hu, B.C. Pan, W.C. Fan, and A. Ignatiev. *Phys. Rev. B* **41**:9692 (1990).

20. A.W. Robinson, D.P. Woodruff, J.S. Somers, A.L.D. Kilcoyne, D.E. Ricken, and A.M. Bradshaw. *Surf. Sci.* **237**:99 (1990).

21. M.J. Ashwin and D.P. Woodruff. *Surf. Sci.* **237**:108 (1990).

22. A.P. Baddorf and D.M. Zehner. *Surf. Sci.* **238**:255 (1990).

23. H. Niehus, R. Spitzl, K. Besocke, and G. Comsa, *Phys. Rev. B* **43**:12619 (1991).

24. M. Bader, A. Puschmann, C. Ocal, and J. Haase. *Phys. Rev. Lett.* **57**:3273 (1986).

25. R. Feidenhans'l, F. Grey, R.L. Johnson, S.G.J. Mochrie, J. Bohr, and M. Nielsen. *Phys. Rev. B* **41**:5420 (1990).

26. S.R. Parkin, H.C. Zeng, M.Y. Zhou, and K.A.R. Mitchell. *Phys. Rev. B* **41**:5432 (1990).

27. J. Wever, D. Wolf, and W. Moritz. *Surf. Sci.* **272**:94 (1992).

28. H. Dürr, Th. Fauster, and R. Schneider. *Surf. Sci.* **244**:237 (1991).

29. G. Kleinle, J. Wintterlin, G. Ertl, R.J. Behm, F. Jona, and W. Moritz. *Surf. Sci.* **225**:171 (1990).

30. R. Feidenhans'l, F. Grey, M. Nielsen, F. Besenbacher, F. Jensen, E. Laegsgaard, I. Steensgaard, K.W. Jacobsen, J.K. Norskov, and R.L. Johnson. *Phys. Rev. Lett.* **65**:2027 (1990).

31. G. Kleinle, V. Penka, R.J. Behm, G. Ertl, and W. Moritz. *Phys. Rev. Lett.* **58**:148 (1987).

32. D.T. VU Grimsby, Y.K. Wu, and K.A.R. Mitchell. *Surf. Sci.* **232**:51 (1990).

33. M.A. Mendez, W. Oed, A. Fricke, L. Hammer, K. Heinz, and K. Müller. *Surf. Sci.* **253**(1-3)99 (1991).

34. Y.K. Wu and K.A.R. Mitchell, *Can. J. Chem.* **67**:1975 (1989).

35. W. Oed, B. Doetsch, L. Hammer, K. Heinz and K. Müller. *Surf. Sci.* **207**:55 (1988).

36. W. Nichtl, N. Bickel, L. Hammer, K. Heinz, and K. Müller. *Surf. Sci.* **188**:L729 (1987).

37. W. Nichtl, L. Hammer, K. Müller, N. Bickel, K. Heinz, K. Christmann, and M. Ehsasi. *Surf. Sci.* **11**:201 (1988).

38. W. Oed, W. Puchta, N. Bickel, K. Heinz, W. Nichtl, and K. Müller. *J. Phys. C* **21**:237 (1988).

39. K. Lehnberger, W. Nichtl-Pecher, W. Oed, K. Heinz, and K. Müller. *Surf. Sci.* **217**:511 (1989).

40. M. Michl, W. Nichtl-Pecher, W. Oed, H. Landskron, K. Heinz, and K. Müller, *Surf. Sci.* **220**:59 (1989).

41. W. Puchta, W. Nichtl, W. Oed, N. Bickel, K. Heinz, and K. Müller. *Phys. Rev. B* **39**:1020 (1989).

42. H. Pfnür, G. Held, M. Lindroos, and D. Menzel. *Surf. Sci.* **220**:43 (1989).

43. M. Lindroos, H. Pfnür, G. Held, and D. Menzel. *Surf. Sci.* **222**:451 (1989).

44. Y. Joly, Y. Gauthier, and R. Baudoing. *Phys. Rev. B* **40**:10119 (1989).

45. P.J. Rous, D. Jentz, D.G. Kelly, R.Q. Hwang, M.A. Van Hove, and G.A. Somorjai. In: S.Y. Tong, M.A. Van Hove, K. Takayanagi, and X.D. Xie, editors, *The Structure of Surfaces III.* Springer-Verlag, Berlin, 1991, p. 432.

46. A. Wander, M.A. Van Hove, and G.A. Somorjai. *Phys. Rev. Lett.,* **67**:626 (1991).

47. R.J. Hamers, Ph. Avouris, and F. Bozso. *J. Vacuum Sci. Technol.* A**6**:508 (1988).

48. F. Shoji, K. Kashikara, K. Sumitomo, and K. Oura. *Surf. Sci.* **242**:422 (1991).

49. B.I. Craig and P.V. Smith. *Surf. Sci.* **266**:L55 (1990).

50. M.J. Bozack, M.J. Dresser, W.J. Choyke, P.A. Taylor, and J.T. Yates. *Surf. Sci.* **184**:L332 (1987).

51. G. Thornton, P.L. Wincott, R. McGrath, I.T. McGovern, F.M. Quinn, D. Norman, and D.D. Vvedensky. *Surf. Sci.* **211/212**, 959 (1989).

52. L. Incoccia, A. Balerna, s. Cramm, C. Kunz, F. Senf, and I. Storjohann. *Surf. Sci.* **189/190**:453 (1987).

53. P.V. Smith and A. Wander. *Surf. Sci.* **219**:77 (1989).

54. Y. Miyamoto and A. Oshiyama. *Phys. Rev. B* **41**:12680 (1990).

55. T. Urano, Y. Uchida, S. Hongo, and T. Kanaji. *Surf. Sci.* **242**:39 (1991).

56. T. Kendelewicz, P. Soukiassian, R.S. List, J.C. Woicik, P. Pianetta, I. Lindau, and W.E. Spicer. *Phys. Rev. B* **37**:7115 (1988).

57. Y. Ling, A.J. Freeman, and B. Delley. *Phys. Rev. B* **39**:10144 (1989).

58. C.M. Wei, H. Huang, S.Y. Tong, G.S. Glander and M.B. Webb, *Phys. Rev. B* **42**:11284 (1990).

59. K. Kobayashi, S. Bluegel, H. Ishida, and K. Terakura. *Surf. Sci.* **242**:349 (1991).

60. H. Huang, S.Y. Tong, W.S. Yang, H.D. Shih, and F. Jona. *Phys. Rev. B* **42**:7483 (1990).

61. J.E. Northrup. *Phys. Rev. Lett.* **53**:683 (1984).

62. A. Kawazu and H. Sakama. *Phys. Rev. B* **37**:2704 (1988).

63. J. Nogami, S.-I. Park, and C.F. Quate. *Surf. Sci.* **203**:L631 (1988).

64. J. Zegenhagen, J.R. Patel, P.E. Freeland, D.M. Chen, J.A. Golovchenko, P. Bedrossian, and J.E. Northrup. *Phys. Rev. B* **39**:1298 (1989).

65. K.M. Conway, J.E. McDonald, C. Norris, E. Vlieg, and J.F. van der Veen. *Surf. Sci.* **215**:555 (1989).

66. H. Huang, C.M. Wei, H. Li, B.P. Tonner, and S.Y. Tong. *Phys. Rev. Lett.* **62**:559 (1989).

67. R.L. Headrick, I.K. Robinson, E. Vlieg, and L.C. Feldman. *Phys. Rev. Lett.* **63**:1253 (1989).

68. H. Huang, S.Y. Tong, J. Quinn, and F. Jona. *Phys. Rev. B* **41**:3276 (1990).

69. T. Urano, M. Kaburagi, S. Hongo, and T. Kanaji. *Appl. Surf. Sci.* **41**:103 (1989).

70. J.R. Patel, J.A. Golovchenko, P.E. Freeland and H.-J. Gossmann. *Phys. Rev. B* **36**:7715 (1987).

71. R.L. Headrick and W.R. Graham. *Phys. Rev. B* **37**:1051 (1988).

72. M. Copel, R.M. Tromp, and U.K. Koehler. *Phys. Rev. B* **37**:10756 (1988).

73. C.H. Patterson and R.P. Messmer. *Phys. Rev. B* **39**:1372 (1989).

74. J.R. Patel, P.E. Freeland, M.S. Hybertsen, D.C. Jacogsen, and J.A. Golovchenko. *Phys. Rev. Lett.* **59**:2180 (1987).

75. J.E. Northrup. *Phys. Rev. B* **37**:8513 (1988).

76. T. Abukawa, C.Y. Park, and S. Kono. *Surf. Sci.* **201**:L513 (1988).

77. T. Takahashi, S. Nakatani, T. Ishikawa, and S. Kikuta. *Surf. Sci.* **191**:L825 (1987).

TABLE 2.7a. Surface Structures of Carbon Monoxide (CO), on Different Substrates

Adsorbed Gas	Surface	Surface Structure	References
CO	Co(0001)	$(\sqrt{3} \times \sqrt{3})R30°$–CO	1
		Hexagonal overlayer	1
CO	Cu(111)	Not adsorbed	2
		$(\sqrt{3} \times \sqrt{3})R30°$	3, 4
		$(\sqrt{7/3} \times \sqrt{7/3})R49.1°$	3, 4
			4
		$(\frac{3}{2} \times \frac{3}{2})$	
CO	Cu/Ni(111)	Disordered	4
CO	Ir(111)	$(\sqrt{3} \times \sqrt{3})R30°$–CO	5–10
		$(2\sqrt{3} \times 2\sqrt{3})R30°$–CO	6–10
CO	Ni(111)	$(\sqrt{3} \times \sqrt{3})R30°$–CO	11–14, 123, 124
		Hexagonal overlayer	14
		(2×2)–CO	15
CO	Pd(111)	$(\sqrt{3} \times \sqrt{3})R30°$–CO	16, 17
		Hexagonal overlayer	16
		$c(4 \times 2)$–CO	17
CO	Pt(111)	$(\sqrt{3} \times \sqrt{3})R30°$–CO	18
		$c(4 \times 2)$–CO	18–21, 134
		Hexagonal overlayer	18
		(2×2)–CO	21
CO	Re(0001)	Not adsorbed	22
		(2×2)–CO	23
		Disordered	24
		$(2 \times \sqrt{3})$	24
CO	Rh(111)	$(\sqrt{3} \times \sqrt{3})R30°$–CO	25, 132
		(2×2)–CO	25, 26, 131
CO	Ru(0001)	$(\sqrt{3} \times \sqrt{3})R30°$–CO	26–28, 130
		(2×2)–CO	26, 27

TABLE 2.7a. (*Continued*)

Adsorbed Gas	Surface	Surface Structure	References
		(1×1)–CO	130
CO	Ti(0001)	(1×1)–CO	29, 30
		(2×2)–CO	30
CO	Th(111)	Disordered	31
		$ThO_2(111)$	31
CO	Au(100)	Disordered	32
CO	Co(100)	c(2×2)–CO	33
		(2×2)–C	33
CO	Cu(100)	c(2×2)–CO	34–36, 129
		Hexagonal overlayer	35–37
		(2×2)–C	34, 38
CO	Mo(100)	Disordered	39
		(1×1)–CO	39–42
		c(2×2)–CO	40, 42, 43
		(4×1)–CO	40
CO	Ni(100)	c(2×2)–CO	44–51
			123–128
		(2×2)–CO	52
		Hexagonal overlayer	48, 50, 51
		(2×2)–C	44
CO	Fe(100)	c(2×2)–CO	53
		c(2×2)–C, O	122
CO	Ir(100)	c(2×2)–CO	54, 55
		(2×2)–CO	54
		(1×1)–CO	56
CO	Pd(100)	Disordered	57
		c(4×2)–CO	57
		c(2×2)–CO	58
		$(2 \times 4)R45°$–CO	59–61
		$(2\sqrt{2} \times 2)R45°$–CO	133
		Hexagonal overlayer	59, 61
CO	Pt(100)	c(4×2)–CO	62–67
		$(3\sqrt{2} \times \sqrt{2})R45°$–CO	63–65, 68
		$(\sqrt{2} \times \sqrt{5})R45°$–CO	65, 69
		(2×4)–CO	70
		(1×3)–CO	70
		(1×1)–CO	62, 66, 67, 71
		c(2×2)–CO	67, 71
CO	Rh(100)	c(2×2)–CO	72
		Hexagonal overlayer	72
		(4×1)–CO	73
CO	Ta(100)	c(3×1)–O	74
CO	Th(100)	Disordered	75
CO	W(100)	Discovered	76
		c(2×2)–CO	76, 77
CO	Cu(110)	One-dimensional order	78
		(2×3)–CO	78
		(2×1)–CO	78
		Hexagonal overlayer	79

TABLE 2.7a. (*Continued*)

Adsorbed Gas	Surface	Surface Structure	References
CO	Cu/Ni(110)	(2 × 1)–CO	80
		(2 × 2)–CO	80
CO	Fe(110)	$\begin{vmatrix} 3 & -2 \\ 0 & 4 \end{vmatrix}$ –CO	81
CO	Ir(110)	(2 × 2)–CO	82, 83
		(4 × 2)–CO	83
CO	Mo(110)	(1 × 1)–CO	84, 85
		c(2 × 2)–CO	86
		Disordered	87
	Mo(211)	Disordered	88
CO	Nb(110)	Disordered	89
		(3 × 1)–O	89
CO	Ni(110)	(1 × 1)–CO	90, 91
		Adsorbed	92
		c(2 × 1)–CO	93–95
		(2 × 1)–CO	92, 96, 97
		c(2 × 2)–CO	95
		(4 × 2)–CO	95
CO	Pd(110)	(5 × 2)–CO	98
		(2 × 1)–CO	98,99
		(4 × 2)–CO	99
		c(2 × 2)–CO	99
CO	Pt(110)	(1 × 1)–CO	100, 101
		(2 × 1)–CO	102
CO	Rh(110)	(2 × 1)–CO	103
		c(2 × 2)–C	103
CO	Ru(1010)	Disordered	104
	Ru(101)	$\begin{vmatrix} 1 & 1 \\ 3 & 0 \end{vmatrix}$ –CO	105
		$\begin{vmatrix} 0 & 1 \\ 2 & 0 \end{vmatrix}$ –C	105
CO	Ta(100)	Disordered	106, 107
		(3 × 1)–O	106, 107
	Ta(211)	Disordered	106, 107
		(3 × 1)–)	107
CO	V(110)	Disordered	106
		(3 × 1)–O	106
CO	W(110)	Disordered	108
		c(9 × 5)–CO	108
		(1 × 1)–CO	109
		c(2 × 2)–CO	109
		(2 × 7)–CO	110
		c(4 × 1)–CO	110
		(3 × 1)–CO	110
		(4 × 1)–CO	110
		(5 × 1)–CO	110, 111
		(2 × 1)–(C + O)	110, 111
		c(9 × 5)–(C + O)	110

TABLE 2.7a. (*Continued*)

Adsorbed Gas	Surface	Surface Structure	References
	W(211)	Disordered	112
		c(6 × 4)–CO	112
		(2 × 1)–CO	112
		c(2 × 4)–CO	112
	W(210)	(2 × 1)–CO	113
		(1 × 1)–CO	113
CO	Cu[2_1(100) + 1_1(111)]	Adsorbed	114
	Cu(S)–[3_3(100) + 1_1(010)]	Not adsorbed	115
	Cu(S)–[4_4(100) + 1_1(010)]	Not adsorbed	115
CO	Ir(S)–[5_5(111) + 2_1(100)]	Disordered	116
CO	Pd[2_2(100) + 1_1(010)]	(1 × 1)–CO	117, 118
		(1 × 2)–CO	117, 118
	Pd[2_1(100) + 1_1(111)]	(2 × 1)–CO	117
		3(1d)–CO	117
	Pd(S)–[7_7(111) + 2_1(110)]	($\sqrt{3}$ × $\sqrt{3}$)R30°–CO	117
		Hexagonal overlayer	117
CO	Pt(S)–[5_5(111) + 2_1(100)]	Disordered	119
	Pt(S)–[7_7(111) + 2_1(110)]	Disordered	119
CO	Re(S)–[14(0001) × (10$\bar{1}$1)]	(2 × 2)–CO	120
		(2 × 1)–C	120
CO	Rh(S)–[1_1(111) + 2_1(110)]	$\begin{vmatrix} 1 & 2 \\ 5 & -1 \end{vmatrix}$–CO	121
		$\begin{vmatrix} 1 & 2 \\ 2 & 0 \end{vmatrix}$–CO	121
		Hexagonal overlayer	121
	Rh(S)–[5(111) + 2_1(100)]	($\sqrt{3}$ × $\sqrt{3}$)R30°–CO	121
		(2 × 2)–CO	121
CO	Fe(001)	c(2 × 2)–CO	122
CO	Ni(001)	c(2 × 2)–CO	123, 124, 126 129
CO	Cu(001)	c(2 × 2)–CO	129

REFERENCES

1. M.E. Bridge, C.M. Comrie, and R.M. Lambert. *Surf. Sci.* **67**:393 (1977).

2. G. Ertl. *Surf. Sci.* **7**:309 (1967).

3. J. Kessler and F. Thieme. *Surf. Sci.* **67**:405 (1977).

4. C. Benndorf, K.H. Gressman, and F. Thieme. *Surf. Sci.* **61**:646 (1976).

5. T. Edmonds and R.C. Pitkethyly. *Surf. Sci.* **17**:450 (1969).

6. V.P. Ivanov, G.K. Boreskov, V.I. Savchenko, W.F. Egelhoff, Jr., and W.H. Weinberg. *J. Catal.* **48**:269 (1977).

7. D.I. Hagen, B.E. Nieuwenhuys, G. Rovda, and G.A. Somorjai. *Surf. Sci.* **57**:632 (1976).

8. J. Küppers and A. Plagge. *J. Vacuum sci. Technol.* **13**:259 (1976).

9. C.M. Comrie and W.H. Weinberg. *J. Vacuum Sci. Technol.* **13**:264 (1976).

10. C.M. Comrie and W.H. Weinberg. *J. Chem. Phys.* **64**:250 (1976).

11. P.H. Holloway and J.B. Hudson. *Surf. Sci.* **43**:141 (1974).

12. H. Conrad, G. Ertl, J. Küppers, and E.E. Latta. *Surf. Sci.* **57**:475 (1976).

13. G. Ertl. *Surf. Sci.* **47**:86 (1975).
14. K. Christmann, O. Schober, and G. Ertl. *J. Chem. Phys.* **60**:4719 (1974).
15. L.H. Germer, E.J. Schneiber, and C.D. Hartman. *Philos. Mag.* **5**:222 (1960).
16. H. Conrad, G. Ertl, J. Koch, and E.E. Latta. *Surf. Sci.* **43**:462 (1974).
17. A.M. Bradshaw and F.M. Hoffman. *Surf. Sci.* **72**:513 (1978).
18. G. Ertl, M. Neumann, and K.M. Streit. *Surf. Sci.* **64**:393 (1977).
19. B. Lang, R.W. Joyner, and G.A. Somorjai, *Surf. Sci.* **30**:454 (1972).
20. S.L. Bernasek and G.A. Somorjai, *J. Chem. Phys.* **60**:4552 (1974).
21. A.E. Morgan and G.A. Somorjai. *J. Chem. Phys.* **51**:3309 (19690.
22. G.J. Dooley and T.W. Haas. *Surf. Sci.* **19**:1 (1970).
23. H.E. Farnsworth and D.M. Zehner. *Surf. Sci.* **17**:7 (1969).
24. M. Housley, M. Ducros, G. Piquard, and A. Cassuto. *Surf. Sci.* **68**:277 (1977).
25. D.G. Castner, B.A. Sexton, and G.A. Somorjai. *Surf. Sci.* **71**:519 (1978).
26. J.T. Grant and T.W. Haas. *Surf. Sci.* **21**:76 (1970).
27. J.C. Fuggle, E. Umbach, P. Feulner, and D. Menzel. *Surf. Sci.* **64**:69 (1977).
28. T.E. Madey and D. Menzel. *Proceedings of the 2nd International Conference on Solid Surfaces*, 1974, p. 229.
29. H.E. Farnsworth, R.E. Schlier, T.H. George, and R.M. Buerger. *J. Appl. Phys.* **29**:1150 (1958).
30. H.D. Shih, F. Jona, D.W. Jepsen, and P.M. Marcus. *J. Vacuum Sci. Technol.* **15**:596 (1978).
31. R. Bastasz, C.A. Colmenares, R.L. Smith, and G.A. Somorjai. *Surf. Sci.* **67**:45 (1977).
32. G. McElhiney and J. Pritchard. *Surf. Sci.* **60**:397 (1976).
33. M. Maglietta and G. Robida. *Surf. Sci.* **71**:495 (1978).
34. J.C. Tracy. *J. Chem. Phys.* **56**(6):2748 (1971).
35. C.R. Brundle and K. Wandelt. *Proceedings of the 7th International Vacuum Congress and 3rd International Conference on Solid Surface.*, Vienna, 1977, p. 1171.
36. M.A. Chesters and J. Pritchard. *Surf. Sci.* **28**:460 (1971).
37. G. Ertl. *Surf. Sci.* **7**:309 (1967).
38. R.E. Schlier and H.E. Farnsworth. *J. Appl. Phys.* **25**:1333 (1954).
39. H.E. Farnsworth and K. Hayek. *Nuovo Cimento Suppl.* **5**:2 (1967).
40. R. Riwan, C. Guillot and J. Paigne. *Surf. Sci.* **47**:183 (1975).
41. C. Guillot, R. Riwan, and J. Lecante. *Surf. Sci.* **59**:581 (1976).
42. C.J. Dooley and T.W. Haas. *J. Chem. Phys.* **52**:993 (1970).
43. J. Perdereau and G.E. Rhead. *Surf. Sci.* **24**:555 (1971).
44. E.G. McRae and C.W. Caldwell. *Surf. Sci.* **57**:63 (1976).
45. R.L. Park and H.E. Farnsworth. *J. Chem. Phys.* **43**:2351 (1965).
46. L.H. Germer. *Adv. Catal.* **13**:191 (1962).
47. J.C. Tracy. *J. Chem. Phys.* **56**(6):2736 (1971).
48. S. Andersson and J.B. Pendry. *Surf. Sci.* **71**:75 (1978).
49. S. Andersson, *Proceedings of the 3rd International Vacuum Congress and 7th Int. Conference on Solid Surfaces*, Vienna, 1977, p. 1019.
50. K. Horn, A.M. Bradshaw, and K. Jacobi. *Surf. Sci.* **72**:719 (1978).
51. R.A. Armstrong. In: G.A. Somorjai, editor, *The Structure and Chemistry of Solid Surfaces*. Wiley, New York, 1969.
52. M. Onchi and H.E. Farnsworth. *Surf. Sci.* **13**:425 (1969).
53. F. Boszo, G. Ertl, M. Grunze, and M. Weiss. *Appl. Surf. Sci.* **1**:103 (1978).
54. G. Brodén and T.N. Rhodin. *Solid State Commun.* **18**:105 (1976).
55. B.E. Nieuwenhuys and G.A. Somorjai. *Surf. Sci. Surf.* **72**:8 (1978).
56. J.J. Lander and J. Morrison. *J. Appl Phys.* **34**:1411 (1963).
57. A.M. Bradshaw and F.M. Hoffman. *Surf. Sci.* **72**:513 (1978).

58. G. Pirug, G. Brodén and H.P. Bonzel. *Proceedings of the 7th International Vacuum Congress and 3rd International Conference on Solid Surfaces*, Vienna, 1977, p. 907.

59. R.L. Park and H.H. Madden. *Surf. Sci.* **11**:188 (1968).

60. H. Conrad, G. Ertl, J. Koch, and E.E. Latta. *Surf. Sci.* **43**:462 (1974).

61. H. Conrad, G. Ertl, J. Küppers, and E.E. Latta. *Surf. Sci.* **65**:235 (1977).

62. B. Lang, R.W. Joyner, and G.A. Somorjai. *Surf. Sci.* **30**:454 (1972).

63. A.E. Morgan and G.A. Somorjai. *J. Chem. Phys.* **51**:3309 (1969).

64. A.E. Morgan and G.A. Somorjai. *Surf. Sci.* **12**:405 (1968).

65. C. Burggraf and A. Mosser. *C.R. Acad. Sci.* **268B**:1167 (1969).

66. G. Kneringer and F.P. Netzer. *Surf. Sci.* **49**:125 (1975).

67. G. Brodén and G. Pirug, and H.P. Bonzel. *Surf. Sci.* **72**:45 (1978).

68. F.P. Netzer and G. Kneringer. *Surf. Sci.* **51**:526 (1975).

69. H.P. Bonzel and G. Pirug. *Surf. Sci.* **62**:45 (1977).

70. A.E. Morgan and G.A. Somorjai. *Trans. Am. Crystallogr. Assoc.* **4**:59 (1968).

71. C.R. Helms, H.P. Bonzel, and S. Kelemen. *J. Chem. Phys.* **65**:1773 (1976).

72. D.G. Castner, B.A. Sexton, and G.A. Somorjai. *Surf. Sci.* **71**:159 (1978).

73. J.J. Lander and J. Morrison. *J. Appl. Phys.* **33**:2089 (1962).

74. M.A. Chesters, B.J. Hopkins, and M.R. Leggett. *Surf. Sci.* **43**:1 (1974).

75. T.N. Taylor, C.A. Colmenares, R.L. Smith, and G.A. Somorjai. *Surf. Sci.* **54**:317 (1976).

76. J. Anderson and P.J. Estrup. *J. Chem. Phys.* **46**:563 (1967).

77. P.J. Estrup. In: G.A. Somorjai, editor, *The Structure and Chemistry of Solid Surfaces*. Wiley, New York, 1969.

78. G. Ertl. *Surf. Sci.* **7**:309 (1967).

79. K. Horn, M. Hussain, and J. Pritchard. *Surf. Sci.* **63**:244 (1977).

80. G. Ertl and J. Küppers. *Surf. Sci.* **24**:104 (1971).

81. G. Gafner and R. Feder. *Surf. Sci.* **57**:37 (1976).

82. B.E. Nieuwenhuys and G.A. Somorjai. *Surf. Sci.* **72**:8 (1978).

83. J.L. Taylor and W.H. Weinberg, *J. Vacuum Sci. Technol.* **15**:590 (1978).

84. K. Hayek and H.E. Farnsworth. *Surf. Sci.* **10**:429 (1956).

85. T.W. Haas and A.G. Jackson. *J. Chem. Phys.* **44**:2121 (1966).

86. A.J. Pignosco and G.E. Pellisier. *Surf. Sci.* **7**:261 (1967).

87. E. Gillet, J.C. Chiarena, and M. Gillet. *Surf. Sci.* **67**:393 (1977).

88. G.J. Dooley and T.W. Haas. *J. Vac. Sci. Technol.* **7**:49 (1970).

89. T.W. Haas, A.G. Jackson, and M.P. Hooker. *J. Chem. Phys.* **46**:3025 (1967).

90. A.U. McRae. *Surf. Sci.* **1**:319 (1964).

91. A.G. Jackson and M.P. Hooker. *Surf. Sci.* **6**:297 (1967).

92. J.E. Demuth and T.N. Rhodin. *Surf. Sci.* **45**:249 (1974).

93. J. Küppers. *Surf. Sci.* **36**:53 (1973).

94. H.H. Madden, J. Küppers, and G. Ertl, *J. Chem. Phys.* **58**:3401 (1973).

95. T.N. Taylor and P.J. Estrup. *J. Vac. Sci. Technol.* **10**:26 (1973).

96. H.H. Madden and G. Ertl. *Surf. Sci.* **35**:211 (1973).

97. H.H. Madden, J. Küppers, and G. Ertl. *J. Vacuum Sci. Technol.* **11**:190 (1974).

98. G. Ertl and P. Rau. *Surf. Sci.* **15**:443 (1969).

99. H. Conrad, G. Ertl, J. Koch, and E.E. Latta. *Surf. Sci.* **43**:462 (1974).

100. H.P. Bonzel and R. Ku. *Surf. Sci.* **33**:91 (1972).

101. R.M. Lambert and C.M. Comrie. *Surf. Sci.* **46**:61 (1974).

102. R.M. Lambert. *Surf. Sci.* **49**:325 (1974).

103. R.A. Marbrow and R.M. Lambert. *Surf. Sci.* **67**:489 (1977).

104. R. Ku, N.A. Gjostein and H.P. Bonzel. *Surf. Sci.* **64**:465 (1977).

105. P.D. Reed, C.M. Comrie, and R.M. Lambert. *Surf. Sci.* **59**:33 (1976).

106. T.W. Haas, A.G. Jackson and M.P. Hooker. *J. Chem. Phys.* **46**:3025 (1967).

107. T.W. Haas. In: G.A. Somorjai, editor, *The Structure and Chemistry of Solid Surfaces.* Wiley, New York, 1969.

108. J.W. May and L.H. Germer. *J. Chem. Phys.* **44**:2895 (1966).

109. J.M. Baker and D.E. Eastman. *J. Vacuum Sci. Technol.* **10**:223 (1973).

110. Ch. Steinbruchel and R. Gomer. *Surf. Sci.* **67**:21 (1977).

111. Ch. Steinbruchel and R. Gomer. *J. Vacuum Sci. Technol.* **14**:484 (1977).

112. C.C. Chang. *J. Electrochem. Soc.* **115**:354 (1968).

113. D.L. Adams and L.H. Germer. *Surf. Sci.* **32**:205 (1972).

114. H. Papp and J. Pritchard. *Surf. Sci.* **53**:371 (1975).

115. J. Perdereau and G.E. Rhead. *Surf. Sci.* **24**:555 (1971).

116. D.I. Hagen, B.E. Nieuwenhuys, G. Rovida, and G.A. Somorjai. *Surf. Sci.* **57**:632 (1976).

117. H. Conrad, G. Ertl, J. Koch, and E.E. Latta. *Surf. Sci.* **43**:462 (1974).

118. A.M. Bradshaw and F.M. Hoffman. *Surf. Sci.* **72**:513 (1978).

119. B. Lang, R.W. Joyner, and G.A. Somorjai. *Surf. Sci.* **30**:454 (1972).

120. M. Housley, R. Ducros, G. Piquard, and A. Cassuto. *Surf. Sci.* **68**:277 (1977).

121. D.G. Castner and G.A. Somorjai. *Surf. Sci.* **83**:60 (1979).

122. F. Jona, K. Legg, H.D. Shih, D.W. Jepsen, and P.M. Marcus. *Phys. Rev. Lett.* **40**:1466 (1978).

123. S.D. Kevan, R.F. Davis, D.H. Rosenblatt, J.G. Tobin, M.G. Mason, D.A. Shirley, C.H. Li, and S.Y. Song. *Phys. Rev. Lett.* **46**:1629 (1981).

124. R.F. Davis, S.D. Kevan, D.H. Rosenblatt, M.G. Mason, J.G. Tobin, and D.A. Shirley. *Phys. Rev. Lett.* **45**:1877 (1980).

125. K. Heinz, E. Lang, and K. Mueller. *Surf. Sci.* **87**:595 (1979).

126. M. Passler, A. Ignatiev, F. Jona, D.W. Jepsen, and P.M. Marcus. *Phys. Rev. Lett.* **43**:360 (1979).

127. L.G. Petersson, S. Kono, N.F.T. Hall, C.S. Fadley, and J.B. Pendry. *Phys. Rev. Lett.* **42**:1545 (1979).

128. S.Y. Tong, A. Maldonado, C.H. Li, and M.A. Van Hove. *Surf. Sci.* **94**:73 (1980).

129. S. Andersson and J.B. Pendry. *J. Phys.* **C13**:3547 (1980).

130. G. Michalk, W. Moritz, H. Pfnur, and D. Menzel. *Surf. Sci.* **129**:92 (1983).

131. M.A. Van Hove, R.J. Koestner, J.C. Frost, and G.A. Somorjai. *Surf. Sci.* **129**:482 (1983).

132. R.J. Koestner, M.A. Van Hove, and G.A. Somorjai. *Surf. Sci.* **107**:439 (1981).

133. R.J. Behm, K. Christmann, G. Ertl, and M.A. Van Hove. *J. Chem. Phys.* **73**:2984 (1980).

134. D.F. Ogletree, M.A. Van Hove, and G.A. Somorjai. *Surf. Sci.* **173**:351 (1986).

TABLE 2.7b. Surface Structures of Nitric Oxide (NO) on Different Substrates

Adsorbed Gas	Surface	Surface Structure	References
NO	Ag(111)	Disordered	1
NO	Ir(111)	(2 × 2)–NO	2
NO	Ni(111)	c(4 × 2)–NO	3
		Hexagonal overlayer	3
		(2 × 2)–O	3
		(6 × 2)–N	3
NO	Pd(111)	c(4 × 2)–NO	4
		(2 × 2)–NO	4
NO	Pt(111)	Disordered	5

TABLE 2.7b. (*Continued*)

Adsorbed Gas	Surface	Surface Structure	References
NO	Rh(111)	c(4 × 2)–NO	6
		(2 × 2)–NO	6
NO	Ir(100)	(1 × 1)–NO	2, 7
NO	Pt(100)	(1 × 1)–NO	8
		c(4 × 2)–NO	9
NO	Rh(100)	c(2 × 2)–NO	10
NO	Ta(100)	c(3 × 1)–O	11
NO	W(100)	(2 × 2)–NO	12
		(4 × 1)–NO	12
		(2 × 2)–O	12
		(4 × 1)–O	12
		(2 × 1)–O	12
NO	Ag(110)	Disordered	13
NO	Ni(110)	(2 × 3)–N	14
		(2 × 1)–O	14
NO	Pt(110)	(1 × 1)–NO	15, 16
NO	Ru(10$\bar{1}$0)	c(4 × 2)–(N + O)	17, 18
		(2 × 1)–(N + O)	17, 18
		(2 × 1)–O	18
		c(4 × 2)–O	18
		c(2 × 6)–O	17
		(7 × 1)–O	17
		c(4 × 8)–O	17
		(2 × 1)–N	18
		c(4 × 2)–N	18
	Ru(101)	Disordered	19
NO	Pt(S)–[11$_{11}$(111) + 1$_1$(1$\bar{1}$1)]	(2 × 2)–NO	20
NO	Rh(2)–[1$_1$(111) + 2$_1$(110)]	Disordered	21
		$\begin{bmatrix} -1 & 1 \\ 0 & 0 \end{bmatrix}$	21
	Rh(S)–[5(111) + 2$_1$(100)]	(2 × 2)–NO	21

REFERENCES

1. P.J. Goddard, J. West, and R.M. Lambert. *Surf. Sci.* **71**:447 (1978).
2. J. Kanski and T.N. Rhodin. *Surf. Sci.* **65**:63 (1977).
3. H. Conrad, G. Ertl, J. Küppers, and E.E. Latta. *Surf. Sci.* **50**:296 (1975).
4. H. Conrad, G. Ertl, J. Küppers, and E.E. Latta. *Surf. Sci.* **54**:235 (1977).
5. C.M. Comrie, W.H. Weinberg, and R.M. Lambert. *Surf. Sci.* **57**:519 (1976).
6. D.G. Castner, B.A. Sexton, and G.A. Somorjai. *Surf. Sci.* **71**:519 (1978).
7. A. Ignatiev, T.N. Rhodin, and S.Y. Tong. *Surf. Sci.* **42**:37 (1974).
8. H.P. Bonzel and G. Pirug. *Surf. Sci.* **62**:45 (1977).
9. H.P. Bonzel, G. Brodén, and G. Pirug. *J. Catal.* **53**:96 (1978).
10. D.G. Castner, B.A. Sexton, and G.A. Somorjai. *Surf. Sci.* **71**:159 (1978).
11. M.A. Chesters, B.J. Hopkins, and M.R. Leggett. *Surf. Sci.* **43**:1 (1974).
12. S. Usami and T. Nakagima. *Proceedings of the 2nd International Conference on Solid Surfaces*, 1974, p. 237.

13. R.A. Marbrow and R.M. Lambert. *Surf. Sci.* **61**:317 (1976).

14. M. Perdereau and J. Oudar. *Surf. Sci.* **20**:80 (1970).

15. R.M. Lambert and C.M. Comrie. *Surf. Sci.* **46**:61 (1974).

16. C.M. Comrie, W.H. Weinberg, and R.M. Lambert. *Surf. Sci.* **57**:619 (1976).

17. T.W. Orent and R.S. Hansen. *Surf. Sci.* **67**:325 (1977).

18. R. Ku, N.A. Gjostein, and H.P. Bonzel. *Surf. Sci.* **64**:465 (1977).

19. P.D. Reed, C.M. Comrie, and R.M. Lambert. *Surf. Sci.* **72**:423 (1978).

20. J. Gland, *Surf. Sci.* **71**:327 (1978).

21. D.G. Castner and G.A. Somorjai. *Surf. Sci.* **83**:60 (1979).

TABLE 2.8. Adsorption Geometries of Benzene, Indicating Average Carbon-Ring Radius, C–C Bond Lengths (Two Values Where Long and Short Bonds Coexist), Metal–Carbon Distances, and Adsorption Sites of C_6H_6 Ring Centers

System	C_6 Radius (nm)	d_{C-C} (nm)	d_{M-C} (nm)	Site
Benzene/surface				
Pd(111)–(3 × 3)–C_6H_6	0.141	0.141 ± 0.010	0.239 ± 0.005	fcc
+ 2CO (1)	± 0.010			hollow
Rh(111)–(3 × 3)–C_6H_6	0.151	0.158 ± 0.015	0.230 ± 0.005	hcp
+ 2CO (2)	± 0.015	0.146 ± 0.015		hollow
Rh(111)–c(2 $\sqrt{3}$ × 4)rect–C_6H_6	0.165	0.181 ± 0.015	0.235 ± 0.005	hcp
+ CO (3)	± 0.015	0.133 ± 0.015		hollow
Pt(111)–(2 $\sqrt{3}$ × 4)rect–$2C_6H_6$	0.172	0.176 ± 0.015	0.225 ± 0.005	Bridge
+ 4CO (4)	± 0.015	0.165 ± 0.015		
Gas				
C_6H_6 molecule	0.1397	0.1397		

REFERENCES

1. H. Ohtani, M.A. Van Hove, and G.A. Somorjai. *J. Phys. Chem.* **92**:3974 (1988).

2. G.S. Blackman, R.F. Lin, M.A. van Hove, and G.A. Somorjai. *Acta Crystallogr. B* **43**:368 (1987).

3. M.A. Van Hove, R.F. Lin, and G.A. Somorjai. *J. Am. Chem. Soc.* **108**:2532 (1986).

4. D.F. Ogletree, M.A. Van Hove, and G.A. Somorjai. *Surf. Sci.* **183**:1 (1987).

TABLE 2.9. Coadsorption Systems

Type of Coadsorbate	LEED Pattern(s) Observed
Intermixed or Ordered	
CO + C_6H_6	c(2 $\sqrt{3}$ × 4)rect, (3 × 3)
CO + C_6H_5F	Streaky c(2 $\sqrt{3}$ × 4)rect, (3 × 3)
CO + C_2H_2	c(4 × 2)
CO + $\equiv CCH_3$	c(4 × 2)
NO + $\equiv CCH_3$	c(4 × 2)
CO + Na	c(4 × 2), ($\sqrt{3}$ × 7)rect, plus several others
CO + $\equiv CCH_2CH_3$	(2 $\sqrt{3}$ × 2 $\sqrt{3}$)R30°

TABLE 2.9. *(Continued)*

Type of Coadsorbate	LEED Pattern(s) Observed
Nonintermixed or disordered	
NO + CO	Disordered
Na + C_2H_2	Disordered
Na + $\equiv CCH_3$	Disordered
Na + C_6H_6	$(\sqrt{3} \times \sqrt{3})R30° + (2\sqrt{3} \times 3)$rect

REFERENCE

1. C.M. Mate, C.T. Kao, and G.A. Somorjai. *Surface Sci.* **206**:145 (1988).

3

THERMODYNAMICS OF SURFACES

3.1 INTRODUCTION

Atoms on a surface of a solid have an environment that differs markedly from that of atoms in the bulk of the solid. They have fewer neighbors than do bulk atoms, and the neighbors of each surface atom may be distributed anisotropically. We will define the thermodynamic properties associated with this surface region separately from bulk thermodynamic properties. We will show that atoms (or molecules) of one type accumulate at the surface in multicomponent systems. This gives rise to spectacular physical and chemical properties (why stainless steel resists corrosion and the action of detergents, for example). We will also show that because of the balance of various forces acting on surface atoms, small liquid drops are curved and thus are liquid interfaces which also may rise when placed in capillaries. These are just some of the surface properties that will be discussed in this chapter on surface thermodynamics.

3.2 DEFINITION OF SURFACE THERMODYNAMIC FUNCTIONS

Consider a large homogeneous crystalline solid that contains N atoms and which is bounded by surface planes. The energy and the entropy of the solid per atom are denoted by $E°$ and $S°$, respectively. The specific surface energy E^s (energy per unit area) is defined by the relation [1]

$$E = NE° + \alpha E^s \tag{3.1}$$

where N is the number of atoms in the solid, E is the total energy of the solid, and α is the surface area [2]. Thus E^s is the excess of the total energy E that the solid has over the value $NE°$, which is the value it would have if the surface were in the same thermodynamic state as the homogeneous interior. Similarly, we can write the total entropy S of the solid as

$$S = NS° + \alpha S^s \tag{3.2}$$

where S^s is the specific surface entropy (entropy per unit area of surface created). The surface work content A^s (energy per unit area) is defined by the equation

$$A^s = E^s - TS^s \tag{3.3}$$

where T is the temperature in degrees Kelvin and the surface free energy G^s (energy per unit area) is defined by the equation

$$G^s = H^s - TS^s \tag{3.4}$$

where H^s is the specific surface enthalpy—that is, the heat absorbed by the system per unit surface area created. The total free energy of a system G can also be expressed as

$$G = NG° + \alpha G^s \tag{3.5}$$

where $G°$ and G^s are the free energy per atom and per unit area of surface, respectively, analogous to the total energy and entropy in Eqs. 3.1 and 3.2. We have thus defined the thermodynamic properties of the surface as excesses of the bulk thermodynamic properties, due to the presence of the surface surrounding the condensed phase.

3.3 WORK NEEDED TO CREATE A SURFACE OF A ONE-COMPONENT SYSTEM: SURFACE TENSION

To increase the surface area of a solid, we have to bring atoms from the bulk of a solid or liquid to the surface and move the atoms that are already on the surface along the surface to accommodate the new surface atoms. Under conditions of equilibrium at constant temperature T and pressure P, the reversible surface work* δW^s required to increase the surface area $\mathcal{Q}$ by an amount $d\mathcal{Q}$ of a one-component system is given by

$$\delta W^s_{T,P} = \gamma \, d\mathcal{Q} \tag{3.6}$$

Here γ is the two-dimensional analogue of the pressure and is called the "surface tension," while the volume change is replaced by the change in surface area. Equation 3.6 can be compared with the reversible work needed to increase the volume of a one-component system at constant pressure, $P \, dV$. Both P and γ have directions: While P is always perpendicular to the surface, γ is always parallel to it.

The pressure P is the force per unit area [dynes/cm^2 (N/m^2)], while the surface tension γ or surface pressure has units of force per unit length [dynes/cm (N/m)]. The customary units of surface tension, N/m or J/m^2, are dimensionally identical. We may consider γ as a pressure along the surface plane that opposes the creation of more surface.

A rough estimate of the magnitude of surface tension can be made by assuming that the surface work is of the same magnitude as the heat of sublimation, since sublimation continually creates a new surface. For many metals the heat of sublimation is in the range of 10^5 cal/mole (4.18×10^5 J/mole). Using a units-conversion table, one obtains $10^5 \times 6.94 \times 10^{-24} = 6.94 \times 10^{-19}$ J/atom $= 6.94 \times 10^{-12}$ erg/atom. For a typical surface concentration of 10^{15} atoms/cm^2 (10^{19} atoms/m^2), the estimated surface tension should be on the order of $6.94 \times 10^{-19} \times 10^{19} \approx 7$ J/m^2 (7000 ergs/cm^2). Actually, for metals [3] there is a good experimental correlation between the heat of sublimation ΔH_{subl} and the surface tension γ:

$$\gamma \approx 0.16 \, \Delta H_{subl} \quad \text{(in practice)} \tag{3.7}$$

The small coefficient is due in part to the fact that it is not necessary to break all the metal–metal bonds in the bulk to create a new surface. The estimate is also too high because it does not take into consideration that there may be a "relaxation" of surface atoms in the freshly created surface. The surface atoms, as a consequence of their less symmetric atomic environment compared to atoms in the bulk, change their equilibrium positions. They usually contract toward the bulk to maximize their bonding with the remaining neighbors. This relaxation lowers the surface tension appreciably.

---

*We use the notation δW to indicate that the work W, unlike the free energy G or other thermodynamic functions, is not independent of the reaction path; that is, it is not a total differential.

The experimental values of the surface tension of several metals are listed in Table 3.1, along with the surface tensions of other liquids and solids that were measured in equilibrium with their own vapor. More comprehensive lists may be found in references [4–7].

In order to estimate the magnitude of the surface pressure or surface tension in terms of the magnitudes of the three-dimensional pressure P, we should consider γ as the pressure distributed over a 1-cm^2 surface a few atomic layers thick: $P = \gamma/d$. Assuming $\gamma = 10^3$ dyne/cm (1 N/m) and that the anisotropic surface environment influences bonding in the top three atomic layers, we may estimate $d = 1$ nm (10

TABLE 3.1. Surface Tension γ of Selected Solids and Liquids

Material	γ (ergs/cm^2)	γ (J/cm^2)	$T(°C)$
W (solid) (1)	2900	2.900	1727
Nb (solid) (1)	2100	2.100	2250
Au (solid) (1)	1410	1.410	1027
Ag (solid) (1)	1140	1.140	907
Ag (liquid) (2)	879	0.879	1100
Fe (solid) (1)	2150	2.150	1400
Fe (liquid) (2)	1880	1.880	1535
Pt (solid) (1)	2340	2.340	1311
Cu (solid) (1)	1670	1.670	1047
Cu (liquid) (2)	1300	1.300	1535
Ni (solid) (1)	1850	1.850	1250
Hg (liquid) (2)	487	0.487	16.5
LiF (solid) (3)	340	0.340	−195
NaCl (solid) (3)	227	0.227	25
KCl (solid) (3)	110	0.110	25
MgO (solid) (3)	1200	1.200	25
CaF_2 (solid) (3)	450	0.450	−195
BaF_2 (solid) (3)	280	0.280	−195
He (liquid) (2)	0.308	3.08×10^{-4}	−270.5
N_2 (liquid) (2)	9.71	9.71×10^{-3}	−195
Ethanol (liquid) (2)	22.75	0.02275	20
Water (2)	72.75	0.07275	20
Benzene (2)	28.88	0.02888	20
n-Octane (2)	21.80	0.02180	20
Carbon tetrachloride (2)	26.95	0.02695	20
Bromine (2)	41.5	0.0415	20
Acetic acid (2)	27.8	0.0278	20
Benzaldehyde (2)	15.5	0.0155	20
Nitrobenzene (2)	25.2	0.0252	20

REFERENCES

1. J.M. Blakely and P.S. Maiya. In J. J. Burke et al., editors, *Surfaces and Interfaces*. Syracuse University Press, Syracuse, N.Y., 1967.

2. A.W. Adamson. In *Physical Chemistry of Surfaces*. John Wiley & Sons (Interscience Division), New York, 1967.

3. G.C. Benson and R.S. Yuen. In E.A. Flood, editor, *The Solid-Gas Interface*. Marcel Dekker, New York, 1967.

Å). Thus, $P = 10^{10}$ dyne/cm^2 (10^9 J/m^2), or about 10^4 atmospheres. From this perspective, the atoms in a metal surface are subjected to very large compressive forces. Other solids or liquids with lower surface tension require less energy to produce a unit area of new surface and can also be much more compressible. The surface tension values may vary by about three orders of magnitude, as can be seen in Table 3.1.

3.3.1 The Surface Free Energy Is Always Positive

The change in the total free energy dG of a one-component system* can be written, with the inclusion of the surface work $\gamma\, d\alpha$, as

$$dG = -S\, dT + V\, dP + \gamma\, d\alpha \tag{3.8}$$

At constant temperature and pressure, Eq. 3.8 reduces to

$$dG_{T,P} = \gamma\, d\alpha \tag{3.9}$$

Since, in our present discussion, the change in the total free energy of the system is due only to the change in the surface free energy, we have, from Eq. 3.5,

$$dG_{T,P} = d\, (G^s\alpha) \tag{3.10}$$

In principle, there are two ways to form a new surface: (1) increasing the surface area by adding new atoms from the bulk and (2) stretching the already existing surface (as if it were a rubber mat) with the number of atoms fixed and thereby

---

*The change in the total free energy dG of a one-component is defined by the equation

$$G = H - TS \tag{a}$$

Thus the free-energy change is given by

$$dG = dH - T\, dS - S\, dT \tag{b}$$

The enthalpy is defined as

$$H = E + PV \tag{c}$$

and therefore the change in enthalpy is given by

$$dH = dE + P\, dV + V\, dP \tag{d}$$

The reversible change in energy dE due to heat absorbed by the system and expansion work by the system is given by

$$dE = T\, dS - P\, dV \tag{e}$$

Substitution of Eqs. (d) and (e) into Eq. (b) yields

$$dG = -S\, dT + V\, dP \tag{f}$$

altering the state of strain (which amounts to changing γ) [8]. We can rewrite Eq. 3.10 to yield

$$dG_{T,P} = \left(\frac{\partial(G^s\,\alpha)}{\partial\alpha}\right)_{T,P} d\alpha = \left[G^s + \alpha\left(\frac{\partial G^s}{\partial\alpha}\right)_{T,P}\right] d\alpha \qquad (3.11)$$

If we create the new surface by adding atoms from the bulk, the specific surface free energy G^s is independent of the surface area: $(\partial G^s/\partial\alpha)_{T,P} = 0$.

Combining Eqs. 3.9 and 3.11, we have

$$dG_{T,P} = G^s\,d\alpha = \gamma\,d\alpha \qquad (3.12)$$

or

$$\gamma = G^s \qquad (3.13)$$

In other words, the surface tension is equal to the specific surface free energy for a one-component system. These terms are frequently used interchangeably in the literature. However, for solids at low temperatures, "cold working" of the material can lead to the formation of a new surface by strain that is not relieved because of negligible mobility of the surface species. (Consider, for example, stretched polymer chains that interlock.)

Creation of a surface always results in a positive free energy of formation. This reluctance of the solid or liquid to form a surface defines many of the interfacial properties of condensed phases. To minimize the surface free energy, solids will form surfaces of the lowest specific surface free energy or surface tension γ, which are usually crystal faces with the closest packing of atoms. Surfaces with high values of γ will always be covered with substances that have lower surface tensions, if possible. This is shown schematically in Figure 3.1. Metals are covered by oxides (often called *complete wetting*) if the metal–gas interfacial energy, γ_{m-g}, is larger than the sum of the oxide–gas (γ_{ox-g}) and oxide–metal (γ_{ox-m}) interfacial energies: $\gamma_{m-g} > \gamma_{ox-g} + \gamma_{ox-m}$. Water will adsorb on and cover the oxide if $\gamma_{ox-g} > \gamma_{H_2O-g} + \gamma_{H_2O-ox}$. Adsorbed water can be displaced or covered by organic molecules with even lower surface tension. Liquids tend to assume a spherical shape to minimize

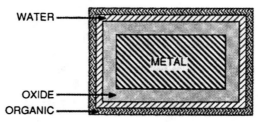

Figure 3.1. Representation of materials of lower surface energy coating materials of higher surface energy, leading to a net reduction of total surface energy ($\gamma_{new\,surface} + \gamma_{interface} < \gamma_{old\,surface}$).

their surface area. This is why curved surfaces play such an important role in surface chemistry. We will continue our discussion of curved surfaces later in the chapter.

3.3.2 Temperature Dependence of the Specific Surface Free Energy

Equation 3.13 holds for most systems in which surface-tension measurements can conveniently be carried out and for the temperature ranges used in these studies. Differentiating Eq. 3.13 as a function of temperature, we can write

$$\left(\frac{\partial G^s}{\partial T}\right)_P = \left(\frac{\partial \gamma}{\partial T}\right)_P = -S^s \qquad (3.14)$$

That is, from the temperature dependence of the surface tension we can obtain the specific surface entropy. A semiempirical equation for predicting the temperature dependence of the surface tension was proposed by van der Waals and Guggenheim [9]:

$$\gamma = \gamma^\circ (1 - T/T_c)^n \qquad (3.15)$$

where T_c is the critical temperature (the temperature at which the condensed phase vanishes) and $\gamma^\circ = \gamma$ at $T = 0$ K. According to Eq. 3.15, the surface tension should vanish at $T = T_c$ as expected, since the interface vanishes at the critical temperature. The exponent n is determined by experiments to be near unity for metals [10] and somewhat larger than unity for many organic liquids [11]. Substitution of Eq. 3.14 into Eq. 3.4 with subsequent rearrangement yields the specific surface enthalpy

$$H^s = G^s + TS^s = \gamma - T\left(\frac{\partial \gamma}{\partial T}\right)_P \qquad (3.16)$$

Thus, at constant pressure the heat absorbed upon the creation of a unit surface area is given by Eq. 3.16. If no volume change is associated with this process, then H^s equals E^s and the specific surface energy is given by the same equation:

$$E^s_{P,V} = \gamma - T\left(\frac{\partial \gamma}{\partial T}\right)_{P,V} \qquad (3.17)$$

Because the surface tension usually decreases with increasing temperature, the derivative $(\partial \gamma / \partial T)_P$ is negative. Therefore, the specific surface energy is somewhat larger than the specific surface free energy G^s (or γ). In theoretical calculations of surface thermodynamic properties, the specific surface energy is given more frequently than the specific surface free energy. However, G^s is determined more readily by experiments.

3.3.3 Surface Heat Capacity

3.3.3.1 Experimental Estimates The temperature derivative of the specific surface enthalpy is the specific surface heat capacity C_P^s:

$$C_P^s = \left(\frac{\partial H^s}{\partial T}\right)_P = T\left(\frac{\partial S^s}{\partial T}\right)_P = -T\left(\frac{\partial^2 \gamma}{\partial T^2}\right)_P \tag{3.18}$$

Equation 3.18 is obtained by differentiating Eq. 3.16 and substituting Eq. 3.14. We can see that the specific surface heat capacity can also be expressed in terms of the temperature derivative of the surface tension. Thus accurate surface-tension measurements as a function of temperature should be good sources of surface-heat-capacity data. Although many of the γ-versus-T curves show marked curvature (i.e., $(\partial S^s/\partial T)_P \neq 0$) instead of a straight line (i.e., $(\partial S^s/\partial T)_P = 0$), the data are not accurate enough to permit computation of reliable surface-heat-capacity values. Such data are more readily available from direct surface-heat-capacity measurements on finely divided powders of large surface/volume ratio (see, for example, references [12, 13]). When the heat capacity of the powder is compared with the heat capacity of large crystallites of the same material, the difference yields the surface heat capacity. Heat-capacity measurements on powdered samples yield larger heat capacities than do measurements on samples with small surface area, as expected. However, quantitative determination of C_P^s is difficult, owing to the uncertainties of surface-area measurements and the difficulties of assessing the role of strain in the surface, which could also affect the surface-heat-capacity values obtained in this manner.

3.3.3.2 Theoretical Estimates

The use of the Debye model (Figure 3.2), which assumes that a solid behaves as a three-dimensional elastic continuum with a frequency distribution $f(\nu) = B\nu^2$, allows accurate prediction of the temperature dependence of the vibrational heat capacity C_V of solids at low temperatures ($C_V \propto T^3$), as well as at high temperatures ($C_V = 3Nk_B$). One may also use the same model with confidence to evaluate the temperature dependence of the surface heat capacity due to vibrations of atoms in the surface.

Let us consider a surface of area $\mathcal{Q}$ at the termination of the bulk lattice in which the atoms have the same properties as in the three-dimensional elastic continuum.

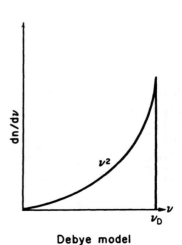

Figure 3.2. Frequency distribution $dn/d\nu$ of lattice vibrations ν assumed by the Debye model.

Debye model

The number of frequencies n in the range ν and $\nu + d\nu$ is given by [14]

$$\frac{dn}{d\nu} = f(\nu) = \frac{2\pi \mathbb{Q}\nu}{C_2} \tag{3.19}$$

where C_2 is the sound velocity in two dimensions.

In general, for an elastic continuum of μ dimensions, the frequency distribution is $f(\nu) \approx \nu^{\mu - 1}$. If N surface atoms still have $3N$ vibrational modes (they are allowed to have out-of-plane vibrations), we have

$$\int_0^{\nu_D} \frac{2\pi \mathbb{Q}}{C_2} \nu \, d\nu = 3N = \frac{\pi \mathbb{Q}\nu_D^2}{C_2} \tag{3.20}$$

Therefore, substituting Eq. 3.20 into 3.19 yields

$$f(\nu) = \frac{6N\nu}{\nu_D^2} \tag{3.21}$$

and the total energy of the surface is given by

$$E = \frac{6Nh}{\nu_D^2} \int_0^{\nu_D} \frac{\nu^2}{\exp{(h\nu/k_B T)} - 1} \, d\nu \tag{3.22}$$

Using the approximation for high temperatures,

$$\exp\left(\frac{h\nu}{k_B T}\right) \approx 1 + \frac{h\nu}{k_B T}, \qquad k_B T \gg h\nu \tag{3.23}$$

we obtain

$$E_{T \to \infty} = \frac{6N k_B T}{\nu_D^2} \frac{\nu_D^2}{2} = 3N k_B T \tag{3.24}$$

and

$$C_V(T \to \infty) = 3N k_B \tag{3.25}$$

Thus both the total energy and the heat capacity for substances yield the same limiting value as that for three-dimensional solids. At low temperatures, following the same procedure used in computing the total energy and heat capacity in three dimensions, we have, after substituting

$$x = \frac{h\nu}{k_B T}, \qquad x_{\max} = \frac{h\nu_D}{k_B T}, \qquad \nu_D = \frac{k_B \Theta_D}{h} \tag{3.26}$$

into Eq. 3.22

$$E = \frac{6Nk_B}{\Theta_D^2} T^3 \int_0^{x_{max}} \frac{x^2 \, dx}{e^x - 1} \tag{3.27}$$

After extending the upper limit to infinity instead of cutting off at x, and using the tabulated value of the pertinent Riemann zeta function (for $s = 3$), we have

$$\int_0^\infty \frac{x^2 \, dx}{e^x - 1} \approx 2.404 \tag{3.28}$$

Thus, substitution of Eq. 3.28 into Eq. 3.27 yields

$$E_{T \to 0} \approx \frac{14.4 N k_B}{\Theta_D^2} T^3 \tag{3.29}$$

and the surface heat capacity is given by

$$C_P^s = \left(\frac{\partial E}{\partial T}\right)_V \approx 43.2 N k_B \left(\frac{T}{\Theta_D}\right)^2 \tag{3.30}$$

We see that, at low temperatures, the surface heat capacity C_P^s is proportional to T^2, as opposed to the T^3 dependence of the bulk-heat capacity. However, the model we have considered here, consisting of a surface layer of atomic thickness, is quite unrealistic; it would be difficult to measure the heat capacity of a single atomic layer. In most cases the solid samples that can be used in experiments are small particles of variable surface/volume ratio or thin films many atomic layers thick. It would therefore be important to consider the heat capacity of such a sample and to see what contribution, if any, the surface makes to the total vibrational heat capacity.

For particles consisting of atoms in the bulk and on the surface, the total number of vibrational modes $3N$ can be expressed after Montroll [15] as

$$3N = \frac{4\pi V}{3C_3} \nu_D^3 + \frac{\pi \alpha}{4C_2} \nu_D^2 \tag{3.31}$$

where C_3 is the speed of sound in three dimensions, ν_D is the maximum frequency, the first term on the right-hand side is the bulk contribution, and the second term is the surface contribution, by use of the elastic continuum model. Solving Eq. 3.31, one obtains, for the maximum frequency ν_D,

$$\nu_D = \left(\frac{9NC_3}{4\pi V}\right)^{1/3} \left[1 - \frac{\pi \alpha}{36C_2 N^{1/3}} \left(\frac{9C_3}{4\pi V}\right)^{2/3} + O(N^{-2/3})\right] \tag{3.32}$$

where $O(N^{-2/3})$ indicates that all other terms proportional to $N^{-2/3}$ were neglected. Now we proceed as before, using the frequency distribution

$$f(\nu) = \frac{4\pi V}{C_3}\nu^2 + \frac{2\pi \mathfrak{A}}{C_2}\nu \tag{3.33}$$

The integration is somewhat more complex than before, since it involves several terms. Also, the method of counting surface modes depends on the boundary conditions—that is, the shape of the thin sample being considered. However, the boundary conditions change only the constants but do not affect the temperature dependence of the different terms. Derivations using different boundary conditions can be found in the literature [16, 17]. Both surface and bulk heat capacities have the same high-temperature limit $3R$. Here we shall only give the heat capacity obtained at low temperature for a rectangular solid:

$$\frac{C_V(T \to 0)}{Nk_B} \approx 234\left(\frac{T}{\Theta_{bulk}}\right)^3 + 50\left(\frac{T}{\Theta_{surface}}\right)^2 N^{-1/3} \tag{3.34}$$

We can see the familiar T^3 and T^2 dependences of the bulk and surface terms, respectively.

It should be noted that the effective Debye temperature $\Theta_{surface}$, which is characteristic of the vibration of surface atoms, may be different from Θ_{bulk}, which characterizes the vibration of bulk atoms. The experimental technique used to obtain $\Theta_{surface}$ will be discussed in the next chapter.

We would like to find the contribution of the surface heat capacity to the total lattice heat capacity at a given temperature for a particle of a given size. Because the surface heat capacity is proportional to the surface area and the bulk term is proportional to the volume, the surface/volume ratio will clearly play an important role in determining the magnitude of the contribution of the surface heat capacity to the total heat capacity. The ratio of the bulk- and surface-heat-capacity terms ξ indicates both the temperature range and the thickness of the specimen for which the surface-heat-capacity contribution will become detectable. The ratio ξ for a cube with sides of length L is approximately given by

$$\xi = \frac{\text{bulk heat capacity}}{\text{surface heat capacity}} \approx 3.6 \times 10^8 L\left(\frac{\Theta_{surface}^2}{\Theta_{bulk}^3}\right)\left(\frac{\rho}{M}\right)^{1/3} T \tag{3.35}$$

where ρ is the density and M is the molar weight of the sample. For a thin film of volume $L \times L \times qL$, where $qL \ll L$, or, for a wire of length L and diameter qL, where again $qL \ll L$, we have

$$\xi \approx 11 \times 10^8 qL\left(\frac{\Theta_{surface}^2}{\Theta_{bulk}^3}\right)\left(\frac{\rho}{M}\right)^{1/3} T \tag{3.36}$$

For most metals, $(\rho/M)^{1/3} \approx 0.5$ cm^{-1} (50 m^{-1}); for platinum, for example, Θ_{bulk} = 234 K and $\Theta_{surface}$ = 110 K; $\xi \approx 5$ for a 1000-Å (100-nm)-thick film. Thus samples in the 10^{-5} to 10^{-4}-cm (10^{-7} to 10^{-6}-m) thickness range should show a detectable contribution of surface heat capacity at temperatures $T < 10$ K.

3.4 THE SURFACE ENERGY AND SURFACE COMPOSITION OF TWO-COMPONENT SYSTEMS

Consider a metal B that is dissolved at low concentrations in another metal A. Metal B may have a tendency to segregate to the surface of A if it forms a strong surface bond for one reason or another. Figure 3.3 shows schematically the various relative energies of vaporization, dissolution, and adsorption of B on A that lead to surface segregation. At any bulk concentration, some of the B atoms will always be at the surface as a result of the surface-bulk equilibrium. By forming a strong surface chemical bond, the surface concentration of B could be greatly enhanced. This occurs if the heat of desorption of B from the surface of A is larger than the heat of vaporization of pure B. Because the binding energy of B on the various surfaces of A may change markedly as a function of crystal orientation, the extent of surface segregation can depend strongly on the surface structure of A.

The energy diagram for the nickel–sulfur system is shown in Figure 3.4. Sulfur segregates to the surface of the metal because of the stronger surface bonds it forms as compared to its bond energies at grain boundaries or in the bulk nickel phase (the heat of solution).

It is also possible that B atoms form a weaker chemical bond with A atoms at the surface compared with the bulk (A). In this circumstance the B atoms will be repelled from the surface, and their surface concentration will be less than expected from their bulk concentration.

It is useful to express the concentration of B in the surface region as compared to its concentration in the bulk. This surface excess, following the *IUPAC Manual of Definitions in Surface Chemistry* [18], is given by

$$\Gamma_B = \frac{n_B}{\alpha} \tag{3.37}$$

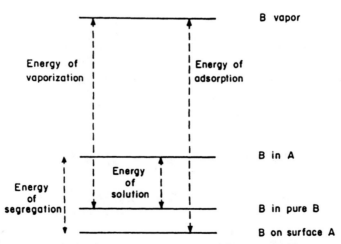

Figure 3.3. Relative energies of vaporization, dissolution, and adsorption of metal B on metal A.

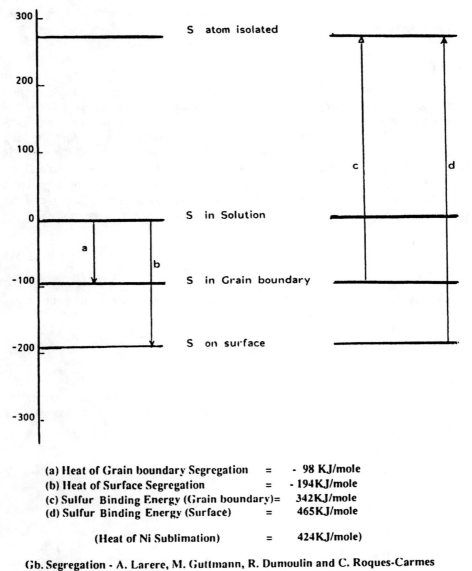

(a) Heat of Grain boundary Segregation   =   - 98 KJ/mole
(b) Heat of Surface Segregation   =   - 194KJ/mole
(c) Sulfur Binding Energy (Grain boundary)= 342KJ/mole
(d) Sulfur Binding Energy (Surface)   =   465KJ/mole

    (Heat of Ni Sublimation)   =   424KJ/mole)

Gb. Segregation - A. Larere, M. Guttmann, R. Dumoulin and C. Roques-Carmes
                Acta Met. (1982)
    Adsorption - J. McCarty and H. Wise, J. Chem. Phys. (1980)
    Solubility - N. Barbouth and J. Oudar, Scripta Met. (1972)

Figure 3.4. Energy diagram for the nickel–sulfur system. Note that sulfur is more strongly bound to the nickel surface than to nickel in the bulk.

Here n_B is the excess number of moles of B atoms in the surface above the number of moles of B atoms in a bulk region that would contain the same number of moles of atoms as that part of the surface. For example, we usually have on the order of 10^{15} atoms in a square-centimeter area of the surface; if 90% of these 9×10^{14} atoms are B atoms, whereas in the bulk only 50% of 10^{15} atoms (5×10^{14}) are B atoms, the surface excess is $n_B = 4 \times 10^{14}$. The term Γ_B has units of surface concentration, since $\mathcal{Q}$ is the surface area.

3.4.1 The Wagner Experiment

It is easy to show how the surface excess changes as we change the surface area, for example, by considering the experiment suggested by Wagner [19] shown in Figure 3.5. A glass tube contains both a liquid and a gaseous phase of a two-component system—for example, Hg and HCl at a constant temperature. When the long axis of the tube is vertical, the interfacial area between the liquid and the vapor is small. When the tube is rotated 90°, so that its long axis is horizontal, the interfacial area increases substantially. As a result, more HCl adsorbs, causing a pressure change. The number of moles of HCl that must be added for a unit increase of the interfacial area may be used as the definition of surface excess concentration:

$$\Gamma_{HCl} = \left(\frac{\partial n_{HCl}}{\partial \mathcal{Q}}\right)_{T,P,V,n_{Hg}} \tag{3.38}$$

Similarly the surface excess concentration of mercury, the other component, is defined as

$$\Gamma_{Hg} = \left(\frac{\partial n_{Hg}}{\partial \mathcal{Q}}\right)_{T,P,V,n_{HCl}} \tag{3.39}$$

The surface tension of a multicomponent system changes as the surface concentrations of its various constituents are altered. Rigorous derivation of the variation of surface tension with surface excess concentration yields

$$\Gamma_i = -\left(\frac{\partial \gamma}{\partial \mu_i}\right)_T \tag{3.40}$$

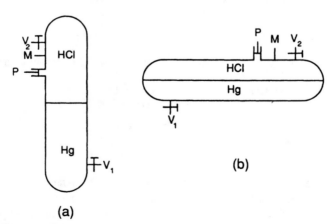

(a)

(b)

Figure 3.5. The Wagner experiment. When the tube of Hg liquid and HCl gas is rotated form a vertical position to a horizontal position, the gas–liquid interface area increases.

where Γ_i is the excess surface concentration of compound i, and $d\mu_i$ is the change in chemical potential associated with the placing of extra atoms of the ith component on the surface. Equation 3.40 is often written in the form

$$d\gamma = -S^s \, dT - \sum_i \Gamma_i \, d\mu_i \qquad (3.41)$$

(known as the Gibbs equation [20]) or, for an isothermal system, as

$$d\gamma = -\sum_i \Gamma_i \, d\mu_1 \qquad (3.42)$$

Let us use the Gibbs equation to predict surface segregation in binary alloy systems.

3.4.2 Surface Segregation in Binary Alloy Systems

For a dilute binary system exhibiting ideal solution behavior,

$$\mu_2^b = \mu_2^{0,b} + RT \ln x_2^b \qquad (3.43)$$

where μ_2^b and $\mu_2^{0,b}$ are the chemical potentials of the second component in the bulk solution and the pure component, respectively, and x_2^b is the mole fraction of the second component in the bulk. Because by definition the mole fraction x_2^b is less than unity, the logarithm term on the right-hand side of Eq. 3.43 must be negative. Thus, the chemical potential is decreasing for a two-component system as compared to the chemical potential of a pure material.

The driving force for surface segregation is the difference in the binding energies between the two metal atoms, A–B, and the binding energies in the pure components, A–A and B–B. The same change in chemical bonding gives rise to a change in the surface tension of the binary system compared to the surface tension for the pure constituents. Assuming ideal solution behavior, Eq. 3.43 can be rewritten to express the equality of the chemical potentials in the bulk and at the surface:

$$\mu_2^b = \mu_2^{0,b} + RT \ln x_2^b = \mu_2^{0,s} + RT \ln x_2^s - \gamma a_2 \qquad (3.44)$$

where a_2 is the surface area covered by 1 mole of the second component of the binary system. For a pure, one-component system, $x_2^s = x_2^b = 1$, and we have

$$\mu_2^{0,s} - \mu_2^{0,b} = \gamma_2 a_2 \qquad (3.45)$$

where γ_2 is the surface tension of pure substance 2. Using Eq. 3.45 to rewrite Eq. 3.44, and assuming that $a_1 = a_2 = a$, then, for a two-component system, we have the equations

$$\gamma a = \gamma_1 a + RT \ln x_1^s - RT \ln x_1^b \qquad (3.46)$$

and

$$\gamma a = \gamma_2 a + RT \ln x_2^s - RT \ln x_2^b \qquad (3.47)$$

This can be rewritten as

$$\frac{x_2^s}{x_1^s} = \frac{x_2^b}{x_1^b} \exp\left\{\frac{(\gamma_1 - \gamma_2)a}{RT}\right\} \tag{3.48}$$

which is the final result for the monolayer ideal-solution model, where γ_2 and γ_1 are the surface tensions of pure components 2 and 1. Thus if $\gamma_2 < \gamma_1$, the surface fraction of component 2 will increase exponentially, resulting in marked surface segregation. From Eq. 3.48, it is clear that the constituent with the lower surface tension will have a higher concentration at the surface.

The surface tension of solids and liquids can be measured by a variety of techniques, or it may be obtained by correlation with other thermodynamic properties. For metals, excellent correlations between γ and the heat of sublimation ΔH_{subl} can be established. Because sublimation and the creation of a unit area of surface are processes that give rise to the ΔH_{subl} and γ, respectively, a correlation between these two parameters is expected if the bond energies can be estimated by the addition of nearest-neighbor bonds. For example, in an fcc solid, each bulk atom has 12 nearest neighbors ($z = 12$), whereas on the (111) surface (the highest atomic-density crystal plane), three of these nearest neighbors are missing. Thus 12 bonds are broken when an atom is moved from the solid into the vapor phase, whereas only three bonds are removed when the surface is created. Converting γ (energy/area) to a molar quantity by multiplying it by a (area/mole), we have the relation

$$\gamma a = \gamma_m = \tfrac{3}{12}\Delta H_{subl} = 0.25\,\Delta H_{subl} \tag{3.49}$$

The experimental data for metals actually lead to the result [3]

$$\gamma_m \approx 0.16\,\Delta H_{subl} \tag{3.50}$$

By using this experimental correlation, we can rewrite Eq. 3.48 to yield

$$\frac{x_2^s}{x_1^s} = \frac{x_2^b}{x_1^b} \exp\left\{\frac{0.16\,(\Delta H_{subl_1} - \Delta H_{subl_2})}{RT}\right\} \tag{3.51}$$

According to Eq. 3.51, the metal constituent with the lower heat of sublimation will accumulate at the surface in excess. This relationship, with small modifications, has been used to predict surface enrichment at alloy surfaces. Note that, for oxide or organic solid surfaces, no simple $\gamma \leftrightarrow \Delta H_{subl}$ correlation exists like the one found for metal surfaces.

Equations 3.48 and 3.51 also predict that the surface composition of ideal solutions should be an exponential function of temperature. Thus, while the bulk composition of multicomponent systems is not much affected by temperature, the surface concentration of the constituents may change markedly. According to these equations, both the surface and bulk compositions should approach the same atom-fraction ratios at high temperatures.

The surface segregation of one of the constituents becomes more pronounced with increases in the difference in surface tensions between the components making up the solution. Surface segregation is expected to be prevalent for metal solutions, since metals have the highest surface tensions. Surface segregation can also be readily detected for oxides and organic solutions, which are systems with smaller surface tensions.

Metallic alloys are not ideal solutions, since they generally have some finite heat of mixing. In the deviation of Eq. 3.48, this heat of mixing was ignored by assuming that the bond energy between unlike atoms $E_{1,2}$ is equal to the average of the bond enthalpies between like atoms, namely by assuming that [21, 22]

$$E_{1,2} = \frac{E_{1,1} + E_{2,2}}{2} \tag{3.52}$$

In the "regular solution approximation," this equality is no longer assumed and the heat of mixing is finite. We define the regular solution parameter Ω as

$$\Omega = N_A z \left(E_{1,2} - \frac{E_{1,1} + E_{2,2}}{2} \right) \tag{3.53}$$

where N_A is Avogadro's number and z is the bulk coordination number. Then the regular solution parameter Ω is directly correlated with the heat of mixing ΔH_m by the relation

$$\Omega = \frac{\Delta H_m}{x_1^b(1 - x_1^b)} \tag{3.54}$$

Therefore, from heat-of-mixing data, which in many cases are readily available, the parameter Ω and the bond energy E_{12} can be estimated.

The surface composition in the regular-solution monolayer approximation is given by

$$\frac{x_2^s}{x_1^s} = \frac{x_2^b}{x_1^b} \exp\left\{ \frac{(\gamma_1 - \gamma_2)a}{RT} \right\} \exp\left\{ \frac{\Omega(l + m)}{RT} \left[(x_1^b)^2 - (x_1^s)^2 \right] + \frac{\Omega l}{RT} \left[(x_2^s)^2 - (x_1^s)^2 \right] \right\} \tag{3.55}$$

where l is the fraction of nearest neighbors to an atom in the plane and m is the fraction of nearest neighbors below the layer containing the atom. For example, for an atom with $z = 12$ nearest neighbors (three above, three below, and six in the same plane), $l = \frac{6}{12} = 0.5$ and $m = \frac{3}{12} = 0.25$. In this approximation, the surface composition becomes a fairly strong function of the heat of mixing, its sign, and its magnitude, in addition to the surface-tension difference and temperature [23].

3.4.3 Surface Composition of Alloys from Model Calculations

Let us show a few examples of how Eq. 3.55 predicts surface segregation. For Au–Ag alloys, the quantity Ω is constant and negative (exothermic heat of mixing) to within 17% throughout the entire composition range [24], which is about as close to regular behavior as found for any metallic alloys. When Eq. 3.55 is applied to the Au–Ag system, the result is shown in Figure 3.6. The calculation was carried out for an fcc(111) face, and surface energy data were used instead of heats of sublimation. The agreement between the experimental data and model calculations is quite good.

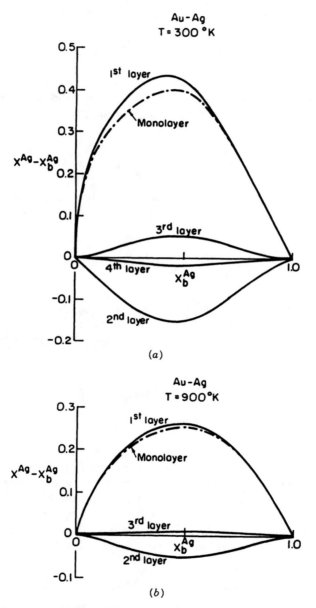

Figure 3.6. Surface segregation in the Au–Ag system. (**a**) Surface excess of Ag as a function of bulk composition at 300 K. (**b**) Surface excess of Ag as a function of bulk composition at 900 K.

Note again that surface composition is strongly temperature-dependent. As the temperature increases, the surface excess concentration of the segregating constituent should diminish exponentially if the models expressed by Eqs. 3.48, 3.51, and 3.55 are obeyed. This effect is readily discernible by comparing Figure 3.6a with Figure 3.6b, where the surface excesses of silver in Au–Ag are plotted at two different temperatures.

The most artificial aspect of this model, which is expressed in Eq. 3.55, is the monolayer approximation, and in fact it is unnecessary to require all layers below the top monolayer to have the bulk composition. Williams and Nason [25] presented a four-layer model in which the top four layers were allowed to have compositions different from the bulk, while the fifth and deeper layers had the bulk composition. The results of the derivation are four coupled equations relating the surface composition of the four layers to T, Ω, ΔH_{subl}, x^b, and the crystal-structure parameters. To demonstrate this model, results of these types of calculations for the Au–Ag system and the liquid Pb–In system are shown in Figures 3.6a, 3.6b, and 3.7, respectively. The surface enrichment diminishes rapidly with depth into the surface, as might be expected for this model, which considers only nearest-neighbor bonding. Furthermore, if, as in Ag–Au, $\Omega < 0$ [which, by Eq. 3.54, implies attractive interactions between unlike atoms], then there is a reversal in enrichment in adjacent layers. That is, Ag enrichment occurs in the first layer, but Ag depletion takes place in the second layer. This represents a tendency toward ordering at the surface of the alloy. For Pb–In, where $\Omega > 0$, the attraction between like atoms is greater on the average than between unlike atoms (endothermic heat of mixing), so that Pb, the component with the lowest surface energy, clusters at the surface. If $\Omega = 0$, the depth distribution collapses to only a single-monolayer type of segregation.

There are irregularities at solid surfaces, namely steps and kinks, at which atoms have fewer nearest neighbors than in the (111) plane of an fcc solid. According to the thermodynamic models discussed so far, surface segregation should be different

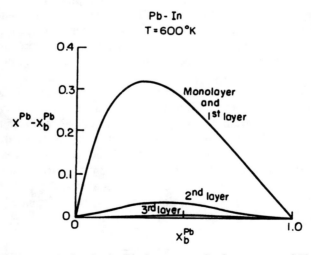

Figure 3.7. Surface segregation in the Pb–In system. Surface excess of Pb as a function of bulk composition at 600 K.

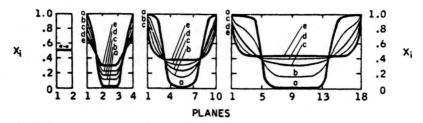

Figure 3.8. Segregation in the Ni–Au system. Calculated composition, X_i, for the various planes of 50% Ni–Au (average composition) thin films with (111) surfaces. Data are shown for film thicknesses of 2–18 planes and various temperatures. Temperature of a miscibility gap is 1100 K [26].

at these sites. These surface irregularities have no analogy in liquids, and they can be of great importance in a variety of surface phenomena, ranging from crystal growth to heterogeneous catalysis. Burton et al. [26] approached this problem using the regular-solution approximation and using appropriate values for l and m, the fraction of in-plane and out-of-plane nearest neighbors, respectively. The lower the coordination of a particular site, the greater the tendency for segregation. Thus, in alloy microclusters, sites at edges and corners are more enriched in the segregating species than are sites on flat terraces.

Burton et al. [26] also studied thin films of various thickness for alloys in which there is a miscibility gap ($\Omega > 0$). Figure 3.8 shows the results of calculations for a 50 atom% Au–Ni alloy. The Au–Ni bulk phase diagram has a miscibility gap below $T_c = 1000$ K. For $T > T_c$ the segregation of gold takes place only in the surface region, with a core that approaches the bulk composition as the film thickness increases. For $T < T_c$ the films exhibit phase separation, with the Au-rich phase accumulating at the surface and the Ni-rich phase accumulating at the center.

This structure, consisting of a film with an outer shell of one phase and an inner region of another, was suggested by Sachtler and Jongpier in their explanation of experimental findings in Cu–Ni alloys [27]. Their "cherry" model suggests surface segregation of the more easily diffusing species for systems where the two components are deposited separately; this species is also usually the more surface-active.

In the theories presented so far, the driving force for segregation has been the fact that the surface is a region of reduced atomic coordination. In solids, there is another driving force for segregation, the reduction of strain. McLean [28] has pointed out that solute atoms that differ in size form the solvent lattice atoms create a strain in the lattice. At a grain boundary, there are open sites where more space is available to the atoms. By migrating to these sites, a solute can reduce the strain energy. McLean [28] used the ideal-solution model and gave an expression identical to Eq. 3.48, except that the argument of the exponential involves a difference between the strain energy caused by the solute atom located at the grain boundary and one located within the bulk. His expression (however, see reference [23]) is

$$\frac{x_2^s}{x_1^s} = \frac{x_2^b}{x_1^b} \exp\left(-\frac{E_s}{RT}\right) \tag{3.56}$$

In Eq. 3.56,

$$E_s = 24\pi \frac{K_{sm}G_{sm}}{3K_{sm}r_2 + 4G_{sm}r_1} r_1 r_2 (r_2 - r_1)^2 \tag{3.57}$$

where K_{sm} is the bulk shear modulus of the solute, G_{sm} is the shear modulus of the solvent, and r_1 and r_2 are the appropriate radii for the solvent and solute, respectively. The atomic radius is frequently obtained from the atomic volume, although it can also be readily obtained from crystallographic data.

Complete treatment of the equilibrium surface composition must involve the minimization of the total free energy of the multicomponent regular solution [23]. To this end, the contributions of atom interactions and surface and solute strain energies must all be included in the calculation. To the first approximation, all these effects can be combined into a unified formalism by setting the heat of segregation ΔH_{segr} equal to the exponent of the right-hand side of Eqs. 3.55 and 3.57:

$$\Delta H_{segr} = (\gamma_1 - \gamma_2) a + \Omega(l + m) [(x_1^b)^2 - (x_2^b)^2] + \Omega l [(x_2^s)^2 - (x_1^s)^2]$$

$$- 24\pi \frac{K_{sm}G_{sm}}{3K_{sm}r_2 + 4G_{sm}r_1} r_1 r_2 (r_2 - r_1)^2 \tag{3.58}$$

3.5 SURFACES WHEN NO BULK PHASE EXISTS: TWO-DIMENSIONAL PHASES

Studies of small particles by Sinfelt [29] and his co-workers have shown that when the particles' sizes become very small and dispersions tend toward unity (that is, when virtually every atom is at the surface), alloy systems exhibit phase diagrams very different from those that characterize bulk systems. For example, microclusters containing Cu and Ru, Cu and Os, or Au and Ni can be produced in any ratio of the two elements, indicating complete miscibility or solid solution behavior. In the bulk phase these elements are completely immiscible. This very different behavior of the surface phases of bimetallic systems finds important applications in the design of catalysts to carry out selective chemical reactions. Morán-López and Falicov [30] developed a theory—using pairwise interactions—of alloy surface segregation that explains this effect. Bimetallic systems remain miscible at lower temperatures in two dimensions than in three dimensions.

It is likely that similar changes occur for two-dimensional mixed oxide phases compared to their bulk phases.

3.5.1 Monomolecular Films

When an insoluble substance of lower surface energy is placed on the surface of another substance of higher surface energy, it will spread until it covers the whole surface provided that the solid–solid interfacial energy is sufficiently low. We can coat a metal pan or a cotton fabric with fluorocarbons to make them non-sticky,

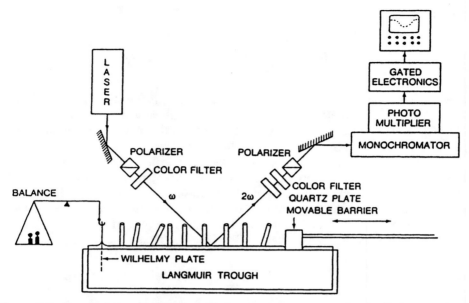

Figure 3.9. Langmuir trough with movable barrier to compress molecules in the monolayer. The balance measures surface tension and the laser optics monitor changes of molecular orientation by second harmonic signal generation at various angles of incidence.

thereby resisting soiling. We can also spread olive oil on the surface of wine or water to slow down oxidation or evaporation.

The thermodynamic driving force for spreading a substance on top of another is the reduction of the overall surface tension or the surface free energy. If the concentration of the lower-surface-tension substance is low, it forms a film one molecule thick. For example, titanium oxide over aluminum oxide or long-chain organic molecules over water were found to form monomolecular films. An ingenious balance was devised by Langmuir [31] to measure changes in interfacial tension as the film spreads or is compressed over liquids. A modern version of Langmuir's device is shown in Figure 3.9. The film is confined on one side by a floating barrier. The surface tension is then measured by a variety of techniques as the barrier is moved, either to compress the molecules in the film even more or to let them spread out. Optical measurements can monitor changes of orientation of long-chain hydrocarbons as the surface pressure is varied.

3.6   METASTABLE SURFACE PHASES

Microporous solids with surface areas in the range of 400 m^2/g or larger can be prepared in several ways. Controlled thermal decomposition of silica gel or aluminum hydroxides to AlO(OH) and then to the γ- or η-phases of alumina produces such high-surface-area materials. The so-called sol–gel method, which uses both aluminate and silicate ions or aluminum and silicon alkyl compounds at a well-controlled hydrogen ion concentration, produces crystalline microporous solids that are the synthetic zeolites. High-surface-area carbons, phosphates, and carbonates are well known and are employed for selective adsorption or to carry out surface reactions. When these microporous materials are heated to high enough temperatures, their pore structures collapse with heat evolution to produce the more stable low-surface-area phases; α-alumina, quartz, or graphite. The exothermic heat of phase transformation clearly indicates that the microporous phases are metastable with respect to the high-density, low-surface-area phases. This is not surprising, since the surface free energy is always positive; thus, any reduction in surface area will lead to the formation of a more stable material. Nevertheless, microporous crystalline materials of high surface area play a dominant role in many applications of surface science, ranging from gas separation to heterogeneous catalysis. They retain their structural integrity as long as they are used below the temperature range of their transformation to the more stable high-density phase.

Rapid heating and cooling of surfaces by the use of laser beams can also lead to the formation of metastable surface phases. Heating and quenching rates on the order of 10^6–10^9 degrees per second have been obtained. In this circumstance, the high-temperature surface structure (including the melt) can be retained as the surface region is cooled so rapidly that diffusion-controlled reequilibration cannot take place. Glassy metal films have been prepared this way. Silicon crystal surfaces with metastable unreconstructed surface structures have also been produced. The surface composition may also be altered by selective evaporation or segregation of some of the constituents as a result of the large temperature gradient between the surface and the bulk.

Another way to produce metastable surface phases is by ion bombardment or ion implantation. Oxygen ion bombardment of the reconstructed (100) face of platinum converts it to the metastable unreconstructed surface structure. By using high-energy (10–10^3 keV) ion beams, the near-surface region can be saturated with the material used for the bombardment so that concentrations much higher than the solubility limit in equilibrium can be obtained. Ion implantation is frequently used to dope semiconductors with high concentrations of electron acceptors or donors whose penetration depth can be precisely controlled by the ion energy.

In all these circumstances, the thermodynamically metastable phases are stabilized by the very slow rates of transformation into the more thermodynamically stable phases. These metastable materials play very important roles in surface chemistry, on account of their very high surface area or unique and unusual surface properties.

3.7 CURVED SURFACES

Solids and liquids will always tend to minimize their surface area in order to decrease the excess surface free energy. For liquids, therefore, the equilibrium surface becomes curved, where the radius of curvature will depend on the pressure difference on the two sides of the interface and on the surface tension. Remember that we have considered the surface tension as a surface pressure exerted tangentially along the surface. Now we will also consider the role of the external and internal pressures that act normal to the interface on the properties of the surface.

In Figure 3.10 we have an equilibrium curved surface (a bubble in this case),

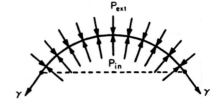

Figure 3.10. Section of a bubble surface with internal and external pressures P_{in} and P_{ext} and surface tension γ.

with internal and external pressures P_{in} and P_{ext}, respectively, and with a surface tension γ. The radius of curvature of the bubble is given by r, and hence its volume is given by $V = \frac{4}{3}\pi r^3$. The force operating normal to the surface, which would change the volume by $dV = 4\pi r^2\, dr$, is given by $4\pi r^2(P_{in} - P_{ext})\, dr$. The force exerted on the bubble surface tangentially is given by $4\pi r^2\gamma$. If there is equilibrium, any change in the external pressure that would, for instance, increase the volume must be counteracted by a corresponding increase in the surface energy because of the extension of the surface by

$$\gamma\, d\mathcal{Q} = 8\pi\gamma r\, dr \tag{3.59}$$

In equilibrium the two changes are equal, and we have

$$4\pi r^2(P_{in} - P_{ext})\, dr = 8\pi r\gamma\, dr \tag{3.60}$$

$$(P_{in} - P_{ext}) = \frac{2\gamma}{r} \tag{3.61}$$

This equation is quite significant in explaining the properties of liquid surfaces and bubbles. First, Eq. 3.61 indicates that, in equilibrium, a pressure difference can be maintained across a curved surface. The pressure inside the liquid drop or gas bubble is higher than the external pressure because of the surface tension. The smaller the droplet or larger the surface tension, the larger the pressure difference that can be maintained. For a flat surface, $r = \infty$, and the pressure difference normal to the interface vanishes.

3.7.1 Capillary Rise

If the liquid is constrained in a capillary with very small radius, the difference between the internal and external pressures can become very large. Because of the large internal pressure, the liquid in the capillary will rise until it is balanced by the hydrostatic pressure of the rising liquid column. In this circumstance Eq. 3.61 becomes

$$\Delta P = \Delta\rho g h = \frac{2\gamma}{r} \tag{3.62}$$

where $\Delta\rho$ is the density difference between the liquid and gas phases, g is the gravitational acceleration, and h is the height of the capillary column. The higher the surface tension and the smaller the capillary radius, the larger is the capillary rise.

This phenomenon plays an important role in the growth of plants and in controlling their height, because the transport of water through capillary rise influences their rate of photosynthesis.

3.7.2 The Vapor Pressure of Curved Surfaces

Let us now consider how the vapor pressure of a droplet depends on its radius of curvature r. In equilibrium, the pressure difference across the droplet is given by Eq. 3.61. If we transfer atoms from, say, the liquid to the surrounding gas phase, there is a small and equal equilibrium displacement on both sides of the interface:

$$dP_{in} - dP_{ext} = d\left(\frac{2\gamma}{r}\right) \tag{3.63}$$

At constant temperature, the free-energy change associated with the transfer of atoms across the interface is given by

$$dG_{in} = \overline{V}_{in}\, dP_{in} \tag{3.64}$$

and

$$dG_{ext} = \overline{V}_{ext}\, dP_{ext} \tag{3.65}$$

In equilibrium, the free-energy changes are equal; that is, dG_{in} equals dG_{ext}, and we have

$$\overline{V}_{in}\, dP_{in} = \overline{V}_{ext}\, dP_{ext} \tag{3.66}$$

Substitution of Eq. 3.63 into Eq. 3.66, after rearrangement, yields

$$\frac{\overline{V}_{ext} - \overline{V}_{in}}{\overline{V}_{in}}\, dP_{ext} = d\left(\frac{2\gamma}{r}\right) \tag{3.67}$$

We can neglect the molar volume of the liquid with respect to the much larger molar volume of the gas ($\overline{V}_{ext} \gg \overline{V}_{in}$). Assuming that the vapor behaves as an ideal gas, we have for the molar volume

$$\overline{V}_{ext} = \frac{RT}{P_{ext}} \tag{3.68}$$

Substituting Eq. 3.68 into Eq. 3.67 yields, after rearrangement,

$$\frac{dP_{ext}}{P_{ext}} = \frac{2\gamma \overline{V}_{in}}{RT}\, d\left(\frac{1}{r}\right) \tag{3.69}$$

We can now integrate Eq. 3.43 between the limits of a flat surface with zero curvature ($r = \infty$, $P = P_0$) and some other state corresponding to a curved surface

$(1/r, P)$ and assume that the molar volume of the liquid remains unchanged along this path. We obtain

$$\ln\left(\frac{P_r}{P_0}\right) = \frac{2\gamma \overline{V}_{in}}{RTr} \qquad (3.70)$$

which is the well-known Kelvin equation for describing the dependence of the vapor pressure of any spherical particle on its size. We can see that, according to Eq. 3.70, small particles have higher vapor pressures than larger ones. Similarly, very small particles of solids have greater solubility than large particles. If we have a distribution of particles of different sizes, we will find that the larger particles will grow at the expense of the smaller ones, as predicted by Eq. 3.70.

Differences in vapor pressure or solubility that depend on particle size (radius of curvature) can only be observed for particles smaller than $r < 10$ nm (100 Å). If we assume representative values for a water droplet [$\gamma = 72.8$ ergs/cm^2 (7.28 × 10^{-2} J/m^2), $\overline{V}_{in} = 18$ cm^3/mole (18 × 10^{-6} m^3/mole)], P_r/P_0 approaches unity rapidly above this radius.

3.7.3   The Contact Angle and Adhesion

The shape of the curved surface, in turn, allows one to determine the surface tension of the liquid when it is in equilibrium with its own vapor or to determine the interfacial tension if the droplet is in contact with a different substance (gas, liquid, or solid). The interfacial tension is determined by measuring the contact angles at the liquid–solid and solid–vapor interfaces. The contact angle is defined in Figure 3.11, which shows a typical liquid–solid interface.

Common experience tells us that the smaller the angle between the liquid and the solid, the more evenly the liquid is spread over, or adheres to, the solid surface, until at $\psi \approx 0°$ complete wetting of the solid surface takes place. If the contact angle is large ($\psi \approx 90°$), the liquid does not readily wet the solid surface. For $\psi > 90°$ the liquid tends to form sphere-shaped droplets on the solid surface that may easily run off; that is, the liquid does not wet the solid surface at all.

For a liquid that rests on a smooth surface with a finite contact angle, one can determine the relationship between the interfacial tensions at the different interfaces from consideration of the balance of surface forces at the line of contact of the three phases (solid, liquid, and gas). Remembering that the interfacial tension always exerts a pressure tangentially along the surface, the surface free-energy balance (a

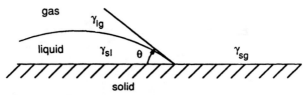

Figure 3.11. Definition of contact angle at the solid–liquid interface. γ_{lg} denotes the liquid–gas interfacial energy, γ_{sl} denotes the solid–liquid interfacial energy, and γ_{sg} denotes the solid–gas interfacial energy.

condition of equilibrium) between the surface forces acting in opposite directions is given by

$$\gamma_{lg} \cos \psi + \gamma_{sl} = \gamma_{sg} \tag{3.71}$$

or

$$\cos \psi = \frac{\gamma_{sg} - \gamma_{sl}}{\gamma_{lg}} \tag{3.72}$$

Here γ_{lg} is the interfacial tension at the liquid–gas interface, and γ_{sg} and γ_{sl} are the interfacial tensions between the solid–gas and solid–liquid interfaces, respectively. Thus, when the solid–liquid–gas interface is at equilibrium, by using the knowledge of γ_{lg} and the contact angle, we can determine the difference $\gamma_{sg} - \gamma_{sl}$ but not their absolute values.

There are extreme cases when Eq. 3.72 does not define the equilibrium position of the line of contact. This happens under conditions of complete wetting, when $\gamma_{sg} > \gamma_{sl} + \gamma_{lg}$, or when the liquid does not wet the solid at all and the gas displaces the liquid completely along the solid surface ($\gamma_{sl} > \gamma_{sg} + \gamma_{lg}$). Therefore, it is more convenient to consider the wetting coefficient

$$k_w = \frac{\gamma_{sg} - \gamma_{sl}}{\gamma_{lg}} \tag{3.73}$$

in describing the wetting ability of a liquid. If $k \geq +1$, the solid is completely wetted by the liquid. For $k = \pm 1$, the wetting is described by Eq. 3.72; and for $k \leq -1$, the solid is not wetted at all.

Because the wetting ability of the liquid at the solid surface is so important in practical problems of adhesion or lubrication, a great deal of work is being carried out to determine the interfacial tensions for different combinations of interfaces [32–34]. For example, it is useful in these studies to determine the energy necessary to separate the solid–liquid interface. In Figure 3.12 the two states are shown before and after separation. We define the reversible work of separation as the difference in free energy between the two states in Figure 3.12. In the process we eliminate the solid–liquid interface and re-form the solid–gas and liquid–gas interfaces. Thus the reversible work per unit area W^s is given by

$$W^s = \gamma_{lg} + \gamma_{sg} - \gamma_{sl} \tag{3.74}$$

Equation 3.74 is called Dupré's equation. Combining Eqs. 3.72 and 3.74, we have

$$W^s = \gamma_{lg}(1 + \cos \psi) \tag{3.75}$$

which is frequently called Young's equation [35]. Harkins and his co-workers have defined the work of adhesion by redefining the second state (in Figure 3.12), in which the solid and liquid phases are separated, to be in a vacuum [36]. The surfaces in their separated state are free of adsorbed molecules. Thus the work of adhesion

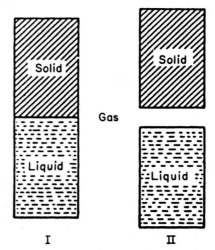

Figure 3.12. Two states to define the reversible work of separation.

is defined as

$$W_A^s = \gamma_{l,0} + \gamma_{s,0} - \gamma_{s,l} \tag{3.76}$$

where $\gamma_{l,0}$ and $\gamma_{s,0}$ are the surface tensions of the liquid and the solid, respectively, in vacuum.

3.7.4 Nucleation

Because the free energy of formation of a surface is always positive, a particle that consists only of surfaces (that is, platelets or droplets of atomic dimensions) would be thermodynamically unstable. This is also apparent from the Kelvin equation [Eq. 3.70], which states that a particle that falls below a certain size will have an increased vapor pressure and will therefore evaporate. There must be a stabilizing influence, however, that allows small particles of atomic dimensions to form and grow a common occurrence in nature. This influence is given by the free energy of formation of the bulk condensed phase. In this process, n moles of vapor are transferred to the liquid phase under isothermal conditions. This work of isothermal compression is given by

$$\Delta G = -nRT \ln \left(\frac{P}{P_{eq}} \right) \tag{3.77}$$

The condensed phase of n moles will grow if the vapor pressure P is larger than the equilibrium vapor pressure P_{eq}, or it will vaporize if $P < P_{eq}$. For small particles, we must take into account both their positive surface free energy, which would impede their growth, and the free energy of formation of the bulk condensed phase, which is negative for any external pressure larger than the equilibrium pressure. For a spherical droplet the total surface free energy is $4\pi r^2 \gamma$. Thus the total free energy

ΔG of a growing particle is

$$\Delta G_{total} = -nRT \ln \left(\frac{P}{P_{eq}} \right) + 4\pi r^2 \gamma \tag{3.78}$$

In order to see the dependence of ΔG on the dimensions of the condensed particle, let us substitute for n the number of moles of condensed vapor, $n = 4\pi r^3/3\overline{V}$, where $\overline{V} = M/\rho$ is the molar volume of the condensed phase, M is its atomic weight, and ρ is the density. We have

$$\Delta G_{total} = -\left(\frac{\frac{4}{3}\pi r^3}{\overline{V}} \right) RT \ln \left(\frac{P}{P_{eq}} \right) + 4\pi r^2 \gamma \tag{3.79}$$

Initially, when the condensed particle is very small, the surface free-energy term must be the larger of the two terms on the right-hand side of Eq. 3.79, and ΔG increases with r^2. In this range of sizes the particles are unstable. Above a critical size, however, the volumetric term becomes larger and dominates, since it increases as r^3 while the surface free-energy term increases only as r^2. Hence the particle of that size or larger grows spontaneously (for $P > P_{eq}$). When ΔG is at a maximum, that is, $(\partial \Delta G_{total}/\partial r) = 0$, the particle reaches the critical size it must have for spontaneous growth to begin:

$$\frac{\partial \Delta G_{total}}{\partial r} = -\frac{4\pi}{\overline{V}} r^2 RT \ln \left(\frac{P}{P_{eq}} \right) + 8\pi\gamma r = 0 \tag{3.80}$$

Hence

$$r_{critical} = \frac{2\gamma \overline{V}}{RT \ln \left(\dfrac{P}{P_{eq}} \right)} \tag{3.81}$$

The value of $r_{critical}$ is about 6–10 Å (0.6–1 nm) for most materials, which indicates that the droplet of critical size contains between 50 and 100 atoms or molecules. Figure 3.13 shows the variation of the total free energy with the radius of the condensed particle.

The free energy of a particle of critical size can be expressed by substituting Eq. 3.81 into Eq. 3.79. We obtain

$$\Delta G_{max} = \tfrac{4}{3}\pi\gamma r_{crit}^2 \tag{3.82}$$

Thus the total free energy for a spherical particle of critical size is one-third of its surface free energy. Figure 3.14 shows the case of water at 0°C and also shows its critical radius, which is about 8 Å (0.8 nm).

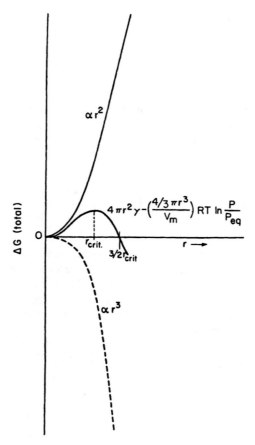

Figure 3.13. Total free-energy change, ΔG_{total}, of a particle as a function of its radius r and the change of its surface and volume free energy as a function of r.

A condensed particle must be larger than a certain critical size for spontaneous growth to occur at pressures $P > P_{\text{eq}}$. Homogeneous nucleation of the condensed phase by simultaneous clustering of many vapor atoms to reach this critical size, however, is very improbable. This is why supersaturated vapor can exist; that is, ambient conditions in which vapor pressures are larger than the equilibrium vapor pressure ($P > P_{\text{eq}}$) of a condensable substance can be established without the formation of the condensed phase. Precipitation in the absence of nuclei is very difficult, and large pressures much higher than the equilibrium vapor pressure ($P \gg P_{\text{eq}}$) must be established before condensation can occur within reasonable experimental times. There is, in general, a long induction period before growth of a liquid or solid phase commences in the absence of stable nuclei. Supersaturated vapor or undercooled liquid (cooled many degrees below its freezing point) are common occurrences in nature and in the laboratory.

Because of the difficulty of homogeneous nucleation, growth of condensed phases generally occurs on solid surfaces already present, such as the walls of the reaction cell or dust particles present in the atmosphere or in interstellar space. The intro-

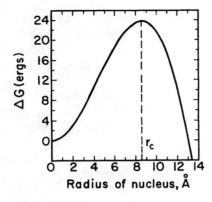

Figure 3.14. Total free-energy change of a water droplet as a function of its radius r.

duction of solid crystallites to induce the formation and growth of the condensed phase is frequently called *seeding* or *heterogeneous nucleation*. It is used to facilitate the growth of crystals or the condensation of water droplets from the atmosphere (rain making).

3.8 THERMODYNAMICS OF ADSORBED MONOLAYERS

3.8.1 Heat of Adsorption

When an atom or molecule strikes a surface and forms a bond with it, heat evolves. One can measure the temperature rise of the solid by calorimetry. In recent years the availability of improved heat sensors has permitted the measurement of temperature changes on the order of 10^{-3} K, a temperature change that should occur when a carbon monoxide monolayer chemisorbs on both sides of a well-characterized platinum single-crystal disk that is 1 cm^2 in surface area in the (111) orientation and 1 mm thick. Such a disk can chemisorb on its two sides 2×10^{15} molecules, which equals 2.2×10^{-9} moles. Taking an average heat of adsorption ΔH_{ads} of 25 kcal/mole (105 kJ/mole), an average Pt heat capacity of 0.032 cal/g/K (134 J/kg/K), and an average density of 21.45 g/cm^3 (21.45×10^3 kg/m^3), we have $\Delta T \approx 1.2 \times 10^{-3}$ K. Measurements such as the one derived above can yield the heats of adsorption of any molecule directly as a function of both surface structure and a wide range of coverages. A single crystal adsorption microcalorimeter was reported recently [37, 38] that uses an ultrathin sample (≈ 2000 Å thick) to minimize its heat capacity. The temperature rise upon adsorption is measured by focusing the infrared radiation from the crystal face onto a sensitive detector. Most calorimetric studies up to now have been carried out on high-surface-area microporous samples with surface structures and surface compositions that have not been well-characterized.

The heat of adsorption is measured more frequently by desorption, by breaking the adsorbate–surface bond. For each molecule–substrate combination, there is an optimum temperature at which the adsorbed molecules are removed at a maximum rate. By rapidly heating the surface (at rates of a few degrees per second) to this optimum temperature, the adsorbed molecules are removed at a maximum rate before their surface concentration is depleted. Working from this optimum temperature, at which the maximum rate of desorption occurs, the activation energy of desorption, which is closely related to ΔH_{ads}, can be calculated. This technique, which is most useful for molecules which adsorb reversibly and do not chemically alter during the temperature rise, is called *temperature-programmed desorption* (TPD) and will be discussed in more detail in the next chapter.

Even though adsorption is an exothermic process, the heat of adsorption is always given with a positive sign, unlike the negative sign used for exothermic events as dictated by the thermodynamic convention. We shall follow the tradition and denote ΔH_{ads} as positive.

3.8.2 Two-Dimensional Phase Approximation

A great deal of exchange takes place among atoms and molecules adsorbed at different surface sites of a solid. The reason for this lies in the low activation-energy values for transport along the surface for the diffusion of atoms or molecules from one step site along a terrace to another. These are frequently one-half or less of the activation energies for bulk diffusion or heats for desorption into the gas phase. As a result, a great deal of movement of molecules occurs along the surface from one site to another by surface diffusion during their residence time there. Therefore, we may assume equilibrium among molecules in the various surface sites in most circumstances. The long residence times τ are forced on the system as a result of the large adsorption energies because

$$\tau = \tau_0 \exp\left(\frac{\Delta H_{ads}}{RT}\right) \tag{3.83}$$

Because of the long residence times of the various gases that are common in the earth's atmosphere (oxygen, water vapor, hydrocarbons), the adsorbed layer may be viewed as a two-dimensional phase that is well-protected from exchange with the gas or the bulk of the condensed phase by large potential-energy barriers, while transport and chemical exchange along the surface is facile. This is shown schematically in Figure 3.15.

Theories that assume equilibrium among atoms at various surface sites and among different adsorbates have been successful in explaining the nature of evaporation [39], crystal growth [40], and adsorption [41], in many systems.

3.8.3 Adsorption Isotherms We will now concern ourselves with the properties of the adsorbed layer of a weakly interacting gas on a solid surface. We can see from Eq. 3.83 that the residence time of weakly interacting gas atoms can be increased markedly by decreasing the temperature at which the experiment is carried out. Assuming a heat of adsorption $\Delta H_{ads} \approx 2$ kcal/mole (8 kJ/mole), and $\tau_0 = 10^{-12}$ sec, the residence time at 300 K is on the order of 10^{-11} sec. At 35 K, however, the residence time is greater than 2 sec! Thus, by judicious choice of the experimental conditions, we can maintain a large concentration of gas atoms σ on the surface, even for small values of ΔH_{ads}. It is not difficult to see that, in addition to the residence time, the surface concentration or surface coverage σ will also depend on the flux of gas atoms F striking the surface per unit area per second. The surface coverage σ (molecules/cm^2) when a large concentration of surface atoms is

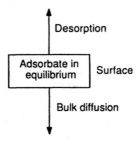

Exceptions :

Low pressure - High surface temperatures studies
(molecular beam - surface scattering)

Exothermic surface reactions ?

Figure 3.15. Adsorbates on the surface which exist as a two-dimensional phase protected by large potential barriers from desorption and from bulk diffusion. Because of their long residence time on the surface, they come to equilibrium among the various surface sites.

still available as adsorption sites (the low-coverage situation) is given by the product [42]

$$\sigma = \tau F \tag{3.84}$$

The gas flux is proportional to the pressure; and, from the kinetic theory of gases, it is given by Eq. 1.2. The residence time is defined in Eq. 3.83. Using these two equations, we can rewrite the surface coverage as

$$\sigma = \frac{N_A P}{\sqrt{2\pi MRT}} \tau_0 \exp\left(\frac{\Delta H_{\text{ads}}}{RT}\right) \tag{3.85}$$

From a knowledge of ΔH_{ads}, P, and T, σ can be estimated. We assume that the adsorbed gas atoms undergo complete thermal accommodation on the surface.

Much of our information about the nature of the adsorbed gas layer comes from studies of the amount of gas adsorbed on a surface σ (the surface coverage) as a function of gas pressure P at a given temperature. The σ–P curves derived from these experiments are called *adsorption isotherms*. Adsorption isotherms are used primarily to determine thermodynamic parameters that characterize the adsorbed layer (heats of adsorption, and the entropy and heat capacity changes associated with the adsorption process) and to determine the surface area of the adsorbing solid. The latter measurement is of great technical importance because of the widespread use of porous solids of high surface area in various industrial processes. The effectiveness of participation by a porous solid in a surface reaction is often proportional to the surface area of the solid. The simplest adsorption isotherm at a constant temperature is obtained from Eq. 3.85, which we can rewrite as

$$\sigma = kP \tag{3.86}$$

where

$$k = N_A (2\pi MRT)^{-1/2} \tau_0 \exp\left(\frac{\Delta H_{ads}}{RT}\right) \qquad (3.87)$$

Thus the coverage is proportional to the first power of the pressure at a given temperature, if the model for adsorption that led to the formulation of Eq. 3.84 is correct. That is, the adsorbed gas atoms do not interact with each other and we have an unlimited number of surface sites at which adsorption can occur. It is also assumed that the adsorption energy ΔH_{ads} is the same for all of the molecules. Equation 3.86 is unlikely to be suitable for describing the overall adsorption process. Nevertheless, it approximates the adsorption isotherms for many real systems at low pressures [10^{-5} torr ($< 1.3 \times 10^{-3}$ Pa)] and at high pressure [10 torr (1.3×10^3 Pa)] at the initial stages of adsorption. The adsorption isotherms of argon on silica gel, which obey Eq. 3.86, are shown in Figure 3.16.

Langmuir derived a different adsorption isotherm by assuming that adsorption is terminated upon completion of a monomolecular adsorbed gas layer [43]. He did this by asserting that any gas molecule that strikes an adsorbed atom must reflect from the surface. All the other assumptions (i.e., homogeneous surface and noninteracting adsorbed species) used to obtain Eq. 3.85 were also maintained. If σ_0 is the surface coverage of a completely covered surface, the concentration of surface sites available for adsorption, after adsorbing σ molecules, is $\sigma_0 - \sigma$. Of the total flux F incident on the surface, a fraction [$(\sigma/\sigma_0)F$] will strike molecules already adsorbed and therefore be reflected. Thus a fraction [$1 - (\sigma/\sigma_0)$]F of the total incident flux will be available for adsorption. Equation 2.84 should thus be modified as

$$\sigma = \left(1 - \frac{\sigma}{\sigma_0}\right) F\tau \qquad (3.88)$$

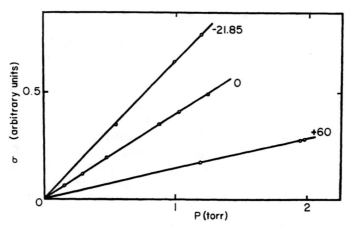

Figure 3.16. Adsorption isotherms of argon on silica gel at $-21.85°C$, $0°C$, and $+60°C$.

which can be rearranged to give

$$\sigma = \frac{\sigma_0 F \tau}{\sigma_0 + F \tau} = \frac{\sigma_0 b P}{\sigma_0 + b P} \tag{3.89}$$

By writing $\theta = \sigma / \sigma_0$, where θ is often called the *degree of coverage*, Eq. 3.89 can be rewritten as

$$\theta = \frac{b' P}{1 + b' P} \tag{3.90}$$

where $b' = b / \sigma_0$. From Eq. 3.90, $b' P$ may be neglected at low pressures in comparison with 1 in the denominator, and Eq. 3.86 is obtained. The adsorption isotherm of ethyl chloride on charcoal, which appears to obey an equation of the form of Eq. 3.89, is shown in Figure 3.17.

Equation 3.89 can be rearranged to give

$$\frac{1}{\sigma} = \frac{1}{bP} + \frac{1}{\sigma_0} \tag{3.91}$$

Therefore, a linear Langmuir plot is obtained by plotting $1/\sigma$ against $1/P$. Such a plot is shown for the adsorption of oxygen, carbon monoxide, and carbon dioxide on silica in Figure 3.18.

There are several other derivations of the Langmuir adsorption isotherm from statistical mechanics and thermodynamics. Although the model is physically unrealistic for describing the adsorption of gases on real surfaces, its successes, just like the success of other adsorption isotherms also based on different simple adsorption models, is due to the relative insensitivity of macroscopic adsorption measurements to the atomic details of the adsorption process. Thus the adsorption isotherm

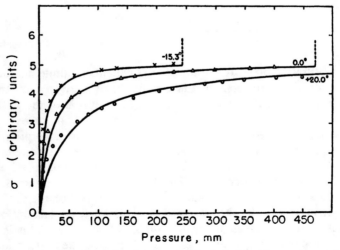

Figure 3.17. Adsorption isotherms of ethyl chloride on charcoal.

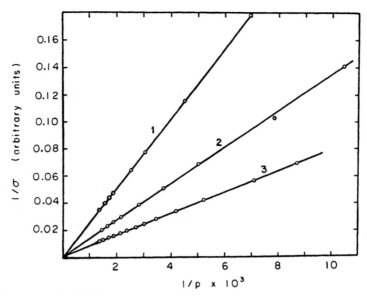

Figure 3.18. Adsorption isotherm of (1) oxygen, (2) carbon monoxide, and (3) carbon dioxide on silica plotted as $1/\sigma$ versus $1/P$.

provides one with useful approximate values of the important adsorption parameters σ and ΔH_{ads} and permits the determination of the surface area.

Another frequently used adsorption model that allows for adsorption in multilayers where gas atoms or molecules adsorb on top of already adsorbed molecules was proposed by Brunauer, Emmett, and Teller [44] (the BET model). With the exception of the assumption that the adsorption process terminates at monolayer coverage, the BET model retains all other assumptions made in deriving the Langmuir adsorption isotherm. The BET model leads to a two-parameter adsorption equation of the form

$$\frac{P}{\sigma(P_0 - P)} = \frac{1}{\sigma_0 c} + \frac{c - 1}{\sigma_0 c}\frac{P}{P_0} \tag{3.92}$$

where P_0 is the saturation pressure of the vapor at which an infinite number of layers can be built up on the surface, and c is a constant at a given temperature. The constant is an exponential function of the heat of adsorption of the first layer and the heat of condensation or liquefaction of the vapor ΔH_L ($c \propto \exp\{(\Delta H_{ads} - \Delta H_L)/RT\}$), which had been assumed to equal the heat of adsorption above the second layer. A plot of $P/[\sigma(P_0 - P)]$ versus P/P_0 should yield a straight line with slope $(c - 1)/\sigma_0 c$ and intercept $1/\sigma_0 c$. The adsorption isotherm of nitrogen on titanium dioxide (anatase), which obeys Eq. 3.92, is shown in Figure 3.19.

The adsorption isotherm yields the amount of gas adsorbed on the surface. Unless the molecular area occupied by the adsorbed gas is known, the adsorption isotherm yields only relative surface areas rather than the absolute values. This is the reason for using only one gas (nitrogen or krypton) to determine the surface areas of different solids. However, Harkins and Jura [45] developed an absolute method of

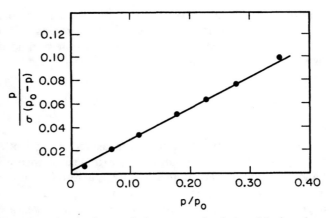

Figure 3.19. The adsorption isotherm of nitrogen on titanium oxide that obeys the Langmuir isotherm.

determining the surface area of nonporous solids by using an immersion calorimeter. If a finely divided crystal is coated with an adsorbed film (gas or liquid) and then is dropped into a liquid inside a calorimeter, the heat of immersion divided by the surface free energy of the liquid gives the surface area of the solid directly. By using the absolute method and the BET or other adsorption isotherms for the same system, the average area occupied by a molecule of a given gas can be obtained. For nitrogen, this value was found to be 16.2 Å^2 (0.162 nm^2) [and for krypton, 25.6 Å^2 (0.256 nm^2)] for a variety of surfaces and was adopted as a standard for surface-area determinations [46].

Several other theoretical models [47–49] have attempted to give a more realistic description than the Langmuir and BET models of the gas–surface interactions that lead to physical adsorption. The variable parameters in these models are the interaction potential, the structure of the adsorbed layer (mobile or localized monolayer of multilayer), and the structure of the surface (homogeneous or heterogeneous, number of nearest neighbors).

3.8.4 Integral and Differential Heats of Adsorption

When a single atom or molecule is adsorbed on a clean surface, the heat of adsorption reflects the strength of the adsorbate–substrate bond. As more molecules adsorb, each one contributes its heat of adsorption to the total (integral) measured value. If the heat of adsorption per molecule is q_{ads}, the integral heat of adsorption, ΔH_{ads}, of N molecules is

$$\Delta H_{\text{ads}} = N q_{\text{ads}} \tag{3.93}$$

As we pack more molecules on the surface, they begin to interact and influence each other's bounding. This adsorbate–adsorbate interaction can be repulsive or attractive. In either case, q_{ads}, which is the result of both adsorbate–substrate and adsorbate–adsorbate interactions, will change. In addition, q_{ads} may also change if the surface contains many adsorption sites. Usually the strongly binding surface sites fill up first with adsorbates. Then the weaker bonding sites become occupied as the coverage increases. Thus, determining the heat of adsorption as a function of coverage can provide information about the adsorbate–substrate and adsorbate–adsorbate interactions separately. If we differentiate ΔH_{ads} with respect to N, we obtain

$$\Delta H_{ads}^{diff} = \left(\frac{d\Delta H_{ads}}{dN}\right)_T = q_{ads} + N\left(\frac{dq_{ads}}{dN}\right)_T \tag{3.94}$$

If the differential heat of adsorption ΔH_{ads} is measured under isothermal conditions, it is commonly called the *isothermal* or *isosteric* heat of adsorption [42].

The integral and differential heats of adsorption are determined by measuring the adsorption isotherms for a given system at different temperatures. From the data, the equilibrium pressures necessary to obtain the same coverages at the different temperatures are determined. From the slope of the plots of $\ln P_{\theta = const}$ versus $1/T$, the differential isosteric heats of adsorption for a given coverage are determined by the use of the Clausius–Clapeyron equation:

$$\left(\frac{d \ln P}{dT}\right)_{\theta = const} = -\frac{\Delta H_{ads}^{diff}}{RT^2} \tag{3.95}$$

In Figure 3.20a the coverage-dependent ΔH_{ads}^{diff} values are shown for several weakly adsorbed gases as a function of coverage.

In recent years, differential heats of adsorption have been obtained using single-crystal metal surfaces with well-characterized structures. One example of the variation of the isosteric heat with coverage on such a surface is shown in Figure 3.20b. The heat of adsorption of strongly chemisorbed CO on the Pd(111) crystal face was determined as a function of CO coverage. At coverages of up to half a monolayer, the heat of adsorption remains high, because neighboring CO molecules do not interfere with each other's bonding to the metal substrate. Above this coverage, however, ΔH_{ads} declines rapidly due to repulsive adsorbate–adsorbate interactions; and at about 80% of a monolayer coverage, ΔH_{ads} drops to one-third of its value at low coverages.

Most chemisorption systems show similar variations of ΔH_{ads} with coverage.

3.8.5 Molecular and Dissociative Adsorption

The magnitude of ΔH_{ads} can tell us whether a diatomic molecule remains intact upon adsorption or dissociates into its constituent atoms. For adsorption as a diatomic molecule, ΔH_{ads} is defined as the energy needed to break the bond between the substrate (M) and the adsorbed molecule (X_2), MX_2:

$$MX_{2\,ads} \xrightarrow{\Delta H_{ads}} M + X_{2\,gas} \tag{3.96}$$

If the diatomic molecule dissociates upon adsorption, the heat of adsorption for this purpose is defined as

$$2MX_{ads} \xrightarrow{\Delta H_{ads}} 2M + X_{2\,gas} \tag{3.97}$$

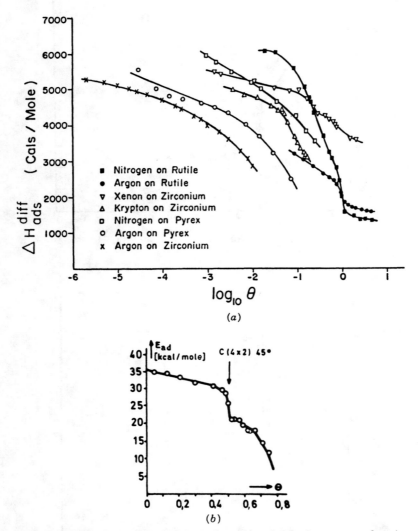

Figure 3.20. (a) Heat of adsorption of several weakly adsorbed gases as a function of coverage. (b) Heat of adsorption for chemisorbed CO on the Pd(111) crystal face as a function of coverage [50].

From the knowledge of ΔH_{ads} (which is always positive) and the binding energy of the gas-phase molecule, the energy of the surface chemical bond is given by

$$\Delta E_{bond} (MX_2) = \Delta H_{ads} \tag{3.98}$$

or

$$\Delta E_{bond} (MX) = \frac{\Delta H_{ads} + \Delta E_{X_2}}{2} \tag{3.99}$$

where ΔE_{X_2} is the dissociation energy of the X_2 gas molecule.

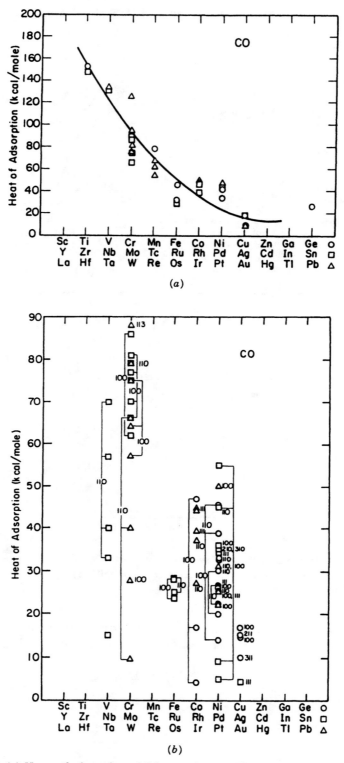

Figure 3.21. (a) Heats of adsorption of CO on polycrystalline transition metal surfaces. (b) Heats of adsorption of CO on various single crystal surfaces of transition metals.

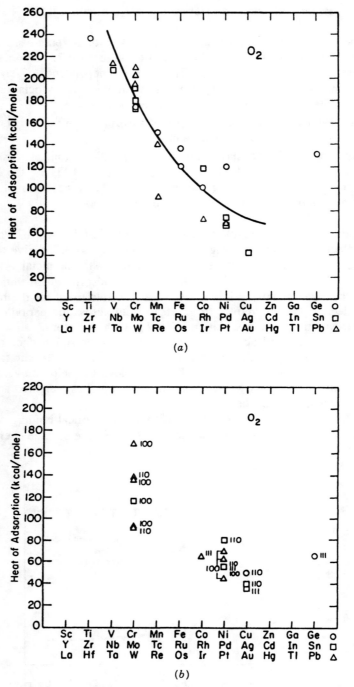

Figure 3.22. (a) Heats of adsorption of O_2 on polycrystalline transition metals surfaces. (b) Heats of adsorption of O_2 on various single crystal surfaces of transition metals.

There are several difficulties in determining ΔH_{ads} reliably:

- The heat of chemisorption changes markedly with coverage θ, as shown in Figure 3.20b for CO on Pd. Thus, the coverage must be known in order to interpret the thermodynamic data. The coverage dependence of ΔH_{ads} can also lead to the circumstance in which adsorbing molecules dissociate at low coverages while remaining intact at high coverages.
- The surface may restructure as a result of adsorption. In this event, the heat of restructuring is part of the experimental ΔH_{ads}.
- The adsorbed molecule may decompose upon adsorption, at least at certain sites and at low coverages, leaving fragments behind. Thus, the heat of adsorption may include the energy needed for bond breaking, and molecular fragments may permanently alter the surface composition.

Nevertheless, it is most useful to obtain an estimate of the heat of adsorption of different molecules on surfaces. Figures 3.21a, 3.22a, and 3.23a give the heats of adsorption of CO, O_2, and H_2, respectively, on polycrystalline metal surfaces. Figures 3.21b, 3.22b, and 3.23b display the heats of adsorption on various crystal surfaces of the same metals. While there is a noticeable trend for the heats of adsorption to decrease as one moves from left to right across the periodic table, this trend is overwhelmed by the variation of ΔH_{ads} with the atomic structure of the surface. Evidently, several adsorption sites on a given crystal surface that exhibit large variations of ΔH_{ads} mask any trends in ΔH_{ads} across the periodic table. These trends are noticeable only for polycrystalline samples that yield an average heat of adsorption compared to ΔH_{ads} values from the different sites weighted according to the relative concentration of these sites.

We will discuss the surface-structure sensitivity of chemical binding of adsorbed atoms and molecules when we review the properties of the surface chemical bond in Chapter 6.

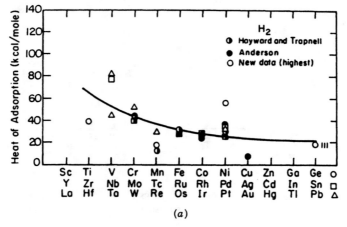

(a)

Figure 3.23. (a) Heats of adsorption of H_2 on polycrystalline transition metal surfaces. (b) Heats of adsorption of H_2 on various single crystal surfaces of transition metals.

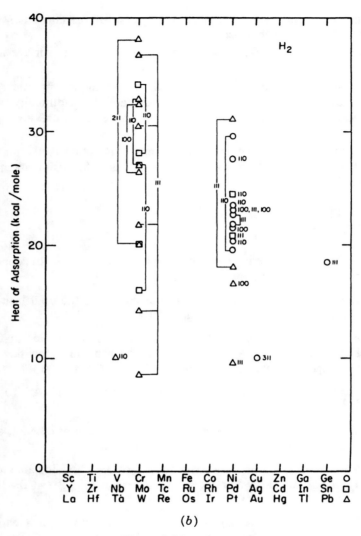

Figure 3.23. (*Continued*)

3.9 SUMMARY AND CONCEPTS

- Surface thermodynamic functions are defined to be separable from the functions that characterize the properties of bulk atoms.
- The surface pressure or surface tension is large, indicating that the surface atoms are subjected to large compressive forces.
- The surface free energy is always positive; it is highest for metals, lower for oxides, and lowest for organic molecules, especially fluorocarbons.
- The surface heat capacity has a different temperature dependence than the three-dimensional heat capacity.

- In multicomponent systems, surface segregation of the constituent with the lowest surface free energy occurs.
- Mixtures of small particles in which every atom is on the surface have very different phase diagrams compared to the bulk phase.
- Liquid systems with curved surfaces exhibit many unique properties (capillary rise, curvature-dependent vapor pressure, contact angle, difficulty of nucleation) because of differences between internal and external pressures at the interface.
- The external pressure–coverage relationships (adsorption isotherms) and the heats of adsorption for the adsorbate layers provide information about bonding at surfaces.

3.10   PROBLEMS

3.1   Calculate the change in energy when two atomically smooth copper plates, each of 10-cm^2 area, are joined. Assume $T = 1047°C$.

3.2   Estimate the surface-heat-capacity contribution to the total heat capacity of a nickel cube with a side length of 100 Å.

3.3   Compute the surface concentration of a Au-Ag solid solution that contains only 5 atom% silver. Assume ideal solution behavior. How does the surface composition change upon heating from 300 K to 700 K?

3.4   Compute the vapor pressure of a water droplet of 10-Å, 10^2-Å, and 10^3-Å diameter.

3.5   Calculate the critical radius and ΔG_{max} for a droplet of ethanol, assuming a vapor pressure $P_{C_2H_5OH} = 50$ torr supercooled to 20°C.

3.6   A water droplet on mercury has a contact of 33.3°. Calculate the interfacial energy at the water–mercury interface. ($\gamma_{Hg-air} = 0.436$ J/m^2, $\gamma_{H_2O-air} = 0.073$ J/m^2).

3.7   Calculate the pressure gradient $P_{in} - P_{ext}$ for water in a capillary of 10^{-4}-cm diameter. How great is the capillary rise if P_{ext} is 1 atm?

3.8   Using silica gel and nitrogen as absorbent, the following adsorption isotherm data were obtained [G. Constabaris et al., Chevron Research Co.]:

P/P_0:	0.055	0.061	0.077	0.094	0.120	0.158	0.177
$\sigma(cm^3/g)$:	131.3	134.3	139.9	148.9	153.5	164.0	169.3

P/P_0:	0.209	0.240	0.270	0.300	0.330	0.352
$\sigma(cm^3/g)$:	176.9	184.5	192.3	200.0	207.7	212.7

Plot the BET function, $P/\sigma (P_0 - P)$ versus P/P_0, compute the slope and the intercept, and obtain the surface area and the parameter c.

*3.9   Compute the vapor pressure of a spherical platinum particle with $r = 10$ Å at 200°C and 800°C using vapor pressure data given in the literature (e.g., see reference [51]).

(b) Small particles of comparable size are frequently used as catalysts in many industrial processes. Could these particles be used continuously for 3 years assuming that the loss of material would occur only by vaporization?

*3.10   When two metals with atoms of different sizes form a solid solution, the strain energy that is produced by mixing small and large atoms modifies the surface composition. Discuss the effects of the strain energy on surface segregation [52–57].

*3.11   A monolayer of n-octodecanol, when spread over water, reduces its evaporation rate markedly [58]. Calculate the change in the loss of water from a 1-km²-surface-area lake by evaporation over a 24-h period by this monolayer. Assume that the film pressure is 0.02 J/m² and that the relative humidity is zero.

*3.12   The Langmuir trough is used to measure surface pressures of monolayers. Describe the operation of the device and the variation of monolayer film behavior with surface pressure for a given system described in the literature.

*3.13   High-surface-area alumina silicates (zeolites) can be produced from solution. Review the literature and discuss how these microporous materials are prepared [59–62].

*3.14   The heat of adsorption of CO declines rapidly at greater than one-half monolayer coverage [50]. Discuss the reasons for this behavior.

*3.15   The hydrogen molecule readily dissociates on nickel but does not dissociate on gold. Show by calculation the thermodynamic reasons for this behavior [63].

**3.16   Review the behavior of micelles and reverse micelles [64]. Name three micelle systems that are used in our everyday life.

REFERENCES

[1]   G.N. Lewis and M. Randall. *Thermodynamics*. McGraw–Hill, New York, 1961. Revised by K.S. Pitzer and L. Brewer.

[2]   R. Defay and I. Prigogine. *Surface Tension and Adsorption*. John Wiley & Sons, New York, 1966.

[3]   S. Overbury, P. Bertrand, and G.A. Somorjai. The Surface Composition of Binary Systems. Prediction of Surface Phase Diagrams of Solid Solutions. *Chem. Rev.* **75**:547 (1975).

[4]   G. Kőrösi and E.S.Z. Kováts. Density and Surface Tension of 83 Organic Liquids. *J. Chem. Eng. Data* **26**:323 (1981).

[5]   S. Blairs. The Surface Tension and Critical Properties of Liquid Metals. *J. Colloid Interface Sci.* **67**:548 (1978).

[6]   J.J. Jasper. The Surface Tension of Pure Liquid Compounds. *J. Phys. Chem. Ref. Data* **1**:841 (1972).

[7]   B.C. Allen. The Surface Tension of Liquid Transition Metals at Their Melting Points. *Trans. Metall. Soc. AIME* **227**:1175 (1963).

[8] J.C. Erickson. Thermodynamics of Surface Phase Systems. V. Contribution to the Thermodynamics of the Solid–Gas Interface. *Surf. Sci.* **14**:221 (1969).

[9] E.A. Guggenheim. The Principle of Corresponding States. *J. Chem. Phys.* **13**:253 (1945).

[10] A.V. Grosse. The Relationship between the Surface Tensions and Energies of Liquid Metals and Their Critical Temperatures. *J. Inorg. Nucl. Chem.* **24**:147 (1962).

[11] A.W. Adamson. *Physical Chemistry of Surfaces*. John Wiley & Sons, New York, 1990.

[12] P. Balk and G.C. Benson. Calorimetric Determination of the Surface Enthalpy of Potassium Chloride. *J. Phys. Chem.* **63**:1009 (1959).

[13] G. Jura and C.W. Garland. The Experimental Determination of the Surface Tension of Magnesium Oxide. *J. Am. Chem. Soc.* **74**:6033 (1952).

[14] T.L. Hill. *Introduction to Statistical Thermodynamics*. Addison–Wesley, Reading, MA, 1962.

[15] E.W. Montroll. Size Effect in Low Temperature Heat Capacities. *J. Chem. Phys.* **18**:183 (1950).

[16] M. Dupuis, R. Mazo, and L. Onsager. Surface Specific Heat of an Isotropic Solid at Low Temperatures. *J. Chem. Phys.* **33**:1452 (1960).

[17] R. Stratton. A Surface Contribution to the Debye Specific Heat. *Philos. Mag.* **44**:519 (1953).

[18] D.H. Everett, editor. Manual of Symbols and Terminology for Physiochemical Quantities and Units. Appendix II. Definitions, Terminology and Symbols in Colloid and Surface Chemistry. Part I. *Pure Appl. Chem.* **31**:577 (1972).

[19] C. Wagner. Phenomenal and Thermodynamic Equations of Adsorption. *Nachrichten der Akademie der Wissenschaften in Göttingen, II. Mathematisch-Physikalische Klasse*, 1973, p. 37.

[20] J.W. Gibbs. *The Collected Works of J.W. Gibbs*. Longmans, Green & Company, London, 1931.

[21] S. Overbury and G.A. Somorjai. Auger Electron Spectroscopy of Alloy Surfaces. *Faraday Discuss. Chem. Soc.* **60**:279 (1975).

[22] J. Blakely and J. Shelton. Equilibrium Adsorption and Segregation. In: J. Blakely, editor, *Surface Physics of Materials*, Volume 1. Academic Press, New York, 1975.

[23] P. Wynblatt and R.C. Ku. Surface Energy and Solute Strain Energy Effects in Surface Segregation. *Surf. Sci.* **65**:511 (1977).

[24] R. Hultgren, P.D. Desai, D.T. Hawkins, M. Gleiser, and K.K. Kelley, editors. *Selected Values of the Thermodynamic Properties of Binary Alloys*. American Society for Metals, Metals Park, Ohio, 1973.

[25] F.L. Williams and D. Nason. Binary Alloy Surface Compositions from Bulk Alloy Thermodynamic Data. *Surf. Sci.* **45**:377 (1974).

[26] J.J. Burton, E. Hyman, and D.A. Fedak. Surface Segregation in Alloys. *J. Catalysis*, **37**:106 (1975).

[27] W.M.H. Sachtler and R. Jongpier. The Surface of Copper–Nickel Alloy Films. II. Phase Equilibrium and Distribution and Their Implications for Work Function, Chemisorption, and Catalysis. *J. Catal.* **4**:665 (1965).

[28] D. McLean. *Grain Boundaries in Metals*. Oxford University Press, London, 1957.

[29] J.H. Sinfelt. Heterogeneous Catalysis: Some Recent Developments. *Science* **195**:641 (1977).

[30] J.L. Morán-López and L.M. Falicov. Segregation and Short-Range Order Properties at the Boundaries of Two-Dimensional Bimetallic Clusters. *Surf. Sci.* **79**:109 (1979).

[31] I. Langmuir. The Constitution and Fundamental Properties of Solids and Liquids. II. Liquids. *J. Am. Chem. Soc.* **39**:1848 (1917).

[32] F.M. Fowkes, editor. *Contact Angle, Wettability, and Adhesion. Advances in Chemistry*, Volume 43. American Chemical Society, Washington, D.C., 1964.

[33] J.J. Bikerman. *Surface Chemistry for Industrial Research*. Academic Press, New York, 1947.

[34] D.J. Alner, editor. *Aspects of Adhesion*. University of London Press, London, 1966.

[35] L. Leger and J.F. Joanny. Liquid Spreading. *Rep. Prog. Phys.* **55**:431 (1992).

[36] W.D. Harkins. *The Physical Chemistry of Surface Films*. Van Nostrand Reinhold, New York, 1952.

[37] C.E. Borroni-Bird and D.A. King. An Ultrahigh Vacuum Single Crystal Adsorption Microcalorimeter. *Rev. Sci. Instrum.* **62**:2177 (1991).

[38] C.E. Borroni-Bird, N. Al-Sharaf, S. Andersson, and D.A. King. Single Crystal Adsorption Microcalorimetry. *Chem. Phys. Lett.* **183**:516 (1991).

[39] J.P. Hirth and A.M. Pound. *Condensation and Evaporation*. Pergamon, Elmsford, NY, 1969.

[40] W.K. Burton, N. Cabrera, and F.C. Frank. The Growth of Crystals and the Equilibrium Structure of Their Surfaces. *Philos. Trans. R. Soc. London* **243A**:299 (1951).

[41] M.W. Roberts and C.S. McKee. *Chemistry of the Metal-Gas Interface*. Oxford University Press, New York, 1978.

[42] J.H. de Boer. *The Dynamical Character of Adsorption*. Oxford University Press, New York, 1968.

[43] I. Langmuir. The Adsorption of Gases on Plane Surfaces of Glass, Mica, and Platinum. *J. Am. Chem. Soc.* **40**:1361 (1918).

[44] S. Brunauer, P.H. Emmett, and E. Teller. Adsorption of Gases in Multimolecular Layers. *J .Am. Chem. Soc.* **60**:309 (1938).

[45] W.D. Harkins and G. Jura. Surfaces of Solids. XII. An Absolute Method for the Determination of the Area of a Finely Divided Crystalline Solid. *J. Am. Chem. Soc.* **66**:1362 (1944).

[46] A.L. McClellan and H.F. Harnsberger. Cross-Sectional Areas of Molecules Adsorbed on Solid Surfaces. *J. Coll. Interface Sci.* **23**:577 (1967).

[47] S. Ross and J.P. Oliver. *On Physical Adsorption*. John Wiley & Sons, New York, 1964.

[48] D.M. Young and A.D. Crowell. *Physical Adsorption of Gases*. Butterworth, Woburn, MA, 1962.

[49] W.M. Champion and G.D. Halsey, Jr. A New Multilayer Isotherm with Reference to Entropy. *J. Am. Chem. Soc.* **76**:974 (1954).

[50] H. Conrad, G. Ertl, J. Koch, and E.E. Latta. Adsorption of CO and Pd Single Crystal Surfaces. *Surf. Sci.* **43**:462 (1974).

[51] L.H. Dreger and J.L. Margrave. Vapor Pressures of Platinum Metals. I. Palladium and Platinum. *J. Phys. Chem.* **64**:1323 (1960).

[52] D.C. Peacock. Tests of Segregation Theory. I. The Ni–5Pt Alloy. *Appl. Surf. Sci.* **26**:306 (1986).

[53] W.M.H. Sachtler. Surface Composition of Alloys. *Appl. Surf. Sci.* **19**:167 (1984).

[54] F.F. Abraham and C.R. Brundle. Surface Segregation in Binary Solid Solutions: A Theoretical and Experimental Perspective. *J. Vacuum Sci. Technol.* **18**:506 (1981).

[55] M.J. Kelley and V. Ponec. Surface Composition of Alloys. *Prog. Surf. Sci.* **11**:139 (1981).

[56] M.J. Kelley, P.W. Gilmour, and D.G. Swarzfager. Strain Effects in Surface Segregation—The Au/Ni System. *J. Vacuum Sci. Technol.* **17**:634 (1980).

[57] F.F. Abraham, N.H. Tsai, and G.M. Pound. Bond and Strain Energy Effects in Surface Segregation. *Scr. Metall.* **13**:307 (1979).

[58] V.K. La Mer, T.W. Healy, and L.A.G. Aylmore. The Transport of Water Through Monolayers of Long-Chain *n*-Paraffinic Alcohols. *J. Colloid Sci.* **19**:673 (1964).

[59] H. van Bekkum, E.M. Flanigen, and J.C. Jansen, editors. *Introduction to Zeolite Science and Practice, Studies in Surface Science and Catalysis*, Volume 58. Elsevier, Amsterdam, 1991.

[60] M.L. Occelli and H.E. Robson, editors. *Zeolite Synthesis. ACS Symposium Series*, Volume 398. American Chemical Society, Washington, D.C., 1989.

[61] P.A. Jacobs and J.A. Martens. *Synthesis of High-Silica Aluminosilicate Zeolites. Studies in Surface Science and Catalysis*, Volume 33. Elsevier, Amsterdam, 1987.

[62] A.B. Stiles. *Catalysts Manufacture: Laboratory and Commercial Preparations. Chemical Industries*, Volume 14. Marcel Dekker, New York, 1983.

[63] K.W. Frese, Jr. Calculation of Surface Binding Energies for Hydrogen, Oxygen, and Carbon Atoms on Metallic Surfaces. *Surf. Sci.* **182**:85 (1987).

[64] K.L. Mittal, editor. *Micellization, Solubilization, and Microemulsions*. Plenum Press, New York, 1977.

4

DYNAMICS AT SURFACES

4.1 INTRODUCTION

In this chapter we shall review the motion of atoms and molecules at surfaces. First we discuss how atoms vibrate about their equilibrium surface sites. Then, the elementary surface processes during the collisions of gas atoms and molecules with surfaces are described. We then discuss several elementary gas–surface interactions: adsorption, surface diffusion, and desorption.

4.2 SURFACE ATOM VIBRATIONS

4.2.1 The Harmonic Oscillator Model

In studying the motion of surface atoms about their equilibrium positions, it is frequently useful to relate it to the motion of model systems. Perhaps the most useful analogy is the one-dimensional harmonic oscillator: a mass on a spring that is fixed on one end. The properties of the one-dimensional harmonic oscillator can be directly related to the properties of surface atoms vibrating about their equilibrium positions. The reason for this model's success is that the displacement of the vibrating atoms is only a small fraction of their distance from their neighbors (interatomic distance), and thus the harmonic approximation remains appropriate for describing many phenomena related to atomic vibration.

As shown in Figure 4.1, the restoring force f that the spring exerts on the displaced mass m is linearly proportional to the amount of stretch or displacement x and opposite in sign to it:

$$f = m \frac{d^2x}{dt^2} = -kx \tag{4.1}$$

where d^2x/dt^2 is the acceleration and k is the constant of proportionality, called the "force constant." r is the length of the spring, at which $x = 0$, and $x = a$ is the distance of maximum displacement of the mass from its equilibrium position at r. Equation 4.1 has a solution of the form

$$x = a \cos \omega t \tag{4.2}$$

where $\omega = \sqrt{k/m}$ is the angular frequency, which depends on the mass m and the "stiffness" of the spring, described by the force constant k; a is the amplitude of oscillation, which is equal to the maximum displacement, and t is time.

The potential-energy change dV of the system as the particle is displaced by dx against the force exerted by the spring is

$$\frac{dV}{dx} = kx \tag{4.3}$$

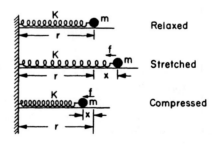

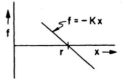

Figure 4.1. One-dimensional harmonic oscillator model.

Using a boundary condition in which the mass, in its equilibrium position, has zero displacement ($x = 0$), we integrate Eq. 4.3 to obtain, for the potential energy V,

$$V = \tfrac{1}{2}kx^2 \qquad (4.4)$$

It is also important to recognize that differentiation of Eq. 4.4 twice yields $d^2V/dx^2 = k$. Thus the force constant of the spring is equal to the curvature of the potential-energy function, which is shown in Figure 4.2.

For an arbitrary (not necessarily harmonic) attractive potential, the potential-energy term can be expanded in a Taylor series about its minimum ($x = 0$) to give

$$V(x) = V|_{x=0} + \frac{dV}{dx}\bigg|_{x=0} x + \frac{1}{2}\frac{d^2V}{dx^2}\bigg|_{x=0} x^2 + \frac{1}{6}\frac{d^3V}{dx^3}\bigg|_{x=0} x^3 + \cdots \qquad (4.5)$$

Because $V_{x+0} = 0$ and the first derivative must also be zero (because the potential energy is at a minimum at $x = 0$), the potential energy is given by the second- and higher-order derivatives in the series. In the "harmonic" approximation, one neglects the higher-order terms, an adequate approximation for our purposes at present. By taking the third derivative into consideration, we can describe the "anharmonicity" of the oscillator. The anharmonic nature of the real crystal potential is responsible for the thermal expansion of solids at elevated temperatures and for the interaction between certain modes of lattice vibration. For surface atoms placed in an anisotropic environment, the inclusion of the anharmonic term in the potential-energy function can be important in describing many of the physical–chemical properties of the atoms.

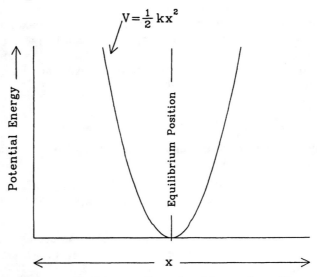

Figure 4.2. Potential-energy V of the one-dimensional harmonic oscillator as a function of the displacement x.

4.2.2 Vibrational Modes of Surface Atoms

A solid may be viewed as comprising many harmonic oscillators, each coupled to its neighbors. The elastic continuum theory has been used to compute the frequency distribution of waves propagating in such a medium [1]. Debye suggested that the frequency distribution—the number of frequencies n—in the frequency range from ν to $\nu + d\nu$ has the form of $dn/d\nu \propto \nu^2$, and that there is a cutoff frequency ν_D at the high-frequency end. Figure 3.2 shows the Debye frequency distribution. The surface vibrational modes may be obtained using this model by considering waves propagating in a semi-infinite isotropic elastic medium that possesses a stress-free surface. Rayleigh [2] found that such a surface generates long-wavelength modes of wave propagation in directions parallel to the surface and which decay exponentially with depth into the bulk. These surface waves are also called *Rayleigh waves*.

Inelastic helium scattering appears to be an excellent technique to determine the vibration spectrum of surface atoms at solid surfaces. Helium atoms with incident energies near thermal energies [≈ 30 meV ($\approx 5 \times 10^{-21}$ J)] can excite the surface atoms upon collision and can also diffract, since their de Broglie wavelength at these energies is less than the interatomic distance [≈ 0.1 nm $= 1$ Å], corresponding to an energy ≈ 0.02 eV (3×10^{-21} J). As long as their energy distribution can be kept narrow, ≈ 0.3 meV ($\approx 5 \times 10^{-23}$ J), the vibrational excitation spectrum of surface atoms can be mapped by monitoring the number of scattered helium atoms as a function of their kinetic energy and scattering angle. The surface-phonon dispersion spectrum (another name for the surface-atom vibration spectrum) has been determined for NaF, Ag, and Si single-crystal surfaces in this manner, for example, and vibration modes that are localized at the surface have been identified. High-resolution electron-energy-loss spectroscopy (HREELS) also provides information on surface-atom vibration, although with a lower energy resolution [≈ 4 meV ($\approx 6 \times 10^{-22}$ J or ≈ 32 cm^{-1})]. The higher-frequency vibrations of surface atoms in ZnO surfaces [3] [≈ 69 meV ($\approx 10\text{-}20$ J or ≈ 552 cm^{-1})] and the higher-frequency vibrations of atoms at steps on platinum surfaces have been detected in this way. The inward relaxation of surface atoms at steps is greater than the relaxation of atoms in close-packed flat surfaces, giving rise to a higher force constant and a higher vibration frequency [4]. On a stepped platinum surface, a 205-cm^{-1} [≈ 25-meV (4.0 $\times$ 10^{-21}-J)] vibration was observed, somewhat above the maximum bulk frequency of 195 cm^{-1} [≈ 24 meV ($\approx 3.8 \times 10^{-21}$ J)].

4.2.3 Surface Mean-Square Displacements

It can be readily shown that the average potential energy of a classical harmonic oscillator is $\frac{1}{2} k_B T$. Thus, from Eq. 4.4 we have

$$\langle V \rangle = \langle \tfrac{1}{2} k x^2 \rangle = \tfrac{1}{2} k \langle x^2 \rangle = \tfrac{1}{2} m \omega^2 \langle x^2 \rangle = \tfrac{1}{2} k_B T \qquad (4.6)$$

where $\langle x^2 \rangle$ is the mean-square displacement. Because the average energy E of a harmonic oscillator is twice the average potential energy V, we can express the mean-square displacement, after rearrangement of Eq. 4.6, as

$$\langle x^2 \rangle = \frac{E}{m \omega^2} \qquad (4.7)$$

By using the high-temperature limit for the total energy of N harmonic oscillators

$$E = 3Nk_bT \qquad (4.8)$$

and equating ω with the Debye cutoff frequency $\omega_D = k_B\Theta_D/h$, where $\omega_D = 2\pi\nu_D$, $\hbar = h/2\pi$, and Θ_D is the Debye temperature, Eq. 4.7 can be rewritten as

$$\langle x^2 \rangle = \frac{3N\hbar^2 T}{mk_B\Theta_D^2} \qquad (4.9)$$

Thus the mean-square displacement of atoms that vibrate as harmonic oscillators is linearly proportional to the temperature at high temperatures ($T > \Theta_D$).

Let us consider the vibration of atoms as represented by a collection of harmonic oscillators. It has been found by experiment [5–13] that the intensity of diffracted low-energy electrons or X-ray beams depends strongly on the temperature of the scattering solid. The intensities of the diffracted beams decrease exponentially with increasing temperature. This is because, at any one instant, many of the vibrating atoms are displaced from their equilibrium positions. Thus the incident electron beam encounters a partially disordered lattice. Those atoms displaced from their equilibrium positions during the scattering process will scatter out of phase, and a portion of the elastically scattered electrons will be found in the background instead of in the diffraction spot. The larger the amplitude of vibration of the surface atoms, the more likely that the back-scattered electrons will be in the background instead of contributing to the diffraction-spot intensity. It can be shown [14] that the intensity I_{hkl} of a diffracted beam (neglecting multiple scattering events) is given by

$$I_{hkl} = |F_{hkl}|^2 \exp\left\{ -\left(\frac{16\pi^2(\cos^2\psi)}{\lambda^2}\right)\langle x^2 \rangle \right\} \qquad (4.10)$$

where the exponential term is the Debye–Waller factor, λ is the electron wavelength, ψ is the angle of incidence with respect to the surface normal, and $|F_{hkl}|^2$ is the scattered intensity caused by the rigid lattice. The constant factor $[16\pi^2(\cos^2\pi)]/\lambda^2$ gives the magnitude of the scattering vector: $|\Delta\vec{k}| = |\vec{k} - \vec{k}_0| = [(4\pi/\lambda)\cos\psi]^2$, and $\langle x^2 \rangle$ is the mean-square displacement component in the direction of $\Delta\vec{k}$. Substituting Eq. 4.9 into Eq. 4.10, we have

$$I_{hkl} = |F_{hkl}|^2 \exp\left\{ -\left[\left(\frac{12N\hbar^2}{mk_B}\right)\left(\frac{\cos\psi}{\lambda}\right)^2\frac{T}{\Theta_D^2}\right] \right\} \qquad (4.11)$$

Thus, according to Eq. 4.11, the intensity of the diffracted beam decreases exponentially with increasing temperature. By measuring the intensity of low-energy electrons diffracted from surface atoms as a function of temperature, the mean-square displacement of surface atoms can be obtained. According to Eq. 4.11, the logarithm of the intensity plotted as a function of temperature T gives a straight line. The root-mean-square (rms) displacement of surface atoms can be calculated from the slope and the surface Debye temperature and by using Eq. 4.11.

The rms displacement of surface atoms in several cubic metals has been determined in this manner. In most cases the mean displacement component perpendicular to the surface plane, $\sqrt{\langle x_\perp^2 \rangle}$, was obtained from low-energy electron-diffraction

TABLE 4.1. Surface and Bulk Root-Mean-Square Displacement Ratios and Debye Temperature for Several Metals

	$\dfrac{\langle u^2 \perp \rangle^{1/2} \text{ (surface)}}{\langle u^2 \rangle^{1/2} \text{ (bulk)}}$	Θ (surface) (K)	Θ (bulk) (K)
Pb (110), (111) (refs. 1, 2)	2.43 (1.84)	37 (49)	90
Bi (0001), (0112) (ref. 2)	2.42	48	116
Pd (100), (111) (ref. 1)	1.95	142	273
Ag (100) (ref. 3), (110) (ref. 3), (111) (ref.4)	2.16 (1.48)	104 (152)	225
Pt (100), (110), (111) (ref. 5)	2.12	110	234
Ni (110) (ref. 6)	1.77	220	390
Ir (100) (ref. 7)	1.63	175	285
Cr (110) (ref. 8)	1.80	333	600
Nb (110) (ref. 9)	2.65	106	281
V (100) (ref.10)	1.52	250	380
Rh (100), (111) (ref. 11)	1.35	260	350

REFERENCES

1. R.M. Goodman, H.H. Farrell, and G.A. Somorjai. *J. Chem. Phys.* **48**:1046 (1968).
2. R.M. Goodman and G.A. Somorjai. *J. Chem. Phys.* **52**:6325 (1970).
3. J.M. Morabito, Jr., R. F. Steiger, and G. A. Somorjai. *Phys. Rev.* **179**:638 (1969).
4. E.R. Jones, J.T. McKinney, and M.B. Webb. *Phys. Rev.* **151**:476 (1966).
5. H.B. Lyon and G.A. Somorjai. *J. Chem. Phys.* **44**:3707 (1966).
6. A.U. McRae. *Surf. Sci.* **2**:522 (1964).
7. R. M. Goodman, Ph.D. dissertation, University of California, Berkeley, 1969.
8. R. Kaplan and G. A. Somorjai, *Solid State Commun.* **9**:505 (1971).
9. D. Tabor and J. Wilson, *Surf. Sci.* **20**:203 (1970).
10. D.J. Chneng, R.F. Wallis, C. Megerle, and G.A. Somorjai. *Phys. Rev.* **B12**:5599 (1975).
11. D.G. Castner. Ph.D. dissertation, University of California, Berkeley, 1979.

experiments (Table 4.1). For many metals, the rms displacement of surface atoms perpendicular to the surface plane is 1.4 to 2 times as large as the bulk value (see also reference [15]). Similar large rms displacements were obtained for different crystal faces of the same solid in most cases. Correspondingly, the calculated surface Debye temperatures are smaller than their bulk values.

The mean-square displacement of surface atoms should be sensitive to changes in the number and type of neighboring atoms. The adsorption on the clean surface of gases that chemically interact with the surface atoms (e.g., oxygen on nickel or tungsten) strongly affects the vibrational amplitude of surface atoms.

4.2.4 Vibrations of Adsorbed Atoms and Molecules

Adsorbed atoms that form chemical bonds with atoms in a solid surface vibrate about their equilibrium position, as is readily detectable by vibration spectroscopies. Usually the frequencies of these vibrations are in the 300- to 800-cm^{-1} [$\approx$37- to 100-meV ($\approx 6.15 \times 10^{-21}$-J)] range, and thus they are easily distinguishable from the vibration spectra of the clean solid surface, which appear at lower frequencies. A good example is that of hydrogen dissociatively chemisorbed on a tungsten surface as hydrogen atoms, shown in Figure 4.3. Hydrogen is thought to occupy sites bridg-

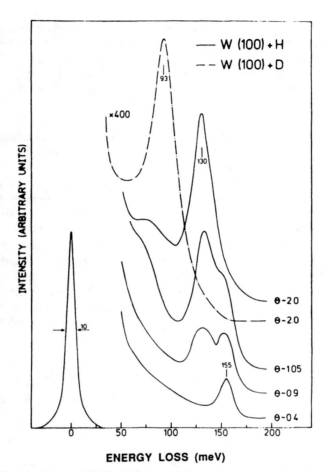

Figure 4.3. Vibrational spectra of hydrogen and deuterium adsorbed on the W(100) surface obtained by HREELS. The curves correspond to the spectrum obtained at different coverages [89] θ.

ing two tungsten atoms, although H atoms that are located at top sites, above the metal atoms, may also be present, especially at higher coverages. Infrared spectroscopy in its various forms (reflection Fourier transform infrared, sum frequency generation) provides higher energy resolution [≈ 2 cm^{-1} $\approx 2.5 \times 10^{-4}$ eV ($\approx 4 \times 10^{-23}$ J)] than HREELS, but electron scattering is usually more sensitive to the presence of small surface concentrations of adsorbed species because of their very high electron-scattering cross sections, compared to photons.

By monitoring the vibration spectra of chemisorbed species as a function of coverage, crystal surface, and temperature, the location and site symmetries of the adsorbed atoms (4-fold, 3-fold, bridge, on-top) can be monitored and variations in site occupancy can be determined.

Adsorbed molecules that form chemical bonds with surface atoms exhibit vibrations of the surface chemical bond as well as within the bonds in the molecule. Carbon monoxide chemisorbed on various metal surfaces is the most frequently studied molecular adsorbate system. CO molecules, which adsorb usually with the CO bond perpendicular to the surface, were found to occupy mainly bridge and on-top sites at lower coverages (up to one-half monolayer), but their chemisorption in a 3-fold site [16] and in a *gem*-dicarbonyl configuration [17] (where two CO molecules are bound to one metal atom) were also observed (Table 4.2). At higher coverages, CO molecules occupy sites of lower symmetry because of repulsion between the adsorbate molecules, a repulsion that also weakens their bonding to the metal. These changes can be monitored by analysis of the vibrational spectrum. In Figure 4.4 the HREELS spectrum of CO on the Rh(111) surface is displayed as a function of coverage. Table 4.3 lists the CO vibrational frequencies associated with chemisorption at surface sites of different symmetry.

The chemisorption of polyatomic molecules provides rich information on the molecular structure of the adsorbed species. The vibrational spectra of benzene and deuterated benzene are displayed in Figure 4.5a. The benzene ring lies parallel to the Rh(111) crystal face. The vibrational spectra of *o*-xylene (C_2H_4) chemisorbed on the Pt(111) crystal face are shown at three different temperatures (Figure 4.5b). At 245 K the molecule lies with its benzene ring parallel to the surface. At 370 K there is hydrogen loss from its methyl side groups, but no change in molecular orientation. At 470 K the molecule is bound to the metal surface through its side-group carbon atoms and the benzene ring is oriented perpendicular to the platinum surface. Using a combination of diffraction (LEED) (to determine bond length) and

TABLE 4.2. CO Adsorption Sites (Determined by LEED) and C–O Stretch Frequencies

Site	1600	1700	1800	1900	2000	2100	Free CO 2200 cm^{-1}
			Hollow	Bridge		Top	
Ni(100) c(2 × 2)–CO					●		
Cu(100) c(2 × 2)–CO					●		
Pd(100) $(2\sqrt{2} \times \sqrt{2})R45°$–2CO				●			
Ru(0001) $(\sqrt{3} \times \sqrt{3})R30°$–CO					●		
Rh(111) $(\sqrt{3} \times \sqrt{3})R30°$–CO					●		
Rh(111) (2 × 2)–3CO				●		●	
Rh(111) c($2\sqrt{3}$ × 4)rect–C_6H_6 + CO	●						
Pt(111) c(4 × 2)–2CO				●			●

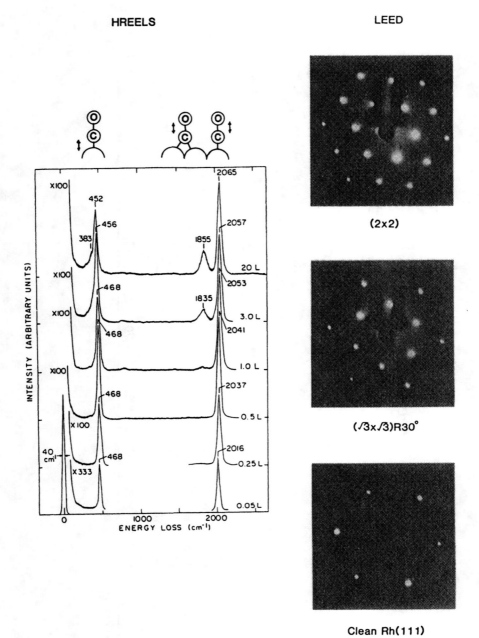

Figure 4.4. Vibrational spectrum of carbon monoxide chemisorbed on the Rh(111) surface obtained by HREELS. The curves correspond to the spectrum obtained at different CO exposures, *L*. The LEED diffraction pattern indicates that the CO monolayer is ordered at the different coverages.

TABLE 4.3. Selected Vibrational Frequencies for Carbon Monoxide on Rhodium[a]

Surface	Technique	ν_{Rh-C} / $\delta_{Rh-C=O}$	Surface Sites			
			$\nu_{3\text{-fold}}$	ν_{bridge}	ν_{atop}	$\nu_{Rh-(CO)_2}$
Rh(111)	ELS	420		1870	2070	2031, 2101
Rh/Al$_2$O$_3$	IR			1870	2070	1990 2020, 2080
Rh/SiO$_2$	IR			1890–1900	2040–2065	(2111)
Evaporated Rh film	Transmission IR	400–575		1852, 1905	2055	1942
Rh/Al$_2$O$_3$	IETS	413, 465, 600		1721	1942	
Rh$_2$(CO)$_8$	Solution IR			1845, 1861		2061, 2086
Rh$_4$(CO)$_{12}$	Solid IR	393, 423, 488		1848		2028–2105
Rh$_6$(CO)$_{16}$	Solid IR	413, 427, 513	1770			2016–2077

[a] All frequency in cm^{-1}.

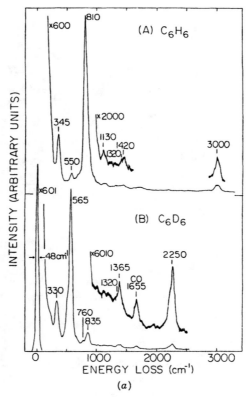

Figure 4.5. (a) Vibrational spectra obtained by HREELS in the specular direction for a saturation coverage of benzene chemisorbed on Rh(111) at 300 K for a well-ordered $c(\sqrt{3} \times 4)$ surface structure: **(A)** C_6H_6; **(B)** C_6D_6 [90]. **(b)** The vibrational spectra of o-xylene at three different temperatures [91].

vibration spectroscopy to (yield the bonding site, the molecular symmetry, and bonding orientation), the structure of the adsorbed molecule can be unambiguously determined.

4.3 ELEMENTARY PROCESSES OF GAS–SURFACE INTERACTION

Even the simplest gas–surface interaction involves several steps, beginning with the collision of the incident atoms or molecules with the surface. As the gas or vapor species nears the surface, it experiences an attractive potential whose range depends on the electronic and atomic structures of the collision partners. The interaction may vary in strength in proportion to the reciprocal of the distance between the collision partners (for example, between incident gas ions and surface ions of opposite charge) and therefore be long-range. It may also be much shorter in range and vary in proportion with the inverse third or sixth power of the distance. The type of interaction (that is, how the force between an approaching atom and the surface changes with the distance between them) may be measured directly by the recently developed

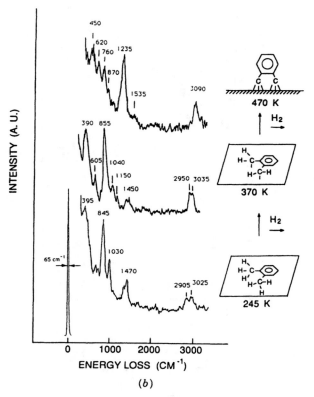

Figure 4.5. (*Continued*)

atomic force microscope (AFM). In the AFM, a tip of atomic dimension can be prepared and brought to the surface in atomic ranges of 0–0.5 nm (0–5 Å) by piezoelectric ceramics that can expand by the application of an external voltage in the range of about 1 volt per angstrom. As the tip is brought to the surface this way, the tip–surface distance can be stabilized by a force that is equal and opposite to the force operating between the tip and the surface. This is usually done by using a spring with a suitable spring constant. The force on the tip, attractive or repulsive, as it approaches the surface can be measured with a sensitivity of 10^{-9} newtons.

A certain fraction of the incident gas atoms are trapped in the attractive potential well, and once trapped they can move along the surface by diffusion. The adsorbed species may desorb from the surface if sufficient energy is imparted to it at a given surface site to overcome the attractive surface forces. The types of interactions that take place between a gas atom or molecule and the surface depend on the energy of the gaseous species (kinetic or translational energy, internal energy, rotation, vibration, or electronic excitation when appropriate), the temperature, and the atomic structure of the solid surface.

During the collision of the adsorbed gas atom with the surface, it exchanges kinetic or translational energy T with the vibrational modes V_s of the surface atoms. The type of energy transfer that takes place in this circumstance is often called the $T \leftrightarrow V_s$ energy exchange. During the collision of gas molecules with the surface,

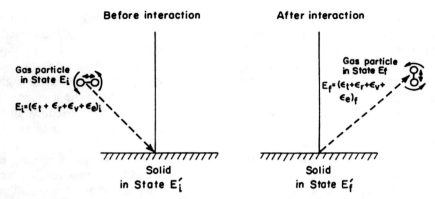

Figure 4.6. Scheme of energy transfer in gas–surface interaction. E_i and E_f indicate the initial and final energy states, respectively. For gas molecules these are the sum of translational (E_t), rotational (E_r), vibrational (E_v), and electronic (E_e) energy states.

they may exchange internal energy, including rotation R or vibration V, for example, with the vibrating surface atoms. In this circumstance there are also $R \leftrightarrow V_s$ and $V \leftrightarrow V_s$ energy-transfer processes. The various energy-transfer processes are depicted schematically in Figure 4.6.

To understand the dynamics of the gas–surface interaction, it is essential to determine how much energy is exchanged between the gas and surface atoms through the various energy-transfer channels. By determining the residence time of the adsorbed atoms or molecules on the surface at the various surface sites available for bonding, we can determine the dependence of energy transfer on residence time and surface structure. Finally, we would like to know the kinetic parameters, rate constants, activation energies, and preexponential factors for each of the elementary surface steps of adsorption, surface migration, and desorption in order to obtain a complete description of the gas–surface energy-transfer process.

Perhaps the most versatile method for studying gas–surface collision dynamics is molecular-beam surface scattering. In an ideal experiment, a well-collimated beam of molecules of uniform and known translational energy and known rotational and vibrational state population strikes a clean surface of well-characterized atomic surface structure and temperature. Some of the molecules are back-reflected after a very short residence time ($1–10^3$ vibrations of the surface collision partners), while others are trapped for much longer times in the attractive surface potential before desorbing. The experimenter measures the amount of translational energy exchanged by detecting the velocity and angle distribution of the scattered molecules by suitable time-of-flight analysis using a mass-spectrometer detector. Rotational and vibrational energy exchange with the surface is measured by appropriate laser techniques (two-photon ionization, for example) that probe the internal energy states of the molecules back-scattered from the surface.

4.3.1 Adsorption. Energy Accommodation Coefficients

When an atom approaches the surface, it may become trapped and only come to rest after traveling some distance along the surface to a suitable site of high binding energy. We call this process *adsorption*. Adsorption will be likely to occur if the incident atom has a kinetic energy that is smaller than the well depth of the attractive surface potential. An incident atom with a higher kinetic energy may first be trapped in the surface potential and then slide along the surface before desorbing again into the gas phase, since it has enough kinetic energy left to escape. At even higher kinetic energies, the incident atom may simply back-reflect from the surface after spending on the order of vibrational times ($\nu = 10^{-12}$ sec) in the proximity of the vibrating surface atoms during the collision, or it may back-reflect without any energy exchange with the surface at all.

All of these processes have been observed in atomic-beam surface-scattering experiments, during which a beam of atoms of well-defined kinetic energy impinges on a single-crystal surface. The kinetic energy distribution of the back-scattered atoms can be detected by using a "correlation chopper" velocity selector [18].

It is customary to define an energy accommodation coefficient

$$\alpha_E = \frac{E_{incident} - E_{scattered}}{E_{incident} - E_{surface}} \tag{4.12}$$

where $E_{surface}$ is the thermal energy of the solid.

For an incident atom, $\alpha_E = 0$ if the kinetic energies of the incident and scattered atoms are the same: $E_{incident} = E_{scattered}$. Conversely, $\alpha_E = 1$ if the kinetic energy of the scattered species is equal to the kinetic energy expected for desorbing from a surface of thermal energy $k_B T_{surface}$, since, in this circumstance, $E_{scattered} = E_{surface}$. For a scattered atom, the energy accommodation coefficient α_E becomes the translational-energy accommodation coefficient,

$$\alpha_T = \frac{T_{incident} - T_{scattered}}{T_{incident} - T_{surface}} \tag{4.13}$$

In Figure 4.7, $T_{scattered}$ is plotted as a function of T_s, the surface temperature for the scattering of inert gases from a graphite surface. The experimental data show surface equilibrium below 300 K and very little energy accommodation at higher thermal energies, $k_B T_s$. This result indicates only partial energy transfer between the incident He, Ar, Kr, and Xe atoms and the graphite surface when the surface is hot.

For a diatomic or a polyatomic molecule, energy transfer to or from the surface involves deexcitation or excitation of vibrational and rotational modes of motion, in addition to changes of kinetic energy. Therefore, we can define α_V and α_R as the accommodation coefficients for vibrational and rotational energy transfer, respectively, during collision with the surface. These coefficients are displayed for NO molecules scattered after a single collision with a platinum crystal surface of (111) orientation in Figure 4.8. Under the conditions of the experiment, the kinetic energy and vibration energy accommodation coefficients are near unity for a wide range of metal surface temperatures, indicating equilibration with the surface. The rotation energy accommodation coefficient, however, is much smaller than unity. It appears that the molecule does not rotate freely on the platinum surface but probably is aligned in the direction of the chemical bond; thus it scatters preferentially in this orientation.

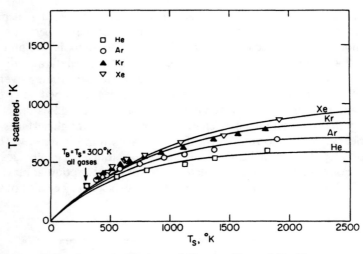

Figure 4.7. Translational energy of scattered He, Ar, Kr, and Xe $T_{scattered}$, as a function of surface temperature of graphite, T_s [92].

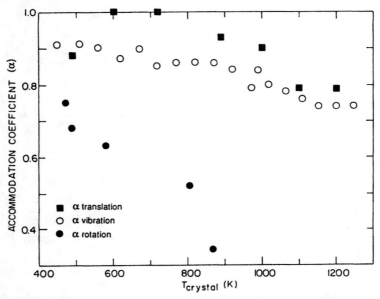

Figure 4.8. Translational, vibrational, and rotational energy accommodation coefficients for NO scattered from Pt(111) as a function of crystal temperature [93].

4.3.2 Sticking Probability

In all experiments designed to determine α_T, α_V, and α_R, the energy contents of the atoms and molecules are scrutinized before and after scattering. Ultimately, we would like to study the scattering species while it is on the surface during the collision process to learn about the details of bonding and local geometries on the molecular level. To this end, time-resolved techniques that take a "snapshot" of the adsorbate–surface complex during collision must be developed. These techniques have not been developed yet. In the absence of these tools, we aim to determine the fraction of incident atoms and molecules that ultimately adsorb and how the adsorbate concentration depends on surface structure, adsorbate coverage θ (θ = number of molecules adsorbed/number of molecules adsorbed in a complete monolayer), and temperature. The sticking probability S, defined as the adsorption rate divided by the collision rate, is greater on more open or rough surfaces (stepped surfaces, for example) than on smooth surfaces, as shown in Figure 4.9. This effect is caused partly by the higher heats of adsorption of atoms and molecules at these surface sites. For diatomic molecules such as H_2 or O_2, adsorption at elevated temperatures (≈ 300 K) can occur only if the molecules dissociate, because of the weak

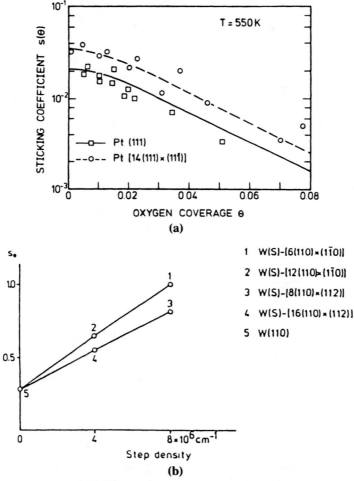

Figure 4.9. (a) The sticking coefficients of O_2 on the flat Pt(111) and stepped Pt[14(111) × (111)] crystal faces as a function of oxygen coverage [94]. (b) The sticking coefficients of nitrogen as a function of step density on various crystal faces of tungsten [95].

TABLE 4.4. Structure Sensitivity of $H_2 \leftrightarrow D_2$ Exchange at Low Pressures ($\approx 10^{-6}$ torr)

Surface	Reaction Probability
Stepped Pt(332)	0.9
Flat Pt(111)	$\approx 10^{-1}$
Defect-free Pt(111)	$\leq 10^{-3}$

molecular bonding. In this circumstance, the adsorption or sticking probability is equal to the dissociation probability. For hydrogen, the probability of $H_2 \leftrightarrow D_2$ exchange upon a single collision of the molecules with the surface yields the dissociation and thus the sticking probability. In Table 4.4, the $H_2 \leftrightarrow D_2$ exchange probabilities are listed for a flat (111) and a stepped crystal face of platinum [19]. For the stepped surface, the $H_2 \leftrightarrow D_2$ exchange probability is near unity; that is, every impinging molecule dissociates. On the (111) metal surface, the reaction probability is lower by at least an order of magnitude. For a defect-free (111) surface, the reaction probability is below the detection limit—that is, $< 10^{-3}$. Thus, defects and more open, rough surfaces can markedly increase the adsorption probability, indicating the structure sensitivity of the adsorption process. The value of S often decreases with increasing coverage or remains unchanged over a wide coverage range. Examples for this type of behavior are also displayed in Figure 4.9. In some cases, however, S increases with coverage, as shown in Figure 4.10 for N_2 adsorption on the (110) face of tungsten.

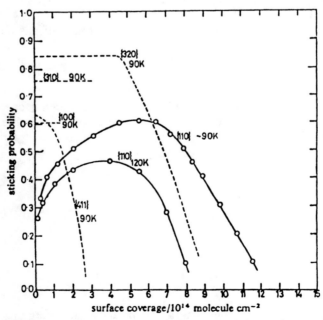

Figure 4.10. Sticking probabilities of N_2 on various crystal faces of tungsten as a function of surface coverage [95].

The sticking probability may decrease with increasing temperature or remain unchanged. When molecules must dissociate in order to adsorb on a surface, under certain experimental conditions S may increase with increasing temperature, indicating that there is an activation energy for adsorption.

4.3.3 Models of Energy Transfer and Adsorption

Several models have been proposed to explain the complex behavior of energy accommodation coefficients and the sticking probability. These models have proved to be useful in interpreting some of the energy-transfer processes during gas–surface collisions. One model, the hard-cube model, assumes that the surface is a flat repulsive wall and that no forces parallel to the surface act on the gas atoms during collision. The gas atoms are also assumed to be lighter than the surface atoms. This model explains qualitatively the angular distribution of argon scattered from silver. Another model, the soft-cube model, assumes a gas–surface attractive potential that can trap the incident atoms. This model qualitatively explains xenon scattering from silver or argon scattering from platinum. More realistic surface potentials and trajectory calculations have been utilized in recent years; these have showed success in interpreting both angular and energy distributions obtained by atomic and molecular-beam surface-scattering experiments [20–33].

The existence of a precursor state, where the incident atom or molecule is temporarily trapped in a shallow attractive potential, has been proposed by Kisliuk [34, 35]. In this precursor state, the molecule may visit several surface sites, one of which permits chemisorption. This model provides enough variables to explain the decrease of the sticking probability with increasing coverage and its complex temperature dependence, as found during the chemisorption of N_2 on W(100), and O_2 on W(110) and Pt(111).

For noninteracting adsorbates [36], the rate of adsorption of gas-phase species, r_a, can be written as

$$r_a = S^{(\alpha)}(\theta)F \tag{4.14}$$

where F is the flux of the adsorbate and $S^{(ga)}(\theta)$ is the sticking coefficient. The superscript α on $S^{(\alpha)}(\theta)$ is 1 for nondissociative and 2 for dissociative adsorption. The form of $S^{(\alpha)}(\theta)$ depends on whether adsorption occurs directly from the gas phase or via a precursor state. For direct adsorption,

$$S^{(\alpha)}(\theta) = S^{(\alpha)}(0)(1 - \theta)^\alpha \tag{4.15}$$

where $S^{(\alpha)}(0)$ is the sticking coefficient at zero coverage. If $S^{(\alpha)}(0)$ is assumed to obey an Arrhenius expression, then

$$S^{(\alpha)}(0) = S_0^\alpha \exp\left(-\frac{E_a}{k_B T}\right) \tag{4.16}$$

where $S_0^{(\alpha)}$ and E_a are the pre-exponential factor and activation energy for adsorption, respectively, and k_B is the Boltzmann constant.

When adsorption is assumed to proceed via a precursor state, the mechanism of adsorption can be represented by [37]

$$A \underset{k_d^*}{\overset{\zeta F_A}{\rightleftharpoons}} A^* \xrightarrow{k_a^*} A_s \tag{4.17}$$

$$A_2 \underset{k_d^*}{\overset{\zeta F_{A_2}}{\rightleftharpoons}} A_2^* \xrightarrow{k_a^*} A_s \tag{4.18}$$

where the precursor species for nondissociative adsorption is denoted by A^* (A_2^* for dissociative adsorption) and the chemisorbed species by A_s, ζ is the trapping probability from the gas phase into the precursor state, k_a^* is the rate constant for adsorption from the precursor state into the chemisorbed state, and k_d^* is the rate constant for desorption from the precursor state. The precursor state can be located over an empty site (an intrinsic precursor) or over an occupied site (an extrinsic precursor). Two different approaches have been used to represent the adsorption rate. The first is based on a continuum description and uses the stationary-state approximation to determine the concentration of precursor species [38, 39]. The second approach is based on a successive-site model of the type first proposed by Kisliuk [34, 35, 40]. Although conceptually different, both approaches have been shown to lead to equivalent forms of the adsorption rate in many cases [38, 41].

If the intrinsic and extrinsic precursors are energetically equivalent and each occupies only a single adsorption site, then the rates of non-dissociative and dissociative adsorption can be written as [37]

$$r_a = \frac{\zeta F_A k_a^* (1 - \theta_A)}{k_d^* + k_a^* (1 - \theta_A)} \tag{4.19}$$

and

$$r_a = \frac{\zeta F_{A_2} k_a^* (1 - \theta_A)^2}{k_d^* + k_a^* (1 - \theta_A)^2} \tag{4.20}$$

Equations 4.19 and 4.20 can be used together with Eq. 4.14 to write expressions for $[S^{(\alpha)}(\theta)/S^{(\alpha)}(0)]$. Thus,

$$\frac{S^{(1)}(\theta)}{S^{(1)}(0)} = \frac{(1 + K)(1 - \theta_A)}{1 + K(1 - \theta_A)} \tag{4.21}$$

and

$$\frac{S^{(2)}(\theta)}{S^{(2)}(0)} = \frac{(1 + K)(1 - \theta_A)^2}{1 + K(1 - \theta_A)^2} \tag{4.22}$$

where $K = (k_a^*/k_d^*)$. A plot of $[S^{(1)}(\theta)/S^{(1)}(0)]$ versus θ is shown in Figure 4.11. When $K \gg 1$, $[S^{(1)}(\theta)/S^{(1)}(0)] = 1$, and when $K \ll 1$, $[S^{(1)}(\theta)/S^{(1)}(0)] = 1 - \theta$. Because the value of K is temperature-dependent, the shape of $[S^{(\alpha)}(\theta)/S^{(\alpha)}(0)]$ versus θ will depend on temperature.

The rate of desorption for a randomly distributed adsorbate in the absence of

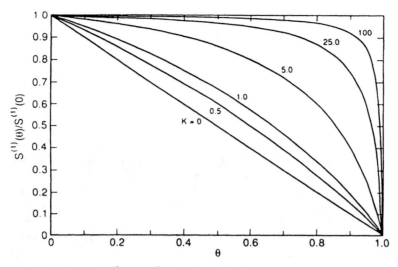

Figure 4.11. Variation of $S^{(1)}(\theta)/S^{(1)}(0)$ with θ for adsorption via precursor mechanism. The precursor parameter K equals $k*a/k*d$ [36].

lateral interactions can be written as

$$r_d = k_d^{(\alpha)}\theta^\alpha \tag{4.23}$$

where $k_d^{(\alpha)}$ is the rate coefficient for desorption. If $k_d^{(\alpha)}$ is assumed to obey an Arrhenius expression, then

$$k_d^{(\alpha)} = A_d^{(\alpha)} \exp\left(-\frac{E_d}{k_BT}\right) \tag{4.24}$$

where $A_d^{(\alpha)}$ and E_d are the preexponential factor and activation energy for desorption, respectively.

Implicit in the formulation of Eq. 4.24 is the assumption that desorption occurs directly from the adsorbed state. This assumption is unnecessarily restrictive because it is conceivable that the adsorbate passes through a weakly bound precursor state before leaving the surface. Making the same assumptions used in deriving Eqs. 4.19 and 4.20, the rates of nonassociate and associative desorption can be written as [37]

$$r_d = \frac{k_d^{(1)}k_d^*\theta_A}{k_a^*(1 - \theta_A) + k_d^*} \tag{4.25}$$

and

$$r_d = \frac{k_d^{(2)}k_d^*\theta_A^2}{k_a^*(1 - \theta_A)^2 + k_d^*} \tag{4.26}$$

In the limit $k_a^*(1 - \theta)^\alpha \ll k_d^*$, Eqs. 4.25 and 4.26 reduce to Eq. 4.23.

Surface reactions can be classified into two generic types. The first includes reactions between two adsorbed species or between an adsorbed species and a vacant site (Langmuir–Hinshelwood processes). For randomly distributed adsorbates on a surface in the absence of adsorbate–adsorbate interactions, the rate of reaction is given by

$$r_r = k_r \theta_A \theta_B \tag{4.27}$$

or

$$r_r = k_r \theta_A \theta_V \tag{4.28}$$

where k_r is the rate coefficient, θ_i is the surface coverage of species i, and θ_V is the fraction of vacant sites. If k_r follows an Arrhenius expression, then

$$k_r = A \exp\left(-\frac{\Delta E^*}{k_B T}\right) \tag{4.29}$$

where A and ΔE^* are the pre-exponential factor and the activation energy for reaction, respectively.

The second class of reactions includes the direct interaction of a gas-phase species with an adsorbed species to form a product which may either remain adsorbed or desorb into the gas phase (Eley–Rideal processes). For such processes, the rate of reaction can be written as

$$r_r = k_r \theta_A P_B \tag{4.30}$$

where P_B is the partial pressure of reactant B. The form of Eq. 4.30 is similar to that for adsorption and thus k_r can be represented as a reactive sticking coefficient, S_0, by the expression

$$k_r = \frac{S_0 a_s}{\sqrt{2\pi m_B k_B T}} \exp\left(-\frac{\Delta E^*}{k_B T}\right) \tag{4.31}$$

where a_s is the area per reaction site, and m_B is the molecular weight of species B.

When lateral interactions become significant, the relationships between the rate of an elementary process and the adsorbate coverage become quite complex and cannot, in general, be written in closed form. An exception to this occurs in the case where a lattice–gas model is used to describe the effects of lateral interactions. In such a model, each adsorbate is assumed to be localized on a two-dimensional array of surface sites, and each site is assumed to be either vacant or occupied by a single adsorbate. A given adsorbate can interact with adsorbates on nearest-neighbor sites, next-nearest-neighbor sites, and so on, but in most variants of the lattice–gas model, only nearest-neighbor interactions are taken into account. Using these assumptions, relationships can be derived between the adsorbate coverage and the rate of adsorption, desorption, and surface reactions [42–44].

The partial energy transfer during gas–surface interactions observed by experiments could be the result of the existence of a precursor state that temporarily traps the atom or molecule in an attractive potential, from which adsorption or desorption subsequently occurs. It would be of great value if the energy states of the trapped atoms or molecules could be studied, in addition to monitoring their energy states before and after collision with the surface. The development of time-resolved surface spectroscopies will likely make this possible in the future. At present, it is difficult to define the precursor state more precisely or on the molecular level.

However, new experimental findings promise the possibility of a molecular-level description of the gas–surface collision complex [45]. One of these is the observation that the dissociation probability of certain molecules—CH_4 [46], N_2, and CO_2—upon collision with a metal surface increases exponentially with increasing kinetic energy above a certain threshold energy, while the dissociation probability of other molecules—CO, for example—is constant over a large range of kinetic energy [47]. For CO_2, increased vibrational excitation instead of increased velocity also increases the dissociation probability. It appears that these polyatomic molecules must be distorted with respect to their gas-phase structure before dissociation can occur and that this distortion is accomplished by high-velocity collision with the surface or by appropriate vibrational excitation.

Another observation that sheds light on the gas–surface collision complex is adsorbate-induced restructuring of surfaces. Chemisorption induces rearrangements of the surface metal atoms around the adsorption site in smooth surfaces, and the rate of restructuring can be coverage-dependent, increasing with increasing coverage [48]. Perhaps the gas–surface collision complex in the precursor state induces local surface rearrangements that create the chemisorption sites.

4.3.4 Surface Diffusion

Surfaces are heterogeneous on the atomic scale. Atoms appear in flat terraces, at steps, and at kinks. There are also surface point defects, vacancies, and adatoms. These various surface sites achieve their equilibrium surface concentrations through an atom-transport process along the surface that we call *surface diffusion*. Adsorbed atoms and molecules reach their equilibrium distribution on the surface in the same way. This view of surface diffusion as a site-to-site hopping process leads to the ''random-walk'' picture, in which the mean-square displacement of the adsorbed particle along the *x*-component of the coordinate is given by

$$\langle \Delta x^2(t) \rangle = \nu t d^2 \tag{4.32}$$

where ν is the frequency of jumps, t is the time, and d^2 is the mean-square jump length. The ratio of the mean-square displacement to the time defines the time-independent self-diffusion constant D:

$$D = \lim_{t \to \infty} \frac{\langle \Delta x^2(t) \rangle}{2bt} \tag{4.33}$$

where b is the dimensionality associated with the diffusion process ($b = 1$ for linear motion, $b = 2$ for motion along a plane or flat surface). This linear time dependence of the mean-square displacement is characteristic of diffusion. The self-diffusion coefficient

$$D = \frac{\nu d^2}{2b} \tag{4.34}$$

is the property of a material that characterizes its atom transport.

The rms distance $\langle \Delta x^2 \rangle^{1/2}$ can be expressed in terms of the diffusion coefficient by substitution of Eq. 4.34 into 4.32 to give, for $b = 2$,

$$\langle x^2 \rangle^{1/2} = (4Dt)^{1/2} \tag{4.35}$$

From measurements of the mean travel distance of diffusing atoms, the diffusion coefficient can be evaluated. Conversely, knowledge of the diffusion coefficient allows one to estimate the rms distance or the time necessary to carry out the diffusion. For example, the diffusion coefficients of silver ions on the surface of silver bromide can be estimated to be 10^{-9} and 10^{-13} cm^2/sec (10^{-13} and 10^{-17} m^2/sec) at 300 and 100 K, respectively. Assuming that an rms distance of 10^{-4} cm (10^{-6} m) is required for silver-particle aggregation (photographic printout) to begin, what durations of light exposure are required? Using Eq. 4.35, we find $t = 5$ sec and $t = 5 \times 10^4$ sec at 300 and 100 K, respectively. The exponential temperature dependence of D is, of course, the reason that silver bromide photography cannot be carried out at low temperatures (much below 300 K) but is done easily at about room temperature. We can also see that, at a slightly elevated temperature (≈ 450 K), the thermal diffusion of silver particles should be rapid enough [$D \approx 3 \times 10^{-7}$ cm^2/sec ($\approx 3 \times 10^{-11}$ m^2/sec)] so that their aggregation will take place rapidly even in the dark ($t \approx 10^{-2}$ sec), in the absence of any photoreaction.

The frequency ν with which an atom with a vibrational frequency ν_0 will escape from a site depends on the height ΔE^* of the potential-energy barrier it has to climb in order to escape:

$$\nu = z\nu_0 \exp\left(-\frac{\Delta E^*}{k_B T}\right) \tag{4.36}$$

where z is the number of equivalent neighboring sites. Equation 4.19 can therefore be rewritten as

$$D = D_0 \exp\left(-\frac{\Delta E^*}{k_B T}\right) \tag{4.37}$$

where $D_0 = \frac{1}{6}\nu_0 d^2, \frac{1}{4}\nu_0 d^2$ for 3-fold or 2-fold symmetry, respectively.

Surface diffusion has so far been discussed in terms of a single surface atom. However, on a real surface many atoms diffuse simultaneously; and in most diffusion experiments the measured diffusion distance after a given diffusion time is an average of the diffusion lengths of a large, statistical number of surface atoms. A statistical thermodynamic treatment in terms of macroscopic parameters leads to the expression

$$D = D_0 \exp\left(-\frac{\Delta E_D^*}{RT}\right) \tag{4.38}$$

where ΔE_D^* is the total activation energy for the overall diffusion process. It is usually assumed that only one diffusion mechanism is involved.

Experimentally, the diffusion coefficient D is obtained by using a relationship between the diffusion rate and concentration gradient, namely, Fick's second law of diffusion in one dimension:

$$\frac{\partial c}{\partial t} = D\frac{\partial^2 c}{\partial x^2} \tag{4.39}$$

where c is the concentration of adatoms, t is the time, and x is the distance along the surface. In most surface-diffusion studies, the surface concentration of diffusing atoms c is measured as a function of distance x along the surface, and Eq. 4.39 is solved by the use of boundary conditions that approximate the experimental geometry. These experiments are by no means easy, and many novel experimental techniques have been applied to study surface diffusion on single crystals.

Diffusion experiments at surfaces are designed to measure self-diffusion or the diffusion of adsorbates. The techniques used [49–55] may provide atomic-scale diffusion data or macroscopic diffusion parameters. The techniques that provide atomic-level information include (a) field ion microscopy, which can be used to observe the surface migration of isolated adatoms or clusters of atoms, (b) field electron microscopy, and (c) scanning tunneling microscopy (for descriptions of the techniques, see references [56–68]. Macroscopic mass transport along the surface can be monitored by the use of radiotracers or by techniques that monitor the restructuring of surfaces as a function of time.

For example, the surface self-diffusion coefficient can be measured by etching a periodic surface profile (e.g., sinusoidal) into a single-crystal surface. The amplitude of the profile is measured as a function of time via the intensity distribution of a laser diffraction pattern generated by the profile itself [52]. The self-diffusion coefficient can be evaluated from the change of the profile amplitude $A(t)$ with time as the surface relaxes into its equilibrium surface structure upon heating:

$$A(t) = A_0 \exp\left(-B\omega^4 t\right), \qquad B = \frac{\gamma\sigma\Omega^2 D}{k_B T}, \qquad \omega = \frac{2\pi}{d} \tag{4.40}$$

where A_0 is the initial profile amplitude, γ is the surface free energy, σ is the number of surface atoms per unit area, Ω is the atomic volume, d is the profile periodicity, and $k_B T$ is the Boltzmann constant multiplied by the absolute temperature. The diffusion coefficient is obtained from the slope of a plot of $\ln A(t)$ versus time.

Laser-induced desorption [69] has been used successfully for adsorbate surface-diffusion studies. Laser-heating a small area on the surface induces thermal desorption of adsorbates, followed by in-diffusion of adsorbate species from the periphery of the heated area. Second and successive laser pulses desorb the molecules that have diffused into the depleted area, and their concentration can be monitored as a function of time and surface coverage.

4.3.5 Mechanisms of Surface Diffusion

The experimentally determined surface self-diffusion constants for face-centered cubic and body-centered cubic metals are plotted in Figures 4.12 and 4.13 as a function of T_m/T, where T_m is the absolute melting temperature. For face-centered cubic metals this dependence is approximated by two functions:

$$D = 740 \exp\left(-\frac{Q_1 T_m}{RT}\right), \qquad 0.77 < \frac{T}{T_m} < 1 \qquad (4.41)$$

and

$$D = 0.014 \exp\left(-\frac{Q_2 T_m}{RT}\right), \qquad \frac{T}{T_m} < 0.77 \qquad (4.42)$$

where $Q_1 = 30$ cal/mole K (125 J/mole K) and $Q_2 = 13$ cal/mole K (54 J/mole K). For body-centered cubic metals the data can be represented by

$$D = 3.2 \times 10^4 \exp\left(-\frac{Q_1 T_m}{RT}\right), \qquad 0.75 < \frac{T}{T_m} < 1 \qquad (4.43)$$

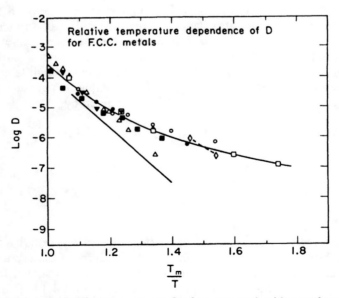

Figure 4.12. Surface self-diffusion constants for face-centered cubic metals as a function of T_m/T, where T_m is the melting temperature [96].

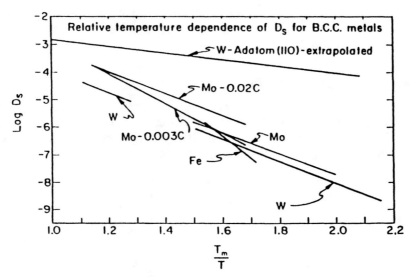

Figure 4.13. Surface self-diffusion constants for body-centered cubic metals as a function of T_m/T, where T_m is the melting temperature [96].

and

$$D = 1.0 \exp\left(-\frac{Q_2 T_m}{RT}\right), \qquad \frac{T}{T_m} < 0.75 \tag{4.44}$$

where $Q_1 = 35$ cal/mole K (146 J/mole K) and $Q_2 = 18.5$ cal/mole K (77 J/mole K). The changes in slope indicate changes in the mechanism of surface diffusion. While, at low temperatures, adatom diffusion or adatom–surface atom exchange appears to be the dominant atom-transport mechanism, the diffusion of surface vacancies created by thermal roughening is likely to be dominant at high temperatures to account for the increased activation energies. For example, copper adatom and vacancy diffusion rates in the Cu(110) crystal face are given by

$$D_{(110)}^{\text{adatom}} = 6.2 \times 10^{-3} \exp\left(-63.5 \text{ kJ} \cdot \text{mole}^{-1}/RT\right) \tag{4.45}$$

and

$$D_{(110)}^{\text{vacancy}} = 2.45 \times 10^{-3} \exp\left(-78.2 \text{ kJ} \cdot \text{mole}^{-1}/RT\right) \tag{4.46}$$

respectively.

Recent single-crystal studies reveal the surface-structure sensitivity and anisotropy of self-diffusion [70, 71]. Depending on the structure of the crystal face, diffusion coefficients may vary by orders of magnitude. This is shown for rhodium adatom diffusion on various rhodium crystal faces in Figure 4.14. Diffusion rates parallel to steps are greater than diffusion rates perpendicular to them.

The collective diffusion of dimers and clusters of atoms has also been observed [52, 54, 55]. The diffusion rates of dimers can be greater (Re_2 on Re) or smaller

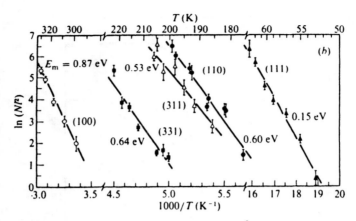

Figure 4.14. Rhodium adatom diffusion coefficients (Nl^2) on different Rh single-crystal planes as a function of reciprocal temperature [97].

(W$_2$ on W) than that of single adatoms. The diffusion rates of platinum clusters on platinum decrease with cluster size.

The presence of coadsorbates can markedly influence surface diffusion. On the one hand, the presence of Bi, Pb, Tl, and S can greatly increase the surface self-diffusion of Cu and Ag. Elements that reduce the melting point of the substrate can cause an increase in surface diffusion rates and a decrease in activation energies for diffusion in general. On the other hand, carbon markedly decreases the diffusion rate of copper.

The surface diffusion D_s of adsorbed atoms often shows a strong dependence on surface concentration. An example is oxygen diffusion on the W(110) surface [72], shown in Figure 4.15. The variation of D_s with oxygen coverage is over two orders of magnitude, with a maximum at $\theta = 0.4$, corresponding to the formation of an ordered chemisorbed oxygen layer.

In Tables 4.5 and 4.6, surface self-diffusion and adsorbate diffusion data are plotted. The diffusion coefficient D_0 and the activation energy ΔE_D^* are given, along with the temperature range of the study.

The migration of atoms or molecules along the surface is one of the most important elementary steps of gas–surface interactions, reactive or nonreactive. It appears that the activation energies for surface self-diffusion are much smaller than the heats of sublimation, thus permitting equilibration of atoms among the various surface sites. All the studies reported indicate that the migration of adsorbates along the surface is also rapid, and that the activation energies for surface diffusion are much smaller than the heats of desorption. Ordering of chemisorbed monolayers is often observed at low temperatures (< 77 K). This ordering would not be possible without high surface-diffusion rates. The two-dimensional phase approximation that assumed equilibrium of adsorbed species among all available surface sites would not be valid without high rates of surface diffusion.

A closely related and frequently observed phenomenon is the "spillover" of adsorbed species. In a multiphase system such as metal islands dispersed on an oxide, it is possible for molecules to adsorb or even react on one of the constituents (the metal, for example) before diffusing over onto the second phase (the oxide in this

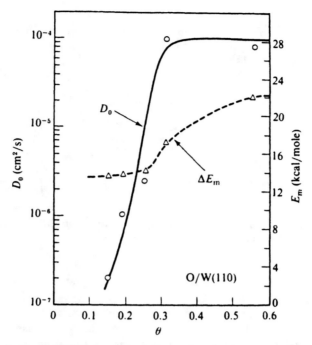

Figure 4.15. Variation in diffusion constant D_s, and activation energy for diffusion, ΔE_D^*, for oxygen atoms on the W(110) surface as a function of oxygen coverage [97].

TABLE 4.5. Activation Energies for Surface Self-Diffusion ΔE^* and Diffusion Constants D_0 for Several Metals

Material	ΔE_D^* (kJ/mole)[a]	D_0 ($10^{-4} \times m^2$/sec)
Ni	158.8	300
Pt	109.1–125	4×10^{-3}
Rh	173.5	$\approx 4 \times 10^{-2}$
Re	217.4	≈ 1.0
W	284–326	0.85
Cu	171–192	≈ 650
Au	146–176	0.37
Ta	188	—
Mo	217–234	≈ 0.8
Fe	249.1	—

[a] N.A. Gjostein. In: J.J. Burke et al., editor *Surfaces and Interfaces*. Syracuse Univeristy Press, Syracuse, NY 1967.

TABLE 4.6. Surface Diffusion Coefficient and Activation Energies of Diffusion for Selected Adsorbate-Substrate Systems

System	Method and Conditions	D_0 $(10^{-4} \times cm^2/sec)$	ΔE_D^* (kJ/mole)	Reference
O/W(110)	FEM			
	$\theta < 0.2$	1×10^{-7}	59	1
	$\theta = 0.56$	1×10^{-4}	92	1
CO/W(110)	FEM			
	α phase	Small	Small	2
	β phase	1×10^{-5}	96	2
Xe/W(110)	FEM	7×10^{-8}	4.6	3
H/Ni(100)	LID	2.5×10^{-3}	14.6	4
CO/Ni(100)	LID, $\theta = 0.4$	0.05	20.5	5
O/Pd(100)	LEED	—	52.3	6
Ni/N–i(110)	Tracer	300	158.8	7

REFERENCES

1. J.R. Chen and R. Gomer. *Surf. Sci.* **79:**413 (1979).
2. J.R. Chen and R. Gomer. *Surf. Sci.* **81:**589 (1979).
3. J.R. Chen and R. Gomer. *Surf. Sci.* **94:**416 (1980).
4. D.R. Mullins, B. Roop, S. A. Costello, and J.M. White. *Surf. Sci.* **186:**67 (1987).
5. B. Roop, S.A. Costello, D.R. Mullins, and J.M. White. *J. Chem. Phys.* **86:**303 (1987).
6. S.L. Chang and P.A. Thiel. *Phys. Rev. Lett.* **59:**296 (1987).
7. J.R. Wolfe and H.W. Weart. In G.A. Somorjai, editor, *The Structure and Chemistry of Solid Surfaces.* John Wiley & Sons, New York, 1969.

case), where they may react with a different adsorbed species or desorb. This phenomenon, which is predicated on rapid surface diffusion, is most important in heterogeneous catalysis.

4.3.6 Desorption

When an adsorbed layer of atoms or molecules is heated by a laser or an electron beam, or by resistively heating the substrate, the surface species may desorb. This is because their surface residence time depends exponentially on temperature ($\tau = \tau_0 \exp (\Delta E/RT)$). If the adsorbate is not resupplied from the gas phase, its surface concentration diminishes rapidly with increasing temperature until the surface becomes clean. This phenomenon is the basis of temperature-programmed desorption (TPD). Figure 4.16 shows the TPD spectra of xenon obtained by starting from several different coverage levels. Maximum desorption occurs at a given temperature. From these data and the known initial coverages, the heat of desorption and a preexponential factor can be determined, if one assumes a certain order (first, second, etc.) for the desorption process.

An analysis of the desorption process, using the temperature where the desorption peak is obtained, is given by Redhead [73]. Assuming that ν and E_{des} are independent of the adsorbate concentration σ and time t, E_{des} can be obtained for zero-, first-,

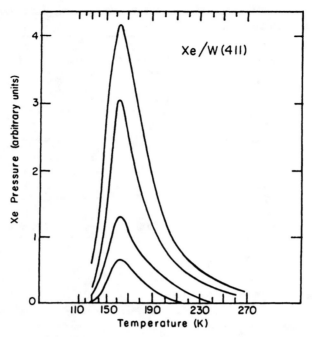

Figure 4.16. Temperature-programmed desorption of xenon from W(411) at various coverages [98].

and second-order desorption, respectively, as

$$\frac{E_0}{R} = \frac{\nu_0}{\sigma\alpha} \exp\left(-\frac{E_0}{RT_p}\right) \tag{4.47}$$

and

$$\frac{E_1}{RT_p^2} = \frac{\nu_1}{\alpha} \exp\left(-\frac{E_1}{RT_p}\right) \tag{4.48}$$

and

$$\frac{E_2}{RT_p^2} = \frac{\nu_2\sigma}{\alpha} \exp\left(-\frac{E_2}{RT_p}\right) \tag{4.49}$$

where T_p is the temperature at which a desorption peak is at the maximum and σ is the initial adsorbate concentration. The subscript 0, 1, or 2 denotes the zeroth-, first-, or second-order desorption processes, respectively. The term α is a constant of proportionality for the increase in temperature with time. The temperature rise is usually in the form $T = T_0 + \alpha t$; that is, the temperature of the sample rises linearly with time. As seen from Eqs. 4.47–4.49, T_p is independent of σ for the first-order process. Alternatively, T_p increases or decreases with σ for the zeroth- or second-order processes, respectively. Equations 4.47–4.49 allow one to determine the activation energy and the preexponential factor, and also to distinguish between ze-roth-, first-, and second-order desorption processes from the measurements of the dependence of the peak temperatures on initial adsorbate concentration.

For a first-order process, the desorption-peak maximum temperature T_p and the width of the peak at half-maximum are independent of the initial coverage. An example of this behavior is shown in Figure 4.16 for xenon on a tungsten crystal surface.

For a second-order desorption process, both the desorption-peak maximum temperature T_p and the half-width change with increasing initial coverage. An example of this type of behavior is the desorption of N_2 from W(100) (Figure 4.17), clearly indicating the dissociative adsorption of nitrogen before desorption under the conditions of the experiment.

A variety of procedures for analyzing desorption spectra have been developed. These are reviewed in detail elsewhere (see references [61, 74–76]).

4.3.7 Surface-Structure Sensitivity of Thermal Desorption

The thermal-desorption spectrum reflects the binding energy of the adsorbed species. Because its value and, therefore, the heats of adsorption may change from site to site on the surface, TPD can be used as a "fingerprint" of the surface structure. Figure 4.18 shows the thermal desorption of hydrogen from flat (111), stepped (557), and kinked (12,9,8) crystal faces of platinum. The flat surface exhibits one broad peak, while the stepped surface shows two peaks. It is relatively straightforward to associate the desorption peak that appears at the higher temperature with hydrogen desorbing from step sites. The peak intensity ratios may yield the ratios of step and terrace atom-site concentrations, while the temperatures that are associated with the

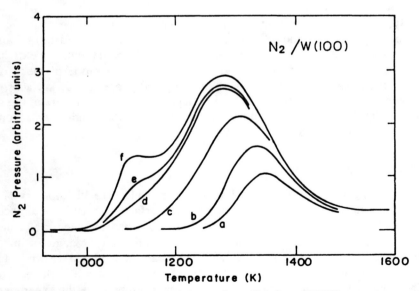

Figure 4.17. Temperature-programmed desorption of N_2 from W(100) at various coverages [98].

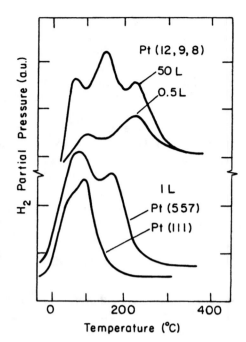

Figure 4.18. Temperature-programmed desorption of hydrogen from the flat Pt(111), stepped Pt(557) and kinked Pt(12, 9, 8) crystal surfaces [99].

maximum desorption rates can be used to calculate the heat of desorption from a given surface site. For example, the kinked platinum surface exhibits three thermal-desorption peaks; the highest temperature peak is due to hydrogen desorbing from the kink sites. Surface defects, steps, and kinks are usually sites of higher heats of adsorption for most adsorbates.

Recently, xenon has been used as a nonreactive probe of surface structure. As long as the surface can be cooled to a low enough temperature to adsorb this inert gas atom, its local interaction with surface sites of different structure yields large enough variations in its heat of adsorption to be used as a probe of the surface structure. As we shall see in the chapter on electrical properties of surfaces, the surface electric dipole varies from site to site, depending on the structure of the site. This electric dipole influences the polarizability and thus the bonding of adsorbed atoms or molecules at that site.

4.3.8 Collision-Induced Desorption

When a beam of argon atoms of high kinetic energy (113–213 kJ/mole = 27–51 kcal/mole) impinges on a methane-covered nickel surface, the methane molecules desorb [77, 78]. As an Ar atom collides with an adsorbed methane molecule, it transfers a certain fraction of its kinetic energy, depending on the angle of impact. The methane collides with the metal surface, and it can rebound and desorb if it has enough kinetic energy to overcome the attractive force that holds it in the adsorbed state. Calculations that treat Ar and CH_4 as hard spheres reproduce the experimental observations well.

4.3.9 Electron-Beam-Induced Desorption

When adsorbed molecules are bombarded with electrons, local heating effects occur that lead to thermal desorption. In addition, there is a small but finite probability that electrons in the chemical bonds that hold the adsorbate to the surface will be excited into a repulsive state, leading to the desorption of that molecule either as a neutral species or as a molecular ion. Desorption of neutral species under electron-beam bombardment is frequently observed in studies of electron–surface interactions. A fraction of the adsorbed molecules will be ionized. These can be detected as positive ions, and the spatial distribution of this ion flux can be imaged on a fluorescent screen. Electron-stimulated desorption ion-angular distribution (ESDIAD) [56, 61, 64, 79–84] is the name of the technique that is used to learn about the site symmetry and orientation of adsorbed molecular species, since the molecular ions are usually emitted in the directions of their chemical bonds with the surface and with an unchanged orientation with respect to the orientation of the molecule when it was adsorbed on the surface.

4.3.10 Photon-Stimulated Desorption

When a molecule chemisorbed on a metal surface (CO on nickel [85], for example) is illuminated by light of appropriate energy, desorption of some of the molecules is observed. Careful studies of the threshold energy for the process and the kinetic energy distribution of the desorbing molecules indicate that local heating by light is the dominant mechanism for desorption. However, weakly bound methyl bromide adsorbed on the LiF(100) surface can be photodissociated into CH_3 and Br, and the kinetic-energy distribution of the desorbing CH_3 fragments clearly indicate that the molecule has photofragmented without the participation of the ionic substrate in the C$-$Br bond-breaking process [86]. Chemisorbed species on semiconductor surfaces often desorb when the surface is illuminated with photons of band-gap energy. Excited electron–hole pairs are created in this manner. These electron–hole pairs cause reduction and oxidation of adsorbed ionic species that subsequently desorb as neutral molecules. For example, water dissociates to H^+ and OH^- species on $SrTiO_3$ or TiO_2 surfaces. Upon irradiation by photons of band-gap energy [≈ 3.1 eV ($\approx 5 \times 10^{-19}$ J)], H atoms and OH radicals are produced, which desorb when appropriately catalyzed by forming H_2 and O_2 molecules [87].

Surface photochemistry leading to the desorption of desired products is a rapidly growing area of surface chemistry.

4.3.11 Ion-Beam-Induced Desorption and Sputtering

Ions have greater masses than electrons; thus their transfer of energy to surface species is much more efficient. Ions incident on surfaces can break chemical bonds and eject atoms, molecules, or molecular clusters. Most of these species are neutral, but the ion impact may also ionize a fraction of these ejected particles. The detection of these ions, called *secondary-ion mass spectroscopy* (SIMS), is an important technique of surface-composition analysis. Ion bombardment is used frequently to remove unwanted molecular or atomic layers of impurities in order to clean a surface efficiently. Ion sputtering is also used to deposit thin films of the bombarded material without much change in composition. Because big chunks of the solid are lifted off the surface by sputtering, the local surface composition or even the local atomic structure may remain unperturbed. Thus, complex multicomponent solids, oxides, carbides, and so on, may be sputter-deposited as thin films under appropriate conditions. The mechanism of ion-induced sputtering has been the subject of intense studies because of the complexity of the high-energy process and its many uses [88]. A more recent view proposes that the bombarding ions deposit most of their energy [10^2–10^3 eV (1.6×10^{-17}–1.6×10^{-16} J)] in the near-surface region in a cluster of atoms that disintegrate to yield the various fragments observed in the mass spectrometer.

4.4 SUMMARY AND CONCEPTS

- The harmonic-oscillator and elastic-continuum models can be used to explain the presence of surface phonons (Rayleigh waves and localized surface modes of vibration) and the larger mean-square displacement of surface atoms compared to that of atoms in the bulk.

- Surface–atom displacements play an important role in inducing the restructuring of surfaces.

- Atomic and molecular-beam surface-scattering studies reveal efficient energy transfer between the translational, vibrational, and rotational energy modes of the incident molecules and the surface atoms.

- Atom transport by surface diffusion is rapid and requires much lower activation energy compared to desorption. Thus, the surface atoms or the adsorbates equilibrate among the various surface sites.

- Desorption by heating and by electron, photon, and ion bombardment can be used to learn about the binding and structure of adsorbates and adsorption sites.

4.5 PROBLEMS

4.1 Using the Debye model, compute the root-mean-square displacement of nickel at 300 K and at its melting point. What is the fractional displacement of the metal atoms relative to the interatomic distance at the melting temperature?

4.2 Calculate the distance traveled by a copper surface atom on a copper surface in 1 h at $T = 300$ K.

4.3 The maximum desorption rate of CO from Pt(111) occurs at 480 K. Assuming first-order desorption kinetics and a heating rate of 30 K/sec, calculate the activation energy for desorption.

4.4 Why does the C≡O stretching frequency change with the site symmetry of the adsorbed molecule?

*4.5 What is the physical meaning of the Debye–Waller factor? How would it differ for low-energy electron diffraction, helium-atom diffraction, and X-ray diffraction from the same solid (see [100–102])?

*4.6 The melting of surfaces [103] as the temperature of the crystal is increased has been theoretically treated in reference [104]. One such experiment is described in reference [105]. How may such events be detected?

*4.7 Adsorption probability or sticking probability of atoms and diatomic molecules have been measured for many transition metal surfaces. These experiments are carried out as a function of temperature and coverage and also as a function of the energy content and angle of incidence of the incident atoms or molecules. Pick a gas–surface system (e.g., H_2/Ni [106]) and describe the results of these experiments. In particular, how does the sticking probability depend on the surface structure and the temperature of the transition metal?

*4.8 Field-ion microscopy may be used to study surface diffusion. Enumerate the various atom diffusion mechanisms that have been proposed for clean transition metal surfaces [107, 108].

*4.9 Laser desorption that "burns a hole" in the adsorbed monolayer is utilized to measure the diffusion rate of adsorbed atoms and molecules. Describe the technique and discuss the results of one recent investigation [109, 110].

*4.10 The dissociation of molecules on metal surfaces may depend on their kinetic energy at the time of impact at the surface. Describe the models that can explain these experimental findings (e.g., see reference [111]).

*4.11 Electron beam impact can induce the desorption of molecules—neutral or ionized—in the direction of their chemical bond to the surface. The technique that takes advantage of this phenomenon to learn about surface bond directionality is called *electron-stimulated desorption ion-angular distribution* (ESDIAD) [56, 61, 64, 79–83]. Describe the bonding of a polyatomic molecule on a transition metal surface as determined by ESDIAD [112].

*4.12 The photon-induced dissociation and desorption of methyl bromide from a LiF surface was monitored [113]. Discuss the evidence that the photon energy was absorbed directly by the molecule adsorbed on the alkali halide surface. Would you expect the same photon-induced dissociation behavior if methyl bromide was chemisorbed on a transition metal surface?

*4.13 The translational-energy accommodation and transfer between a monatomic gas and a metal surface depends on both the translational energy, E_T, of the incident atom and the temperature of the solid, $E_{surface}$. Describe the nature of energy transfer in the two extremes when (a) $E_T \gg E_{surface}$ and (b) $E_T \ll E_{surface}$.

*4.14 Electron-beam incidence induces the desorption of adsorbed molecules either as neutral or as ionized species [114–116]. Discuss the mechanisms of this process, and indicate the structural information that can be obtained and the cross sections for the desorption processes.

*4.15 Ion-beam sputtering is used to etch away materials and to deposit thin films of controlled composition. Discuss what is known about the mechanisms of ion sputtering, and give two examples of its useful applications [117–119].

*4.16   Bound states of inert gas atoms have been detected by atomic beam scattering studies that monitored the energy and angular distribution of scattered atoms from LiF crystal surfaces. Describe the experiments and results of these studies [120, 121] (see also references [122, 123]).

**4.17   The theory of desorption of atoms and molecules by temperature-programmed desorption has been reviewed in references [61, 74–76, 124–126]. Discuss the assumptions made in deriving the first- and second-order desorption rates and their correlation to the temperature of the maximum desorption rates. How does the magnitude of the preexponential factor reflect the assumption of (a) a mobile adsorbate layer or (b) an immobile adsorbate layer?

**4.18   The elementary steps of the silver halide photographic process are discussed in reference [127]. The surface diffusion rates of silver atoms on a single silver crystal surface were found to increase by orders of magnitude in the presence of chemisorbed sulfur. Could sulfur-induced sensitization of the photographic process be related to this observation?

**4.19   The vibrational spectra of carbon monoxide chemisorbed on various metal surfaces are a function of coverage and temperature [128]. Review the experimental techniques that were employed and discuss the surface structure sensitivity, the coverage and temperature dependencies of the adsorbed molecule.

**4.20   The interactions of ions with surfaces have many important applications that include sputter cleaning, ion implantation, ion etching, and surface chemical analysis (SIMS and ISS) [56, 61, 64, 84, 117, 129–133]. Discuss each of these processes and by reviewing recent papers in the literature describe a case history of the application of each of these processes. Theories of high-energy ion–surface interactions have been proposed. Discuss the various models that explain the phenomena.

REFERENCES

[1]   L.D. Landau and E.M. Lifshitz. *Theory of Elasticity*. Pergamon, Elmsford, NY, 1970.

[2]   L. Rayleigh. On Waves Propagated Along the Plane Surface of an Elastic Solid. *Proc. London Math. Soc.* **17**:4 (1885).

[3]   H. Ibach. Low Energy Electron Spectroscopy—a Tool for Studies of Surface Vibrations. *J. Vacuum Sci. Technol.* **9**:713 (1972).

[4]   S. Lehwald, F. Wolf, H. Ibach, B.M. Hall, and D.L. Mills. Surface Vibrations on Ni(110): The Role of Surface Stress. *Surf. Sci.* **192**:131 (1987).

[5]   S. Thevuthasan and W.N. Unertl. Depth Dependence of the Anomalous Thermal Behavior of Cu(110). *Appl. Phys. A* **51**:216 (1980).

[6]   U. Breuer, H.P. Bonzel, K.C. Prince, and R. Lipowsky. LEED Investigation of

Temperature Dependent Surface Order on Pb Single Crystal Surfaces. *Surf. Sci.* **223**:258 (1989).

[7] Y. Darici, J. Marcano, H. Min, and P.A. Montano. LEED Measurements of Fe Epitaxially Grown on Cu(001). *Surf. Sci.* **182**:477 (1987).

[8] F.C.M.J.M. van Delft, M.J. Koster van Groos, R.A.G. de Graaf, A.D. van Langfeld, and B.E. Nieuwehuys. Determination of Surface Debye Temperatures by LEED. *Surf. Sci.* **189**:695 (1987).

[9] P.R. Watson and J. Mischenko III. Measurement of the Effective Debye Temperature for the Ti(0001) Surface. *Surf. Sci.* **186**:184 (1987).

[10] M.A. Vasiliev and S.D. Gorodetsky. Composition and Dynamical Characteristics of the Surface of Single Crystals FeNi$_3$ (100), (110), (111). *Surf. Sci.* **171**:543 (1986).

[11] N.J. Wu, V. Kumykov, and A. Ignatiev. Vibrational Properties of the Graphite (0001) Surface. *Surf. Sci.* **163**:51 (1985).

[12] S. Mróz and A. Mróz. LEED Measurements of the Surface Debye Temperature for the (110) Nickel Face. *Surf. Sci.* **109**:444 (1981).

[13] S.D. Bader. LEED Debye–Waller Analysis of Vibrational Hybridization for p(2 × 2)–O/Pd(111). *Surf. Sci.* **99**:392 (1980).

[14] H.B. Lyon and G.A. Somorjai. Surface Debye Temperatures of the (100), (111), and (110) Faces of Platinum. *J. Chem. Phys.* **44**:3707 (1966).

[15] D.P. Jackson. Approximate Calculation of Surface Debye Temperatures. *Surf. Sci.* **43**:431 (1974).

[16] H. Ohtani, M.A. Van Hove, and G.A. Somorjai. Leed Intensity Analysis of the Surface Structures of Pd(111) and of CO Adsorbed on Pd(111) in a $(\sqrt{3} \times \sqrt{3})R30°$ Arrangement. *Surf. Sci.* **187**:372 (1987).

[17] J.T. Yates, Jr. and K. Kolasinski. Infrared Spectroscopic Investigation of the Rhodium *gem*-Dicarbonyl Surface Species. *J. Chem. Phys.* **79**:1026 (1983).

[18] G. Comsa, R. David, and B.J. Schumacher. Magnetically Suspended Cross-Correlation Chopper in Molecular Beam Experiments. *Rev. Sci. Instrum.* **52**:789 (1981).

[19] T.H. Lin and G.A. Somorjai. Angular and Velocity Distributions of HD Molecules Produced by the H$_2$–D$_2$ Reaction on the Stepped Pt(557) Surface. *J. Chem. Phys.* **81**:704 (1984).

[20] A.E. Depristo and A. Kara. Molecular-Surface Scattering and Dynamics. In: I. Prigogine and S.A. Rice, editors, *Advances in Chemical Physics*, Volume 77. John Wiley & Sons, New York, 1990.

[21] S.T. Ceyer. Translational and Collision-Induced Activation of CH$_4$ on Ni(111): Phenomena Connecting Ultra-High-Vacuum Surface Science to High-Pressure Heterogeneous Catalysis. *Langmuir* **6**:82 (1990).

[22] D.J. Doren and J.C. Tully. Dynamics of Precursor-Mediated Chemisorption. *J. Chem. Phys.* **94**:8428 (1991).

[23] M. Head-Gordon, J.C. Tully, C.T. Rettner, C.B. Mullins, and D.J. Auerbach. On the Nature of Trapping and the Desorption at High Surface Temperatures. Theory and Experiments for the Ar–Pt(111) System. *J. Chem. Phys.* **94**:1516 (1991).

[24] S.I. Ionov and R.B. Bernstein. Hard-Cube Analysis of the Steric Effect in Molecule–Surface Scattering. *J. Chem. Phys.* **92**: 680 (1990).

[25] J.C. Tully. Washboard Model of Gas–Surface Scattering. *J. Chem. Phys.* **92**:680 (1990).

[26] C.T. Rettner and D.J. Auerbach. Probing the Dynamics of Gas–Surface Interactions with High Energy Molecular Beams. *Comments At. Mol. Phys.* **20**:153 (1989).

[27] J.A. Barker and D.J. Auerbach. Gas–Surface Interactions and Dynamics; Thermal Energy Atomic and Molecular Beam Studies. *Surf. Sci. Rep.* **4**:1 (1985).

[28] M.P. D'Evelyn and R.J. Madix. Reactive Scattering from Solid Surfaces. *Surf. Sci. Rep.* **3**:413 (1984).

[29] C. Steinbrückel. Gas–Surface Scattering Distributions According to the Hard-Spheroid Model. *Surf. Sci.* **115**:247 (1982).

[30] J. Lorenzen and L.M. Raff. A Comparison of Detailed Lattice Model Gas–Solid Theory with Molecular Beam Data: Scattered Velocity Distributions for an Ar/W System. *J. Chem. Phys.* **74**:3929 (1981).

[31] E.K. Grimmelmann, J.C. Tully, and M.J. Cardillo. Hard-Cube Model Analysis of Gas–Surface Energy Accommodation. *J. Chem. Phys.* **72**:1039 (1980).

[32] R.M. Logan and J.C. Keck. Classical Theory for the Interaction of Gas Atoms with Solid Surfaces. *J. Chem. Phys.* **49**:860 (1968).

[33] R.M. Logan and R.E. Stickney. Simple Classical Model for the Scattering of Gas Atoms from a Solid Surface. *J. Chem. Phys.* **44**:195 (1966).

[34] P. Kisliuk. The Sticking Probabilities of Gases Chemisorbed on the Surfaces of Solids. II. *J. Phys. Chem. Solids* **5**:78 (1958).

[35] P. Kisliuk. The Sticking Probabilities of Gases Chemisorbed on the Surfaces of Solids. *J. Phys. Chem. Solids* **3**:95 (1957).

[36] S.J. Lombardo and A.T. Bell. A Review of Theoretical Models of Adsorption, Diffusion, Desorption, and Reaction of Gases on Metal Surfaces. *Surf. Sci. Rep.* **13**:1 (1991).

[37] W.H. Weinberg. Precursor Intermediates and Precursor-Mediated Surface Reactions. General Concepts, Direct Observations and Indirect Manifestations. In: M. Grunze and H.J. Kreuzer, editors, *Workshop on Interface Phenomena*, *Springer Series in Surface Sciences*, Volume 8. Springer-Verlag, New York, 1987.

[38] A. Cassuto and D.A. King. Rate Expressions for Adsorption and Desorption Kinetics with Precursor States and Lateral Interactions. *Surf. Sci.* **102**:388 (1981).

[39] R. Gorte and L.D. Schmidt. Desorption Kinetics with Precursor Intermediates. *Surf. Sci.* **76**:559 (1978).

[40] D.A. King. The Influence of Weakly Bound Intermediates on Thermal Desorption Kinetics. *Surf. Sci.* **64**:43 (1977).

[41] K. Schönhammer. On the Kisliuk Model for Adsorption and Desorption Kinetics. *Surf. Sci.* **83**:L633 (1979).

[42] V.P. Zhdanov. Effect of the Lateral Interaction of Adsorbed Molecules on Preexponential Factor of the Desorption Rate Constant. *Surf. Sci.* **111**:L662 (1981).

[43] V.P. Zhdanov. Lattice–Gas Model for Description of the Adsorbed Molecule of Two Kinds. *Surf. Sci.* **111**:63 (1981).

[44] V.P Zhdanov. Lattice–Gas Model of Bimolecular Reaction on Surface. *Surf. Sci.* **102**:L35 (1981).

[45] J.C. Tully and M.J. Cardillo. Dynamics of Molecular Motion at Single-Crystal Surfaces. *Sci.* **223**:445 (1984).

[46] M.B. Lee, Q.Y. Yang, and S.T. Ceyer. Dynamics of the Activated Dissociative Chemisorption of CH_4 and Implication for the Pressure Gap in Catalysis: A Molecular Beam High-Resolution Electron Energy Loss Study. *J. Chem. Phys.* **87**:2724 (1987).

[47] S.L. Tang, J.D. Beckerle, and S.T. Ceyer. Effect of Translational Energy on the Molecular Chemisorption of CO on Ni(111): Implications for the Dynamics of the Chemisorption Process. *J. Chem. Phys.* **84**:6488 (1986).

[48] R.D. Levine and G.A. Somorjai. Kinetic Model for Cooperative Dissociative Che-
      misorption and Catalytic Activity via Surface Restructuring. *Surf. Sci.* **232**:407
      (1990).

[49] J.C. Dunphy, P. Sautet, D.F. Ogletree, O. Dabbousi, and M.B. Salmerón. Scanning
      Tunneling Microscopy Study of Surface Diffusion of Sulfur on Re(0001). *Phys. Rev.
      B*, 1993.

[50] G. Ehrlich. Direct Observations of the Surface Diffusion of Atoms and Clusters. *Surf.
      Sci.* **246**:1 (1991).

[51] T.T. Tsong. Experimental Studies of the Behavior of Single Adsorbed Atoms on
      Solid Surfaces. *Rep. Prog. Phys.* **51**:759 (1988).

[52] H.P. Bonzel. Surface Diffusion on Metals. In: H. Mehrer, editor, *Diffusion in Metals
      and Alloys*. Springer-Verlag, New York, 1990.

[53] R. Gomer. Diffusion of Adsorbates on Metal Surfaces. *Rep. Prog. Phys.* **53**:917
      (1990).

[54] R.J. Borg and G.J. Dienes. *An Introduction to Solid State Diffusion*. Academic Press,
      New York, 1988.

[55] G. Ehrlich and K. Stolt. Surface Diffusion. *Annu. Rev. Phys. Chem.* **31**:603 (1980).

[56] J.B. Hudson. *Surface Science: An Introduction*. Butterworth-Heinemann, Boston,
      1992.

[57] L.E.C. van de Leemput and H. van Kempen. Scanning Tunneling Microscopy. *Rep.
      Prog. Phys.* **55**:1165 (1992).

[58] M. Tsukada, K. Kobayashi, N. Isshiki, and H. Kagashima. First-Principles Theory
      of Scanning Tunneling Microscopy. *Surf. Sci. Rep.* **13**:265 (1991).

[59] F. Ogletree and M. Salmerón. Scanning Tunneling Microscopy and the Atomic Struc-
      ture of Solid Surfaces. *Prog. Solid State Chem.* **20**:235 (1990).

[60] P.K. Hansma, V.B. Elings, O. Marti, and C.E. Bracker. Scanning Tunneling Mi-
      croscopy and Atomic Force Microscopy: Application to Biology and Technology.
      Science **242**:157 (1988).

[61] D.P. Woodruff and T.A. Delchar. *Modern Techniques of Surface Science. Cam-
      bridge Solid State Science Series*. Cambridge University Press, New York, 1986.

[62] R.P.H. Gasser. *An Introduction to Chemisorption and Catalysis by Metals*. Oxford
      University Press, New York, 1985.

[63] J.A. Panitz. High-Field Techniques. In: R.L. Park and M.G. Lagally, editors, *Solid
      State Physics: Surfaces. Methods of Experimental Physics*, Volume 22. Academic
      Press, New York, 1985.

[64] M.J. Higatsberger. Solid Surfaces Analysis. In: C. Marton, editor, *Advances in Elec-
      tronics and Electron Physics*, Volume 56. Academic Press, New York, 1981.

[65] G. Ehrlich. Wandering Surface Atoms and the Field Ion Microscope. *Phys. Today*
      June: 44 (1981).

[66] R. Gomer. Recent Applications of Field Emission Microscopy. In: R. Vanselow,
      editor, *Chemistry and Physics of Solid Surfaces*, Volume 2. CRC Press, Cleveland,
      OH, 1979.

[67] G.A. Somorjai and M.A. Van Hove. Adsorbed Monolayers on Solid Surfaces. In:
      J.D. Dunitz, J.B. Goodenough, P. Hemmerich, J.A. Albers, C.K. Jørgensen, J.B.
      Neilands, D. Reinen, and R.J.P. Williams, editors, *Structure and Bonding*, Volume
      38. Springer-Verlag, New York, 1979.

[68] M. Prutton. *Surface Physics*. Oxford University Press, New York, 1975.

[69] R.B. Hall. Pulsed-Laser-Induced Desorption Studies of Kinetics of Surface Reac-
      tions. *J. Phys. Chem.* **91**:1007 (1987).

[70] E. Preuss, N. Freyer, and H.P. Bonzel. Surface Self-Diffusion on Pt(110): Directional Dependence and Influence of Surface-Energy Anisotropy. *Appl. Phys. A* **41**:137 (1986).

[71] G. Ayrault and G. Ehrlich. Surface Self-Diffusion on an FCC Crystal: An Atomic View. *J. Chem. Phys.* **60**:281 (1974).

[72] J.R. Chen and R. Gomer. Mobility of Oxygen on the (110) Plane of Tungsten. *Surf. Sci.* **79**:413 (1979).

[73] P.A. Redhead. Thermal Desorption of Gases. *Vacuum* **12**:203 (1962).

[74] J.T. Yates. The Thermal Desorption of Adsorbed Species. In: R.L. Park and M.G. Lagally, editors, *Solid State Physics: Surfaces, Methods of Experimental Physics*, Volume 22. Academic Press, New York, 1985.

[75] R.J. Madix. The Application of Flash Desorption Spectroscopy to Chemical Reactions on Surfaces; Temperature Programmed Reaction Spectroscopy. In: R. Vanselow, editor, *Chemistry and Physics of Solid Surfaces*, Volume 2. CRC Press, Cleveland, OH, 1979.

[76] J.L. Beeby. The Theory of Desorption. In: R. Vanselow, editor, *Critical Reviews in Solid State and Material Sciences*, Volume 7. CRC Press, Cleveland, OH, 1977.

[77] J.D. Beckerle, A.D. Johnson, and S.T. Ceyer. Collision-Induced Desorption of Physisorbed CH_4 from Ni(111): Experiments and Simulations. *J. Chem. Phys.* **93**:4047 (1990).

[78] J.D. Beckerle, A.D. Johnson, Q.Y. Yang, and S.T. Ceyer. Collision-Induced Dissociative Chemisorption of CH_4 on Ni(111) by Inert Gas Atoms: The Mechanism for Chemistry with a Hammer. *J. Chem. Phys.* **91**:5756 (1989).

[79] R.D. Ramsier and J.T. Yates, Jr. Electron Stimulated Desorption: Principles and Applications. *Surf. Sci. Rep.* **12**:243 (1991).

[80] R.H. Stulen. Recent Developments in Angle-Resolved Electron-Stimulated Desorption of Ions from Surfaces. *Prog. Surf. Sci.* **32**:1 (1989).

[81] T.E. Madey and R. Stockbauer. Experimental Methods in Electron-and Photon-Stimulated Desorption. In: R.L. Park and M.G. Lagally, editors, *Solid State Physics: Surfaces, Methods of Experimental Physics*, Volume 22. Academic Press, New York, 1985.

[82] M.L. Knotek. Stimulated Desorption. *Rep. Prog. Phys.* **47**:1499 (1984).

[83] M.L. Knotek. Stimulated Desorption from Surfaces. *Phys. Today* **September**:24 (1984).

[84] D.P. Woodruff, G.C. Wang, and T.M. Lu. Surface Structure and Order–Disorder Phenomena. In: D.A. King and D.P. Woodruff, editors, *Adsorption at Solid Surfaces. The Chemical Physics of Solid Surfaces and Heterogeneous Catalysis*, Volume 2. Elsevier, New York, 1983.

[85] B. Roop, S.A. Costello, D.R. Mullins, and J.M. White. Coverage-Dependent Diffusion of CO on Ni(100). *J. Chem. Phys.* **86**:3003 (1987).

[86] I. Harrison, J.C. Polanyi, and P.A. Young. Photochemistry of Adsorbed Molecules. III. Photodissociation and Photodesorption of CH_3Br Adsorbed on LiF(001). *J. Phys. Chem.* **88**:6100 (1984).

[87] F.T. Wagner and G.A. Somorjai. Photocatalytic and Photoelectrochemical Hydrogen Production on $SrTiO_3$ Single Crystals. *J. Am. Chem. Soc.* **102**:5494 (1980).

[88] E. Taglauer, W. Heiland, and J. Onsgaard. Ion Beam Induced Desorption of Surface Layers. *Nucl. Instrum. Methods* **168**:571 (1980).

[89] H. Froitzheim, H. Ibach, and S. Lehwald. Surface Sites of H on W(100). *Phys. Rev. Lett.* **36**:1549 (1976).

[90] B.E. Koel and G.A. Somorjai. Vibrational Spectroscopy Using HREELS of Benzene Adsorbed on the Rh(111) Crystal Surface. *J. Electron Spectrosc. Relat. Phenom.* **29**:287 (1983).

[91] D.E. Wilk, C.D. Stanners, Y.R. Shen, and G.A. Somorjai. The Structure and Thermal Decomposition of *Para-* and *Ortho-*Xylene on Pt(111). A HREELS, LEED, and TPD Study. *Surf. Sci.* **280**:298 (1993).

[92] W.J. Siekhaus, J.A. Schwarz, and D.R. Olander. A Modulated Molecular Beam Study of the Energy of Simple Gases Scattered from Pyrolytic Graphite. *Surf. Sci.* **33**:445 (1972).

[93] M. Asscher, W.L. Guthrie, T.H. Lin, and G.A. Somorjai. Energy Redistribution among Internal States of Nitric Oxide Molecules upon Scattering from Pt(111) Crystal Surface. *J. Chem. Phys.* **78**:6992 (1983).

[94] H. Hopster, H. Ibach, and G. Comsa. Catalytic Oxidation of Carbon Monoxide on Stepped Platinum 111 Surfaces. *J. Catal.* **46**:37 (1977).

[95] R. Raval, M.A. Harrison, and D.A. King. Nitrogen Adsorption on Metals. In: D.A. King and D.P. Woodruff, editors, *Chemisorption Systems, Part B. The Chemical Physics of Solid Surfaces and Heterogeneous Catalysis*, Volume 3. Elsevier, New York, 1990.

[96] N.A. Gjostein. Surface Self-Diffusion in FCC and BCC Metals: A Comparison of Theory and Experiment. In: J.J. Burke, N. Reed, and V. Weiss, editors, *Surfaces and Interfaces I: Chemical and Physical Characteristics*. Syracuse University Press, Syracuse, NY, 1967.

[97] A. Zangwill. *Physics at Surfaces*. Cambridge University Press, New York, 1988.

[98] D.A. King. Kinetics of Adsorption, Desorption, and Migration at Single-Crystal Metal Surfaces. In: R. Vanselow, editor, *CRC Critical Reviews in Solid State and Materials Sciences*, Volume 7. CRC Press, Cleveland, OH, 1978, p. 163.

[99] G.A. Somorjai. *Chemistry in Two Dimensions: Surfaces*. Cornell University Press, Ithaca, NY, 1981.

[100] G. Comsa, G.H. Comsa, and J.K. Fremery. Peculiarities of the Helium-Beam Scattering on Metal Surfaces. *Z. Naturforsch.* **29A**:189 (1974).

[101] W.H. Weinberg. Atomic Helium Scattering and Diffraction from Solid Surfaces. *J. Phys. C* **5**:2098 (1972).

[102] J.L. Beeby. The Debye–Waller Factor in Atom–Surface Scattering. *Jpn. J. Appl. Phys. suppl. 2, part 2: Proceedings of the 2nd International Conference on Solid Surfaces (Kyoto)*, 1974, p. 537.

[103] J.F. van der Veen, B. Pluis, and A.W. Denier van der Gon. Surface Melting. In: R. Vanselow and R.F. Howe, editors, *Chemistry and Physics of Solid Surfaces VII, Springer Series in Surface Sciences*, Volume 10. Springer-Verlag, Berlin, 1988.

[104] P. Stolze, J.K. Nørskov, and U. Landman. Disordering and Melting of Aluminum Surfaces. *Phys. Rev. Lett.* **61**:440 (1988).

[105] P. von Blanckenhagen, W. Schommers, and V. Voegele. Summary Abstract: Temperature Dependence of the Structure of the Al(110) Surface. *J. Vacuum Sci. Technol. A* **5**:649 (1987).

[106] K.D. Rendulic, A. Winkler, and H. Karner. Adsorption Kinetics of H_2/Ni and Its Dependence on Surface Structure, Surface Impurities, Gas Temperature, and Angle of Incidence. *J. Vacuum Sci. Technol. A* **5**:488 (1987).

[107] G.L. Kellogg and P.J. Feibelman. Surface Self Diffusion on Pt(001) by an Atomic Exchange Mechanism. *Phys. Rev. Lett.* **64**:3143 (1990).

[108] C. Chen and T.T. Tsong. Displacement Distribution and Atomic Jump Direction in Diffusion of Ir Atoms on the Ir(001) Surface. *Phys. Rev. Lett.* **64**:3147 (1990).

[109] J.L. Brand, M.V. Arena, A.A. Deckert, and S.M. George. Surface Diffusion of n-Alkanes on Ru(001). *J. Chem. Phys.* **92**:5136 (1990).

[110] E.G. Seebauer, A.C.F. Kong, and L.D. Schmidt. Surface Diffusion of Hydrogen and CO on Rh(111): Laser-Induced Thermal Desorption Studies. *J. Chem. Phys.* **88**:6597 (1988).

[111] S.T. Ceyer, J.D. Beckerle, M.B. Lee, S.L. Tang, Q.Y. Yang, and M.A. Hines. Effect of Translational and Vibrational Energy on Adsorption: The Dynamics of Molecular and Dissociative Chemisorption. *J. Vacuum Sci. Technol. A* **5**:501 (1987).

[112] M.D. Alvey, J.T. Yates, Jr., and K.J. Uram. The Direct Observation of Hindered Rotation of a Chemisorbed Molecule: PF_3 on Ni(111). *J. Chem. Phys.* **87**:7221 (1987).

[113] E.B.D. Bourdon, J.P. Cowin, I. Harrison, J.C. Polanyi, J. Segner, C.D. Stanners, and P.A. Young. UV Photodissociation and Photodesorption of Adsorbed Molecules. I. CH_3 on LiF(001). *J. Phys. Chem.* **88**:6100 (1984).

[114] D. Menzel and R. Gomer. Desorption from Metal Surfaces by Low-Energy Electrons. *J. Chem. Phys.* **41**:3311 (1964).

[115] P.J. Feibelman and M.L. Knotek. Reinterpretation of Electron-Stimulated Desorption Data from Chemisorption Systems. *Phys. Rev. B* **18**:6531 (1978).

[116] J.T. Yates, Jr., J.N. Russell, Jr., and S.M. Gates. Kinetic and Spectroscopic Investigations of Surface Chemical Processes.. In: R. Vanselow and R. Howe, editors, *Chemistry and Physics of Solid Surfaces VI. Springer Series in Surface Sciences*, Volume 5. Springer-Verlag, Berlin, 1986.

[117] N. Winograd and B.J. Garrison. Surface Structure and Reaction Studies by Ion–Solid Collisions. In: A.W. Czanderna and D.M. Hercules, editors, *Ion Spectroscopies for Surface Analysis. Methods of Surface Characterization*, Volume 2. Plenum Press, New York, 1991.

[118] J.M. Poate, K.N. Tu, and J.W. Mayer, editors. *Thin Films—Interdiffusion and Reactions*. John Wiley & Sons, New York, 1978.

[119] G.K. Wehner and G.S. Anderson. The Nature of Physical Sputtering. In: L.I. Maissel and R. Glang, editors, *Handbook of Thin Film Technology*. McGraw–Hill, New York, 1970.

[120] J.A. Meyers and D.R. Frankl. Selective Adsorption of ^4He on Clean LiF(001) Surfaces. *Surf. Sci.* **51**:61 (1975).

[121] A. Tsuchida. The (0,0) Fourier Component of the Interaction Potential of He/LiF(001). *Surf. Sci.* **52**:685 (1975).

[122] G. Lilienkamp and J.P. Toennies. The Observation of One-Phonon Assisted Selective Desorption and Adsorption of He Atoms in Defined Vibrational Levels on a LiF(001) Single Crystal Surface. *J. Chem. Phys.* **78**:5210 (1983).

[123] S.T. Ceyer and G.A. Somorjai. Surface Scattering. *Annu. Rev. Phys. Chem.* **28**:477 (1977).

[124] A.M. de Jong and J.W. Niemantverdriet. Thermal Desorption Analysis: Comparative Test of Ten Commonly Applied Procedures. *Surf. Sci.* **233**:355 (1990).

[125] E. Seebauer, A.C.F. Kong, and L.D. Schmidt. The Coverage Dependence of the Pre-Exponential Factor for Desorption. *Surf. Sci.* **193**:417 (1988).

[126] D.A. King. Thermal Desorption from Metal Surfaces: A Review. *Surf. Sci.* **47**:384 (1975).

[127] J.F. Hamilton. The Photographic Process. *Prog. Solid State Chem.* **8**:167 (1973).

[128] J.C. Campuzano. The Adsorption of Carbon Monoxide by the Transition Metals. In: D.A. King and D.P. Woodruff, editors, *Chemisorption Systems, Part A. The Chem-*

ical Physics of Solid Surfaces and Heterogeneous Catalysis, Volume 3. Elsevier, New York, 1990.

[129]   A.W. Czanderna. Overview of Ion Spectroscopies for Surface Compositional Analysis. In: A.W. Czanderna and D.M. Hercules, editors, *Ion Spectroscopies for Surface Analysis. Methods of Surface Characterization*, Volume 2. Plenum Press, New York, 1991.

[130]   C.J. Powell, D.M. Hercules, and A.W. Czanderna. Comparison of SIMS, SNMS, ISS, RBS, AES, and XPS Methods for Surface Compositional Analysis. In: A.W. Czanderna and D.M. Hercules, editors, *Ion Spectroscopies for Surface Analysis. Methods of Surface Characterization*, Volume 2. Plenum Press, New York, 1991.

[131]   W. Heiland and E. Taglauer. Ion Scattering and Secondary-Ion Mass Spectrometry. In: R.L. Park and M.G. Lagally, editors, *Solid State Physics: Surfaces. Methods of Experimental Physics*, Volume 22. Academic Press, New York, 1985.

[132]   H.W. Werner. Introduction to Secondary Ion Mass Spectrometry (SIMS). In: L. Fiermanns, J. Vennik, and V. Dekeyser, editors, *NATO Advanced Study Institutes Series: Series B, Physics*, Volume 32. Plenum Press, New York, 1978.

[133]   J.A. McHugh. Secondary Ion Mass Spectrometry. In: A.W. Czanderna, editor, *Methods of Surface Analysis. Methods and Phenomena: Their Applications in Science and Technology*, Volume 1. Elsevier, New York, 1975.

5

ELECTRICAL PROPERTIES OF SURFACES

5.1   THE SURFACE ELECTRON POTENTIAL

The electrostatic electron potential V^s in a surface region of thickness x (the region most affected by the presence of an interface) may be divided into three parts:

$$V^s(x) = V_{core}(x) + V_{exchange}(x) + V_{dipole}(x) \tag{5.1}$$

The $V_{core}(x)$ term represents the potential between the core electrons and valence electrons, while $V_{exchange}(x)$ is the exchange potential between the valence electrons. In the solid each electron lowers its energy by pushing the other electrons of like spin aside because of the Pauli exclusion principle (exchange interaction) or by having electrons of either spin avoid each other to minimize their repulsive (Coulomb) interaction (correlation interaction). This produces the exchange-correlation hole, a deficit of electronic charge that surrounds each electron. These potentials, along with those of the ion cores, are responsible for keeping the electrons in the solid in spite of their high concentration. $V_{core}(x)$ is not likely to change, whether the atom is at the surface or in the bulk, because of the localized nature of the core electrons. Therefore, we may neglect it for purposes of this discussion. However, $V_{exchange}(x)$ and $V_{dipole}(x)$ are very much the properties of the surface atoms. $V_{dipole}(x)$ is specific to the surface and is closely connected with the space charge that normally accumulates at the surface.

Perhaps the two most frequently measured electrical properties in surface science are the surface space charge potential V_{dipole} and the related work function ϕ.

5.2 THE SURFACE SPACE CHARGE

5.2.1 The Surface Space Charge at the Solid–Vacuum Interface

Consider an atomically smooth slab of solid in an ultrahigh-vacuum chamber. The ion cores of its atoms may be viewed as being smeared out to produce a uniform density of positive charge; electrons are bound to this uniform charge by electrostatic forces. This is the so-called jellium model that has been successful in reproducing many of the surface electrical properties of metals. At the surface the electrons are bound only toward the solid, and thus they spill out by tunneling into the vacuum, with the charge density dropping exponentially away from the surface (Figure 5.1). The electrons that accumulate on the outer edges of the solid–vacuum boundary leave a partial positive charge behind. This charge separation leads to the formation of the surface space charge. For a metal, the free-electron (those electrons that are free to hop from atom to atom in an applied potential) concentration is so high (about one electron per atom) that spatial extent of the space charge is limited to the topmost layer of atoms at the surface, since the atoms below the surface are effectively screened by the large free-electron density. For insulators or semiconductors, the

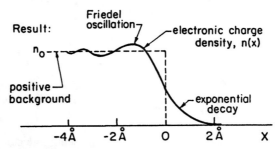

Figure 5.1. Schematic representation of the electronic charge density distribution at a metal surface.

space charge may extend many tens or hundreds of atomic layers into the solid, as we shall see shortly.

A more realistic model will take full account of the atomic nature of the surface and yield charge densities and electronic potentials similar to those obtained by the jellium model. In this circumstance, however, the charge density on the solid side of the surface exhibits fluctuations that are often called *Friedel oscillations* and which are due to the screening by the free electrons (Figure 5.1). The amplitude of this oscillation is a sensitive function of the electron density, as are the height and extent of the surface space-charge potential.

Both the height of the surface space-charge potential barrier V_s and its distance of penetration into the bulk d depend on the concentration of mobile charge carriers in the surface region. In order to discuss the properties of a surface space charge, let us consider an n-type semiconductor with a bulk carrier concentration n_e^{bulk}. In order to calculate the properties of the space-charge layer as a function of its charge density ρ_e, consider a homogeneous one-dimensional solid in thermal equilibrium (Figure 5.2). The potential at any point is only a function of the distance x from the surface (where $x = 0$) and is determined by the Poison equation:

$$\frac{d^2V}{dx^2} = -\frac{\rho_e(x)}{\epsilon\epsilon_0} \tag{5.2}$$

where ϵ is the dielectric constant in the solid and ϵ_0 is the permittivity of free space, a constant. In our model of the space-charge layer, $\rho_e(x) = eN_D^+$, where N_D^+ is the

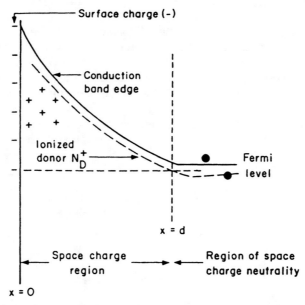

Figure 5.2. Scheme of space-charge buildup at an n-type semiconductor surface upon adsorption of electron acceptor molecules.

concentration if ionized donors. Integrating twice, one obtains

$$V(x) = -\frac{e}{2\epsilon\epsilon_0} N_D^+ (x - d)^2 \qquad (5.3)$$

At $x = d$, $V(x) = 0$; that is, d defines the distance at which the electrostatic potential due to the charge imbalance in the space-charge layer becomes zero and the electron concentration attains its bulk value again. At the surface $(x = 0)$,

$$V_s = \frac{e}{2\epsilon\epsilon_0} N_D^+ d^2 \qquad (5.4)$$

where V_s is the height of the space-charge potential at the interface. Assuming that all the electrons of concentration n_e^{bulk} are removed from the space-charge region and trapped at the surface, leaving behind an equal static positive charge, we have

$$en_e^{bulk}d \approx eN_D^+ d \qquad (5.5)$$

Substitution of Eq. 5.5 into Eq. 5.4 and subsequent rearrangement yield

$$d = \left[\frac{2\epsilon\epsilon_0 V_s}{en_e^{bulk}}\right]^{1/2} \qquad (5.6)$$

This distance is called the *Debye length*. It measures the penetration depth of the electrostatic surface effects.

Thus the higher the free-carrier concentration in the material, the smaller the penetration depth of the applied field into the medium. For electron concentrations of 10^{22} cm^{-3} (10^{28} m^{-3}) or larger, the space charge is restricted to distances on the order of one atomic layer or less, because the large free-carrier density screens the solid from the penetration of the electrostatic field caused by the charge imbalance. For most metals, almost every atom contributes one free valence electron. Because the atomic density for most solids is on the order of 10^{22} cm^{-3} (10^{28} m^{-3}), the free-carrier concentration in metals is in the range of 10^{20}–10^{22} cm^{-3} (10^{26}–10^{28} m^{-3}). Thus V_s and d are small. For semiconductors or insulators, however, typical free-carrier concentrations at room temperatures are in the range of 10^{10}–10^{16} cm^{-3} (10^{16}–10^{22} m^{-3}). Therefore, at the surfaces of these materials, there is a space-charge barrier of appreciable height and penetration depth that could extend over thousands of atomic layers into the bulk. This is the reason for the sensitivity of semiconductor devices to ambient changes that affect the space-charge barrier height.

5.2.2 Surface Space Charge at the Solid–Liquid Interface

So far we have only considered the properties of the surface space charge at the solid–vacuum interface. Let us now immerse the solid into a liquid. The molecules in the liquid adsorb onto the solid surface and become polarized as they respond to the electrical field at the interface to produce an electrochemical double layer. They may also line up in preferential bonding directions if they possess a permanent dipole moment. In this circumstance an electrical field is induced on the liquid side of the solid–liquid interface, with marked consequences for molecule transport or charge transport to and from the interface. The charge layer that forms on the liquid side of the solid–liquid interface is often called the *Helmholtz layer*, and it plays an important role in affecting electrochemical changes associated with reduction or oxidation of charged species when the solid is used as an electrode.

The presence of charged species at the solid–liquid interface helps to stabilize colloids. These are small particles, 10^4–10^5 Å (10^3–10^4 nm) in diameter, that carry the same charge and thus exhibit repulsive electrostatic interaction. Because all particles in a colloid system are of the same size, they are thermodynamically stable because their solubilities (or vapor pressures) are the same according to the Kelvin equation. Their surface charge provides extra stability because of the strong repulsion. Milk, blood, paint, and latex are examples of important colloid systems in biology and in the chemical technologies. Mechanical agitation can strip off the charged protein coating of a milk colloid, causing coagulation (whipping cream). The elimination of the repulsive charge by other means (adsorption or drying) can readily destabilize these systems. One important application of the surface space charge is copying in xerography. In this process a charged organic or inorganic insulator is selectively illuminated by using the mirror reflection of the printed page, thereby removing the charge only at the illuminated sites. By pressing an insulating surface (paper) against the charged surface, its charge is transferred and is then fixed by colored polymer particles that melt when heated at their sites of adsorption (that is, where the charges were on the paper).

5.3 THE WORK FUNCTION

The work function is defined as the minimum potential the most loosely bound valence electrons in the solid must overcome in order to be ejected into the vacuum outside the solid with zero kinetic energy at absolute zero temperature. A schematic energy-level diagram showing the work function, using the free-electron gas model of a metal, is depicted in Figure 5.3. The potential energy of the electrons at the top of the valence band is often called the Fermi energy E_F. Thus the work function can be defined as

$$e\phi = eV_{\text{exchange}} + eV_{\text{dipole}} - E_F \tag{5.7}$$

where V_{exchange} is a bulk property that depends on the bulk electron density. The term V_{dipole} is due to the surface space-charge potential that must be overcome by the electrons in the solid in order to exit into the vacuum. The term V_{dipole} is responsible

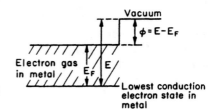

Figure 5.3. Energy level diagram to define the work function.

TABLE 5.1. Work Functions Measured from Different Crystal Faces of Tungsten, Molybdenum, and Tantalum

	Work Function					
	Tungsten (ref. 1)		Molybdenum (ref. 2)		Tantalum (ref. 2)	
Crystal Face	eV	10^{-19} J	eV	10^{-19} J	eV	10^{-19}
(110)	4.68	7.50	5.00	8.01	4.80	7.69
(112)	4.69	7.51	4.55	7.29	—	—
(111)	4.39	7.03	4.10	6.57	4.00	6.41
(001)	4.56	7.31	4.40	7.05	4.15	6.65
(116)	4.39	7.03	—	—	—	—

REFERENCES

1. M. Kaminsky, *Atomic and Ionic Impact Phenomena on Metal Surfaces*. Academic Press, New York, 1965.
2. O.D. Protopopov et al. *Sov. Phys. Solid State* (English translation), **8**:909 (1966).

for having each crystal surface exhibit a different work function. The work functions measured for various crystal faces of several metals are listed in Table 5.1. Calculations using the jellium model result in work functions very similar to those listed in Table 5.1.

5.3.1 Effect of Surface Roughness on Work Function

As the atomic density at the surface decreases, it becomes rougher on the atomic scale. The valence electrons spill out into the vacuum as before, but they smooth out the roughness in the positive charge distribution. The result is an electrostatic dipole oriented opposite to the spill-out dipole. As a consequence, the net dipole is reduced relative to its value at the higher-atomic-density, smoother surface, yielding a lower work function.

Changes in work function with increasing step density (roughness) are shown for stepped platinum and gold surfaces in Figure 5.4. As can be seen, the work function decreases linearly with increasing step density. The induced dipole moment μ due to steps can be calculated from the work-function change because $\Delta\phi = 4\pi N_s \mu$, where N_s is the step density. In Figure 5.4, note that platinum has a steeper slope for the work-function change with step density than does gold, indicating a larger dipole per step atom (0.6 Debye per step atom = 2×10^{-30} C · m per step atom) than on the gold stepped surface (0.27 Debye per step atom = 0.9×10^{-30} C · m per step atom). A step site on a tungsten surface has a 0.30-Debye dipole on the average, while at a tungsten adatom on the surface there is a dipole moment as large as 1.0 Debye, 3.3×10^{-30} C · m).

5.3.2 Change of Work Function with Particle Size

The work function of a solid is equivalent to its ionization potential. It is always lower than the ionization potential of the single atoms that make up the solid.* These

*This is because the remaining electrons partially screen the positively charged "hole" left behind.

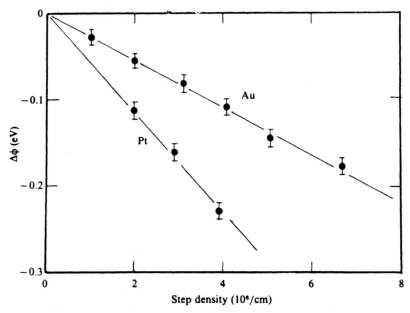

Figure 5.4. The work function of stepped gold and platinum surfaces as a function of step density (roughness) [6, 52].

values are listed for alkali atoms and solids in Table 5.2. It is important to find out how the single-atom ionization potential approaches the value of the work function as the atoms aggregate into clusters. It might even be possible to determine the ionization potential of clusters of ever-increasing size. Such experiments are in progress in several laboratories. By producing atom clusters of different sizes by laser evaporation or by ion bombardment, the ionization threshold of these clusters can then be probed using photoionization and employing variable photon energies. At present, the existing data indicate that the ionization potential decreases with increasing particle size in an oscillatory manner rather than smoothly [1]. This effect is shown for iron clusters in Figure 5.5. Whether these variations are due to changes in cluster structure or to other effects will have to be elucidated by future studies.

TABLE 5.2. Work Functions and Ionization Potentials of Alkali Metals[a]

Element	Work Function		Ionization Potential	
	ϕ (eV)	ϕ ($10^{-19} \times$ J/mole)	V_{ion} (eV)	V_{ion} ($10^{-19} \times$ J/mole)
Lithium	2.9	4.6	5.392	8.639
Sodium	2.75	4.41	5.139	8.234
Potassium	2.30	3.69	4.341	6.955
Rubidium	2.16	3.46	4.177	6.692
Cesium	1.81	3.03	3.894	6.239

[a] Data from *CRC Handbook of Chemistry and Physics*, 64th edition. CRC Press, Boca Raton, FL, 1983.

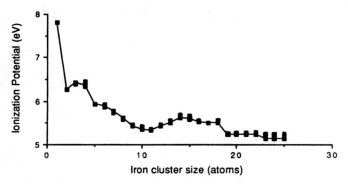

Figure 5.5. The ionization potential for iron clusters as a function of cluster size [1].

5.4 ADSORPTION-INDUCED CHARGE TRANSFER AT SURFACES: METALS AND INSULATORS

The surface space charge or surface dipole presents an electric field that influences atoms or molecules that may adsorb. Even inert gas atoms such as argon or xenon respond to this field upon adsorption because of their polarizability. The magnitude of the polarizability depends on the volume occupied by the electrons, and it increases with increasing atomic volume. The charge distribution of these atoms will be modified slightly in a way that lowers the work function of the adsorbing metal. Typical values of metal-work-function changes upon adsorption of a monolayer of inert gas atoms are shown in Table 5.3. The work-function change is also site-dependent, as shown for xenon at a stepped Pd(810) surface in Figure 5.6. Because the work-function change upon adsorption is greater at steps, it provides information about surface structure. Changes in work function can often be calibrated to monitor the coverage of adsorbates. The relationship between work function and surface composition is complicated, however, because of the adsorbate–adsorbate interaction, which also influences adsorption-induced changes of the surface dipole.

The opposite situation from weak interaction of inert gases with the surface space charge is surface ionization, when the adsorbate is ionized by the substrate. This typically occurs in alkali-metal adsorption on transition-metal surfaces. In the more usual situation with chemisorbed molecules, only partial charge transfer occurs to or from the substrate to the molecule. If the negative pole of the molecule points toward the vacuum, the induced electric fields cause an increase in the work function. Table 5.4 lists the work-function changes obtained by the chemisorption of several molecules on rhodium.

Let us discuss the model that is commonly used to interpret the chemisorption-induced changes in work function, along with some of the experimental results.

Chemisorption of an atom or a molecule leads to charge transfer to or from the metal. This charge transfer is larger than it is for inert gas adsorption. The charge transfer results in a larger change of work function, with the magnitude of the change depending on the nature of the adsorbate–substrate bond and on the coverage. If we consider well-separated chemisorbed species with surface concentration σ and po-

TABLE 5.3. Sign of Work-Function Change, $\Delta\phi$, Upon Adsorption at 300 K

System	Sign of Work-Function Change, $\Delta\phi$	Reference
O/Ag(111)	Positive	1
O/Cu(100)	Positive	2
O/Cu(110)	Positive	2
O/Cu(111)	Positive	2
O/Ni(110)	Positive	3
CO/Cu(100)	Negative	4
CO/Mn	Positive	5
CO/Ni(111)	Positive	6
H/Mo(001)	Positive	7
H/Mo(011)	Positive	7
H/Mo(111)	Positive	7
H/Ni(110)	Positive	8
H/W(100)	Positive	9
Cs/Ni(100)	Negative	10
Cs/W(100)	Negative	11
K/Ni(110)	Negative	12
Xe/Pd(100)	Negative	13
Substitute/Pt(111)	Negative	14
Aromatic molecules/Pt(100)	Negative	12
Cl/W(100)	Positive	15
Cl/$_\beta$Ti	Positive	16
Ba/W(100)	Negative	17
CH_4/W(110)	Positive	18
Na/W(110)	Negative	19
Li/W(110)	Negative	20

REFERENCES

1. A.W. Dweydari and C.B.H. Mee. *Phys. Status Solidi A* **17**:247 (1973).

2. T.A. Delchar. *Surf. Sci.* **27**:11 (1971).

3. J. Küppers. *Vacuum* **21**:393 (1971).

4. J.C. Tracy. *J. Chem. Phys.* **56**:2748 (1972).

5. G.H. Hall and C.H.B. Mee. *Phys. Status Solidi A* **12**:509 (1972).

6. K. Christmann, Ol Schober, and G. Ertl. *J. Chem. Phys.* **60**:4719 (1974).

7. E. Chrzanowski. *Acta Phys. Pol.* **A44**:711 (1973).

8. T.N. Taylor and P. J. Estrup. *J. Vacuum Sci. Technol.* **11**:244 (1974).

9. C.A. Papageorgopoulos and J.M. Chen. *Surf. Sci.* **39**:283 (1973).

10. C.A. Papageorgopoulos and J.M. Chen. *Surf. Sci.* **52**:40 (1975).

11. T.J. Lee, B.H. Blott, and B.J. Hopkins. *J. Phys. F* **1**:309 (1971).

12. R.L. Gerlach and T.N. Rhodin. *Surf. Sci.* **19**:403 (1970).

13. P.W. Palmberg. *Surf. Sci.* **25**:598 (1971).

14. J.L. Gland and G.A. Somorjai. *Surf. Sci.* **41**:387 (1974).

15. D.L. Fehrs. *Surf. Sci.* **17**:298 (1969).

16. J.R. Anderson and N. Thompson. *Surf. Sci.* **28**:84 (1971).

17. Yu.S. Vedula, Yu.M. Konoplev, V.K. Medvedev, A.G. Naumovets, T.P. Smereka, and A.G. Fedorous. *Materials of the 3rd International Conference on Thermionic Electrical Power Generation*, Jülich, Germany 1972.

18. S. Hellwig and J.H. Block. *Surf. Sci.* **19**:523 (1972).

19. E.V. Klimenko and V.K. Medvedev. *Sov. Phys. Solid State* **10**:1562 (1969); *Fiz. Tverd. Tela* **10**:1986 (1968).

20. V.K. Medvedev and T.P. Smereka. *Sov. Phys. Sol. Stat.* **16**:1046 (1974); *Fiz. Tverd. Tela* **16**:1599 (1974).

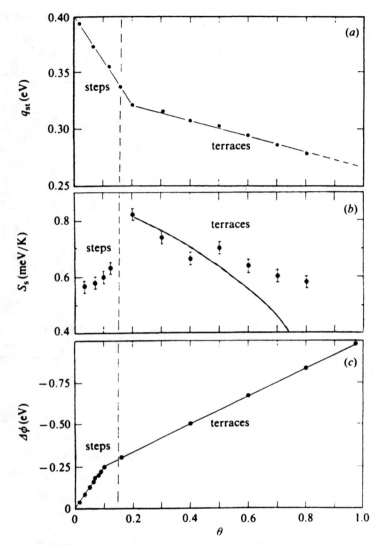

Figure 5.6. The enthalpy change (a), the entropy (b), and the work function change (c) for xenon adsorption as a function of xenon coverage on the Pd(810) stepped crystal surface [53].

larizability α, then the work-function change $\Delta\phi$ is given by the Helmholtz equation:

$$\Delta\phi = 4\pi e\mu\sigma \qquad (5.8)$$

where μ is the dipole moment induced by the adsorbate that localizes a fraction of the surface space charge in the form of a charge density that screens the field due to the charged adsorbate. This screening charge and the charge on the adsorbate form a dipole $\mu = qa$, where a is the separation between the adsorbate and the screening

TABLE 5.4. Work-Function Changes Relative to the Clean Rh(111) Surface for Various Ordered Structures, with and without Coadsorbed CO

Coadsorbate	Number of CO Molecules per Coadsorbate		
	0	1	2
C_6D_6	$(2\sqrt{3} \times 3)$rect	$c(2\sqrt{3} \times 4)$rect	(3×3)
	-1.36 eV	-0.64 eV	-0.26 eV
	-2.18×10^{-19} J	-1.03×10^{-19} J	-0.42×10^{-19} J
C_6D_5F	$(\sqrt{19} \times \sqrt{19})R23.4°$	$c(2\sqrt{3} \times 4)$rect	(3×3)
	-1.24 eV	-0.61 eV	-0.24 eV
	-1.99×10^{-19} J	-0.98×10^{-19} J	-0.38×10^{-19} J
$\equiv CCH_3$	Disordered	$c(4 \times 2)$	—
	-1.23 eV	-0.32 eV	—
	-1.97×10^{-19} J	-0.51×10^{-19} J	—

charge and q is the unit charge. Equation 5.8 can be written as

$$\Delta\phi = -3.76 \times 10^{-5}\mu\sigma \qquad (5.9)$$

where $\Delta\phi$ is in eV, μ is in Debyes, and N is the number of adsorbate atoms per cm^2.

Dipoles of like orientation cause depolarization, which shows up at higher coverages that modify the Helmholtz equation according to the point-depolarization model developed by Topping:

$$\Delta\phi = -\frac{4\pi e\mu\sigma}{1 + 9\alpha\sigma^{3/2}} \qquad (5.10)$$

where μ is the initial dipole moment observed at low coverages and α is the polarizability. Equation 5.10 is given in SI units by

$$\Delta\phi = -\frac{e\sigma\mu}{\epsilon_0\left[1 + \left(\dfrac{9}{4\pi}\right)\alpha\sigma^{3/2}\right]} \qquad (5.11)$$

where ϵ_0 is the vacuum permittivity.

Carbon monoxide increases the work function of rhodium upon chemisorption. Figure 5.7 shows the work-function change as a function of coverage θ. A good fit can be obtained for $\theta \leq 0.33$ (shown by the dashed curve in Figure 5.7) with $\mu_{CO} = -0.2$ Debye (-0.67×10^{-30} C $\cdot$ m) and $\alpha_{CO} = 0.34 \times 10^{-28}$ m^3.

For $\theta_{CO} > 0.33$, $\Delta\phi$ increases dramatically until reaching a value of $+1.05$ eV (1.7×10^{-19} J) and $\theta_{CO} = 0.75$, which is near the saturation coverage. Above $\theta_{CO} = 0.33$, the CO molecules begin to occupy bridge sites (bridging two rhodium atoms) in addition to top sites. The large increase of $\Delta\phi$ above one-third monolayer coverage is attributed to the fact that bridge-bonded CO has a larger surface-dipole moment than does top-site-bonded CO. Thus, changes in bonding that occur with changes in surface concentration have a strong effect on the work-function change.

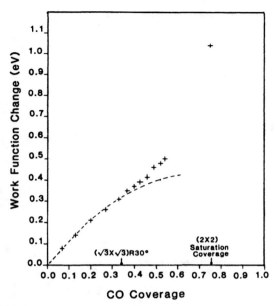

Figure 5.7. The change of work function of Rh(111) upon CO adsorption as a function of coverage [54].

Figure 5.8 shows the work-function change as associated with sodium chemisorption on the (111) crystal face of rhodium. The 5.4-eV (8.7×10^{-19}-J) work function of the transition metal decreases rapidly to 2.5 eV at $\theta_{Na} \approx 0.2$. Using Eq. 5.11, we obtain a surface-dipole moment of $\mu_{Na} = +5.1$ Debye (1.7×10^{-30} C · m) and polarizability $\alpha_{Na} = 2.9 \times 10^{-28}$ m^3 for low sodium coverages. Similar values have been observed for the chemisorption of alkali adatoms on other transition-metal surfaces as well.

The large charge transfer and the high heat of adsorption [≈ 60 kcal/mole (≈ 251 kJ/mole)] associated with the initial stages of alkali-metal chemisorption indicate ionization of the alkali atoms. At higher coverages of alkali metal, however, further work-function change becomes minimal, and the heat of adsorption declines to that of the heat of sublimation of sodium [23 kcal/mole (96 kJ/mole)]. Detailed surface studies on several alkali-metal–transition-metal systems reveal that, at above 20% coverage, the repulsive interaction between the dipoles created by the presence of the alkali ions leads to depolarization and neutralization until a metallic alkali atomic layer is produced.

The chemisorption of organic molecules on transition metals usually reduces their work function. According to the available experimental data, the chemisorption of ethylene in the form of ethylidyne reduces the work function by about -1.2 eV (-1.9×10^{-19} J), corresponding to the formation of a surface dipole of $+0.9$ Debye (3×10^{-30} C · m). Benzene chemisorption reduces the work function by -1.4 eV (-2.2×10^{-19} J), corresponding to a dipole of ($+2.0$ Debye) (6.7×10^{-30} C · m).

Gas adsorption on insulator or semiconductor surfaces can cause very large changes in the height of the surface space-charge potential and its Debye length. As

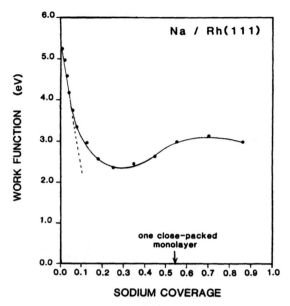

Figure 5.8. The change of work function of Rh(111) upon adsorption of sodium as a function of Na coverage [54].

a result, when used as thin-film adsorbers, the electrical conductivity of these surfaces can be markedly altered. This effect provides a way to detect minute amounts of gases or vapors, such as methane in coal mines or smoke induced by fires. Tin oxide (SnO_2) appears to be the semiconductor of choice for many gas-sensor applications because of its chemical resilience [2, 3]. The thickness of the SnO_2 film controls its sensitivity to adsorbates by adsorption-induced charge transfer, which shows up in changes in electrical conductivity.

The co-deposition of transition metals can enhance chemical reactivity and further increase the sensitivity of the semiconductor detectors. Chemisorption-induced changes in surface electrical properties promise to be important in the chemical analysis of blood and in other biochemical applications.

Adsorption-induced charge transfer also markedly influences the heat of adsorption. A molecule that transfers more charges to or from a surface adsorbs more strongly. This gives rise to the possibility of separating mixtures of molecules (gases or liquids) by virtue of stronger adsorption of those with larger charge transfer. Usually molecules with lower ionization potentials or larger electron affinities are likely to transfer more charge and thus are likely to adsorb more strongly. The rate of charge transfer dn_s/dt to or from the molecule is related to the height of the surface space charge V_s, because the electrons must flow over the top of the potential energy barrier:

$$\frac{dn_s}{dt} \approx \exp\left(-\frac{eV_s}{k_B T}\right) \tag{5.12}$$

Because V_s is proportional to n_s^2 (assuming a parabolic drop-off of the surface barrier height toward the bulk), the rate of charge transfer can be expressed as

$$\frac{dn_s}{dt} \approx \exp\left(-n_s^2\right) \tag{5.13}$$

This type of rate law frequently describes the adsorption of oxygen on semiconductor surfaces and gives straight-line plots of current versus logarithm of time.

The magnitude of charge transfer may be one electron per 10^2 adsorbed molecules; nevertheless, even this magnitude leads to the preferential adsorption of O_2 from air (an N_2–O_2 mixture) because oxygen forms stronger charge-transfer bonds than does nitrogen. On surfaces that do not exhibit charge transfer (microporous alumina-silicate molecular sieves, for example), N_2 would adsorb more strongly than O_2 because of its larger polarizability.

5.4.1 Charge Transfer at the Solid–Solid Interface

When two different metal surfaces are brought into contact, the surface space charges that were present at their interfaces with a vacuum will be modified. The electrons from the metal of lower work function will flow into the other metal until an interface potential develops that opposes further electron flow. This is called the *contact potential* and is related to the work-function difference of the two metals. The contact potential depends not only on the materials that make up the solid–solid interface but also on the temperature. This temperature dependence is used in thermocouple applications, where the reference junction is held at one temperature while the other junction is in contact with the sample. The temperature difference induces a potential (called the *Seebeck effect*), because of electron flow from the hot to the cold junction, that can be calibrated to measure the temperature. Conversely, the application of an external potential between the two junctions can give rise to a temperature difference (Peltier effect) that can be used for heat removal (refrigeration).

Metal–semiconductor contacts play important roles in the technologies of electronic circuitry. Because electrons flow from the material with the lower work function to the material with the higher work function, a blocking contact (Schottky barrier) is produced that inhibits the further flow of electrons in one direction while aiding the flow of electrons in the other direction. Often, however, ohmic metal–semiconductor contacts are needed that permit charge transport across the interface in both directions without a barrier. This can often be accomplished by forming interface compounds between the metal and the semiconductor that can eliminate the formation of the interface space charge. For silicon technology, nickel or cobalt silicides can serve as ohmic contacts between the metal and the semiconductor.

Recently, work-function changes at semiconductor–metal interfaces have been used to image the oxide–metal interface on the atomic scale by using the scanning tunneling microscope (STM). Figure 5.9 shows a picture of the TiO_2–Rh system prepared by depositing titanium oxide islands on the transition-metal surface. The tunnel current I in the STM varies exponentially with both work function ϕ and distance d: $I \approx \exp\left(-\phi d/2\right)$. By periodically changing the tunneling tip-surface distance by oscillating the tip, the derivative dI/dd can be measured; this derivative is proportional to the barrier height, which is related to ϕ. In this way an atomic resolution of the structure of oxide–metal interfaces can be obtained.

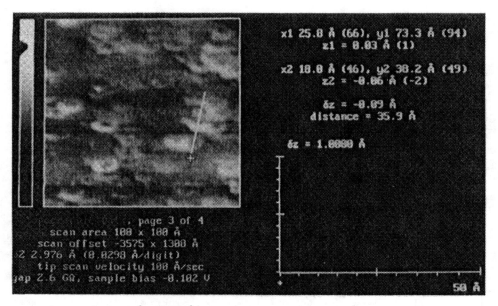

Figure 5.9. A $100\,\text{Å} \times 100\,\text{Å}$ STM image of 0.9 monolayer of TiO_2 on rhodium.

5.4.2 Gas-Phase Ion Production by Surface Ionization: Emission of Positive and Negative Ions

Consider an atom of ionization potential V_{ion} adsorbed on a metal surface of work function ϕ. If the atom is in thermal equilibrium with the solid, it may vaporize as a neutral atom from the surface after acquiring thermal energy equal to its heat of desorption from the metal ΔE_{des}. The desorption energy ΔE_{des}^{+} necessary to vaporize it as a positive ion, on the other hand, can be estimated by [4]

$$\Delta E_{des}^{+} \approx \Delta E_{des} + V_{ion} - \phi \tag{5.14}$$

The value of ΔE_{des}^{+} is obtained by summing the energies needed to vaporize a neutral atom, ionize it in the vapor phase, and then return the electron to the metal surface. If $V_{ion} - \phi$ is positive, the surface atoms are likely to desorb as neutral species, since $\Delta E_{des} \leq \Delta E_{des}^{+}$. However, for systems in which the metal work function is greater than the ionization potential of the adsorbing atom (i.e., if $V_{ion} - \phi \leq 0$), the vaporization of ionic species will occur preferentially. Thus, for studies of surface ionization, high-work-function metals (W, Pt) and adsorbates with low ionization potentials (Cs, Rb, K) are used.

The degree of ionization—the ratio of ion flux j_+ to the flux of neutral atoms j_0 desorbing from the metal surface—is given by the Saha-Langmuir equation [5]:

$$\frac{j_+}{j_0} = \frac{g_+}{g_0} \exp\left[-\frac{e(V_{ion} - \phi)}{k_B T} \right] \tag{5.15}$$

TABLE 5.5. Calculated Values for the Degree of Ionization for Different Alkali Metals on a Tungsten Surface at Different Temperatures

T (K)	Li (V_{ion} = 5.40 eV = 8.65 × 10^{-19} J)	Na (V_{ion} = 5.12 eV = 8.20 × 10^{-19} J)	K (V_{ion} = 4.32 eV = 6.92 × 10^{-19} J)	Rb (V_{ion} = 4.10 eV = 6.57 × 10^{-19} J)	Cs (V_{ion} = 3.88 eV = 6.22 × 10^{-19} J)
1000	1.8×10^{-5}	5.0×10^{-4}	6.3	103.9	790.0
1500	5.5×10^{-4}	5.0×10^{-3}	2.2	35.8	72.0
2000	3.0×10^{-3}	1.5×10^{-2}	1.6	11.4	19.9
2500	8.4×10^{-3}	3.2×10^{-2}	1.3	7.0	9.8

where g_+/g_0 is the ratio of the statistical weights of the ionic and atomic states. Table 5.5 shows the calculated values for the degree of ionization for different alkali metals on a tungsten surface at different temperatures. We can see that elements with small ionization potentials (Cs and Rb) yield predominantly ion fluxes, while for elements with large ionization potentials, such as lithium and sodium fluxes, the neutral species predominate.

Equation 5.15 has been verified by experiments using several different metal surfaces. Deviations from the predicted ion flux are due to the presence of impurities on the metal surface that may change its work function, and the fact that thermal equilibrium may not be completely established between the adsorbate and the surface within its residence time on the metal [4]. This latter effect can give rise to a partial reflection of the incident vapor atoms as neutral species, thereby reducing the ion flux to below a value predicted by Eq. 5.15. The surface temperature in surface-ionization experiments should be high enough so that thermal desorption of the adsorbed species can take place rapidly. Otherwise, accumulation of the adsorbate on the surface would impede the surface-ionization reaction by reducing the concentration of surface sites on which ionization can take place and by decreasing the work function of the clean surface.

Because a metal surface is heterogeneous, there are local variations of the work function along the crystal surface. For a polycrystalline substrate that exposes many crystal faces, the work function changes from crystal face to crystal face. Therefore, it is advantageous in surface-ionization experiments to establish conditions that allow surface diffusion of the adsorbed species to occur. In this way the ionization probability may be increased.

Alkali metals are not the only alkali species that have been ionized by surface ionization. Alkali halides (NaCl, LiF, etc.) and alkali-earth metal atoms (Ba, Mg, etc.) have also been ionized by this method. Tungsten and platinum surfaces are used most frequently in these studies.

The emission of negative ions has also been observed under conditions of surface ionization. If the electron affinity S_e of a negative ion is defined by the reaction $A^- \rightarrow A + e$, the desorption energy of negative ions, ΔE_{des}^- can be estimated by [4]

$$\Delta E_{des}^- \approx \Delta E_{des} - S_e + \phi \qquad (5.16)$$

Here we form the negative ion by vaporizing a neutral atom and an electron from the surface and then combining them in the vapor. The electron affinities of several

TABLE 5.6. Electron Affinities of Several Elements Which Exhibit the Largest Positive Values

	F	Cl	Br	I	O	S
Electron affinity (eV):	3.6	3.7	3.5	3.2	3.1 (2.3)	2.4
Electron affinity (10^{-19} J):	5.77	5.93	5.61	5.13	4.97 (3.69)	3.85

elements that exhibit the largest positive values of S_e are shown in Table 5.6. For many elements, however, the electron affinity is negative; these elements are not likely candidates for negative surface ionization. If $(S_e - \phi) > 0$—that is, if the electron affinity is greater than the work function (which is always positive)—the atoms adsorbed on the metal surface are most likely to desorb as negative ions. The degree of ionization is given by

$$\frac{j_-}{j_0} = \frac{g_-}{g_0} \exp\left[-\frac{e(-S_e + \phi)}{k_B T} \right] \tag{5.17}$$

which is similar to Eq. 5.15. Negative-ion emission requires metal surfaces with relatively low work functions. This negative surface-ionization process has been studied to a lesser extent that positive-ion emission. These studies should be somewhat more difficult to carry out, because the negative-ion flux and the flux of electrons that may be emitted thermally from the surface at the same time would have to be separated and identified.

5.5 SURFACE ELECTRON DENSITY OF STATES

When X-ray photons of energy $h\nu$ impinge on a solid, electrons are emitted from those occupied electronic states where the electron-binding energies E_B are less than the energy of the incident photons minus the work function. The kinetic energy E_{kin} of the emitted so-called photoelectrons is related to the binding energy of electrons in the states they occupied by

$$E_{kin} = h\nu - E_B - \phi \tag{5.18}$$

If the ejection probability is the same for all electron states, the intensity distribution of photoemitted electrons as a function of their kinetic energy provides a true image of the occupied electronic density of states (number of electrons with a given binding energy in the range E_B to $E_B + \Delta E$). Figure 5.10 shows the electron density of states for nickel and copper determined in this way.

Because the X-ray photons may eject electrons from a depth of over 10 atomic layers, mainly the bulk density of states is obtained in this way. The electron density of states for surface atoms should be different from that in the bulk because the bonding environment for surface atoms is different in their number of nearest neighbors, relaxation or reconstruction, and anisotropy of bonding that could give rise to

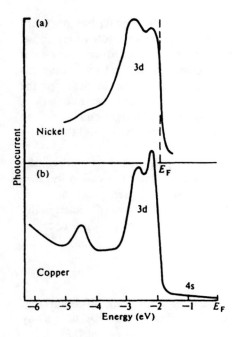

Figure 5.10. Electron density of states for nickel and copper [55].

new electronic states, called *surface states*. It turns out that by observing the photo-emitted electrons at near-grazing exit angles or by using lower-energy photons in the ultraviolet range [≈ 22 eV] (3.5×10^{-18} J) (ultraviolet photoelectron spectroscopy), the photoemission becomes much more surface-sensitive [6]. Experiments of this type yield the surface electron density of states, which has been found to be narrower than for the bulk, both by experiment and by calculation. In addition, new features, electrons that are localized at surface atoms in bound states, appear due to surface states. Often electrons in these states participate in bonding with adsorbed atoms or molecules. These measurements yield the electronic structure of the surface atoms. Many of the physical and chemical properties of surfaces depend on this property. These include electron transport along the surface, and bonding of atoms and molecules are controlled by the surface electronic structure.

5.6 ELECTRON EXCITATION AT SURFACES

When an electron in an atom or molecule is excited into an unoccupied bound state, it leaves an electron vacancy, or hole, behind. The electron–hole pair thus created exhibits a Coulomb attraction that is modified by the screening of all the other electrons in the system. The same phenomenon occurs when an electron is excited—by light (or electron beam) of appropriate energy—into a bound state above the Fermi level in an semiconductor. The electron–hole pair created in this circumstance is called an *exciton*, and its attractive Coulomb interaction is screened by the static dielectric constant of the solid. There is a finite probability that the exciton may migrate from atom to atom (or molecule to molecule) through the solid before deexcitation, the destruction of the electron–hole pair by recombination, occurs. Exciton hopping (that is, the correlated migration of the electron–hole pair) has been observed in molecular solids. In semiconductors, the electrons and holes may move independently under the influence of an applied potential that overcomes their screened Coulomb attraction. When electron–hole pairs are created by photoexcitation, photocurrents can be observed that are proportional to the intensity of the photon flux. The deexcitation process leading to electron–hole recombination, whether excited by photons or electrons, is often associated with light emission. The light emission is used in such applications as television screens or light-emitting diodes.

In a metal, the superposition of many electron–hole pairs leads to a wave-like disturbance of the charge density at the surface. This disturbance is called the *surface plasmon*. Its frequency is related to the bulk plasma frequency ω_b: as $\omega_s = \omega_b / \sqrt{2}$. The existence of both surface and bulk plasma excitation was detected under conditions of electron-beam or photon excitation, and their corresponding energies are in the range of 5–20 eV (8–32×10^{-19} J).

5.6.1   Thermal Emission of Electrons from Surfaces

When a metal or an oxide filament is heated in vacuum, electrons boil off its surface. These electrons can be collected on a positive charged plate a short distance away or can be focused by charged plates. This phenomenon is often called *thermionic emission* and is often used to produce electron beams. The electrons that require the least amount of thermal energy to overcome their binding energy in the solid and evaporate are at the top of the valence band. The energy distribution of these electrons can be approximated by a Boltzmann distribution.

$$f(E) \approx \exp\left(-\frac{\phi}{k_B T}\right) \qquad (5.19)$$

where ϕ is the work function.* One can compute the flux at energy $E > \phi + E_F$ of electrons leaving the metal at any one temperature. This value will give us the current density j (A/cm^2):

$$j\left(\frac{\text{A}}{\text{cm}^2}\right) = en(E)v_z \qquad (5.20)$$

---

*Electrons in a solid obey Fermi–Dirac statistics,

$$f(E) = \frac{1}{1 + \exp\left[(E - E_F)/k_B T\right]}$$

where $f(E)$ gives the probability that a state of energy E will be occupied in thermal equilibrium. E_F is the chemical potential (or Fermi level) and is defined as the energy of the topmost filled electron state at absolute zero temperature. For the high-energy tail of the distribution, we have $(E - E_F) > k_B T$. Since, under these conditions, the exponential term is dominant, the unity in the denominator can be neglected, and we have essentially the Boltzmann distribution,

$$f(E) = \exp\left[-(E - E_F)/k_B T\right]$$

TABLE 5.7. Thermionic Work Functions of Several Metals

Metal	Temperature (K) for 1.3×10^{-5} Pa Vapor Pressure	Richardson Constants $(10^{-4}$ A$/$m^{2}K$^2)$	Φ (eV)	Φ $(10^{-19}$ J$)$
Cs	273	160	1.81	2.90
Ba	580	60	2.11	3.38
Ni	1270	60	4.1	6.57
Pt	1650	170	5.4	8.65
Mo	1970	55	4.15	6.65
C	2030	48	4.35	6.97
Ta	2370	60	4.10	6.57
W	2520	80	4.54	7.27
Th	1800		2.7	4.33
Zr	1800		3.1	4.97

where $n(E)$ is the concentration of high-energy electrons, v_z is their velocity normal to the surface, and e is the unit charge. After integration over the Boltzmann distribution (treating the electrons in the metal as an electron gas) between the limits of E and ∞ in the z direction and between $-\infty$ and ∞ in the x and y directions, we have

$$j\left(\frac{\text{A}}{\text{cm}^2}\right) = \frac{1}{2}\, eN_0 \left(\frac{2k_BT}{\pi m}\right)^{1/2} \exp\left(-\frac{\phi}{k_BT}\right) \tag{5.21}$$

where m is the electron mass and N_0 is the density of states that gives the number of electron states per unit volume. It is given by $N_0 = 2(2\pi m k_B T/h^2)^{3/2}$. Substitution of N_0 into Eq. 5.21 gives

$$j\left(\frac{\text{A}}{\text{cm}^2}\right) = AT^2 \exp\left(-\frac{\phi}{k_BT}\right) \tag{5.22}$$

where $A = 4\pi emk_b^2/h^3 = 120$ A$/$cm^{2}K^2. This is the well-known Richardson–Dushman equation.

The electron flux leaving the surface increases with increasing temperature and decreasing work function. Thermionic emission is the method used most frequently to produce electron beams.* Table 5.7 gives the thermionic work function of several materials. Barium and its compounds (oxide and silicate) and cesium are used most frequently as cold cathodes, since large electron currents may be obtained from their surfaces even at low temperatures because of their work functions.

---

*More in-depth discussions of the uses of thermionic emission and the experimental variations in cathode emission are given in references [7–11].

5.7 ELECTRON EMISSION FROM SURFACES BY INCIDENT ELECTRON OR PHOTON BEAMS

Electron emission from surfaces induced by an electron or photon beam is one of the most successful means of learning about structure, composition, and bonding at surfaces on the atomic level. Electrons can be emitted readily from a solid by incident photons or electrons with energies of the emitted electron greater than the work function (a few electron volts). When electrons are emitted at low kinetic energies, they come only from surface atoms, because electrons emitted from atomic layers under the topmost layer lose their energy by collisions before exiting the surface. Figure 5.11 shows the mean free path of electrons for inelastic scattering in solids as a function of the kinetic energy of the emitted electron. The curve shown in Figure 5.11, which is often called the *universal curve* (because it is applicable to most solids), exhibits a broad minimum in the energy range between 10 and 500 eV ($1.6 \times 10^{-18} \rightarrow 8 \times 10^{-17}$ J), with the corresponding mean free paths on the order of 4–20 Å (0.4–2 nm).

Let a monochromatic beam of electrons of energy E_p (the primary electron beam) strike a solid surface. A typical plot of the number of scattered electrons $N(E)$ as a function of their kinetic energy E is shown in Figure 5.12. The $N(E)$ versus E curve shows a broad peak at low energies, due to secondary electrons created as a result of inelastic collisions between the incident electrons and the electrons bound to the solid. Thus one incident electron may cause the emission of several low-energy electrons from the solid.

Some of the electrons elastically back-scatter with energy E_p. These electrons will back-diffract from the surface if their de Broglie wavelength,

$$\lambda(\text{nm}) = \sqrt{\frac{1.5}{E(\text{eV})}} \qquad \lambda(\text{Å}) = \sqrt{\frac{150}{E(\text{eV})}}$$

is smaller than or equal to the interatomic distance. This occurs in the 10–500 eV [1.6–80.1×10^{-18} J] range, corresponding to $\lambda = 3.9$ Å (0.39 nm) and 0.64 Å (0.064 nm), respectively.

Low-energy electron diffraction (LEED) has proved to be a powerful tool for providing information about periodic surface structures. Small energy losses [in the meV (10^{-22} J) range] by the incident electrons provide the energy to excite vibrations and produce vibrational spectra of adsorbed atoms and molecules (high-resolution electron-energy-loss spectroscopy). Energy losses in the 1–20 eV (1.6–32×10^{-18} J) range are caused by electronic excitations such as plasma excitation or electron excitation of adsorbed species. Higher-energy electrons cause electron emission from inner shells of surface atoms. The deexcitation processes that follow lead to Auger electron emission and X-ray fluorescence. Both of these processes provide information about surface composition, since the energies of the emitted electrons or photons identify the emitting atom.

Photon beams incident on surfaces induce vibrational excitation of adsorbed molecules at low energies [in the meV ($\approx 10^{-22}$ J) range]. Photoemission of electrons from the valence band [the 5–30 eV (8–48×10^{-19} J) range] yields the surface density of electronic states. Photoemission of electrons from inner shells (30–104

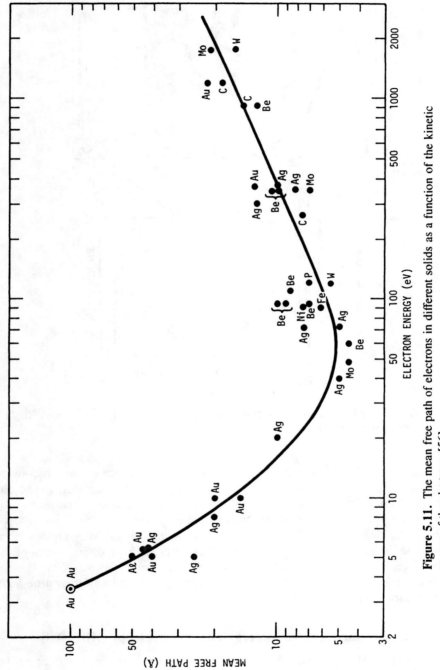

Figure 5.11. The mean free path of electrons in different solids as a function of the kinetic energy of the electrons [56].

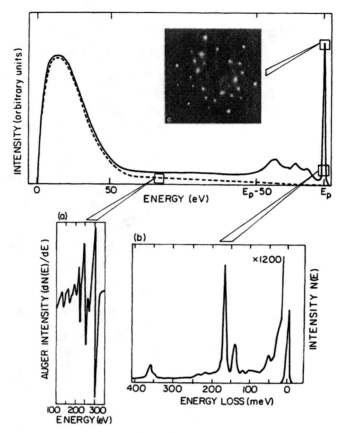

Figure 5.12. Energy distribution of scattered electrons from an ethylene-covered Rh(111) crystal surface at 300 K [57].

eV $\approx 4.8 \times 10^{-18} = 1.6 \times 10^{-15}$ J) is used in photoelectron spectroscopy to identify the surface composition and the oxidation states of surface atoms. The diffraction of X-ray photoelectrons yields information about surface structure.

Many of the processes that occur during electron (photon)–surface interactions form the foundation of the various techniques of surface analysis. They are described briefly in the list of techniques discussed in Chapter 1. Here we discuss three of the spectroscopies—high-resolution electron-energy-loss spectroscopy, X-ray photoelectron spectroscopy, and Auger electron spectroscopy—because of their prominent roles in surface chemistry.

5.7.1 High-Resolution Electron-Energy-Loss Spectroscopy (HREELS)

In HREELS [12–16], an electron beam of 5–20 eV ($\approx 10^{-19}$ J) energy strikes a solid surface. The back-reflected electrons' energies can be measured with an energy resolution of about 5 meV (40 cm^{-1} $\approx 8 \times 10^{-22}$ J), which is about an order of magnitude better than the energy resolution used in other electron spectroscopies. (This is the reason for the name, although photon spectroscopies have much higher energy resolutions.) This highly monochromatic beam, upon incidence, excites the various chemical bonds (M—H, M—O, C—H, or C—C, where M is the substrate atom). The frequency modes of these chemical bonds are in the 500–2500 cm^{-1} (1–5 × 10^{-20} J) range. A typical vibrational spectrum of an organic molecule, o-xylene, on the Rh(111) crystal face is shown in Figure 4.5. The electrons are back-reflected from the surface with energies equal to $E_{reflected} = E_{incident} - E_{vibration}$, and they are detected by a suitable energy analyzer. Using HREELS, not only is hydrogen readily detectable at coverages much lower than a monolayer, but also isotope shifts due to different masses of H and D can be observed (Figure 5.13). Adsorbed species with chemical bonds perpendicular to the surface are more readily detectable than adsorbed species with chemical bonds parallel to the surface. The surface sensitivity of this technique is so high ($\approx 1\%$ of a monolayer) that the structure of the molecules

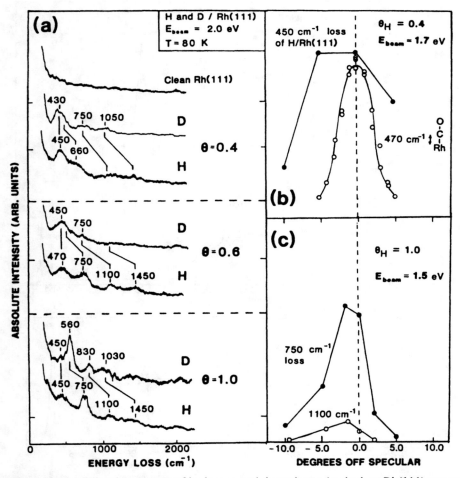

Figure 5.13. (a) Vibration spectra of hydrogen and deuterium adsorbed on Rh(111) systems at different coverages at 80 K. The data were taken by HREELS. (b) The angular distribution of the 450-cm^{-1} and 470-cm^{-1} loss peaks. (c) The angular distribution of the 750-cm^{-1} and 1100-cm^{-1} loss peaks [58].

adsorbed at the different adsorption sites can be monitored as they fill up the various sites with increasing coverages.

5.7.2   X-Ray Photoelectron Spectroscopy (XPS)

XPS [13, 14, 17–25] provides information about elemental surface composition. The principle of photoelectron spectroscopy is the excitation of electrons in an atom or molecule by means of X-rays into vacuum. The ejected photoelectrons have a kinetic energy E_{kin} equal to

$$E_{kin} \approx h\upsilon - E_B \qquad (5.23)$$

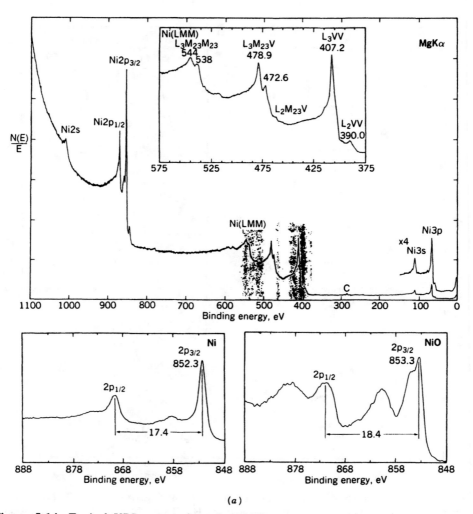

Figure 5.14. Typical XPS spectra from three different transition metals [59]: **(a)** nickel (nickel oxide is shown for comparison); **(b)** copper (copper oxide is shown for comparison); **(c)** zirconium (carbon and oxygen impurity peaks are also shown).

where $h\nu$ is the energy of the incident X-rays and E_B is the binding energy of the ejected electron. The X-ray source [26] often consists of an anode of material, usually Al or Mg. The energies of these lines are 1253.6 eV (2×10^{-16} J) for Mg, with a full width at half-maximum (FWHM) of 0.7 eV (10^{-10} J), and 1486.6 eV (2.4×10^{-16} J) for Al, with an FWHM of 0.85 eV (1.4×10^{-19} J). Typical XPS spectra are shown in Figure 5.14.

Another frequently used photon source is synchrotron radiation. When high-energy electrons are accelerated to energies of 1 to 6 GeV (1.6×10^{-14} to 10^{-12} J), electromagnetic radiation is emitted in the 10 to 10^4 eV (1.5×10^{-18} to 1.6×10^{-15} J) energy range. Continuous radiation in this energy range, which has intensities more than five orders of magnitude higher than a conventional X-ray tube, provides a powerful probe of the electronic structure of atoms and molecules.

Equation 5.23 gives a highly simplified relationship between the kinetic energy

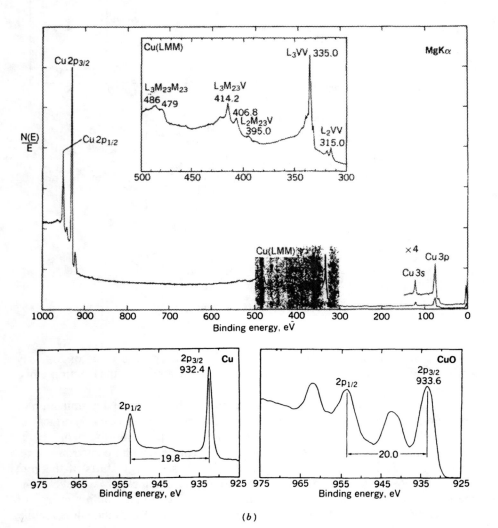

(b)

Figure 5.14. (*Continued*)

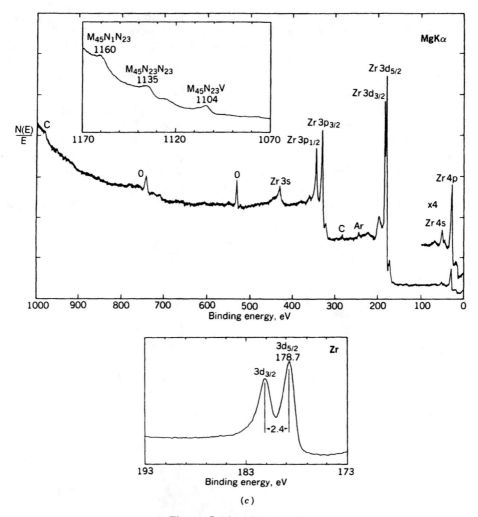

Figure 5.14. (*Continued*)

E_{kin} of the emitted photoelectrons and their binding energy. The value of E_{kin} may be modified by several atomic parameters that are associated with the electron emission process.

One of the most important additional applications of XPS is the determination of the oxidation state of elements at the surface. The electronic binding energies for inner-shell electrons shift as a result of changes in the chemical environment. An example of these shifts can be seen in nitrogen, indicating the photoelectron energy for various chemical environments (Figure 5.15). These energy shifts of the core electrons are closely related to charge transfer in the outer electronic level. The charge redistribution of valence electrons induces changes in the binding energy of the core electrons, so that information on the valence state of the element is readily obtainable. A loss of negative charge (oxidation) is usually accompanied by an increase in the binding energy E_B of the core electrons.

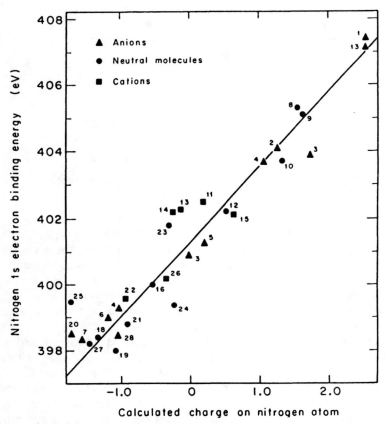

Figure 5.15. The binding energy of nitrogen $1s$ electrons as a function of the calculated charge on the nitrogen atom in different chemical environments [57].

The surface sensitivity of photoelectron spectroscopy is increased by collecting the emitted electrons that emerge at small angles to the surface plane, as was mentioned before. These electrons must travel a longer distance in the solid, and therefore they are more likely to be absorbed unless they are generated at the surface or in the near surface region.

5.7.3 Auger Electron Spectroscopy (AES)

Auger electron spectroscopy [13, 14, 19–21, 27–38] is suitable for studying the composition of solid and liquid surfaces. Its sensitivity is about 1% of a monolayer, and it may be used with relative ease, compared with several other techniques of electron spectroscopy. When an energetic beam of electrons or X-rays [1000–5000 eV (1.6–8 × 10^{-16} J)] strikes the atoms of a material, electrons which have binding energies less than the incident beam energy may be ejected from the inner atomic level. By this process a singly ionized, excited atom is created. The electron vacancy thus formed is filled by deexcitation of electrons from other electron energy states. The energy released in the resulting electronic transition can be transferred by electrostatic interaction to still another electron in the same atom or in a different atom.

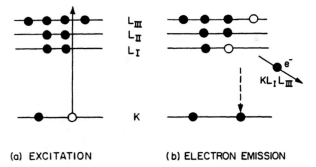

(a) EXCITATION   (b) ELECTRON EMISSION

Figure 5.16. Scheme of the Auger electron emission process.

If this electron has a binding energy that is less than the energy transferred to it from the deexcitation of the previous process (which involves filling the deep-lying electron vacancy), it will then be ejected into vacuum, leaving behind a doubly ionized atom. The electron ejected as a result of the deexcitation process is called an *Auger electron*, and its energy is primarily a function of the energy-level separations in the atom. Thus measurement of the Auger electron energy identifies the atom it comes from. These processes are schematically displayed in Figure 5.16.

Most Auger spectroscopy studies of surfaces are carried out for qualitative as well as quantitative surface chemical analysis [39]. Typical Auger spectra from alloy surfaces are shown in Figure 5.17. While the raw experimental data yield the electron intensity as a function of its energy (I versus eV), it is usually displayed as the second derivative of intensity d^2I/dV^2 as a function of electron energy eV. In this way the Auger peaks are readily separated from the background, due to other electron-loss processes that take place simultaneously.

5.8 FIELD ELECTRON EMISSION

When a potential V is applied between a metal tip and a plate, a large electrical field E can be generated at the tip because of its small curvature ($E \approx V/r$). Thus, for $V = 10^3$ volt and $r = 10^{-4}$ cm (10^{-6} m), an electrical field of 10^7 volt/cm ($\approx 10^9$ V/m) can be obtained, large enough to cause field electron emission [13, 14, 19, 21, 30, 40–42]. The applied potential reduces the barrier height, which is represented by the work function in the absence of the electrical field, according to the Fowler–Nordheim equation

$$I = AV^2 \exp\left(-\frac{b\phi^{3/2}}{V}\right) \qquad (5.24)$$

where A and b are constants for a given material. One way to observe field electron emission is by using the field electron microscope. A fine tip of radius of 10^2 nm (10^3 Å) is produced, and a potential of 1–5 kV is applied between it and a plate. (The tip is negatively charged.) The plate is covered with a phosphor that emits light in proportion to the incident electron flux. The space between the tip and the detector

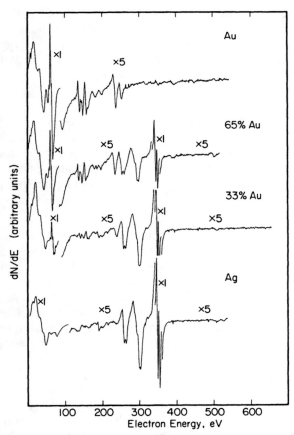

Figure 5.17. Typical AES spectra from pure gold and silver and their alloys [60].

plate is evacuated so that the electron mean free path in the partial vacuum is larger than the size of the apparatus.

5.9 FIELD IONIZATION

The large electrical field at the tip of a field electron microscope can be used to ionize gas atoms that approach it or adsorb on it [13, 14, 19, 21, 30, 40, 42, 43]. If the tip is positively charged in order to repel the positive ions formed, field ionization of an approaching atom occurs at a critical distance X [about 4–8 Å (0.4–0.8 nm)] from the tip, defined by $X \approx (V_{\text{ion}} - \phi)/E$, where V_{ion} is the ionization potential of the gas atom. The ionization probability depends strongly on the local field variations induced by the atomic structure of the surface. Therefore, field ionization microscopy (FIM) can be used to image the surface. Protruding atoms (adatoms, atoms at steps or kinks) have lower local work functions and therefore generate more ionization than atoms embedded in close-packed atomic planes. Thus, protruding atoms produce individual bright spots on the screen. The imaging of the atomic surface structure of the tip by ions occurs with very little uncertainty in the location of the site from which the ions are emitted, because the ions move very little tangential to the tip surface, especially at low temperatures ($T \approx 21$ K is often used for that reason). This small lateral movement allows a spatial resolution of 2–3 Å (0.2–0.3 nm). Small-radius tips are needed to produce the large field required for ionization, but small-radius tips also permit the immense magnification of this microscope, about 10^7-fold.

5.10   ELECTRON TUNNELING

When a sharp tip is brought within 5–10 Å (0.5–1.0 nm) of a surface and a small (≈ 10 volt) potential is applied between the two, quantum tunneling of electrons occurs. This effect produces a tunneling current I, given by the formula

$$I = CV \exp(-AVwx) \tag{5.25}$$

where C and A are constants, w is the tunnel barrier height, and x is the distance between the tip and the sample surface. (The formula is for one-dimensional tunneling and is an approximation for the tunneling tip.) The tunneling current changes exponentially with the tip-surface distance and can be stabilized by an electronic control unit. The tip can be moved along and perpendicular to the surface by piezoelectric ceramics that extend or contract by the application of suitable voltages (about 1 Å per volt). Measurement of the tunnel current as a function of atomic-scale displacement of the tip along the surface permits imaging of the surface structure.

5.10.1   The Scanning Tunneling Microscope (STM)

Electron tunneling and tip motion are the basis of the STM [14, 44–51] (Figure 5.18). Macroscopic tips always have certain whiskers of atomic dimensions due to clusters of atoms that act like minitips. The minitip that happens to be closest to the surface will draw all the tunnel current and will provide the spatial localization required for high-resolution imaging. Typical values for tunnel currents vary from 10

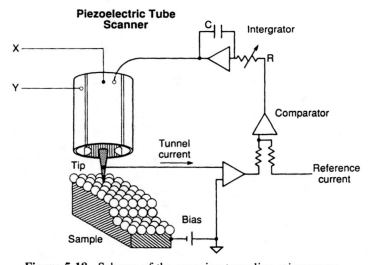

Figure 5.18. Scheme of the scanning tunneling microscope.

pA to 10 nA. An STM picture of a graphite surface at atomic resolution is shown in Figure 2.5.

The STM may be operated by moving it along the surface in the constant-height mode or in the constant-current mode (topographic mode). In the constant-height mode, the height of the tip is kept constant, and the variation in current is measured (the closer the distance, the greater the current crossing the gap). In the topographic mode, the tunneling current is kept constant by moving the tip in and out, always at a constant distance from the surface. By oscillating the tip the tunnel current can be modulated (dI/dx) to image the local barrier height. Investigations of the applied voltage dependence of the tunnel current should provide information about the local electron density of states.

Electron tunneling spectroscopy applied in a different experimental configuration can yield the vibrational structure of adsorbates. For example, by adsorbing a monolayer of molecules at an aluminum oxide–lead interface, the vibrational spectrum of benzoic acid was obtained by plotting d^2V/dI^2, the second derivative of the applied voltage with respect to the tunnel current, versus the applied voltage V. The result is shown in Figure 5.19. The experiment was performed at 4.2 K.

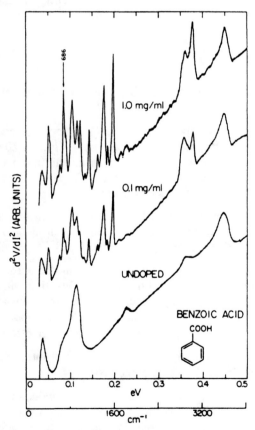

Figure 5.19. Electron tunneling spectra from aluminum oxide–lead junctions with benzoic acid adsorbate. Acid solution strength are indicated. The junction spectrum without the adsorbate is included for comparison [61].

5.11 SUMMARY AND CONCEPTS

- The discontinuity and change of dimensionality (from three to two dimensions) at solid–vacuum, solid–gas, and solid–liquid interfaces gives rise to electron redistribution. These effects result in surface space charges, surface electronic states, and work functions that are altered by changes of surface structure and adsorption.
- Charge transfer at the interface may control (a) electron transport near the surface and (b) the nature of adsorption.
- Surface ionization of adsorbates can take place under appropriate circumstances.
- Excited surface-atom vibrations induced by incident electrons have many applications in surface science.
- Electron emission from the valence band induced thermally by photons or by other electrons also has many applications in surface studies.
- The emission of inner-shell electrons from surface atoms is used for chemical analysis and determination of oxidation states.
- Electric-field-induced electron emission or tunneling and ionization of atoms at sharp tips are used to image surface atoms.

5.12 PROBLEMS

5.1 Milk and latex paint are two colloid systems. Describe the reasons for their stability [62].

5.2 When potassium is deposited on the (111) crystal face of rhodium, the work function of the metal decreases markedly. At 20% monolayer coverage, the work function change attains its minimum value of -1.8 eV. If the Rh—K interatomic distance is 1.2 Å, what is the charge transfer at the adsorption site of the alkali metal?

5.3 (a) Calculate the electron flux from a hot tungsten filament at 2200 K. (b) At what temperature would you need to operate a BaO cathode to obtain the same current? Assume $\phi_{BaO} = 1.1$ eV and $A_{BaO} = 3.5$ mA/cm^{2}K^2.

*5.4 Low-energy (5–15 eV) incident electrons can excite the vibrational modes of surface atoms and adsorbed molecules. Describe what has been learned about the (a) bonding of hydrogen atoms on rhodium and tungsten metal surfaces [63–65] and (b) the bonding of CO on rhodium and nickel [66, 67]. Discuss the meaning of dipole and impact electron scattering.

**5.5 What is the Helmholtz layer at the solid–liquid interface? Discuss the experimental evidence for its existence [68–71].

**5.6 The contact potential that develops at semiconductor–oxide/metal interfaces dramatically alters the transport of the electrons in the two directions (oxide to metal and metal to oxide). Describe how the current varies as a function in the two directions. Discuss different applications of this phenomenon [72].

**5.7   The distribution of the space charge can be calculated along the surface of an electrode of a given shape by solving the Poisson equation using appropriate boundary conditions. Review the literature and discuss how the shape of an electrode alters the space charge layer [73].

**5.8   Gas adsorption changes the electrical conductivity of oxide semiconductor thin films [2,3]. How is this phenomenon utilized to detect methane or carbon monoxide? How would charge transfer to the substrate affect the heat of adsorption of an organic molecule, and how would it affect that of oxygen and nitrogen? Could you use this phenomenon for the separation of O_2 and N_2 in air?

**5.9   Discuss the principles of Fourier transform infrared spectroscopy and non-linear laser optics [74, 75], and of second harmonic and sum frequency generation. Compare the relative advantages of both techniques for the study of the bonding, orientation, and location of adsorbed molecules either on metal or on insulator surfaces.

**5.10   Xerography is one of the dominant methods of producing copies of written material. Describe its principle of operation [76, 77]. How are colored copies made?

REFERENCES

[1]   R.L. Whetten, D.M. Cox, D.J. Trevor, and A. Kaldor. Advances in Research on Clusters of Transition Metal Atoms. *Surf. Sci.* **156**:8 (1985).

[2]   M.J. Madou and S.R. Morrison. *Chemical Sensing with Solid State Devices.* Academic Press, New York, 1989.

[3]   T. Takeuchi. Oxygen Sensors. *Sensors and Actuators* **14**:109 (1988).

[4]   M. Kaminsky. *Atomic and Ionic Impact Phenomena on Metal Surfaces.* Academic Press, New York, 1965.

[5]   M.D. Scheer and J. Fine. Positive and Negative Self-Surface Ionization of Tungsten and Rhenium. *J. Chem. Phys.* **46**:3998 (1967).

[6]   A. Zangwill. *Physics at Surfaces.* Cambridge University Press, New York, 1988.

[7]   S. Dushman. *Scientific Foundations of Vacuum Technique*, 2nd edition. John Wiley & Sons, New York, 1962. Revised by the General Electric Research Staff, J.M. Lafferty, editor.

[8]   L.M. Field. High Current Electron Guns. *Rev. Mod. Phys.* **18**:353 (1946).

[9]   J.P. Blewett. The Properties of Oxide-Coated Cathodes. I. *J. Appl. Phys.* **10**:668 (1939).

[10]   J.P. Blewett. The Properties of Oxide-Coated Cathodes. II. *J. Appl. Phys.* **10**:831 (1939).

[11]   S. Dushman. Thermionic Emission. *Rev. Mod. Phys.* **2**:281 (1930).

[12]   W. Ho. High Resolution Electron Energy Loss Spectroscopy. In: B.W. Rossiter and R.C. Baetzold, editors, *Investigations of Surfaces and Interfaces, Part A. Physical Methods of Chemistry*, Volume 9A, 2nd edition. John Wiley & Sons, New York, 1993.

[13]   D.P. Woodruff and T.A. Delchar. *Modern Techniques of Surface Science. Cambridge Solid State Science Series.* Cambridge University Press, New York, 1986.

[14] J.B. Hudson. *Surface Science: An Introduction*. Butterworth-Heinemann, New York, 1992.

[15] J.L. Erskine. High Resolution Electron Energy Loss Spectroscopy. In: J.E. Greene, editor, *Critical Reviews in Solid State and Material Sciences*, Volume 13. CRC Press, Boca Raton, FL, 1987.

[16] J.J. Pireaux, P.A. Thiry, R. Sporken, and R. Caudano. Analysis of Semiconductors and Insulators by High Resolution Electron Energy Loss Spectroscopy—Prospects for Quantification. *Surf. Interface Anal.* **15**:189 (1990).

[17] J.F. Moulder, W.F. Stickle, P.E. Sobol, and K.D. Bomben. *Handbook of X-Ray Photoelectron Spectroscopy*. Perkin Elmer, Eden Prairie, MN, 1992.

[18] M.H. Kibel. X-Ray Photoelectron Spectroscopy. In: D.J. O'Conner, B.A. Sexton, and R. St. C. Smart, editors, *Surface Analysis Methods in Materials Science. Springer Series in Surface Sciences*, Volume 23. Springer-Verlag, Berlin, 1992.

[19] M.J. Higatsberger. Solid Surfaces Analysis. In: C. Marton, editor, *Advances in Electronics and Electron Physics*, Volume 56. Academic Press, New York, 1981.

[20] R.L. Park. Core-Level Spectroscopies. In: R.L. Park and M.G. Lagally, editors, *Solid State Physics: Surfaces. Methods of Experimental Physics*, Volume 22. Academic Press, New York, 1985.

[21] M. Prutton. *Surface Physics*. Oxford University Press, New York, 1975.

[22] D.T. Clark. Structure, Bonding, and Reactivity of Polymer Surfaces Studies by Means of ESCA. In: R. Vanselow, editor, *Chemistry and Physics of Solid Surfaces*, Volume 2. CRC Press, Cleveland, OH, 1979.

[23] D. Briggs. Analytical Applications of XPS. In: C.R. Brundle and A.D. Baker, editors, *Electron Spectroscopy: Theory, Techniques and Applications*, Volume 3. Academic Press, New York, 1979.

[24] C.D. Wagner, W.M. Riggs, L.E. Davis, J.F. Moulder, and G.E. Muilenberg, editors. *Handbook of X-Ray Photoelectron Spectroscopy*. Perkin Elmer, Eden Prairie, MN, 1978.

[25] W.M. Riggs and M.J. Parker. Surface Analysis by X-Ray Photoelectron Spectroscopy. In: A.W. Czanderna, editor, *Methods of Surface Analysis. Methods and Phenomena: Their Applications in Science and Technology*, Volume 1. Elsevier, New York, 1975.

[26] P.K. Ghosh. *Introduction to Photoelectron Spectroscopy. Chemical Analysis*, Volume 67. John Wiley & Sons, New York, 1983.

[27] R. Browning. Auger Spectroscopy and Scanning Auger Microscopy. In: D.J. O'Conner, B.A. Sexton, and R. St. C. Smart, editors, *Surface Analysis Methods in Materials Science. Springer Series in Surface Sciences*, Volume 23. Springer-Verlag, Berlin, 1992.

[28] L.E. Davis, N.C. MacDonald, P.W. Palmberg, G.E. Riach, and R.E. Weber. *Handbook of Auger Electron Spectroscopy*. Perkin Elmer, Eden Prairie, MN, 1978.

[29] B.V. King. Sputter Depth Profiling. In: D.J. O'Conner, B.A. Sexton, and R. St. C. Smart, editors, *Surface Analysis Methods in Materials Science. Springer Series in Surface Sciences*, Volume 23. Springer-Verlag, Berlin, 1992.

[30] R.P.H. Gasser. *An Introduction to Chemisorption and Catalysis by Metals*. Oxford University Press, New York, 1985.

[31] R. Weissmann and K. Müller. Auger Electron Spectroscopy—A Local Probe for Solid Surfaces. *Surf. Sci. Rep.* **1**:251 (1981).

[32] G. Ertl and J. Küpper. *Low Energy Electrons and Surface Chemistry*. VCH, New York, 1985.

[33] J. Kirschner. Polarized Electrons at Surfaces. In: G. Höhler, editor, *Springer Tracts in Modern Physics*, Volume 106. Springer-Verlag, New York, 1985.

[34] G.E. McGuire and P.H. Holloway. Applications of Auger Spectroscopy in Materials Analysis. In: C.R. Brundle and A.D. Baker, editors, *Electron Spectroscopy: Theory, Techniques and Applications*, Volume 4. Academic Press, New York, 1981.

[35] J.C. Fuggle. High Resolution Auger Spectroscopy of Solids and Surfaces. In: C.R. Brundle and A.D. Baker, editors, *Electron Spectroscopy: Theory, Techniques and Applications*, Volume 4. Academic Press, New York, 1981.

[36] C.C. Chang. Analytical Auger Electron Spectroscopy. In: P.F. Kane and G.B. Larrabee, editors, *Characterization of Solid Surfaces*. Plenum Press, New York, 1974.

[37] P.M. Hall and J.M. Morabito. Compositional Depth Profiling by Auger Electron Spectroscopy. In: R. Vanselow, editor, *Chemistry and Physics of Solid Surfaces*, Volume 2. CRC Press, Cleveland, OH, 1979.

[38] A. Joshi, L.E. Davis, and P.W. Palmberg. Auger Electron Spectroscopy. In: A.W. Czanderna, editor, *Methods of Surface Analysis, Methods and Phenomena: Their Applications in Science and Technology*, Volume 1. Elsevier, New York, 1975.

[39] J.M. Slaughter, W. Weber, G. Güntherot, and C.M. Falco. Quantitative Auger and XPS Analysis of Thin Films. *MRS Bull.* **December:**39 (1992).

[40] G.A. Somorjai and M.A. Van Hove. Adsorbed Monolayers on Solid Surfaces. In: J.D. Dunitz, J.B. Goodenough, P. Hemmerich, J.A. Albers, C.K. Jørgensen, J.B. Neilands, D. Reinen, and R.J.P. Williams, editors, *Structure and Bonding*, Volume 38. Springer-Verlag, New York, 1979.

[41] R. Gomer. Recent Applications of Field Emission Microscopy. In: R. Vanselow, editor, *Chemistry and Physics of Solid Surfaces*, Volume 2. CRC Press, Cleveland, OH, 1979.

[42] J.A. Panitz. High-Field Techniques. In: R.L. Park and M.G. Lagally, editors, *Solid State Physics: Surfaces. Methods of Experimental Physics*, Volume 22. Academic Press, New York, 1985.

[43] G. Ehrlich. Wandering Surface Atoms and the Field Ion Microscope. *Phys. Today* **June:**44 (1981).

[44] A.L. de Lozanne. Scanning Tunneling Microscopy. In: B.W. Rossiter and R.C. Baetzold, editors, *Investigations of Surfaces and Interfaces, Part A. Physical Methods of Chemistry*, Volume 9A, 2nd edition, John Wiley & Sons, New York, 1993.

[45] B.A. Sexton. Scanning Tunneling Microscopy. In: D.J. O'Conner, B.A. Sexton, and R. St. C. Smart, editors, *Surface Analysis Methods in Materials Science. Springer Series in Surface Sciences*, Volume 23. Springer-Verlag, Berlin, 1992.

[46] H.J. Günterodt and R. Wiesendanger, editors. Scanning Tunneling Microscopy I: General Principles and Applications to Clean and Adsorbate Covered Surfaces. *Springer Series in Surface Sciences*, Volume 20. Springer-Verlag, Berlin, 1992.

[47] P.K. Hansma, V.B. Elings, O. Marti, and C.E. Bracker. Scanning Tunneling Microscopy and Atomic Force Microscopy: Application to Biology and Technology. *Science* **242:**157 (1988).

[48] L.E.C. van de Leemput and H. van Kempen. Scanning Tunneling Microscopy. *Rep. Prog. Phys.* **55:**1165 (1992).

[49] M. Tsukada, K. Kobayashi, N. Isshiki, and H. Kagashima. First-Principles Theory of Scanning Tunneling Microscopy. *Surf. Sci. Rep.* **13:**265 (1991).

[50] F. Ogletree and M. Salmerón. Scanning Tunneling Microscopy and the Atomic Structure of Solid Surfaces. *Prog. Solid State Chem.* **20:**235 (1990).

[51] G. Binnig and H. Rohrer. Scanning Tunneling Microscopy—from Birth to Adolescence. *Rev. Mod. Phys.* **59:**615 (1987).

[52] K. Besocke, B. Krahl-Urban, and H. Wagner. Dipole Moments Associated with Edge Atoms; a Comparative Study on Stepped Pt, Au, and W Surfaces. *Surf. Sci.* **68**:39 (1977).

[53] R. Miranda, S. Daiser, K. Wandelt, and G. Ertl. Thermodynamics of Xenon Adsorption on Pd(s)[8(100) × (110)]: From Steps to Multilayers. *Surf. Sci.* **131**:61 (1983).

[54] C.M. Mate, C.T. Kao, and G.A. Somorjai. Carbon Monoxide Induced Ordering of Adsorbates on the Rh(111) Crystal Surface: Importance of Surface Dipole Moments. *Surf. Sci.* **206**:145 (1988).

[55] J.B. Pendry. Electron Emission from Solids. In: B. Feuerbacher, B. Fitton, and R.F. Willis, editors, *Photoemission and the Electronic Properties of Surfaces*. Wiley–Interscience, New York, 1978.

[56] G.A. Somorjai. *Chemistry in Two Dimensions: Surfaces*. Cornell University Press, Ithaca, NY, 1981.

[57] S.R. Bare and G.A. Somorjai. Surface Chemistry. In: R.A. Meyers, editor, *Encyclopedia of Physical Science and Technology*. Academic Press, New York, 1987.

[58] C.M. Mate. A Molecular Surface Science Study of the Structure of Adsorbates on Surfaces: Importance to Lubrication. Ph.D. thesis, University of California, Berkeley, 1986.

[59] C.D. Wagner, W.M. Riggs, L.E. Davis, J.F. Moulder, and G.E. Muilenberg, editors. *Handbook of X-Ray Photoelectron Spectroscopy*. Perkin Elmer, Eden Prairie, MN, 1978.

[60] S.H. Overbury and G.A. Somorjai. The Surface Composition of the Silver–Gold System by Auger Electron Spectroscopy. *Surf. Sci.* **55**:209 (1976).

[61] J.D. Langan and P.K. Hansma. Can the Concentration of Surface Species be Measured with Inelastic Electron Tunneling? *Surf. Sci.* **52**:211 (1975).

[62] P.C. Hiemenz. *Principles of Colloid and Surface Chemistry*, 2nd edition. Marcel Dekker, New York, 1986.

[63] C.M. Mate and G.A. Somorjai. Delocalized Quantum Nature of Hydrogen Adsorbed on the Rh(111) Crystal Surface. *Phys. Rev. B* **34**:7417 (1986).

[64] R.F. Willis, W. Ho, and E.W. Plummer. Vibrational Excitation of Hydrogenic Modes on Tungsten by Angle Dependent Electron Energy Loss Spectrometry. *Surf. Sci.* **80**:593 (1979).

[65] H. Froitzheim, H. Ibach, and S. Lehwald. Surface Sites of H on W(100). *Phys. Rev. Lett.* **36**:1549 (1976).

[66] C.M. Mate and G.A. Somorjai. Carbon Monoxide Induced Ordering of Benzene on Pt(111) and Rh(111) Crystal Surfaces. *Surf. Sci.* **160**:542 (1985).

[67] W. Erley, H. Wagner, and H. Ibach. Adsorption Sites and Long Range Order—Vibrational Spectra for CO on Ni(111). *Surf. Sci.* **80**:612 (1979).

[68] J.K. Sass, N.V. Richardson, H. Neff, and D.K. Roe. Towards Model Systems for the Helmholtz Layer: Coadsorption of Water and Bromine on Cu(100). *Chem. Phys. Lett.* **73**:209 (1980).

[69] W.N. Hansen, C.L. Wang, and T.W. Humphries. Electrode Emersion and the Double Layer. *J. Electroanal. Chem. Interfacial Electrochem.* **90**:137 (1978).

[70] W.J. Anderson and W.N. Hansen. Observing the Electrochemical Interphase via Electrode Conductance. *Electroanal. Chem. Interfacial Electrochem.* **43**:329 (1973).

[71] R.S. Perkins and T.N. Andersen. Potentials of Zero Charge of Electrodes. In: J. O'M. Bockris and B.E. Conway, editors, *Modern Aspects of Electrochemistry*, Volume 5. Plenum Press, New York, 1969.

[72] C.A. Wert and R.M. Thomson. *Physics of Solids*, 2nd edition. McGraw–Hill, New York, 1970.

[73] R.N. Adams. *Electrochemistry at Solid Electrodes*. Marcel Dekker, New York, 1969.

[74] Y.R. Shen. *The Principles of Non-Linear Optics*. John Wiley & Sons, New York, 1984.

[75] A. Yariv. *Quantum Electronics*, 3rd edition. John Wiley & Sons, New York, 1989.

[76] D.M. Burland and L.B. Schein. Physics of Electrophotography. *Phys. Today* **May**:46 (1986).

[77] E.M. Williams. *The Physics and Technology of Xerographic Processes*. John Wiley & Sons, New York, 1984.

6

THE SURFACE CHEMICAL BOND

6.1 INTRODUCTION

We define the formation of a surface chemical bond to be adsorption accompanied by charge transfer and charge redistribution between the adsorbate and the substrate, producing strong bonds of covalent or ionic character. Heats of adsorption on the order of 63 kJ/mole (15 kcal/mole) or larger would certainly indicate the formation of a chemical bond, leading to long surface residence times τ [$\tau = \tau_0 \exp (\Delta H_{ads}/RT)$], even at elevated temperatures, compared to τ_0 ($\tau_0 \approx 10^{-12}$ sec) related to vibrational times for surface atoms.

Both surface atoms and adsorbates must participate to form the surface chemical bond. In order to determine the nature of the bond, the heat of adsorption is measured as a function of the pertinent variables. These include trends across the periodic table, variations of bond energies with adsorbate size, molecular structure and coverage, and substrate structure. Changes in the electronic and atomic structure of the bonding partners are determined and compared with their electronic and atomic (or molecular) structure before they formed the surface bond.

When a molecule from the gas phase adsorbs on a surface by forming a chemical bond, the process is similar to a stoichiometric reaction. The product—the adsorbate—may resemble the gas-phase reactant; it may have greatly rearranged its bonding and, thus, its molecular structure; or it may have even dissociated on the surface. The other reactant—the surface—may have undergone similar changes: Its structure may be altered only slightly in the presence of the adsorbed molecule, or the surface atoms may have moved to new equilibrium positions by displacement perpendicular or parallel to the surface. Thus the surface may completely restructure as the adsorbate bonds form.

Some experimental techniques [e.g., low-energy electron diffraction (LEED)–surface crystallography] can detect the structural changes that occur on both sides of the surface chemical bond. However, most currently used techniques are only capable of detecting the structural changes that occur on the adsorbate side (e.g., infrared spectroscopy) or on the substrate side (e.g., electron microscopy). As a result, we often gain only incomplete information about the surface chemical bond, leading to a one-sided "molecule-centric" or "surface-centric" view of the adsorbate–surface compound that is produced.

The rest of this chapter reviews what is known about the nature of the surface chemical bond. It will become clear that a combination of techniques, which yield diverse information on the atomic, molecular, and electronic structure of the adsorbate–substrate compound, are needed to obtain a complete physical–chemical picture of bonding at surfaces and interfaces. We will summarize the information available and present the current models of the surface chemical bond, along with the unique properties of these bonds that have been uncovered by surface-science studies.

6.2 BONDING TRENDS ACROSS THE PERIODIC TABLE

Most available data of the heats of adsorption concern atoms (e.g., hydrogen, nitrogen, oxygen, and potassium) and small molecules (e.g., N_2, O_2, CO, and CO_2) on transition-metal surfaces. Some of the data are displayed in Figures 3.21–3.23. The heat of adsorption generally increases from right to left in the periodic table. This trend has been explained by the chemisorption model developed by Nørskov [1] using the effective-medium theory, whereby the interaction of the adsorbate with the metal is primarily determined by the so-called one-electron energy term. This term is defined as due to changes in the charge distribution of the atom or molecule when it is taken from a homogeneous electron gas (uniform charge) onto the surface. The main difference in a transition-metal surface is due to the d-electrons that lie in a band around the Fermi level. These d-states can interact with the adsorbate states, and thus hybridization can occur, giving rise to bonding and antibonding shifts. The d-electron contribution to the bonding is proportional to $(1 - f_d)$, where f_d is the degree of filling of the d-band. The d-electron contribution to the surface chemical bond depends on the degree of filling of the antibonding states (since the bonding states are already filled). Thus, early transition metals with fewer d-electrons form stronger chemical bonds. This effect is shown in Figure 6.1 for hydrogen and oxygen adsorbed on the $3d$ transition metals. The effective-medium theory provides good agreement between experimental data and calculated energies. Similar trends and agreement are found for the heats of adsorption of CO and N_2 as well.

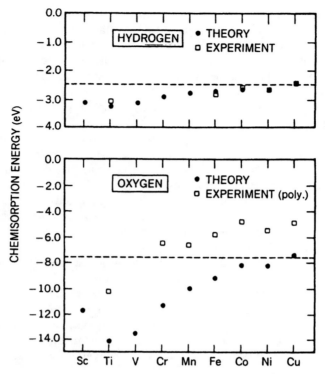

Figure 6.1. Chemisorption energies of hydrogen and oxygen on transition metals across the periodic table that were calculated using the effective-medium theory and also measured on polycrystalline surfaces [23].

It is important to note that the heats of adsorption of oxygen and hydrogen correlate well with the heats of formation of the corresponding oxides and hydrides (per metal atom) as shown by Toyoshima and Somorjai [2] (Figure 6.2). When the heat of adsorption of carbon monoxide is plotted as a function of the heat of formation of the corresponding oxide (per metal atom), two straight lines are obtained (Figure 6.3). The metals that chemisorb less strongly than iron would not readily dissociate CO, whereas those that chemisorb more strongly would dissociate the molecule, as proposed by Joyner and Roberts [3].

6.3   CLUSTER-LIKE BONDING OF MOLECULAR ADSORBATES

When ethylene chemisorbs at ≈ 300 K on the (111) crystal faces of various transition metals (Pt, Rh, Pd), it chemically rearranges to form the molecule–surface compound shown in Figure 6.4. Its structure is determined by LEED–surface crystallography and is very similar to those of the multinuclear organometallic complexes listed in Figure 6.5. The rearranged ethylene, which has also lost a hydrogen, is called *ethylidyne* and belongs to the alkylidyne group (species of the formula C_nH_{2n-1}), a common substituent in surface and in organometallic chemistry (Figures

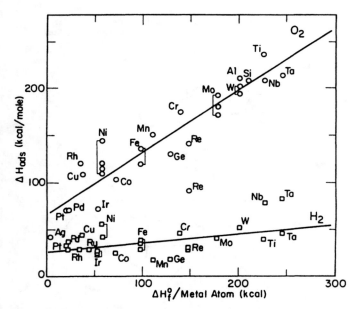

Figure 6.2. Heats of adsorption of O_2 and H_2 on various transition metals as a function of the heats of formation of the corresponding oxides and hydrides (per metal atom) [2].

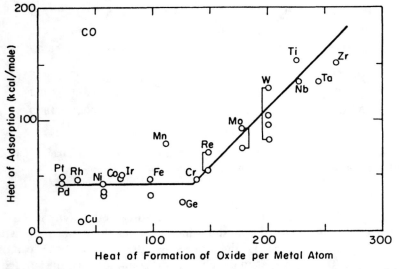

Figure 6.3. Heats of adsorption of CO on various transition metals as a function of the heats of formation of the corresponding oxides (per metal atom) [2].

Different ethylidyne species: bond distances and angles
(r_C = carbon covalent radius; r_M = bulk metal atomic radius)

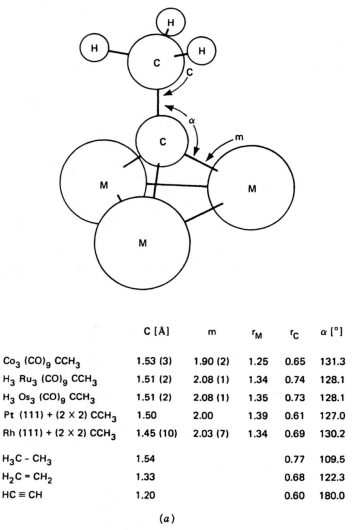

	C [Å]	m	r_M	r_C	α [°]
$Co_3 (CO)_9 CCH_3$	1.53 (3)	1.90 (2)	1.25	0.65	131.3
$H_3 Ru_3 (CO)_9 CCH_3$	1.51 (2)	2.08 (1)	1.34	0.74	128.1
$H_3 Os_3 (CO)_9 CCH_3$	1.51 (2)	2.08 (1)	1.35	0.73	128.1
Pt (111) + (2 × 2) CCH_3	1.50	2.00	1.39	0.61	127.0
Rh (111) + (2 × 2) CCH_3	1.45 (10)	2.03 (7)	1.34	0.69	130.2
$H_3C - CH_3$	1.54			0.77	109.5
$H_2C = CH_2$	1.33			0.68	122.3
$HC \equiv CH$	1.20			0.60	180.0

(a)

Figure 6.4. Alkylidyne structure of (a) ethylene [24] and (b) other alkenes on transition metal surfaces [24] and in (c) organometallic clusters [25].

6.4b and 6.4c). The vibrational spectrum of chemisorbed ethylidyne is nearly identical to that in the organometallic cluster which contains three metal atoms (Figure 6.5a). The C—C bond distance is slightly less than the single carbon—carbon bond length of 1.54 Å (0.154 nm), as in the cluster compounds. Thus, the surface chemical bond of chemisorbed ethylene can, as a first approximation, be viewed as a cluster-like bond that contains at least three metal atoms. The C—C bond order

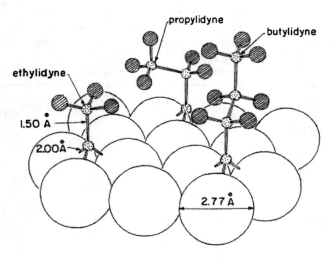

Pt (III) + ethylidyne, propylidyne and butylidyne

(b)

top site   bridge bonded

3-fold coordinated   4-fold coordinated
                           "butterfly"

(c)

Figure 6.4. (*Continued*)

present in gaseous ethylene is reduced from two to nearly one upon chemisorption. This reduction in bond orders of alkenes and alkynes upon chemisorption on metal surfaces is commonly observed, indicating charge transfer from the molecules into the metal. In fact, the metal work function usually decreases when organic molecules are adsorbed, further proving the direction and magnitude of the charge transfer as the chemisorption bonds form.

There are many chemisorbed organic groups whose surface bonding can be viewed as identical to that of organometallic clusters. Figures 6.5 and 6.6 show the equiv-

Structure			Vibrational Frequencies (cm^{-1})		
Surfaces	Proposed Surface Geometry	Cluster Analogue	Characteristic Dipole–Active Modes	Surface	Cluster
Rh(111) Pt(111) Pd(111) Ru(001) Rh(100)	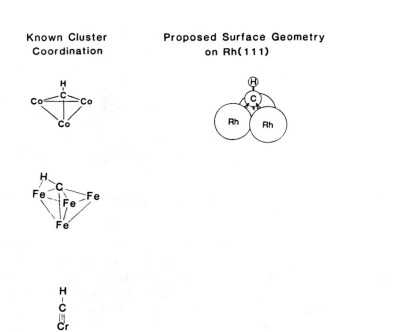		ν_s(MC) ν(CC δ_s(CH$_3$) ν_s(CH$_3$)	435 1121 1337 2880	401 1163 1356 2888
Rh(111) Ru(001) W(110)			ν_s(MC) ν(CH)	~740 2930	715 3041
gas phase			ν(CC) δ_s(CH$_3$) ν_s(CH$_3$)		995 1379 2954

(*a*)

Known Cluster
Coordination

Proposed Surface Geometry
on Rh(111)

(*b*)

Figure 6.5. (**a**) Vibrational spectra and (**b**) structure of methylidyne on surfaces and in organometallic clusters [25].

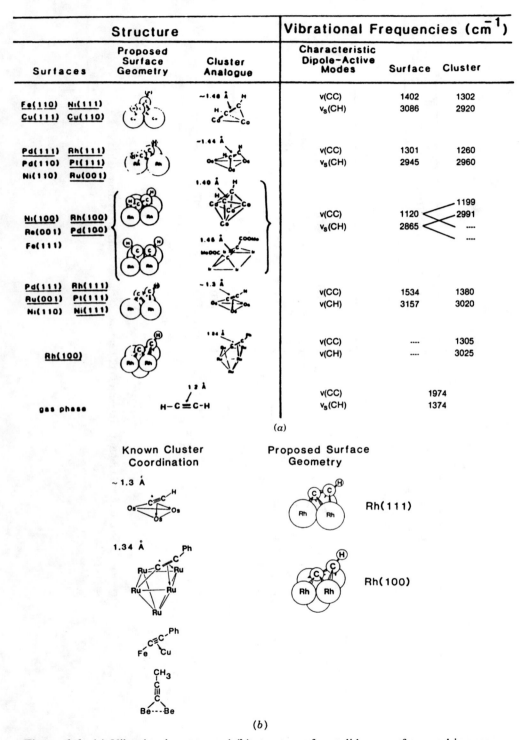

Figure 6.6. **(a)** Vibrational spectra and **(b)** structure of acetylide on surfaces and in organometallic clusters [25].

Structure			Vibrational Frequencies (cm^{-1})		
Surfaces	Proposed Surface Geometry	Cluster Analogue	Characteristic Dipole-Active Modes	Surface	Cluster
Rh(111) + CO			ν_s(CH)	3000	3098
Pt(III) Rh(111)			γ(CH)	776	817
Pd(111) Pd(100)			ν(CC)	1420	{ 1396, 1373
Ni(111) Pt(110)					
Ni(100) Ni(110)			ν_s(CH)	3059	
			γ(CH)	670	
			ν(CC)	1479	

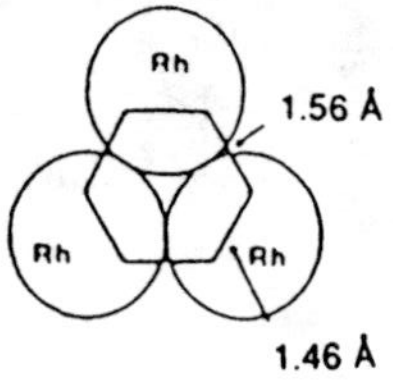

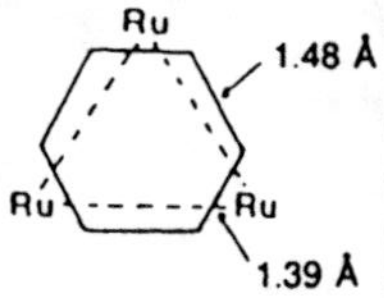

1.56 Å   1.46 Å   1.48 Å   1.39 Å   Rh   Ru   1.41 Å

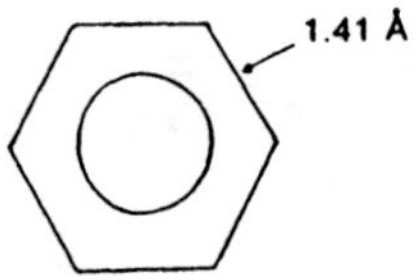

gas phase

Figure 6.7. Structure and vibrational spectrum of benzene on surfaces and in organometallic clusters [25].

alent bonding arrangements of methylidyne ($-CH$) and acetylide ($-C_2H$) groups, respectively, on surfaces and in organometallic clusters.

Benzene usually chemisorbs on metals with its ring parallel to the surface (although it may adsorb in a different configuration when it loses hydrogen). Because of charge transfer to the metal, $C-C$ bond elongations occur with respect to the gas-phase configuration, with periodic distortions of the $C-C$ distance that reflect the symmetry of the adsorption site (Figure 6.7). The ring may even bend (see Chapter 2), with two of the opposing carbon atoms closer to the metal surface than the other four carbon atoms. Distortions and elongations of $C-C$ bonds are also found when benzene is bound to clusters of metal atoms in organometallic complexes. Thus the cluster-like bonding model appears to be valid for chemisorbed benzene as well.

The bonding picture of adsorbed molecules becomes more complicated if there are more bonding sites available on the same molecule. For example, pyridine (C_5H_5N) may bind through the lone electron pairs of its nitrogen or through the π electrons of its carbon ring. Thus, depending on the metal, the binding geometry of the substrate, the temperature, or the adsorbate coverage, the molecule may be tilted with respect to the substrate surface, its ring may be parallel with it, or it may be upright with bonding solely through the nitrogen. Partial dehydrogenation can also occur (Figure 6.8).

It is too simplistic to consider that only the nearest-neighbor metal atoms of the substrate participate in the bonding. There is evidence that the atoms at next-nearest-neighbor sites change their location when chemisorption occurs, moving either closer or further away from the chemisorption bonds. This effect will be discussed in Section 6.5.

6.4   THE CARBON MONOXIDE CHEMISORPTION BOND

Carbon monoxide chemisorbed on various transition-metal surfaces is the most intensively studied of all adsorption systems. Thus it provides a model of how surface-science studies, using a combination of techniques, reveal the nature of the surface chemical bond in detail.

For instance, ultraviolet photoemission studies [using 40.8 eV ($\approx 6.5 \times 10^{-18}$ J) photons] compared the energy distribution of photoelectrons from molecular CO, CO chemisorbed on the (100) crystal face of iridium, and iridium carbonyl, $Ir_4(CO)_{12}$ [4]. The spectra reveal the concentration of electrons in the various occupied states. For the CO molecule, these are the 5σ molecular orbital, which has electrons with the lowest binding energy, and the 1π and 4σ orbitals, whose electrons can still be emitted using the applied photon energy. The photoelectron spectra from the chemisorbed CO and from the transition-metal-carbonyl cluster are very similar, indicating that the electron-energy distribution in their chemical bonds is nearly identical (Figure 6.9). Thus the surface chemical bond of CO on iridium is cluster-like. Comparing these spectra to that of molecular CO indicates that the binding energy of the 5σ electrons increases and that these states mix with the 1π states.

The 2π molecular orbitals of gas-phase CO are unoccupied. Molecular-orbital calculations for CO chemisorbed on Ni(100) indicate that the 5σ orbital electrons of CO interact with the $3d_{z^2}$ orbital of nickel and that there is back-donation to the

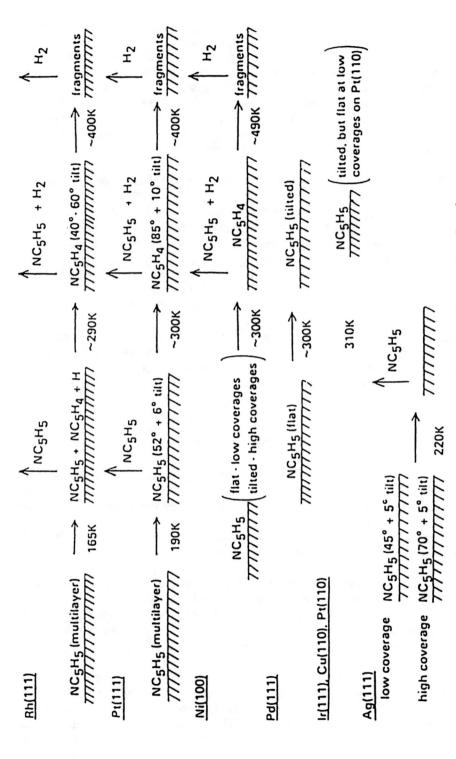

Figure 6.8. Structure and orientation of chemisorbed pyridine as a function of temperature and coverage on different transition metal surfaces [26].

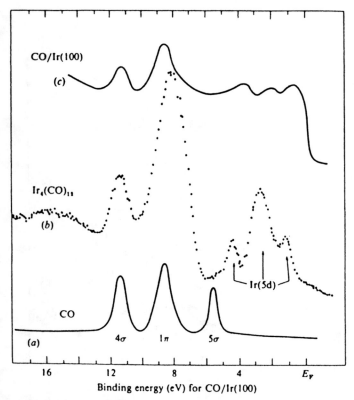

Figure 6.9. Photoelectron spectra [4] of **(a)** CO in the gas phase, **(b)** the $Ir_4(CO)_{12}$ cluster, and **(c)** CO chemisorbed on Ir(100).

antibonding 2π orbital from the filled density of electron states of the metal (Figure 6.10). These calculations have been validated by observed photoelectron spectra. Electron transfer from the filled CO molecular orbitals to the metal and from the metal to the unfilled molecular orbitals explains why the work-function change accompanying CO adsorption on most transition metals is small in magnitude, even though a strong chemisorption bond is produced.

Vibrational-spectroscopy studies indicate that the CO stretching frequency is sensitive to the symmetry of the CO adsorption site (see Table 6.1). Adsorption on a top site leaves CO with a high frequency of vibration, although about 200 cm^{-1} (4 $\times$ 10^{-21} J) lower than in the gas phase. Chemisorption in a threefold site lowers the CO vibration frequency the most, nearly to that of a C—O single bond in an alcohol or ether, for example.

Techniques that are sensitive to the orientation of the chemisorbed molecules (NEXAFS, for example) clearly show that the C=O bond, when chemisorbed on the nickel (100) surface, is perpendicular to that surface.

Higher coverages of CO lead to repulsive interactions between the coadsorbed molecules. These higher coverages (a) lower the average heat of adsorption (as shown in Figure 3.20) and (b) push the CO molecules into new adsorption sites (as shown in Figure 2.31) to maximize the distance between them. When CO is coadsorbed

Figure 6.10. Electron density contours of CO chemisorbed on the Ni(100) surface [27].

with benzene, there is an attractive CO—benzene interaction that causes (a) rotation of the benzene molecule with respect to its orientation on the metal surface without CO and (b) relocation of CO into threefold sites that it would not occupy if the benzene were not present.

Coadsorption of CO with alkali metals on transition-metal surfaces increases the heat of adsorption of the molecule.

Thus, coadsorption can change both the strength of the CO chemical bond and its chemisorption site.

6.5 ADSORBATE-INDUCED RESTRUCTURING. THE FLEXIBLE SURFACE

The chemisorption of an atom or a molecule often induces rearrangement of the substrate atoms around the adsorption site. For example, the chemisorption of carbon atoms on the nickel (100) surface occurs at the fourfold site. The nearest-neighbor nickel atoms are displaced away from the carbon, permitting it to move more into the metal surface and bond to the metal atom in the second layer [5, 6]. A small in-plane rotation of the surface nickel atoms around the carbon, shown in Figure

TABLE 6.1. CO-Induced Ordered Structures on Rh(111) Surface

Coadsorbate	LEED Structure	Number of Rh Surface Atoms per Unit Cell	Number of Coadsorbates per Unit Cell	Number of CO Molecules per Unit Cell	C–O Stretching Frequenies (cm^{-1})
—	$(\sqrt{3} \times \sqrt{3})R30°$	3	0	1	2010 (top)
	(2×2)	4	0	3	2060 (top)
					1855 (bridge)
Ethylidyne ($\equiv CCH_3$)	$c(4 \times 2)$	4	1	1	1790 (hcp hollow)
Propylidyne ($\equiv CCH_2CH_3$)	$(2\sqrt{3} \times 2\sqrt{3})$ $R30°$	12	3	1	1750
Acetylene (C_2H_2)	$c(4 \times 2)$	4	1	1	1725
Fluorobenzene (C_6H_5F)	(3×3)	9	1	2	1720
	$c(2\sqrt{3} \times 4)$rect	8	1	1	1670
Benzene (C_6H_6)	(3×3)	9	1	2	1700 (hcp hollow[a])
	$c(2\sqrt{3} \times 4)$rect	8	1	1	1655 (hcp hollow)
Sodium (Na)	$(\sqrt{3} \times 7)$rect	14	4	7	1695
	$c(4 \times 2)$	4	1	1	1410

[a] hcp hollow means that one second-layer metal atom lies below the threefold hollow site—in contrast to a fcc hollow, where no second-layer atom lies below the threefold site.

2.18, relieves the stress that would have been caused by the shortened distance between the nearest-neighbor and next-nearest-neighbor metal atoms. This massive local restructuring around the chemisorption site weakens the metal—metal bonds at the surface (an endothermic process). The formation of the strong metal—carbon bonds (an exothermic process) provides the energy needed for the restructuring of the substrate.

Another example of adsorbate-induced restructuring is sulfur chemisorption on the Fe(110) crystal face (Figure 2.19). In this case the nearest-neighbor iron atoms move closer to the sulfur chemisorption site, forming an Fe_4S-like cluster [7]. Again the weakening of the metal—metal bonds nearest to the next-nearest-neighbor bonds is more than offset by the formation of the four strong Fe—S bonds. In Figure 6.11a we show the restructuring of the Cu(110) surface induced by chemisorbed oxygen as monitored by scanning tunneling microscopy (STM).

Chemisorption-induced restructuring can be very well seen using a small metal tip and field ion microscopy. In Figure 6.11b the field ion microscope picture of a rhodium tip is shown when clean and after exposure to carbon monoxide at 420 K at low pressures ($\approx 10^{-4}$ Pa) [8]. The metal tip has been completely reshaped as a result of CO chemisorption. The tip becomes faceted and rougher, the step density is reduced, and extended low-Miller-index terraces are formed.

Rough surfaces that are also chemically active (as will be discussed later) appear to be flexible. The uncovered surface atoms move toward the bulk and to new equilibrium positions. The more open the surface, the larger the movement and the more flexible the surface atoms are. Upon chemisorption these surfaces restructure more readily. It is perhaps instructive to divide surfaces according to their flexibility as shown in Figure 6.12. Close-packed surfaces, like the face-centered cubic (fcc) (111)

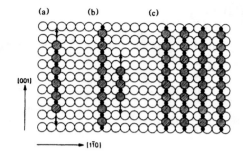

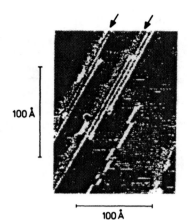

(2x1)-O/Cu(110)

STM image of (2×1)O nuclei in different growth phases: two nuclei two and three rows wide, respectively, at the upper edge of steps along [001] and a single-row nucleus on the flat terrace. Step edges are marked by arrows.

(a)

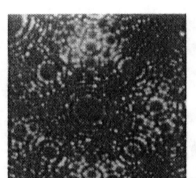

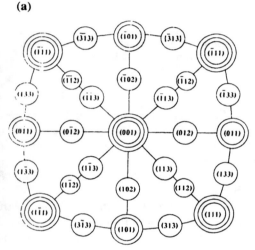

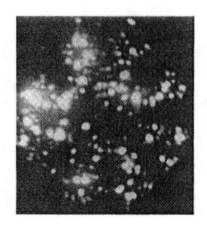

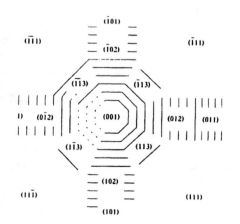

(b)

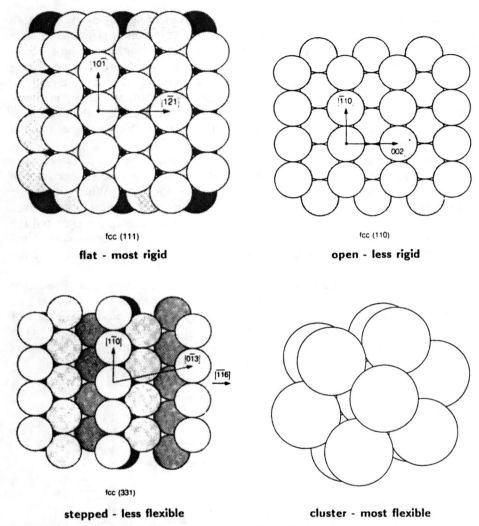

fcc (111)

flat - most rigid

fcc (110)

open - less rigid

fcc (331)

stepped - less flexible

cluster - most flexible

Figure 6.12. Models of surfaces divided according to their atom coordination. Atoms in the close-packed (111) surfaces of fcc metals have the highest coordination, their relaxation is small, and chemisorption-induced restructuring is most difficult. These we call *rigid surfaces*. Clusters have the lowest coordination accompanied by large relaxation and thermodynamically favorable chemisorption-induced restructuring; these are the most flexible. The more open fcc (110) surface and stepped surfaces show intermediate flexibility [29].

Figure 6.11. (a) Oxygen-chemisorption-induced restructuring of the Cu(110) surface monitored by scanning tunneling microscopy [28]. (b) Field ion micrographs (image gas: Ne; T = 85 K) of a (001)-oriented Rh tip before (*top left*) and after reaction with 10^{-4} Pa CO during 30 min at 420 K (*bottom left*); stereographic projections at the right demonstrate the change in the morphology from nearly hemispherical to polygonal (scheme at the bottom right indicates the coarsening of the crystal and the dissolution of a number of crystallographic planes due to the reaction with CO) [8].

crystal faces are fairly rigid because of the large number of nearest neighbors (high coordination); the atoms stay close to their bulk-like equilibrium positions in spite of the anisotropy of the surface environment. Upon chemisorption these surfaces may restructure; however, the thermodynamic driving force for such restructuring is not large. Clusters of atoms are perhaps the most flexible; the atoms are ready to relocate because of the low coordination of atoms at each surface site. Upon chemisorption, massive restructuring of these clusters may occur because the thermodynamic stability of the chemisorption bonds readily offsets the weakening of the few metal—metal bonds.

Substrate restructuring occurs during the chemisorption of molecules as well. The metal surface atoms that are "relaxed" by moving inward when the surface is clean move outward during the formation of the chemisorption bond. When ethylidyne forms on the platinum (111) surface, in addition to outward movement of the surface atoms, the nearest-neighbor metal atoms move toward the adsorption site, and the next-nearest-neighbor Pt atom moves inward, causing a slight corrugation of the surface, while the Pt atom underneath the adsorption site moves down, away from the carbon atom (Figure 6.13).

Adsorption-induced restructuring can occur on the chemisorption time scale ($\approx 10^{-15}$ sec for charge transfer or $\approx 10^{-12}$ sec for vibrational times). There is evidence, however, that adsorbate-induced restructuring can occur on the time scale of catalytic reactions (seconds). CO oxidation to CO_2 or ammonia reacting with NO to produce N_2 and H_2O show oscillatory behavior under certain circumstances of temperature and reactant partial pressures. The reaction rate alternates periodically be-

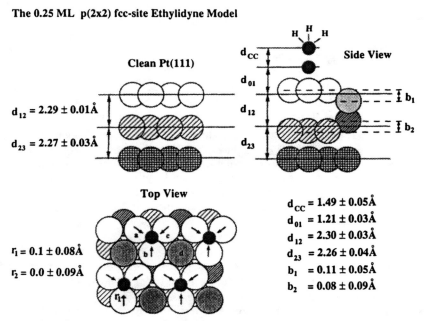

Figure 6.13. Ethylene chemisorption restructures the Pt(111) surface. The Pt atoms move inward around the bonding site, the next-nearest-neighbor metal atom moves downward, and the Rh atom in the second layer moves upward [30].

tween two values. One reason for the oscillation is the periodic restructuring of the surface. In this circumstance the sticking probability of one of the reactants is greater on one type of surface structure, while the sticking probability of the other reactant is greater on the surface structure of the other type. Thus, the reaction rate alternates between the two branches of the reaction, one taking place on the CO-covered or NO-covered metal surface, and the other taking place on the oxygen-covered or ammonia covered metal surface.

Adsorbate-induced restructuring can occur on even longer time scales (hours), involving massive restructuring of the surface by atom transport. For example, sulfur restructures the (111) crystal face of nickel until the metal surface assumes the (100) orientation. Alumina restructures iron through the formation of an iron-aluminate phase to produce (111) crystal faces during ammonia synthesis, regardless of the original crystallite orientation (see Chapter 7). In this circumstance the chemisorption-induced restructuring can be viewed as the initial phase of a solid-state reaction whose kinetics are controlled by atom transport (by diffusion).

Restructuring occurs in order to maximize the bonding and stability of the adsorbate–substrate complex. Thus it is driven by thermodynamic forces and is most likely to occur when the stronger adsorbate–substrate bonds that form compensate for the weakening of bonds between the substrate atoms, an inevitable accompaniment to the chemisorption-induced restructuring process.

6.6 THERMAL ACTIVATION OF BOND BREAKING

When molecules adsorb on a solid surface of low enough temperature (say 20–25 K), they maintain their gas-phase-like structure and remain chemically intact even on the most reactive metal surfaces. As the temperature is increased, either chemical rearrangement of the adsorbed molecule or bond breaking occurs at a certain temperature or narrow temperature range. Each adsorbate–substrate system has a characteristic temperature of bond activation. As the temperature is increased further, another bond-breaking or molecular rearrangement occurs; and sequential bond scission continues at characteristic temperatures until the molecule breaks up into its atomic constituents, which then desorb or diffuse into the bulk. An example, ethylene chemisorption on the platinum (111) surface, is shown in Figure 6.4. The thermal-desorption spectrum indicates sequential hydrogen evolution, while the vibrational spectra taken in the different temperature ranges (Figure 6.14) indicate that molecular rearrangements and chemical bond breaking occur simultaneously, as follows:

$$2C_2H_4(ads) \xrightarrow{300K} 2 -C_2H_3 + H_2 \uparrow \xrightarrow{410K} -C_2H + 2 -CH + \tfrac{3}{2}H_2 \uparrow \xrightarrow{>900K}$$

$$4C + \tfrac{3}{2}H_2 \uparrow$$

In Figure 2.30 the bond-breaking sequences for ethylene and benzene chemisorbed on the Rh(111) surface are compared. At low temperatures the structures of the chemisorbed molecules are different. As the temperature is increased, benzene appears to break into three short-lived acetylene molecules, which become C_2H spe-

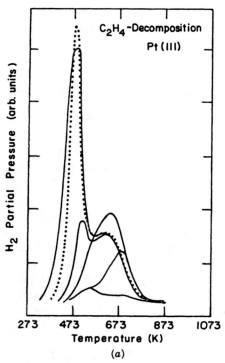

Figure 6.14. (a) Temperature-programmed desorption of hydrogen from the thermal decomposition of chemisorbed ethylene on Pt(111) [31]. (b) Proposed surface reaction mechanisms to account for the sequential decomposition [32, 33]. Asterisk: denote data taken from reference [33]; daggers denote data taken from reference [32].

cies after hydrogen desorption. At higher temperatures, the molecular fragments produced from ethylene and benzene on the metal surface are the same. The molecular fragmentation sequences for three C_3 hydrocarbons—propadiene, propyne (methylacetylene), and propylene—are shown in Figure 6.15a, and those for *o*- and *p*-xylene are shown in Figure 6.15b.

From the examples above, it is clear that molecular rearrangement or bond breaking on the surface has to be "activated" by increasing the temperature. Perhaps the first experimental observation of this phenomenon was the activated dissociation of dinitrogen (N_2) on iron surfaces, a phenomenon that gave rise to the suggestion of "physisorption" to "chemisorption" transition. Lennard-Jones [9] modeled this transition by a one-dimensional potential-energy curve-crossing diagram that is a simplified reaction coordinate for dissociative chemisorption. A typical diagram is shown in Figure 6.16. By using data from a combination of experiments, one can construct the more complex potential-energy diagram for CO_2 and H_2 formation from CO and H_2O (the so-called water–gas shift reaction) shown in Figure 6.17a or for the dehydrogenation of ethyl amine to acetonitrile in Figure 6.17b.

The molecular mechanisms that give rise to the breakup and reactions of the adsorbate–substrate cluster at a well-defined temperature are not clear, although they

Conversion of Ethylene to Ethylidyne (CCH₃)

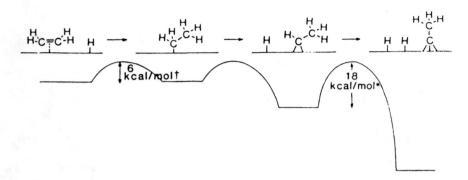

Fragmentation of Ethylidyne
to Vinylidene (CCH₂) and Acetylide (CCH)

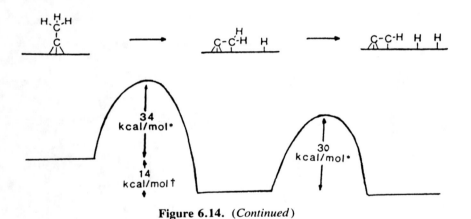

Figure 6.14. (*Continued*)

are a unique property of the surface chemical bond. Nevertheless, their characteristics, dependence on the substrate structure, and adsorbate coverage are well documented and are described below.

6.7 SURFACE-STRUCTURE SENSITIVITY OF BOND BREAKING

Molecules dissociate on more open and atomically rough surfaces at lower temperatures than on flat, close-packed surfaces of low Miller indices. For example, ethylene dissociates on a nickel (111) crystal face at around 250 K (Figure 6.18). On a stepped nickel surface, however, dissociation occurs at below 130 K. Rough surfaces are much more chemically active than flat surfaces at a given temperature. Using a mixed H_2/D_2 molecular beam, the probability of H—H bond breaking was

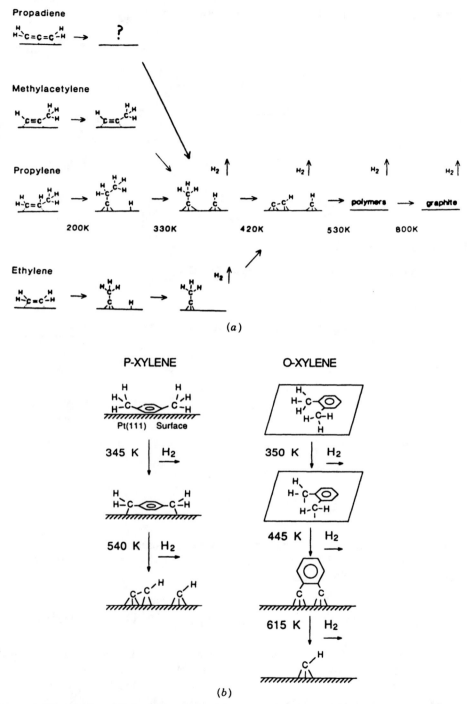

Figure 6.15. (a) Thermal fragmentation of C_3 hydrocarbons, propadiene, methylacetylene, propylene. Comparison with ethylene [34]. (b) Thermal fragmentation pathways for p-xylene and o-xylene [35].

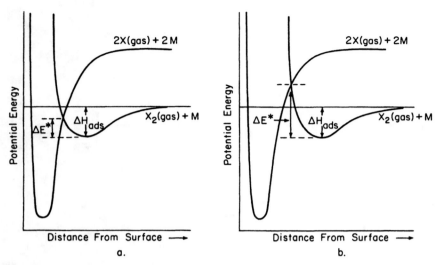

Figure 6.16. One-dimensional potential-energy diagrams showing the possible transition from molecular physisorption to dissociative chemisorption.

studied upon a single collision with a platinum surface by detecting the appearance of HD [10]. On a well-ordered Pt(111) close-packed surface, the reaction probability was below the detection limit of 10^{-3}. On a stepped metal surface, however, the reaction probability was near unity (Chapter 4). Likewise, the dissociative chemisorption of N_2 on the open (111) crystal face of body-centered cubic (bcc) iron was much more probable than on the close-packed (110) crystal face.

Thermal-desorption studies clearly indicate that adsorbed atoms and molecules have higher heats of adsorption at defect sites on the surface. This effect is demonstrated in Figure 4.18, where the thermal desorption of hydrogen is shown for flat, stepped, and kinked crystal faces of platinum. The flat metal surface shows a low-temperature desorption peak. The stepped surface exhibits two desorption peaks; the higher-temperature peak can be readily assigned to desorption from the steps, corresponding to a higher heat of desorption from this site. The kinked surface shows three desorption peaks, with the highest temperature peak corresponding to desorption from the kink sites.

This sensitivity of bonding to surface structure leads to sequential filling of adsorption sites as the coverage is increased, with the sites of highest adsorption energy filling first. This is shown for CO adsorption on a stepped platinum surface in Figure 6.19. At low coverages the step sites are covered with CO because of their high adsorption energy. As the CO coverage is increased, CO fills the terrace sites after all the step sites are covered; two thermal-desorption peaks appear, with the lower-temperature peak indicating the weaker bonding.

Defect sites (steps or kinks) and rough, low-packing-density surfaces have higher charge densities near the Fermi level. This is shown by lower work functions and the higher densities of filled electronic states detected by photoemission studies. These rough surfaces restructure more readily when clean, as described in Chapter 2. They are likely to participate in more massive adsorbate-induced restructuring

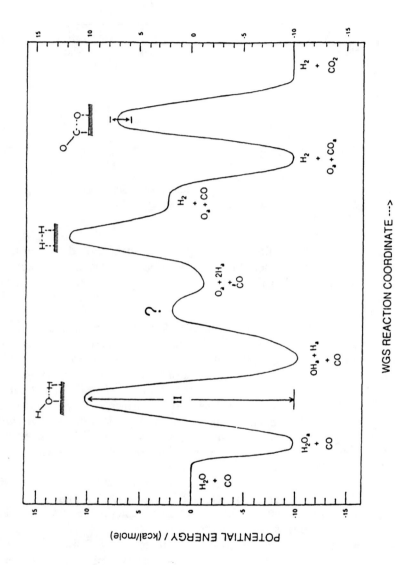

POTENTIAL ENERGY / (kcal/mole)

WGS REACTION COORDINATE --->

(a)

422

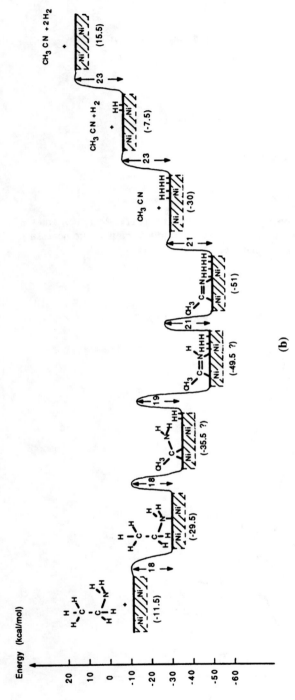

Figure 6.17. (a) Complex potential-energy diagram for the water–gas shift reaction (H_2O + CO → CO_2 + H_2) on copper surfaces constructed by using the available experimental data [36]. (b) Potential energy diagram for the dehydrogenation of ethyl amine to acetontrile on Ni(111) [37].

(b)

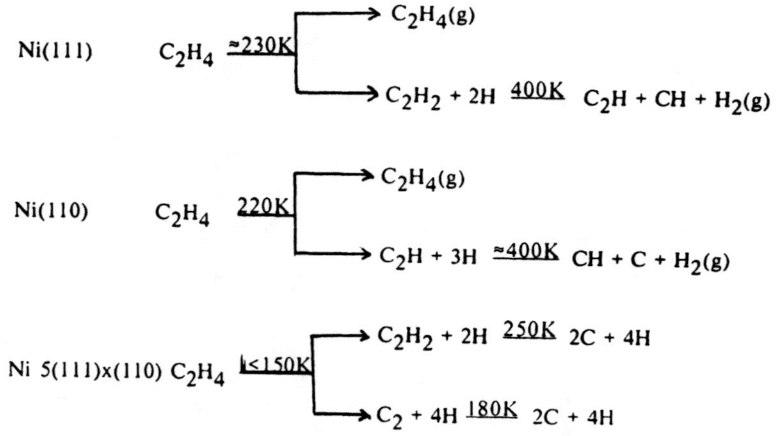

Ni(111)
C2H4
≈230K
C2H4(g)
C2H2 + 2H
400K
C2H + CH + H2(g)
Ni(110)
C2H4
220K
C2H4(g)
C2H + 3H
≈400K
CH + C + H2(g)
Ni 5(111)x(110) C2H4
≤150K
C2H2 + 2H
250K
2C + 4H
C2 + 4H
180K
2C + 4H

Figure 6.18. Sequential thermal decomposition of ethylene on the (111), the (110), and the stepped 5(111) × (110) crystal faces of nickel. Note the much lower temperature necessary to dissociate the organic molecule on the stepped metal surface [38].

CLEAN Pt (533)
CO PARTIAL PRESSURE (MASS 28 SIGNAL)
CO EXPOSURE
3.6 L
2.4 L
1.2 L
0.9 L
0.6 L
0.3 L
400
600
TEMPERATURE (K)

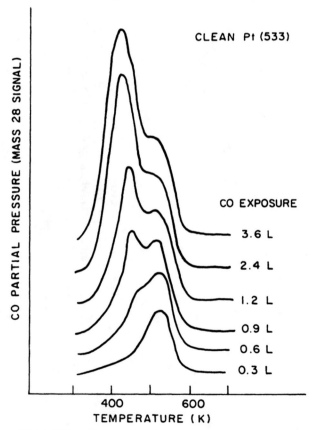

Figure 6.19. Sequential filling of step sites then terrace sites on the stepped Pt(533) surface during CO chemisorption [24].

processes. All these factors can contribute to the enhanced reactivity and bond strength of rough surfaces that lead to the marked surface-structure sensitivity of the adsorbate bond.

6.8 COVERAGE DEPENDENCE OF BONDING AND COADSORPTION

The heat of chemisorption per atom or per molecule declines with increasing coverage for most chemisorption systems. This is shown for potassium on a rhodium (111) crystal face and for CO on a palladium (100) face, respectively, in Figures 6.20 and 3.20b. At low coverages, potassium is strongly bound to the transition metal as it transfers electrons to it to become positively charged. With increasing coverage, adsorbate–adsorbate interaction causes repulsion among the charged species, leading to depolarization and much weakened adsorption bonds until the heat of adsorption becomes equal to the heat of sublimation of metallic potassium. Carbon monoxide chemisorbs with its $C—O$ bond perpendicular to the metal surface, occupying on-top and bridge sites, as shown in Figure 2.31, until about one-half monolayer coverage is reached. The heat of adsorption stays relatively constant with coverage in this coverage range, indicating that very little adsorbate–adsorbate interaction is influencing the bonding of the molecule to the metal. At higher coverages, however, the molecules strongly repel each other, forcing the on-top-site CO molecules to relocate to maximize adsorbate–adsorbate distances (see Figure 2.31), and ΔH_{ads} declines rapidly until it reaches about one-third of its value at low CO coverages.

Thus, increasing coverage of chemisorbed species not only leads to sequential filling of binding sites (the stronger binding sites filling first), as shown in the previous section, but can also weaken the adsorbate–substrate bonds markedly. This effect of coverage influences the surface residence times of adsorbates and subsequently their behavior during chemisorption and surface chemical reactions.

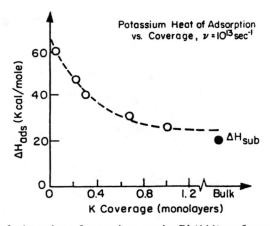

Figure 6.20. Heat of adsorption of potassium on the Rh(111) surface as a function of coverage [39].

6.8.1 Coadsorption

The coadsorption of two different species can lead to either attractive or repulsive adsorbate–adsorbate interaction. The coadsorption of ethylene and carbon monoxide demonstrates the attractive interaction that can occur in the adsorbed layer. CO and C_2H_4 chemisorbed together on the Rh(111) crystal face form the structure shown in Figure 6.21. There are two different molecules per unit cell, indicating attraction among the molecular species of different type. Ethylene adsorption decreases the work function of rhodium (as shown in Figure 6.22), whereas CO increases the work function of rhodium (Figure 6.23) upon chemisorption. Thus, C_2H_4 is an electron donor whereas CO is an electron acceptor on the transition metal, resulting in an attractive donor–acceptor interaction among the two types of adsorbates.

The ordering of one adsorbate by the coadsorption of another through donor–acceptor interaction is commonly observed, as shown for several coadsorbed systems listed in Table 6.1. For Rh(111) the magnitude of the adsorbate–adsorbate attractive interaction is about an order of magnitude weaker [about 4–6 kcal/mole (17–25 kJ/mole)] than most adsorbate–substrate chemisorption bonds [about 30–60 kcal/mole (125–250 kJ/mole)]. Repulsive interaction between two donor or two acceptor coadsorbed molecules leads to separation of the adsorbates by island for-

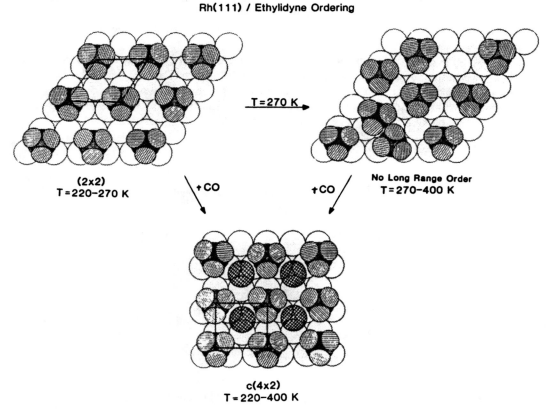

Figure 6.21. Coadsorption of ethylene and carbon monoxide on the Rh(111) surface [25].

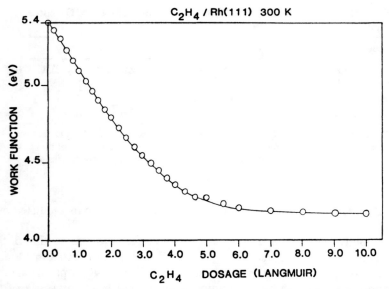

Figure 6.22. Decrease of rhodium work function upon chemisorption of ethylene on the (111) surface [40].

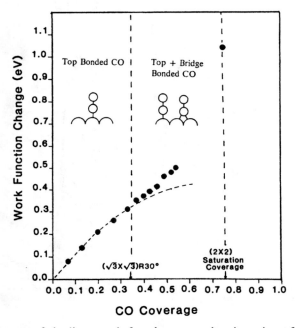

Figure 6.23. Increase of rhodium work function upon chemisorption of carbon monoxide on the (111) surface [40].

**TABLE 6.2. Combinations of Adsorbates with Similarly Oriented
Dipole Coadsorbed on the Rh(111) Surface**

Coadsorbates	LEED Patterns Observed
CO + NO	Disordered or compressed (2 × 2)–3CO [26]
Na + C₂H₂	Disordered
Na + ≡CCH₃	Disordered
Na + C₆H₆	$(\sqrt{3} \times \sqrt{3})R30°$ + $(2\sqrt{3} \times 3)$rect[a]

[a] Because the $(\sqrt{3} \times \sqrt{3})R30°$ and $(2\sqrt{3} \times 3)$rect are observed for Na and benzene,
respectively, adosrbed alone on Rh(111), the observation of a mixture of these two
LEED structures implies that these two coadsorbates segragate on the surface.

mation or disorder in the adsorbed layer. Several of these systems are listed in Table
6.2.

Strong attractive interaction among adsorbates can lead to dissociation of the molecular species. This is observed during the coadsorption of potassium (donor) and
CO (acceptor) on several transition-metal surfaces. Thermal-desorption data indicate
CO desorbing at much higher temperatures than normal in the presence of the adsorbed alkali metal, often showing a 17-kcal/mole (71-kJ/mole) increase in its heat
of adsorption (Figure 6.24). The CO stretching frequency decreases with increasing
dipole moment of coadsorbed donors (Figure 6.25). Isotope-labeling studies (using
$^{12}C^{18}O$ and $^{13}C^{16}O$) indicate scrambling of the two isotopic species in the presence

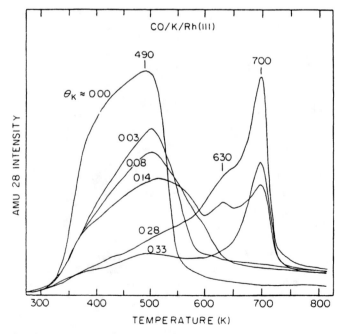

Figure 6.24. Large shift of CO thermal desorption to higher temperature upon potassium
coadsorption [41].

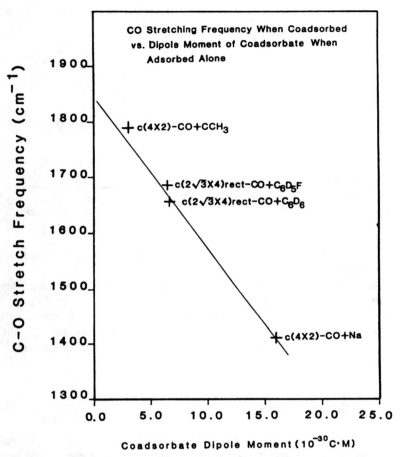

Figure 6.25. Decrease in chemisorbed CO stretching frequency on Rh(111) with increasing dipole movement of coadsorbed donors [40].

of potassium, signaling molecular dissociation, while no dissociation is apparent in the absence of potassium on rhodium. Up to three CO molecules dissociate per potassium atom at an alkali metal coverage of 20% of a monolayer, as shown in Figure 6.26.

Repulsive interaction is also observed with the coadsorption of potassium and ammonia. Both species are electron donors to transition metals. On iron, a 4-kcal/mole (17-kJ/mole) decrease in the heat of chemisorption of NH_3 is observed due to coadsorbed potassium.

Alkali metals are often used as additives during catalytic reactions. They are "bonding modifiers"; that is, they influence the bonding and thus the reactivity of the coadsorbed molecules. Potassium is a promoter in CO hydrogenation reactions where CO dissociation is desired and is one of the elementary reaction steps. The alkali metal also reduces the hydrogen chemisorption capacity of the transition metal. Potassium is a promoter in ammonia synthesis for the opposite reason, because it weakens the NH_3 product molecule bonding to the metal, thereby reducing its sur-

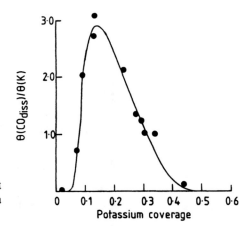

Figure 6.26. Number of CO molecules that dissociate per potassium atom on Rh(111) as a function of potassium coverage [42].

face concentration, which would block important reaction sites. It also aids in the dissociation of dinitrogen (see Chapter 7).

Halogen species can also be important bonding modifiers, because they are powerful electron acceptors. Indeed, they are used as promoters in several catalytic processes (for example, ethylene oxidation to ethylene oxide over silver, or during partial oxidation of methane). Nevertheless, their molecular and atomic chemisorption behavior has been studied less and therefore is not as well understood as the role of coadsorbed alkali-metal ions.

6.9   WEAK SURFACE BONDS

A gas atom or molecule approaching a surface "feels" an attractive potential. The nature of the gas–surface interaction determines the depth of the potential well (it is deep for chemisorption) and the range of the interaction. We may call the adsorbate–surface interaction weak if it leads to heats of adsorption of less than 10 kcal/mole (42 kJ/mole). This usually means that the adsorbate–adsorbate and adsorbate–substrate interactions are of the same order of magnitude. Therefore, the influence of the substrate atomic surface structure on the adsorption site is considerably weaker in this case than it is for chemisorption and more strongly influenced by coverage (i.e., adsorbate–adsorbate interaction).

In the absence of strong attractive interactions induced by charge transfer between adsorbates and surface atoms, weak attractive interactions can be induced in several ways. When a gas atom or molecule with no permanent dipole moment approaches the surface of a metal in which the conduction electrons constitute a mobile, fluctuating electron gas, the surface charge induces a dipole in the approaching species. This attractive induced dipole–surface-charge interaction is similar to that of a gas molecule with a permanent dipole, and the potential energy of interaction is of the form

$$V_{L-J} = -\frac{C}{r^3} \tag{6.1}$$

where C is a constant.

According to the model developed by Lennard-Jones [9] for spherically symmetrical atoms, C is given by $C = mc^2\chi/N_A$, where m is the electronic mass, c is the velocity of light, N_A is Avogadro's number, and χ is the diamagnetic susceptibility of the gas atom. The value of C is on the order of 10^2 kcal$\cdot$Å^3/mole when r is given in angstroms. In Table 6.3 the surface-interaction energies of several monatomic and diatomic gases are listed at a distance of closest approach of 4 Å (0.4 nm). The interaction potential between metal surfaces and approaching gas atoms has the same range, as given in Eq. 6.1, as was shown by Bardeen [11] and Margenau and Pollard [12], and the constant of proportionality in these cases is of the same magnitude as that derived by Lennard-Jones. It should be noted that, in using these models of gas–surface interactions, the gas atom is assumed to interact with the surface as a whole instead of with individual surface atoms. Recently [13] there has been experimental evidence that in some cases the interaction potential between a metal surface and organic molecules of different types varies inversely as the square of the distance: $V \propto r^{-2}$.

For certain types of gas–surface interactions, it may be useful to view the interaction as between the gas atom and a single surface atom. Weak attractive interaction between a pair of atoms can be due to dispersion forces (London [14, 15]) that represent the interaction of induced fluctuating charge distributions. In addition, molecules that possess permanent dipoles can further polarize each other (Debye [16, 17]) and can have dipole–dipole interactions (Keesom [18, 19]). All these pairwise interaction potentials fall off inversely as the sixth power of the distance.

The dispersion force is due to induced dipole interaction between atoms or molecules through electron density fluctuations. According to London, the potential energy of interaction V_{London} is given by

$$V_{London} = -\frac{C'}{r^6} \qquad (6.2)$$

where, using an approximate model, C' is given by

$$C' = \frac{3h\nu_1\nu_2}{2(\nu_1 + \nu_2)}\alpha_1\alpha_2 \qquad (6.3)$$

TABLE 6.3. Values of the Constants of the Lennard-Jones and London Interaction Potentials and the Interaction Energies at the Distance of Closest Approach of 4 Å

Atom or Molecule	C (kcal$\cdot$Å^3/ mole)	C (J$\cdot$nm^3/ mole)	C' (kcal$\cdot$Å^6/ mole)	C' (J$\cdot$nm^6/ mole)	V_{L-J} (kcal/mole) ($r = 4$ Å)	V_{L-J} (kJ/mole) ($r = 0.4$ nm)	V_{London} (kcal/mole) ($r = 4$ Å)	V_{London} (kJ/mole) ($r = 0.4$ nm)
Nθ	129	539	67	0.280	2.0	8.4	0.016	0.067
Ar	352	1471	802	3.352	5.5	23.0	0.19	0.79
H$_2$	76	318	176	0.736	1.2	5.0	0.04	0.17
N$_2$	144	602	919	0.341	2.2	9.2	0.22	0.92
CO$_2$	362	1513	1872	7.825	5.6	23.4	0.46	1.92

Here α_1 and α_2 are the polarizabilities of the interacting species, ν_1 and ν_2 are their characteristic frequencies of oscillation (oscillator strengths), and h is Planck's constant. The value of C' can be calculated to be on the order of 10^3 kcal Å^6 (4.18 J $\cdot$ nm^6).

Because the values of ν_1 and ν_2 are not easily available, one seeks other ways to estimate the dispersion constant C' from readily measurable molecular properties. One very good approximate expression, which was developed by Slater and Kirkwood [20], can be written as

$$C' \left(\frac{\text{kJ} \cdot \text{nm}^6}{\text{mole}} \right) = 1.52 \times 10^{-3} \frac{\alpha_1 \alpha_2}{\left(\dfrac{\alpha_1}{n_1} \right)^{1/2} + \left(\dfrac{\alpha_2}{n_2} \right)^{1/2}} \tag{6.4}$$

where α_1 and α_2 are the polarizabilities in units of Å^3, and n_1 and n_2 are the number of electrons in the outer shells of the molecules. In Table 6.3 the London interaction energies are also listed for a radius of $r = 4$ Å (0.4 nm), along with the dispersion constants for several pairs of like atoms. (For interaction between like species, $\alpha_1 = \alpha_2$ and $n_1 = n_2$.) It can be seen that, due to its short range, V_{London} is a much weaker attractive potential at that distance when compared with the $V \propto r^{-3}$.

Molecules that possess permanent dipole moments can further polarize each other, giving, for the mutual attractive potential energy V_{Debye},

$$V_{\text{Debye}} = - \frac{\alpha_1 \mu_2^2 + \alpha_2 \mu_1^2}{r^6} \tag{6.5}$$

where μ_1 and μ_2 are the dipole moments of the interacting molecules. Direct interaction of two different molecules with permanent dipoles without additional polarization yields

$$V_{\text{Keesom}} = - \frac{2}{3 k_B T} \cdot \frac{\mu_1^2 \mu_2^2}{r^6} \tag{6.6}$$

Both V_{Debye} and V_{Keesom} are orientation-averaged expressions, and Eq. 6.6 is restricted to gases in thermal equilibrium. The dispersion interaction V_{London} is appreciably larger than these other two effects (except for the most polar molecules, such as water, for which V_{Keesom} is somewhat larger). Table 6.4 lists the average polarizabilities of several atoms and molecules; Table 6.5 lists the dipole moments of several molecules. There are also many other types of interactions (for example, the ion–induced dipole interaction, which varies as r^{-4}), but they are likely to be less important in gas–surface interactions and will not be discussed here.

It has been shown [21] that the dispersion interaction between pairs of atoms is additive. Calculations show [22] that a large long-range attractive interaction may result from the simultaneous dispersion interaction of many atoms. For example, the attractive potential energy of interaction between two flat plates V in vacuum, due to the summation of the pairwise dispersion forces, varies inversely with the square of the distance: $V \propto r^{-2}$.

TABLE 6.4. Average Polarizabilities for Several Atoms and Molecules

Atom or Molecule	$\bar{\alpha}$ (Å^3)	Atom or Molecule	α (Å^3)
Neon (Ne)	0.39	Hydrogen sulfide (H$_2$S)	3.78
Argon (Ar)	1.63	Ammonia (NH$_3$)	2.26
Krypton (Kr)	2.46	Nitrous oxide (N$_2$O)	3.00
Xenon (Xe)	4.00	Methane (CH$_4$)	2.60
Hydrogen (H$_2$)	0.79	Ethane (C$_2$H$_6$)	4.47
Nitrogen (N$_2$)	1.76	Ethylene (C$_2$H$_4$)	4.26
Oxygen (O$_2$)	1.60	Benzene (C$_6$H$_6$)	10.32
Carbon monoxide (CO)	1.95	Acetone (CH$_3$COCH$_3$)	6.33
Carbon dioxide (CO$_2$)	2.65		

TABLE 6.5. Dipole Moments of Several Molecules

Molecule	μ (Debye[a])
H$_2$O	1.84
H$_2$S	0.89
NO	0.16
CO	0.12
N$_2$O	0.166
HF	1.91
HCl	1.08
NH$_3$	1.45
CH$_3$OH	1.68
CH$_3$CHO	2.72
(CH$_3$)$_2$CO	2.9

[a] 1 Debye = 1×10^{-18} esu.

6.9.1 Phase Transformations in the Weakly Adsorbed Layer

A great deal of information can be obtained about the structural changes that occur in weakly adsorbed layers from adsorption–isotherm measurements (amount adsorbed versus pressure). The most commonly studied systems are inert gases adsorbed on graphite or on metal surfaces. At low temperatures the adsorbed atoms form ordered structures (identified by LEED on crystal surfaces), which may be viewed as two-dimensional condensation and evidence of the existence of gas–solid equilibrium. As the temperature is increased, the adsorption isotherm changes. Figure 6.27 shows the adsorption isotherms of krypton on the (0001) crystal face of graphite in the 79.2- to 88.5-K temperature range. The dashed line indicates the possible phase diagram. At around 85 K there is a first-order phase transformation that is associated with the onset of disorder in the adsorbed monolayer, indicating the formation of a liquid-like film. Above this temperature, therefore, a solid–liquid equilibrium exists.

When the coverage increases at a given temperature (one that is below the temperature at which the liquid-like film forms), the surface structure of the adsorbed atoms changes due to repulsive adsorbate–adsorbate interactions. This effect is also

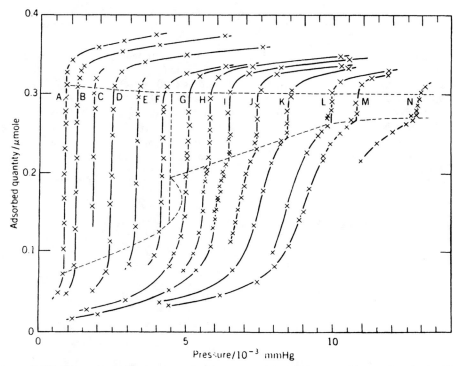

Figure 6.27. The adsorption isotherms of krypton on the (0001) graphite surface in the 79.2- to 88.5-K temperature range: A, 79.24; B, 80.54; C, 81.77; D, 82.83; E, 83.84; F, 84.69; G, 85.33; H, 85.74; I, 86.12; J, 86.58; K, 87.08; L, 87.61; M, 87.81; N, 88.46 K [43].

implied by the rapid decline of the heat of adsorption with increasing coverage, as shown for several weakly adsorbed systems in Figure 6.28. Incommensurate surface structures usually form because the substrate structure (periodicity) has little influence on the ordering behavior of these systems. At low enough temperatures, multilayers can be produced, and their properties (number of layers, heats of interaction between the layers) can be analyzed using the BET isotherm analysis (see Chapter 3).

Total surface-area measurements are a useful application of weakly adsorbed monolayers, because their interaction energies are largely independent of the chemical composition of the substrates. Gas separation is another application of weak adsorption. Oxygen (O_2) and nitrogen (N_2) can be separated from air by selective adsorption, because the polarizability of N_2 and therefore its heat of adsorption is greater than that of O_2. The gas chromatograph operates on the principle of small differences in heats of adsorption of molecules. The difference changes the residence times of adsorbates (called *retention times* in this circumstance) on a column, thus separating them by delaying their arrival at the detector. Separation of macromolecules can be achieved at the solid–liquid interface (liquid-phase chromatography), where the diffusion rates of different molecules are influenced by their somewhat different binding at the interface.

Weakly adsorbing surfaces can be prepared that either preferentially adsorb (hydrophilic interface) or repel (hydrophobic) water. Hydroxylated silica surfaces and

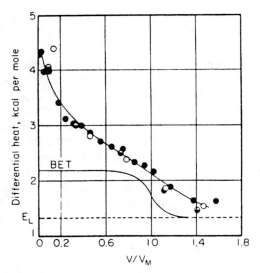

Figure 6.28. Differential heat of adsorption of nitrogen on carbon black at 78.5 K [44].

fluorocarbons behave in this way, respectively, and high-surface-area molecular sieves have been developed with these properties. Such interfaces can be used to separate organic and aqueous phases of solutions.

Weakly adsorbing insulator or semiconductor surfaces that operate by charge transfer (such as SnO_2, as shown in Chapter 5) can be used as detectors or for

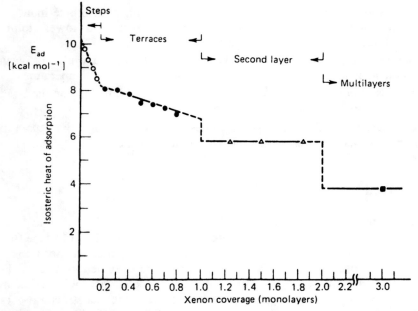

Figure 6.29. Heat of xenon adsorption on the stepped palladium 8(100) × (110) surface as a function of coverage [45].

separating gases. Their principles of operation are different from those of surfaces that form polarizability bonds, because they respond to small differences in ionization potentials or electron affinities.

The adsorption of even weakly bound atoms shows variations due to the structure of transition metals. Xenon exhibits easily detectable changes in heat of adsorption on palladium crystal surfaces (as shown in Figure 6.29), depending on location (at a step site or on a terrace site). The higher heat of adsorption at the terrace sites is not surprising, since the lower work function at defect sites indicates larger electric fields at these sites, thus influencing the bonding of the highly polarizable xenon atom. The weak adsorption of the chemically passive xenon can then be used to learn about the atomic heterogeneity of the metal surface structure in a noninvasive manner, using relatively simple thermal desorption studies. As Figure 6.29 indicates, there are differences in the heat of adsorption of xenon in the first and second monolayers, and even between the second layer and the multilayers that adsorb over it. Thus, the effect of weak polarizability bonding at the metal surface can influence the bonding of at least two layers of adsorbates.

6.10 SUMMARY AND CONCEPTS

- The formation of the surface chemical bond is accompanied by charge redistribution in the adsorbate and the substrate that may also change the structures of both.
- Bond energies for a given atom or molecule adsorbed on transition metals increase from right to left in the periodic table.
- Molecular adsorbates exhibit bonding and structure that are similar to those in cluster compounds (multinuclear organometallic clusters, for example).
- The adsorbate bond is surface-structure-sensitive, and adsorbate-induced surface restructuring frequently occurs. Rough surfaces (with lower atomic coordination) restructure more readily.
- Bond breaking in adsorbates requires thermal activation and usually occurs in several steps at well-defined temperatures with increasing temperature.
- Changes of coverage and coadsorption have marked influence on the bonding, location, and orientation of adsorbates.
- Weak surface bonds play important roles in gas chromatography separation of molecular mixtures and in gas separation and detection.

6.11 PROBLEMS

⋆⋆6.1   Ethylidyne restructures the Rh(111) crystal face [30], sulfur restructures the Fe(110) face [7], and carbon restructures the Ni(100) face [6, 46]. The surface metal atoms move into new equilibrium positions upon chemisorption in different ways, and there is evidence of restructuring even in the second substrate layer under the surface. Review the available data and point out the important electronic and structural parameters that influence the nature and magnitude of chemisorption-induced surface restructuring.

**6.2 Adsorbate-induced restructuring of surfaces could explain the formation of cluster-like bonding of adsorbates on metal surfaces. Discuss how the strength of the chemisorption bond is likely to influence the restructuring of metal surfaces.

**6.3 One of the unique features of the surface chemical bond is how sequential bond breaking occurs in the adsorbed monolayer as the temperature increases. Find two examples in the surface-science literature for the thermal activation of bond scission. Describe how sharp the transition is (its temperature range), its substrate structure dependence, and its coverage dependence. Speculate on the reasons for breaking 400-kJ/mole chemical bonds by merely increasing the temperature a few degrees in a given temperature range.

**6.4 The heat of adsorption of CO is determined as a function of coverage by several research groups using single-crystal metal surfaces [47]. Review and describe the experimental procedures of how such experiments are carried out.

**6.5 Surface defects, steps, and kinks dissociate molecular bonds more readily and exhibit higher heats of adsorption for the chemisorbed atoms or the molecular fragments. Find two examples of such chemical behavior and discuss the possible relationship between the electronic structure and the atomic structure of the defect and its reactivity to break adsorbate bonds.

**6.6 Low-energy electron diffraction studies of rare gases on copper have detected several ordered surface structures that form as a function of coverage [48, 49]. Explain how changes in bonding give rise to alterations of surface structure and the two-dimensional phase diagram that has been constructed.

**6.7 The binding energy of potassium is strongly coverage-dependent. When it is coadsorbed with CO, it markedly strengthens the CO bond to the transition-metal substrate. When it is coadsorbed with ammonia, it decreases the heat of adsorption of the molecule on iron surfaces. Explain the reasons for these intriguing properties [50–52] of adsorbed potassium for altering the chemical bonding in the monolayer.

**6.8 Inert gases decrease the work function of transition metals [53–55]. Although their bonding is weak, they exhibit detectable surface-structure sensitivity. This has been particularly well demonstrated for the adsorption of xenon. Review the available literature and discuss the nature of inert gas bonding to metal substrates that gives rise to these effects.

**6.9 The adsorption isotherm of xenon on graphite has been measured at different temperatures [56–60]. Review the experimental results and discuss the surface phases of xenon that were detected. Would you expect krypton to behave similarly on the same substrate? Explain.

**6.10 When ethylene chemisorbs on the (111) face of rhodium, it lies with its C=C bond parallel to the surface at low temperatures, forms ethylidyne (C_2H_3-) at 300 K, and dissociates to C_2H- and $CH-$ groups at 410

K. Find organometallic multinuclear cluster compounds with similar organic species attached and discuss their bonding behavior (bond distances, binding sites, and bond angles) [61].

**6.11 The heat of adsorption of carbon monoxide varies across the periodic table. There is a great deal of data available to demonstrate this, especially for transition metals [47]. The heat of adsorption per CO molecule also varies markedly with coverage, especially above one-half monolayer for most metals. CO may also occupy top, bridge, threefold, and other sites where its binding energy is different at each site. Review the available data, discuss the trends, and comment on the effects of the changing electronic structure of the substrate and the variation of the atomic structure of a given substrate on the binding of CO.

REFERENCES

[1]   J.K. Nørskov. Covalent Effects in the Effective-Medium Theory of Chemical Binding: Hydrogen Heats of Solution in the 3d Metals. *Phys. Rev. B* **26**:2875 (1982).

[2]   I. Toyoshima and G.A. Somorjai. Heats of Chemisorption of O_2, H_2, CO, CO_2, and N_2, on Polycrystalline and Single Crystal Transition Metal Surfaces. *Catal. Rev. Sci. Eng.* **19**:105 (1979).

[3]   R.W. Joyner and M.W. Roberts. Auger Electron Spectroscopy Studies of Clean Polycrystalline Gold and the Adsorption of Mercury on Gold. *J. Chem. Soc. Faraday Trans. I* **69**:1242 (1973).

[4]   E.W. Plummer, W.R. Salaneck, and J.S. Miller. Photoelectron Spectra of Transition-Metal Carbonyl Complexes: Comparison with the Spectra of Adsorbed CO. *Phys. Rev. B* **18**:1673 (1978).

[5]   Y. Gauthier, R. Baudoing-Savois, K. Heinz, and H. Landskron. Structure Determination of p4g Ni(100)–(2 × 2)C by LEED. *Surf. Sci.* **251**:493 (1991).

[6]   J.H. Onuferko, D.P. Woodruff, and B.W. Holland. LEED Structure Analysis of the Ni{100}(2 × 2)C(p4g) Structure; a Case of Adsorbate-Induced Substrate Distortion. *Surf. Sci.* **87**:357 (1979).

[7]   H.D. Shih, F. Jona, D.W. Jepsen, and P.M. Marcus. Metal-Surface Reconstruction Induced by Adsorbate: Fe(110)p(2 × 2)–S. *Phys. Rev. Lett.* **46**:731 (1981).

[8]   N. Kruse and A. Gaussmann. Changes in the Morphology of Rh Field Emitter Tips due to the Reaction with Carbon Monoxide. *Surf. Sci.* **266**:51 (1992).

[9]   J.E. Lennard-Jones. Processes of Adsorption and Diffusion on Solid Surfaces. *Trans. Faraday Soc.* **28**:333 (1932).

[10] T.H. Lin and G.A. Somorjai. Angular and Velocity Distributions of HD Molecules Produced by the H_2-D_2 Reaction on the Stepped Pt(557) Surface. *J. Chem. Phys.* **81**:704 (1984).

[11] J. Bardeen. The Image and van der Waals Forces at a Metallic Surfaces. *Phys. Rev.* **58**:727 (1940).

[12] H. Margenau and W.G. Pollard. The Forces between Neutral Molecules and Metallic Surfaces. *Phys. Rev.* **60**:128 (1941).

[13] D. Lando and L.J. Slutsky. Surface van der Waals Forces. *J. Chem. Phys.* **52**:1510 (1970).

[14] F. London. The General Theory of Molecular Forces. *Trans. Faraday Soc.* **33**:8 (1937).

[15] F. London. Zur Theorie und Systematik der Molekularkräfte. *Z. Phys.* **63**:245 (1930).

[16] P. Debye. Molekularkräfte und Ihre Elektrische Deutung. *Physikalische Zeitschrift,* **22**:302 (1921).

[17] P. Debye. Die van der Waalsschen Kohäsionskräfte. *Phys. Z.* **21**:178 (1920).

[18] W.H. Keesom. Die van der Waalsschen Kohäsionskräfte. Berichtigung. *Phys. Z.* **22**:643 (1921).

[19] W.H. Keesom. Die van der Waalsschen Kohäsionskräfte. *Phys. Z.* **22**:129 (1921).

[20] J.C. Slater and J.G. Kirkwood. Van der Waals Forces in Gases. *Phys. Rev.* **37**:682 (1931).

[21] F. London. Über einige Eigenschaften und Anwendungen der Molekularkräfte. *Z. Phys. Chem. Abt. B* **11**:222 (1930).

[22] A.D. Crowell. Surface Forces and the Solid–Gas Interface. In: E.A. Flood, editor, *The Solid–Gas Interface,* Volume 1. Marcel Dekker, New York, 1967.

[23] J.K. Nørskov. Chemisorption on Metal Surfaces. *Rep. Prog. Phys.* **53**:1253 (1990).

[24] G.A. Somorjai and B.E. Bent. The Structure of Adsorbed Monolayers. The Surface Chemical Bond. *Prog. Colloid Polym. Sci.* **70**:38 (1985).

[25] B.E. Bent. Bonding and Reactivity of Unsaturated Hydrocarbons on Transition Metal Surfaces: Spectroscopic and Kinetic Studies of Platinum and Rhodium Single Crystal Surfaces. Ph.D. thesis, University of California, Berkeley, 1986.

[26] C.M. Mate, G.A. Somorjai, H.W.K. Tom, X.D. Zhu, and Y.R. Shen. Vibrational and Electronic Spectroscopy of Pyridine and Benzene Adsorbed on the Rh(111) Crystal Face. *J. Chem. Phys.* **88**:441 (1988).

[27] E. Wimmer, C.L. Fu, and A.J. Freeman. Catalytic Promotion and Poisoning: All-Electron Local-Density-Functional Theory of CO on Ni(001) Surfaces Coadsorbed with K or S. *Phys. Rev. Lett.* **55**: 2618 (1985).

[28] D.J. Coulman, J. Winterlin, R.J. Behm, and G. Ertl. Novel Mechanism for the Formation of Chemisorption Phases: the $(2 \times 2)O-Cu(110)$ "Added Row" Reconstruction. *Phys. Rev. Lett.* **64**:1761 (1990).

[29] G.A. Somorjai. Directions of Theoretical and Experimental Investigations into the Mechanisms of Heterogeneous Catalysis. *Catal. Lett.* **9**:311 (1991).

[30] A. Wander, M.A. Van Hove, and G.A. Somorjai. Molecule-Induced Displacive Reconstruction in a Substrate Surface: Ethylidyne Adsorbed on Rh(111) Studied by Low-Energy-Electron Diffraction. *Phys. Rev. Lett.* **67**:626 (1991).

[31] S.M. Davis, F. Zaera, B.E. Gordon, and G.A. Somorjai. Radiotracer and Thermal Desorption Studies of Dehydrogenation and Atmospheric Hydrogenation of Organic Fragments Obtained from [^{14}C]Ethylene Chemisorbed over Pt(111) Surfaces. *J. Catal.* **92**:240 (1985).

[32] D. Godbey, F. Zaera, R. Yeates, and G.A. Somorjai. Hydrogenation of Chemisorbed Ethylene on Clean, Hydrogen, and Ethylidyne Covered Platinum (111) Crystal Surfaces. *Surf. Sci.* **167**:150 (1986).

[33] D.B. Kang and A.B. Anderson. Adsorption and Structural Rearrangements of Acetylene and Ethylene on Pt(111); Theoretical Study. *Surf. Sci.* **155**:639 (1985).

[34] B.E. Bent, C.M. Mate, J.E. Crowell, B.E. Koch, and G.A. Somorjai. Bonding and Thermal Decomposition of Propylene, Propadiene, and Methyl Acetylene on the Rh(111) Single-Crystal Surface. *J. Phys. Chem.* **91**:1493 (1987).

[35] D.E. Wilk, C.D. Stanners, Y.R. Shen, and G.A. Somorjai. The Structure and Thermal Decomposition of *Para*- and *Ortho*-Xylene on Pt(111). A HREELS, LEED, and TPD Study. *Surf. Sci.* (1993).

[36] J. Nakamura, J.M. Campbell, and C.T. Campbell. Kinetics and Mechanism of the

Water–Gas Shift Reaction Catalysed by the Clean and Cs-promoted Cu(110) Surface: A Comparison with Cu(111). *J. Chem. Soc. Faraday Trans.* **86**:2725 (1990).

[37] D.E. Gardin and G.A. Somorjai. The Vibrational Spectra (HREELS) and Thermal Decomposition (TPD) of Methylamine (CH$_3$NH$_2$) and Ethylamine (C$_2$H$_5$NH$_2$) on Ni(111). *J. Phys. Chem.* **96**:9424 (1992).

[38] G.A. Somorjai, C.M. Kim, and C. Knight. Building of Complex Catalysts on Single-Crystal Surfaces. In: D.J. Dwyer and F.M. Hoffmann, editors, *Surface Science of Catalysis: In Situ Probes and Reaction Kinetics. ACS Symposium Series*, Volume 482. American Chemical Society, Washington, D.C., 1992.

[39] J.E. Crowell. Chemical Modification of Surfaces: The Effect of Potassium on the Chemisorption of Molecules on Transition Metal Crystal Surfaces. Ph.D. thesis, University of California, Berkeley, 1984.

[40] C.M. Mate, C.T. Kao, and G.A. Somorjai. Carbon Monoxide Induced Ordering of Adsorbates on the Rh(111) Crystal Surface: Importance of Surface Dipole Moments. *Surf. Sci.* **206**:145 (1988).

[41] J.E. Crowell and G.A. Somorjai. The Effect of Potassium on the Chemisorption of Carbon Monoxide on the Rh(111) Crystal Face. *Appl. Surf. Sci.* **19**:73 (1984).

[42] J.E. Crowell, W.T. Tysoe, and G.A. Somorjai. Potassium Coadsorption Induced Dissociation of CO on the Rh(111) Crystal Face: An Isotope Mixing Study. *J. Phys. Chem.* **89**:1598 (1985).

[43] Y. Larher. Triple Point of the First Monomolecular Layer of Krypton Adsorbed on the Cleavage Face of Graphite. *J. Chem. Soc. Faraday Trans. I* **70**:320 (1974).

[44] A.W. Adamson. *Physical Chemistry of Surfaces*. John Wiley & Sons, New York, 1990.

[45] R. Miranda, S. Daiser, K. Wandelt, and G. Ertl. Thermodynamics of Xenon Adsorption on Pd(s) [8(100) × (110)]: From Steps to Multilayers. *Surf. Sci.* **131**:61 (1983).

[46] A. Atrei, U. Bardi, M. Maglietta, G. Rovida, M. Torrini, and E. Zanazzi. SEELFS Study of Ni(001)(2 × 2)C p4g Structure. *Surf. Sci.* **211/212**:93 (1989).

[47] J.C. Campuzano. The Adsorption of Carbon Monoxide by the Transition Metals. In: D.A. King and D.P. Woodruff, editors, *Chemisorption Systems, Part A. The Chemical Physics of Solid Surfaces and Heterogeneous Catalysis*, Volume 3. Elsevier, New York, 1990.

[48] A. Glachant, M. Jaubert, M. Bienfait, and G. Baoto. Monolayer Adsorption of Kr and Xe on Metal Surfaces: Structures and Uniaxial Phase Transitions on Cu(110). *Surf. Sci.* **115**:219 (1981).

[49] U. Bardi, A. Glachant, and M. Bienfait. Phase Transitions on Stepped and Disordered Surfaces: Xe Adsorbed on Cu and NaCl Single Crystal Surfaces. *Surf. Sci.* **97**:137 (1980).

[50] M.P. Kiskinova. *Poisoning and Promotion in Catalysis Based on Surface Science Concepts and Experiments. Studies in Surface Science and Catalysis*, Volume 70. Elsevier, Amsterdam, 1992.

[51] G.E. Rhead. On the Variation of Work Function with Coverage for Alkali-Metal Adsorption. *Surf. Sci.* **203**:L663 (1988).

[52] H.P. Bonzel. Alkali-Metal-Affected Adsorption of Molecules on Metal Surfaces. *Surf. Sci. Rep.* **8**:43 (1987).

[53] K. Christmann and J.E. Demuth. Interaction of Inert Gases with a Nickel (100) Surface. I. Adsorption of Xenon. *Surf. Sci.* **120**:291 (1982).

[54] Y.C. Chen, J.E. Cunningham, and C.P. Flynn. Dependence of Rare-Gas–Adsorbate Dipole Moment on Substrate Work Function. *Phys. Rev. B* **30**:7317 (1984).

[55] C.P. Flynn and Y.C. Chen. Work Function and Bonding Energy of Rare-Gas Atoms Adsorbed on Metals. *Phys. Rev. Lett.* **46**:447 (1981).

[56] H. Hong, C.J. Peters, A. Mak, R.J. Birgeneau, P.M. Horn, and H. Suematsu. Synchrotron X-Ray Study of the Structures and Phase Transitions of Monolayer Xenon on Single-Crystal Graphite. *Phys. Rev. B* **40**:4797 (1989).

[57] J. Suzanne, J.P. Coulomb, and M. Bienfait. Two-Dimensional Phase Transition in Xenon Submonolayer Films Adsorbed on (0001) Graphite. *Surf. Sci.* **47**:204 (1976).

[58] J. Suzanne, J.P. Coulomb, and M. Bienfait. Transition Bidimensionelle du Premier Ordre; Cas du Xénon Adsorbé sur la Face (0001) du Graphite. *Surf. Sci.* **44**:141 (1974).

[59] J. Suzanne, J.P. Coulomb, and M. Bienfait. Auger Electron Spectroscopy and LEED Studies of Adsorption Isotherms: Xenon on (0001) Graphite. *Surf. Sci.* **40**:414 (1973).

[60] A. Thomy, X. Duval, and J. Regnier. Two-Dimensional Phase Transitions as Displayed by Adsorption Isotherms on Graphite and Other Lamellar Solids. *Surf. Sci. Rep.* **1**:1 (1981).

[61] E.L. Muetterties, T.N. Rhodin, E. Band, C.F. Brucker, and W.R. Pretzer. Clusters and Surfaces. *Chem. Rev.* **79**:91 (1979).

7

CATALYSIS BY SURFACES

442

7.1 INTRODUCTION

In a surface catalytic process, the reaction occurs repeatedly by a sequence of elementary steps that includes adsorption, surface diffusion, the chemical rearrangements (bond breaking, bond forming, molecular rearrangement) of the adsorbed reaction intermediates and the desorption of the products.

Catalytic reactions play all important roles in our life. Most biological reactions that build the human body, as well as the reactions that control the functioning of the brain and other vital organs, are catalytic. Photosynthesis and the majority of chemical processes that are utilized in chemical technology are also catalytic reactions. These range from oil refining and the production of chemicals by hydrogenation, dehydrogenation, partial oxidation, and organic molecular rearrangements (isomerization, cyclization), to ammonia synthesis and fermentation. The chemical bonds that form during these processes that can turn over repeatedly are very different from those that form during stoichiometric reactions that characterize the formation of the chemisorption bond.

7.1.1 Brief History of Surface Catalysis

In 1814 Kirchhoff reported that acids aid the hydrolysis of starch to glucose. The oxidation of hydrogen by air over platinum was observed by H. Davy (1817) and E. Davy (1820) as well as by Döbereiner (1823), who constructed a "tinderbox" to produce flame when a small dose of hydrogen generated by the reaction of zinc and hydrochloric acid reacts with air in the presence of platinum. His device sold handily in the early part of the 19th century when matches were not yet available. Platinum was also found to aid the oxidation of CO and ethanol (Döbereiner).

Faraday was the first to carry out experiments to explore why platinum facilitates the oxidation reactions of different molecules. He found that ethylene adsorption deactivates the platinum surface temporarily while the adsorption of sulfur deactivates platinum permanently. He measured the rate of hydrogen oxidation, suggested a mechanism, and observed its deactivation and regeneration. Thus, Faraday was the first scientist who studied catalytic reactions. In 1836 Berzelius [1, 2] defined the phenomenon and called it *catalysis* and suggested the existence of a "catalytic force" associated with the action of catalysts.

Catalyst-based technologies were introduced in the second half of the 19th century. The Deacon process ($2HCl + \frac{1}{2}O_2 + \xrightarrow{CuCl_2} H_2O + Cl_2$) was discovered in 1860, and the oxidation of SO_2 to SO_3 by platinum was discovered by Messel in 1875. Mond introduced the nickel-catalyzed reaction of methane with steam ($CH_4 + H_2O \xrightarrow{Ni} CO + 3H_2$). In the early 20th century, Ostwald developed the process of ammonia oxidation ($2NH_3 + \frac{5}{2}O_2 \xrightarrow{Pt} 2NO + 3H_2O$) to form nitric oxide, the precursor to nitric acid manufacture (1902); and in 1902, Sebatier developed a process for the hydrogenation of ethylene ($C_2H_4 + H_2 \xrightarrow{Ni} C_2H_6$). In 1905, Ipatieff used the catalytic action of clays to carry out different organic reactions: dehydrogenation, isomerization, hydrogenation, and polymerization.

Better understanding of thermodynamics established the limits of reaction rates in catalyzed reactions. A catalyst can bring a reaction closer to equilibrium but cannot produce molecules in excess of equilibrium concentrations. The ammonia synthesis from N_2 and H_2 became the reaction to provide the testing ground for both catalysis science and technology. The quality of the catalyst could be tested based on how closely chemical equilibrium could be attained. High-pressure reactors were designed to shift the chemical equilibrium during catalyzed ammonia production.

Catalyzed reactions of carbon monoxide and hydrogen were utilized to produce methanol ($CO + 2H_2 \xrightarrow{ZnO, Cr_2O_3} CH_3OH$) in 1923 and higher-molecular-weight liquid hydrocarbons by 1930. The production of motor fuels became one of the chief aims of catalysis during the 1930–1950 period. The cracking of long-chain hydrocarbons to produce lower-molecular-weight products was achieved over oxide catalysts composed mostly of alumina and silica. Acid-catalyzed alkylation reactions provided high-octane fuel and important organic molecules.

In the meantime, catalysis science was developed (1915–1940) through the efforts of Langmuir (sticking probability, adsorption isotherm, dissociative adsorption, role of monolayers), Emmett (surface area measurements, kinetics of ammonia synthe-

sis), Taylor (active sites, activated adsorption), Bonhoeffer, Rideal, Roberts, Polanyi, Farkas (kinetics and molecular mechanisms of ethylene hydrogenation, *ortho-
para* hydrogen conversion, isotope exchange, intermediate compound theories), and
many others.

The discovery of abundant and inexpensive oil in Arabia in the early 1950s focused the development of catalytic processes to convert petroleum crude to fuels and
chemicals. Oil and oil-derived intermediates (ethylene, propylene) became the dominant feedstocks.

Platinum (metal)- and acid (oxide)-catalyzed processes were developed to convert
petroleum to high-octane fuels. Hydrodesulfurization catalysis removed sulfur from
the crude to prevent catalyst deactivation. The discovery of microporous crystalline
alumina silicates (zeolites) provided more selective and active catalysts for many
reactions, including cracking, hydrocracking, alkylation, isomerization, and oligomerization. Catalysts that polymerize ethylene, propylene, and other molecules were
discovered. A new generation of bimetallic catalysts that were dispersed on high-
surface-area ($100–400$ m^2/g) oxides was synthesized.

The energy crisis in the early 1970s renewed interest in chemicals and fuels,
producing technologies using feedstocks other than crude oil. Intensive research was
carried out utilizing coal, shale, and natural gas to develop new technologies and to
improve on the activity and selectivity of older catalyst-based processes. Increasing
concern about environmental quality led to the development of the catalytic converter for automobiles and to other, nitrogen-oxide-reducing catalysts.

Modern surface science developed during the same period and has been applied
intensively to explore the working of catalysts on the molecular level, to characterize
the active surface, and to aid the development of new catalysts for new chemical
reactions. Indeed, surface science provided the means to explore the molecular structure and mechanisms of elementary reaction steps and to provide for rational design
for modification of catalyst activity and selectivity. This was carried out usually by
altering the structure of the surface and by using coadsorbed additives as bonding
modifiers for reaction intermediates on the surface.

In this chapter we describe the important macroscopic and molecular concepts of
surface catalysis that emerged from studies of recent decades. Then we shall review
what is known about a few important catalytic reactions that provide case histories
of the state of modern surface science of catalysis and of catalytic science.

7.2   CATALYTIC ACTION

One of the major functions of a catalyst is to aid in rapidly achieving chemical
equilibrium for certain chemical reactions.

Two of the simpler, although important, reactions that demonstrate this type of
catalytic action are the formation of water from oxygen and hydrogen ($\frac{1}{2}O_2 + H_2 \rightarrow
H_2O$) and the formation of ammonia from hydrogen and nitrogen ($3H_2 + N_2 \rightarrow
2NH_3$). Water has a standard free energy of formation $\Delta G_{298}^0 = -58$ kcal/mole
(232 kJ/mole). Yet O_2 and H_2 gas mixtures may be stored indefinitely in a glass
bulb without showing signs of any chemical reaction. Just by dropping a high-surface-area platinum gauze into the mixture, the reaction occurs instantaneously and
explosively—as demonstrated to the delight of freshman chemistry students in the

introductory chemistry courses. The reason for this striking effect can be explained as follows. H_2 and O_2 have large activation energies for several of the elementary steps for the reaction in the gas phase. First, one of the diatomic molecules must be dissociated. Dissociation energies are very large compared with thermal energies, RT (103 kcal/mole (412 kJ/mole) for H_2 and 117 kcal/mole (468 kJ/mole) for oxygen [3]). The subsequent atom–molecule reactions ($H + O_2$ or $H_2 + O$) still require an activation energy of about 10 kcal/mole (40 kJ/mole) [4]. Thus the gas-phase reaction is very improbable under any circumstances. In the presence of a properly structured platinum surface, however, both molecules dissociate to atoms with zero activation energies ($2Pt + H_2 \rightarrow 2Pt-H$, or $2Pt + O_2 \rightarrow 2Pt-O$) [5, 6], as shown by low-pressure surface studies. In addition, the atom–atom or atom–molecule reactions that subsequently take place on the surface have very low or no activation energies in contrast to that in the gas phase [5]. Thus the surface catalytic action involves its ability to atomize the large-binding-energy diatomic molecules by forming chemisorbed atomic intermediates and to lower the activation energy for the reaction on the surface that follows.

Similarly, the synthesis of ammonia from dinitrogen and hydrogen ($N_2 + 3H_2 \rightarrow 2NH_3$) required the "activation" of the $N-N$ bond to dissociate the molecule. The nitrogen atoms that form then must react with hydrogen atoms or molecules to produce NH_3. The very large dissociation energy of N_2 ($\Delta E = 280$ kcal/mole or 1120 kJ/mole) makes it virtually impossible for this reaction to occur in the gas phase. On an iron surface, however, N_2 dissociates on a properly structured surface [the (111) crystal face, for example] with a small activation energy (3 kcal or 12 kJ/mole). This is the key initiation step for the catalytic reaction. Iron also readily atomizes the hydrogen molecules. The chemisorbed nitrogen atoms then react with hydrogen atoms on the surface to produce NH, NH_2, and finally NH_3 molecules that desorb into the gas phase.

7.2.1 Kinetic Expressions

Catalysis is a kinetic phenomenon; we would like to carry out the same reaction with an optimum rate over and over again using the same catalyst surface. Therefore, in the sequence of elementary reactions leading to the formation of the product molecule, the rate of each step must be of steady state. Let us define the catalytic reaction turnover frequency, $\mathfrak{J}$, as the number of product molecules formed per second. Its inverse, $1/\mathfrak{J}$, yields the turnover time, the time necessary to form a product molecule. By dividing the turnover frequency by the catalyst surface area, $\mathfrak{A}$, we obtain the specific turnover rate, $\mathfrak{R}$ (molecules/cm^2/sec) $= \mathfrak{J}/\mathfrak{A}$ ($\mathfrak{R}$ often called the turnover frequency also, in the literature). This type of analysis assumes that every surface site is active. Although the number of catalytically active sites could be much smaller (usually uncertain) than the total number of available surface sites, the specific rate defined this way gives a conservative lower limit of the catalytic turnover rate. If we multiply $\mathfrak{R}$ by the total reaction time, δt, we obtain the turnover number, the number of product molecules formed per surface site. A turnover number of one corresponds to a stoichiometric reaction. Because of the experimental uncertainties, the turnover number must be on the order of 10^2 or larger for the reaction to qualify as catalytic.

While the turnover number provides a figure of merit for the activity of the catalyst sites, the reaction probability RP reveals the overall efficiency of the catalytic process under the reaction conditions. The reaction probability is defined as

$$RP = \frac{\text{rate of formation of product molecules}}{\text{rate of incidence of reactant molecules}} \qquad (7.1)$$

RP can be readily obtained by dividing $\Re$ by the rate of molecular incidence F which is obtained from the kinetic theory expression $F = P/(2\pi MRT)^{1/2}$.

The specific catalytic reaction rate $\Re$ can often be expressed as the product of the rate constant k and a reactant pressure (or concentration)-dependent term

$$\Re = k \times f(P_i) \qquad (7.2)$$

where P_i is the partial pressure of the reactants. The rate constant for the overall catalytic reaction may contain the rate constants of many of the elementary reaction steps that precede the rate-determining step. Because the slowest rate-reaction step may change as the reaction conditions vary (temperature, pressure, relative surface concentrations of reactants, catalyst structure), k may also change to reflect the changing reaction mechanism. Nevertheless, k can be defined using the Arrhenius expression

$$k = A \exp\left(-\frac{\Delta E^*}{RT}\right) \qquad (7.3)$$

where A is the temperature-independent preexponential factor and ΔE^* is the apparent activation energy measured under the catalytic reaction conditions.

Ranges of turnover rates for hydrocarbon reactions are shown in Figure 7.1. Turnover rates between 10^{-4} and 100 are used in the various technologies, and thus the temperature employed is adjusted to obtain the desired rates. The more complex isomerization, cyclization, dehydrocyclization, and hydrogenolysis reactions have activation energies ΔE^* in the range of 35–45 kcal/mole (140–180 kJ/mole); and thus according to the Arrhenius expression for the rate constant k, $k = A \exp(-\Delta E^*/RT)$, high temperatures are required to carry them out at the desired rates. Hydrogenation reactions have activation energies of 6–12 kcal/mole (24–28 kJ/mole) and therefore may be performed at high rates at 300 K or below. Thus, there are at least two classes of reactions distinguishable by their very different activation energies that may be carried out at high and at low temperature, respectively, under very different experimental conditions.

The rates of surface catalyzed reactions are usually measured by monitoring the concentrations of reactants and products as a function of time under steady-state conditions. Such studies tell us relatively little about the elementary surface reaction steps. Dynamic methods that alter the flow of reactants or introduce pulses of isotopically labeled reacting species have been useful to distinguish between reacting intermediates and adsorbed spectator species on surfaces. These investigations are carried out by following changes of the concentrations of adsorbates beginning when changes in flow rate commence, as a function of time, and by monitoring the time-dependent changes in the concentrations of isotopically labeled product molecules.

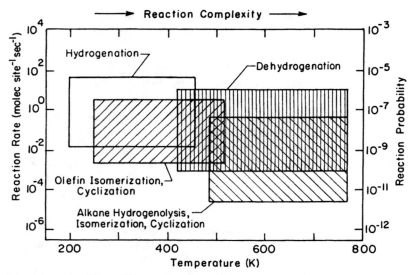

Figure 7.1. Block diagram of hydrocarbon conversion over platinum catalysts showing the approximate range of reaction rates and temperature ranges that are most commonly studied.

7.2.2 Selective Catalysis

A good catalyst is also selective and permits the formation of only one type of product when reactions may occur along several reaction paths. CO and H_2 react to produce methane (CH_4) exclusively when nickel is used as a catalyst, whereas only methanol (CH_3OH) is formed when the catalyst is copper and zinc oxide. The reaction of n-hexane in the presence of excess hydrogen can produce benzene, cyclic molecules, branched isomers, or shorter-chain species as shown in Figure 7.2. A selective catalyst will produce only one of these products.

In more general terms, catalyzed reactions involve either (a) successive kinetic steps leading to the final product or (b) alternative, simultaneous reaction paths yielding two or more products. The former reaction scheme may be represented by

$$A \xrightarrow{R_1} B \xrightarrow{R_2} C \qquad (7.4)$$

and a good example is the stepwise dehydrogenation of cyclohexane to cyclohexene and then to benzene. When two or more parallel reaction paths are operative as is the case during n-hexane conversion, the reaction scheme is

$$A \begin{array}{c} \xrightarrow{R_1} B \\ \xrightarrow{R_2} C \\ \xrightarrow{R_3} X \end{array}$$

We define the fractional catalytic selectivity, S_j, as the fraction of reacting mol-

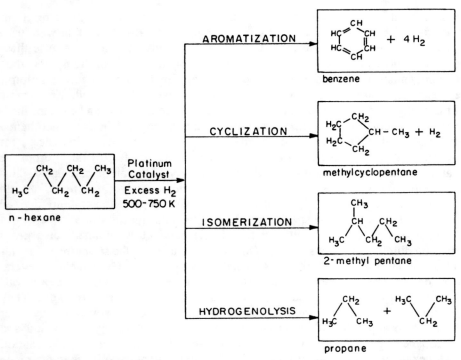

Figure 7.2. Various organic molecules that can all be produced by the catalyzed reactions of *n*-hexane [188].

ecules which are converted along a specified pathway

$$S_j = \frac{R_j}{\sum_{i=1}^{n} R_i} \qquad (7.5)$$

An additional possibility is provided by competitive parallel reactions

$$A \xrightarrow{R_1} B \qquad (7.6)$$
$$X \xrightarrow{R_2} Y \qquad (7.7)$$

Here the ratio of rates, R_1/R_2, defines the kinetic selectivity. The activity (rate) and the selectivity are the key parameters of any catalytic reaction.

7.2.3 Tabulated Kinetic Parameters for Catalytic Reactions

A great deal of kinetic information has been obtained for different types of catalyzed hydrocarbon reactions carried out over metal catalyst surfaces. These reactions include dehydrogenation, hydrogenation, hydrogenolysis and cracking, ring opening, dehydrocyclization, and isomerization. The kinetic parameters for these reactions are listed in Tables 7.1 to 7.41. In these tables the catalyst systems that were used are listed together with the temperature range of the investigation. Because these reactions are always carried out in the presence of hydrogen, both the hydrocarbon and hydrogen concentrations (in molecules$/cm^3$) are tabulated in a logarithmic form. These exponents are also displayed whenever they were determined. From these data the changes of the reaction rates with reactant concentrations can be determined. The rate of reaction at a given temperature, in the range used in the experimental study, is also calculated and listed, together with the apparent activation energy for the reaction, ΔE^*, and the logarithm of the preexponential factor, $\ln A$. From the rate and reactant concentrations a reaction probability (RP) can be calculated. This is also displayed in Tables 7.1 to 7.41 for the various catalytic reactions, as $-\ln$ RP. Fractional selectivities, S, are also supplied when reported. These are defined as the ratio of the rate of the specific reaction to the total reaction rate.

There is a great deal of scatter in the kinetic parameters obtained for a given reaction on different catalyst systems. This is expected, since the structure and bonding characteristics of the different metal catalysts vary widely. Nevertheless, several conclusions may be reached from the inspection of the data. The reaction probabilities are very low under the conditions where these reactions were carried out. They range from 10^{-8} to 10^{-5} for hydrogenation to 10^{-12} to 10^{-8} for most of the other reactions. The apparent activation energies are the lowest for hydrogenation and cyclopropane ring opening, 9–15 kcal/mole (36–60 kJ/mole). For dehydrogenation of cyclohexane and for the hydrogenolysis of C_4 to C_6 alkanes, ΔE^* is in the range 16–25 kcal/mole (64–100 kJ/mole). For most of the other reactions, which include (a) hydrogenolysis (the most frequently studied reaction) of ethane, propane, and other alkanes, (b) cracking of olefins and benzene, (c) dehydrogenation of alkanes, and (d) isomerization of C_5 to C_6 hydrocarbons, the apparent activation energies are in the range 25–50 kcal/mole (100–200 kJ/mole).

The kinetic information displayed in Tables 7.1 to 7.41 can be useful in establishing the optimum reaction conditions and catalyst systems. It is hoped that reliable kinetic parameters will become available for many other important catalyzed hydrocarbon reactions in the near future.

7.3 CATALYST PREPARATION, DEACTIVATION, AND REGENERATION

7.3.1 Catalyst Preparation

The higher the active surface area of the catalyst, the greater the number of product molecules produced per unit time. Therefore, much of the art and science of catalyst preparation deals with high-surface-area materials. Usually materials with 100- to 400-m^2/g surface area are prepared from alumina, silica, or carbon; and more recently other oxides (Mg, Zr, Ti, V oxides), phosphates, sulfides, or carbonates have been used. These are prepared in such a way that they are often crystalline with well-defined microstructures and behave as active components of the catalyst system in spite of their accepted name "supports." Transition-metal ions or atoms are then deposited in the micropores, which are then heated and reduced to produce small metal particles 10–10^2 Å in size with virtually all the atoms located on the surface (unity dispersion). The surface structure of the metal particles can often be controlled by the method of preparation. Usually more than one metal component is used, with bimetallic systems being the most popular in recent years. Frequently another oxide (e.g., TiO_2) is dispersed on the high-surface-area oxide (alumina) to impart unique catalytic properties as well. Additives [7] that are usually electron donors (alkali metals) or electron acceptors (halogens) are adsorbed on the metal or on the oxide to act as bonding modifiers for the coadsorbed reactants (see Chapter 4). This complex and intricately fabricated catalyst system is used for hundreds or thousands of hours and often millions of turnovers to produce the desired molecules at high rates and selectivity before their deactivation.

7.3.2 Catalyst Deactivation

Catalysts live long and active lives, but they do not last forever. The type of supported metal catalysts that are used in petroleum refining produces in the range of 200–800 barrels of products per pound of catalyst (1 barrel = 42 gallons). Once the catalyst is deactivated, it is either regenerated or replaced. There can be many reasons for the deactivation. At the operating temperatures some of the reactant hydrocarbons may completely decompose and deposit a thick layer of inactive carbon on the catalyst surface (coke). For many catalysts the deactivation is slow enough that they are used in steady-state operation. The liquid or gaseous reactants are passed through the catalyst with a well-defined "space velocity" that is normally measured as the weight hourly space velocity (WHSV)—that is, the pound of liquids or gas passed over the unit weight of catalyst per hour. For other active catalysts, deactivation is so rapid that they are used in a cyclic fashion; the reactors "swing" between running the catalytic reactions and regenerating. Thus understanding the causes of deactivation and developing new catalysts that are more resistant to "poisoning" are constant concerns of the catalytic chemist.

Many of the catalyst poisons act by blocking active surface sites. In addition, poisons may change the atomic surface structure in a way that reduces the catalytic activity. Sulfur, for example, is known to change the surface structure of nickel [8]. By forming chemical bonds of different strengths on the different crystal planes, it provides a thermodynamic driving force for the restructuring of the metal particles. Sometimes the rate of deactivation of metal catalysts by small concentrations of sulfur can indeed be dramatic. The automobile catalytic converter necessitated the removal of tetraethyl-lead from gasoline, one of the best antiknocking agents, because it readily poisoned the Pt–Pd catalyst by depositing lead sulfate on the noble-metal surfaces. One of the major causes of deactivation in crude oil cracking catalysts is the deposition on the catalyst surface of metallic impurities that are present as compounds in the reactant mixture. Vanadium- and titanium-containing organometallic compounds decompose and not only deactivate the catalyst surface but often plug the pores of the high-surface-area supports, thereby impeding the reactant–catalyst contact during petroleum refining.

A freshly prepared catalyst may not exhibit optimum catalytic activity upon its first introduction into the reactant stream. There may be efficient but undesirable side reactions that need to be eliminated. For this purpose a small amount of "poison" is often added to the reaction mixture or introduced in the form of pretreatment. Thus deactivating impurities may also be used, in small quantities, to improve the selectivity of the working catalyst.

7.3.3   Catalyst Regeneration

The regeneration treatment of the catalyst depends on the causes of deactivation. Most frequently, carbon deposition is the primary source of deactivation during hydrocarbon conversion reactions. In this circumstance, heating the spent catalyst in air or in oxygen burns off the carbon. The heat generated in this exothermic combustion reaction can be used beneficially in the overall catalytic process. Sintering of catalyst particles due to exposure to high temperatures for extended periods leads to loss of surface area. Oxygen can often oxidize the metal component of the catalyst to alter the shape and size of the metal particles. Metal oxides have lower surface energy than metals, and therefore oxidation could lead to better "wetting" of the high-surface-area oxide support. Subsequent reduction of the metal oxides in hydrogen may lead to redispersion of the metal constituent as small particles with increased total surface area. Additives such as chlorine that may form volatile metal halides can also help the redispersion of some of the catalyst components.

At high enough temperatures the micropores of the high-surface-area catalyst may collapse by sintering or melting. It is therefore essential that the materials chemistry be understood and that compounds with the proper surface and bulk thermodynamic properties be chosen to maintain their thermal stability under diverse (oxidizing or reducing) reaction conditions.

The removal of impurities that deposit from the reactant mixture poses particular challenge. Sulfur, arsenic, phosphorous, and vanadium are often deposited during oil refining. The reader is referred to publications that deal with these special problems of catalyst deactivation and regeneration (e.g., see references [9, 10]).

7.4   METAL CATALYSIS

Transition metals and their compounds, oxides, sulfides, and carbides are uniquely active as catalysts, and they are used in most surface catalytic processes. The effective-medium theory of the surface chemical bond (Chapter 6) emphasizes the dominant contribution of d-electrons to bonding of atoms and molecules at surfaces. Other theories [11] also point out that d-electron metals in which the d-bond is mixed with the s and p electronic states provide a large concentration of low-energy electronic states and electron vacancy states. This is ideal for catalysis because of the multiplicity of degenerate electronic states that can readily donate or accept electrons to and from adsorbed species. Those surface sites where the degenerate electronic states have the highest concentrations are most active in breaking and forming chemical bonds. These electronic states have high charge fluctuation probability (configurational and spin fluctuations) especially when the density of electron vacancy or hole states is high.

7.4.1   Trends Across the Periodic Table

One prediction of these theoretical models is that the heat of chemisorption of atoms should increase from right to left in the periodic table. This trend is well-documented in Chapter 6, and there is good agreement between experiments and theory. Thus, one of the important functions of transition metals in catalytic reactions is to atomize diatomic molecules and then to supply the atoms to other reactants and reaction intermediates. H_2, O_2, N_2, and CO are the diatomic molecules of importance, in order of increasing bond energy. The strength of bonding of hydrogen, carbon, nitrogen, and oxygen atoms on transition-metal surfaces provides the thermodynamic driving force for the atomization and for the release of atoms for reactions with other molecules. If the surface bonds are too strong, the reaction intermediates block the adsorption of new reactant molecules because of their long surface residence times and the reaction stops. For too weak adsorbate–surface bonds, the necessary bond-scission processes may be absent. Hence the catalytic reaction will not occur. A good catalyst is thought to be able to form chemical bonds of intermediate strength. These bonds should be strong enough to induce bond dissociation in the reactant molecules. However, the bond should not be too strong, thereby ensuring only short residence times for the surface intermediates and rapid desorption of the product molecules so that the reaction can proceed with a large turnover number.

These considerations are strikingly demonstrated by the volcano-shaped pattern of variation of catalytic activity as shown schematically in Figure 7.3. While the heat of adsorption is steadily decreasing from left to right, the catalytic reaction rates peak at the group VIII metals in the periodic table. Figure 7.3 shows the pattern of variation of catalytic reaction rates across the series of transition metals Re, Os, Ir, Pt, and Au for the hydrogenolysis of the C—C bond in ethane, the C—N bond in methylamine, and the C—Cl bond in methyl chloride.

The influence of the electronic structure of surface atoms show up not only in producing the volcano-shaped trends of transition metal catalytic activity across the periodic table, but also in producing the structure sensitivity of certain catalytic reactions on a given transition metal. A catalytic reaction is defined as *structure-sensitive* if the rate changes markedly as the particle size of the catalyst is changed. Reaction studies on single crystals revealed the importance of steps of atomic height and of kinks in the steps in increasing reaction rates for H_2/D_2 exchange, for dehydrogenation and hydrogenolysis. Theoretical studies indicate large changes in the local density of electronic states at the surface defect sites that correlate with changes in catalytic activity.

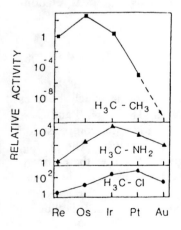

Figure 7.3. Catalytic activities of transition metals across the periodic table [188] for the hydrogenolysis of the C–C bond in ethane, the C–N bond in methylamine, and the C–Cl bond in methyl chloride.

7.4.2   Some Frequently Used Concepts of Metal Catalysis

During the operation of complex catalyst systems, several macroscopic experimental parameters have been uncovered that provide useful practical information about the nature of the catalyst or the catalyzed surface reaction. A catalytic reaction is defined to be *structure-sensitive* if the rate changes markedly as the particle size of the catalyst changes [12]. Conversely, the reaction is *structure-insensitive* on a given catalyst if its rate is not influenced appreciably by changing the dispersion of the particles under the usual experimental conditions. In Table 7.42 we list several reactions that belong to these two classes. Clearly, variations of particle size give rise to changes of atomic surface structure. The relative concentrations of atoms in steps, kinks, and terraces are altered. Nevertheless, no quantitative correlation has been made to date between variations of macroscopic particle size and the atomic surface structure.

During the development of mechanistic interpretations of catalytic reactions using the macroscopic rate equations that were determined by experiments, two types of reaction models found general acceptance. In one of them the rate-determining surface reaction step involves interaction between two atoms or molecules, both in the adsorbed state. This reaction model is called the *Langmuir–Hinshelwood* mechanism [13, 14]. In the other the rate-determining reaction step involves a chemical reaction between a molecule from the gas phase and one in the adsorbed state. This is called the *Rideal–Eley* mechanism [15]. Most reactions have rate equations that fit the first of these two mechanisms. Recently, the oxidation of CO has been identified by molecular-scale studies as obeying the Langmuir–Hinshelwood reaction mechanism [16]. However, correlation of these reaction mechanisms (suggested by inspection of the macroscopic rate equations) with molecular-level studies of the elementary surface reactions remains one of the future challenges of catalysis.

During studies of a given catalyzed reaction over catalysts that were prepared in different ways, an interesting phenomenon was found, called the *compensation effect* [17]. Using the Arrhenius expression for the rate constant, both the preexponential factor and the activation energy for the reaction were found to have varied greatly from catalyst to catalyst. However, they varied in such a way as to compensate each other, so that the rate constant (or the reaction rate under the same conditions of pressure and temperature) remained almost constant. For example, for the methanation reaction (that is, the hydrogenation of CO), the following empirical relationship was found to hold between A and ΔE^*:

$$\ln A = \alpha + \frac{\Delta E^*}{R\Theta} \tag{7.8}$$

where α is a constant and Θ is called the *isokinetic temperature*, at which the rates on all the catalysts are equal. For the methanation reaction [18], $\alpha \approx 0$ and $\Theta = 436$ K. Thus $\ln A_{CH_4} \approx 1.1 \Delta E^*$ kcal/mole. Figure 7.4 shows the compensation effect for the methanation reaction for eight different metal catalysts. The $\ln A_{CH_4}$ versus ΔE^* plots yield a straight-line relationship. Figure 7.5 shows the compensation effect for the hydrogenolysis reactions whose rates are displayed in Figure 7.3.

The compensation effect has been rationalized in a variety of ways. It is thought

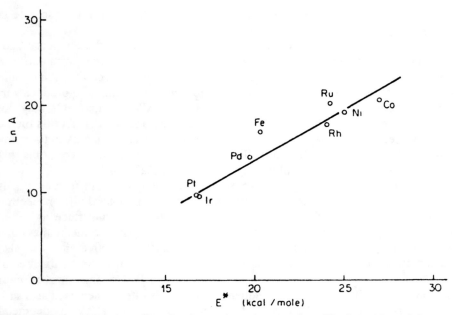

Figure 7.4. Compensation effect for the methanation reaction. The logarithm of the preexpontial factor is plotted againt the apparent activation energy, ΔE^*, for this reaction over several transition-metal catalysts [18].

that one catalyst may have a large concentration of active sites where the reaction requires a high activation energy, while the other catalyst, which is prepared differently, has a small concentration of active sites that have low activation energies for the same surface reaction. An atomic-level explanation of the compensation effect remains the task of scientists in the future.

During most reactions, the surface of the active metal catalyst is covered with a *strongly chemisorbed overlayer* that remained tenaciously bound to the surface for 10^2–10^6 turnovers. During hydrocarbon reactions, this is a carbonaceous overlayer with a composition of about $(H/C) \approx 1$, during ammonia syntheses it is chemisorbed nitrogen, and during hydrodesulfurization it is a mixture of sulfur and carbon. It is believed that this overlayer may play a role in restructuring the surface to create new active sites and in altering the bonding of reactants, intermediates, and prod-

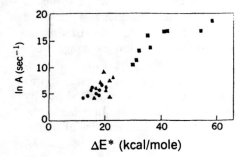

Figure 7.5. Compensation effect for the hydrogenolysis reaction. The logarithm of the preexponential factor is plotted against the apparent activation energy, ΔE^*, for this reaction over several transition-metal catalysts. The squares, triangles, and circles represent values for ethane, methylamine, and methyl chloride hydrogenolysis, respectively [188].

ucts. However, more experimental evidence is needed before the precise role of these strongly held surface deposits can be identified.

Structure modifiers and *bonding modifiers* are often introduced as important additives when formulating the complex catalyst systems. Structural promoters can change the surface structure that is often the key to catalyst selectivity. Aluminum oxide facilitates the restructuring of iron in the presence of nitrogen to produce surfaces that are most active during ammonia synthesis. Alloy components may not participate in the reaction chemistry but modify structure and site distribution on the catalyst surface. Site blocking could improve selectivity as has been proven for many working catalyst systems. Sulfur and silicon or other strongly adsorbed atoms that seek out certain active sites can block undesirable side reactions.

Bonding modifiers are employed to weaken or strengthen the chemisorption bonds of reactants and products. Strong electron donors (such as potassium) or electron acceptors (such as chlorine) that are coadsorbed on the catalyst surface are often used for this purpose. Alloying may create new active sites (mixed metal sites) that can greatly modify activity and selectivity. New catalytically active sites can also be created at the interface between the metal and the high-surface-area oxide support. In this circumstance the catalyst exhibits the so-called strong metal-support interaction (SMSI). Titanium oxide frequently shows this effect when used as a support for catalysis by transition metals. Often the sites created at the oxide–metal interface are much more active than the sites on the transition metal.

7.5 CATALYSIS BY IONS AT SURFACES. ACID–BASE CATALYSIS

Most surface reactions and the formation of surface intermediates involve charge transfer—either an electron transfer or a proton transfer. These processes are often viewed as modified acid–base reactions. It is common to refer to an oxide catalyst as acidic or basic according to its ability to donate or accept electrons or protons [19].

The electron transfer capability of a catalyst is expressed according to the Lewis definition. A surface site capable of receiving a pair of electrons from the adsorbate is a Lewis acid. A site having a free pair of electrons that can be transferred to the adsorbate is a Lewis base. The acidity of metal ions of equal radius increases with the increasing charge of the metal ions: $Na^+ < Ca^{2+} < Y^{3+} < Th^{4+}$. The strength of the Lewis acidity is measured by determining the binding energies of the charge-transfer complexes that form by this type of electron-transfer process.

The proton-transfer capability of a catalyst is expressed according to the Brønsted definition. A surface site capable of losing a proton to the adsorbate is a Brønsted acid. A site that can accept a proton from the adsorbed species is a Brønsted base. The Brønsted acidity of the catalyst is usually determined by ion exchange from solution (surface proton is substituted by alkali ions Li^+, Na^+, etc.) or by the adsorption of weak acids or bases, such as phenol and pyridine, or the surface. In this way the proton-transfer ability of the surface can be titrated. The Brønsted acidity for oxides has also been related to the metal—oxygen bond energies. In general, the acidity increases with an increase of charge on the metal ion. In the series of oxides Na_2O, CaO, MgO, Ag_2O, BeO, Al_2O_3, CdO, ZnO, SnO, H_2O, B_2O_3, FeO, SiO_2,

Ca_2O_3, Fe_2O_3, P_4O_6, SnO_2, GeO_2, TiO_2, SO_2, N_2O_5, and Cl_2O_7, those that are to the left of water are bases, and those to the right are acids [19].

7.5.1 Acid Catalysis in Solutions

In aqueous solutions of acids the hydronium ion H_3O^+ has been identified as a proton donor. Undissociated acids in high concentrations of acid solutions, HF for example, can also act as proton donors. The stronger the acid, the more active the catalyst. This acid strength is related to the dissociation constant of the acid, K_{HA}:

$$K_{HA} = \frac{a_{H^+}a_{A^-}}{a_{HA}} \tag{7.9}$$

where a_{H^+}, and a_{A^-}, and a_{HA} are the activities of the ionic and undissociated species, respectively. The rate of acid-catalyzed reactions can be represented by the Brønsted relation

$$\ln k = \alpha \ln K_{HA} + \text{const.} \tag{7.10}$$

where k is the second-order rate constant for the reaction and α is a constant with a value between 0 and 1. The Brønsted relation is one of the so-called free energy relations encountered in physical organic chemistry that relate the kinetic parameters of a reaction (the rate constant, for example) to an equilibrium constant.

In concentrated strong acid solutions there can be many different proton donors. A useful function, the Hammett acidity function measures the tendency of the solution to donate a proton to a neutral base, B (e.g., a neutral organic molecule). The protonation of the base can be expressed as

$$H^+ + B \rightleftharpoons BH^+ \tag{7.11}$$

The dissociation equilibrium constant of BH^+ can be written as

$$K_a = \frac{a_{H^+}a_B}{a_{BH^+}} = \frac{a_{H^+}\gamma_B}{\gamma_{BH^+}} \cdot \frac{C_B}{C_{BH^+}} \tag{7.12}$$

where γ_B, γ_{BH^+}, and C_B, C_{BH^+} are the activity coefficients and the concentrations of the neutral and the protonated base, respectively. C_B/C_{BH^+} can be measured experimentally. The negative logarithm of K_a defines the Hammett acidity function H_0 as

$$-\log K_a = pK_a = H_0 + \log \frac{C_{BH^+}}{C_B} \tag{7.13}$$

The H_0 values for aqueous solutions for several strong acids are given in Figure 7.6.

There are acids with values of the Hammett acidity function of -20 or less; these are called *superacids* because the H_0 value for pure H_2SO_4 is only -12. These

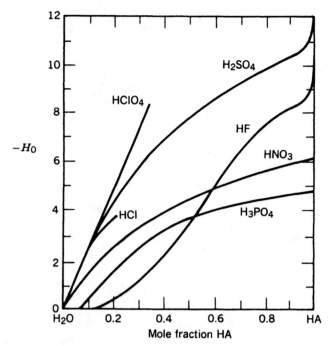

Figure 7.6. The Hammett acidity function, H_0, for several acids as a function of their mole fraction in aqueous solution [189].

extremely strong acids can be formed from the combination of Lewis and Brønsted acids. These include:

Brønsted Acid	Lewis Acid
HF	BF_3
HF	SbF_5
FSO_3H	SbF_5

The great proton donor strength of superacids is due to the stabilization of the protonated forms of the Brønsted acid in an ion pair; $H_2F^+SbF_6^-$ for example.

7.5.2 Solid Acids

There are solid acids with Hammett acidity functions that are greater than sulfuric acid and similar to that of the superacids. Many of these are alumina silicates (commonly called *zeolites*), which are among the most common minerals in nature. The acid strength of these materials often depends on the Si/Al ratio. The aluminum ion, having one less valence electron than the silicon ion, has high electron affinity, thereby stabilizing a proton (Brønsted acid) near the AlO_4^- tetrahedra by weakening the O—H bond in the hydrogen-form zeolite:

$$
\begin{array}{ccccccc}
 & \overset{+}{H} & & & \overset{+}{H} & & \\
 & | & & & | & & \\
O \quad O & O \quad O & O \quad O \\
\diagdown \diagup \diagdown \diagup \diagdown & \diagup \diagdown \diagup \diagdown \diagup \diagdown & \diagup \diagdown \diagup \\
Si \quad Al^- \quad Si \quad Al^- \quad Si \\
\diagup \diagdown \diagup \diagdown \diagup \diagdown \diagup \diagdown \diagup \diagdown \\
O \quad O \quad O \quad O \quad O \quad O \quad O \quad O \quad O
\end{array}
$$

When this material is heated to high temperatures, water is driven off and coordinately unsaturated Al^{3+} ions are formed; these are strong electron-acceptor Lewis acid sites.

Very high internal surface area zeolites ($10^2 m^2/g$) can be synthesized with controlled pore sizes of 8–20 Å and controlled acidity [(Si/Al) ratio]. These find applications in the cracking and isomerization of hydrocarbons that occur in a shape-selective manner as a result of the uniform pore structure and are the largest-volume catalysts utilized in petroleum refining at present [20]. They are also the first of the "high-technology" catalysts where the chemical activity is tailored by atomic-scale study and control of the internal surface structure and composition.

7.5.3 Carbenium Ion Reactions

Hydrocarbons may be viewed as weak bases that can be protonated by strong acids to form carbenium ions. For an olefin this reaction may be written as

$$
R-CH{=}CH-R' + H^+ \rightleftharpoons R-CH_2-\overset{+}{C}H-R' \tag{7.14}
$$

Tertiary carbenium ions are more stable than secondary ions, which are more stable than primary ions:

$$
R_3C-\overset{+}{\underset{\underset{CR_3''}{|}}{C}}- CR_3' > R_3C-\overset{+}{C}H-CR_2' > R_3C-CH_2-\overset{+}{C}H_2 \tag{7.15}
$$

Isomerization, carbon–carbon bond scission (cracking), and carbon–carbon bond formation (alkylation) are among the most important hydrocarbon conversion reactions catalyzed by acids. Zeolites are often used to carry out these reactions during the refining of petroleum. Some of the zeolites are particularly active to convert olefins and cycloparaffins to paraffins and aromatics to produce jet fuel and gasoline.

7.6 MOST FREQUENTLY USED CATALYST MATERIALS

It may be instructive to review how widely catalysts are applied in the various technologies and to identify some of the most frequently used materials. There are three major areas of catalyst application at present [21]: automotive [22, 23], fossil-fuel refining, and production of chemicals. Table 7.43 lists the chemical processes that are the largest users of heterogeneous catalysts and the catalyst systems that are employed most frequently at present.

The automotive industry uses mostly noble metals (platinum, rhodium, and palladium) for catalytic control of car emissions: unburned hydrocarbons, CO, and NO. These highly dispersed metals are supported on oxide surfaces, and the catalyst system is specially prepared to be active at the high space velocities of the exhaust gases and over a wide temperature range. In petroleum refining, zeolites are most widely used for cracking of hydrocarbon in the presence of hydrogen. The important hydrodesulfurization process uses mostly sulfides of molybdenum and cobalt on an alumina support. The "reforming" reactions to produce cyclic and aromatic molecules and isomers from alkanes to improve the octane number are carried out mostly over platinum or platinum-containing bimetallic catalysts, such as Pt–Re and Pt–Sn. Sulfuric and hydrofluoric acids are the catalysts for alkylation. In the chemical technologies, steam reforming of natural gas (mostly methane) to produce hydrogen and CO is an important large-volume catalytic process. The purified natural gas is reacted with steam to form CO and H_2, mostly over supported nickel catalyst. The water–gas shift reaction ($CO + H_2O \rightarrow CO_2 + H_2$) is then employed to produce more hydrogen. The most frequently used catalyst for this purpose is iron-based. Methanol is produced from CO and H_2, and ammonia is produced from H_2 and N_2. Copper oxide and zinc oxide are also used for the shift reaction, as well as for the production of methanol from CO and H_2. Nickel is the catalyst for methanation from CO and H_2, and iron is the major catalyst for the ammonia synthesis.

Catalytic hydrogenation processes primarily use nickel and palladium as catalysts. Hydrogenation of nitrile groups to amines and various edible and inedible oils for the preparation of margarine, salad oils, and stearine are some of the major applications. Selective hydrogenation of olefins is also an important catalytic process. Among the larger-volume oxidation reactions, the oxidation of ammonia to nitric oxide to produce nitric acid uses noble metals: Pt, Pt–Rh, and Pt–Pd–Rh. The oxidation of SO_2 to SO_3 to produce sulfuric acid uses mostly vanadium oxide as catalyst. Ammoxidation, which makes acrylonitrile from propylene, oxygen and ammonia uses bismuth and molybdenum oxides as catalysts. Oxychlorination to make vinyl chloride from acetylene and HCl uses copper chloride as a catalyst. Polymerization reactions of ethylene and propylene are catalyzed by titanium trichloride, aluminum alkyls, chrome oxide on silica, and peresters. While these are the catalysts that are used in the largest quantity, many other highly selective catalysts serve as the basis of entire chemical technologies. In fact, the value of a very selective catalyst that aids a complex chemical transformation and the production of precious lifesaving pharmaceuticals is without compare.

Most of the catalysts employed in the chemical technologies are heterogeneous. The chemical reaction takes place on surfaces, and the reactants are introduced as gases or liquids. Homogeneous catalysts, which are frequently metalloorganic molecules or clusters of molecules, also find wide and important applications in the chemical technologies [24]. Some of the important homogeneously catalyzed processes are listed in Table 7.44. Carbonylation, which involves the addition of CO and H_2 to a C_n olefin to produce a C_{n+1} acid, aldehyde, or alcohol, uses rhodium and cobalt complexes. Cobalt, copper, and palladium ions are used for the oxidation of ethylene to acetaldehyde and to acetic acid. Cobalt(II) acetate is used mostly for alkane oxidation to acids, especially butane. The air oxidation of cyclohexane to cyclohexanone and cyclohexanol is also carried out mostly with cobalt salts. Further oxidation to adipic acid uses copper(II) and vanadium(V) salts as catalysts. The hydrocyanation of butadiene to adiponitrile uses zero-valent nickel complexes. Polymerization technologies also frequently use homogeneous catalysts. The manufacture of polyethylene terephthalate uses antimony salts, and the copolymerization of ethylene and propylene to produce rubber uses alkylvanadium compounds.

7.7 SURFACE-SCIENCE APPROACH TO CATALYTIC CHEMISTRY

The purpose of surface-science studies of heterogeneous catalyst systems is to understand how they work on the atomic scale. One aims to identify the active sites where bond breaking and rearrangement take place and to detect surface intermediates that form. Studies are conducted to determine how the atomic surface structure and surface composition determine activity and selectivity. Once such an atomic-scale understanding is obtained, more active and selective catalysts can be designed, or one might find substitutes for precious-metal catalysts that are not readily available. Working catalyst systems have complicated structures, however, that do not lend themselves easily to atomic-scale investigations. The large-surface-area internal pore structure of the support hides the metal particles and makes it difficult to study their structure, oxidation state, and composition, which determine both activity and selectivity. Characterization of these complex but practical catalyst systems are the aims of many laboratories.

There is a different approach to the study of catalyst systems, which I would like to call the model system or synthetic approach [25]. It is similar to the technique used by synthetic organic chemists to prepare complex organic molecules by linking the smaller segments one by one until the final product is obtained. The catalyst particle is viewed as composed of single-crystal surfaces, as shown in Figure 7.7. Each surface has different reactivity, and the product distribution reflects the chemistry of the different surface sites. One may start with the simplest single-crystal surface [e.g., the (111) crystal face of platinum] and examine its reactivity. It is expected that much of the chemistry of the dispersed catalyst system would be absent on such a homogeneous crystal surface. Then high-Miller-index crystal faces are prepared to expose surface irregularities, steps, and kinks of known structure and concentration, and their catalytic behavior is tested and compared with the activity of the dispersed supported catalyst under identical experimental conditions. If there are still differences, the surface composition is changed systematically or other variables are introduced until the chemistries of the model system and the working catalyst become identical. This approach is described by the following sequence:

Studies of the structure of crystal surfaces when clean
and in the presence of chemisorbed reactants and
products. Chemisorption, structure, and bonding studies
of reactants and products at low pressures ($\leq 10^{-4}$ torr)

$\updownarrow$

Surface reactions on external surfaces (small area
(≈ 1 cm^2) crystals, foils, thin films, deposited particles)
at high pressures (10^3 to 10^5 torr)

$\updownarrow$

Reactions on dispersed (high surface area) catalysts at
high pressures (10^3 to 10^5 torr)

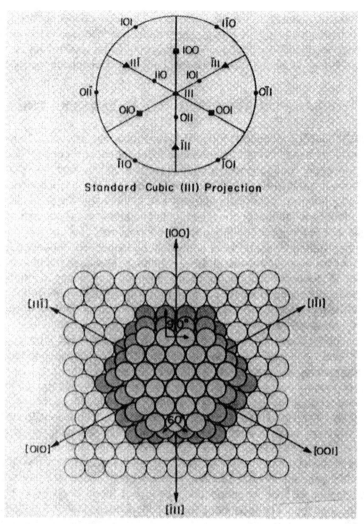

Figure 7.7. Catalyst particle viewed as a crystallite, composed of well-defined atomic planes, steps, and kink sites.

Investigations in the first step define the surface structure and composition on the atomic scale and the chemical bonding of adsorbates. Studies in the second phase reveal many of the elementary surface reaction steps and the dynamics of surface reactions. Combined studies in the second and third steps establish the similarities and differences between the model systems and the dispersed catalysts under practical reaction conditions.

The advantage of using small-area (1 cm²) catalyst samples is that their surface structure and composition can be prepared with uniformity and can be characterized by the many available surface diagnostic techniques. However, the small catalyst area that must be used in studies of this type necessitated the development of new instrumentation, which will be described next.

7.7.1 Techniques to Characterize and Study the Reactivity of Small-Area Catalyst Surfaces

7.7.1.1 High-Pressure Reactors In our synthetic approach to catalytic reaction studies, it is imperative that we determine the surface composition and surface structure in the same chamber where the reactions are carried out, without exposing the crystal surface to the ambient atmosphere. This necessitates the combined use of an ultrahigh-vacuum enclosure, where the surface characterization is to be carried out, and a high-pressure isolation cell, where the catalytic studies are performed. Such an apparatus is shown in Figure 7.8. The small-surface-area (approximately 1 cm^2) catalyst is placed in the middle of the chamber, which can be evacuated to 10^{-9} torr. The surface is characterized by LEED and AES and by other desired surface diagnostic techniques. Then the lower part of the high-pressure isolation cell is lifted to enclose the sample in a 0.5-liter volume that is sealed by a copper gasket (approximately 2000 psi pressure is needed to provide a leak-free seal). The isolation chamber can be pressurized to 100 atm if desired and is connected to a gas chro-

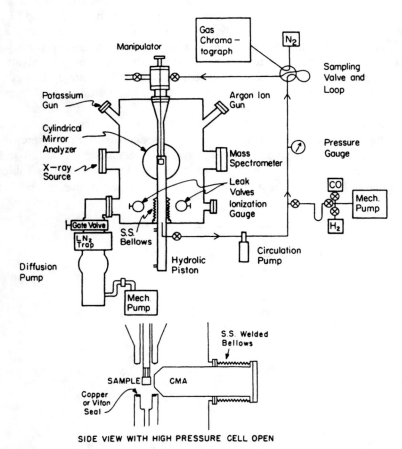

Figure 7.8. Schematic representation of one type of apparatus capable of carrying out catalytic-reaction-rate studies on single-crystal surfaces of low surface area at high pressures (atmospheres) and also to perform surface characterization in ultrahigh vacuum.

matograph that detects the product distribution as a function of time and surface temperature. The sample may be heated resistively, both at high pressure or in ultrahigh vacuum. After the reaction study, the isolation chamber is evacuated and opened, and the catalytic surface is again analyzed by the various surface-diagnostic techniques. Ion-bombardment cleaning of the surface or means to introduce controlled amounts of surface additives by vaporization are also available. The reaction at high pressures may be studied in the batch or the flow mode.

There are many different designs available for combined high-pressure reaction studies and ultrahigh-vacuum surface science investigations. Transfer rods that move the sample from the environmental cells to the UHV chamber and reaction cells that permit liquid-phase or gas-phase reaction studies have been described in the literature.

7.7.1.2 Comparison of the Reactivities of Small- and Large-Surface-Area Catalysts

It is essential to test the high-pressure chamber to make sure that the measured reaction rates using the small-surface-area sample can be readily compared to reaction rates obtained on large-surface-area catalysts. This comparison has been made using the ring opening of cyclopropane [26] and the hydrogenation of carbon monoxide [27] as test reactions. Table 7.45 shows the turnover numbers and the activation energies obtained for the ring opening of cyclopropane to form propane on small-area single-crystal platinum and on dispersed platinum catalysts under identical experimental conditions. The agreement is indeed excellent. This is a structure-insensitive reaction at high pressures that lends itself well to such correlative studies. For structure-sensitive reactions, marked differences are found, with the single-crystal catalyst being much more active in general. Similarly, excellent agreements among rates, activation energies, and the product distribution were obtained for the hydrogenation of carbon monoxide over polycrystalline rhodium foils and dispersed, silica-supported rhodium catalyst particles. This is shown in Table 7.46. Figures 7.9 and 7.10 show the agreement reached between studies of the same re-

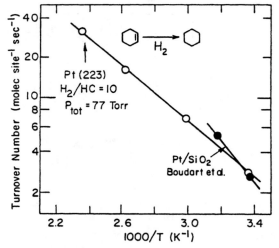

Figure 7.9. Arrhenius plot of the rate of cyclohexene hydrogenation to cyclohexane on Pt(111) crystal surfaces and on platinum particles dispersed on silica. Both the rates and activation energies are similar [190].

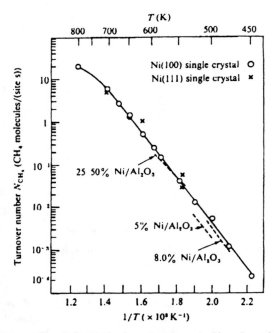

Figure 7.10. Arrhenius plot of the rate of methane production from hydrogen and carbon monoxide over Ni(111) and (100) compared to the production over supported Ni/Al$_2$O$_3$ dispersed catalyst. Both the rates and the activation energies are the same [191].

actions (benzene hydrogenation over Pt and CO hydrogenation over Ni) over low-surface-area model single crystal and high-surface-area dispersed catalysts.

7.8 CASE HISTORIES OF SURFACE CATALYSIS

Surface-science studies succeeded to identify many of the molecular ingredients of surface catalyzed reactions. Each catalyst system that is responsible for carrying out important chemical reactions with high turnover rate (activity) and selectivity has unique structural features and composition. In order to demonstrate how these systems operate, we shall review what is known about (a) ammonia synthesis catalyzed by iron, (b) the selective hydrogenation of carbon monoxide to various hydrocarbons, and (c) platinum-catalyzed conversion of hydrocarbons to various selected products.

7.8.1 Ammonia Syntheses

7.8.1.1 Thermodynamics and Kinetics The reaction of nitrogen and hydrogen to produce ammonia, $N_2 + 3H_2 \rightarrow 2NH_3$, is somewhat exothermic. The free energy of ammonia formation as a function of temperature is shown in Figure 7.11. The reaction is carried out over iron catalyst that is "promoted" by adding alumina and potassium most frequently. The reaction temperature is around 400°C, and total pressures utilized are in the range of 150–300 atm.

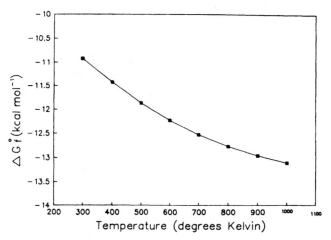

Figure 7.11. The free energy of ammonia formation as a function of temperature [33].

7.8.1.1.1 Kinetics From the experimental data, the observed dependencies of the rate on N_2 and H_2 pressures several rate laws have been proposed; the best known is perhaps the one by Temkin [28, 29]. An extension of this rate law by Nielsen [30, 31] yields

$$\frac{dP_{NH_3}}{dt} = \frac{k(P_{N_2}K_a - P_{NH_3}^2/P_{H_2}^3)}{(1 + K_3 P_{NH_3}/P_{H_2}^w)^{2\alpha}} \tag{7.16}$$

where $w = 1.5$ and $\alpha = 0.75$. k, K_a, and K_3 are constants. The rate of ammonia formation depends in a rather complex manner on the partial pressures of N_2, H_2, and NH_3 mostly because of the possibility of a back-reaction. Far from equilibrium this may be neglected, and in this circumstance the rate depends only on the nitrogen pressure. This conclusion indicates that the rate-limiting step is the dissociative adsorption of nitrogen on the catalyst surface—a conclusion that is shared by most of the practitioners.

Other important rate equations that are applicable in a variety of experimental conditions have been proposed by Ozaki, Taylor, and Boudart [32].

The net activation energy for the reaction is 76 kJ/mole (e.g., see references [33–36]), which is in excellent agreement with the 81-kJ/mole value determined using single-crystal iron surfaces [37].

7.8.1.2 Catalyst Preparation The industrial catalyst is prepared by the reduction of iron oxide, Fe_3O_4 (94 wt%). It is in the shape of small porous particles with a surface area in the range of 10–15 m^2/g. Additives that improve its performance include Al_2O_3 (2.3 wt%), K_2O (0.8 wt%), and often CaO (1.7 wt%), MgO (0.5 wt%), and SiO_2 (0.4 wt%). Al, Mg, Ca, and Si oxides stabilize the pore structure and the surface structure of the iron catalyst K_2O, although decreases the iron surface area somewhat still greatly increases the ammonia yield at 613 K from 0.2 mol% to 0.34 mol%.

7.8.1.3 Activity for Ammonia Synthesis Using Transition Metals Across the Periodic Table   There are two factors that are all important in determining the ammonia synthesis rate. One is the N_2 dissociative sticking probability. N_2 dissociation turns out to be rate-limiting, and at low conversions the total rate of the reaction equals the dissociation rate of N_2. The other factor is the nitrogen atom chemisorption energy. Chemisorbed atomic nitrogen is by far the most stable reaction intermediate. Therefore, the surface is mainly covered by nitrogen atoms up to 90% of a monolayer; and the number of free sites on the surface where the nitrogen can adsorb is proportional to $(1 - \theta_N)$, where θ_N is the nitrogen coverage.

Using a kinetic model that was reported by A. Nielsen [30], the ammonia formation rate can be calculated as a function of the number of *d*-electrons in the transition metals. The results are shown in Figure 7.12. It produces a volcano curve similar to that observed experimentally by Ozaki and Aika, who have plotted the variation of the activity of various transition metals for the ammonia synthesis reaction as a function of the degree of filling of the *d*-band (Figure 7.13). The calculated results, and those found by experiments, overlap very well indeed. On the right side of the maximum in the volcano curve, the ammonia production decreases because the rate of N_2 dissociation drops as a consequence of the increase in the activation energy for dissociation. To the left of the tip of the volcano, the dissociation rate increases; but since the nitrogen chemisorption bond also increases in strength, the number of surface sites where the nitrogen molecule can dissociate decreases so fast that the overall rate decreases.

7.8.1.4 Surface Science of Ammonia Synthesis

7.8.1.4.1 Structure Sensitivity of Ammonia Synthesis   An ultrahigh vacuum chamber equipped with a high-pressure cell was developed to study the ammonia synthe-

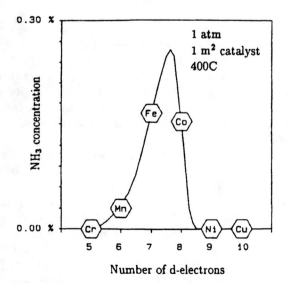

Figure 7.12. The calculated ammonia concentration for a fixed set of reaction conditions as a function of the number of *d*-electrons [71].

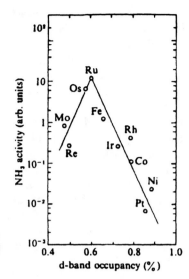

Figure 7.13. The activity of various transition metals for ammonia synthesis as a function of the degree of filling of the *d*-band [66].

sis reaction on iron single-crystal surfaces. A single crystal is enclosed in a high-pressure cell which constitutes part of a microbatch reactor. High pressures of gases—15 atm of hydrogen and 5 atm of nitrogen, for example—are introduced and the sample is heated to reaction temperatures, 600–700 K. The ammonia production is monitored using a selective photoionization detector with such photon energy that it ionizes ammonia and not nitrogen or hydrogen. After the reaction is completed, the reaction loop is evacuated and the cell opened, returning the sample to the ultrahigh vacuum environment where surface characterization is performed by Auger electron spectroscopy, low-energy electron diffraction, and temperature-programmed desorption.

In Figure 7.14 the rates of ammonia synthesis are shown over five iron crystal orientations. The Fe(111) and Fe(211) surfaces are by far the most active in ammonia synthesis and they are followed in reactivity by Fe(100), Fe(210), and Fe(110) [38]. Schematic representations of the idealized unit cells for these surfaces are shown in Figure 7.15. There are two possible reasons for the high activity of the (111) and (211) faces compared to the other (210), (100), and (110) orientations: their exceptionally high surface roughness or the presence of unique active sites the other crystal faces may not possess.

The (111) surface can be considered a rough surface, since it exposes second- and third-layer atoms to reactant gases in contrast to the (110) surface which only exposes first-layer atoms. Work functions are related to the roughness of a surface [39], and it is useful to quantify the corrugation of a plane in this way. Open faces, such as the (111) surface, have lower work functions than do close-packed faces, such as the (110) surface. The work functions of all the iron faces are not currently available but they are for tungsten [40], another body-centered cubic (bcc) metal which also shows structure sensitivity for ammonia decomposition [41]. The order of decreasing work function (ϕ) is as follows: $\phi_{110} > \phi_{211} > \phi_{100} > \phi_{111} > \phi_{210}$. However, the order of decreasing work function from crystal face to crystal face does not correlate with variations of catalytic activity.

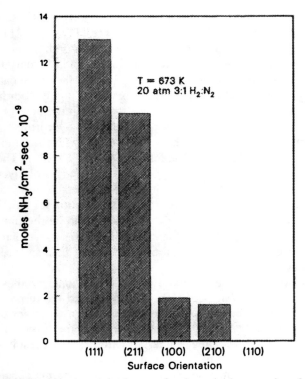

Figure 7.14. Rates of ammonia synthesis over five iron single-crystal surfaces with different orientations: (111), (211), (100), (210), and (110) [38].

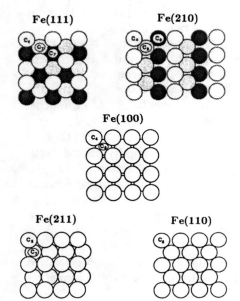

Figure 7.15. Schematic representations of the idealized surface structures of the (111), (211), (100), (210), and (110) orientation of iron single crystals. The coordination of each surface atom is indicated [38].

The second possible explanation for the structure sensitivity of ammonia synthesis rate of iron involves the nature of the active sites. The (111) and (211) faces of iron are the only surfaces which expose C_7 sites (iron atoms with seven nearest neighbors) to the reactant gases. Theoretical work by Falicov and Somorjai [11] has suggested that highly coordinated surface atoms would show increased catalytic activity due to low-energy charge fluctuations in the d-bands of highly coordinated surface atoms. Examination of the results suggests that the latter argument of active sites is the key to the structure sensitivity of ammonia synthesis over iron.

The reaction rates, in Figure 7.14, show that the (211) face is almost as active as the (111) plane of iron, while Fe(210) is less active than Fe(100). The Fe(210) and Fe(111) faces are open faces which expose second- and third-layer atoms. The Fe(211) face is more close-packed, but it exposes C_7 sites. If either surface roughness or a low work function were the important consideration for an active ammonia synthesis catalyst, then the Fe(210) would be expected to be the most active face. However, in marked contrast, Fe(111) and Fe(211) faces are much more active, indicating that the presence of C_7 sites is more important than surface roughness in an ammonia synthesis catalyst.

The idea of C_7 sites being the most active site in ammonia synthesis on iron has been suggested in the past. Dumesic et al. [42] found that the turnover number for ammonia synthesis was lower on small iron particles than on larger ones. Pretreatment of an Fe/MgO catalyst with ammonia enhanced the turnover number over small iron particles, but did not affect the larger particles. This result was explained by noting that the concentration of C_7 sites would be expected to be higher on the smaller iron particles and that restructuring induced by ammonia enhanced the number of these sites on the catalyst.

7.8.1.4.2 Kinetics of Dissociative Nitrogen Adsorption Because this step is rate-determining for ammonia synthesis, considerable effort has been expended on its detailed investigation. It has turned out to be of great complexity so that, even now, complete understanding of the underlying microscopic dynamics is still lacking, although there exists general agreement about the experimental findings.

In Figure 7.16, the variation in the relative surface concentration of N_{ad} (as monitored by Auger electron spectroscopy) with N_2 exposure at elevated temperatures for the Fe(110), (100), and (111) surfaces [43, 44] is shown. The slopes of these curves yield the sticking coefficients for dissociative chemisorption which are obviously very small and depend markedly on the surface orientation. More specifically, the initial sticking coefficient (at 683 K) changes from 7×10^{-8} to 2×10^{-7} to 4×10^{-6} in the sequence Fe(110) < Fe(100) < Fe(111); that is, the (111) plane is about two orders of magnitude more active than the most densely packed (110) plane. This sequence of activity toward dissociative nitrogen adsorption at low pressures ($< 10^{-4}$ torr) is in agreement with that found for the rate of ammonia production at high pressure (20 atm) described in the previous section. Moreover, the sticking coefficients are approximately of the same orders of magnitude as the reaction probabilities derived from the high-pressure work. This remarkable result demonstrates that kinetic parameters derived from well-defined single-crystal surfaces are obviously transferable over the "pressure gap" and it confirms that the dissociative nitrogen adsorption is indeed the rate-limiting step, since the rate of NH_3 formation equals that of dissociative nitrogen adsorption.

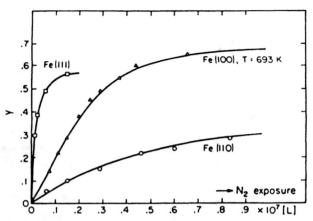

Figure 7.16. Variation of the relative surface concentration, y, of atomic nitrogen as a function of N_2 exposure [33]. 1 L (Langmuir) $= 10^{-6}$ torr-sec.

Similar conclusions had already been reached many years ago by Emmett and Brunauer [45], who measured the uptake of nitrogen by commercial catalysts and concluded likewise that the sticking coefficient is only on the order of 10^{-6}.

The sticking coefficient can be formulated in terms of the usual Arrhenius equation for a rate constant, $s = A \exp(-\Delta E^*/RT)$, with the preexponential A and activation energy ΔE^* as parameters. Measurements at different temperatures revealed that the differences between the three crystal planes can essentially be traced back to differences in the net activation energy E^* for the overall process $N_2 \rightarrow 2N_{ad}$, which in the limit of zero coverage was found to be about 27 kJ/mole for Fe(110), about 21 kJ/mole for Fe(100), and about 0 kJ/mole for Fe(111). These activation energies increase continuously with increasing coverage, in qualitative agreement with previous measurements using supported Fe catalysts [46].

7.8.1.4.3 Effects of Aluminum Oxide in Restructuring Iron Single-Crystal Surfaces for Ammonia Synthesis

The initial rate of ammonia synthesis has been determined over the clean Fe(111), Fe(100), and Fe(110) surfaces with and without aluminum oxide. The addition of aluminum oxide to the (110), (100), and (111) faces of iron decreases the rate of ammonia synthesis in direct proportion to the amount of surface covered [47]. This suggests that the promoter effect of aluminum oxide involves reaction with iron which cannot be achieved by simply depositing aluminum oxide on an iron catalyst.

Remembering that the industrial catalyst is prepared by fusion of 2–3% by weight of aluminum oxide and potassium with iron oxide (Fe_3O_4), experiments were performed in which Al_xO_y/Fe single-crystal surfaces were pretreated in an oxidizing environment prior to ammonia synthesis. These experiments were carried out by depositing about 2 ml of Al_xO_y on Fe(111), Fe(100), and Fe(110) surfaces and then treating them in varying amounts of water vapor at 723 K in order to oxidize the iron and to induce an interaction between iron oxide and aluminum oxide. After removal of the water vapor, high pressures of nitrogen and hydrogen were added to determine the rates of ammonia synthesis. The rate of ammonia synthesis over

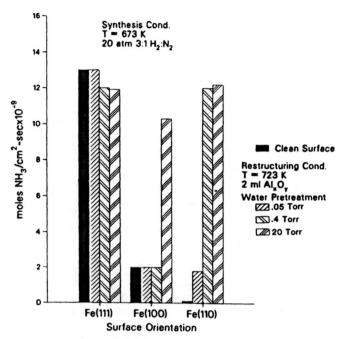

Figure 7.17. Rates of ammonia synthesis over clean iron single crystals and water-induced restructured Al_xO_y/Fe surfaces. Restructuring conditions are given in the figure [38].

Al_xO_y/Fe surfaces pretreated with water vapor prior to ammonia synthesis is shown in Figure 7.17. The initially inactive $Al_xO_y/Fe(110)$ surface restructures and becomes as active as the Fe(100) surface after a 0.05-torr water vapor treatment and as active as the Fe(111) surface after a 20-torr water-vapor pretreatment. This is about a 400-fold increase in the rate of ammonia synthesis compared with clean Fe(110) [37]. The activity of the $Al_xO_y/Fe(100)$ surface can also be enhanced to that of the highly active Fe(111) surface by utilizing a 20-torr water-vapor pretreatment, and this high activity is maintained indefinitely as in the case for the restructured $Al_xO_y/Fe(110)$. Little change in the activity of the Fe(111) surface is seen experimentally when it is treated in water vapor in the presence of Al_xO_y.

The activity of the Fe(110) and Fe(100) surfaces for ammonia synthesis can also be enhanced to the level of Fe(111) by water-vapor pretreatments in the absence of aluminum oxide, but in this circumstance the enhancement in activity is only transient. Figure 7.18 shows the rate of ammonia synthesis as a function of reaction time for restructured Fe(110) and $Al_xO_y/Fe(110)$ surfaces. Both surfaces have an initial activity similar to that of the clean Fe(111) surface. The restructured $Al_xO_y/Fe(110)$ surface maintains this activity for over 4 hr while the restructured Fe(110) surface loses its activity for ammonia synthesis within 1 hr of reaction.

7.8.1.4.4 Characterization of the Restructured Surfaces The observation that the $Al_xO_y/Fe(110)$ and $Al_xO_y/Fe(100)$ become as active as the Fe(111) surface for ammonia synthesis suggests that new crystal orientations are being created upon restructuring the $Al_xO_y/Fe(110)$ and $Al_xO_y/Fe(100)$ surfaces in water vapor. A sug-

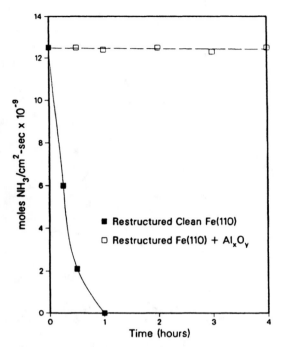

Figure 7.18. Deactivation of the restructured Fe(110) surface occurs within 1 hr while the restructured Al_xO_y/Fe(110) surface maintains its activity under ammonia synthesis conditions [38].

gested increase in the surface area cannot account for the enhancement in rate, since it has been shown that about 40% less carbon monoxide adsorbs on restructured Al_xO_y/Fe(110) and Al_xO_y/Fe(100) relative to the clean respective surfaces [48] (i.e., the iron surface area actually decreases).

Electron spectroscopies, low-energy electron diffraction, temperature-programmed desorption (TPD), and scanning electron microscopy (SEM) have been used to characterize the restructured surfaces. SEM micrographs for restructured Fe(110) and Al_xO_y/Fe(110) surfaces following a 20-torr water-vapor pretreatment show that the surfaces seem to be completely recrystallized. Auger electron spectroscopy finds that only about 5% of the iron surface is covered by aluminum oxide, and sputtering the surface with argon ions reveals aluminum oxide beneath the iron surface.

TPD of ammonia from iron single-crystal surfaces following high-pressure ammonia synthesis proves to be a sensitive probe of the new surface binding sites formed upon restructuring. Ammonia TPD spectra for the four clean surfaces are shown in Figure 7.19. Each surface shows distinct desorption sites. The Fe(110) surface displays one desorption peak (β_3) with a peak maximum at 658 K. Two desorption peaks are seen for the Fe(100) surface (β_2 and β_3) at 556 K and 661 K. The Fe(111) surface exhibits three desorption peaks (β_1, β_2, and β_3) with peak maxima at 495 K, 568 K, and 676 K, and the Fe(211) plane has two desorption peaks (β_2 and β_3) at 570 K and 676 K. PD spectra for the Al_xO_y/Fe(110), Al_xO_y/Fe(100), and Al_xO_y/Fe(111) surfaces restructured in 20 torr of water vapor are shown in

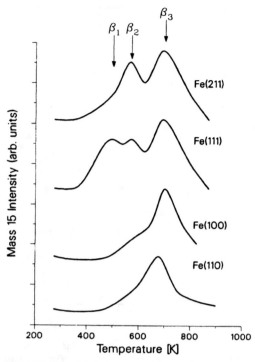

Figure 7.19. Ammonia TPD after high-pressure ammonia synthesis. The low-temperature peaks exhibited by Fe(111) and Fe(211) (β_1 and β_2) are attributed to the presence of C_7 sites [38].

Figure 7.20. A new desorption peak, β_2, develops on the restructured $Al_xO_y/Fe(110)$ surface, and an increase in the β_2 peak occurs on the restructured $Al_xO_y/Fe(100)$ surface. The β_2 peaks from the restructured $Al_xO_y/Fe(110)$ and $Al_xO_y/Fe(100)$ surfaces grow in the same temperature range as the Fe(111) and Fe(211) β_2 peaks.

The ammonia TPD results point toward the formation of surface orientations which contain C_7 sites during water-vapor-induced restructuring. The growth of the β_2 peaks upon restructuring of the Fe(110) and Fe(100) surfaces suggests that the surfaces change orientation upon water-vapor treatment. The β_2 peaks also reside in the same temperature range as the Fe(111) β_2 peak. It seems likely that the TPD peaks in this temperature range act as a signature for the C_7 sites because the Fe(211) surface which contains C_7 sites is highly active in the ammonia synthesis reaction and also exhibits a β_2 peak after ammonia synthesis, with a peak maximum at 570 K. These results suggest that surface orientations which contain C_7 sites, such as the Fe(111) and Fe(211) planes, are formed during the reconstruction of clean and Al_xO_y-treated iron surfaces, but only in the presence of Al_xO_y does the active restructured surface remain stable under the ammonia synthesis conditions.

With the addition of Al_xO_y, the mobility of the iron is increased and restructuring can occur at lower pressure of water vapor. The SEM micrographs suggest that iron forms crystallites on top of the restructured $Al_xO_y/Fe(110)$ surface [as opposed to the uniform appearance of the restructured clean Fe(110) surface]. AES finds little

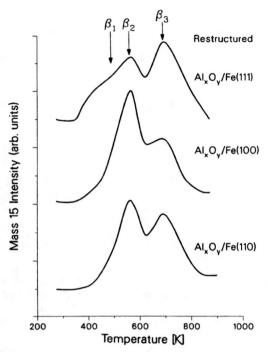

Figure 7.20. Ammonia TPD following ammonia synthesis from restructured $Al_xO_y/Fe(100)$ surfaces exhibit low-temperature peaks similar to those of Fe(111) and Fe(211). Thus, restructuring by water vapor creates active C_7 sites [38].

Al_xO_y on the surface, suggesting that the iron has diffused through the Al_xO_y islands, covering them. These findings can be explained by considering wetting properties and the minimization of the free energy for the iron oxide–aluminum oxide system. The formation of iron aluminate (i.e., $FeAl_2O_4$) in the presence of an oxygen source was also postulated [49] on the basis of microelectron diffraction data.

The formation of an iron aluminate during reconstruction of the iron surface may be responsible for the stability of the restructured Al_xO_y/Fe surfaces. The presence of iron aluminate has been postulated from XPS studies on $Fe-Al_2O_3$ and $Fe_3O_4-Al_2O_3$ systems [50, 51] as well as in numerous studies on the industrial ammonia synthesis catalyst [52–54]. The low coverages of Al_xO_y on the restructured surfaces suggest that $FeAl_2O_4$ plays the role of support on which iron surfaces grow with (111) orientation that is most active in ammonia synthesis. This is supported by the fact that ion sputtering the restructured surfaces reveal subsurface Al_xO_y. This model of the role of alumina as a structure modifier of iron for ammonia synthesis is shown in Figure 7.21.

7.8.1.4.5 *Effect of Potassium on the Dissociative Chemisorption of Nitrogen on Iron Single-Crystal Surfaces in UHV*

The rate-determining ammonia synthesis reaction is widely accepted to be the dissociation of nitrogen [32, 55–57]. Consequently the direct interaction between nitrogen and iron has been studied [43, 44] together with the addition of submonolayer amounts of potassium [56, 58]. All the

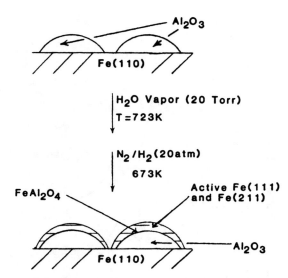

Figure 7.21. Scheme of the restructuring process of iron induced by water vapor and the presence of aluminum oxide. The oxidation of iron permits the migration of the metal on top of the aluminum oxide. The formation of $FeAl_2O_4$ may facilitate this process. Upon reduction in nitrogen and hydrogen, iron is left in active and stable (111) orientation for ammonia synthesis on top of $FeAl_2O_4$.

work that will be referred to in this section was carried out in a UHV chamber, which therefore limits the pressure range to lie between 10^{-4} torr and 10^{-10} torr.

Using both iron single crystals and polycrystalline foils, the sticking probability of molecular nitrogen on iron was found to be on the order of 10^{-7}. This result reveals why, in addition to thermodynamic considerations, ammonia synthesis from the elements is favored at high reactant gas pressures. Because the sticking probability of dissociating nitrogen is so low on iron, higher pressures of nitrogen enhance the kinetics of the rate-limiting step in ammonia synthesis. The structure sensitivity of the reaction is also revealed in the nitrogen chemisorption studies. It was found that the Fe(111) surface dissociatively chemisorbed nitrogen 20 times faster than the Fe(100) surface and 60 times faster than the Fe(110) surface. This agrees well with the structure sensitivity of ammonia synthesis and adds more credence to dissociative chemisorption being the rate-limiting step. The addition of submonolayer amounts of elemental potassium has dramatic effects on the nitrogen chemisorption properties of the (110), (100), and (111) faces of iron.

The effect of potassium on the initial sticking coefficient (S_0) of nitrogen on a Fe(100) surface is shown in Figure 7.22. For clean Fe(100), S_0 is 2×10^{-7}, but with the addition of potassium S_0 increases almost linearly, until at a potassium concentration of 1.5×10^{14} potassium atoms per cm^2, where S_0 maximizes at a value of 3.9×10^{-5}, a factor of $280 \times$ enhancement is seen. Higher coverages of potassium start to decrease S_0, presumably due to the blocking of iron sites by potassium which would otherwise dissociatively chemisorb nitrogen. The maximum increase in S_0, due to potassium adsorption, on Fe(111) is about a factor of 10 (S_0 = 4×10^{-5}) at a potassium concentration of 2×10^{14} K atoms per cm^2. The potassium-induced enhancement of S_0 on the Fe(110) surface is greater than that

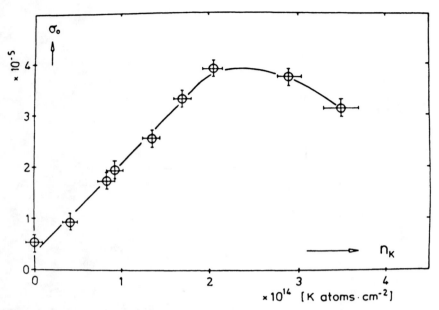

Figure 7.22. Variation of the initial sticking coefficient of N_2, σ_0, with the addition of potassium to Fe(100) at 430 K. The N_2 sticking coefficient can be enhanced by a factor of 280 relative to clean Fe(100) [56].

observed on either Fe(111) or Fe(100), so that the differences in activities for nitrogen dissociation seen on the clean surfaces is much smaller in the presence of potassium.

The mechanism by which potassium promotes nitrogen chemisorption is usually attributed to the lowering of the surface work function in the vicinity of a potassium ion. This effect is greatest at sufficiently low coverages (<0.15 ML) where the potassium–iron bond has strong ionic character, so that the local ionization potential of the surface iron atoms is greatest. This allows for more electron density to be transferred to the nitrogen $2\pi^*$ antibonding orbitals from the surface. This phenomenon increases the adsorption energy of molecular nitrogen and simultaneously lowers the activation energy for dissociation. For example, on the Fe(100) surface the addition of 1.5×10^{14} K atoms per cm^2 decreases the work function by about 1.8 eV and increases the rate of nitrogen dissociation by more than a factor of 200. This enhancement in rate is accompanied by an increase in the adsorption energy of nitrogen on Fe(100) by 11.5 kcal/mole, which decreases the activation barrier for dissociation, in the presence of potassium, from 2.5 kcal/mole to about 0 kcal/mole.

7.8.1.4.6 Temperature-Programmed Desorption Studies of Ammonia from Iron Surfaces in the Presence of Potassium The TPD of ammonia from clean Fe(111) and K/Fe(111) is shown in Figure 7.23 [59]. Ammonia desorbs through a wide temperature range, resulting in a broad peak with a maximum rate of desorption occurring at around 300 K. With the addition of 0.1 ML of potassium, the temperature of the peak maximum is reduced by about 40 K. Assuming first-order desorption for ammonia, the 40-K decrease corresponds to a 2.4-kcal/mole drop in the

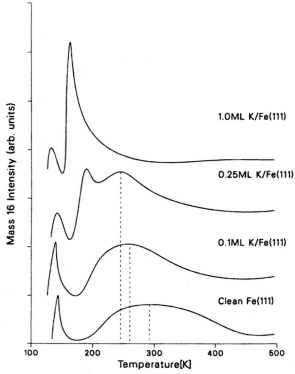

Figure 7.23. Ammonia TPD from clean Fe(111) and K/Fe(111) surfaces. The desorption temperature of ammonia from Fe(111) is lowered in the presence of potassium. Thus potassium lowers the adsorption energy of ammonia on the iron surface [59].

adsorption energy of ammonia on iron in the presence of 0.1 ML potassium. The peak maximum continuously shifts to lower temperature with increasing amounts of coadsorbed potassium. At a coverage of 0.25 ML a new desorption peak appears at about 189 K. Increasing coverages of potassium now increase the intensity of the new peak (it also shifts to lower temperatures) and decreases the intensity of the original ammonia desorption peak. At a potassium coverage of about 1.0 ML, only a weakly bound ammonia species is present, with a maximum rate of desorption occurring at 164 K. This observation of decreasing adsorption energy for ammonia with the coadsorption of potassium on iron is similar to what is found for ammonia desorption from nickel and ruthenium with adsorbed sodium [60, 61].

7.8.1.4.7 Effects of Potassium on Ammonia Synthesis Kinetics Extensive research has been completed in which the effects of potassium on ammonia synthesis over iron single-crystal surfaces of (111), (100), and (110) orientations [59] have been determined. The apparent order of ammonia and hydrogen for ammonia synthesis over iron and K/Fe surfaces has been determined in addition to the effect of potassium on the apparent activation energy (E_a) for the reaction. In all the experiments, potassium was coadsorbed with oxygen because only about 0.15 ML of potassium coadsorbed with oxygen is stable under ammonia synthesis conditions (20 atm total pressure: 3 to 1 H_2 to N_2: $T = 673$ K) [47, 59, 62]. It has been shown that the addition of 0.15 ML of potassium to Fe(111) and Fe(100) increases the ammonia partial pressure dependence from -0.60 for the clean iron surfaces to -0.35 for the 0.15 ML K/Fe(111) and 0.15 ML K/Fe(100) surfaces under high-pressure ammonia synthesis conditions (Figure 7.24). The apparent order in hydrogen partial pressure has been found to decrease from 0.76 for clean Fe(111) to 0.44 for the 0.15 ML K/Fe(111) surface (Figure 7.25). The Fe(110) is inactive for ammonia synthesis under these conditions with or without potassium. These changes in both the apparent order of hydrogen and ammonia pressure dependence occurring with no change in the activation energy suggests that potassium does not change the elementary steps of ammonia synthesis (Figure 7.26). The data show that the promotional effect of potassium is enhanced as the reaction conversion increases (i.e., increasing ammonia partial pressure.

These results are consistent with earlier literature [63, 64] in which the effects of potassium on doubly promoted (aluminum oxide and potassium) catalysts were studied. It was shown that the turnover number for ammonia synthesis is roughly the same over singly (aluminum oxide) and double promoted iron when 1 atm reactant

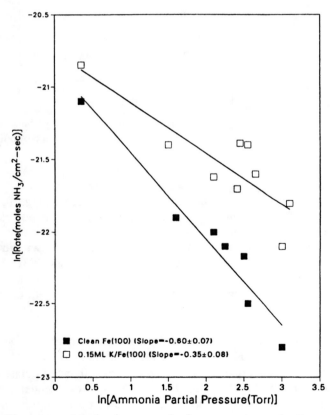

Figure 7.24. The apparent order in ammonia for ammonia synthesis over Fe(100) and K/Fe(100) surfaces. The order in ammonia becomes less negative when potassium is present. The same values were found for Fe(111) and K/Fe(111) surfaces [59].

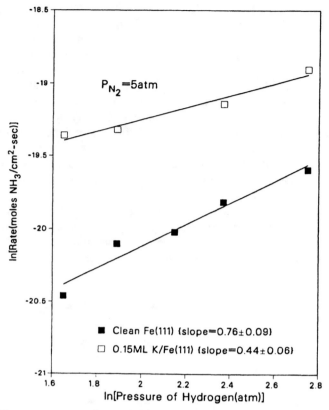

Figure 7.25. The apparent reaction order in hydrogen for ammonia synthesis over Fe(111) and K/Fe(111) surfaces. The order in hydrogen decreases in the presence of potassium [59].

pressure of nitrogen and hydrogen is used [64]. This implies that at low-pressure conditions, the gas-phase ammonia concentration is not high enough for potassium to exert a promoter effect. As higher reactant pressures are achieved (95–200 atm), the promoter effect of potassium becomes significant. It was found that doubly promoted catalysts became increasingly more active than catalysts without potassium when the concentration of ammonia in the gas phase increased [63]. This implies that potassium makes the apparent reaction order dependence in ammonia partial pressure less negative over commercial catalysts, in agreement with the single-crystal work.

7.8.1.4.8 Effects of Potassium on the Adsorption of Ammonia on Iron Under Ammonia Synthesis Conditions The changes in the apparent reaction order dependence in ammonia partial pressure suggest that to elucidate the effects of potassium on both iron single crystals and the industrial catalyst, it is necessary to understand the readsorption of gas-phase ammonia on the catalyst surface during ammonia synthesis; The fact that the rate of ammonia synthesis is negative order in ammonia synthesis. Once adsorbed, the ammonia has a certain residence time (τ) on the catalyst which is determined by its adsorption energy (ΔH_{ads}) on iron [$\tau \propto \tau_0 \exp (\Delta H_{ads}/RT)$]

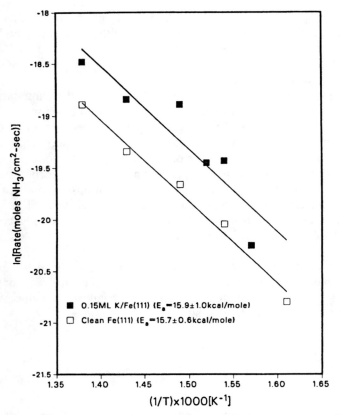

Figure 7.26. The activation energy for ammonia synthesis on Fe(111) and K/Fe(111). Within experimental error there is no change, suggesting that potassium does not change the reaction mechanism of ammonia synthesis [59].

[65]. During this residence on the catalyst, ammonia can either diffuse on the surface or decompose to atomic nitrogen and hydrogen [28, 29, 32]. In both cases the species produced by ammonia might reside on surface sites that would otherwise dissociatively chemisorb gas-phase nitrogen and thereby decrease the rate of ammonia synthesis [28, 29, 32, 66]. The promoter effect of potassium then involves lowering the adsorption energy of the adsorbed ammonia so that the concentration of adsorbed ammonia is decreased. This is supported by the TPD results, which show that ammonia desorption from Fe(111) shifts to lower temperatures when potassium is adsorbed on the surfaces. Even at a 0.1-ML coverage of potassium (coverage roughly equivalent to that stable under ammonia synthesis conditions), the adsorption energy of ammonia is decreased by 2.4 kcal/mole. Thus, the residence time for the adsorbed ammonia is reduced and more of the active sites are available for the dissociation of nitrogen. At higher coverages of potassium, the adsorption energy of ammonia decreases to an even greater extent, but these coverages could not be maintained under ammonia synthesis conditions. There also seems to be an additional adsorption site for ammonia when adsorbed on iron at high coverages of potassium as indicated by the TPD results. The development of a new desorption peak

with coverages of potassium greater than 0.25 ML might result from ammonia molecules interacting directly with potassium atoms, with the negative end of the ammonia dipole interacting with the potassium ion on the iron surface [60]. This interaction appears to be weak, since at a potassium coverage of 1 ML, ammonia desorbs from the surface at 164 K.

Additional experimental evidence supporting the notion that ammonia blocks active sites comes from the post-reaction Auger data. Within experimental error, there is no change in the intensity of the nitrogen Auger peak between a Fe surface and a K/Fe surface after a high-pressure ammonia synthesis reaction. This suggests that potassium does not change the coverage of atomic nitrogen, but instead the presence of potassium helps to inhibit the readsorption or promote the desorption of molecular ammonia on the catalyst. High-pressure reaction conditions are probably needed to stabilize this ammonia product on the iron surface at 673 K, so it will not be present in the ultrahigh vacuum environment. Thus, only the more strongly bound atomic nitrogen will be detected by AES in UHV.

7.8.1.5 Mechanism and Kinetics of Ammonia Synthesis
If all the experimental evidence presented in the preceding sections is put together, the reaction scheme for the catalytic synthesis of ammonia on iron-based catalysts can unequivocally be formulated in terms of the following steps:

$$H_2 + * \rightleftharpoons 2H_{ad}$$

$$N_2 + * \rightleftharpoons N_{2,ad}$$

$$N_{2,ad} + * \rightleftharpoons 2N_{ad}$$

$$N_{ad} + H_{ad} \rightleftharpoons NH_{ad}$$

$$NH_{ad} + H_{ad} \rightleftharpoons NH_{2,ad}$$

$$NH_{2,ad} + H_{ad} \rightleftharpoons NH_{3,ad}$$

$$NH_{3,ad} \rightleftharpoons NH_3 \uparrow$$

where * denotes schematically an ensemble of atoms forming an adsorption site.

The progress of the reaction may be rationalized in terms of its energy profile as reproduced in Figure 7.27.

Attempts at theoretical modeling of the kinetics along these lines were recently performed independently by two groups: Bowker et al. [67, 68] and Stolze and Nørskov [69–74]. The latter group starts with the experimentally well-established fact that dissociation of adsorbed nitrogen is rate-limiting. The overall rate can then be calculated from the rate of this step and the equilibrium constants of all the other steps. This reduces the number of input parameters significantly. The adsorption-desorption equilibria are treated with the approximation of competitive Langmuir-type adsorption and by evaluation of the partition functions for the gaseous and adsorbed species. The data for the potassium-promoted Fe(111) surface were used for the rate of dissociative nitrogen adsorption and are also representative of the other crystal planes of the promoted catalyst, as outlined above. The active area of the commercial catalyst was assumed to equal that derived from selective carbon monoxide chemisorption as a well-established standard procedure. A particular

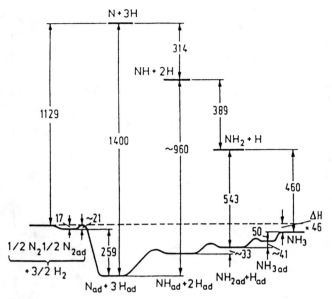

Figure 7.27. Schematic energy profile of the progress of ammonia synthesis on iron (in kJ/mole) [33].

strength of this model is the fact that experimental data from single-crystal studies (such as TPD traces) are reproduced well with the same set of parameters and the same model as used for the determination of the rate under "real" conditions. Comparison of the resulting yields against those determined experimentally with a commercial catalyst yielded general agreement to within a factor better than 2. In Figure 7.28, a compilation of data over a wide range of conditions is presented that demonstrate this almost-too-perfect agreement.

A general conclusion from these models based on single-crystal data is that the most abundant surface species under practical synthesis conditions will be adsorbed atomic nitrogen ($>90\%$), despite the fact that its formation is the rate-limiting step of the overall reaction.

7.8.2 Hydrogenation of Carbon Monoxide

7.8.2.1 Thermodynamics Using relatively easily available small molecules (e.g., CO, H_2, CO_2, and H_2O), all of the smaller or larger hydrocarbon molecules could be synthesized by reactions that are thermodynamically feasible. For example, the standard free energies of three of the four reactions that produce methane from these molecules are negative:

$$CO + 3H_2 \rightleftharpoons CH_4 + H_2O \qquad \Delta G^0_{298} = -33.4 \text{ kcal/mole}$$

$$4CO + 2H_2O \rightleftharpoons CH_4 + 3CO_2 \qquad \Delta G^0_{298} = -54.1 \text{ kcal/mole}$$

$$CO_2 + H_2 \rightleftharpoons CH_4 + 2H_2O \qquad \Delta G^0_{298} = -26.2 \text{ kcal/mole}$$

$$CO_2 + 2H_2O \rightleftharpoons CH_4 + 2O_2 \qquad \Delta G^0_{298} = +19.1 \text{ kcal/mole}$$

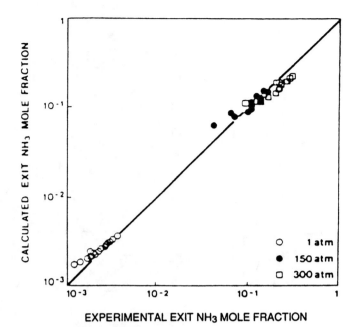

Figure 7.28. Comparison of calculated and measured ammonia production over commercial iron-based catalysts for a broad range of temperatures, pressures, N/H ratios, and gas flows [192].

The only reaction that is thermodynamically uphill is the one between CO_2 and H_2O, which produces oxygen as well. This type of reaction is the basis of photosynthesis, which requires external energy and has played an important role in the evolution of this planet. The other three reactions can be readily carried out using transition metals and their compounds as catalysts.

The usual sources of CO, CO_2, and H_2 are coal or natural gas, which is mostly methane (≈ 72 mole %). The gasification of coal using steam at high temperatures produces predominantly carbon monoxide and hydrogen, a gas mixture that is appropriately called "water gas" or "synthesis gas" ("syn gas"):

$$Coal + H_2O \rightleftharpoons CO + H_2 \qquad G^0_{298} = +88 \text{ kJ/mole} \qquad (7.17)$$

At lower gasification temperatures (≈ 900 K) in the presence of appropriate catalysts (CaO and K_2O, for example) the gasification produces almost exclusively CO_2 and H_2:

$$Coal + 2H_2O \overset{CaO, K_2O}{\rightleftharpoons} CO_2 + 2H_2 \qquad \Delta G^0_{298} = +60 \text{ kJ/mole} \qquad (7.18)$$

Both of these reactions are endothermic, and the heat needed to carry them out is often obtained by combustion of some of the coal.

The reaction of steam with methane is another method of obtaining syn gas:

$$CH_4 + H_2O \rightleftharpoons CO + 3H_2 \qquad \Delta G^0_{298} = +136 \text{ kJ/mole} \qquad (7.19)$$

This reaction [75] is often called "steam reforming." Once CO, CO_2, and H_2 are obtained, their molar ratio can be adjusted to the desired value using the water–gas shift reaction

$$CO + H_2O \rightleftharpoons CO_2 + H_2 \qquad \Delta G^0_{298} = -28 \text{ kJ/mole} \qquad (7.20)$$

This is a virtually thermoneutral reaction that can readily be catalyzed by copper oxide at low temperatures (≈ 600 K) and by potassium promoted iron oxide at elevated temperatures (≈ 1000 K).

In recent years the reaction of CH_4 with CO_2 instead of steam was also used to produce syn gas. This is also an endothermic reaction, but in some circumstances there are advantages in using carbon dioxide as an oxidizing agent.

One of the most promising new methods of producing syn gas from methane is by partial oxidation with oxygen directly (e.g., see reference [76]):

$$CH_4 + \tfrac{1}{2}O_2 \rightleftharpoons CO + H_2$$

$$\Delta G^0_{298} = -20.7 \text{ kcal/mole}, \quad \Delta H^0_{298} = -8.5 \text{ kcal/mole} \qquad (7.21)$$

This is an exothermic reaction that produces H_2 and CO in the 2:1 ratio, which is desirable for the synthesis of many hydrocarbons. By the use of appropriate catalysts, this reaction may be carried out at much lower temperatures than the reactions 7.17 and 7.19.

Using various ratios of carbon monoxide and hydrogen, the production of hydrocarbons of different types is thermodynamically feasible.

Let us consider the formation of alkanes according to the reaction

$$(n + 1)H_2 + 2nCO \rightleftharpoons C_nH_{2n+2} + nCO_2 \qquad (7.22)$$

$$(2n + 1)H_2 + nCO \rightleftharpoons C_nH_{2n+2} + nH_2O \qquad (7.23)$$

Both reactions are thermodynamically feasible, although reaction 7.22 has a somewhat lower negative free energy of formation. Thus the by-product of alkane formation is either CO_2 or H_2O. These reactions are not independent but are related through the water–gas shift reaction. Catalysts that carry out the water–gas shift readily (iron, for example) may produce alkanes and both CO_2 and H_2O, depending on the reaction conditions. Other catalysts that are poor for catalyzing the water–gas shift may produce alkanes and mostly water or mostly CO_2. Catalysts that produce alkanes and CO_2 are often more desirable, because less hydrogen is used up in this circumstance. Hydrogen is generally the costlier of the two reactants. Let us write only one of these reactions for the formation of alkanes, alkenes, and alcohols, which are also produced from CO and H_2, and compare their free energies for formation:

$$(2n + 1)H_2 + nCO \rightleftharpoons C_nH_{2n+2} + nH_2O \qquad (7.24)$$

$$2nH_2 + nCO \rightleftharpoons C_nH_{2n} + nH_2O \qquad (7.25)$$

$$2nH_2 + nCO \rightleftharpoons C_nH_{2n+1}OH + (n - 1)H_2O \qquad (7.26)$$

The reactions that produce higher-molecular-weight hydrocarbon from CO and H_2 are often called *Fischer–Tropsch reaction processes*, named after their discoverers. The standard free energies of formation of the various products, as a function of temperature, are shown in Figures 7.29, 7.30, and 7.31. Because these are all exothermic reactions, low temperatures favor the formation of the products. At present, however, none of the known catalysts for the hydrogenation of CO can produce hydrocarbons at high enough rates to approach the concentrations that are predicted from thermodynamic equilibrium consideration. In fact, the reaction rates are orders of magnitude lower than the maximum rates calculable at equilibrium. Thus these thermodynamic calculations provide only guidance and boundary conditions of the product distribution that may be produced under various experimental conditions. Because a slow surface reaction step (or perhaps several steps that have large activation energies) controls the rate of the reaction as well as the product distribution (they have low turnover frequencies, 10^{-6} to 10 molecules/surface-atom/sec), higher temperatures (in the range 500–700 K) are usually employed to optimize the rates of formation of the products.

Another reaction that appears to play an important role in the synthesis of hydrocarbons from CO and H_2 is the disproportionation of carbon monoxide:

$$2CO \rightleftharpoons C + CO_2 \qquad \Delta G^0_{298} = -116 \text{ kJ/mole} \qquad (7.27)$$

There is a great deal of experimental evidence, which is presented later in the section on methanation, that the hydrogenation of the active form of carbon that

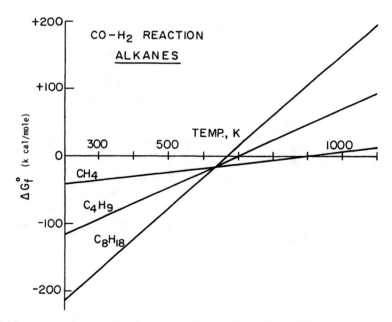

Figure 7.29. Free energies of formation of alkanes from CO and H_2 as a function of temperature.

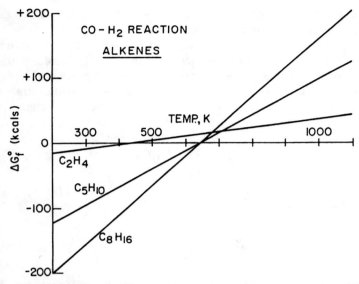

Figure 7.30. Free energies of formation of alkenes from CO and H_2 as a function of temperature.

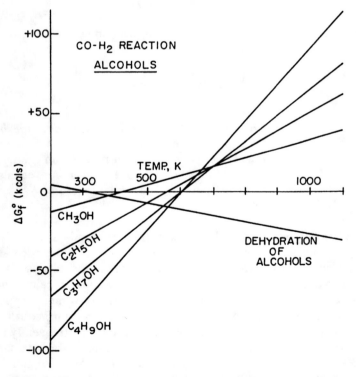

Figure 7.31. Free energies of formation of alcohols from CO and H_2 as a function of temperature.

deposits as a result of the reaction in Eq. 7.27—frequently called the *Boudouard reaction*—leads to the formation of hydrocarbons.

According to the Le Chatelier principle, high pressures favor the association reactions, which are accompanied by a decrease in the number of moles in the reaction mixture as the product molecules are formed. Thus the formation of higher-molecular-weight products is more favorable at high pressures. To demonstrate this [27], let us consider the reaction $aA + bB \rightleftharpoons cC + dD$. The equilibrium constant in terms of partial fugacities is

$$K_f = \frac{(f_C)^c (f_D)^d}{(f_A)^a (f_B)^b} \tag{7.28}$$

In terms of partial pressures, this becomes

$$K_P = \frac{(Px_C)^c (Px_D)^d}{(Px_A)^a (Px_B)^b} \tag{7.29}$$

where P is the total pressure and x_A, x_B, and so on, are the mole fractions. It follows that $K_f = K_p K_\gamma$, where $\gamma = f/p$ and

$$K_\gamma = \frac{(\gamma x_C)^c (\gamma x_D)^d}{(\gamma x_A)^a (\gamma x_B)^b} \tag{7.30}$$

The approximation $K_\gamma \approx 1$ for Fischer–Tropsch reaction conditions (less than 100 atm) yields

$$\frac{(x_C)^c (x_D)^d}{(x_A)^a (x_B)^b} = K_f P^{-\Delta n} \tag{7.31}$$

where $-\Delta n = a + b - c - d$. Thus, associative reactions, where $(a + b)$ is larger than $(c + d)$, are favored by a pressure increase. For all of the Fischer–Tropsch reactions, $(a + b)$ is larger than $(c + d)$, as a general rule. As an example, for the reaction $3H_2 + CO \rightleftharpoons CH_4 + H_2O$ ΔG_f at 730 K and 1 atm equals -11.42 kcal/mole and K equals 3.68×10^3. At 10^{-4} torr total pressure, $K_f P^2 = 3.68 \times 10^3 \times 1.78 \times 10^{-14} = 6.4 \times 10^{-11}$.

7.8.2.2 Catalyst Preparation

Nickel is used most frequently as a catalyst to produce methane. It is usually deposited on high-surface-area oxides such as $\gamma\text{-}Al_2O_3$ and TiO_2. Potassium is used frequently as a promoter in the catalyst formulation. Copper, copper oxide, and zinc oxide together are used to produce methanol selectively. Small particles of the mixed oxides are used usually without support.

Iron and its compounds (carbide, nitride), as well as ruthenium, cobalt, rhodium, and molybdenum compounds (sulfide, carbide), are used most frequently to produce high-molecular-weight hydrocarbons. Iron can be prepared as a high-surface-area catalyst (≈ 300 m^2/g) even without using a microporous oxide support. $\gamma\text{-}Al_2O_3$, TiO_2, and silica are frequently used as supports of the dispersed transition-metal particles. Recently zeolites, as well as thorium oxide and lanthanum oxide, have been employed with success as catalyst supports and as active catalyst components that alter the product distribution.

Potassium is used most frequently as a promoter to increase the molecular weight of the organic products and reduce the hydrogenation rate during the reactions that leads to the formation of unsaturated hydrocarbon. Bimetallic systems are utilized frequently to change the product distribution during CO hydrogenation. Copper and manganese, rhodium, platinum, palladium with iron, and two catalytically active transition metals with various combinations (Fe–Co, Fe–Ru) are used to alter product selectivity.

7.8.2.3 *Methanation. Kinetics, Surface Science, Mechanisms* One of the main products of the hydrogenation of CO is methane. It is produced almost exclusively over nickel, while it forms together with higher-molecular-weight hydrocarbons over many other transition-metal surfaces. Vannice [77] has determined the relative activity of various transition metals for methanation at 1 atm total pressure under conditions in which most other hydrocarbon molecules are not likely to form because of thermodynamic limitation. The order of decreasing activity is Ru > Fe > Ni > Co > Rh > Pd > Pt > Ir. The activation energy for methanation from CO and H_2 is in the range 23–25 kcal/mole for ruthenium, iron, nickel, cobalt, and rhodium metals, for which this has been determined. The nearly identical activation energies indicate that the mechanism of methanation is likely to be similar. Recent studies in several laboratories clearly show that the dominant mechanism involves the dissociation of CO followed by the hydrogenation of the surface carbon atoms to methane. The adsorbed oxygen is removed from the surface as CO_2 by reaction with another CO molecule. The net process by which the active surface carbon that is to be hydrogenated forms is often described as the disproportionation of CO:

$$2CO \rightleftharpoons C + CO_2 \tag{7.32}$$

This is called the *Boudouard reaction*. This mechanism has been confirmed in several ways. The formation of a carbonaceous overlayer has been detected on polycrystalline rhodium [27] and on iron and nickel surfaces [78, 79] in CO–H_2 mixtures in the temperature range 500–700 K. After pumping out the reaction mixture and introducing hydrogen, methane is produced at the same rate as in the presence of water gas. Ventrcek et al. [80] and Rabo et al. [81] have been able to titrate the amount of surface carbon by quantitative measurement of the amount of CO_2 evolved ($2CO \rightleftharpoons C + CO_2$) over nickel, ruthenium, and cobalt catalyst surfaces, respectively. Rabo et al. [81] have introduced pulses of H_2 after forming the surface carbon to produce predominantly methane. Biloen et al. [82] have deposited the active surface carbon on nickel, cobalt, and ruthenium by dissociating labeled ^{13}CO. The isotopically labeled carbon layer is readily hydrogenated subsequently in the presence of a CO–H_2 mixture to yield labeled $^{13}CH_4$.

The active surface carbon that forms from the dissociation of CO maintains its activity to produce methane only in a rather narrow temperature range. Above 700 K the carbon layer becomes graphitized and loses its reactivity with hydrogen. At temperatures below 450 K, the dissociation rate of CO to produce the active carbon is too slow to produce the active surface carbon in high enough concentrations. The temperature dependence of the nature of the CO and carbon chemical bonds introduces a narrow range of conditions for the production of methane. The fact that the dissociation of molecules on surfaces is an activated process is well established by many studies of the formation of surface chemical bonds.

However, changes in the chemical activity of the surface carbon that forms are less well established. The unique hydrogenation activity of the carbon that forms upon the dissociation of carbon monoxide on transition-metal surfaces in the range 450–700 K indicates the formation of active carbon–metal bonds that deserve further experimental scrutiny. The formation of reactive carbene or carbyne species is not unlikely, because these active metal–carbon bonds can yield the hydrogenation activity that was detected. Araki and Ponec [79] have compared the catalytic activity of nickel and nickel–copper alloys for methanation. Upon the addition of less than 10 atom % of copper, the activity drastically decreased. Their results indicate that more than one nickel atom is involved in forming the strong and active metal–carbon bond that yields methane by direct hydrogenation. Because ethylidyne molecules were detected on the Pt(111) crystal face upon adsorption of C_2H_4 and C_2H_2, where the strongly bound carbon is in a threefold site [83], a similar location for the carbon or CH fragments on nickel surfaces [84], which would bind them to three or four nickel atoms, seems likely. The active carbon is metastable with respect to the formation of graphite, however. Heating to above 700 K produces a stable graphite surface layer that is unreactive with hydrogen. Once the graphitic carbon is formed, the catalyst loses its activity for the formation of hydrocarbons of any type. The rate of methane formation is usually positive order (often first order) in hydrogen pressure and negative order in CO pressure.

While the hydrogenation of the active surface carbon that forms from CO dissociation appears to be the predominant mechanism of CH_4 formation, it is not the only mechanism that produces methane. Poutsma et al. [85] have detected the formation of CH_4 over palladium surfaces that do not readily dissociate carbon monoxide. They also observed methane formation over nickel surfaces at 300 K under conditions in which only molecular carbon monoxide appears to be present on the catalyst surfaces [81]. Vannice [86] also reported the formation of methane over platinum, palladium, and iridium surfaces, and independent experiments indicate the absence of carbon monoxide dissociation over these transition-metal catalysts in most cases. It appears that the direct hydrogenation of molecular carbon monoxide can also occur but that this reaction has a much lower rate than methane formation via the hydrogenation of the active carbon that is produced from the dissociation of carbon monoxide in the appropriate temperature range.

Another mechanism, proposed by Pichler [87] and Emmettt [88, 89], involves the direct hydrogenation of molecular carbon monoxide to an enol species, followed by dehydration and further hydrogenation to produce methane. It is likely that this mechanism provides an additional reaction channel that may compete over certain transition-metal catalysts with CH_4 formation via the dissociation of carbon monoxide. Recent studies of methane formation over molybdenum indicate positive CO and H_2 pressure dependencies of the reaction rate

$$R_{CH_4} \propto (P_{H_2})^{1.0}(P_{CO})^{0.5} \tag{7.33}$$

This could be an indication of the formation of an enol intermediate.

The chemistry of C_{Ni} formed from CO disproportionation and from other carbon sources has been further investigated by Rabo et al. [90]. It was found that the C_{Ni} reacts readily with water. Upon injecting a pulse of steam at 600 K over freshly prepared C_{Ni}, this species rapidly reacted with water to form equimolar CO_2 and CH_4 according to the equation

$$2C_{Ni} + 2H_2O \overset{600\,K}{\rightleftharpoons} CO_2 + CH_4 \qquad (7.34)$$

These experimental results are consistent with the thermodynamics of the reaction between carbon and water. At low temperatures they favor the formation of methane, in contrast to the same reaction occurring at high temperatures where the product is $CO + H_2$. The fresh C_{Ni} species reacts readily with both H_2 and H_2O, whereas this species aged at higher temperatures is rendered substantially inert to both. The reaction of C_{Ni} with H_2O, similar to the reaction of C_{Ni} with H_2, is rapid at about 600 K, reaching 90% conversion of the C_{Ni} layer in a few minutes.

The formation of C_{Ni} from CO, according to the Boudouard reaction, is exothermic. The reaction of C_{Ni}^{CO} with H_2O is also exothermic. This latter observation is in contrast to the reaction of graphite and water, which is calculated to be endothermic at about 600 K by about 3 kcal/mole. The exothermic nature of the reaction between C_{Ni} and H_2O indicates a higher-energy state for C_{Ni} relative to graphite.

An interesting change in the kinetics of methanation was observed by Castner et al. [91] when the reaction rates were monitored over clean rhodium and oxidized rhodium surfaces in $CO-H_2$ and CO_2-H_2 gas mixtures. The rates obtained at 600 K, the activation energies, and the preexponential factors for methanation are listed in Table 7.47. The turnover frequencies are much greater and the activation energies are much lower over the preoxidized metal surface. It appears that the oxidized metal surface is not only a better catalyst, but the mechanism of methanation is very different, as indicated by the large change in the kinetic parameters. The activation energy for methane formation is in the 12- to 15-kcal range over the oxidized surface and also when CO_2-H_2 gas mixtures are used for the reaction instead of CO and H_2, in contrast with the 24-kcal activation energy for this reaction on clean metal surfaces. High-resolution electron spectroscopy studies revealed that CO_2 dissociates on the clean rhodium surface to CO and O, and thus the molecule may act as an oxidizing agent on the clean metal surface. It is then likely that the CO_2-H_2 reaction occurs on a partially oxidized rhodium surface, and for this reason it exhibits similar kinetics to the $CO-H_2$ reaction on the oxidized metal surface.

Surface-science studies using nickel single-crystal surfaces revealed that the methanation reaction is surface-structure-insensitive. Both the (111) and (100) crystal faces yield the same reaction rates over a wide temperature range. These specific rates are also the same as those found for alumina-supported nickel, further proving the structure insensitivity of the process. This is also the case for the reaction over ruthenium, rhodium, molybdenum, and iron.

7.8.2.4 *Promotion of the Rates of C—O Bond Hydrogenation by the Oxide–Metal Interface*   CO hydrogenation catalysis has benefited greatly from the rediscovery of the unique catalytic behavior of oxide–metal interfaces first observed by Schwab [92]. The effect is commonly referred to as *strong metal-support interaction*, or SMSI (see also reference [93]). Tauster et al. [94, 95] reported large enhancement

in the CO hydrogenation rates for transition-metal catalysts when supported on high-surface-area titanium oxide. This effect is clearly shown in Figure 7.32, where the rate of methane formation from CO and H_2 is compared for different nickel catalysts. These include unsupported nickel and nickel deposited on silica, alumina, and titanium oxide. As can be seen, the nickel deposited on titanium oxide is orders of magnitude more active for CO hydrogenation than the pure, unsupported nickel catalyst. Subsequent studies of catalyst activation involving reduction and reoxidation using H_2 and O_2, respectively, indicated that the catalyst is activated by optimizing the oxide–metal interface area. Because the same catalytic behavior can be obtained by depositing the metal on the oxide support or by deposition of oxide islands on the transition metal, the oxide–metal periphery area is implicated as the active site responsible for the increased reaction rates. A typical reaction rate behavior exhibits a maximum with increasing oxide coverage over a transition-metal catalyst as shown in Figure 7.33 for CO_2 hydrogenation over TiO_2 on Rh. The oxide alone is inactive while the metal is active for methane formation. At about 50% of a monolayer of oxide coverage, which corresponds to the optimum oxide–metal interface area, the reaction rate exhibits a maximum.

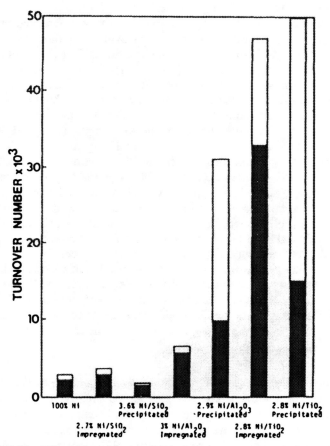

Figure 7.32. Effect of support on CO hydrogenation over Ni catalysts [193].

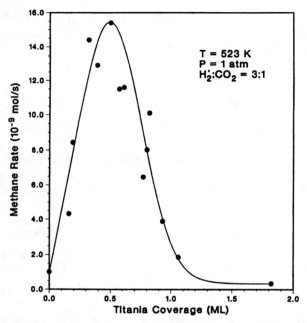

Figure 7.33. CO_2 hydrogenation rate over the rhodium–titanium oxide catalyst as a function of oxide coverage of the metal [194].

This large oxide–metal interface catalysis effect is observed for several transition metals (including Ni, Rh, Co, and Fe) and for several oxides (including TiO_2, La_2O_3, Nb_2O_5, and Ta_2O_5). In addition to CO activation, other molecules that have CO bonds (CO_2, acetone, alcohols) are also activated for hydrogenation.

This oxide–metal interface activation phenomenon is under intense investigation in many laboratories because several new catalyst systems have been reported based on SMSI [96–126]. Recent studies indicate a correlation between the Lewis acidity of high oxidation state transition metal oxides (utilized as supports) and the enhancement of the reaction rates: The stronger the Lewis acidity of the oxide, the greater the activity of the oxide-metal interface catalyst. The scanning tunneling microscope (STM) has been able to image the oxide–metal interface on the atomic scale. It is suggested that the periodic restructuring of metal atoms that drives the catalytic reaction can occur at faster rates at the oxide–metal interface, leading to enhanced catalytic activity. It is hoped that STM experiments that are performed while the reaction is occurring can investigate the dynamic changes of surface structure at the oxide–metal interface and elsewhere on the surface of the active catalyst.

If the predominant reaction mechanism involves CO dissociation (as appears to be the case over nickel and most other transition-metal catalysts), methane formation may be expressed by writing the following elementary surface reaction steps:

$$CO \rightleftharpoons C + O$$

$$CO + O \rightleftharpoons CO_2$$

$$C + H \rightleftharpoons CH + 3H \rightleftharpoons CH_2 + 2H \rightleftharpoons CH_3 + H \rightleftharpoons CH_4 \uparrow$$

All of the species are reacting in their adsorbed states. If enol species form as reaction intermediates as suggested for CO hydrogenation over molybdenum, the elementary surface reaction sequence may be expressed as follows:

$$CO + 2H \rightleftharpoons CHOH$$

$$CHOH + H \rightleftharpoons CH + H_2O \uparrow$$

$$CH + 3H \rightleftharpoons CH_2 + 2H \rightleftharpoons CH_3 + H \rightleftharpoons CH_4 \uparrow$$

7.8.2.5 Methanol Production. Kinetics, Surface Science, and Mechanisms

Methanol production from CO, CO_2, and H_2 is an industrial process that yields about 3×10^6 kg per day. The relevant thermodynamic parameters for the two reactions are [127]

$$CO + 2H_2 \rightleftharpoons CH_3OH \qquad \Delta H_{600}^0 = -100.5 \text{ kJ/mole}$$
$$\Delta G_{600}^0 = +45.4 \text{ kJ/mole}$$
$$CO_2 + 3H_2 \rightleftharpoons CH_3OH + H_2O \qquad \Delta H_{600}^0 = -61.6 \text{ kJ/mole}$$
$$\Delta G_{600}^0 = +61.8 \text{ kJ/mole}$$

Over copper-based catalysts the water–gas shift reaction may also occur in the presence of the three reacting molecules [127].

The activation energy for the reaction is about 64 kJ/mole, and it is usually carried out in the 530- to 580-K temperature range and a few atmospheres of total pressure [127].

The first catalyst utilized was a mixed ZnO/Cr_2O_3 oxide catalyst that operated at high temperatures (700 K) and at high pressures (100 atm). The catalyst presently used is Cu/ZnO with Al or Cr promoters that operates at much lower temperatures and atmospheric pressures [127].

There are continuing questions about the oxidation state of copper during the reaction. The observation that indicates that metallic copper plays an important role during the reaction is that reaction rate appears to be proportional to the metallic copper surface area over Al-promoted Cu/ZnO catalysts. However, over the binary Cu/ZnO catalyst no correlation between the rate of methanol formation and copper surface area is found. The presence of partially oxidized copper on the catalyst surface was detected by recent EXAFS studies and by temperature-programmed reduction [127]. Surface-science studies using Cu(310) crystal surfaces covered with ZnO islands indicate the presence of strongly chemisorbed oxygen on the metal and that oxygen from zinc oxide can readily spill over to the copper.

There is evidence that the hydrogenation of both CO and CO_2 can produce methanol, depending on catalyst formulation and the reaction conditions. Over a cesium-promoted Cu/ZnO catalyst, methanol forms from CO and H_2 without the presence of CO_2. In this circumstance CO is the primary reactant. Isotope labeling studies using ^{14}C-labeled [14]CO_2 in a reactant mixture of $H_2/$[12]$CO/$[14]CO_2 over $Cu/ZnO/Al_2O_3$ catalysts detected only ^{14}C-labeled methanol, implicating CO_2 as the primary reactant. It appears that over most Cu/ZnO catalysts both CO and CO_2 may hydrogenate to produce methanol and that CO hydrogenation is retarded by CO_2, but the reverse is not true [127].

The presence of H_2O accelerates methanol production at low concentrations, and the use of D_2O yields $CH_2DOH(D)$. H_2O is a reaction inhibitor at high concentrations. These results implicate water as a possible reactant and point to the importance of the water–gas shift reaction during the synthesis of methanol [127].

Surface-science studies using copper single-crystal surfaces of (110) and (310) orientation onto which ZnO islands had been deposited indicate that CO and CO_2 chemisorption can be used to identify the metal and the oxide sites, respectively. Methanol chemisorption produces both formate and methoxy species. The concentration of formate is enhanced by the presence of ZnO–copper interfaces, implicating these species as a reaction intermediate.

Palladium dispersed on silica or on other supports (La_2O_3, ZrO_2, etc.) can also form methanol selectively [127]. Surface-science investigations produced methanol on the Pd(110) crystal face without the presence of any oxide near atmospheric pressures and at 550 K. The activation energy was 74 kJ/mole.

Because the catalysts, copper, palladium, and zinc oxide do not dissociate CO, the hydrogenation of molecular CO is one of the likely mechanisms for methanol formation:

$$CO + 2H \rightleftharpoons CH-OH + H \rightleftharpoons CH_2-OH + H \rightleftharpoons CH_3OH \qquad (7.35)$$

Carbon dioxide may dissociate to CO and O or it may also hydrogenate to produce a formate intermediate, $CO_2 + H \rightleftharpoons HCOO$. Further hydrogenation leads to the formation of methanol and water by a series of elementary reaction steps that are yet to be investigated.

7.8.2.6 Production of Higher-Molecular-Weight Hydrocarbons. Kinetics, Surface Science, and Mechanisms

The hydrogenation of carbon monoxide over iron, cobalt, and ruthenium surfaces produces a mixture of hydrocarbons with a wide range of molecular-weight distribution. Most of the hydrocarbons produced are normal paraffins; however, olefins and alcohols in smaller concentrations are also obtained.

The wide product distribution indicates that a polymerization mechanism may be operative. Some of the reaction intermediates serve as chain initiators; then the chain propagation proceeds rapidly until termination by hydrogen occurs before the molecule desorbs from the catalyst surface. The distribution of reaction products, which has been shown to follow a Schulz–Flory [128, 129] distribution of molecular weights frequently encountered in polymerization processes, is given by

$$M(P) = (\ln \alpha)^2 P \alpha^P \qquad (7.36)$$

where $M(P)$ is the weight fraction of hydrocarbons containing P carbon atoms. The chain-growth probability factor is defined as

$$\alpha = R_P/(R_P + R_T) \qquad (7.37)$$

where R_P and R_T are the rate of propagation and termination, respectively. Equation 7.36 can be expressed in logarithmic form as

$$\ln \frac{M(P)}{P} = \ln (\ln^2 \alpha) + P \ln \alpha \qquad (7.38)$$

A plot of ln $(M(P)/P)$ versus P yields a value of α from either the slope or the ordinate intercept. Agreement between the slope and intercept is used as a criterion of the soundness of Schulz–Flory fit. A typical product distribution of CO hydrogenation that reflects the polymerization kinetics is shown in Figure 7.34.

Dwyer and Somorjai [130] have studied the Fischer–Tropsch reaction using a polycrystalline iron foil of 1-cm^2 surface area; and the reaction was carried out with a hydrogen/carbon monoxide ratio of 3:1, at 6 atm and 600 K. At the low conversions (below 1%) obtained under these conditions, the products are primarily methane and ethylene, with trace amounts of other α-olefins up to C_5. This product distribution is compared with that obtained from pilot-plant studies over iron catalysts under industrial conditions and at high conversions (85%). Under industrial conditions, high-molecular-weight paraffins are obtained in large concentrations. In order to simulate the experimental conditions that exist at high conversions, Dwyer and Somorjai [130] have added ethylene to the synthesis gas, since C_2H_4 was one of the products detected. The fate of ethylene was then monitored as a function of reaction time. The majority of the ethylene was hydrogenated to ethane. However, about 10% of the added ethylene was converted to other hydrocarbons. The conversion of ethylene to other hydrocarbons had a significant impact on the product distribution of the CO—H$_2$ reaction. The relative amount of C_3 to C_5 hydrocarbons increased due to the presence of ethylene in the synthesis gas. The influence of ethylene concentration on the product distribution was investigated by varying the partial pressure of ethylene between 2 and 150 torr, while the H$_2$/CO ratio was held constant at 3:1 and at a total pressure of 6 atm. As the initial ethylene partial pressure is increased, the relative amount of methane in the product distribution decreased, although the amount of methane formed remains largely unchanged. The C_{5+} fraction, however, increases with increasing ethylene in an almost linear fashion. The C_3 and C_4 fractions increase to limiting values.

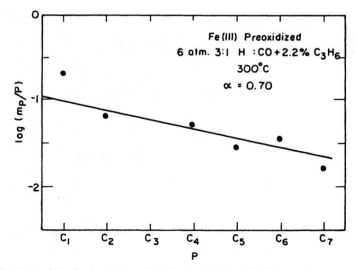

Figure 7.34. Plot of the hydrocarbon distribution in the CO–H$_2$ reaction over iron that reflects the polymerization kinetics [130].

Experiments in which propylene was added to synthesis gas produced results similar to those when ethylene was added. Propylene seems to produce larger molecules than did the same amount of added ethylene. By adding small concentrations of propylene, it is possible to obtain the product distribution found under high-conversion conditions.

Considerable work has been published concerning the incorporation of radioactive-isotope-labeled olefins in hydrocarbons during Fischer–Tropsch reactions. The pioneering work of Kummer and Emmett [89] and of Hall et al. [88] suggested that ethylene acted as a chain initiator over iron catalysts. The same results were obtained over cobalt catalyst by Eidus et al. [131].

It has long been suspected that α-olefins are the primary products of Fischer–Tropsch synthesis, although they are thermodynamically unstable under the reaction conditions. It appears that readsorption and subsequent secondary reactions of α-olefins occur readily under Fischer–Tropsch conditions. Readsorption and secondary reaction of these olefins may be a major reaction pathway leading to the growth of hydrocarbon molecules during Fischer–Tropsch synthesis. In standard flow reactors with large-surface-area catalysts, it is expected that at the leading edge of the bed, the product distribution will be similar to that obtained at low conversions. As these initial reaction products proceed along the bed, they will be readsorbed and undergo secondary reactions, leading to higher-molecular-weight products. As a result of the changing product distribution along the catalyst bed, the surface composition of the catalyst is also likely to change. The presence of readsorption as an important reaction step should permit one to devise ways of controlling the product distribution. Various additives to the reactant mixture, changing the size and geometry of the catalyst bed, and mixing of catalysts are among the experimental variables that may be used to tailor product distribution in the Fischer–Tropsch reaction.

We may then write the formation of high-molecular-weight hydrocarbons over iron or ruthenium surfaces as a two-step process, starting with olefin production:

$$2CO + 4H_2 \underset{550-650\,K}{\overset{Fe,\,6\,atm}{\rightleftharpoons}} C_2H_4(C_3H_6) + 2H_2O \qquad (7.39)$$

This is followed by the readsorption of olefins that induces polymerization:

$$CO + H_2 \underset{550-650\,K}{\overset{Fe,\,C_2H_4(C_3H_6)}{\rightleftharpoons}} C_{5-9}H_{12-20} + 2H_2O \qquad (7.40)$$

Recent studies of the $CO-H_2$ reaction on ruthenium surfaces have also shown the importance of readsorption on the metal catalyst surface. The presence of a multiple-step reaction that proceeds via the readsorption of the initial products provides opportunities for altering the product distribution by using several different catalysts simultaneously in the reaction mixture. By physical mixing of two catalysts, for example, experimental conditions can be realized where the olefins readsorb and further react on the other catalyst instead of on the iron catalyst surface. This way the product distribution can be changed to obtain molecules that are more desirable than the saturated straight-chain hydrocarbons. Chang and Silvestri [132] and Lechthaler and co-workers [133] have reported on a process that converts CO and

H$_2$ to aromatic molecules or to high-octane-number gasoline. First, methanol and olefins are produced by the catalytic reactions of CO and H$_2$, as discussed above. Then, using a zeolite shape-selective catalyst that is introduced along with the ruthenium or other metal catalyst in the same reaction chamber, methanol and the olefins are converted to aromatic molecules, cycloparaffins, and paraffins. The mechanism involves the dehydration of methanol to dimethyl ether. The light olefins that also form are alkylated by methanol and by the dimethyl ether [134] to produce higher-molecular-weight olefins and then the final cyclic and aromatic products.

The formation of aromatic molecules from CO and H$_2$ over ThO$_2$ surfaces has been reported at higher temperatures [87], whereas C$_4$ isomers were produced at lower temperatures and high pressures over the same catalyst (isosynthesis). However, the mechanism of this reaction has not yet been subjected to detailed scientific scrutiny.

During CO hydrogenation over transition-metal surfaces, the presence of potassium usually increases the rate and also selectivities for C$_{2+}$ hydrocarbon production as expected if the CO dissociation rate is increased. As noted earlier, potassium adsorbed on transition metals exists in largely ionic states; this results from the transfer of valence charge density into the metal d-band, which reduces the metal work function. This charge transfer has a profound influence on the adsorption behavior of CO a revealed by thermal desorption and vibrational spectroscopy studies using platinum, nickel, and ruthenium single-crystal surfaces. In all cases, the CO desorption temperature is increased by 100–200 K in the presence of potassium [135, 136] reflecting a 5- to 12-kcal/mole increase in the heat of molecular CO chemisorption. In addition, the CO bond is weakened substantially as compared to adsorption on clean metal surfaces. Figure 7.35 illustrates the HREELS spectra for CO coadsorbed with potassium at several coverages on the hexagonal (111) platinum surface. With increasing coverage, there is a continued shift of the CO stretching frequencies from 1875 and 2120 cm^{-1} to 1565 cm^{-1}. These shifts correlate with a change in bonding from mostly top sites to bridge sites and a decrease in CO bond order from 2.0 to 1.5. This dramatic bond weakening reflects enhanced population of the CO 2π* antibonding orbital as a result of the increased density of metal electronic states in the presence of potassium. One should anticipate that the weakened CO bond and strengthened metal–carbon bond would facilitate CO dissociation. This was demonstrated by Campbell and Goodman [137] using nickel (100) where the CO dissociation rate was increased fourfold at potassium coverage of 10% of a monolayer. The activation energy for CO dissociation was also lowered from about 23 to 10 kcal/mole.

It is now clearly established that potassium chemisorbed on transition metals functions as an unusually powerful donor. This increases the density of surface electron states available for back-bonding with certain adsorbates, if they possess orbitals with energy and symmetry that correlate near the Fermi energy of the metal. Examples of such orbitals would be the 2π* of CO and N$_2$. The important general consequences of this interaction include increased heat of molecular adsorption and increased dissociation probability.

On the clean rhodium (111) surface, CO stays molecularly adsorbed at low pressures while it dissociates in the presence of potassium [138, 139]. This can be studied by the adsorption of a mixture of ^{12}C^{18}O and ^{13}C^{16}O and detecting ^{13}C^{18}O and ^{12}C^{16}O, the products of scrambling, which clearly identify the dissociation of mo-

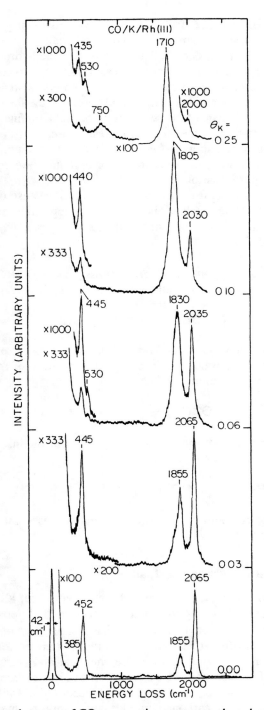

Figure 7.35. Vibrational spectra of CO at saturation coverage when chemisorbed on Rh(111) at 300 K as a function of preadsorbed potassium coverage [139].

lecular CO on the metal surface. In Figure 6.26, we show that three CO molecules may dissociate per potassium atom at a potassium coverage where maximum charge transfer to the transition metal occurs. The reduction of the hydrogenation ability of the catalyst induced by potassium is another reason for the markedly altered product distribution. Manganese oxides, when used as promoters, also increase the olefin selectivities. Thorium oxide and lanthanum oxide promote the formation of branched hydrocarbons and also enhance the selectivity for light olefins. Alkali-promoted molybdenum proves to be an excellent, sulfur-resistant catalyst.

When titania is used as a support for cobalt iron or ruthenium, very active catalysts are prepared, indicating the importance of certain oxide–metal interfaces as active sites for CO hydrogenation.

Zeolites as co-catalysts shift the product distribution because their acid sites can carry out secondary reactions such as alkylation, cracking, oligomerization, and isomerization. These reactions can be important in shifting selectivities toward high-octane gasoline or olefins.

7.8.2.7 *Formation of Oxygenated Hydrocarbons from CO and H_2 and Organic Molecules*

The carbonylation of methanol produces acetic acid:

$$CH_3OH + CO \rightleftharpoons CH_3COOH \tag{7.41}$$

This reaction is carried out over rhodium carbonyls as catalyst using HI as a promoter. Acetic anhydride is produced from the carbonylation of methylacetate over lithium-iodide-promoted rhodium catalyst:

$$CH_3COOCH_3 + CO \rightleftharpoons (CH_3CO)_2O \tag{7.42}$$

The hydroformylation reaction produces aldehydes from olefins, CO and H_2. For example,

$$CH_2{=}CH_2 + CO + H_2 \rightleftharpoons H_3C{-}CH_2{-}CHO \tag{7.43}$$

Rhodium, cobalt, and ruthenium are the most frequently used catalysts to carry out this family of reactions.

7.8.3 Hydrocarbon Conversion on Platinum

7.8.3.1 *Introduction*

Platinum is one of the most versatile, all-purpose, heterogeneous metal catalysts. It is employed under reducing conditions (in the presence of excess hydrogen) for the conversion of aliphatic straight-chain hydrocarbons to aromatic molecules (dehydrocyclization) and to branched molecules (isomerization), and for hydrogenation on a large scale in the chemical and petroleum-refining industries [140–142]. It is also used as an oxidation catalyst for ammonia oxidation, an important step in the process of producing fertilizers [143, 144]. Platinum is the catalyst for the oxidation of carbon monoxide and unburned hydrocarbons in the control of car emissions [145, 146]. Platinum is perhaps the most widely used and most active electrode for catalyzed reactions in electrochemical cells [147–150]. Its chemical stability in both oxidizing and reducing conditions makes this metal an ideal catalyst in many applications. Mined mostly in South Africa and in Russia, platinum, along with rhodium (which occurs as an impurity in platinum ores), is very rare and therefore expensive. Its regeneration and recovery must be an important part of any technology that uses this metal.

For this reason it is of considerable importance to scrutinize the catalytic activity of platinum on the atomic scale, to learn what makes this metal so versatile as a catalyst and so selective for important catalytic transformations after suitable preparation. Once the elements of catalytic activity are revealed, it should be possible to use this metal more economically or perhaps to find ways to synthesize new catalyst systems to substitute for this excellent but rare catalyst.

Let us concentrate on the atomic-scale study of the platinum surface under the reducing conditions used during hydrocarbon conversion reactions. In this circumstance $H-H$, $C-H$, and $C-C$ bond-breaking processes are essential. In Figure 7.36 the various hydrocarbon conversion reactions of interest are listed. Dehydrogenation involves $C-H$ bond breaking only, while hydrogenolysis necessitates the breaking of $C-C$ bonds. Dehydrocyclization must involve the complex process of dehydrogenation and ring closure.

Figure 7.36 shows several reactions that are all catalyzed by platinum. The simpler hydrogenation and dehydrogenation reactions have turnover frequencies in the

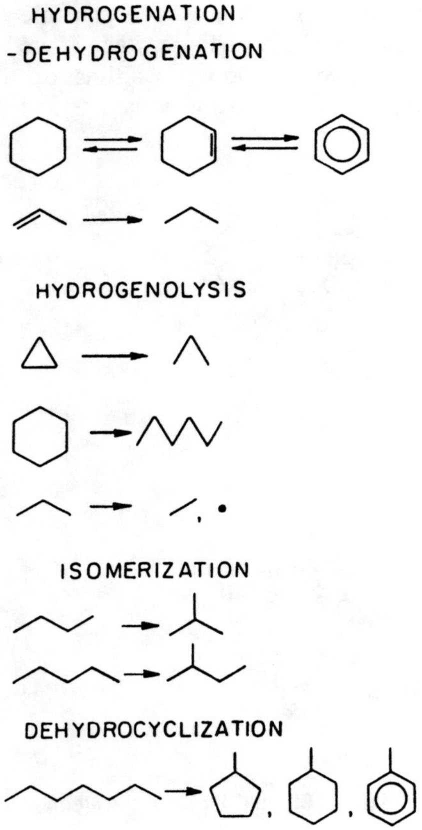

Figure 7.36. Several competing hydrocarbon reactions that occur on platinum catalyst surfaces.

range 0.1 to 10 sec^{-1} under the usual conditions of 400 to 600 K, atmospheric pressures of reactant and excess hydrogen) that are employed in the chemical industry [151, 152]. However, platinum is really noted for being an excellent catalyst for the more complex reactions of dehydrocyclization (e.g., *n*-heptane to toluene) and isomerization (for *n*-pentane to 2-methylbutane) that have turnover frequencies of about 10^{-4} to 10^{-2} sec^{-1} under experimental conditions similar to those used to carry out the more facile reactions [153, 154]. One of the key questions in the molecular-scale study of the hydrocarbon catalysis of platinum is how this metal selectively catalyzes the complex, low-turnover frequency reactions while blocking the simpler, high-rate dehydrogenation and hydrogenation reactions and the slower, but unwanted, hydrogenolysis reaction. This happens after suitable preparation of the platinum catalyst prior to exposure to the reaction mixture.

Various crystal faces of platinum single crystals 1 mm thick with surface areas of about 1 cm^2 serve as excellent model catalysts. These samples can be prepared with quite uniform and ordered surface structures that can be analyzed by the various surface-science techniques. Low-energy electron diffraction was particularly useful for the determination of the structure of single-crystal surfaces. The flat surfaces where each platinum atom is surrounded by six and four nearest neighbors, respectively, are the two closest-packed platinum crystal faces of the highest atomic density (Figure 7.37). Stepped crystal faces can also be prepared easily; these faces display close-packed terraces several atoms in width, which are separated by atomic steps one atom in height. The lowered coordination of the step atoms is responsible for the unique chemical activity that is often displayed at these surfaces sites. There can be kinks in the steps, and atoms at these ledges have even lower coordination. The structure and concentration of steps and kinks, along with the structure and

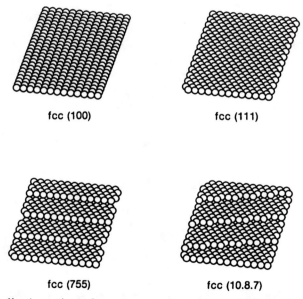

fcc (100)   fcc (111)

fcc (755)   fcc (10.8.7)

Figure 7.37. Idealized atomic surface structures for the flat Pt(100) and Pt(111), the stepped Pt(755), and the kinked Pt(10, 8, 7) surfaces.

width of the terraces, can be varied by cutting the platinum single crystals along different crystals planes and then by appropriately polishing and etching them to remove the surface damage introduced by the mechanics of surface preparation.

7.8.3.2 Structure Sensitivity of Hydrocarbon Conversion Reactions on Platinum Surfaces

How does the reaction rate depend on the atomic structure of the platinum catalyst surface? To answer this question, reaction rate studies using flat, stepped, and kinked single-crystal surfaces with variable surface structure were very useful indeed. For the important aromatization reactions of *n*-hexane to benzene and *n*-heptane to toluene, it was discovered that the hexagonal platinum surface where each surface atom is surrounded by six nearest neighbors is three to seven times more active than the platinum surface with the square unit cell [155, 156]. Aromatization reaction rates increase further on stepped and kinked platinum surfaces. Maximum aromatization activity is achieved on stepped surfaces with terraces about five atoms wide with hexagonal orientation, as indicated by reaction rate studies over more than 10 different crystal surfaces with varied terrace orientation and step and kink concentrations (Figure 7.38).

The reactivity pattern displayed by platinum crystal surfaces for alkane isomerization reactions is completely different from that for aromatization. Studies revealed that maximum rates and selectivity (rate of desired reaction/total rate) for butane isomerization reactions are obtained on the flat crystal face with the square unit cell. Isomerization rates for this surface are four to seven times higher than those for the hexagonal surface. Isomerization rates are increased to only a small extent by surface irregularities (steps and kinks) on the platinum surfaces (Figure 7.39).

For the undesirable hydrogenolysis reactions that require C—C bond scission,

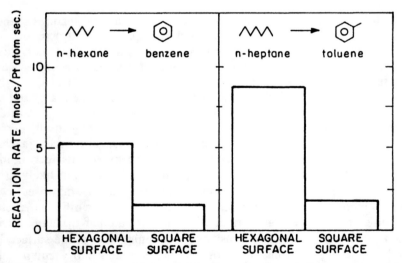

Figure 7.38. The structure sensitivity of dehydrocyclization of alkanes to aromatic hydrocarbons. The bar graphs compare reaction rates for *n*-hexane and *n*-heptane catalyzed at 573 K and atmospheric pressure over the two flat platinum single-crystal faces with different atomic structure. The Pt surface with a hexagonal atomic arrangement is several times more active than the surface with a square unit cell over a wide range of reaction conditions [155].

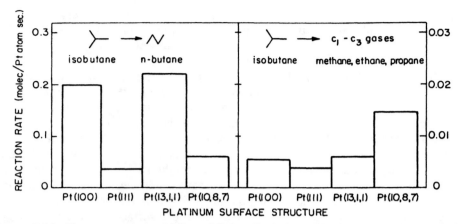

Figure 7.39. The structure sensitivity of light alkane isomerization and hydrogenolysis. Shown here are the reaction rates of isobutane catalyzed at 570 K and atmospheric pressure over four platinum surfaces shown in Figure 7.37. Isomerization is favored over Pt surfaces that have a square atomic arrangement. Hydrogenolysis rates are maximized when kink sites are present in high concentrations on the platinum surface [155].

the two flat surfaces with highest atomic density exhibit very similar reaction rates. However, the distribution of hydrogenolysis products varies sharply over these two surfaces. The hexagonal surface displays high selectivity for scission of the terminal C—C bonds, whereas the surface with a square unit cell always prefers cleavage of C—C bonds located in the center of the reactant molecule. The hydrogenolysis rates increase markedly (three- to fivefold) when kinks are present in high concentrations on the platinum surfaces.

Because different reactions are sensitive to different structural features of the catalyst surface, we must prepare the catalyst with the appropriate structure to obtain maximum activity and selectivity. The terrace structure, the step or kink concentrations, or a combination of these structural features is needed to achieve optimum reaction rates for a given reaction. Studies indicate that H—H and C—H bond-breaking processes are more facile on stepped surfaces than on the flat crystal faces, while C—C bond scission is aided by kink sites that appear to be the most active for breaking any of the chemical bonds that are available during the hydrocarbon conversion reactions. Because molecular rearrangement must also occur, in addition to bond breaking, it is not surprising that the terrace structure exerts such an important influence on the reaction path that the adsorbed molecules are likely to take. The difference in chemical behavior of terrace, step, and ledge atoms arises not only from their different structural environment but also from their different electronic charge densities that result from variation of the local atomic structure. Electron spectroscopy studies reveal altered density of electronic states at the surface irregularities; there are higher probabilities of electron emission into vacuum at these sites (lower work function), indicating the redistribution of electrons [157].

One of the important attributes of transition-metal surfaces is that they atomize diatomic molecules with large binding energy (H_2 or O_2) by forming strong M—H or M—O bonds and hold the atoms in high surface concentrations so that they are readily available during the surface reaction (see Section 7.2). The hydrogen atom surface concentration is especially important in permitting catalyzed hydrocarbon conversion reactions to proceed unimpeded. The presence of excess hydrogen facilitates removal of product molecules and also inhibits catalyst deactivation. For this reason the reforming reaction of organic molecules is always catalyzed in the presence of excess hydrogen.

7.8.3.3 Carbonaceous Overlayers

What is the composition of the working platinum catalyst surface? When the surface is examined after carrying out any one of the hydrocarbon conversion reactions, it is always covered by a near-monolayer amount of carbonaceous deposit.

In order to determine the surface residence time of the carbonaceous deposit, the platinum surface was dosed by the ^{14}C-labeled organic molecules under the reaction conditions. Carbon-14 is a β-particle emitter. The β-particle detector was used to monitor its surface concentration as a function of time during the catalytic reaction. The hydrogen content of the adsorbed organic layer is determined by detecting the amount of desorbing hydrogen with a mass spectrometer. These investigations reveal that the residence time of the adsorbed carbonaceous layer depends on its hydrogen content, which in turn depends on the reaction temperature (Figure 7.40).

Although the amount of deposit does not change much with temperature, the composition does; it becomes much poorer in hydrogen as the reaction temperature is increased. The adsorption reversibility decreases markedly with increasing temperature as the carbonaceous deposit becomes more hydrogen-deficient. As long as

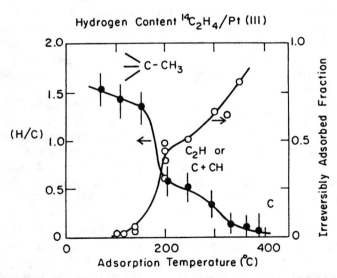

Figure 7.40. Carbon-14-labeled ethylene (or other alkenes) was chemisorbed as a function of temperature on the flat Pt(111) crystal face. The (H/C) ratio of the adsorbed species was determined from hydrogen thermal desorption. The amount of preadsorbed alkene that could not be removed by subsequent treatment in 1 atm of hydrogen represents the irreversibly adsorbed fraction. The adsorption reversibility decreases markedly with increasing adsorption temperature as the surface species become more hydrogen-deficient. The irreversibly adsorbed species have long residence times, on the order of days [195].

the composition is about $C_nH_{1.5n}$ and below 450 K, the organic deposit can be removed readily in hydrogen with increasing reaction temperatures ($>$450 K), it converts to an irreversible adsorbed deposit with a composition of $C_{2n}H_n$ that can no longer be readily removed (hydrogenated) in the presence of excess hydrogen [158].

Nevertheless, the catalytic reaction proceeds readily in the presence of this active carbonaceous deposit [158, 159]. Above 750 K this active carbon layer is converted to a graphitic layer that deactivates the metal surface, and all chemical activity for any hydrocarbon conversion reaction ceases. Hydrogen exchange studies indicate rapid exchange between the hydrogen atoms in the adsorbing reactant molecules and the hydrogen in the active but irreversibly adsorbed deposit. Only the carbon atoms in this layer do not exchange. Thus, one important property of the carbonaceous deposit is its ability to store and exchange hydrogen [158–160].

The structure of the adsorbed hydrocarbon monolayers was submitted to detailed studies by LEED and HREELS [161]. In the temperature range of 300–400 K the adsorbed alkenes form alkylidyne molecules that are shown in Chapter 6. The $C-C$ bond closest to the metal is perpendicular to the surface plane, and its 1.5-Å length corresponds to a single bond. The carbon atom that bonds the molecule to the metal is located in a threefold site equidistant 2.0 Å from the nearest metallic neighbors [162]. This bond is appreciably shorter than the covalent metal–carbon bond (2.2 Å) and is indicative of multiple metal–carbon bonds of the carbene or carbyne type. Although this layer is ordered, on being heated to about 100°C it disorders and hydrogen evolution is detectable by a mass spectrometer that is attached to the system. As the molecules dehydrogenate, the disordered layer is composed of CH_2-, C_2H-, and CH-type fragments that can be identified by HREELS [161]. Only after being heated to about 400°C do the fragments lose all their hydrogen and the graphite overlayer forms. These sequential bond-breaking processes, which occur as a function of temperature, are perhaps the most important and unique characteristics of the surface chemical bond (Chapter 6). Although the surface remains active in the presence of organic fragments of C_2H stoichiometry, it loses all activity when the graphite monolayer forms.

How is it possible that the hydrocarbon conversion reaction exhibits great sensitivity to the surface structure of platinum, while under the reaction conditions the metal surface is covered with a near-monolayer of carbonaceous deposit? In fact, often more than a monolayer amount of carbon-containing deposit is present, as indicated by surface-science measurements. Recent scanning tunneling microscopy studies that were carried out at high hydrocarbon and hydrogen pressures (atm) and hydrocarbon reaction temperatures indicate that CH_2, C_2H, and CH fragments are mobile on the surface; they move around by surface diffusion in the presence of coadsorbed molecular reactants. While they do not desorb, their mobility makes the active metal sites on the surface available to the molecular reactants. When the carbonaceous species polymerize at higher temperatures to form a graphite deposit, they lose their mobility and deactivate the metal surface by permanently blocking the active sites.

In order to determine how much of the platinum surface is exposed and remains uncovered, the adsorption and subsequent thermal desorption of carbon monoxide was utilized. This molecule, although readily adsorbed on the metal surface at 300 K at low pressures, does not adsorb on the carbonaceous deposit. The results indicate that up to 10–15% of the surface remains uncovered metal sites decreases slowly with increasing reaction temperature. The structure of these uncovered metal islands is not very different from the structure of the initially clean metal surface during some of the organic reactions.

As a result of catalyzed hydrocarbon conversion reaction studies on platinum crystal surfaces, a model for the working platinum reforming catalyst could be proposed [163] and is shown in Figure 7.41. Between 80% and 95% of the catalyst surface is covered with an irreversibly adsorbed carbonaceous deposit that stays on the surface for times much longer than the reaction turnover time. The structure of this carbonaceous deposit varies continuously from two-dimensional to three-dimensional with increasing reaction temperature. There are platinum patches that are not covered by this deposit. These metal sites can accept the reactant molecules which then compress the carbonaceous deposit by surface diffusion to free up the active sites where the reactions occur. Upon desorption of the products, the carbonaceous species may diffuse back to cover the metal sites. The adsorption of new reactant molecules repeats the process; compression of the carbonaceous deposit by surface diffusion, reaction at the metal sites and product desorption. There is evidence that the carbonaceous deposit participates in some of the reactions by hydrogen transfer by providing sites for rearrangement and desorption while remaining inactive in other reactions; its chemical role requires further exploration.

7.8.3.4 Catalysis in the Presence of a Strongly Adsorbed Overlayer
Reactions of this type do not occur directly on the metal surface and therefore are usually structure-insensitive [164]. In fact, the role of the metal could be reduced to providing atoms, hydrogen for example, via the dissociation of diatomic molecules. The metal is usually covered by strongly adsorbed overlayers and thus the incoming reactants (other than hydrogen) cannot form strong metal-adsorbate bonds. An example of

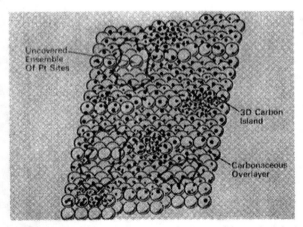

Figure 7.41. Model for the working structure and composition of a platinum dehydrocyclization catalyst. Most of the surface is continuously covered by a strongly bound carbonaceous deposit whose structure varies from two-dimensional to three-dimensional with increasing reaction temperature. Uncovered patches or ensembles of platinum surface sites always exist in the presence of this carbonaceous deposit. Bond breaking and chemical rearrangement in reacting hydrocarbon molecules take place readily at these uncovered sites [158].

this type of reaction is the hydrogenation of ethylene [165]. This facile reaction occurs at 300 K and at atmospheric pressures on many transition metal surfaces. It has been the subject of investigations of many researchers [166–172]. Table 7.48 shows that hydrogenation occurs equally well on platinum crystals, films, foils, and supported particles, indicating that the reaction is structure-insensitive [173]. When the clean metal surfaces are exposed to ethylene, a strongly adsorbed overlayer of ethylidyne (C_2H_3) forms. This molecule, shown in Figure 2.26 along with its vibration spectrum (Figure 2.25), is obtained by high-resolution electron-energy-loss spectroscopy (HREELS) [174]. The kinetics of ethylene hydrogenation and those of ethylidyne have been studied extensively over the (111) faces of rhodium and platinum, and the rates of these processes are displayed in Figure 7.42. Ethylene hydrogenation occurs at a rate six orders of magnitude higher than the rehydrogenation of the strongly adsorbed ethylidyne [165]. Even the deuteration of the methyl group of ethylidyne occurs very slowly. Studies using ^{14}C labeling of ethylidyne and vibrational spectroscopy confirm these findings.

The (111) faces of platinum of rhodium are instantly covered with a monolayer of ethylidyne during ethylene hydrogenation because reaction rates are nearly identical over initially clean surfaces and surfaces precovered with ethylene. Vibrational spectroscopy studies confirm that the adsorbed monolayer structure on these surfaces following hydrogenation is ethylidyne. Thus ethylene hydrogenation occurs rapidly on the C_2H_3-covered surfaces. The packing of the ethylidyne on the overlayer does not permit C_2H_4 adsorption directly on the metal surface, as proven by exchange studies with C_2H_4 and C_2D_4. On the other hand, thermal desorption studies have

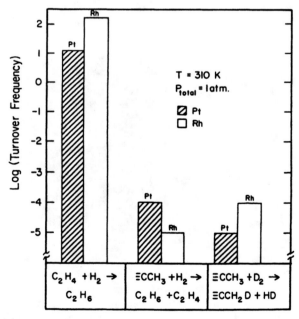

Figure 7.42. Turnover rates for ethylene hydrogenation, the rehydrogenation of ethylidyne, and the deuteration of the methyl- group of ethylidyne on platinum and rhodium crystal surfaces [190]. Note that ethylene hydrogenation rates are orders of magnitude faster than the rate of removal of chemisorbed ethylidyne.

shown that H_2 (D_2) can be dissociated and readsorbed on the ethylidyne-covered surfaces up to about one-fourth monolayer coverage.

One reaction model that explains these results has been proposed [165]. The hydrogen atom is transferred to the ethylene molecule that is weakly adsorbed on top of the ethylidyne and in the second layer perhaps by forming an ethylidene intermediate. This model of hydrogen transfer from hydrocarbons to ethylene was first proposed by Thomson and Webb [175]. This mechanism is of the Eley–Rideal type and is characterized by low activation energy and structure insensitivity.

Another reaction model involves the compression of the ethylidyne overlayer at high pressure of ethylene. Because of repulsive adsorbate–adsorbate (ethylene–ethylidyne) interaction, and the expected small activation energy of ethylidyne surface diffusion, ethylene could adsorb on the metal in the small hole created near the compressed ethylidyne. Compression of this type has been detected by STM upon the adsorption of hydrocarbons on platinum and the coadsorption of CO and sulfur on both platinum and rhenium surfaces [218].

However, there are other mechanisms of C_2H_4 for hydrogenation that studies have uncovered [165]. At higher temperatures, the rate of rehydrogenation of C_2H_3 is significant and the bare metal becomes available, in part, for C_2H_4 hydrogenation. During the electrochemical hydrogenation of C_2H_4, the platinum surface is covered with a layer of hydrogen atoms (hydride) that react rapidly with the approaching C_2H_4 and do not permit the formation of ethylidyne. The complexity of surface reactions cannot be underestimated.

Nevertheless, ethylene hydrogenation may provide an example of reactions of weakly adsorbed molecules or high coverages in the second layer, an important class of catalytic reactions that could occur at low temperatures or high pressures. It should be noted that most catalyzed biochemically important reactions occur at 300 K and at high turnover rates, virtually excluding the possibility of forming strong chemical bonds with the enzyme catalyst surface by the adsorbed reactants or reaction intermediates.

These types of structure-insensitive reactions may be compared with homogeneous catalytic reactions that are facile, occurring at lower temperatures, and include hydrogenation or hydroformylation. Because the metal plays secondary roles in this process, high coordination sites are not needed to carry out the reaction. It is hoped that future studies will reveal the possible correlation between homogeneous catalytic reactions and heterogenous reactions of this type.

The organic overlayer may also serve as a template to orient or align the reactants. LEED surface crystallography and HREELS studies of the structure of these monolayers indicate that their structural integrity is preserved at temperatures as high as 400 K; thus their presence only allows us to carry out various specific reactions below this temperature. Above 400 K, fragmentation to small organic CH and C_2H groups occurs (Figure 2.30). While at low temperatures [176], benzene and ethylene maintain their molecular identify on the platinum and rhodium crystal surfaces, above 400 K the fragments are the same small organic moieties. Thus catalysis that requires an organic template to properly line up the reactant molecules can be carried out only below 400 K.

7.8.3.5 Structure Modifiers

7.8.3.5.1 Site Blocking by Sulfur Let us consider the interaction of coadsorbed sulfur with thiophene which occurs during the hydrodesulfurization of thiophene on the molybdenum (100) crystal surface [177]. This gentle reaction removes the sulfur from the molecule as H_2S in the presence of hydrogen, leaving behind the C_4 species that readily hydrogenates to butadiene, butenes, and butane without fragmentation. Molybdenum metal strongly adsorbs and decomposes thiophene and butenes as shown by surface studies, and thus the clean surface cannot be an active catalyst. MoS_2 is a layer compound, and its basal plane holds thiophene so weakly that its thermal desorption occurs at 165 K [178]. Thus this surface is not chemically active. The active molybdenum surface contains about one-half monolayer of strongly adsorbed sulfur. These atoms block the metal sites where thiophene decomposition would occur. Studies using ^{35}S labeling indicate that these sulfur atoms remain permanently on the metal surface during the catalytic reactions. The sulfur atom that is removed from the thiophene molecule occupies sites of weaker bonding where hydrogenation to H_2S and subsequent desorption occurs, while the C_4 species becomes partly hydrogenated and desorbs.

Thus the blockage of certain adsorption sites on the surface of early transition metal attenuates the strong bonding and permits the catalytic reaction to occur.

The hydrogenolysis of organic molecules over platinum is frequently an undesirable reaction that leads to the production of lower-molecular-weight products. Kink sites on transition metal surfaces are especially active for the C—C bond-breaking reaction [179]. While their surface concentration is no more than about 5% of the total number of metal sites, they may account for 90% of hydrogenolysis activity. These hydrogenolysis sites can often be poisoned by the chemisorption of controlled amounts of sulfur (produced by H_2S decomposition) that bonds more strongly to kink sites as compared to terrace sites. In this way, the hydrogenolysis reaction can be poisoned selectively as the kink sites are blocked and rendered inactive.

7.8.3.5.2 Ensemble Effect in Alloy Catalysis and the Creation of New Sites by Alloys

As compared to pure platinum, bimetallic alloys such as platinum–rhenium and platinum–gold frequently exhibit superior activity, selectivity, and deactivation resistance while catalyzing reforming reactions. The influence of gold on hydrocarbon conversion catalysis by platinum was recently studied by evaporating gold onto platinum single-crystal surfaces [180]. At low temperatures, gold forms epitaxial overlayers on platinum, but upon heating it dissolves to form an alloy in the near surface region. This Pt–Au alloy displays markedly different activity and selectivity for the conversion of *n*-hexane as shown in Figure 7.43. Isomerization activities increase substantially as compared to those for clean platinum, whereas the aromatization and hydrogenolysis rates decrease exponentially with increasing gold surface concentration. This remarkable change in catalytic behavior can be explained by a change in the geometric distribution of platinum sites that are present in the (111) alloy surface. Substitution of gold atoms dilutes the surface platinum atoms such that the high-coordination threefold platinum sites are eliminated much faster than the twofold bridge and single-atom top sites. This change in the distribution of the available reaction sites is frequently called the *ensemble effect* [180]. As a result of this effect, catalyzed reactions that involve adsorption and rearrangement at threefold sites are eliminated, whereas reactions that require one or two atoms sites are attenuated to a much lesser extent. Although minor changes in electronic structure may also occur at the alloy surface sites, most of the reaction results can be explained by this high-coordination-site elimination model. Similar results revealing pro-

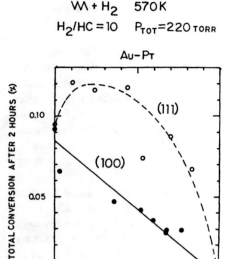

$\wedge\wedge + H_2$ 570 K

$H_2/HC = 10$ $P_{TOT} = 220$ TORR

Au-Pt

Figure 7.43. Rates of formation of various products from *n*-hexane conversion as a function of fractional gold coverage for gold–platinum alloys that were prepared by vaporizing gold onto platinum (111) and platinum (100) crystal surfaces, respectively [180, 190].

nounced changes in catalytic behavior with alloy composition were reviewed by Ponec [181] and Sinfelt [182]. For a variety of hydrocarbon reactions, catalyzed over metal films and high-area-supported catalysts, in most cases the geometrical ensemble effect is decisive in controlling the reaction selectivity.

The effect of alloying is also surface-structure-sensitive, as shown by studies where gold was the alloying constituent in the Pt(100) crystal face instead of the Pt(111) surface [183]. The (100) surface has a square unit cell that contains fourfold bridge and top sites, and unlike the (111) surface it does not have threefold sites. When this surface is alloyed with gold, all reaction rates decline in proportion to the concentration of inactive gold on the Pt(100) surface when *n*-hexane was used as a reactant. This is shown in Figure 7.43. Thus the enhancement of the isomerization activity requires a presence of threefold sites. When gold is used as an alloying agent, there are three types of threefold sites available. One contains only platinum atoms, whereas the other two mixed Pt–Au sites contain one and two atoms, respectively. Thus alloying produces new mixed metal sites with catalytic behavior that can modify the selectivity. Figure 7.43 clearly indicates that the high isomerization rate of *n*-hexane is sustained until the surface was covered up to two-thirds monolayer gold [183]. Thus all three threefold sites are active for isomerization. The mixed Pt–Au sites are then responsible for the enhanced isomerization activity of the Pt–Au alloy that exhibits markedly higher rates than the pure platinum (111) crystal surface.

Boudart and co-workers [184] have shown a 50-fold increase in the rate of H_2/O_2 reaction to produce water over Pd–Au alloys. Such large effects cannot be explained by site-blocking ensemble effects. The new sites that are created by alloying have unique structure and bonding. In fact, a new catalyst is created with structural and bonding properties that are not derived from the structural and bonding properties of the pure alloy constituents.

7.8.3.6 The Building of Improved Platinum and Other Metal Catalysts The atomic-scale ingredients of selective hydrocarbon catalysis by platinum have been identified and a model of the working catalyst has been constructed. Attention now turns toward building improved catalyst systems. Additives are being used to alter the surface structure beneficially, to reduce the amount of carbon deposit, or to slow down its conversion to the inactive graphitic form. Bimetallic or multimetallic platinum catalyst systems have been developed by the addition of one or more other transition metals (Re, Pd, Ir, or Au) that can be operated at higher reaction temperatures to obtain higher reaction rates [185]. They show slower rates of deactivation (have longer lifetimes) and can also be more selective for a given chemical reaction (dehydrocyclization or isomerization) than the one-component catalyst [185].

One of the major challenges in preparing scientifically tailored, high-technology metal catalysts is to deposit the metal particles with the specific surface structure needed to obtain optimum reaction selectively. The structure of the support and its chemical interaction with the metal are utilized to achieve this goal. Deposition of ordered platinum monolayers on sulfides or oxides with well-defined substrate structure is one important approach in this direction. Zeolites, aluminosilicates that are available with variable but well-defined pore structure and Al/Si ratio, could perhaps provide the structural definition that was obtained on the low-surface-area single-crystal catalysts without sacrificing the availability of high surface area [186]. There are attempts to prepare metal catalyst particles with uniform size and equal distances of separation by using microelectronic circuitry fabrication technology. Using electron beam lithography (or perhaps X-ray lithography in the future), metal particles in the size range of $10^2 \text{Å} - 10^3 \text{Å}$ can be deposited in ordered arrays on silica or alumina substrates. The reactivity and the stability of these metal "nanocluster" arrays are under investigation in my laboratory. Strong chemical interaction between the metal particles and the support induces charge transfer toward or away from the metal that again could beneficially alter its catalytic properties [187]. Other additives are being investigated that increase catalytic activity by decreasing the surface residence times required for the reaction and product desorption, thereby reducing the amount of platinum required in conventional reforming catalysts. Identification of new, less expensive catalyst materials with platinum-like chemical activity and selectivity is another important direction of research for catalysis science. Many transition-metal materials are in short supply worldwide and are not readily available in the United States.

As combined surface-science and catalytic-reaction studies develop working models for catalysts of many types, the building of new high-technology catalysts, using this molecular-level understanding, will become more frequent. This transition from art to catalysis science can come none too soon. The rising cost of petroleum necessitates the use of new fuel sources (natural gas, coal, shale, tar sand) and the use of new feedstocks for chemicals (methane, $CO + H_2$, coal liquids). The fuel and chemical technologies based on these new feedstocks require the development of an entirely new generation of catalysts. Ultimately, our fuels and chemicals must be produced form the most stable and abundant molecules we live with on our planet, including CO_2, H_2O, N_2 and O_2. To build the catalytic chemistry starting from these species is a considerable challenge that will be met by catalysis science in the future.

7.9 SUMMARY AND CONCEPTS

- Surface catalysis aims to carry out the same reaction repeatedly at high rates (activity) and selectivity.
- The catalytic process can be characterized by its kinetic parameters (rate constant, preexponential factor, activation energy, reactant pressure dependencies, reaction probability).
- The preparation, activation, deactivation, and regeneration of high-surface-area catalyst materials are dominant concerns of surface catalysis.
- Catalysis by transition-metal surfaces exhibit trends across the periodic table whereby metals that form chemical bonds of intermediate strength have the highest activities.
- Important catalytic reaction concepts include: structure sensitivity and insensitivity of reactions, mechanistic classifications (Langmuir–Hinshelwood, Eley–Rideal), the compensation effect, the presence of strongly chemisorbed overlayer, and the roles of structure and bonding modifier additives (promoters).
- Acid–base catalysis produces mostly carbenium ions by electron or by proton transfer. Among the solid acids, microporous, crystalline alumina silicates (zeolites) are utilized most frequently.
- Surface-science studies of catalysis employ mostly small-surface-area (1 cm^2) crystal surfaces or model catalysts that are well-characterized on the atomic scale. Promoters are deposited on such a surface with known concentration and composition.
- The state and accomplishments of catalysis science are demonstrated through discussions of the ammonia synthesis, carbon monoxide hydrogenation, and hydrocarbon conversion over platinum.

7.10 PROBLEMS

7.1 Calculate the reaction probability of a catalytic reaction that has a turnover rate of 10^{-3} molecules/surface site/sec at 1 atm.

*7.2 The hydrogenation of carbon monoxide and carbon dioxide to methane can be described by a series of elementary reaction steps [194, 196] that are given below:

$$(1) \quad CO_{(g)} + S \rightleftharpoons CO_{(a)}$$

$$(2) \quad H_2 + 2S \rightleftharpoons 2H_{(a)}$$

$$(3) \quad 2H_{(a)} + CO_{(a)} \rightleftharpoons H_2CO_{(a)} + 2S$$

$$(4) \quad H_2CO_{(a)} + S \rightleftharpoons CH_{2(a)} + O_{(a)}$$

$$(5) \quad O_{(a)} + H_{(a)} \rightleftharpoons HO_{(a)} + S$$

$$(6) \quad HO_{(a)} + H_{(a)} \rightleftharpoons H_2O_{(g)} + 2S$$

$$(7) \quad CH_{2(a)} + H_{(a)} \rightleftharpoons CH_{3(a)} + S$$

$$(8) \quad CH_{(a)} + H_{(a)} \rightleftharpoons CH_{4(a)} + 2S$$

Write the rate expression that gives the CO and H_2 pressure dependencies of the reaction rate assuming that (a) step 3 or (b) step 4 is rate-determining.

**7.3   The determination of the equilibrium constant of ammonia formation N_2 + $3H_2 \rightleftharpoons 2NH_3$ has been performed by Haber and Nerst. Using different catalysts they obtained different results. Review the literature [197] on these studies and describe the outcome of this important debate in the history of catalysis.

**7.4   Search the literature to find the important surface catalyzed processes that are used to convert crude oil to gasoline and describe them in sequence of application in the refining technology [198, 199].

**7.5   The partial oxidation of ethylene to ethylene oxide is an important chemical reaction in the chemical technology [198, 200, 201]. Describe the process, the catalyst that is employed, and the nature of the catalyst promoters.

**7.6   Acrylonitrile (CH_2CHCN) is produced from propylene, ammonia, and oxygen over a mixed oxide catalyst [198, 199, 202]. Describe the process.

**7.7   The hydrogenation of nitriles (R-CN) to amines (R-NH_2) is carried out using Raney nickel as a catalyst. Describe what Raney nickel is and describe the process [198].

**7.8   Microporous, crystalline oxides (alumina, silicates, phosphates, etc.) are used as catalyst is in the petroleum and in the chemical technologies in large volume to carry out cracking, isomerization, alkylation, and many other important hydrocarbon conversion reactions [198, 199, 203]. Discuss the structure of these so-called "zeolites" that have one-dimensional and two-dimensional micropores. How can the acidity of the catalysts be altered? How do their acid strengths compare with concentration H_2SO_4 and HF?

**7.9   The catalytic reduction of nitrogen oxides, NO_x, that are produced by combustion of fuels at high temperatures ($\approx 1800°C$ during electric power generation) is one of the important environmental catalytic problems. Review the process that uses ammonia or small hydrocarbons as reducing agents, and list the catalysts that are employed [204, 205].

**7.10   The "three-way" catalytic converter used in automobiles catalyzes the oxidation of unburned hydrocarbons and CO while reducing simultaneously NO to N_2 [198, 206]. Describe the process.

**7.11   The water–gas–shift reaction [207] is utilized to produce hydrogen by the reaction of CO and H_2O. Describe the process.

**7.12   The oxidation of CO to CO_2 and the reduction of NO by NH_3 are complex catalyzed surface reactions that have two or more branches depending on the composition of the reactant mixture and the temperature. There are periodic oscillations in the reaction rates including instabilities [208]. Review the literature describing the ratio rate oscillations for these two processes and discuss the experimental conditions that give rise to this phenomenon.

**7.13 The removal of sulfur from organosulfur compounds is an important catalytic reaction during petroleum refining [198, 199]. A test reaction for this process is the hydrodesulfurization of thiophene to butenes. Describe the process [209]. The removal of nitrogen from organonitrogen compounds is equally important. Describe the process [210].

**7.14 The polymerization of ethylene over chromium compounds is responsible for the production of much of the polyethylene that is produced [211]. Describe the process.

**7.15 The catalyzed gasification (using steam) of carbon solids (coals, chars, organic solid waste, graphite) to H_2, CO_2, and CO is utilized to convert these materials to gaseous fuels (coal gasification) [23, 212]. Describe the process.

**7.16 The conversion of methane by oxydehydrogenation to ethane, ethylene, oxygen-containing organic molecules, or CO and H_2 are important reactions that are at the frontier of catalyst research. Methane, which is the most abundant fraction of natural gas, is an increasingly significant source of fuels as the supply of crude oil diminishes. Review the processes [213–217].

REFERENCES

[1] J. Berzelius. *Jahres-Bericht über die Fortschritte der Physichen Wissenschaften*, Volume 15, H. Laupp, Tübingen, 1836.

[2] P. Emmett. Fifty Years of Progress in Surface Science. *CRC Crit. Rev. Solid State Sci.* **4**:127 (1974).

[3] A. Gaydon. *Dissociation Energies and Spectra of Diatomic Molecules*. Chapman and Hall, London, 1953.

[4] G.W. Koeppl. Best *Ab Initio* Surface Transition State Theory Rate Constants for the $D+H_2$ and $H+D_2$ Reactions. *J. Chem. Phys.* **59**:3425 (1973).

[5] M. Salmerón, R.J. Gale, and G.A. Somorjai. A Modulated Molecular Beam Study of the Mechanism of the H_2–D_2 Exchange Reaction on Pt(111) and Pt(332) Crystal Surfaces. *J. Chem. Phys.* **70**:2807 (1979).

[6] H. Bonzel, G. Brodén, and G. Pirug. Structure Sensitivity of NO Adsorption on a Smooth and Stepped Pt(100) Surface. *J. Catal.* **53**:96 (1978).

[7] M.P. Kiskinova. Electronegative Additives and Poisoning in Catalysis. *Surf. Sci. Rep.* **8**:359 (1988).

[8] G.A. Somorjai. On the Mechanism of Sulfur Poisoning of Platinum Catalysts. *J. Catal.* **27**:453 (1972).

[9] S. Bhatia, J. Beltramini, and D.D. Do. Deactivation of Zeolite Catalysts. *Catal. Rev. Sci. Eng.* **31**:431 (1989/1990).

[10] E.E. Petersen and A.T. Bell, editors. *Catalyst Deactivation. Chemical Industries*, Volume 30. Marcel Dekker, New York, 1987.

[11] L. Falicov and G.A. Somorjai. Correlation between Catalytic Activity and Bonding and Coordination Number of Atoms and Molecules on Transition Metal Surfaces: Theory and Experimental Evidence. *Proc. Natl. Acad. Sci. USA* **82**:2207 (1985).

[12] M. Boudart. Catalysis by Supported Metals. *Adv. Catal. Relat. Subj.* **20**:153 (1969).

[13] H.P. Bonzel and R. Ku. Mechanisms of the Catalytic Carbon Monoxide Oxidation on Pt(110). *Surf. Sci.* **33**:91 (1972).

[14] I. Langmuir. The Mechanism of the Catalytic Action of Platinum in the Reactions $2CO + O_2 \rightleftharpoons 2CO_2$ and $2H_2 + O_2 \rightleftharpoons 2H_2O$. *Trans. Faraday Soc.* **17**:621 (1922).

[15] D.D. Eley. Catalysis, an Art Becoming a Science. *Chem. Ind.* (1), **1976**:12 (1976).

[16] T. Engel and G. Ertl. A Molecular Beam Investigation of the Catalytic Oxidation of CO on Pd(111). *J. Chem. Phys.* **69**:1267 (1978).

[17] A.K. Galway. Compensation Effect in Heterogeneous Catalysis. *Adv. Catal.* **26**:247 (1977).

[18] M.A. Vannice. The Catalytic Synthesis of Hydrocarbons from H_2/CO Mixtures over the Group VIII Metals. II. The Kinetics of the Methanation Reaction over Supported Metals. *J. Catal.* **37**:462 (1975).

[19] O.V. Krylov. *Catalysis by Non-Metals*. Academic Press, New York, 1970.

[20] P.B. Venuto and E.T. Habib. Catalyst-Feedstock-Engineering Interactions in Fluid Catalytic Cracking. *Catal. Rev. Sci. Eng.* **18**:1 (1978).

[21] D.P. Burke. Catalysts I. A $600-Million Market in Cars and Refineries. *Chem. Week* **124**:42 (1979).

[22] K.C. Taylor. Automobile Catalytic Converters. In: J.R. Anderson and M. Boudart, editors, *Catalysis: Science and Technology*, Volume 5. Springer-Verlag, New York, 1984.

[23] J.A. Cusumano, R.A. Dalla Betta, and R.B. Levy. *Catalysis in Coal Conversion*. Academic Press, New York, 1978.

[24] G.W. Parshall. Industrial Applications of Homogeneous Catalysis. A Review. *J. Mol. Catal.* **4**:243 (1978).

[25] G.A. Somorjai. Active Sites in Heterogeneous Catalysis. *Adv. Catal.* **26**:1 (1977).

[26] D.R. Kahn, E.E. Petersen, and G.A. Somorjai. The Hydrogenolysis of Cyclopropane on a Platinum Stepped Single Crystal at Atmospheric Pressure. *J. Catal.* **34**:294 (1974).

[27] B.A. Sexton and G.A. Somorjai. The Hydrogenation of CO and CO_2 on Polycrystalline Rhodium: Correlation of Surface Composition, Kinetics, and Product Distributions. *J. Catal.* **46**:167 (1977).

[28] M. Temkin and V. Pyzhev. *Acta Physicochim. URSS* **12**:327 (1940).

[29] M. Temkin and V. Pyzhev. Kinetics of the Synthesis of Ammonia on Promoted Iron Catalysts. *J. Phys. Chem. (USSR)* **13**:851 (1939).

[30] A. Nielsen. Review of Ammonia Catalysis. *Catal. Rev.* **4**:1 (1970).

[31] A. Nielsen. *An Investigation on Promoted Iron Catalysts for the Synthesis of Ammonia*. J. Gjellerups Forlag, Copenhagen, 1968.

[32] A. Ozaki, H. Taylor, and M. Boudart. Kinetics and Mechanism of the Ammonia Synthesis. *Proc. R. Soc. London* **A258**:47 (1960).

[33] G. Ertl. Elementary Steps in Ammonia Synthesis: The Surface Science Approach. In: J.R. Jennings, editor, *Catalytic Ammonia Synthesis: Fundamentals and Practice, Fundamental and Applied Catalysis*. Plenum Press, New York, 1991.

[34] J.W. Geuss and K.C. Waugh. Chemisorption at More Elevated Pressures on Indus-

trial Ammonia Synthesis Catalysts. In: J.R. Jennings, editor, *Catalytic Ammonia Synthesis: Fundamentals and Practice, Fundamental and Applied Catalysis*. Plenum Press, New York, 1991.

[35] S.R. Tennison. Alternative Noniron Catalysts. In: J.R. Jennings, editor, *Catalytic Ammonia Synthesis: Fundamentals and Practice. Fundamental and Applied Catalysis*. Plenum Press, New York, 1991.

[36] C.A. Vancini. *Synthesis of Ammonia*. Macmillan Press, London, 1971. Translated by L. Pirt.

[37] N.D. Spencer, R.C. Schoonmaker, and G.A. Somorjai. Iron Single Crystals as Ammonia Synthesis Catalysts. Effect of Surface Structure on Catalyst Activity. *J. Catal.* **74**:129 (1982).

[38] D.R. Strongin, J. Carrazza, S.R. Bare, and G.A. Somorjai. The Importance of C_7 Sites and Surface Roughness in the Ammonia Synthesis Reaction over Iron. *J. Catal.* **103**:213 (1987).

[39] R. Smoluchowsky. Anisotropy of the Electronic Work Function of Metals. *Phys. Rev.* **60**:661 (1941).

[40] J. Hölzl and F.K. Schulte. Work Function of Metals. In: G. Höhler, editor, *Solid Surface Physics. Springer Tracts in Modern Physics*, Volume 85. Springer-Verlag, Berlin, 1979.

[41] J. McAllister and R.S. Hansen. Catalytic Decomposition of Ammonia on Tungsten (100), (110), and (111) Crystal Faces. *J. Chem. Phys.* **59**:414 (1973).

[42] J.A. Dumesic, H. Topsøe, and M. Boudart. Surface, Catalytic, and Magnetic Properties of Small Iron Particles. III. Nitrogen Induced Surface Reconstruction. *J. Catal.* **37**:513 (1975).

[43] F. Bozso, G. Ertl, and M. Weiss. Interaction of Nitrogen with Iron Surfaces. II. Fe(110). *J. Catal.* **50**:519 (1977).

[44] F. Bozso, G. Ertl, and M. Weiss. Interaction of Nitrogen with Iron Surfaces. I. Fe(100) and Fe(111). *J. Catal.* **49**:18 (1977).

[45] P.H. Emmett and S. Brunauer. The Adsorption of Nitrogen by Iron Synthetic Ammonia Catalysts. *J. Am. Chem. Soc.* **56**:35 (1934).

[46] J.J. Scholten, P. Zwietering, J.A. Konvalinka, and J.H. de Boer. Chemisorption of Nitrogen on Iron Catalysts in Connection with Ammonia Synthesis. Part 1. The Kinetics of the Adsorption and Desorption of Nitrogen. *Trans. Faraday Soc.* **55**:2166 (1959).

[47] S.R. Bare, D.R. Strongin, and G.A. Somorjai. Ammonia Synthesis over Iron Single-Crystal Catalysts: The Effects of Alumina and Potassium. *J. Phys. Chem.* **90**:4726 (1986).

[48] D.R. Strongin, S.R. Bare, and G.A. Somorjai. The Effects of Aluminum Oxide in Restructuring Iron Single Crystal Surfaces for Ammonia Synthesis. *J. Catal.* **103**:289 (1987).

[49] I. Sushumna and E. Ruckenstein. Role of Physical and Chemical Interactions in the Behavior of Supported Metal Catalysts: Iron on Alumina—A Case Study. *J. Catal.* **94**:239 (1985).

[50] E. Paparazzo. XPS Analysis of Iron Aluminum Oxide Systems. *Appl. Surf. Sci.* **25**:1 (1986).

[51] E. Paparazzo, J.L. Dormann, and D. Fiorani. X-Ray Photoemission Study of Fe-Al_2O_3 Granular Thin Films. *Phys. Rev. B* **28**:1154 (1983).

[52] W.S. Borghard and M. Boudart. The Textural Promotion of Metallic Iron by Alumina. *J. Catal.* **80**:194 (1983).

[53] H. Ludwiczek, A. Preisinger, A. Fischer, R. Hosemann, A. Schönfeld, and W. Vo-

gel. Structure, Formation, and Stability of Paracrystalline Ammonia Catalysts. *J. Catal.* **51**:326 (1978).

[54] G. Fagherazzi, F. Galante, F. Garbassi, and N. Pernicone. Structural Study of the Al_2O_3-Promoted Ammonia Synthesis Catalyst. II. Reduced State. *J. Catal.* **26**:344 (1972).

[55] W.G. Frankenburg. The Catalytic Synthesis of Ammonia from Nitrogen and Hydrogen. In: P.H. Emmett, editor, *Hydrogenation and Dehydrogenation. Catalysis*, Volume 3. Reinhold, New York, 1955.

[56] G. Ertl, S.B. Lee, and M. Weiss. Adsorption of Nitrogen on Potassium Promoted Fe(111) and (100) Surfaces. *Surf. Sci.* **114**:527 (1982).

[57] G. Ertl. Surface Science and Catalysis—Studies on the Mechanism of Ammonia Synthesis: The P.H. Emmett Award Address. *Catal. Rev. Sci. Eng.* **21**:201 (1980).

[58] Z. Paál, G. Ertl, and S.B. Lee. Interactions of Potassium, Oxygen, and Nitrogen with Polycrystalline Iron Surfaces. *Appl. Surf. Sci.* **8**:231 (1981).

[59] D.R. Strongin and G.A. Somorjai. The Effects of Potassium on Ammonia Synthesis over Iron Single Crystal Surfaces. *J. Catal.* **109**:51 (1988).

[60] T.E. Madey and C. Benndorf. Influence of Surface Additives (Na and O) on the Adsorption and Structure of NH_3 on Ni(110). *Surf. Sci.* **152/153**:587 (1985).

[61] C. Benndorf and T.E. Madey. Interaction of NH_3 with Adsorbed Oxygen and Sodium on Ru(001): Evidence for Both Local and Long Range Interactions. *Chem. Phys. Lett.* **101**:59 (1983).

[62] D.R. Strongin and G.A. Somorjai. On the Rate Enhancement of Ammonia Synthesis over Iron Single Crystals by Coadsorption of Aluminum Oxide with Potassium. *Catal. Lett.* **1**:61 (1988).

[63] K. Altenburg, H. Bosch, J.G. van Ommen, and P.J. Gellings. The Role of Potassium as a Promoter in Iron Catalysts for Ammonia Synthesis. *J. Catal.* **66**:326 (1980).

[64] S. Khammouma. Ph.D. thesis, Stanford University, Palo Alto, CA, 1972.

[65] G.A. Somorjai. *Chemistry in Two Dimensions: Surfaces.* Cornell University Press, Ithaca, NY, 1981.

[66] A. Ozaki and K. Aika. Catalytic Activation of Dinitrogen. In: J.R. Anderson and M. Boudart, editors, *Catalysis: Science and Technology*, Volume 1. Springer-Verlag, Berlin, 1981.

[67] M. Bowker, I. Parker, and K.C. Waugh. Extrapolation of the Kinetics of Model Ammonia Synthesis to Industrially Relevant Temperatures and Pressures. *Appl. Catal.* **14**:101 (1985).

[68] M. Bowker, I. Parker, and K.C. Waugh. The Application of Surface Kinetic Data to the Industrial Synthesis of Ammonia. *Surf. Sci.* **197**:L223 (1988).

[69] P. Stolze and J.K. Nørskov. Comment on "The Application of Surface Kinetic Data to the Industrial Synthesis of Ammonia" by M. Bowker, I. Parker, and K.C. Waugh. *Surf. Sci.* **197**:L230 (1988).

[70] P. Stolze and J.K. Nørskov. An Interpretation of the High-Pressure Kinetics of Ammonia Synthesis Based on a Microscopic Model. *J. Catal.* **110**:1 (1988).

[71] J.K. Nørskov and P. Stolze. Theoretical Aspects of Surface Reactions. *Surf. Sci.* **189/190**:91 (1987).

[72] P. Stolze. Surface Science as the Basis for the Understanding of the Catalytic Synthesis of Ammonia. *Phys. Scr.* **36**:824 (1987).

[73] P. Stolze and J.K. Nørskov. A Description of the High-Pressure Ammonia Synthesis Reaction Based on Surface Science. *J. Vacuum Sci. Technol.* A **5**:581 (1987).

[74] P. Stolze and J.K. Nørskov. Bridging the "Pressure Gap" between Ultrahigh-Vacuum Surface Physics and High-Pressure Catalysis. *Phys. Rev. Lett.* **55**:2502 (1985).

[75] J.R. Rostrup-Nielsen. Catalytic Steam Reforming. In: J.R. Anderson and M. Boudart, editors, *Catalysis Science and Technology*, Volume 5. Springer-Verlag, New York, 1984.

[76] D.A. Hickman and L.D. Schmidt. Production of Syngas by Direct Catalytic Oxidation of Methane. *Science* **259**:343 (1993).

[77] M.A. Vannice. The Catalytic Synthesis of Hydrocarbons from H_2/CO Mixtures over the Group VIII Metals. I. The Specific Activities and Product Distributions of Supported Metals. *J. Catal.* **37**:449 (1975).

[78] D.J. Dwyer and G.A. Somorjai. Hydrogenation of CO and CO_2 over Iron Foils: Correlations of Rate, Product Distribution, and Surface Composition. *J. Catal.* **52**:291 (1978).

[79] M. Araki and V. Ponec. Methanation of Carbon Monoxide on Nickel and Nickel Copper Alloys. *J. Catal.* **44**:439 (1976).

[80] P.R. Ventrcek, B.J. Wood, and H. Wise. The Role of Surface Carbon in Catalytic Methanation. *J. Catal.* **43**:363 (1976).

[81] J.A. Rabo, A.P. Risch, and M.L. Poutsma. Reactions of Carbon Monoxide and Hydrogen on Co, Ni, Ru, and Pd Metals. *J. Catal.* **53**:295 (1978).

[82] P. Biloen, J.L. Helle, and W.M.H. Sachtler. Incorporation of Surface Carbon into Hydrocarbons during Fischer–Tropsch Synthesis: Mechanistic Implications. *J. Catal.* **58**:95 (1979).

[83] L.L. Kesmodel, L.H. Dubois, and G.A. Somorjai. LEED Analysis of Acetylene and Ethylene Chemisorption on the Pt(111) Surface: Evidence for Ethylidyne Formation. *J. Chem. Phys.* **70**:2180 (1979).

[84] J.E. Demuth and H. Ibach. Identification of CH Species on Ni(111) by High Resolution Electron Energy Loss Spectroscopy. *Surf. Sci.* **78**:L238 (1978).

[85] M.L. Poutsma, L.F. Elek, P. Ibarbia, H. Risch, and J.A. Rabo. Selective Formation of Methanol from Synthesis Gas over Palladium Catalysts. *J. Catal.* **52**:157 (1978).

[86] M.A. Vannice. The Catalytic Synthesis from Carbon Monoxide and Hydrogen. *Catal. Rev. Sci. Eng.* **14**:153 (1976).

[87] H. Pichler. Twenty-five Years of Synthesis of Gasoline by Catalytic Conversion of Carbon Monoxide and Hydrogen. *Adv. Catal. Relat. Subj.* **4**:271 (1952).

[88] W.K. Hall, R.J. Kokes, and P.H. Emmett. Mechanism Studies of the Fischer–Tropsch Synthesis: The Incorporation of Radioactive Ethylene, Propionaldehyde and Propanol. *J. Am. Chem. Soc.* **82**:1027 (1960).

[89] J.T. Kummer and P.H. Emmett. Fischer–Tropsch Synthesis Mechanism Studies. The Addition of Radioactive Alcohols to the Synthesis Gas. *J. Am. Chem. Soc.* **75**:5177 (1953).

[90] J.A. Rabo, L.F. Elek, and J.N. Francis. The Reaction of Surface Carbon with Water on Ni and Co Catalysts; a New Process for the Production of Concentrated Methane from Dilute Carbon Monoxide Waste Streams. In: T. Seiyama and K. Tanabe, editors, *New Horizons in Catalysis. Studies in Surface Science and Catalysis*, Volume 7A. Elsevier, New York, 1981, p. 490.

[91] D.G. Castner, R. Blackadar, and G.A. Somorjai. CO Hydrogenation on Clean and Oxidized Rhodium Foil and Single Crystal Catalysts. Correlations of Catalyst Activity, Selectivity, and Surface Composition. *J. Catal.* **66**:257 (1980).

[92] G.M. Schwab. Metal Electrons and Catalysis. *Trans. Faraday Soc.* **42**:689 (1946).

[93] S.J. Tauster. Strong Metal Support Interactions. *Acc. Chem. Res.* **20**:389 (1987).

[94] S.J. Tauster, S.C. Fung, R.T.K. Baker, and J.A. Horsley. Strong Interactions in Supported Metal Catalysts. *Science* **211**:1121 (1981).

[95] S.J. Tauster, S.C. Fung, and R.L. Garten. Strong Metal Support Interactions. Group 8 Noble Metals Supported on TiO_2. *J. Am. Chem. Soc.* **100**:170 (1978).

[96] V. Andera. Investigation of the Rh/TiO_2 System by XPS and XAES. *Appl. Surf. Science* **51**:1 (1991).

[97] J.P. Belzunegui, J.M. Rojo, and J. Sanz. ^1H NMR Procedure to Estimate the Extent of Metal Surface Covered by TiO_x Overlayers in Reduced Rh/TiO_2 Catalysts. *J. Phys. Chem.* **95**:3463 (1991).

[98] A. Dauscher, L. Hilaire, F. le Normand, G. Maire, G. Mueller, and A. Vasquez. Characterization by XPS and XAS of Supported Pt/TiO_2–CeO_2 Catalysts. *Surf. Interface Anal.* **16**:341 (1990).

[99] A.N. Murty, M. Seamster, A.N. Thorpe, and R.T. Obermyer. NMR Evidence of Metal–Support Interaction in Syngas Conversion Catalyst Co–TiO_2. *J. Appl. Phys.* **67**:5847 (1990).

[100] G. Munuera, A.R. González-Elipe, and J.P. Espinós. XPS Study of Phase Mobility in Ni/TiO_2 Systems. *Surf. Sci.* **211/212**:1113 (1989).

[101] M.G. Cattania, F. Parmagiani, and V. Ragaini. A Study of Ruthenium Catalysts on Oxide Supports. *Surf. Sci.* **211/212**:1097 (1989).

[102] F. Boccuzzi, A. Chiorino, and G. Ghiotti. IR Study of the CO Adsorption on Pt/ZnO Samples. Evidence for a PtZn Phase Formation in the SMSI State. *Surf. Sci.* **209**:77 (1989).

[103] D. Gazzoli, G. Minelli, and M. Valigi. The FeO_x–TiO_2 System: An X-Ray Diffraction, Thermogravimetric and Magnetic Susceptibility Study. *Mater. Chem. Phys.* **21**:93 (1989).

[104] K. Matsuo and K. Nakano. Performances of Platinum Metal Catalysts Supported on Titania Coated Silica Prepared from Metal Organics. *Appl. Surf. Sci.* **33/34**:269 (1988).

[105] A.D. Logan, E.J. Braunschweig, A.K. Datye, and D.J. Smith. Direct Observation of the Surfaces of Small Metal Crystallites; Rhodium Supported on TiO_2. *Langmuir* **4**:827 (1988).

[106] K. Tamura, U. Bardi, and Y. Nihei. An Investigation by Angular Resolved X-Ray Photoelectron Spectroscopy of Strong Metal–Support Interaction (SMSI) in the Pt/TiO_2 System. *Surf. Sci.* **197**:L281 (1988).

[107] G. Munuera, A.R. González-Elipe, J.P. Espinós, J.C. Conesa, J. Soria, and J. Sanz. Hydrogen-Induced TiO_x Migration onto Metallic Rh in Real Rh/TiO_2 Catalysts. *J. Phys. Chem.* **91**:6625 (1987).

[108] A. Katrib, C. Petit, P. Legare, L. Hilaire, and G. Maire. An Investigation of Metal-Support Interaction in Bimetallic Pt–Mo Catalysts Deposited on Silica and Alumina. *Surf. Sci.* **189/190**:886 (1987).

[109] A.B. Anderson and D.Q. Dowd. Co-Adsorption on Pt(111) Doped with TiO, FeO, ZnO, and Fe, and Pt Ad-Atoms. Molecular Orbital Study of Co-Dopant Interactions. *J. Phys. Chem.* **91**:869 (1987).

[110] C. Ocal and S. Ferrer. A New CO Adsorption State on Thermally Treated Pt/TiO_2 Model Catalysts. *Surf. Sci.* **178**:850 (1986).

[111] C. Ocal and S. Ferrer. The Strong Metal–Support Interaction (SMSI) in Pt–TiO_2 Model Catalysts. A New CO Adsorption State on Pt-Ti Atoms. *J. Chem. Phys.* **89**:4974 (1985).

[112] J. Cunningham and G.H. Al-Sayyed. Metal–Support Interaction in Benzene Hydrogenation over Pt–TiO_2. Influence of O_2 and UV. *Surf. Sci.* **169**:289 (1986).

[113] J. Sanz and J.M. Rojo. NMR Study of Hydrogen Adsorption on Rh/TiO_2. *J. Phys. Chem.* **89**:4974 (1985).

[114] H. Orita, S. Naito, and K. Tamuru. Nature of SMSI Effect on CO + H_2 Reaction over Supported Rh Catalysts. *J. Phys. Chem.* **89**:3066 (1985).

[115] S. Takatani and Y.W. Chung. Effect of High Temperature Reduction on the Surface Composition of and CO Chemisorption on Ni/TiO_2. *Appl. Surf. Sci.* **19**:341 (1984).

[116] H.R. Sadeghi and V.E. Henrich. Rh on TiO_2. Model Catalyst Studies of the Strong Metal-Support Interaction. *Appl. Surf. Sci.* **19**:330 (1984).

[117] K. Tanaka, K. Miyahara, and I. Toyoshima. Adsorption of CO_2 on TiO_2 and Pt/TiO_2 Studied by X-Ray Photoelectron Spectroscopy and Auger Electron Spectroscopy. *J. Phys. Chem.* **88**:3504 (1984).

[118] J.C. Conesa, J. Soria, P. Malet, G. Munuera, and J. Sanz. Magnetic Resonance Studies of Hydrogen-Reduced Rh/TiO_2 Catalysts. *J. Phys. Chem.* **88**:2986 (1984).

[119] D.N. Belton, Y.M. Sun, and J.M. White. Thin-Film Models of Strong Metal–Support Interaction Catalysts. Platinum on Oxidized Titanium. *J. Phys. Chem.* **88**:1690 (1984).

[120] T. Pannaparayil, M. Oskooie-Tabrizi, C. Lo, L.N. Mulay, G.A. Melson, and V.U.S. Rao. Magnetic and Mössbauer Study of Metal–Zeolite Interaction in Catalysts. *J. Appl. Phys.* **55**:2601 (1984).

[121] J.E.E. Baglin, G.J. Clark, and J.F. Ziegler. Catalyst Metal–Support Interactions. Rutherford Backscattering Spectrometry Applied to Discontinuous Films. *Nucl. Instrum. Methods Phys. Res.* **218**:445 (1983).

[122] T. Huizinga, H.F.J. van't Blik, J.C. Vis, and R. Prins. XPS Investigations of Pt and Rh Supported on γ-Al_2O_3 and TiO_2. *Surf. Sci.* **135**:580 (1983).

[123] B.H. Chen and J.M. White. Properties of Platinum Supported on Oxides of Titanium. *J. Phys. Chem.* **86**:3534 (1982).

[124] H. Poppa and F. Soria. The Interaction of CO and O_2 with Thin Islands of Pd. *Surf. Sci.* **115**:L105 (1982).

[125] M. Valigi, D. Gazzoli, and D. Cordischi. A Structural, Thermogravimetric and ESR Study of the RhO_x–TiO_2 System. *J. Mater. Sci.* **17**:1277 (1982).

[126] J.A. Horsley. A Molecular Orbital Study of Strong Metal–Support Interaction between Platinum and Titanium Dioxide. *J. Am. Chem. Soc.* **101**:2870 (1979).

[127] R.G. Herman. Classical and Non-Classical Route for Alcohol Synthesis. In: L. Guczi, editor, *New Trends in CO Activation. Studies in Surface Science and Catalysis*, Volume 64. Elsevier, Amsterdam, 1991.

[128] P.J. Flory. *Principles of Polymer Chemistry.* Cornell University Press, Ithaca, NY, 1953.

[129] G.V. Schulz. Über die Kinetik der Kettenpolymerisationen. V. Der Einfluß verschiedener Reaktionsarten auf die Polymolekularität. *Z. Phys. Chem. Abt. B* **43**:25 (1939).

[130] D.J. Dwyer and G.A. Somorjai. The Role of Readsorption in Determining the Product Distribution During CO Hydrogenation over Fe Single Crystals. *J. Catal.* **56**:249 (1979).

[131] Y.T. Eidus, N.D. Zelinski, and N.I. Ershov. *Dokl. Akad. Nauk USSR* **60**:599 (1948).

[132] C.D. Chang and A.J. Silvestri. The Conversion of Methanol and Other O-Compounds to Hydrocarbons over Zeolite Catalysts. *J. Catal.* **47**:249 (1977).

[133] S.L. Meisel, J.P. McCullough, C.H. Lechthaler, and P.B. Weisz. Gasoline from Methanol in One Step. *Chemtech* **6**:86 (1976).

[134] W.W. Kaeding and S.A. Butter. Production of Chemicals from Methanol. I. Low Molecular Weight Olefins. *J. Catal.* **61**:155 (1980).

[135] J.K. Nørskov. Adsorbate–Adsorbate Interactions and Surface Reactivity. In: H.P. Bonzel, A.M. Bradshaw, and G. Ertl, editors, *Physics and Chemistry of Alkali Metal Adsorption. Materials Science Monographs*, Volume 57. Elsevier, Amsterdam, 1989.

[136] J.E. Crowell, E.L. Garfunkel, and G.A. Somorjai. The Coadsorption of Potassium and CO on the Pt(111) Crystal Surface: A TDS, HREELS, and UPS Study. *Surf. Sci.* **121**:303 (1982).

[137] C.T. Campbell and D.W. Goodman. A Surface Science Investigation of the Role of Potassium Promoters in Nickel Catalysts for CO Hydrogenation. *Surf. Sci.* **123**:413 (1982).

[138] J.E. Crowell, W.T. Tysoe, and G.A. Somorjai. Potassium Coadsorption Induced Dissociation of CO on the Rh(111) Crystal Face: An Isotope Mixing Study. *J. Phys. Chem.* **89**:1598 (1985).

[139] J.E. Crowell and G.A. Somorjai. The Effect of Potassium on the Chemisorption of Carbon Monoxide on the Rh(111) Crystal Face. *Appl. Surf. Sci.* **19**:73 (1984).

[140] J.H. Sinfelt. Heterogeneous Catalysis by Metals. *Prog. Solid State Chem.* **10**:55 (1975).

[141] F.G. Ciapetta and D.N. Wallace. Catalytic Naphtha Reforming. *Catal. Rev.* **5**:67 (1972).

[142] G.A. Mills, H. Heineman, T.H. Milliken, and K.G. Oblad. Catalytic Mechanism. *Ind. Eng. Chem.* **45**:135 (1953).

[143] N.I. Il'chenko and G.I. Golodets. Catalytic Oxidation of Ammonia. I. Reaction Kinetics and Mechanism. *J. Catal.* **39**:57 (1975).

[144] J.K. Dixon and J.E. Longfield. Hydrocarbon Oxidation. In: P.H. Emmett, editor, *Oxidation, Hydration, Dehydration and Cracking Catalysts. Catalysis*, Volume 7. Reinhold, New York, 1960.

[145] C.C. Chang and L.L. Hegedus. Surface Reactions of NO, CO, and O_2 near the Stoichiometric Point. I. Pt-Alumina. *J. Catal.* **57**:361 (1976).

[146] M. Shelef. Nitric Oxide: Surface Reactions and Removal from Auto Exhaust. *Catal. Rev. Sci. Eng.* **11**:1 (1975).

[147] E. Segal, R.J. Madon, and M. Boudart. Catalytic Hydrogenation of Cyclohexane. I. Vapor-Phase Reaction on Supported Platinum. *J. Catal.* **52**:45 (1978).

[148] P. Stonehart and P.M. Ross. The Commonality of Surface Processes in Electrocatalysis and Gas-Phase Heterogeneous Catalysis. *Catal. Rev. Sci. Eng.* **12**:1 (1975).

[149] J.C. Schlatter and M. Boudart. Hydrogenation of Ethylene on Supported Platinum. *J. Catal.* **24**:482 (1972).

[150] A.J. Appleby. Electrocatalysis and Fuel Cells. *Catal. Rev.* **4**:221 (1970).

[151] J. Haro, R. Gómez, and J.M. Ferreira. The Role of Palladium in Dehydrogenation of Cyclohexane over Pt-Pd/Al_2O_3 Bimetallic Catalysts. *J. Catal.* **45**:326 (1976).

[152] M. Kraft and H. Spindler. Studies on Some Relations between Structural and Catalytical Properties of Alumina-Supported Platinum. In: *Proceedings of the 4th International Congress on Catalysis. Preprints of the Papers*, 1960, paper #69.

[153] C. Leclerq, L. Leclerq, and R. Maurel. Hydrogenolysis of Saturated Hydrocarbons. III. Selectivity in Hydrogenolysis of Various Aliphatic Hydrocarbons on Platinum/Alumina. *J. Catal.* **50**:87 (1977).

[154] J.R. Anderson and N.R. Avery. The Isomerization of Aliphatic Hydrocarbons on Evaporated Films of Platinum and Palladium. *J. Catal.* **5**:446 (1966).

[155] G.A. Somorjai. Surface Science View of Catalysis: The Past, Present, and Future. In: *8th International Congress on Catalysis. Proceedings. Volume 1: Plenary Lectures*, 1984.

[156] S.M. Davis, F. Zaera, and G.A. Somorjai. Surface Structure and Temperature Dependence of *n*-Hexane Skeletal Rearrangement Reactions Catalyzed over Platinum Single Crystal Surfaces: Marked Structure Sensitivity of Aromatization. *J. Catal.* **85**:206 (1984).

[157] K. Besocke, B. Krahl-Urban, and H. Wagner. Dipole Moments Associated with Edge Atoms; a Comparative Study on Stepped Pt, Au, and W Surfaces. *Surf. Sci.* **68**:39 (1977).

[158] G.A. Somorjai and F. Zaera. Heterogeneous Catalysis on the Molecular Scale. *J. Phys. Chem.* **86**:3070 (1982).

[159] M. Salmerón and G.A. Somorjai. Desorption, Decomposition, and Deuterium Exchange Reactions of Unsaturated Hydrocarbons (Ethylene, Acetylene, Propylene, and Butenes) on the Pt(111) Crystal Face. *J. Phys. Chem.* **86**:341 (1982).

[160] S.M. Davis and G.A. Somorjai. Molecular Ingredients of Heterogeneous Catalysis: New Horizons for High Technology. *Bull. Soc. Chim. France* **3**:(3), 271 (1984).

[161] G.A. Somorjai, J.E. Crowell, R.J. Koestner, L.H. Dubois, and M.A. Van Hove. The Study of the Structure of Adsorbed Molecules on Solid Surfaces by High Resolution Electron Energy Loss Spectroscopy and Low Energy Electron Diffraction. In: K. Fuwa, editor, *Recent Advances in Analytical Spectroscopy*. Pergamon Press, Oxford, 1982.

[162] L.H. Dubois, D.G. Castner, and G.A. Somorjai. A Low Energy Electron Diffraction (LEED), High Resolution Electron Energy Loss (ELS), and Thermal Desorption Mass Spectrometry (TDS) Study. *J. Chem. Phys.* **72**:5234 (1980).

[163] S.M. Davis, F. Zaera, and G.A. Somorjai. The Reactivity and Composition of Strongly Adsorbed Carbonaceous Deposits on Platinum. Model of the Working Hydrocarbon Conversion Catalyst. *J. Catal.* **77**:439 (1982).

[164] G.A. Somorjai. Surface Science and Catalysis. *Philos. Trans. R. Soc. London* **A318**:81 (1986).

[165] A. Wieckowski, S. Rosasco, G. Salaita, A. Hubbard, B. Bent, F. Zaera, D. Godbey, and G.A. Somorjai. Comparison of Gas-Phase and Electrochemical Hydrogenation of Ethylene at Platinum Surfaces. *J. Am. Chem. Soc.* **107**:5910 (1985).

[166] A. Farkas, L. Farkas, and E.K. Rideal. Experiments on Heavy Hydrogen. IV. The Hydrogenation and Exchange Reaction of Ethylene with Heavy Hydrogen. *Proc. R. Soc. London* **A146**:630 (1934).

[167] D.D. Eley and J.L. Tuck. On the Microthermoconductivity Method for the Estimation of *Para*-Hydrogen and Deuterium. *Trans. Faraday Soc.* **32**:1425 (1936).

[168] O. Beek. Hydrogenation Catalysts. *Discuss. Faraday Soc.* **8**:118 (1950).

[169] G.H. Twigg and E.K. Rideal. The Exchange Reaction between Ethylene and Deuterium on a Nickel Catalyst. *Proc. R. Soc. London* **A171**:55 (1939).

[170] J. Horiuti and M. Polanyi. Exchange Reactions of Hydrogen on Metallic Catalysts. *Trans. Faraday Soc.* **30**:1164 (1934).

[171] R.W. Roberts. A Study of the Adsorption and Decomposition of Hydrocarbons on Clean Iridium Surfaces. *J. Phys. Chem.* **67**:2035 (1963).

[172] J. Horiuti and K. Miyahara. *Hydrogenation of Ethylene on Metallic Catalysts*. National Standard Reference Data Series 13, National Bureau of Standards, 1968.

[173] F. Zaera and G.A. Somorjai. Hydrogenation of Ethylene over Platinum (111) Single-Crystal Surfaces. *J. Am. Chem. Soc.* **106**:2288 (1984).

[174] R.J. Koestner, M.A. Van Hove, and G.A. Somorjai. A LEED Crystallography Study of the (2×2)–C_2H_3 Structure Obtained after Ethylene Adsorption on Rh(111). *Surf. Sci.* **121**:321 (1982).

[175] S.J. Thomson and G. Webb. Catalytic Hydrogenation of Olefins on Metals, a New Interpretation. *J. Chem. Soc. Chem. Commun.* **13**:526 (1976).

[176] B.E. Koel, J.E. Crowell, B. Bent, C.M. Mate, and G.A. Somorjai. Thermal Decomposition of Benzene on the Rh(111) Crystal Surface. *J. Phys. Chem.* **90**:2949 (1986).

[177] A.J. Gellman, D. Neiman, and G.A. Somorjai. Catalytic Hydrodesulfurization over

the Mo(100) Single Crystal Surface. I. Kinetics and Overall Mechanism. *J. Catal.* **107**:92 (1987).

[178]   M.H. Farias, A.J. Gellman, G.A. Somorjai, R.R. Chianelli, and K.S. Liang. The CO and Coadsorption and Reactions of Sulfur, Hydrogen, and Oxygen on Clean and Sulfided Mo(100) and on MoS_2 Crystal Faces. *Surf. Sci.* **140**:181 (1984).

[179]   W.D. Gillespie, R.K. Herz, E.E. Petersen, and G.A. Somorjai. The Structure Sensitivity of *n*-Heptane Dehydrocyclization and Hydrogenolysis Catalyzed by Platinum Single Crystals at Atmospheric Pressure. *J. Catal.* **70**:147 (1981).

[180]   J.W.A. Sachtler and G.A. Somorjai. Influence of Ensemble Size on CO Chemisorption and Catalytic *n*-Hexane Conversion by Au-Pt(111) Bimetallic Single-Crystal Surfaces. *J. Catal.* **81**:77 (1983).

[181]   V. Ponec. Catalysis by Alloys in Hydrocarbon Reactions. *Adv. Catal.* **32**:149 (1983).

[182]   J.H. Sinfelt. *Bimetallic Catalysts: Discoveries, Concepts, and Applications.* John Wiley & Sons, New York, 1983.

[183]   R.C. Yeates and G.A. Somorjai. Surface Structure Sensitivity of Alloy Catalysis: Catalytic Conversion of *n*-Hexane over Au–Pt(111) and Au–Pt(100) Alloy Crystal Surfaces. *J. Catal.* **103**:208 (1987).

[184]   Y.L. Lam, J. Criado, and M. Boudart. Enhancement by Inactive Gold of the Rate of the H_2–O_2 Reaction on Active Palladium: a Ligand Effect. *Nouv. J. Chim.* **1**:461 (1977).

[185]   G.A. Somorjai. Building and Characterization of Catalysts on Single Crystal Surfaces. *Catal. Lett.* **15**:25 (1992).

[186]   J.A. Rabo, editor. *Zeolite Chemistry and Catalysis. ACS Monograph*, Volume 171. American Chemical Society, Washington, D.C., 1976.

[187]   Z. Karpiński, Z. Zhang, and W.M.H. Sachtler. Probing Palladium-Cobalt/NaY Catalysts by Neopentane Conversion. *Catal. Lett.* **13**:123 (1992).

[188]   J.H. Sinfelt. Catalytic Hydrogenolysis on Metals. *Catal. Lett.* **9**:159 (1991).

[189]   B.C. Gates. *Catalytic Chemistry.* John Wiley & Sons, New York, 1992.

[190]   G.A. Somorjai and J. Carrazza. Structure Sensitivity of Catalytic Reactions. *Ind. Eng. Chem. Fundam.* **25**:63 (1986).

[191]   D.W. Goodman. Model Catalytic Studies over Metal Single Crystals. *Acc. Chem. Res.* **17**:194 (1984).

[192]   J.K. Nørskov, P. Stoltze, and U. Nielsen. The Reactivity of Metal Surfaces. *Catal. Lett.* **9**:173 (1991).

[193]   C.H. Bartholomew, R.B. Pannell, and J.L. Butler. Support and Crystallite Size Effects in CO Hydrogenation on Nickel. *J. Catal.* **65**:335 (1980).

[194]   K.J. Williams, A.B. Boffa, M. Salmerón, A.T. Bell, and G.A. Somorjai. The Kinetics of CO_2 Hydrogenation on a Rh Foil Promoted by Titania Overlayers. *Catal. Lett.* **9**:415 (1991).

[195]   S.M. Davis, F. Zaera, B.E. Gordon, and G.A. Somorjai. Radiotracer and Thermal Desorption Studies of Dehydrogenation and Atmospheric Hydrogenation of Organic Fragments Obtained from [^{14}C]Ethylene Chemisorbed over Pt(111) Surfaces. *J. Catal.* **92**:240 (1985).

[196]   K.J. Williams. Titania Promotion of Rhodium Model Catalysts: A Surface Science and Reaction Study. Ph.D. thesis, University of California, Berkeley, 1991.

[197]   S.A. Topham. The History of the Catalytic Synthesis of Ammonia. In: J.R. Anderson and M. Boudart, editors, *Catalysis: Science and Technology*, Volume 7. Springer-Verlag, Berlin, 1985.

[198] C.N. Satterfield. *Heterogeneous Catalysis in Practice*. McGraw–Hill Chemical Engineering Series. McGraw–Hill, New York, 1980.

[199] B.C. Gates, J.R. Katzer, and G.C.A. Schuit. *Chemistry of Catalytic Processes*. McGraw–Hill Chemical Engineering Series. McGraw–Hill, New York, 1979.

[200] R.A. van Santen and H.P.C.E. Kuipers. The Mechanism of Ethylene Epoxidation. *Adv. Catal.* **35**:265 (1987).

[201] J.M. Berty. Ethylene Oxide Synthesis. In: B.E. Leach, editor, *Applied Industrial Catalysis*, Volume 1. Academic Press, New York, 1983.

[202] R.K. Grasselly and J.D. Burrington. Selective Oxidation and Ammoxidation of Propylene by Heterogeneous Catalysis. *Adv. Catal.* **30**:133 (1981).

[203] J. Dwyer and P.J. O'Malley. Relation between Acidic and Catalytic Properties of Zeolites. In: S. Kaliaguine, editor, *Keynotes in Energy-Related Catalysis. Studies in Surface Science and Catalysis*, Volume 35. Elsevier, Amsterdam, 1988.

[204] M. Iwamoto. Catalytic Decomposition of Nitrogen Monoxide. In: M. Misano, Y. Morooka, and S. Kimura, editors, *Future Opportunities in Catalytic and Separation Technology. Studies in Surface Science and Catalysis*, Volume 54. Elsevier, Amsterdam, 1990.

[205] H. Bosch and F. Janssen. Catalytic Reduction of Nitrogen Oxides: A Review on the Fundamentals and Technology. *Catal. Today* **2**:369 (1988).

[206] J.E. McEvoy, editor. *Catalysts for the Control of Automobile Pollutants. Advances in Chemistry*, Volume 143. American Chemical Society, Washington, D.C., 1975.

[207] D.S. Newsome. The Water–Gas Shift Reaction. *Catal. Rev. Sci. Eng.* **21**:275 (1980).

[208] G. Ertl. Oscillatory Catalytic Reactions at Single-Crystal Surfaces. *Adv. Catal.* **37**:213 (1990).

[209] P. Grange. Catalytic Dehydrosulfurization. *Catal. Rev. Sci. Eng.* **21**:135 (1980).

[210] T.C. Ho. Hydrodenitrogenation Catalysis. *Catal. Rev. Sci. Eng.* **30**:117 (1988).

[211] M.P. McDaniel. Supported Chromium Catalysts for Ethylene Polymerization. *Adv. Catal.* **33**:48 (1985).

[212] E.E. Donath. History of Catalysis in Coal Liquifaction. In: J.R. Anderson and M. Boudart, editors, *Catalysis: Science and Technology*, Volume 3. Springer-Verlag, Berlin, 1982.

[213] J. Raskó, G.A. Somorjai, and H. Heinemann. Catalytic Low-Temperature Oxydehydrogenation of Methane to Higher Hydrocarbons with Very High Selectivity at 8–12% Conversion. *Appl. Catal. A* **84**:57 (1992).

[214] J. Raskó, P. Pereira, G.A. Somorjai, and H. Heinemann. The Catalytic Low Temperature Oxydehydrogenation of Methane: Temperature Dependence, Carbon Balance and Effects of Catalyst Composition. *Catal. Lett.* **9**:395 (1991).

[215] J.C. Mackie. Partial Oxidation of Methane: The Role of Gas Phase Reactions. *Catal. Rev. Sci. Eng.* **33**:169 (1991).

[216] Y. Amenomiya, V.I. Birss, M. Goledzinowski, J. Galuszka, and A.R. Sanger. Conversion of Methane by Oxidative Coupling. *Catal. Rev. Sci. Eng.* **32**:163 (1990).

[217] J.H. Lunsford. The Catalytic Conversion of Methane to Higher Hydrocarbons. *Catal. Today* **6**:235 (1990).

[218] Surface Science as a Basis for Understanding Heterogeneous Catalysis. In: R.W. Joyner and R.A. van Santen, editors, *Elementary Reaction Steps in Heterogeneous Catalysis*. Kluwer Academic Publishers, Dordrecht, The Netherlands, in press, (1993).

TABLE 7.1. Kinetic Parameters for Ethane Hydrogenolysis Over Metal Catalysts

Catalyst	T (°C)	$\log_{10}$(HC)	$\log_{10}$(H$_2$)	ΔE^* (kcal/mole)	$\log_{10} A$	x	y	Rate (molecules/cm^2·sec) At 250°C	$-\log_{10}$ RP	Ref.[a]
SiO$_2$ Supported										
5% CO	220–260	17.87	18.69	30	25.48	1.0	−0.8	9×10^{12}	9.0	1
5% Ni	180–220	17.87	18.69	41	31.69	1.0	−2.4	4×10^{14}	7.4	1
5% Cu	290–330	17.87	18.69	21	—	1.0	−0.4	—	—	2
1% Ru	170–270	17.87	18.69	33	29.5	—	—	3×10^{15}	6.5	3
5% Ru	180–210	17.87	18.69	32	28.11	0.8	−1.3	6×10^{14}	7.2	1
0.1% Rh	190–250	17.87	18.69	42	31.4	0.8	−2.1	1×10^{14}	7.9	4
0.3% Rh	190–250	17.87	18.69	42	31.4	0.8	−2.1	1×10^{14}	7.9	4
1% Rh	190–250	17.87	18.69	42	31.9	0.8	−2.1	3×10^{14}	7.5	4
5% Rh	190–225	17.87	18.69	42	31.76	0.8	−2.2	2×10^{14}	7.7	1
10% Rh	190–250	17.87	18.69	42	31.8	0.8	−2.1	2×10^{14}	7.6	4
5% Pd	345–375	17.87	18.69	58	33.57	0.9	−2.5	2×10^{9}	12.6	1
5% Re	230–265	17.87	18.69	31	26.26	0.5	0.3	2×10^{13}	8.6	1
5% Os	125–160	17.87	18.69	35	30.85	0.6	−1.2	2×10^{16}	5.7	1
5% Ir	180–210	17.87	18.69	36	28.72	0.7	−1.6	5×10^{13}	8.2	1
5% Pt	345–385	17.87	18.69	54	31.77	0.9	−2.5	2×10^{9}	12.7	1
1% Rh–0.13% Cu	170–270	17.87	18.69	~33	28.5	—	—	3×10^{14}	7.5	3
1% Ru–0.32% Cu	170–270	17.87	18.69	~33	28.1	—	—	1×10^{14}	7.8	3
1% Ru–0.63% Cu	170–270	17.87	18.69	~33	26.3	—	—	2×10^{12}	9.6	3
1% Os–0.10% Cu	170–270	17.87	18.69	~35	29.1	—	—	1×10^{14}	7.8	3
1% Os–0.17% cu	170–270	17.87	18.69	~35	28.2	—	—	2×10^{13}	8.6	3
Unsupported								At 250°C		
Ni powder	225–270	17.87	18.69	43	31.3	1.0	−2.1	2×10^{13}	8.6	5
Ni powder	165–275	17.51	18.81	40	26.9	—	—	2×10^{10}	11.3	6
Ni film	254–273	17.10	18.16	58	35.8	—	—	5×10^{11}	9.4	7
Ru powder	160–200	17.87	18.69	32	28.4	—	—	1×10^{15}	6.9	8
Ru powder	140–300	17.51	18.51	28	25.1	0.9	−0.9	3×10^{13}	8.1	9
Rh powder	190–250	17.87	18.59	39	29.3	0.9	−1.6	1×10^{13}	8.8	4
Rh powder	125–175	17.51	18.51	36	30.9	0.8	−2.1	9×10^{15}	5.6	9

	Temp (°C)									Ref.
Pd film	270–360	17.10	18.16	50	31.9	1.0	—	1×10^{11}	10.0	10
W film	170–180	17.10	18.16	27	26.4	—	—	1×10^{15}	6.0	7
Ir powder	235–300	17.51	18.51	40	28.4	—	—	7×10^{11}	9.7	9
Pt film	270–340	17.10	18.16	57	34.2	1.0	—	3×10^{10}	10.6	10
Unsupported Alloys										
Ni–Cu powder								At 250°C		
($X_{Cu} = 0.062$)	310–340	17.87	18.69	51	31.6	0.9	−1.3	2×10^{10}	11.7	5
($X_{Cu} = 0.103$)	325–355	17.87	18.69	51	31.1	1.0	−1.0	6×10^{9}	12.2	5
($X_{Cu} = 0.315$)	355–395	17.87	18.69	50	29.9	1.0	−1.3	8×10^{8}	13.0	5
($X_{Cu} = 0.633$)	380–440	17.87	18.69	48	28.5	0.8	−1.2	2×10^{8}	13.7	5
Ru–Cu powder										
($X_{Cu} = 0.018$)	280–350	17.87	18.69	~25	~22	—	—	1×10^{12}	9.9	8
($X_{Cu} = 0.03$)	280–350	17.87	18.69	~25	~22	—	—	6×10^{11}	10.2	8
($X_{Cu} = 0.05$)	350–430	17.87	18.69	~25	21	—	—	6×10^{10}	11.2	8
Other Supported and Unsupported Systems								At 180°C		
0.5% Ru/Al₂O₃	160–220	19.17	19.60	—	—	1.0	−2.0	3×10^{12}	10.7	11
Ir film	80–205	14.7	14.85	~23	26	—	—	9×10^{14}	4.0	12
								At 300°C		
Ni powder	315	18.21	0.0	38	25.7	—	—	2×10^{11}	11.0	13
5% Ni/SiO₂	300	18.82	19.26	—	—	—	—	5×10^{14}	8.2	14
3.7% Ni/SiO₂Al₂O₃	300	18.82	19.26	—	—	—	—	2×10^{13}	9.6	11
2.5% Ni/Na-Y	300	18.82	19.26	—	—	—	—	9×10^{12}	9.9	11
2.4% Ni/Li-Y	300	18.82	19.26	—	—	—	—	7×10^{12}	10.0	11
2.3% Ni/Ca-Y	300	18.82	19.26	—	—	—	—	7×10^{12}	10.0	11
1.2% Ni/Mg-Y	300	18.82	19.26	—	—	—	—	8×10^{12}	10.0	11
2% Pt/Al₂O₃	300	18.39	19.35	—	—	—	—	8×10^{9}	12.6	15
								At 350°C		
Co powder	310–360	17.51	18.81	27	20.3	—	—	8×10^{10}	10.7	6
8% Ni/SiO₂–Al₂O₃	350	18.0 (pulse)	19.39	—	—	—	—	3×10^{14}	—	16

527

TABLE 7.1. (Continued)

Catalyst	T (°C)	$\log_{10}$(HC)	$\log_{10}$(H$_2$)	ΔE^* (kcal/mole)	$\log_{10} A$	x	y	Rate (molecules/cm$^2 \cdot$ sec)	$-\log_{10}$ RP	Ref.[a]
								At 350°C		
8% Ni/SiO$_2$–Al$_2$O$_3$ +0.73% Na$^+$	350	18.0 (pulse)	19.39	—	—	—	—	2×10^{14}	—	16
8% Ni/SiO$_2$–Al$_2$O$_3$ +2.4% Na$^+$	350	18.0 (pulse)	19.39	—	—	—	—	1×10^{14}	—	16
Ni–Cu films										
($X_{Cu} = 0.0$)	310–350	16.56	17.64	28	23.2	—	—	3×10^{13}	7.1	17
($X_{Cu} = 0.04$)	310–350	16.56	17.64	22	20.9	—	—	2×10^{13}	7.3	17
($X_{Cu} = 0.18$)	310–350	16.56	17.64	22	20.4	—	—	5×10^{12}	7.9	17
($X_{Cu} = 0.54$)	310–350	16.56	17.64	18	18.8	—	—	3×10^{12}	8.1	17
($X_{Cu} = 0.82$)	310–350	16.56	17.64	21	19.0	—	—	5×10^{11}	8.9	17
($X_{Cu} = 0.96$)	310–350	16.56	17.64	16	16.7	—	—	1×10^{11}	9.5	17
3.1% Pt/SiO$_2$ oxidized at 200°C	205–350	18.21	18.21	—	—	—	—	2×10^{14}	8.1	18
3.6% Pt/SiO$_2$ oxidized at 500°C	205–350	18.21	18.21	—	—	—	—	3×10^{13}	8.8	18
1.8% Mo–4.4% Pt/SiO$_2$	205–350	18.21	18.21	33	27.0	1.0	−1.0	3×10^{15}	6.8	18
Pt powder	370–460	17.51	18.81	53	26.9	0.9	−1.9	3×10^{8}	13.1	19
Pt film										
Clean	300–410	17.02	18.02	53	30.6	—	—	1×10^{12}	9.1	20
Steady state	300–410	17.02	18.02	36	24.7	—	—	1×10^{12}	9.1	20
								At 390°C		
3.8% Pt/Na–Y	390	18.03	19.01	—	—	—	—	8×10^{13}	8.2	21
10.8% Pt/Nh$_4$–Y	390	18.03	19.01	—	—	—	—	1×10^{14}	8.1	21
2.6% Pt/SiO$_2$	390	18.03	19.01	—	—	—	—	2×10^{13}	8.8	21

[a]References for Tables 7.1 to 7.41 are listed at the end of Table 7.41.

TABLE 7.2. Kinetic Parameters for Propane Hydrogenolysis Over Metal Catalysts

Catalyst	T (°C)	$\log_{10}$(HC)	$\log_{10}$(H$_2$)	ΔE^* (kcal/mole)	$\log_{10}$ A	x	y	Rate (molecules/cm^2 · sec)	$-\log_{10}$ RP	Ref.[a]
0.5% Ru/Al$_2$O$_3$	140–170	18.77	19.32	36	31.9	1.0	−1.5	3×10^{13} ($S_{C_2} = 0.89$)	9.3	11
								At 150°C		
Pd–Rh powder										
($X_{Rh} = 0.0$)	150	17.81	18.11	—	—	—	—	6×10^{9}	12.0	22
($X_{Rh} = 0.2$)	150	17.81	18.11	—	—	—	—	1×10^{11}	10.6	22
($X_{Rh} = 0.8$)	150	17.81	18.11	—	—	—	—	2×10^{11}	10.4	22
($X_{Rh} = 1.0$)	150	17.81	18.11	—	—	—	—	2×10^{11}	10.4	22
Pd–Pt powder										
($X_{Pt} = 0.0$)	150	17.81	18.11	—	—	—	—	6×10^{9}	12.0	22
($X_{Pt} = 0.2$)	150	17.81	18.11	—	—	—	—	3×10^{10}	11.3	22
($X_{Pt} = 0.4$)	150	17.81	18.11	—	—	—	—	1×10^{11}	10.8	22
($X_{Pt} = 0.8$)	150	17.81	18.11	—	—	—	—	1×10^{11}	10.7	22
($X_{Pt} = 1.0$)	150	17.81	18.11	—	—	—	—	3×10^{11}	10.3	22
								At 250°C		
6.7% Fe/Al$_2$O$_3$–K$_2$O	270–340	18.79	19.57	31	23.2	1.2	2.0	2×10^{10} ($S_{C_2} = 0.06$)	12.4	23
6.3% Co/SiO$_2$	230–270	18.79	19.57	22	22.8	0.8	−0.7	5×10^{13} ($S_{C_2} = 0.2$)	9.1	23
7% Co/SiO$_2$	245–265	18.79	19.57	34	28.6	0.7	−1.0	3×10^{14} ($S_{C_2} = 0.19$)	8.3	23
4.1% Ni/SiO$_2$	250–300	18.79	19.57	42	30.1	0.7	−1.6	3×10^{12} ($S_{C_2} = 0.53$)	10.1	23
16.7% Ni/SiC	280–305	18.79	19.57	52	34.6	0.9	−2.4	1×10^{13} ($S_{C_2} = 0.56$)	9.8	23

TABLE 7.2. (*Continued*)

Catalyst	T (°C)	$\log_{10}$(HC)	$\log_{10}$(H$_2$)	ΔE^* (kcal/mole)	$\log_{10}$ A	x	y	Rate (molecules/cm^2 · sec)	$-\log_{10}$ RP	Ref.[a]
								At 250°C		
15% Ni/Mg-SiC	250–280	18.79	19.57	46	32.1	0.7	−2.3	1×10^{13} ($S_{C_2} = 0.5$)	9.8	23
Ni powder	250	18.21	18.55	39	27.4	—	—	2×10^{11}	10.9	13
75% Ni/Al$_2$O$_3$	210–250	18.19	19.03	49	35.6	1.0	−1 to −2.3	2×10^{15}	6.9	24
Ni powder Clean	200–270	17.51	18.51	33	27.1	0.9	0 to −0.6	9×10^{11} ($S_{C_2} = 0.65$)	9.5	25
Steady state	200–270	17.51	18.51	—	—	0.9	0 to −0.6	4×10^{9} ($S_{C_2} = 1.0$)	11.9	25
Ni film	217–267	17.10	18.16	31	26.4	—	—	3×10^{13}	7.6	7
0.5% Ru/Al$_2$O$_3$	150–180	18.79	19.57	37	32.2	0.8	−2.0	7×10^{16}	5.9	23
W film	180–190	17.10	18.16	18	21.7	—	—	2×10^{14} ($S_{C_2} = 0.99$)	6.8	7
								At 300°C		
Ni powder	315	18.21	0	36	24.8	—	—	1×10^{11}	11.0	13
2% Pt/Al$_2$O$_3$	300	18.39	19.35	—	—	—	—	1×10^{12}	10.3	15
								At 360°C		
Pt powder	370–440	17.51	18.81	24	19.2	1.0	−1.6	9×10^{10}	10.5	19
Pt film Clean	360	17.02	18.02	—	—	—	—	1×10^{13}	8.0	20
Steady state	360	17.02	18.02	—	—	—	—	5×10^{12}	8.3	20

[a]References for Tables 7.1 to 7.41 are listed at the end of Table 7.41.

TABLE 7.3. Kinetic Parameters for Cyclopropane Ring Opening Over Metal Catalysts

Catalyst	T (°C)	$\log_{10}$(HC)	$\log_{10}$(H$_2$)	ΔE^* (kcal/mole)	$\log_{10} A$	Rate (molecules/cm$^2 \cdot$ sec) At 25°C	$-\log_{10}$ RP	Ref.[a]
Rh film	-78	17.51	18.51	—	—	8×10^{13}	7.6	26
Supported and Unsupported Systems								
35% Co/Kieselguhr	25	19.12	19.05	~ 12	~ 21	1×10^{12}	11.0	27
14–40% Ni/Al$_2$O$_3$	25	19.12	19.05	11	22.1	6×10^{13}	9.3	27
Ni film	-46 to 0	$17.51^{-0.1}$	$18.51^{0.6}$ (D$_2$)	8	19.3	5×10^{13}	7.8	26
10% Mo/Al$_2$O$_3$	25	19.12	19.05	—	—	2×10^{12}	10.8	27
10% Ru/SiO$_2$	0–80	17.87	18.69	12	21.1	2×10^{12}	9.5	28
0.5–5% Rh/Al$_2$O$_3$	25	19.12	19.05	—	—	4×10^{14}	8.5	27
5% Rh/C	25	19.12	19.05	—	—	3×10^{14}	8.6	27
0.36% Rh/SiO$_2$	-35 to -10	17.87	18.69	11	23.4	2×10^{15}	6.5	28
Pd Powder	25	19.12	19.05	—	—	2×10^{12}	10.8	27
5% Pd/Al$_2$O$_3$	25	19.12	19.05	—	—	1×10^{14}	9.1	27
10% Pd/C	25	19.12	19.05	—	—	4×10^{13}	9.5	27
10% Pd/SiO$_2$	-10 to -25	17.87	18.69	16	25.5	7×10^{13}	8.0	28
Pd film	-46 to -8	$17.51^{-0.9}$	$18.51^{0.1}$ (D$_2$)	15	25.3	5×10^{14}	6.8	26
10% Os/SiO$_2$	0 to 60	17.87	18.69	13	22.2	5×10^{12}	9.1	28
10% Ir/SiO$_2$	0 to 30	17.87	18.69	13	22.4	8×10^{12}	8.9	28
Pt powder	25	19.12	19.05	—	—	5×10^{13}	9.4	27
0.5–5% Pt/Al$_2$O$_3$	25	19.12	19.05	—	—	7×10^{13}	9.2	27
Pt(557)	75	18.64	19.35	12	25.8	7×10^{16}	5.8	29
0.3–2%, Pt/η-Al$_2$O$_3$	75	18.64	19.35	8.5	23.1	8×10^{16}	5.7	30
0.3–0.6% Pt/γ-Al$_2$O$_3$	75	18.64	19.35	9	23.9	2×10^{17}	5.3	31
7.1–81% Pt/SiO$_2$	-10 to -21	18.22	19.36	10	22.3	1×10^{15}	7.2	32
0.6% Pt/SiO$_2$	-20 to -30	17.87	18.69	11	22.6	4×10^{14}	7.2	28
Pt film	-78 to -23	$17.51^{-0.2}$	$18.51^{0.2}$ (D$_2$)	11	24.6	3×10^{16}	4.9	26
3.8% Pt/Na-Y	22	17.51	18.98	—	—	1.4×10^{15}	6.4	21
3.5% Pt/NH$_4$-Y	22	17.51	18.98	—	—	1.8×10^{15}	6.2	21
1.1% Pt/Ce-Y	22	17.51	18.98	—	—	4×10^{15}	5.9	21

TABLE 7.3. *(Continued)*

Catalyst	T (°C)	$\log_{10}$(HC)	$\log_{10}$(H$_2$)	ΔE^* (kcal/mole)	$\log_{10} A$	Rate (molecules/cm^2 · sec)	$-\log_{10}$ RP	Ref.[a]
Supported and Unsupported Systems								
						At 25°C		
1.7% Pt/Al$_2$O$_3$	22	17.51	18.98	—	—	1.3×10^{13}	8.4	21
2.6% Pt/SiO$_2$	22	17.51	18.98	—	—	4×10^{13}	7.9	21
Pt powder	24	17.09	19.37	—	—	2×10^{13}	7.8	33
0.6–4.8% Pt/Al$_2$O$_3$	24	$17.09^{0.6}$	$19.37^{0.0}$	—	—	3×10^{13}	7.7	33
						At 140°C		
Fe film	50 to 150	$17.51^{0.0}$	$18.51^{1.0}$ (D$_2$)	23	24.4	2×10^{12}	9.2	26
8% Ni/SiO$_2$-Al$_2$O$_3$	140	18.8 (pulse)	19.3	—	—	1.5×10^{15}	—	16
8% Ni/SiO$_2$-Al$_2$O$_3$ + 0.44% Na$^+$	140	18.8 (pulse)	19.3	—	—	4×10^{14}	—	16
8% Ni/Si$_2$-Al$_2$O$_3$ + 2.4% Na$^+$	140	18.8 (pulse)	19.3	—	—	5×10^{13}	—	16

[a]References for Tables 7.1 to 7.41 are listed at the end of Table 7.41.

TABLE 7.4. Kinetic Parameters for Cyclopropane Hydrogenolysis Over Metal Catalysts

Catalyst	T (°C)	$\log_{10}$(HC)	$\log_{10}$(H$_2$)	ΔE^* (kcal/mole)	$\log_{10} A$	Rate (molecules/cm^2 · sec)	$-\log_{10}$ RP	Ref.[a]
						At 25°C		
35% Co/Kieselguhr	25	19.12	19.05	—	—	5×10^{11}	11.4	27
14–40% Ni/Al$_2$O$_3$	25	19.12	19.05	—	—	8×10^{13}	9.2	27
Ni (film)	−46 to 0	17.51	18.51 (D$_2$)	~8	~19	1.5×10^{13}	8.4	26
10% Mo/Al$_2$O$_3$	25	19.12	19.05	—	—	1×10^{12}	11.1	27
10% Ru/SiO$_2$	0 to 80	17.87	19.69	12	20.3	4×10^{11}	10.3	28
5% Rh/C	25	19.12	19.05	—	—	8×10^{12}	10.2	27
0.5% Rh/Al$_2$O$_3$	25	19.12	19.05	—	—	4×10^{13}	9.5	27
10% Os/SiO$_2$	0 to 60	17.87	18.69	13	21.4	8×10^{11}	9.9	28
						At 140°C		
Fe (film)	50 to 150	17.51	18.51 (D$_2$)	23	24.4	2×10^{12}	9.2	26
8% Ni/SiO$_2$-Al$_2$O$_3$	140	18.7 (pulse)	19.3	—	—	2×10^{14}	—	16
8% Ni/SiO$_2$-Al$_2$O$_3$ + 0.44% Na$^+$	140	18.7 (pulse)	19.3	—	—	1×10^{14}	—	16

[a] References for Tables 7.1 to 7.41 are listed at the end of Table 7.41.

TABLE 7.5. Kinetic Parameters for *n*-Butane Hydrogenolysis Over Metal Catalysts

Catalyst	T (°C)	$\log_{10}$(HC)	$\log_{10}$(H$_2$)	ΔE^* (kcal/mole)	$\log_{10} A$	Rate (molecules/cm^2 · sec)	$-\log_{10}$ RP	S_{C_3}	S_{C_2}	Ref.[a]
						At 150°C				
0.5% Ru/Al$_2$O$_3$	150	$18.35^{0.9}$	$19.35^{-1.35}$	48	39.7	1×10^{15}	7.3	—	—	34
Rh powder	80 to 130	$17.51^{0.5}$	$18.51^{-1.3}$	29	28.9	1×10^{14}	7.4	~0.6	~0.3	9
W film	144 to 164	17.10	18.16	7	16.4	6×10^{12}	8.2	—	—	7
						At 250°C				
Ni film	188 to 209	17.10	18.16	34	28.5	2×10^{14}	6.6	—	—	7
Re/Al$_2$O$_3$	240	18.39	19.35	24	24.0	1×10^{14}	8.2	—	—	35
Os/Al$_2$O$_3$	240	18.39	19.35	34	30.3	1×10^{16}	6.2	—	—	35
Ir/Al$_2$O$_3$	240	18.39	19.35	43	33.8	8×10^{15}	6.4	—	—	35
Ir powder	180 to 230	$17.51^{0.4}$	$18.51^{-0.3}$	24	22.9	9×10^{12}	8.5	≤0.05	~0.15	9
Pt/Al$_2$O$_3$	240	18.39	19.35	34	25.3	1.5×10^{11}	11.1	—	—	35
Pt powder	370 to 460	$17.51^{1.0}$	$18.81^{-0.6}$	23	19.4	7×10^{9}	11.6	~0.65	~0.33	19
Pt film	300 to 400	17.02	18.02	22	20.5	2×10^{11}	9.6	—	—	20
Pt-Re/Al$_2$O$_3$-bimetallic	240	18.39	19.35	36	27.4	3×10^{12}	9.9	—	—	35
($X_{Re} = 0.25$)										
($X_{Re} = 0.55$)	240	18.39	19.35	40	30.1	3×10^{13}	9.9	—	—	35
($X_{Re} = 0.75$)	240	18.39	19.35	43	31.6	6×10^{13}	8.6	—	—	35
						At 300°C				
Ni powder	315	18.2	0.0	44	27.8	1.3×10^{11}	11.0	—	—	13
Pd film	276 to 310	$17.10^{-0.3}$	$18.16^{0.0}$	38	27.4	1×10^{13}	8.0	~0.85	~0.11	10
Pt/Al$_2$O$_3$	300	18.39	19.35	—	—	1×10^{13}	9.3	—	—	36
2% Pt/Al$_2$O$_3$	300	18.39	19.35	—	—	1×10^{13}	9.3	—	—	15
Pt film	256 to 300	$17.10^{0.7}$	$18.16^{1.4}$	21	21.0	1×10^{13}	8.0	~0.6	~0.3	10
Pt(111) film	320	17.10	18.16	~20	—	~6×10^{12}	8.2	~0.6	~0.3	10
Pt(100) film	300	17.10	18.16	—	—	2×10^{13}	7.7	~0.55	~0.4	10

[a]References for Tables 7.1 to 7.41 are listed at the end of Table 7.41.

TABLE 7.6. Kinetic Parameters for Isobutane Hydrogenolysis Over Metal Catalysts

Catalyst	T (°C)	$\log_{10}$(HC)	$\log_{10}$(H$_2$)	ΔE^* (kcal/mole)	$\log_{10}$ A	Rate (molecules/cm^2·sec)	$-\log_{10}$ RP	S_{C_1}	S_{C_2}	Ref.[a]
0.5% Ru/Al$_2$O$_3$	150	$18.35^{0.7}$	$19.35^{-0.7}$	36	32.8	2×10^{14}	At 150°C 8.0	—	—	34
							At 250°C			
Ni film	200–220	17.10	18.16	30	26.8	2×10^{14}	6.7	—	—	7
Pd film	270–310	$17.10^{-0.2}$	$18.16^{0.1}$	21	21.0	2×10^{12}	8.7	~0.94	~0.02	10
Ta film	200–295	16.51	17.51	~3	~14	~3×10^{12}	7.9	~0.6	~0.2	37
W film	150–195	16.51	17.51	~12	~19	~4×10^{13}	6.8	~0.7	~0.2	37
Re film	200–270	16.51	17.51	~5	~15	~2×10^{12}	8.1	~0.4	~0.3	37
Pt film	265–299	$17.10^{0.5}$	$18.16^{-1.4}$	21	20.7	1×10^{12}	9.0	~0.8	~0.04	10
							At 300°C			
Pd film	265–310	17.10	18.16	21	20.8	7×10^{12}	8.1	~0.96	~0.02	38
Re–Au films										
(X_{Re} = 0.71)	290–390	16.51	17.51	~16	~17	~3×10^{10}	9.9	0.0	~0.8	37
(X_{Re} = 0.51)	350–410	16.51	17.51	~3	~12	~7×10^{10}	9.5	~0.2	~0.25	37
Pt/Al$_2$O$_3$	300	18.39	19.35	—	—	8×10^{12}	9.4	—	—	36
Pt powder	370–460	$17.51^{1.0}$	$18.81^{-1.7}$	26	19.9	1×10^{10}	11.4	~0.6	~0.2	19
Pt film	265–299	$17.10^{0.5}$	$18.16^{-1.4}$	21	20.7	6×10^{12}	8.3	~0.8	~0.04	10
Pt(111) film	294–305	$17.10^{0.5}$	$18.16^{-1.4}$	19	18.7	3×10^{11}	9.5	~1.0	0.0	10
Pt(100) film	299	$17.10^{0.5}$	$18.16^{-1.4}$	—	—	7×10^{12}	8.2	~0.6	~0.6	10
							At indicated T (°C)			
Pt/Na-Y (10 Å)	320	18.19	19.36	—	—	2×10^{12}	8.8	—	—	39
Pt/Na-Y (15–20 Å)	320	18.19	19.36	—	—	4×10^{12}	9.5	—	—	39
Pt film	360	17.02	18.02	—	—	2×10^{13}	7.8	—	—	20

[a]References for Tables 7.1 to 7.42 are listed at the end of Table 7.41.

TABLE 7.7. Kinetic Parameters for Methylcyclopropane Ring Opening Over Metal Catalysts

Catalyst	T (°C)	$\log_{10}$(HC)	$\log_{10}$(H$_2$)	ΔE^* (kcal/mole)	$\log_{10}$ A	Rate (molecules/cm$^2\cdot$ sec)	$-\log_{10}$ RF	S_{C_4}	S_{iC_4}	Ref.[a]
						At indicated T (°C)				
Ni film	−46	17.51	18.51	—	—	2×10^{13}	8.3	0.23	0.77	26
Pd film	−23	17.51	18.51	—	—	2×10^{13}	8.1	0.15	0.85	26
Pt film	−64	17.51	18.51	—	—	1×10^{13}	8.4	0.02	0.98	26
						At 0°C				
7.1% Pt/SiO$_2$	0	18.21	19.36	9	22	$0.3\text{–}1.2 \times 10^{15}$	7.1	0.05	0.95	40
27% Pt/SiO$_2$	0	18.21	19.36	9	22	$3\text{–}9 \times 10^{14}$	7.2	0.06	0.94	40
81% Pt/SiO$_2$	0	18.21	19.36	9	22	$3\text{–}9 \times 10^{14}$	7.2	0.03	0.97	40
27% Pt/SiO$_2$	0	18.21	19.36	10	22.3	2×10^{14}	7.8	0.06	0.94	32
0.5% Pt/SiO$_2$	0	17.51	18.21	—	—	6×10^{13}	7.7	0.04	0.96	41
0.5% Pt/SiO$_2 \cdot$ Al$_2$O$_3$ (BF)	0	17.51	18.21	—	—	6×10^{14}	6.6	1.0	0.0	41
2.5% Pt/SiO$_2$	0	17.51	18.21	—	—	2×10^{13}	8.2	0.04	0.96	41
						At 24°C				
Pt powder	24	17.09	19.37	—	—	4×10^{12}	8.4	0.05	0.95	33
0.3% Pt/Al$_2$O$_3$	24	17.09	19.37	—	—	3×10^{13}	7.5	0.13	0.97	33
4.8% Pt/Al$_2$O$_3$	24	17.09	19.37	—	—	2×10^{13}	7.7	0.13	0.87	33
3% Pt/SiO $\cdot$ Al$_2$O$_3$ (BF)	24	17.09	19.37	—	—	2×10^{14}	6.7	0.12	0.88	33

[a] References for Tables 7.1 to 7.41 are listed at the end of Table 7.41.

TABLE 7.8. Kinetic Parameters for *n*-Pentane Hydrogenolysis Over Metal Catalysts

Catalyst	T (°C)	$\log_{10}$(HC)	$\log_{10}$(H₂)	ΔE^* (kcal/mole)	$\log_{10} A$	Rate (molecules/cm²·sec)	$-\log_{10}$ RP	S_{C_4}	S_{C_1}	Ref.[a]
						At 100°C				
0.5% Rh/Al₂O₃	100	17.51	18.51	—	—	3×10^{11}	9.9	~0.12	~0.88	42
1.9% Rh/Al₂O₃	100	17.51	18.51	—	—	2×10^{11}	10.1	~0.15	~0.85	42
8.7% Rh/Al₂O₃	100	17.51	18.51	—	—	9×10^{10}	10.4	~0.60	~0.40	42
12.4% Rh/Al₂O₃	100	17.51	18.51	—	—	7×10^{10}	10.6	~0.66	~0.34	42
						At 185°C				
Rh powder	130–170	17.51	18.51	28	28.2	9×10^{14}	6.4	—	—	9
5.5% Rh/Al₂O₃	110–140	$17.51^{1.0}$	$18.51^{-1.6}$	27 / 17	25.1 for C₃ + C₂ / 21.0 for C₄ + C₁	7×10^{12}	8.5	~0.8	~0.2	43
Os/Al₂O₃	185	18.35	19.35	—	—	2×10^{13}	9.0	—	—	44
0.6% Re/Al₂O₃	185	18.35	19.35	—	—	4×10^{12}	9.6	~0.14	~0.20	44
0.6% Ir/Al₂O₃	185	18.35	19.35	—	—	1×10^{12}	10.2	0.1	0.84	44
0.6% Re-Ir/Al₂O₃-bimetallic										
(X_{Re} = 0.88)	185	18.35	19.35	—	—	6×10^{12}	9.4	—	—	44
(X_{Re} = 0.45)	185	18.35	19.35	—	—	8×10^{12}	9.3	~0.22	~0.56	44
(X_{Re} = 0.10)	185	18.35	19.35	—	—	4×10^{12}	9.6	—	—	44
						At 250°C				
5% Fe/SiO₂	375–450	$18.79^{0.5}$	$20.09^{-1.6}$	23	22.8	2×10^{11}	9.5	0.0	~0.08	45
5% Co/SiO₂	275–375	$18.79^{0.9}$	$20.09^{-1.5}$	31	27.2	2×10^{14}	8.4	~0.12	~0.1	45
5% Ni/SiO₂	275–375	$18.79^{0.9}$	$20.09^{-1.6}$	31	27.4	3×10^{14}	8.2	~0.66	~0.16	45
5% Ru/SiO₂	230–290	$18.79^{0.9}$	$20.09^{-1.6}$	29	29.4	2×10^{17}	5.4	—	—	45
Ru/C	250–325	$18.79^{1.0}$	$20.09^{-1.5}$	37	32.1	5×10^{16}	6.0	~0.06	~0.14	45
5% Rh/SiO₂	250–325	$18.79^{1.0}$	$20.09^{-1.3}$	30	28.0	3×10^{15}	7.2	~0.04	~0.28	45
Rh/C	290–375	$18.79^{1.0}$	$20.09^{-1.3}$	31	27.8	7×10^{14}	7.8	~0.08	~0.30	45
5% Pd/SiO₂	370–480	$18.79^{0.9}$	$20.09^{-1.4}$	49	31.5	1×10^{11}	11.6	~0.62	~0.36	45
Pd/Al₂O₃	250	18.35	19.35	—	—	8×10^{9}	11.6	—	—	44
Re film	190–330	16.51	17.51	~0	~12	$\sim 7 \times 10^{11}$	8.5	~0.56	~0.4	37
0.25% Ir/Al₂O₃	275–360	$18.79^{1.0}$	$20.09^{-1.5}$	~32	27.3	9×10^{13}	8.7	~0.08	~0.38	45
0.25% Ir + Cu/Al₂O₃-bimetallic	250	18.35	19.35	—	—	4×10^{14}	7.6	—	—	44

TABLE 7.8. (Continued)

Catalyst	T (°C)	$\log_{10}$(HC)	$\log_{10}$(H$_2$)	ΔE^* (kcal/mole)	$\log_{10} A$	Rate (molecules/cm$^2 \cdot$ sec)	$-\log_{10}$ RP	S_{C_4}	S_{C_1}	Ref.[a]
							At 250°C			
($X_{Cu} = 0.20$)	250	18.35	19.35	—	—	4×10^{11}	8.6	—	—	44
($X_{Cu} = 0.50$)	250	18.35	19.35	—	—	3×10^{12}	9.8	—	—	44
($X_{Cu} = 0.80$)	250	18.35	19.35	—	—	4×10^{11}	10.6	—	—	44
Ir powder	210–270	17.51	18.51	25	22.7	2×10^{12}	10.2	—	—	9
5% Pt/SiO$_2$	350–500	$18.79^{0.7}$	$20.09^{-1.4}$	28	27.0	2×10^{15}	7.3	0.44	0.56	45
Pt/Al$_2$O$_3$	250	18.35	19.35	—	—	4×10^{10}	11.6	—	—	44
0.33% Ru/Al$_2$O$_3$	450	18.61	19.31	~27	~22	7×10^{13}	8.6	—	—	46
0.86% Ru/Al$_2$O$_3$	490	18.61	19.31	—	—	1×10^{14}	8.4	—	—	46
Rh film	320	16.51	17.51	—	—	~1×10^{11}	9.3	100% C$_1$		47
Rh–Cu films										
($X_{Cu} = 0.06$)	330	16.51	17.51	—	—	~9×10^{10}	9.4	—	—	47
($X_{Cu} = 0.92$)	310	16.51	17.51	—	—	~2×10^{10}	10.1	—	—	47
($X_{Cu} = 1.0$)	310	16.51	17.51	—	—	~4×10^{9}	10.8	—	—	47
Rh–Ag film										
($X_{Ag} = 0.19$)	320	16.51	17.51	~20	~18	~2×10^{10}	10.1	—	—	47
Rh–Sn film										
($X_{Sn} = 0.17$)	280	16.51	17.51	—	—	~1×10^{11}	9.2	—	—	47
Rh–Au film										
($X_{Au} = 0.23$)	320	16.51	17.51	29	20	~4×10^{9}	10.8	—	—	47
Re–Au films										
($X_{Au} = 0.14$)	280	16.51	17.51	~5	—	~3×10^{11}	8.8	~0.5	~0.5	37
(($X_{Au} = 0.68$)	290	16.51	17.51	~15	—	~2×10^{10}	10.1	~0.0	~0.64	37
($X_{Au} = 0.93$)	290	16.51	17.51	~14	—	~2×10^{10}	10.1	~0.0	~0.98	37
16% Pt/SiO$_2$ ($D \approx 0.3$)	312	18.39	19.34	~28	~21	~3×10^{10}	11.8	~0.14	~0.72	48
16% Pt-Au/SiO$_2$ ($D \approx 0.3$)-bimetallic										
($X_{Au} = 0.975$)	370	18.39	19.34	~39	~22	~3×10^{8}	13.8	~0.5	~0.5	48
($X_{Au} = 0.875$)	350	18.39	19.34	~45	~25	~2×10^{9}	13.0	~0.4	~0.56	48
1.3% Pt/SiO$_2$ (14 Å)	283	$18.03^{0.4}$	$19.05^{-0.8}$	—	—	2×10^{12}	9.6	—	—	49
8.2% Pt/SiO$_2$ (195 Å)	283	$18.03^{0.4}$	$19.05^{-0.8}$	—	—	4×10^{11}	10.3	—	—	49

[a]References for Tables 7.1 to 7.41 are listed at the end of Table 7.41.

TABLE 7.9. Kinetic Parameters for Isopentane Hydrogenolysis Over Metal Catalysts

Catalyst	T (°C)	$\log_{10}$(HC)	$\log_{10}$(H$_2$)	ΔE^* (kcal/mole)	$\log_{10}$ A	Rate (molecules/cm^2 · sec)	$-\log_{10}$ RP	S_{iC_4}	S_{nC_1}	Ref.[a]
0.5% Ru/Al$_2$O$_3$	110	18.35	19.35	43	37.5	1×10^{13}	9.1	~0.78	~0.1	34
Pd film	310	17.10	18.16	—	—	2×10^{12}	8.7	~0.32	~0.5	10
2% Pt/Al$_2$O$_3$	300	18.39	19.35	—	—	2×10^{13}	9.0	—	—	15
Pt/Al$_2$O$_3$	300	18.39	19.35	—	—	2×10^{13}	8.9	~0.76	—	36
Pt film	278	17.10	18.16	—	—	3×10^{12}	8.6	—	~0.38	10

[a]References for Tables 7.1 to 7.41 are listed at the end of Table 7.41.

TABLE 7.10. Kinetic Parameters for Neopentane Hydrogenolysis Over Metal Catalysts

Catalyst	T (°C)	$\log_{10}$(HC)	$\log_{10}$(H_2)	ΔE^* (kcal/mole)	$\log_{10}$ A	Rate (molecules/cm² · sec)	$\log_{10}$ RP	S_{iC_4}	S_{C_3}	Ref.[a]
							At 150°C			
0.5% Ru/Al₂O₃	125–155	$18.35^{0.9}$	$19.35^{-0.9}$	43	35.7	4×10^{13}	8.6	~0.16	~0.08	34
Rh powder	110–150	17.51	18.51	40	33.4	8×10^{12}	8.5	—	—	9
							At 250°C			
Ni film	220–265	17.10	18.16	32	26.3	1×10^{13}	8.0	—	—	7
5% Ru/SiO₂	160–180	18.35	19.35	36	30	1×10^{15}	7.2	~1.0	—	50
5% Rh/SiO₂	170–200	18.35	19.35	53	37	8×10^{14}	7.3	~1.0	—	50
5.5% Rh/Al₂O₃	160–200	$17.51^{1.0}$	$18.51^{-1.5}$	20	22.6	2×10^{14}	8.2	—	—	43
W film	200–220	17.10	18.16	11	17.5	9×10^{12}	8.0	—	—	7
10% Os/SiO₂	130–180	18.35	19.35	32	28.4	1×10^{15}	7.1	~1.0	—	50
10% Ir/SiO₂	180–200	18.35	19.35	46	33	6×10^{13}	8.4	~1.0	—	50
1% Pt/spheron	310–370	18.35	19.35	59	33	2×10^{8}	13.9	~1.0	—	50
3% Pt/Na-Y (10 Å)	200–240	18.07	19.37	37	27.6	2×10^{12}	9.6	~0.8	~0.08	51
3% Pt/Ca-Y (10 Å)	200–240	18.07	19.37	35	27.3	6×10^{12}	9.2	~0.6	~0.2	51
3% Pt/LaY (10 Å)	200–240	18.07	19.37	33	26.9	1.6×10^{13}	8.7	~0.5	~0.34	51
2% Pt/SiO₂ (12 Å)	260–300	18.07	19.37	35	26.4	7×10^{11}	10.1	~0.5	~0.1	51
0.9% Pt/SiO₂ (40 Å)	280–320	18.07	19.37	28	23.8	1.5×10^{12}	9.8	~0.6	~0.1	51
0.9% Pt/SiO₂ (70 Å)	280–320	18.07	19.37	25	22.5	1.3×10^{12}	9.8	~0.5	~0.2	51
2.5% Pt/SiO₂	190–360	$17.91^{1.0}$	$19.39^{1.0 \text{ to } -0.3}$	35	28.1	4×10^{13}	8.2	—	—	52
6.3% Pt-Mo/SiO₂ ($X_{Mo} = 0.47$)	190–360	$17.91^{1.0}$	$19.39^{1.0-0}$	16	22.4	5×10^{15}	6.1	—	—	52
6.8% Pt-W/SiO₂ ($X_W = 0.6$)	190–360	$17.91^{1.0}$	$19.39^{1.0-0}$	21	23.6	7×10^{14}	6.9	—	—	52
Pt film	240–295	17.10	18.16	21	20.8	1×10^{12}	8.9	~0.7	—	10

						At indicated T (°C)				
Pt–Pd films										
(X_{Pd} = 0.0)	250	16.34	17.34	—	—	9×10^{10}	9.3	~0.96	—	53
(X_{Pd} = 0.2)	250	16.34	17.34	—	—	2×10^{10}	9.9	~0.6	~0.4	53
(X_{Pd} = 0.3)	250	16.34	17.34	—	—	2×10^{11}	8.9	~0.9	~0.1	53
(X_{Pd} = 0.54)	250	16.34	17.34	—	—	1×10^{12}	8.1	~0.5	~0.3	53
(X_{Pd} = 0.8)	250	16.34	17.34	—	—	2×10^{11}	8.9	~0.7	~0.3	53
(X_{Pd} = 1.0)	250	16.34	17.34	—	—	2×10^{12}	8.0	~0.7	~0.2	53
Pd film	310	17.10	18.16	—	—	9×10^{12}	8.0	~0.7	~0.2	10
Pd film	265–310	17.10	18.16	—	~0	3×10^{12}	8.5	—	—	38
2% Pt/Al$_2$O$_3$	300	18.39	19.35	—	—	1×10^{13}	9.2	—	—	15
Pt/Al$_2$O$_3$	300	18.39	19.35	—	—	8×10^{12}	9.4	—	—	36
1% Pt/C	370	18.35	19.35	—	—	5×10^{13}	8.5	~1.0	—	54
6% Pt–6% Fe/C	370	18.35	19.35	—	—	1.5×10^{12}	10.0	~1.0	—	54
0.6% Pt/γ-Al$_2$O$_3$	307	18.35	19.35	—	—	3×10^{12}	9.8	~1.0	—	55
2% Pt/γ-Al$_2$O$_3$	307	18.35	19.35	—	—	4×10^{12}	9.7	~1.0	—	55
4.3% Pt/SiO$_2$	307	18.35	19.35	—	—	2×10^{13}	8.9	~1.0	—	55
1% Pt/spheron	307	18.35	19.35	—	—	6×10^{10}	11.4	~1.0	—	55
2% Pt/η-Al$_2$O$_3$	307	18.35	19.35	—	—	7×10^{12}	9.4	~1.0	—	55
Pt powder	307	18.35	19.35	—	—	1.2×10^{11}	11.1	~1.0	—	55
5% Pt/Ca-Y (D = 1.0)	272	18.35	19.35	—	—	1.8×10^{13}	8.9	~1.0	—	56
2% Pt/η-Al$_2$O$_3$ (D = 0.64)	272	18.35	19.35	—	—	4×10^{11}	10.7	~1.0	—	56
2% Pt/η-Al$_2$O$_3$ (D = 0.08)	272	18.35	19.35	—	—	6×10^{11}	10.4	~1.0	—	56
Pt film										
Clean	360	17.02	18.02	—	—	7×10^{13}	7.0	—	—	20
Steady state	360	17.02	18.02	—	—	4×10^{12}	8.3	—	—	20
Au powder	450	18.35	19.35	51	20	$\sim 10^{5}$	17	—	—	50

^{a}References for Tables 7.1 to 7.41 are listed at the end of Table 7.41.

TABLE 7.11. Kinetic Parameters for Cyclopentane Ring Opening and Hydrogenolysis Over Metal Catalysts

Catalyst	T (°C)	$\log_{10}$ (HC)	$\log_{10}$ (H$_2$)	ΔE^* (kcal/mole)	$\log_{10}$ A	Rate (molecules/cm$^2 \cdot$ sec)	$-\log_{10}$ RP	S_{C_5}	Ref.[a]
						At 250°C			
Re/Al$_2$O$_3$	240	18.39	19.35	12	17.9	8×10^{12}	9.4	~0.5	35
Os/Al$_2$O$_3$	240	18.39	19.35	18	21.8	3×10^{14}	7.9	—	35
Ir/Al$_2$O$_3$	240	18.39	19.35	28	27.6	8×10^{15}	6.4	—	35
Pt/Al$_2$O$_3$	240	18.39	19.35	35	27.7	1.4×10^{13}	9.1	~1.0	35
Pt-Re/Al$_2$O$_3$-bimetallic									
(X_{Re} = 0.25)	240	18.39	19.35	40	29.6	8×10^{12}	9.4	—	35
(X_{Re} = 0.50)	240	18.39	19.35	26	24.2	2×10^{13}	9.0	—	35
(X_{Re} = 0.75)	240	18.39	19.35	23	24.8	1×10^{15}	7.1	—	35
						At indicated T (°C)			
Rh powder	150	17.51	18.51	18	21.2	9×10^{11}	9.4	—	9
1% Pd/Al$_2$O$_3$									
(D = 0.3–0.8)	290	18.51	19.33	40	28.4	9×10^{12}	9.4	1.0	57
0.25% Pd/Al$_2$O$_3$									
(D = 0.8–1.0)	290	18.51	19.33	40	28.6	1.4×10^{13}	9.2	1.0	57
0.25–8% Pd/Al$_2$O$_3$	290	18.51	19.33	—	—	$9{-}13 \times 10^{12}$	9.3	1.0	58
0.7% Pd/SiO$_2$									
(D = 0.04)	290	18.51	19.33	—	—	3×10^{12}	9.9	1.0	58
1.3% Pd/SiO$_2$									
(D = 0.32)	290	18.51	19.33	—	—	2×10^{13}	9.1	1.0	58
Pd film	300	16.41	17.41 (D$_2$)	~28	~23	~1×10^{12}	8.3	~0.95	59
Pd–Au films									
(X_{Au} = 0.21)	330	16.41	17.41 (D$_2$)	—	—	~2×10^{11}	9.0	~1.0	59
(X_{Au} = 0.34)	310	16.41	17.41 (D$_2$)	—	—	~9×10^{10}	9.3	~1.0	59
(X_{Au} = 0.86)	400	16.41	17.41 (D$_2$)	—	—	~3×10^{12}	7.8	~0.98	59
(X_{Au} = 1.0)	500	16.41	17.41 (D$_2$)	—	—	~4×10^{11}	8.7	~0.85	59
2% Pt/SiO$_2$									
(D = 0.5–1.0)	~300	~16.2	~17.8	—	—	$6{-}9 \times 10^{12}$	7.1	—	60

[a]References for Tables 7.1 to 7.41 are listed at the end of Table 7.41.

TABLE 7.12. Kinetic Parameters for n-Hexane Hydrogenolysis Over Metal Catalysts

Catalyst	T (°C)	$\log_{10}$ (HC)	$\log_{10}$ (H$_2$)	ΔE^* (kcal/mole)	Rate (molecules/cm^2 · sec)	$-\log_{10}$ RP At indicated T (°C)	S_{C_5}	S_{C_4}	S_{C_3}	Ref.[a]
9% Ni/SiO$_2$ ($D = 0.15$)	330	18.69	19.29	—	$\sim 5 \times 10^9$	12.8	97% C$_1$			61
Ni/SiO$_2$-Al$_2$O$_3$ ($D = 0.15$)	330	18.69	19.29	—	$\sim 2 \times 10^7$	15.3	—	—	—	61
Ni/SiO$_2$-Al$_2$O$_3$ ($D = 0.15$)	350	18.69	19.29	—	$\sim 1 \times 10^8$	14.5	—	—	—	61
Ni/SiO$_2$-Al$_2$O$_3$ ($D = 0.15$)	275	18.69	19.29	—	$\sim 2 \times 10^8$	14.3	—	—	—	61
Ni/Li-Y ($D = 0.3$)	300	18.42	19.34	—	$\sim 6 \times 10^{12}$	9.5	~ 0.3	~ 0.2	~ 0.1	14
Ni-Cu powders										
($X_{Cu} = 0.0$)	330	18.12	19.36	44	1×10^{14}	7.8	~ 0.4	~ 0.2	~ 0.1	62
($X_{Cu} = 0.05$)	330	18.12	19.36	44	3×10^{11}	10.5	~ 0.5	~ 0.2	~ 0.1	62
($X_{Cu} = 0.23$)	330	18.12	19.36	47	3×10^{11}	10.5	~ 0.7	~ 0.2	—	62
($X_{Cu} = 0.47$)	330	18.12	19.36	55	$\sim 5 \times 10^{10}$	11.2	~ 0.4	~ 0.2	~ 0.4	62
Cu/SiO$_2$	325	17.74	18.69	~ 18	$\sim 2 \times 10^{14}$	7.2	~ 0.1	~ 0.1	—	63
Rh/SiO$_2$	175	17.74	18.69	~ 22	$\sim 4 \times 10^{14}$	6.9	~ 0.5	~ 0.4	~ 0.1	63
Rh-Cu/SiO$_2$										
($X_{Cu} = 0.85$)	300	17.74	18.69	~ 18	$\sim 2 \times 10^{13}$	8.2	~ 0.3	~ 0.3	~ 0.2	63
Rh-Cu films										
($X_{Cu} = 0.0$)	240	16.51	17.51	~ 12	$\sim 3 \times 10^{11}$	8.9	All C$_1$			47
($X_{Cu} = 0.03$)	280	16.51	17.51	~ 23	$\sim 1 \times 10^{10}$	10.3	67% C$_1$			47
($X_{Cu} = 0.58$)	270	16.51	17.51	~ 15	$\sim 9 \times 10^9$	10.4	All C$_1$			47
($X_{Cu} = 0.92$)	280	16.51	17.51	~ 20	$\sim 1 \times 10^{10}$	10.2	All C$_1$			47
Rh-Cu films										
($X_{Cu} = 1.0$)	310	16.51	17.51	~ 10	$\sim 2 \times 10^{10}$	10.0	All C$_1$			47
Rh-Sn films										
($X_{Sn} = 0.03$)	290	16.51	17.51	~ 20	$\sim 3 \times 10^{10}$	9.9	All C$_1$			47
($X_{Sn} = 0.36$)	320	16.51	17.51	~ 14	$\sim 2 \times 10^{10}$	10.0	All C$_1$			47
Rh-Au film										
($X_{Au} = 0.87$)	300	16.51	17.51	~ 20	$\sim 1 \times 10^{10}$	10.3	All C$_1$			47

TABLE 7.12. (*Continued*)

Catalyst	T (°C)	$\log_{10}$ (HC)	$\log_{10}$ (H$_2$)	ΔE^* (kcal/mole)	Rate (molecules/cm$^2 \cdot$ sec)	$-\log_{10}$ RP	S_{C_5}	S_{C_4}	S_{C_3}	Ref.[a]
						At indicated T (°C)		All C$_1$		
Rh–Ag film										
($X_{\text{Ag}} = 0.26$)	280	16.51	17.51	~16	~1 × 10^{10}	10.3	—	—	—	47
10% Pd/Al$_2$O$_3$ ($D = 0.3$)	350	16.99	19.39	—	~4 × 10^{11}	9.2	~0.6	~0.1	~0.1	64
10% Pd–Au/Al$_2$O$_3$–bimetallic ($D = 0.3$)										
($X_{\text{Au}} = 0.5$)	350	16.99	19.39	—	~9 × 10^{10}	9.9	~0.7	~0.1	~0.1	64
($X_{\text{Au}} = 0.65$)	350	16.99	19.39	—	~2 × 10^{9}	11.5	~0.5	~0.2	~0.3	64
Ta film	370	16.51	17.51	—	~5 × 10^{10}	9.6	—	—	—	37
Ta–Au film										
($X_{\text{Au}} = 0.47$)	360	16.51	17.51	—	~4 × 10^{10}	9.7	—	—	—	37
Re film	180	16.51	17.51	~0	~3 × 10^{12}	7.9	~0.4	~0.2	~0.2	37
Re–Au film										
($X_{\text{Au}} = 0.21$)	340	16.51	17.51	—	~3 × 10^{11}	8.9	~0.1	~0.2	~0.3	37
Ir–Au films										
($X_{\text{Au}} = 0.0$)	240	16.44	17.44	—	~6 × 10^{11}	8.5	~0.3	~0.3	~0.2	65
($X_{\text{Au}} = 0.86$)	360	16.44	17.44	—	~2 × 10^{10}	10.0	—	—	~0.1	65
($X_{\text{Au}} = 0.94$)	400	16.44	17.44	—	~1 × 10^{10}	10.3	—	—	~0.2	65
0.2% Pt/Al$_2$O$_3$ (10 Å)	300	18.3	19.35	—	9 × 10^{12}	9.2	~0.3	~0.3	~0.3	66

Sample										Ref.
0.6% Pt/Al$_2$O$_3$ (10 Å)	300	18.3	19.35	—	7×10^{12}	9.3	~0.3	~0.3	~0.3	66
1.7% Pt/Al$_2$O$_3$ (10 Å)	300	18.3	19.35	—	4×10^{12}	9.5	~0.3	~0.3	~0.3	66
4.7% Pt/Al$_2$O$_3$ (16 Å)	300	18.3	19.35	—	4×10^{12}	9.5	~0.3	~0.3	~0.3	66
9.5% Pt/Al$_2$O$_3$ (55 Å)	300	18.3	19.35	—	4×10^{12}	9.5	~0.3	~0.3	~0.3	66
16% Pt/SiO$_2$ ($D = 0.3$)	300	18.14	19.37	~45	$\sim 1 \times 10^{10}$	12.0	~0.3	~0.3	~0.4	48
16% Pt-Au/SiO$_2$-bimetallic ($D = 0.3$)										
($X_{Au} = 0.875$)	370	18.14	19.37	—	$\sim 6 \times 10^{8}$	13.2	~0.3	~0.3	~0.3	48
($X_{Au} = 0.92$)	320	18.14	19.37	—	$\sim 2 \times 10^{8}$	13.7	~0.4	—	~0.1	48
($X_{Au} = 0.99$)	390	18.14	19.37	—	$\sim 7 \times 10^{8}$	13.1	~0.2	~0.4	~0.3	48
Pt-Cu/SiO$_2$-bimetallic										
($X_{Cu} = 0.0$)	300	18.12	19.36	30–35	$\sim 6 \times 10^{11}$	10.2	~0.5	~0.3	~0.3	67
($X_{Cu} = 0.15$)	300	18.12	19.36	30–35	$\sim 4 \times 10^{11}$	10.3	~0.5	~0.3	~0.3	67
($X_{Cu} = 0.25$)	300	18.12	19.36	30–35	$\sim 2 \times 10^{11}$	10.6	—	—	—	67
($X_{Cu} = 0.85$)	300	18.12	19.36	30–35	$\sim 4 \times 10^{10}$	11.3	~0.5	~0.2	~0.4	67
($X_{Cu} = 1.0$)	300	18.12	19.36	30–35	$\sim 5 \times 10^{10}$	11.2	~0.4	~0.2	~0.4	67
Au/SiO$_2$	300	17.74	18.69	—	6×10^{14}	6.8	~0.4	~0.4	~0.3	63
Sn/SiO$_2$	300	17.74	18.69	~18	1×10^{12}	9.4	100% C$_1$			63
Pt-Sn/SiO$_2$										
($X_{Sn} = 0.05$)	340	17.74	18.69	—	9×10^{12}	8.6	~0.9	—	~0.1	63
Pt-Au/SiO$_2$										
($X_{Au} = 0.95$)	316	17.74	18.69	~28	7×10^{12}	8.7	~0.2	~0.3	~0.4	63

[a]References for Tables 7.1 to 7.41 are listed at the end of Table 7.41.

TABLE 7.13. Kinetic Parameters for 2-Methylpentane Hydrogenolysis Over Metal Catalysts

Catalyst	T (°C)	$\log_{10}$ (HC)	$\log_{10}$ (H$_2$)	Rate (molecules/cm^2 · sec)	$-\log_{10}$ RP	S_{iC_5}	S_{nC_5}	S_{iC_4}	Ref.[a]
10% Pd/Al$_2$O$_3$ ($D = 0.3$)	350	17.11	19.39	$\sim 3 \times 10^{11}$	9.5	~ 0.3	~ 0.45	~ 0.05	64
10% Pd-Au/Al$_2$O$_3$-bimetallic ($D = 0.3$)									
($X_{Au} = 0.5$)	350	17.11	19.39	$\sim 1 \times 10^{10}$	10.9	~ 0.4	~ 0.4	~ 0.1	64
($X_{Au} = 0.65$)	350	17.11	19.39	$\sim 6 \times 10^{9}$	11.2	~ 0.3	~ 0.4	~ 0.1	64
Pt/Al$_2$O$_3$	300	18.39	19.35	3×10^{13}	8.7	—	—	—	36
Pt film (20 Å)	273	17.47	18.47	5×10^{12}	7.8	~ 0.3	~ 0.2	~ 0.3	68
Pt film (38 Å)	273	17.47	18.47	2×10^{12}	8.2	~ 0.25	~ 0.15	~ 0.3	68
Pt film (58 Å)	273	17.47	18.47	1.3×10^{12}	8.4	~ 0.2	~ 0.15	~ 0.3	68
Pt film (>100 Å)	273	17.47	18.47	5×10^{11}	8.6	~ 0.1	~ 0.15	~ 0.4	68

[a]References for Tables 7.1 to 7.41 are listed at the end of Table 7.41.

TABLE 7.14. Kinetic Parameters for 3-Methylpentane Hydrogenolysis Over Metal Catalysts

Catalyst	T (°C)	$\log_{10}$ (HC)	$\log_{10}$ (H$_2$)	Rate (molecules/cm$^2\cdot$sec)	$-\log_{10}$ RP	S_{iC_5}	S_{nC_5}	S_{nC_4}	Ref.[a]
Re film	295	16.51	17.51	$\sim 2 \times 10^{11}$	9.0	—	—	—	37
Re–Au film									
($X_{Au} = 0.27$)	330	16.51	17.51	$\sim 2 \times 10^{11}$	9.1	—	—	—	37
Ir–Au film									
($X_{Au} = 0.0$)	400	16.44	17.44	$\sim 4 \times 10^{11}$	8.7	—	—	—	69
($X_{Au} = 0.4$)	400	16.44	17.44	$\sim 1 \times 10^{11}$	9.3	—	—	—	69
($X_{Au} = 0.8$)	400	16.44	17.44	$\sim 1 \times 10^{11}$	9.3	—	—	—	69
($X_{Au} = 1.0$)	420	16.44	17.44	$\sim 1 \times 10^{11}$	9.3	—	—	—	69
Pt/Al$_2$O$_3$	300	18.39	19.35	4×10^{13}	8.6	—	—	—	36
Pt film (20 Å)	273	17.47	18.47	1.5×10^{13}	8.1	~0.65	~0.2	~0.2	68
Pt film (58 Å)	273	17.47	18.47	9×10^{12}	8.4	~0.6	~0.15	~0.3	68
Pt film (>100 Å)	273	17.47	18.47	6×10^{12}	8.5	~0.4	~0.1	~0.4	68

[a]References for Tables 7.1 to 7.41 are listed at the end of Table 7.41.

TABLE 7.15. Kinetic Parameters for Cyclohexane Hydrogenolysis Over Metal Catalysts

Catalyst	T (°C)	$\log_{10}$ (HC)	$\log_{10}$ (H$_2$)	ΔE^* (kcal/mole)	Rate (molecules/cm^2 · sec)	$-\log_{10}$ RP	%C	Ref.[a]
Ni powder	200–300	17.47	18.47 (D$_2$)	21	$\sim 4 \times 10^{12}$	8.7	—	70
1% Ru/SiO$_2$	316	18.62	19.31	—	6×10^{13}	8.7	≥80	3
Ru powder ($D = 6 \times 10^{-4}$)	316	18.62	19.31	—	$\sim 1.5 \times 10^{14}$	8.3	≥90	71
5% Ru/SiO$_2$ ($D = 0.24$)	316	18.62	19.31	—	$\sim 8 \times 10^{13}$	8.6	≥90	71
1% Ru/SiO$_2$ ($D = 0.41$)	316	18.62	19.31	—	$\sim 6 \times 10^{13}$	8.7	≥90	71
0.1% Ru/SiO$_2$ ($D = 0.0$)	316	18.62	19.31	—	$\sim 2 \times 10^{13}$	9.2	≥90	71
Ru–Cu powder								
($X_{Cu} = 0.0$)	316	18.62	19.31		$\sim 5 \times 10^{13}$	8.8	≥90	8
($X_{Cu} = 0.005$)	316	18.62	19.31		$\sim 2 \times 10^{12}$	10.2	≥90	8
($X_{Cu} = 0.011$)	316	18.62	19.31		$\sim 4 \times 10^{11}$	10.9	≥90	8
1% Os/SiO$_2$	316	18.62	19.31		1.2×10^{14}	8.4	≥80	3
1.3% Os–Cu/SiO$_2$								
($X_{Cu} = 0.5$)	316	18.62	19.31		$\sim 1 \times 10^{13}$	9.5	≥80	3
1.6% Os–Cu/SiO$_2$								
($X_{Cu} = 0.67$)	316	18.62	19.31	—	$\sim 1 \times 10^{13}$	9.5	≥80	3
Pt(111)	300	17.69	18.51	~25	2.2×10^{12} n-C$_6$	9.2	—	72
					1.7×10^{12} C$_1$–C$_3$	9.3	—	
Pt(557)	300	17.69	18.51	~25	1.0×10^{12} n-C$_6$	9.5	—	72
					1.0×10^{12} C$_1$–C$_3$	9.5	—	
Pt(10, 8, 7)	300	17.69	18.51	~25	1.2×10^{12} n-C$_6$	9.5	—	72
					1.0×10^{12} C$_1$–C$_3$	9.5	—	
Pt(25, 10, 7)	300	17.69	18.51	~25	1.0×10^{12} n-C$_6$	9.5	—	72
					7×10^{11} C$_1$–C$_3$	9.7	—	

[a]References for Tables 7.1 to 7.41 are listed at the end of Table 7.41.

TABLE 7.16. Kinetic Parameters for Methylcyclopentane Ring Opening Over Metal Catalysts

Catalyst	T (°C)	$\log_{10}$ (HC)	$\log_{10}$ (H$_2$)	Rate (molecules/cm^2 · sec)	$-\log_{10}$ RP	S_{nC_6}	S_{2MP}	S_{3MP}	Ref.[a]
Ni-Cu powders									
(X_{Cu} = 0.0)	280	18.12	19.36	6×10^{14}	7.2	0.08	0.38	0.54	73
(X_{Cu} = 0.05)	280	18.12	19.36	$\sim 2 \times 10^{12}$	9.7	0.06	0.56	0.38	73
1% Rh-Co/SiO$_2$									
(X_{Co} = 0.5)	260	18.95	19.79	$\sim 6 \times 10^{13}$	9.0	0.20	0.52	0.28	74
Ir-Au films									
(X_{Au} = 0.0)	335	16.44	17.44	$\sim 5 \times 10^{10}$	9.6	—	—	—	69
(X_{Au} = 0.73)	370	16.44	17.44	$\sim 6 \times 10^{10}$	9.5	—	—	—	69
0.3% Pt/Al$_2$O$_3$									
(D = 0.9)	470	19.94	20.63	$\sim 4 \times 10^{14}$	9.2	0.67	0.20	0.13	75
0.7% Pt/SiO$_2$									
(D = 0.9)	485	19.92	$20.61^{0.7}$	$\sim 9 \times 10^{14}$	8.8	0.38	0.40	0.22	76
0.65% Pt/Al$_2$O$_3$									
(D = 0.9)	485	19.92	$20.61^{0.7}$	$\sim 2 \times 10^{15}$	8.5	—	—	—	76
10% Pt/Al$_2$O$_3$									
(D = 0.3)	230	19.12	18.17	$\sim 2 \times 10^{11}$	11.7	—	0.61	0.39	77
10% Pt/Al$_2$O$_3$									
(D = 0.3)	230	18.69	19.29	$\sim 1 \times 10^{12}$	10.5	—	0.77	0.23	77
Pt film (20 Å)	273	17.47	18.47	2.4×10^{13}	7.9	0.31	0.51	0.18	68
Pt film (40 Å)	273	17.47	18.47	1.4×10^{13}	8.2	0.18	0.60	0.22	68
Pt film (>100 Å)	273	17.47	18.47	8×10^{12}	8.4	0.02	0.64	0.34	68

[a] References for Tables 7.1 to 7.41 are listed at the end of Table 7.41.

TABLE 7.17. Kinetic Parameters for Methylcyclopentane Hydrogenolysis Over Metal Catalysts

Catalyst	T (°C)	$\log_{10}$ (HC)	$\log_{10}$ (H_2)	Rate (molecules/cm² · sec)	$-\log_{10}$ RP	%C_1	Ref.[a]
Ni–Cu powders							
($X_{Cu} = 0.0$)	280	18.12	19.36	$\sim 4 \times 10^{14}$	7.4	59	73
($X_{Cu} = 0.05$)	280	18.12	19.36	$\sim 1 \times 10^{12}$	9.9	59	73
1% Rh–Co/SiO₂							
($X_{Cu} = 0.5$)	260	18.95	19.79	$\sim 3 \times 10^{12}$	10.3	—	74
Rh–Cu films							
($X_{Cu} = 0.0$)	310	16.51	17.51	$\sim 2 \times 10^{11}$	9.0	75	47
($X_{Cu} = 0.03$)	350	16.51	17.51	$\sim 4 \times 10^{10}$	9.7	66	47
($X_{Cu} = 0.06$)	375	16.51	17.51	$\sim 4 \times 10^{10}$	9.7	70	47
($X_{Cu} = 0.95$)	300	16.51	17.51	$\sim 2 \times 10^{10}$	10.0	100	47
($X_{Cu} = 1.0$)	200	16.51	17.51	$\sim 4 \times 10^{9}$	10.7	—	47
Re–Au films							
($X_{Au} = 0.0$)	295	16.51	17.51	$\sim 1 \times 10^{11}$	9.2	—	37
($X_{Au} = 0.22$)	340	16.51	17.51	$\sim 6 \times 10^{11}$	8.6	—	37
($X_{Au} = 0.47$)	375	16.51	17.51	$\sim 1 \times 10^{11}$	9.3	—	37
Ir–Au films							
($X_{Au} = 0.0$)	335	16.44	17.44	$\sim 9 \times 10^{10}$	9.2	—	69
($X_{Au} = 0.73$)	370	16.44	17.44	$\sim 4 \times 10^{10}$	9.6	—	69
0.3% Pt/Al₂O₃							
($D = 0.9$)	470	19.94	20.63	$\sim 4 \times 10^{13}$	10.2	—	75
Pt film (20 Å)	273	17.47	18.47	2×10^{12}	9.0	—	68
Pt film (40 Å)	273	17.47	18.47	4×10^{11}	9.7	—	68
Pt film (>100 Å)	273	17.47	18.47	4×10^{11}	9.7	—	68

[a]References for Tables 7.1 to 7.41 are listed at the end of Table 7.41.

TABLE 7.18. Kinetic Parameters for Benzene Hydrogenolysis Over Metal Catalysts

Catalyst	T (°C)	$\log_{10}$ (HC)	$\log_{10}$ (H$_2$)	ΔE^* (kcal/mole)	$\log_{10}$ A	Rate (molecules/cm$^2 \cdot$ sec)	$-\log_{10}$ RP At 227°C	Ref.[a]
1% Ru/SiO$_2$	130–170	16.8 (pulse)	19.6	30	27.3	2×10^{14}	6.4	78
1% Ru/Al$_2$O$_3$	145–190	16.8 (pulse)	19.6	30	27.4	2×10^{14}	6.3	78
1% Tc/SiO$_2$	170–235	16.8 (pulse)	19.6	29	25.3	5×10^{12}	7.9	78
1% Tc/Al$_2$O$_3$	170–235	16.8 (pulse)	19.6	29	25.6	1×10^{13}	7.7	78
1% Re/SiO$_2$	205–235	16.8 (pulse)	19.6	32	26.2	2×10^{12}	8.4	78
1% Re/Al$_2$O$_3$	200–250	16.8 (pulse)	19.6	32	25.7	7×10^{11}	8.8	78

[a] References for Tables 7.1 to 7.41 are listed at the end of Table 7.41.

TABLE 7.19. Kinetic Parameters for *n*-Heptane Hydrogenolysis Over Metal Catalysts

Catalyst	T (°C)	$\log_{10}$ (HC)	$\log_{10}$ (H$_2$)	ΔE^* (kcal/mole)	$\log_{10} A$	Rate (molecules/cm$^2 \cdot$ sec)	$-\log_{10}$ RP	S_{C_6}	S_{C_5}	S_{C_4}	Ref.[a]
							At 205°C				
Pd powder	250–350	18.61	19.31	~28	~22.4	5×10^9	12.7	0.92	0.08	—	79
Rh powder	100–125	18.61	19.31	~35	~31.1	1.5×10^{15}	7.2	0.82	0.10	0.06	79
Ru powder	80–115	18.61	19.31	~35	~32.1	1.7×10^{16}	6.2	~0.5	~0.2	~0.25	79
Ir powder	130–200	18.61	19.31	~30	~28.8	1.5×10^{15}	7.2	~0.3	~0.3	~0.3	79
Pt powder	250–350	18.61	19.31	~27	~23.0	5×10^{10}	11.7	~0.3	~0.2	~0.3	79
							At indicated T (°C)				
0.3% Pt/Al$_2$O$_3$ ($D = 0.9$)	470	19.94	20.63	—	—	$\sim 2 \times 10^{14}$	9.4	0.15	—	—	74
Pt/Al$_2$O$_3$	300	18.39	19.35	—	—	6×10^{12}	9.4	—	—	—	36
Pt(111)	300	17.69	19.19	—	—	7×10^{12}	8.6	~0.3	~0.3	~0.3	72
Pt(557)	300	17.69	19.19	—	—	6×10^{12}	8.7	~0.3	~0.3	~0.3	72
Pt(10, 8, 7)	300	17.69	19.19	—	—	1.2×10^{13}	8.4	~0.3	~0.3	~0.3	72
Pt(25, 10, 7)	300	17.69	19.19	—	—	2×10^{13}	8.2	~0.3	~0.3	~0.3	72
1% Pt/Al$_2$O$_3$ ($D = 0.8$)	420	14.99	—	21	~19	$\sim 2 \times 10^{12}$	6.5	All C$_1$			80

[a]References for Tables 7.1 to 7.41 are listed at the end of Table 7.41.

TABLE 7.20. Kinetic Parameters for Toluene Hydrodealkylation and Hydrogenolysis Over Metal Catalysts

Catalyst	T (°C)	$\log_{10}$ (HC)	$\log_{10}$ (H$_2$)	ΔE^* (kcal/mole)	$\log_{10}$ A	Rate (molecules/cm^2·sec)	$-\log_{10}$ RP	S_{Bz}	Ref.[a]
						At 380°C			
5% Ni/Al$_2$O$_3$	290	$18.39^{0.3}$	$18.95^{-0.2}$	31	25.2	5×10^{14}	7.5	0.94	81
1% Ru/Al$_2$O$_3$	380	$18.39^{0.2}$	$18.95^{1.0}$	33	24.4	3×10^{13}	8.8	0.82	81
Ru/Al$_2$O$_3$	~380	~18.3	~19.3	29	21.9	2×10^{12}	9.9	0.80	82
1% Rh/Al$_2$O$_3$	320	$18.39^{0.2}$	$18.95^{0.2}$	32	25.1	2×10^{14}	8.0	0.98	81
Rh/Al$_2$O$_3$	~380	~18.3	~19.3	30	24.1	1×10^{14}	8.1	0.96	82
1% Pd/Al$_2$O$_3$	450	$18.39^{0.5}$	$18.95^{-0.4}$	39	25.0	1.1×10^{12}	10.2	1.0	81
Pd/Al$_2$O$_3$	~380	~18.3	~19.3	37	24.6	2×10^{12}	9.9	0.99	82
10% Re/Al$_2$O$_3$	450	$18.39^{-0.2}$	$18.95^{1.7}$	33	22.7	5×10^{11}	10.6	0.85	81
2% Os/Al$_2$O$_3$	380	$18.39^{0.0}$	$18.95^{1.2}$	25	22.1	4×10^{13}	8.6	0.93	81
2% Ir/Al$_2$O$_3$	340	$18.39^{0.2}$	$18.95^{0.5}$	28	23.3	1×10^{14}	8.2	0.95	81
2% Pt/Al$_2$O$_3$	380	$18.39^{0.5}$	$18.95^{-0.1}$	34	24.0	5×10^{12}	9.5	0.96	81
Pt/Al$_2$O$_3$	~380	~18.3	~19.3	33	22.5	3×10^{11}	10.7	0.98	82
						At indicated T (°C)			
0.6% Rh/γ-Al$_2$O$_3$	475	18.35	19.35	13	17.7	9×10^{13}	8.3	0.9	83
0.6% Rh/α-Al$_2$O$_3$	475	18.35	19.35	11	16.8	4×10^{13}	8.6	0.9	83
0.6% Rh/glass	475	18.35	19.35	13	17.9	1.3×10^{14}	8.1	0.9	83
0.6% Rh/SiO$_2$-Al$_2$O$_3$	475	18.35	19.35	10	16.0	1.3×10^{13}	9.1	0.9	83
0.6% Rh/SiO$_2$	475	18.35	19.35	12	17.2	5×10^{13}	8.5	0.9	83
0.6% Rh/MgO	475	18.35	19.35	31	20.6	4×10^{11}	10.6	0.9	83
0.6% Rh/ZnO	475	18.35	19.35	38	24.1	1.3×10^{13}	9.1	0.9	83
0.6% Rh/α-Cr$_2$O$_3$	475	18.35	19.35	19	19.2	5×10^{13}	8.5	0.9	83
10% Pt/glass	76	18.29	19.36	8	18.7	5×10^{13}	8.4	1.0	84

[a]References for Tables 7.1 to 7.41 are listed at the end of Table 7.41.

TABLE 7.21. Kinetic Parameters for Other Hydrogenolysis Reactions Over Metal Catalysts

Hydrocarbon	Catalyst	T (°C)	$\log_{10}$ (HC)	$\log_{10}$ (H$_2$)	ΔE^* (kcal/mole)	Rate (molecules/cm^2 · sec)	$-\log_{10}$ RP	Selectivity	Ref.[a]
(cyclobutane)	Ni film	85–200	17.21	18.29	8	—	—	10% nC$_5$, 90% iC$_5$	85
	Pd film	200	17.21	18.29	19	4×10^{10}	10.5	28% nC$_5$, 72% iC$_5$	85
	Pt film	50–150	17.21	18.29	17	—	—	33% nC$_5$, 67% iC$_5$	85
(neopentane)	Pt/Al$_2$O$_3$	300	18.39	19.35	—	2×10^{13}	8.9	—	36
	3% Pt/La-Y (10 Å)	200	18.07	19.37	12	$\sim 3 \times 10^{12}$	9.4	—	51
	3% Pt/Ca-Y (10 Å)	200	18.07	19.37	—	$\sim 2 \times 10^{12}$	9.6	—	51
	3% Pt/Na-Y (10 Å)	200	18.07	19.37	18	$\sim 1.5 \times 10^{12}$	9.7	—	51
(branched alkane)	Pt/Al$_2$O$_3$	300	18.39	19.35	—	3×10^{13}	8.7	—	36
(branched alkane)	Pt/Al$_2$O$_3$	300	18.39	19.35	—	1×10^{13}	9.1	—	36
(branched alkane)	Pt/Al$_2$O$_3$	300	18.39	19.35	—	2×10^{13}	8.9	—	36
(branched alkane)	Pt/Al$_2$O$_3$	300	18.39	19.35	—	2×10^{13}	8.8	—	36
(branched alkane)	Pt/Al$_2$O$_3$	300	18.39	19.35	—	5×10^{13}	8.5	—	36
(branched alkane)	Pt/Al$_2$O$_3$	300	18.39	19.35	—	6×10^{13}	9.4	—	36

Compound	Catalyst							Ref.
(2-methylbutane structure)	Pt/Al$_2$O$_3$	300	18.39	19.35	—	2×10^{13}	8.8	—	36
(2,2-dimethylbutane structure)	Pt/Al$_2$O$_3$	300	18.39	19.35	—	2×10^{13}	8.9	—	36
(methylcyclopentane structure)	Rh–Cu films ($X_{Cu} = 0.0$)	260	16.51	17.51	—	$\sim 4 \times 10^{11}$	8.9	13% ring opening	47
	($X_{Cu} = 0.88$)	250	16.51	17.51	—	$\sim 5 \times 10^{10}$	9.8	8% ring opening	47
(toluene structure)	10% Pt/glass	76	18.29	19.36	8	5×10^{13}	8.4	$S_{Bz} = 1.0$	84
(p-xylene structure)	10% Pt/glass	76	18.29	19.36	17	3×10^{13}	8.7	—	84
(mesitylene structure)	10% Pt/glass	76	18.29	19.36	28	4×10^{12}	9.4	—	84

^{a}References for Tables 7.1 to 7.41 are listed at the end of Table 7.41.

TABLE 7.22. Kinetic Parameters for Cracking Reactions Over Nickel Powder (ref. 13[a])

Hydrocarbon	T (°C)	$\log_{10}$ (HC)	ΔE^* (kcal/mole)	$\log_{10} A$	Rate (molecules/cm^2 · sec) At 300°C	$-\log_{10}$ RP
△	315	18.21	38	25.5	1.3×10^{11}	11.2
	315	18.21	37	24.6	4×10^{10}	11.7
C_2H_4	315	18.21	40	25.6	3×10^{10}	11.9
C_2H_2	315	18.21	38	25.0	4×10^{10}	11.6
	315	18.21	44	27.2	3×10^{10}	11.6
⬡	315	18.21	57	30.5	8×10^{8}	13.2

[a]Reference 13 is listed at the end of Table 7.41.

556

TABLE 7.23. Kinetic Parameters for Ethylene Hydrogenation Over Metal Catalysts

Catalyst	T (°C)	$\log_{10}$ (HC)	$\log_{10}$ (H$_2$)	ΔE^* (kcal/mole)	$\log_{10}$ A	Rate (molecules/cm^2 · sec)	$-\log_{10}$ RP	Ref.[a]
						At 0°C		
Cr film	0	18.12	18.12	—	—	$\sim 3 \times 10^{13}$	8.7	86
Fe film	0	18.12	18.12	—	—	$\sim 6 \times 10^{14}$	7.4	86
Ni film	0	18.12	18.12	—	—	$\sim 2 \times 10^{15}$	6.9	86
Ni film	0 to 72	17.91	17.91	10.5	22.5	1.4×10^{14}	7.8	87
N–Cu films								
($X_{Cu} = 0.18$)	-12 to 25	17.91	17.91	8.6	22.0	1.4×10^{15}	6.8	87
($X_{Cu} = 0.6$)	-11 to 23	17.91	17.91	10.5	23.2	7×10^{14}	7.1	87
($X_{Cu} = 0.98$)	20 to 69	17.91	17.91	9.8	22.3	3×10^{14}	7.4	87
Tc film	0	18.12	18.12	—	—	$\sim 3 \times 10^{13}$	8.7	86
Rh film	0	18.12	18.12	—	—	$\sim 3 \times 10^{17}$	4.7	86
Pd film	0	18.12	18.12	—	—	$\sim 3 \times 10^{16}$	5.7	86
Pd foil	0	17.51	17.51	—	—	4×10^{15}	6.0	88
Pd foil ($\theta_{H_2C_2} = 1.0$)	0	17.51	17.51	—	—	8×10^{14}	6.7	88
Ta film	0	18.12	18.12	—	—	$\sim 3 \times 10^{13}$	8.7	86
W film	0	18.12	18.12	—	—	$\sim 3 \times 10^{13}$	8.7	86
Pt film	0	18.12	18.12	—	—	$\sim 1 \times 10^{16}$	6.2	86
						At 25°C		
Ni(100)	25	17.17^{0}	18.17	—	—	$\leq 2 \times 10^{13}$	≤ 7.9	89
Ni(110)	25	17.17^{0}	18.17	—	—	3×10^{14}	6.8	89
Ni(111)	25	17.17^{0}	18.17	—	—	1×10^{15}	6.3	89
Ni(111) crystallites	25	17.17^{0}	18.17	—	—	1.5×10^{15}	6.1	89
0.05% Pt/SiO$_2$	-70 to 100	17.87	18.69	9	22.0	3×10^{15}	6.5	90
0.05% Pt/SiO$_2$ + Al$_2$O$_3$	-70 to 100	17.87	18.69	9	22.2	4×10^{15}	6.3	90
0.05% Pt/SiO$_2$ preoxidized	-70 to 100	17.87	18.69	9	22.4	7×10^{15}	6.1	90
						At indicated T (°C)		
Ni–Cu film								
($X_{Cu} = 0.03$)	-40	$19.21^{0.0}$	$19.21^{1.0}$	—	—	9×10^{11}	11.3	91
Ni(100)	90	18.35	19.35	~ 4	~ 16	4×10^{13}	8.8	92

TABLE 7.23. (*Continued*)

Catalyst	T (°C)	$\log_{10}$ (HC)	$\log_{10}$ (H$_2$)	ΔE^* (kcal/mole)	$\log_{10}$ A	Rate (molecules/cm^2·sec) At indicated T (°C)	$-\log_{10}$ RP	Ref.[a]
Ni(110)	90	18.35	19.35	~6	~17	2×10^{13}	9.1	92
Ni(111)	90	18.35	19.35	~8	~18	3×10^{13}	8.9	92
Ni(321)	90	18.35	19.35	~5	~17	9×10^{13}	8.5	92
Ni ribbon	200	18.09	19.37	~12	~18	$\sim6 \times 10^{12}$	9.4	93
Ni ribbon	200	18.09	19.37	—	—	$\sim9 \times 10^{14}$	7.2	93
		(+1.3 ppm O$_2$)						
Ni ribbon	310	18.09	19.37	—	—	$\sim2 \times 10^{15}$	6.9	94
		(+22.6 ppm O$_2$)						
($X_{Au} = 0.16$)	180	$17.91^{0.0}$	$17.91^{1.0}$	4.2	16.9	8×10^{14}	7.1	87
($X_{Au} = 0.55$)	180	$17.91^{0.0}$	$17.91^{1.0}$	3.6	16.4	5×10^{14}	7.3	87
($X_{Au} = 1.0$)	486	$17.91^{0.0}$	$17.91^{1.0}$	—	—	6×10^{12}	9.2	87
Ni–Pd films								
($X_{Pd} = 0.0$)	−100	17.68	18.13	—	—	$\sim4 \times 10^{12}$	9.2	95
($X_{Pd} = 0.2$)	−100	17.68	18.13	—	—	$\sim3 \times 10^{13}$	8.3	95
($X_{Pd} = 0.5$)	−100	17.68	18.13	—	—	$\sim5 \times 10^{13}$	8.1	95
($X_{Pd} = 0.75$)	−100	17.68	18.13	—	—	$\sim4 \times 10^{13}$	8.2	95
($X_{Pd} = 1.0$)	−100	17.68	18.13	—	—	$\sim2 \times 10^{13}$	8.5	95
Cu film	150	$17.91^{1.0}$	$17.91^{1.0}$	12	20.4	2×10^{14}	7.6	87
Pt wire	120	17.93	18.57	10	23.1	4×10^{17}	4.5	96
0.05% Pt/SiO$_2$ + Al$_2$O$_3$	100	17.87	18.69	17–20	—	2×10^{16}	5.7	97
0.05% Pt/SiO$_2$	100	17.87	18.69	17–20	—	1×10^{15}	7.0	97
0.05% Pt/SiO$_2$	40	17.87	18.69	16	24.3	1.5×10^{13}	8.8	98
0.54% Pt/Na-Y	−84	17.87	18.69	—	—	8×10^{12}	9.1	56
0.59% Pt/Ca-Y	−84	17.87	18.69	—	—	4×10^{13}	8.4	56
0.6% Pt/Mg-Y	−84	17.87	18.69	—	—	4×10^{13}	8.4	56
0.5% Pt/La-Y	−84	17.87	18.69	—	—	3×10^{13}	8.5	56
0.53% Pt/SiO$_2$	−84	17.87	18.69	—	—	1×10^{13}	9.0	56
1.2–12% Pt/SiO$_2$	−80	17.79	19.38	10–12	24–26	$2\text{–}8 \times 10^{12}$	9.0	99

[a]References for Tables 7.1 to 7.41 are listed at the end of Table 7.41.

TABLE 7.24. Kinetic Parameters for Hydrogenation Reactions of Terminal Olefins

Hydrocarbon	Catalyst	T (°C)	$\log_{10}$ (HC)	$\log_{10}$ (H$_2$)	ΔE^* (kcal/mole)	$\log_{10} A$	Rate (molecules/cm^2 · sec)	$-\log_{10}$ RP	Ref.[a]
C$_3$H$_6$	7–81% Pt/SiO$_2$	−57	18.09	19.37	~10	~24.2	6–15×10^{13}	7.9	32
C$_4$H$_8$	Pt(223)	25	17.36	18.36	—	—	1.3×10^{16}	5.2	100
C$_4$H$_8$	3.2% Ni/mordenite								
	($D = 0.2$)	80	18.91	19.22	~12	~22.2	$\sim 7 \times 10^{14}$	8.2	101
C$_5$H$_{10}$	10^{-2}% Au/SiO$_2$	120	16.62	19.39	—	—	4×10^{11}	8.8	102
C$_5$H$_{10}$	7×10^{-2}% Au/SiO$_2$	120	16.62	19.39	—	—	8×10^{10}	9.5	102
C$_5$H$_{10}$	0.25% Au/SiO$_2$	120	16.62	19.39	—	—	2×10^{10}	10.2	102
C$_5$H$_{10}$	1% Au/SiO$_2$	120	16.62	19.39	—	—	8×10^8	11.6	102
C$_5$H$_{10}$	0.1–1% Au/Al$_2$O$_3$	120	16.62	19.39	—	—	4×10^8	11.9	102

[a]References for Tables 7.1 to 7.41 are listed at the end of Table 7.41.

TABLE 7.25. Kinetic Parameters for Benzene Hydrogenation Over Metal Catalysts

Catalyst	T (°C)	$\log_{10}$ (HC)	$\log_{10}$ (H$_2$)	ΔE^* (kcal/mole)	$\log_{10}$ A	Rate (molecules/cm$^2 \cdot$ sec) At 25°C	$-\log_{10}$ RP	Ref.[a]
4.3–79% Ni/SiO$_2$	60	18.58	19.31	14	22.1	4–9×10^{11}	10.4	103
Ni–Al powders								
(X_{Al} = 0.44)	27–38	18.49	19.33	13	21.6	1×10^{12}	10.3	104
(X_{Al} = 0.58)	27–38	18.49	19.33	13	21.3	7×10^{11}	10.5	104
(X_{Al} = 0.79)	27–38	18.49	19.33	12	20.4	5×10^{11}	10.6	104
4.3–52% Ni/SiO$_2$								
(15–51 Å)	25	18.37	19.29	—	—	1–1.7×10^{13}	9.0	105
10% Ni/SiO$_2$	30	$18.36^{0.2}$	19.29	—	—	$\sim 2 \times 10^{13}$	8.9	106
Ni–Ti–Al powders ($\sim$ 50% by wt. Al)								
(X_{Ti} = 0.02)	27–38	17.67	19.38	12	20.8	1×10^{12}	9.5	107
(X_{Ti} = 0.07)	27–38	17.67	19.38	13	21.6	1×10^{12}	9.5	107
(X_{Ti} = 0.17)	27–38	17.67	19.38	13	21.5	1×10^{12}	9.5	107
Ni–V–Al powders ($\sim$ 50% by wt. Al)								
(X_V = 0.03)	27–38	17.67	19.38	12	20.9	1×10^{12}	9.5	107
(X_V = 0.14)	27–38	17.67	19.38	13	21.5	1×10^{12}	9.5	107
Ni–Cr–Al powders (λ50% by wt. Al)								
(X_{Cr} = 0.03)	27–38	17.67	19.38	13	21.6	1×10^{12}	9.5	107
(X_{Cr} = 0.11)	27–38	17.67	19.38	12	20.7	9×10^{11}	9.6	107
Ni–Fe–Al powders ($\sim$ 50% by wt. Al)								
(X_{Fe} = 0.1)	27–38	17.67	19.38	13	21.5	1×10^{12}	9.5	107
(X_{Fe} = 0.41)	27–38	17.67	19.38	12	20.3	4×10^{11}	9.9	107
(X_{Fe} = 0.61)	27–38	17.67	19.38	12	20.1	2×10^{11}	10.2	107
Ni–Co–Al powders ($\sim$ 50% by wt. Al)								
(X_{Co} = 0.10)	27–38	17.67	19.38	12	20.8	1×10^{12}	9.5	107
(X_{Co} = 0.40)	27–38	17.67	19.38	12	20.6	7×10^{11}	9.7	107
Ni–Mo–Al powders ($\sim$ 50 by wt. Al)								
(X_{Mo} = 0.02)	27–38	17.67	19.38	11	20.1	1×10^{12}	9.5	107
(X_{Mo} = 0.37)	27–38	17.67	19.38	12	20.9	1×10^{12}	9.5	107

Catalyst						At 100°C		Ref.
5.4–75% Ni/Al₂O₃	0–70	17.47	18.61	10	18.6	3×10^{11}	9.8	108
0.6% Re/Al₂O₃	30	17.78	19.38	—	—	$\sim 1 \times 10^{13}$	8.6	44
Os/Al₂O₃	30	17.78	19.38	—	—	$\sim 1 \times 10^{15}$	6.6	44
0.6% Ir/Al₂O₃	30	17.78	19.38	—	—	$\sim 4 \times 10^{13}$	8.0	44
0.6% Ir–Re/Al₂O₃-bimetallic								
($X_{Re} = 0.1$)	30	17.78	19.38	—	—	$\sim 6 \times 10^{13}$	7.9	44
($X_{Re} = 0.45$)	30	17.78	19.38	—	—	$\sim 6 \times 10^{13}$	7.9	44
($X_{Re} = 0.86$)	30	17.78	19.38	—	—	$\sim 4 \times 10^{13}$	8.0	44
Pt/Na-Y (20 Å)	25	$18.36^{0.0}$	19.35	11	21.9	8×10^{13}	8.3	109
Pt/Na-Y (10 Å)	25	$18.36^{0.0}$	19.35	11	21.6	4×10^{13}	8.6	109
3% Pt/Al₂O₃	25	$18.36^{0.0}$	19.35	—	24.3	5×10^{13}	8.5	109
10% Pt/glass	63	18.29	19.36	18	21.0	1.5×10^{11}	10.6	84
0.2–10% Pt/Al₂O₃	50	18.26	19.36	10		8×10^{13}	8.2	110
2.0–2.8% Pt/SiO₂ fired at 100–500°C	25	17.87	19.37	—	—	$3–5 \times 10^{13}$	8.1	111
2.0% Pt/SiO₂ fired at 600°C	25	17.87	19.37	—	—	$\leq 5 \times 10^{11}$	≤ 10	111
10% Co/SiO₂	70–175	17.99	19.09	14	16.5	1.1×10^{13}	8.8	112
Ni/SiO₂	70–140	18.99	20.14	14	22.6	3×10^{14}	8.4	113
Ni powder	100	$18.44^{0.0}$	19.34	—	—	4×10^{13}	8.6	114
10–85% Ni/Cr₂O₃·Al₂O₃	100	$18.44^{0.0}$	19.34	—	—	4×10^{13}	8.6	114
7–75% Ni/Al₂O₃	100	$18.44^{0.0}$	19.34	—	—	4×10^{13}	8.6	114
9–20% Ni/SiO₂	100	$18.44^{0.0}$	19.34	—	—	4×10^{13}	8.6	114
3–70% Ni/Cr₂O₃	100	$18.44^{0.0}$	19.34	—	—	5×10^{13}	8.5	114
11% Ni/SiO₂	60	18.36	$19.29^{0.6}$	13	21.7	3×10^{14}	7.7	115
10% Ni/SiO₂	70–175	17.99	19.09	6	21.4	1.5×10^{13}	8.7	112
1–5% Ni/SiO₂	80–160	$17.99^{0.2}$	$19.09^{1.5}$	6	16.7	1.6×10^{13}	8.7	116
10% Ni/SiO₂	80–160	$17.99^{0.2}$	$19.09^{1.5}$	14	21.2	1.1×10^{13}	8.9	116
1% Ni/SiO₂·Al₂O₃	80–160	$17.99^{0.2}$	$19.09^{1.5}$	—	—	$\leq 10^{11}$	—	116
5% Ni/SiO₂·Al₂O₃	80–160	$17.99^{0.2}$	$19.09^{1.5}$	8	17.1	3×10^{12}	9.4	116
10% Ni/SiO₂·Al₂O₃	80–160	$17.99^{0.2}$	$19.09^{1.5}$	14	21.0	7×10^{12}	9.0	116

561

TABLE 7.25. (*Continued*)

Catalyst	T (°C)	$\log_{10}$ (HC)	$\log_{10}$ (H$_2$)	ΔE^* (kcal/mole)	$\log_{10}$ A	Rate (molecules/cm^2 · sec) At 100°C	$-\log_{10}$ RP	Ref.[a]
1% Tc/SiO$_2$	110–160	16.8 (pulse)	19.6	11	18.4	9×10^{11}	8.7	78
1% Tc/Al$_2$O$_3$	110–170	16.8 (pulse)	19.6	9	17.7	5×10^{12}	8.0	78
1% Ru/SiO$_2$	95–130	16.8 (pulse)	19.6	7	18.2	1×10^{14}	6.7	78
1% Ru/Al$_2$O$_3$	95–145	16.8 (pulse)	19.6	8	18.7	1×10^{14}	6.7	78
Pd/Al$_2$O$_3$	70–140	18.99	20.14	14	22.8	4×10^{13}	9.2	113
Pd/Al$_2$O$_3$	100	17.78	19.38	—	—	4×10^{13}	8.0	44
0.5–2.5% Pd/SiO$_2$ (reduced at 300°C)	100	17.67	19.38	—	—	2×10^{13}	8.2	117
0.5–2.5% Pd/SiO$_2$ (reduced at 450°C)	100	17.67	19.38	—	—	7×10^{12}	8.6	117
1% Pd/SiO$_2$	105–150	16.8 (pulse)	19.6	9.5	18.2	4×10^{12}	8.1	78
1% Pd/Al$_2$O$_3$	105–160	16.8 (pulse)	19.6	9.5	17.5	2×10^{12}	8.4	78
Pd film	80–250	16.33	17.54 (D$_2$)	5.1	17.4	3×10^{14}	5.7	118
Pd–Au films (X$_{Au}$ = 0.17)	80–250	16.33	17.54 (D$_2$)	4.3	16.2	5×10^{13}	6.5	118
(X$_{Au}$ = 0.38)	80–250	16.33	17.54 (D$_2$)	6.8	17.2	2×10^{13}	6.9	118
1% Re/SiO$_2$	130–190	16.8 (pulse)	19.6	11	17.9	3×10^{11}	9.2	78
1% Re/Al$_2$O$_3$	140–185	16.8 (pulse)	19.6	11	17.5	1×10^{11}	9.7	78
Os/Al$_2$O$_3$	100	18.09	19.37	—	—	1.7×10^{14}	7.7	35
Ir/Al$_2$O$_3$	100	18.09	19.37	—	—	8×10^{14}	7.0	35

Catalyst								
0.25% Ir/Al$_2$O$_3$	100	17.78	19.38	—	—	5×10^{14}	6.9	44
0.25% Ir + Cu/Al$_2$O$_3$-bimetallic								
(X$_{Cu}$ = 0.2)	100	17.78	19.38	—	—	3×10^{14}	7.2	44
(X$_{Cu}$ = 0.5)	100	17.78	19.38	—	—	1×10^{14}	7.6	44
(X$_{Cu}$ = 0.8)	100	17.78	19.38	—	—	5×10^{13}	7.9	44
Pt/Al$_2$O$_3$	70–140	18.99	20.14	13	22.4	7×10^{14}	8.0	113
Pt-Ir/Al$_2$O$_3$								
(X$_{Ir}$ = 0.5)	105	18.55^0	19.30	—	—	7×10^{13}	8.5	119
0.5% Pt/Al$_2$O$_3$	40–90	18.26	19.23	10	20.2	2×10^{14}	7.8	120
0.4% Pt-0.5% Sn/Al$_2$O$_3$ (coimpreg.)	40–90	18.26	19.23	10	19.2	2×10^{13}	8.8	120
0.4% Pt-0.7% Sn/Al$_2$O$_3$ (Pt(II)Cl$_2$(SnCl$_3$)$_2$)$^{2-}$	40–90	18.26	19.23	10	20.0	1.5×10^{14}	7.9	120
Pt/Al$_2$O$_3$	100	18.09	19.37	—	—	5×10^{14}	7.2	35
Pt-Re/Al$_2$O$_3$-bimetallic								
(X$_{Re}$ = 0.0)	100	18.09	19.37	—	—	5×10^{13}	8.3	35
(X$_{Re}$ = 0.2)	100	18.09	19.37	—	—	6×10^{13}	8.2	35
(X$_{Re}$ = 0.5)	100	18.09	19.37	—	—	3×10^{13}	8.5	35
(X$_{Re}$ = 0.6)	100	18.09	19.37	—	—	7×10^{13}	8.1	35
10% Pt/SiO$_2$	70–175	17.99	19.09	6	21.4	1.5×10^{13}	8.7	112
Pt/Al$_2$O$_3$	100	17.78	19.38	—	—	5×10^{14}	6.9	44
1% Pt/SiO$_2$	100–150	16.8 (pulse)	19.6	7	18.2	1×10^{14}	6.7	78
1% Pt/Al$_2$O$_3$	110–165	16.8 (pulse)	19.6	8	18.3	4×10^{13}	7.1	78
						At indicated T (°C)		
3% Ni/Na-Y	150	18.16	19.37	—	—	5×10^{12}	9.3	14
3% Ni/Li-Y	150	18.16	19.37	—	—	5×10^{12}	9.3	14
3% Ni/Ca-Y	150	18.16	19.37	—	—	4×10^{12}	9.4	14
3% Ni/Mg-Y	150	18.16	19.37	—	—	4×10^{12}	9.4	14
5% Ni/SiO$_2$	150	18.16	19.37	—	—	1.6×10^{14}	7.8	14
5% Ni/SiO$_2$-Al$_2$O$_3$	150	18.16	19.37	—	—	1.3×10^{13}	8.9	14

TABLE 7.25. (*Continued*)

Catalyst	T (°C)	$\log_{10}$ (HC)	$\log_{10}$ (H$_2$)	ΔE^* (kcal/mole)	$\log_{10}$ A	Rate (molecules/cm^2 · sec) At indicated T (°C)	$-\log_{10}$ RP	Ref.[a]
Ni/Cr$_2$O$_3$	170	18.09	19.35	11	19.2	6×10^{13}	8.1	121
8% Ni/SiO$_2$-Al$_2$O$_3$	200	~18 (pulse)	19.3	—	—	8×10^{14}	—	16
8% Ni/SiO$_2$-Al$_2$O$_3$ + 0.44% Na$^+$	200	~18 (pulse)	19.3	—	—	1.1×10^{15}	—	16
8% Ni/SiO$_2$-Al$_2$O$_3$ + 2.4% Na$^+$	200	~18 (pulse)	19.3	—	—	3×10^{14}	—	16
Ni film	150	17.27	19.02^{1-2}	12	20.4	2×10^{14}	6.9	122
Ni–Cu film ($X_{Cu} = 0.15$–1.0)	150	17.27	19.02^{1-2}	25	26.3	4×10^{13}	7.6	122
Ni(100)	170	17.17	18.17	—	—	1.5×10^{13}	7.8	89
Ni(110)	170	17.17	18.17	—	—	2.5×10^{13}	7.6	89
Ni(111)	170	17.17	18.17	—	—	1.6×10^{13}	7.8	89
Ni(111) crystallites	170	17.17	19.36	11	18.6	1.5×10^{13}	7.8	89
9.1% Ru/Na-A	80	18.62	19.31	—	—	3×10^{12}	10.0	123
14–24% Ru/Na-X	80	18.62	19.31	—	—	2×10^{13}	9.2	123
4–5% Ru/Na-Y	80	18.62	19.31	—	—	4×10^{13}	8.9	123
0.9% Ru/Na-L	80	18.62	19.31	—	—	1×10^{14}	8.5	123
4.3% Ru/Na-L	80	18.62	19.31	—	—	1.4×10^{13}	9.3	123
2.4% Ru/mordenite	80	18.62	19.31	—	—	5×10^{13}	8.8	123
0.25–8% Pd/Al$_2$O$_3$	140	18.26	19.36	—	—	1–1.4×10^{14}	8.0	58
0.7–1.3% Pd/SiO$_2$	140	18.26	19.36	—	—	4–6×10^{13}	8.3	58
0.2–1% Pd/Al$_2$O$_3$	140	18.26	19.36	—	—	8×10^{12}	9.2	124
2.1% Pd/Al$_2$O$_3$	150	$18.07^{0.0}$	19.37	—	—	3×10^{13}	8.0	125
2.2% Pd/Al$_2$O$_3$	150	17.85	19.39	—	—	1.5–5×10^{15}	5.6	126
0.1% Pt/Al$_2$O$_3$ ($D = 0.9$)	85	18.87	19.24	—	—	$\sim 1 \times 10^{14}$	8.7	127
0.1% Pt/Al$_2$O$_3$ ($D = 0.9$) + 0.3% SO$_4^{2-}$	85	18.87	19.24	—	—	$\sim 6 \times 10^{13}$	8.9	127

0.1% Pt/Al$_2$O$_3$ (D = 0.9) + 0.6% SO$_4^{2-}$	85	18.87	19.24	—	~2 × 10^{12}	10.4	127
0.7% Pt/SiO$_2$-Al$_2$O$_3$ (D = 0.1–0.5)	120	18.44	19.34	—	4 × 10^{15}	6.6	126
0.7% Pt/Al$_2$O$_3$ (D = 0.2–0.8)	120	18.44	19.34	—	5 × 10^{15}	6.6	128
0.4% Pt/Al$_2$O$_3$ (D = 0.8)	150	18.55$^{0.0}$	19.32	—	~9 × 10^{14}	7.4	129
0.6% Pt/Al$_2$O$_3$ (D = 0.9)	150	18.55	19.32	—	~1 × 10^{14}	8.4	130
0.6% Pt/Al$_2$O$_3$ + 0.2% Na$^+$ (D = 0.9)	150	18.55	19.32	—	~7 × 10^{13}	8.6	130
0.6% Pt/Al$_2$O$_3$ + 0.6% Na$^+$ (D = 0.9)	150	18.55	19.32	—	~5 × 10^{13}	8.7	130
0.2–1.0% Pt/Al$_2$O$_3$	50	18.26	19.23	—	5 × 10^{13}	8.4	124
1% Pt–Pd/Al$_2$O$_3$-bimetallic (X_{Pd} = 0.2–0.8)	50	18.26	19.23	—	1.5 × 10^{13}	8.9	
3.9% Pt/SiO$_2$	20	18.21	18.99	—	3 × 10^{13}	8.6	131
5% Pt–5% Mo/SiO$_2$	20	18.21	18.99	—	2 × 10^{14}	7.8	131

[a]References for Tables 7.1 to 7.41 are listed at the end of Table 7.41.

TABLE 7.26. Kinetic Parameters for Other Hydrogenation Reactions Catalyzed by Metals

Hydrocarbon	Catalyst	T (°C)	$\log_{10}$ (HC)	$\log_{10}$ (H$_2$)	ΔE^* (kcal/mole)	$\log_{10}$ A	Rate (molecules/cm^2 · sec)	$-\log_{10}$ RP	Selectivity	Ref.[a]
H$_2$C=C=CH$_2$	Co/SiO$_2$	100	18.21	$18.51^{1.0}$	9	~20	~1 × 10^{15}	7.2	~75% C$_3$H$_6$	132
	Ni/SiO$_2$	100	18.21	$18.51^{1.0}$	11	~22	~3 × 10^{15}	6.7	~100% C$_3$H$_6$	132
	Ru/SiO$_2$	100	18.21	$18.51^{1.0}$	5	~18	~1 × 10^{15}	7.2	~75% C$_3$H$_6$	132
	Rh/SiO$_2$	100	18.21	$18.51^{1.0}$	8	~21	~2 × 10^{16}	5.9	~75% C$_3$H$_6$	132
	Pd/SiO$_2$	100	18.21	$18.51^{1.0}$	8	~22	~1 × 10^{17}	5.2	~75% C$_3$H$_6$	132
	Ir/SiO$_2$	100	18.21	$18.51^{1.0}$	9	~20	~3 × 10^{14}	7.7	~75% C$_3$H$_6$	132
	Pt/SiO$_2$	100	18.21	$18.51^{1.0}$	16	~25	~1 × 10^{16}	6.2	~75% C$_3$H$_6$	132
	3.2% Ni/mordenite (D = 0.2)	80	18.91	19.22	—	—	~2 × 10^{14}	8.5	—	101
	3.2% Ni/mordenite (D = 0.2)	80	18.91	19.22	—	—	~3 × 10^{14}	8.3	—	101
	4.3–79% Ni/SiO$_2$	100	18.79	19.26	1	—	7–15 × 10^{13}	8.5	33–45% 1-C$_4$H$_8$ 67–55% trans-2-C$_4$H$_8$	103
H$_3$CC≡CCH$_3$	Pd–Au powders						At 25°C			
	(X_{Au} = 0.0)	−100 to 10	17.46	17.86	13	26.7	1.7 × 10^{17}	4.2	—	133
	(X_{Au} = 0.25)	60 to 120	17.46	17.86	14	24.4	1.6 × 10^{14}	7.2	—	133
	(X_{Au} = 0.40)	110 to 170	17.46	17.86	13	23.0	3 × 10^{13}	7.9	—	133
	(X_{Au} = 0.85)	200 to 260	17.46	17.86	14	22.0	6 × 10^{11}	9.6	—	133

Sample									Ref.[a]
(X_{Au} = 0.94)	300 to 350	17.46	17.86	14	20.8	4×10^{10}	10.8	—	133
0.1–2% Pt/SiO$_2$	20	~17.39$^{0.0}$	19.39	—	—	1.5×10^{16}	5.6	—	134
0.3–1.2% Pt/Al$_2$O$_3$	20	~17.39$^{0.0}$	19.39	—	—	1.0×10^{16}	5.8	—	134
0.6–3.8% Pd/SiO$_2$	0	17.39$^{0.0}$	18.41$^{0.5}$	9.3	22.4	1.1×10^{15}	6.2	—	135
0.5% Pt/Ca-Y	303	20.07	20.84	—	—	$\sim 1 \times 10^{16}$	7.9	—	136
0.4–3.7% Pt/SiO$_2$	22	17.61$^{0.0}$	18.39$^{0.5-0.8}$	8	21.5	4×10^{15}	5.9	—	137
Pt(223)	25	17.36	18.36	5	19.3	4×10^{15}	5.6	—	100
4–39% Ni/SiO$_2$	90	18.55	19.32	15	21.8	$3\text{–}6 \times 10^{12}$	9.6	—	103
51% Ni/SiO$_2$	90	18.55	19.32	—	—	9×10^{12}	9.4	—	103
79% Ni/SiO$_2$	90	18.55	19.32	—	—	1.1×10^{13}	9.3	—	103
10% Pt/glass	63	18.29	19.36	17	23.2	1.5×10^{12}	9.9	—	84
Pd–Au films									
(X_{Au} = 0.0)	100	16.40	17.54			1.8×10^{14}	5.9	68% *trans*	118
(X_{Au} = 0.07)	100	16.40	17.54			1.5×10^{14}	6.0	59% *trans*	118
(X_{Au} = 0.18)	100	16.40	17.54			8×10^{13}	6.3	60% *trans*	118
(X_{Au} = 0.37)	100	16.40	17.54			3×10^{13}	6.7	63% *trans*	118
10% Pt/glass	76	18.29	19.36	28	29.0	4×10^{11}	10.4	—	84
10% Pt/glass	76	18.29	19.36	31	29.4	6×10^{10}	11.3	—	84

[a] References for Tables 7.1 to 7.41 are listed at the end of Table 7.41.

567

TABLE 7.27. Kinetic Parameters for Cyclohexane Dehydrogenation to Benzene Over Metal Catalysts

Catalyst	$T(°C)$	$\log_{10}(HC)$	$\log_{10}(H_2)$	ΔE^* (kcal/mole)	$\log_{10} A$	Rate (molecules/cm^2·sec)	$-\log_{10}$ RP	Ref.[a]
Ni–Cu powders								
($X_{Cu} = 0.0$)	316	18.62	19.31	—	—	$\sim 2 \times 10^{14}$	8.2	5
($X_{Cu} = 0.2–0.9$)	316	18.62	19.31	—	—	$\sim 4 \times 10^{14}$	7.9	5
($X_{Cu} = 1.0$)	316	18.62	19.31	—	—	$\sim 4 \times 10^{12}$	8.9	5
Ni powder	300	17.47	18.47 (D_2)	21	20.7	6×10^{12}	8.5	70
1% Cu/SiO$_2$	316	18.62	19.31	—	—	$\sim 1 \times 10^{12}$	10.5	3
0.81% Pd/Al$_2$O$_3$	225	19.05	19.12	16	21.3	$\sim 2 \times 10^{14}$	8.6	138
0.2–1% Pd/Al$_2$O$_3$	190	18.21	18.46	16	20.8	2×10^{13}	8.7	139
1% Ru/SiO$_2$	316	18.62	19.31	—	—	$\sim 2 \times 10^{14}$	8.2	3
Ru powder	316	18.62	19.31	—	—	$\sim 5 \times 10^{13}$	8.8	71
5% Ru/SiO$_2$	316	18.62	19.31	—	—	$\sim 7 \times 10^{13}$	8.6	71
0.1% Ru/SiO$_2$	316	18.62	19.31	—	—	$\sim 1.1 \times 10^{14}$	8.4	71
1% Ru–0.66% Cu/SiO$_2$	316	18.62	19.31	—	—	$\sim 2 \times 10^{14}$	8.2	71
Ru–Cu powders								
($X_{Cu} = 0.0$)	316	18.62	19.31	—	—	$\sim 2 \times 10^{13}$	9.2	8
($X_{Cu} = 0.005$)	316	18.62	19.31	—	—	$\sim 9 \times 10^{12}$	9.5	8
($X_{Cu} = 0.011$)	316	18.62	19.31	—	—	$\sim 1 \times 10^{13}$	9.5	8
1% Os/SiO$_2$	316	18.62	19.31	—	—	$\sim 2 \times 10^{14}$	8.2	3
1% Os–0.33% Cu/SiO$_2$	316	18.62	19.31	—	—	$\sim 2 \times 10^{14}$	8.2	3
0.06% Pt/Al$_2$O$_3$ ($D = 0.8$)								
Initial	150	19.39	0.0	19	22.2	$\sim 3 \times 10^{12}$	9.6	140
Steady state	150	19.39	0.0	—	—	$\sim 2 \times 10^{12}$	9.8	140
0.55% Pt/Al$_2$O$_3$	350	18.79	19.27	—	—	4×10^{15}	7.0	141
0.1–2% Pt/Al$_2$O$_3$	300	18.65	19.30	—	—	4×10^{15}	6.8	142
2% Pt/Al$_2$O$_3$ ($D = 0.8$)	315	18.61	19.31	—	—	$\sim 3 \times 10^{15}$	7.0	143
Pt/Al$_2$O$_3$	250	18.55	19.30	17	23.3	1.9×10^{16}	6.1	119
Pt–Ir/Al$_2$O$_3$	250	18.55	19.30	17	23.2	1.4×10^{16}	6.2	119

16% Pt/C (D = 0.2)	271	18.35	—	20	~21.7	~5 × 10^{13}	8.5	144
0.2–0.66% Pt/SiO$_2$Al$_2$O$_3$ (D = 0.9)								
(13% Al$_2$O$_3$)	230	18.28	19.36	—	—	~2 × 10^{13}	8.8	145
(1.9% Al$_2$O$_3$)	230	18.28	19.36	—	—	~8 × 10^{13}	8.2	145
(0.5% Al$_2$O$_3$)	230	18.28	19.36	—	—	~1 × 10^{14}	8.1	145
0.2–1% Pt/Al$_2$O$_3$	190	18.21	18.46	16	22.5	1.0 × 10^{15}	7.0	139
0.2% Pt–0.8% Pd/Al$_2$O$_3$	190	18.21	18.46	15	21.5	3 × 10^{14}	7.6	139
0.4% Pt–0.6% Pd/Al$_2$O$_3$	190	18.21	18.46	15	21.3	1.8 × 10^{14}	7.8	139
0.8% Pt–0.2% Pd/Al$_2$O$_3$	190	18.21	18.46	15	21.4	2 × 10^{14}	7.7	139
Pt(111)	260	17.69	18.51	~18	~23	1.2 × 10^{16}	5.5	72
Pt(557)	260	17.69	18.51	~19	~23	5 × 10^{15}	5.8	72
Pt(25,10,7)	260	17.69	18.51	~18	~23	2 × 10^{16}	5.2	72
Pt powder	300	17.47	18.47 (D$_2$)	19	20.7	3 × 10^{13}	7.8	70
2–2.4% Pt/SiO$_2$ (D = 0.05–0.98)	300	16.21	17.84	19	~21	1.3 × 10^{14}	5.9	60
2% Pt/SiO$_2$ (D = 0.98, preoxidized)	300	16.21	17.84	—	—	6 × 10^{14}	5.3	60
2% Pt/SiO$_2$ (D = 0.5, preoxidized)	300	16.21	17.84	—	—	1.7 × 10^{14}	5.8	60
Dehydrogenation to Cyclohexene								
Pt(111)	300	17.69	18.51	—	—	1.0 × 10^{13}	8.5	72
Pt(557)	300	17.69	18.51	—	—	5 × 10^{13}	7.8	72
Pt(25,10,7)	300	17.69	18.51	—	—	3 × 10^{12}	9.0	72

[a]References for Tables 7.1 to 7.41 are listed at the end of Table 7.41.

TABLE 7.28. Kinetic Parameters for Other Dehydrogenation Reactions Catalyzed by Metals

Hydrocarbon	Catalyst	$T(°C)$	$\log_{10}(HC)$	$\log_{10}(H_2)$	ΔE^* (kcal/mole)	$\log_{10} A$	Rate (molecules/cm² · sec)	$-\log_{10}$ RP	S_D	Ref.[a]
	Pt–Au powders									
	($X_{Au} = 0.0$)	360	$17.99^{1.0}$	$19.67^{-1.1}$	30	23.8	4×10^{13}	9.4	—	146
	($X_{Au} = 0.86$)	360	$17.99^{1.0}$	$19.67^{-0.5}$	27	21.1	8×10^{11}	10.0	—	146
	($X_{Au} = 0.93$)	360	17.99	19.67	—	—	3×10^{11}	10.5	—	146
	($X_{Au} = 0.975$)	360	$17.99^{1.0}$	$19.67^{-0.5}$	27	20.4	1.5×10^{11}	10.8	—	146
	($X_{Au} = 0.995$)	360	$17.99^{1.0}$	$19.67^{-0.5}$	27	19.7	3×10^{10}	11.5	—	146
	($X_{Au} = 1.0$)	360	17.99	19.67	—	—	4×10^{9}	12.4	—	146
	W film	425	16.51	17.51	—	—	$\sim 2 \times 10^{12}$	8.1	~0.9	37
	Re–Au films									
	($X_{Au} = 0.29$)	330	16.51	17.51	~13	~16	$\sim 2 \times 10^{11}$	9.1	~0.7	37
	($X_{Au} = 0.49$)	350	16.51	17.51	—	—	$\sim 5 \times 10^{10}$	9.7	~0.6	37
	0.33–0.86% Ru/Al₂O₃	490	18.61	19.31	~27	~22.2	$2\text{-}3 \times 10^{14}$	8.0	—	46
	1.3% Pt/C (20 Å)	460	18.89	19.23	—	—	2×10^{16}	6.4	—	147
	2.8% Pt/C (50 Å)	460	18.89	19.23	—	—	7×10^{15}	6.9	—	147
	Pt(223)	150	17.36	18.36	9	~19.4	6×10^{14}	6.4	~0.01	100
	16% Pt/C ($D = 0.2$)	263	18.35	—	15	~19.0	$\sim 9 \times 10^{12}$	9.2	—	144
	Re/Al₂O₃	310	19.39	0.0	—	—	3×10^{12}	10.7	—	35
	Os/Al₂O₃	310	19.39	0.0	—	—	3×10^{12}	10.7	—	35
	Ir/Al₂O₃	310	19.39	0.0	—	—	4×10^{12}	10.5	—	35
	Pt/Al₂O₃	310	19.39	0.0	—	—	8×10^{13}	9.2	—	35
	Pt-Re/Al₂O₃-bimetallic									
	($X_{Re} = 0.2$)	310	19.39	0.0	—	—	5×10^{13}	9.4	—	35
	($X_{Re} = 0.75$)	310	19.39	0.0	—	—	2×10^{13}	9.8	—	35

[a]References for Tables 7.1 to 7.41 are listed at the end of Table 7.41.

TABLE 7.29. Kinetic Parameters for *n*-Butane Isomerization Over Metal Catalysts

Catalyst	$T(°C)$	$\log_{10}(HC)$	$\log_{10}(H_2)$	ΔE^* (kcal/mole)	$\log_{10} A$	Rate (molecules/cm^2 · sec)	$-\log_{10}$ RP	S_I	Ref.[a]
Pd film	300	17.10	18.16	~38	25.4	1×10^{11}	10.0	~0.01	10
Rh powder	130	17.51	18.51	—	—	$\sim 2 \times 10^{11}$	10.1	~0.02	9
Ir powder	200	17.51	18.51	—	—	$\sim 2 \times 10^{10}$	11.1	~0.02	9
2% Pt/Al$_2$O$_3$	300	18.39	19.35	—	—	1.1×10^{13}	9.3	~0.49	15
Pt/Al$_2$O$_3$	300	18.39	19.35	—	—	3×10^{11}	10.9	~0.03	36
Pt powder	370	17.51$^{1.0}$	18.81$^{-1.7}$	24	19.2	1.2×10^{11}	10.3	~0.15	19
Pt film	300	17.10	18.16	~21	20.1	1.4×10^{12}	8.9	~0.11	10
Pt(111) film	320	17.10	18.16	—	—	1.7×10^{12}	8.8	~0.12	10
Pt(100) film	300	17.10	18.16	—	—	4×10^{12}	8.4	~0.16	10
Pt film	360	17.02	18.02	22	20.2	4×10^{12}	8.3	~0.38	20

[a]References for Tables 7.1 to 7.42 are list ed at the end of Table 7.41.

TABLE 7.30. Kinetic Parameters for Isobutane Isomerization Over Metal Catalysts

Catalyst	T (°C)	$\log_{10}$(HC)	$\log_{10}$(H$_2$)	ΔE^* (kcal/mole)	$\log_{10} A$	Rate (molecules/cm$^2 \cdot$ sec)	$-\log_{10}$ RP	S_I	Ref.[a]
Pd film	300	17.10	18.16	~21	19.5	4×10^{11}	9.4	~0.03	10
Pd film	310	17.10	18.16	21	19.3	3×10^{11}	9.5	~0.05	38
Ta film	200	16.51	17.51	~5	~14	~1×10^{11}	9.4	~0.05	37
W film	155	16.51	17.51	~12	~18	~4×10^{11}	8.8	~0.12	37
Re film	200	16.51	17.51	—	—	~2×10^{11}	9.1	~0.17	37
Pt-Al$_2$O$_3$	300	18.39	19.35	—	—	5×10^{11}	10.6	~0.06	36
Pt powder	375	17.51$^{0.9}$	18.81$^{-2.0}$	29	21.3	4×10^{11}	9.8	~0.65	19
Pt film	300	17.10	18.16	~21	20.7	6×10^{12}	8.2	~0.50	10
Pt film	300	17.10	18.16	~21	20.5	2×10^{12}	8.3	~0.67	10
Pt(111) film	300	17.10	18.16	19	19.4	2×10^{12}	8.7	~0.83	10
Pt(100) film	300	17.10	18.16	—	—	9×10^{12}	8.0	~0.59	10
Pt film									
Clean	360	17.02	18.02	—	—	3×10^{13}	7.5	~0.64	20
Steady state	360	17.02	18.02	—	—	~3×10^{12}	8.5	~0.32	20

[a]References for Tables 7.1 to 7.41 are listed at the end of Table 7.41.

TABLE 7.31. Kinetic Parameters for *n*-Pentane Isomerization Over Metal Catalysts

Catalyst	T (°C)	$\log_{10}$(HC)	$\log_{10}$(H$_2$)	ΔE^* (kcal/mole)	$\log_{10} A$	Rate (molecules/cm²·sec)	$-\log_{10}$ RP	S_I	Ref.[a]
5% Pd/SiO$_2$	400	$19.09^{0.9}$	$20.39^{-1.4}$	49	29.0	1.5×10^{13}	9.8	~0.26	45
Pd/Al$_2$O$_3$	250	18.35	19.35	—	—	$\sim 2 \times 10^9$	13.0	~0.10	44
Rh film	270	16.51	17.51	—	—	$\sim 8 \times 10^9$	12.4	~0.01	47
0.25% Ir/Al$_2$O$_3$	250	18.35	19.35	—	—	$\leq 4 \times 10^{11}$	10.7	$\leq$0.01	44
0.25% Ir + Cu/Al$_2$O$_3$									
($X_{Cu} = 0.4$)	250	18.35	19.35	—	—	$\sim 3 \times 10^{11}$	10.8	~0.06	44
($X_{Cu} = 0.6$)	250	18.35	19.35	—	—	$\sim 2 \times 10^{11}$	11.0	~0.15	44
($X_{Cu} = 0.9$)	250	18.35	19.35	—	—	$\sim 1 \times 10^{11}$	11.3	~0.30	44
5% Pt/SiO$_2$	400	$19.09^{1.0}$	$20.39^{-2.7}$	53	31.1	1.0×10^{14}	9.0	~0.15	45
Pt/C	400	$19.09^{1.0}$	$20.39^{-2.2}$	57	32.2	7×10^{13}	9.1	~0.15	45
0.6% Pt/Al$_2$O$_3$(BF) (D = 0.9)	450	18.79	19.27	—	—	$\sim 2 \times 10^{15}$	7.4	—	130
0.6% Pt-0. % Na$^+$/Al$_2$O$_3$(BF) (D = 0.9)	450	18.79	19.27	—	—	$\sim 2 \times 10^{14}$	8.4	—	130
0.6% Pt-02.5 % Na$^+$/Al$_2$O$_3$ (D = 0.9)	450	18.79	19.27	—	—	$\sim 5 \times 10^{13}$	9.0	—	130
16% Pt/SiO$_2$ (D = 0.3)	312	18.39	19.34	—	—	$\sim 8 \times 10^{10}$	11.4	~0.67	48
16% Pt-Au/SiO$_2$-bimetallic (D = 0.3)									
($X_{Au} = 0.875$)	350	18.39	19.34	—	—	$\sim 5 \times 10^9$	12.6	~0.26	48
($X_{Au} = 0.975$)	370	18.39	19.34	~34	~22	$\sim 1 \times 10^{10}$	12.3	~0.96	48
($X_{Au} = 0.99$)	343	18.39	19.34	~23	~17	$\sim 2 \times 10^9$	13.0	1.0	48
0.3% Pt/SiO$_2$ (D = 0.56)	283	$18.03^{0.6}$	$19.05^{-1.8}$	—	—	8×10^{11}	10.0	~2.9	49
8.2% Pt/SiO$_2$ (D = 0.07)	283	$18.03^{0.6}$	$19.05^{-1.8}$	—	—	1×10^{12}	9.9	~0.74	49

[a]References for Tables 7.1 to 7.41 are listed at the end of Table 7.41.

TABLE 7.32. Kinetic Parameters for Neopentane Isomerization Over Metal Catalysts

Catalyst	T (°C)	$\log_{10}$(HC)	$\log_{10}$(H$_2$)	ΔE^* (kcal/mole)	$\log_{10}$ A	Rate (molecules/cm^2 · sec)	$-\log_{10}$ RP	S_I	Ref.[a]
10% Ir/SiO$_2$	200	18.35	19.35	50	34	1.1×10^{11}	7.2	~0.13	50
Pt/Al$_2$O$_3$	300	18.39	19.35	—	—	5×10^{11}	10.6	~0.05	36
1% Pt/spheron	300	18.35	19.35	49	~31	~2×10^{12}	10.0	~0.97	50
0.6% Pt/Al$_2$O$_3$	307	18.35	19.35	—	—	4×10^{12}	9.6	~0.36	55
2% Pt/Al$_2$O$_3$	307	18.35	19.35	—	—	8×10^{12}	9.3	~0.29	55
4.3% Pt/SiO$_2$	307	18.35	19.35	—	—	6×10^{12}	9.5	~0.23	55
1% Pt/spheron	307	18.35	19.35	—	—	1.7×10^{12}	10.0	~0.96	55
Pt powder	307	18.35	19.35	—	—	1.1×10^{12}	10.2	~0.90	55
6% Pt–6% Fe/C	370	18.35	19.35	—	—	4×10^{12}	9.6	~0.71	54
1% Pt/C	370	18.35	19.35	—	—	1.3×10^{14}	8.1	~0.72	54
3% Pt/Na-Y (10 Å)	225	18.07	19.37	~37	~26	3×10^{10}	11.5	~0.11	51
3% Pt/Ca-Y (10 Å)	225	18.07	19.37	~35	~25	3×10^{10}	11.5	~0.03	51
0.9% Pt/SiO$_2$ (40 Å)	315	18.07	19.37	~28	~24	2.6×10^{13}	8.5	~0.47	51
0.9% Pt/SiO$_2$ (70 Å)	315	18.07	19.37	~25	~23	1.8×10^{13}	8.7	~0.58	51
2% Pt/SiO$_2$ (12 Å)	315	18.07	19.37	~35	~26	2.2×10^{13}	8.6	~0.40	51
2.5% Pt/SiO$_2$	300	17.91$^{1.0}$	19.39$^{1.0-0.1}$	~34	27.4	2.6×10^{14}	7.4	~0.22	52
2.8% Pt–4% W/SiO$_2$	300	17.91$^{1.0}$	19.39$^{1.0-0.1}$	—	—	9×10^{13}	7.8	~0.02	52
Pt film	275	17.10	18.16	—	—	~5×10^{12}	8.3	~0.76	38
Pt film	275	17.10	18.16	~21	21.0	5×10^{12}	8.3	~0.59	10
Au powder	450	18.35	19.35	48	~25	~4×10^{10}	10.6	~0.99	50

[a] References for Tables 7.1 to 7.41 are listed at the end of Table 7.41.

TABLE 7.33. Kinetic Parameters for *n*-Hexane Isomerization Over Metal Catalysts

Catalyst	T (°C)	$\log_{10}$(HC)	$\log_{10}$(H$_2$)	ΔE^* (kcal/mole)	Rate (molecules/cm$^2 \cdot$ sec)	$-\log_{10}$ RP	S_I	S_{2MP}	S_{3MP}	Ref[a]
5% Ni/SiO$_2$-Al$_2$O$_3$ (BF) (D = 0.3)	330	18.69	19.29	—	$\sim 2 \times 10^8$	14.2	0.86	—	—	61
7% Ni/SiO$_2$-Al$_2$O$_3$(BF) (D = 0.3)	350	18.69	19,29	—	$\sim 4 \times 10^9$	12.9	0.95	—	—	61
16% Ni/SiO$_2$-Al$_2$O$_3$(BF) (D = 0.3)	275	18.69	19.29	—	$\sim 1 \times 10^8$	14.5	0.19	—	—	61
3% Ni/Li-Y (BF) (D = 0.3)	300	18.42	19.34	—	$\sim 4 \times 10^{11}$	10.6	$\sim$0.07	0.66	0.34	14
Ni-Cu powders										
(X_{Cu} = 0.0)	330	18.12	19.36	$\sim$44	$\sim 6 \times 10^{12}$	9.2	$\sim$0.04	0.7	0.3	62
(X_{Cu} = 0.05)	330	18.12	19.36	$\sim$44	$\sim 2 \times 10^{10}$	11.2	$\sim$0.08	0.66	0.34	62
(X_{Cu} = 0.47)	330	18.12	19.36	$\sim$55	$\sim 7 \times 10^{10}$	11.1	$\sim$0.61	0.78	0.22	62
Mo/MoO$_2$	330	18.69	19.29	—	—	—	$\sim$0.98	—	—	61
Rh-Cu/SiO$_2$-bimetallic										
(X_{Cu} = 0.0)	175	17.74	18.69	—	$\sim 4 \times 10^{13}$	8.0	$\sim$0.08	0.9	0.1	63
(X_{Cu} = 0.85)	300	17.74	18.69	—	$\sim 4 \times 10^{12}$	9.0	$\sim$0.08	0.66	0.33	63
(X_{Cu} = 1.0)	325	16.99	19.39	—	$\sim 2 \times 10^{12}$	9.3	$\leq$0.01	0.66	0.33	63
10% Pd/Al$_2$O$_3$ (D = 0.3)	350	16.99	19.39	—	$\sim 6 \times 10^{10}$	10.0	$\sim$0.06	0.6	0.4	64
10% Pd-Au/Al$_2$O$_3$ (D = 0.3)- bimetallic										
(X_{Au} = 0.5)	350	16.99	19.39	—	$\sim 1.3 \times 10^{10}$	10.7	$\sim$0.09	0.64	0.36	64
(X_{Au} = 0.35)	350	16.99	19.39	—	$\sim 2 \times 10^8$	12.5	$\sim$0.04	0.6	0.4	64
Pd/Na-Y (BF)	350	18.69	19.29	—	—	—	$\sim$0.91	—	—	61
Re film	180	16.51	17.51	0	$\sim 7 \times 10^{10}$	9.5	$\sim$0.03	1.0	—	37
0.5% Pt/Al$_2$O$_3$ (D = 0.2–0.6)	490	19.67	20.27	—	$1\text{-}1.5 \times 10^{15}$	8.3	$\sim$0.65	0.67	0.33	148
10% Pt/Al$_2$O$_3$ (D = 0.2–0.6)	490	19.67	20.27	—	$1\text{-}1.5 \times 10^{15}$	8.3	$\sim$0.65	0.67	0.33	148
Pt/Al$_2$O$_3$	330	18.69	19.29	—	—	—	0.93	—	—	61

TABLE 7.33. Continued

Catalyst	T (°C)	$\log_{10}$(HC)	$\log_{10}$(H$_2$)	ΔE^* (kcal/mole)	Rate (molecules/cm^2 · sec)	$-\log_{10}$ RP	S_I	S_{2MP}	S_{3MP}	Ref[a]
0.2–9.5% Pt/Al$_2$O$_3$ (D = 0.2–1.0)	300	18.3	19.35	—	6–8×10^{12}	9.2	0.3–0.6	0.74	0.26	66
Pt–Au powder										
(X_{Au} = 0.0)	360	17.77	19.67	—	$\sim 4 \times 10^{11}$	10.0	—	1.0	—	146
(X_{Au} = 0.86)	360	17.77	19.67	—	$\sim 2 \times 10^{7}$	14.3	—	1.0	—	146
(X_{Au} = 0.93)	360	17.77	19.67	—	$\sim 1 \times 10^{6}$	15.6	—	1.0	—	146
16% Pt/SiO$_2$ (D = 0.3)	295	18.14	19.37	45	$\sim 2 \times 10^{10}$	11.7	0.58	0.63	0.37	48
16% Pt–Au/SiO$_2$ (D = 0.03)- bimetallic										
(X_{Au} = 0.875)	370	18.14	19.37	—	$\sim 3 \times 10^{9}$	12.5	0.50	0.23	0.77	48
(X_{Au} = 0.92)	320	18.14	19.37	—	$\sim 2 \times 10^{9}$	12.7	0.90	0.41	0.59	48
(X_{Au} = 0.99)	390	18.14	19.37	—	$\sim 5 \times 10^{9}$	12.3	0.76	0.41	0.59	48
Pt–Cu/SiO$_2$-bimetallic										
(X_{Cu} = 0.0)	300	18.12	19.36	~60	$\sim 2 \times 10^{12}$	9.6	~-0.63	0.66	0.33	67
(X_{Cu} = 0.39)	300	18.12	19.36	~60	$\sim 8 \times 10^{10}$	11.0	~-0.30	—	—	67
(X_{Cu} = 0.98)	300	18.12	19.36	~60	$\sim 1 \times 10^{10}$	11.9	~-0.08	0.6	0.4	67
Pt–Sn/SiO$_2$										
(X_{Sn} = 0.0)	300	17.74	18.69	—	1.1×10^{15}	6.5	0.63	0.65	0.35	63
(X_{Sn} = 0.05)	340	17.74	18.69	—	1.7×10^{12}	9.3	0.12	0.84	0.16	63
Pt–Au/SiO$_2$										
(X_{Au} = 0.5)	320	17.74	18.69	—	5×10^{13}	7.9	0.50	0.55	0.45	63
(X_{Au} = 0.95)	320	17.74	18.69	—	8×10^{11}	9.7	0.08	0.93	0.07	63
Pt film (15 Å)	273	17.47	18.47	—	2.4×10^{13}	7.9	0.84	—	—	68
Pt film (20 Å)	273	17.47	18.47	—	1.8×10^{13}	8.0	0.89	—	—	68
Pt film (38 Å)	273	17.47	18.47	—	8×10^{12}	8.4	0.94	—	—	68
Pt film (58 Å)	273	17.47	18.47	—	4×10^{12}	8.7	0.92	—	—	68
Pt film (>100 Å)	273	17.47	18.47	—	1.3×10^{12}	8.2	0.78	—	—	68

[a]References for Tables 7.1 to 7.41 are listed at the end of Table 7.41.

TABLE 7.34. Kinetic Parameters for 2-Methylpentane Isomerization Over Metal Catalysts

Catalyst	T (°C)	$\log_{10}$ (HC)	$\log_{10}$ (H$_2$)	Rate (molecules/cm^2 · sec)	$-\log_{10}$ RP	S_i	S_{nC_6}	S_{3MP}	Ref.[a]
10% Pd/Al$_2$O$_3$ (D = 0.3)	350	17.11	19.39	$\sim 1 \times 10^{11}$	9.9	0.16	~0.45	~0.55	64
10% Pd–Au/Al$_2$O$_3$ (D = 0.3)- bimetallic									
(X_{Au} = 0.5)	350	17.11	19.39	$\sim 3 \times 10^{9}$	11.4	0.20	~0.5	~0.5	64
(X_{Au} = 0.65)	350	17.11	19.39	$\sim 4 \times 10^{9}$	11.3	0.25	~0.5	~0.5	64
Pt/Al$_2$O$_3$	300	18.39	19.35	1.0×10^{11}	9.2	0.23	—	—	36
Pt film (20 Å)	273	17.47	18.47	5×10^{14}	6.6	~0.94	—	—	68
Pt film (38 Å)	273	17.47	18.47	4×10^{14}	6.7	~0.96	—	—	68
Pt film (58 Å)	273	17.47	18.47	1.5×10^{14}	7.1	~0.95	—	—	68
Pt film (>100 Å)	273	17.47	18.47	6×10^{13}	7.5	~0.92	—	—	68

[a]References for Tables 7.1 to 7.41 are listed at the end of Table 7.41.

TABLE 7.35. Kinetic Parameters for 3-Methylpentane Isomerization Over Metal Catalysts

Catalyst	T (°C)	$\log_{10}$ (HC)	$\log_{10}$ (H$_2$)	Rate (molecules/cm^2 · sec)	$-\log_{10}$ RP	S_i	S_{nC_6}	S_{2MP}	Ref.[a]
Ir–Au films									
(X_{Au} = 0.0)	400	16.44	17.44	$\sim 8 \times 10^{10}$	9.4	~0.12	0.40	0.6	69
(X_{Au} = 0.6)	400	16.44	17.44	$\sim 4 \times 10^{10}$	9.7	~0.06	—	1.0	69
(X_{Au} = 0.19)	400	16.44	17.44	$\sim 1 \times 10^{10}$	10.3	~0.30	—	1.0	69
Pt/Al$_2$O$_3$	300	18.39	19.35	4×10^{13}	8.6	0.42	—	—	36
Pt film (20 Å)	273	17.47	18.47	7×10^{13}	7.5	0.83	—	—	68
Pt film (40 Å)	273	17.47	18.47	3×10^{14}	6.8	0.96	—	—	68
Pt film (58 Å)	273	17.47	18.47	2×10^{14}	7.0	0.96	—	—	68
Pt film (>100 Å)	273	17.47	18.47	7×10^{13}	7.5	0.92	—	—	68

[a]References for Tables 7.1 to 7.41 are listed at the end of Table 7,.41.

TABLE 7.36. Kinetic Parameters for *n*-Heptane Isomerization Over Metal Catalysts

Catalyst	T (°C)	$\log_{10}$(HC)	$\log_{10}$(H_2)	Rate (molecules/cm^3 · sec)	$-\log_{10}$ RP	S_I	Ref.[a]
Ru powder	88	18.61	19.31	$\sim 1 \times 10^{10}$	12.4	~0.07	79
Rh powder	113	18.61	19.31	$\sim 2 \times 10^{10}$	12.1	~0.07	79
Pd powder	300	18.61	19.31	$\sim 4 \times 10^{10}$	11.8	~0.06	79
Ir powder	125	18.61	19.31	$\sim 4 \times 10^{11}$	10.8	~0.13	79
Pt powder	275	18.61	19.31	$\sim 2 \times 10^{12}$	10.1	~0.47	79
0.3% Pt/Al$_2$O$_3$-(BF) ($D = 0.9$)	471	19.94	20.63	$\sim 1 \times 10^{13}$	8.7	0.82	75
Pt/Al$_2$O$_3$	300	18.39	19.35	$\sim 6 \times 10^{11}$	10.4	~0.05	36

[a]References for Tables 7.1 to 7.41 are listed at the end of Table 7.41.

TABLE 7.37 Kinetic Parameters for Other Isomerization Reactions Catalyzed by Metals

Hydrocarbon	Catalyst	T (°C)	$\log_{10}$ (HC)	$\log_{10}$ (H$_2$)	Rate (molecules/cm^2 · sec)	$-\log_{10}$ RP	S_1	Ref.[a]
dbm								
(structure)	3.2% Ni/mordenite ($D = 0.2$)	80	18.91	19.22	$\sim 2 \times 10^{14}$	8.5	~ 0.2 36% *cis*	101
	Pt(223)	25	17.36	18.36	2×10^{15}	6.0	~ 0.15 60% *cis*	100
					$\Delta E \sim 12$ kcal/mole			
(structure)	2% Pt/Al$_2$O$_3$	300	18.39	19.35	1.6×10^{13}	9.0	0.48	15
	Pt film	278	17.10	18.16	7×10^{11}	9.1	0.27	10
(structure)	Pt/Al$_2$O$_3$	300	18.39	19.35	3×10^{11}	10.7	0.02	36
	3% Pt/La-Y(BF)	200	18.07	19.37	$\sim 1.5 \times 10^{13}$	8.7	~ 0.06	51
	3% Pt/CaY(BF)	200	18.07	19.37	$\sim 3 \times 10^{12}$	9.4	~ 0.61	51
	3% Pt/Na-Y(BF)	200	18.07	19.37	$\sim 8 \times 10^{11}$	10.0	~ 0.35	51
(structure)	Pt/Al$_2$O$_3$	300	18.39	19.35	5×10^{11}	10.5	~ 0.02	36
(structure)	Pt/Al$_2$O$_3$	300	18.39	19.35	7×10^{11}	10.3	~ 0.06	36
(structure)	Pt/Al$_2$O$_3$	300	18.39	19.35	3×10^{13}	8.7	~ 0.47	36
(structure)	Pt/Al$_2$O$_3$	300	18.39	19.35	3×10^{13}	8.7	~ 0.52	36
(structure)	Pt/Al$_2$O$_3$	300	18.39	19.35	3×10^{13}	8.7	~ 0.37	36
(structure)	Pt/Al$_2$O$_3$	300	18.39	19.35	2×10^{11}	10.9	~ 0.01	36
cis (structure)	4.8% Pd/C	73	17.12	18.03	8×10^{10}	10.6	$\Delta E^* = 18$ kcal/mole	149
Epimerization (structure)	19% Pd/C	73	17.12	18.03	2×10^{10}	10.6	$\Delta E^* = 20$ kcal/mole	149

[a] References for Tables 7.1 to 7.41 are listed at the end of Table 7.41.

TABLE 7.38. Kinetic Parameters for *n*-Pentane Dehydrocyclization Over Metal Catalysts

Catalyst	T (°C)	$\log_{10}$(HC)	$\log_{10}$(H$_2$)	Rate (molecules/cm^2 · sec.)	$-\log_{10}$ RP	S_C	Ref.[a]
Rh film	320	16.51	17.51	$\sim 1 \times 10^{10}$	10.4	~0.07	47
Rh–Cu films							
($X_{Cu} = 0.06$)	330	16.51	17.51	$\sim 2 \times 10^{10}$	10.1	~0.15	47
($X_{Cu} = 0.92$)	310	16.51	17.51	$\sim 4 \times 10^{10}$	9.8	~0.78	47
Rh–Au films							
($X_{Au} = 0.23$)	320	16.51	17.51	$\sim 4 \times 10^{9}$	10.8	~0.5	47
Rh–Sn films							
($X_{Sn} = 0.03$)	260	16.51	17.51	$\sim 6 \times 10^{10}$	9.6	~0.55	47
($X_{Sn} = 0.17$)	280	16.51	17.51	$\sim 8 \times 10^{10}$	9.5	~0.34	47
Pd/Al$_2$O$_3$	250	18.35	19.35	$\sim 3 \times 10^{9}$	12.7	~0.33	44
0.25% Ir$^+$Cu/Al$_2$O$_3$							
($X_{Cu} = 0.5$)	250	18.35	19.35	$\sim 6 \times 10^{10}$	11.4	~0.02	44
($X_{Cu} = 0.9$)	250	18.35	19.35	$\sim 3 \times 10^{10}$	11.7	~0.08	44
Pt/Al$_2$O$_3$	250	18.35	19.35	$\sim 3 \times 10^{11}$	10.7	~0.50	44
16% Pt/SiO$_2$ ($D = 0.3$)	312	18.39	19.34	$\sim 1 \times 10^{10}$	12.2	~0.08	48
16% Pt-Au/SiO$_2$ ($D = 0.3$)-bimetallic							
($X_{Au} = 0.875$)	350	18.39	19.34	$\sim 8 \times 10^{9}$	12.3	~0.61	48

[a]References for Tables 7.1 to 7.41 are listed at the end of Table 7.41.

TABLE 7.39 Kinetic Parameters for *n*-Hexane Dehydrocyclization Over Metal Catalysts

Catalyst	T(°C)	$\log_{10}$(HC)	$\log_{10}$(H$_2$)	Rate (molecules/cm^2 · sec)	$-\log_{10}$ RP	S_C	S_{MCP}	S_{Bz}	Ref.[a]
Ni–Cu powders									
(X_{Cu} = 0.0)	330	18.12	19.36	~1.5 × 10^{12}	9.8	~0.014	~0.35	~0.65	62
(X_{Cu} = 0.05)	330	18.12	19.36	~4 × 10^{9}	12.3	~0.01	~0.5	~0.5	62
(X_{Cu} = 0.23)	330	18.12	19.36	~8 × 10^{9}	12.0	~0.03	~0.75	~0.25	62
(X_{Cu} = 0.47)	330	18.12	19.36	~9 × 10^{9}	12.0	~0.07	~0.8	~0.2	62
Rh–Cu/SiO$_2$-bimetallic									
(X_{Cu} = 0.85)	300	17.74	18.69	~1 × 10^{13}	8.6	~0.24	~0.1	~0.9	63
(X_{cu} = 1.0)	325	17.74	18.69	~2 × 10^{13}	8.3	~0.08	~0	~1.0	63
Rh film	240	17.74	18.69	~1 × 10^{11}	10.6	~0.25	—	—	47
Rh–Cu films									
(X_{Cu} = 0.03)	280	17.74	18.69	~5 × 10^{10}	10.9	~0.77	—	—	47
(X_{Cu} = 0.58)	270	17.74	18.69	~1 × 10^{10}	11.6	~0.60	—	—	47
(X_{Cu} = 0.92)	280	17.74	18.69	~9 × 10^{9}	11.6	~0.41	—	—	47
Rh–Sn films									
(X_{Sn} = 0.03)	290	17.74	18.69	~4 × 10^{10}	11.0	~0.60	—	—	47
(X_{Sn} = 0.36)	320	17.74	18.69	~8 × 10^{10}	10.7	~0.84	—	—	47
Rh–Au film	300	17.74	18.69	~2 × 10^{10}	11.3	~0.75	—	—	47
(X_{Au} = 0.87)									
Rh–Ag film	280	17.74	18.69	~1 × 10^{10}	11.6	~0.50	—	—	47
(X_{Ag} = 0.26)									
10% Pd/Al$_2$O$_3$	350	16.99	19.39	~2 × 10^{11}	9.5	~0.22	~0.65	~0.35	64
(D = 0.3)									
10% Pd–Au/Al$_2$O$_3$	350	16.99	19.39	~5 × 10^{10}	10.1	~0.21	~0.6	~0.4	64
(D = 0.3)-bimetallic									
(X_{Au} = 0.5)									
(X_{Au} = 0.65)	350	16.99	19.39	~1.3 × 10^{9}	11.7	~0.10	~0.6	~0.4	64
Re film	300	16.51	17.51	~5 × 10^{10}	9.6	~0.14	—	—	37

TABLE 7.39. (*Continued*)

Catalyst	T (°C)	$\log_{10}$(HC)	$\log_{10}$(H$_2$)	Rate (molecules/cm$^2 \cdot$ sec)	$-\log_{10}$ RP	S_C	S_{MCP}	S_{Bz}	Ref.[a]
Re–Au film (X_{Au} = 0.79)	355	16.51	17.51	$\sim 1 \times 10^{11}$	9.3	~ 0.64	—	—	37
W film	370	16.51	17.51	$\sim 1.4 \times 10^{11}$	9.2	~ 1.0	—	—	37
W–Au film (X_{Au} = 0.33)	410	16.51	17.51	$\sim 2 \times 10^{11}$	9.0	~ 1.0	—	—	37
Ta film	375	16.51	17.51	$\sim 4 \times 10^{11}$	8.7	~ 0.9	—	—	37
Ta–Au film (X_{Au} = 0.47)	365	16.51	17.51	$\sim 5 \times 10^{11}$	8.6	~ 0.9	—	—	37
Ir–Au films (ΔE^* = 10–15 kcal/mole) (X_{Au} = 0.0)	390	16.44	17.44	$\sim 8 \times 10^{10}$	9.4	~ 0.33	—	—	65
(X_{Au} = 0.86)	360	16.44	17.44	$\sim 3 \times 10^{10}$	9.8	~ 0.65	—	—	65
(X_{Au} = 0.94)	400	16.44	17.44	$\sim 5 \times 10^{10}$	9.6	~ 0.81	—	—	65
0.5% Pt/Al$_2$O$_3$ (D = 0.2–0.65)	490	19.67	20.27	2–3×10^{14}	9.0	≤ 0.3	~ 0.5	~ 0.5	148
10% Pt/Al$_2$O$_3$ (D = 0.2–0.65)	490	19.67	20.27	2–3×10^{14}	9.0	≤ 0.3	~ 0.5	~ 0.5	148
0.2% Pt/Al$_2$O$_3$ (D = 1.0)	300	18.3	19.35	9×10^{12}	9.2	~ 0.34	1.0	—	66

Catalyst								Ref.[a]
1.7% Pt/Al₂O₃	300	18.3	19.35	1.6×10^{12}	9.9	~-0.13	1.0	—	66
9.5% Pt/Al₂O₃	300	18.3	19.35	8×10^{11}	10.2	~-0.07	1.0	—	66
(D = 0.2)									
Pt-Cu/SiO₂-bimetallic									
(ΔE* = 30–40 kcal/mole)									
(X_Cu = 0.0)	300	18.12	19.36	$\sim 3 \times 10^{11}$	10.5	~-0.1	~0.9	~0.1	67
(X_Cu = 0.39)	300	18.12	19.36	$\sim 1 \times 10^{11}$	11.0	~-0.4	—	—	67
(X_Cu = 0.71)	300	18.12	19.36	$\sim 5 \times 10^{10}$	11.3	~-0.45	—	—	67
(X_Cu = 0.98)	300	18.12	19.36	$\sim 4 \times 10^{10}$	11.4	~-0.33	~0.6	~0.4	67
16% Pt/SiO₂	295	18.14	19.37	$\sim 4 \times 10^{9}$	12.4	~-0.12	0.78	0.22	48
(D = 0.3)									
16% Pt-Au/SiO₂-bimetallic									
(D = 0.3)									
(X_Au = 0.875)	370	18.14	19.37	$\sim 2 \times 10^{9}$	12.7	~-0.40	0.88	0.12	48
(X_Au = 0.99)	390	18.14	19.37	$\sim 9 \times 10^{8}$	13.0	~-0.14	0.77	0.23	48
Pt/SiO₂	300	17.74	18.69	4×10^{13}	8.0	~-0.02	0.95	0.05	63
Pt-Sn/SiO₂	340	17.74	18.69	4×10^{12}	9.0	~-0.25	0.86	0.14	63
(X_Sn = 0.05)									
Pt-Au/SiO₂-bimetallic	316	17.74	18.69	1×10^{13}	8.6	~-0.10	0.5	0.5	63
(X_Au = 0.5)									
(X_Au = 0.95)	320	17.74	18.69	2×10^{12}	9.3	~-0.24	1.0	—	63

[a]References for Tables 7.1 to 7.41 are listed at the end of Table 7.41.

TABLE 7.40. Kinetic Parameters for Other Dehydrocyclization Reactions Catalyzed by Metals

Hydrocarbon	Catalyst	T (°C)	$\log_{10}$(HC)	$\log_{10}$(H$_2$)	Rate (molecules/cm^2 · sec)	$-\log_{10}$ RP	S_c	Ref.[a]
	10% Pd/Al$_2$O$_3$ (D = 0.3)	350	17.11	19.39	~3 × 10^{11}	9.5	~0.41	64
	10% Pd-Au/Al$_2$O$_3$- (D = 0.3) bimetallic	350	17.11	19.35	~7 × 10^{9}	11.0	~0.38	64
	(X_{Au} = 0.5)							
	(X_{Au} = 0.65)	350	17.11	19.39	~6 × 10^{9}	11.1	~0.42	64
	Pt/Al$_2$O$_3$	300	18.39	19.35	4 × 10^{12}	9.6	0.10	36
	Re film	380	16.51	17.51	~5 × 10^{10}	9.6	~0.28	37
	Re–Au film							
	(X_{Au} = 0.27)	330	16.51	17.51	~8 × 10^{10}	9.4	~0.28	37
	Ir–Au films							
	(X_{Au} = 0.0)	400	16.44	17.44	~3 × 10^{11}	8.8	~0.38	69
	(X_{Au} = 0.4)	400	16.44	17.44	~5 × 10^{11}	8.6	~0.80	69
	(X_{Au} = 0.81)	400	16.44	17.44	~3 × 10^{11}	8.8	~0.66	69
	Pt/Al$_2$O$_3$	300	18.39	19.35	~1 × 10^{13}	9.2	0.11	36

584

Catalyst						Ref.
Pd powder	300	18.61	19.31	~2 × 10^10	12.1	~0.03	79
0.3% Pt/Al₂O₃ (D = 0.9)	471	19.94	20.63	~7 × 10^13	9.9	~0.04	75
Pt powder	275	18.61	19.31	~8 × 10^11	10.5	~0.16	79
Pt/Al₂O₃	300	18.39	19.35	6 × 10^12	9.4	~0.45	36
Pt(111)	300	17.69	19.19	1.8 × 10^13 ΔE* ~ 44 kcal/mole	8.2	~0.1–0.2	72
Pt(557)	300	17.69	19.19	2 × 10^13 ΔE* ~ 34 kcal/mole	8.2	~0.1–0.2	72
Pt(10,8,7)	300	17.69	19.19	2 × 10^13	8.2	~0.1–0.2	72
Pt(25,10,7)	300	17.69	19.19	1.0 × 10^13 ΔE* ~ 34 kcal/mole	8.5	~0.1–0.2	72
1% Pt/Al₂O₃ (D = 0.8)	420	14.99	—	~1 × 10^15 to Tol ΔE* ~ 10 kcal/mole	3.8	~0.83	80
1% Pt/Al₂O₃ (D = 0.8)	420	14.99	—	~2 × 10^14 to Bz ΔE* ~ 21 kcal/mole	4.5	~0.16	80
Pt/Al₂O₃	300	18.39	19.35	6 × 10^12	9.4	0.11	36
Pt/Al₂O₃	300	18.39	19.35	6 × 10^15	9.2	0.17	36

^{a} References for Tables 7.1 to 7.41 are listed at the end of Table 7.41.

TABLE 7.41. Kinetic Parameters for Hydro- and Dehydroisomerization Reactions Catalyzed by Metals

Hydrocarbon	Catalyst	T (°C)	$\log_{10}$(HC)	$\log_{10}$(H_2)	Rate (molecules/$cm^2 \cdot$ sec)	$-\log_{10}$ RP	S_{DI}	Ref.
⬡	0.5% Pt/Ca-Y (BF) ($D = 0.8$)	303	19.96	$20.81^{-0.5}$	$\sim 4 \times 10^{14}$ ($\Delta E^* \sim$ 30 kcal/mole)	9.2	~ 0.04	136
⬡	0.5% Pt/Ca-Y (BF) ($D = 0.8$)	288	20.07	$20.84^{-0.3}$	$\sim 2 \times 10^{14}$ ($\Delta E^* \sim$ 33 kcal/mole)	9.6	1.0 (all MCP)	150
⬠	Ni–Cu powder ($X_{Cu} = 0.05$)	280	18.12	19.36	8×10^{10}	11.1	~ 0.03	73
	Rh film	310	16.51	17.51	$\sim 1 \times 10^{10}$	10.3	~ 0.06	47
	Rh–Cu film ($X_{Cu} = 0.03$)	350	16.51	17.51	$\sim 1 \times 10^{10}$	10.3	~ 0.27	47
	($X_{Cu} = 0.06$)	375	16.51	17.51	$\sim 5 \times 10^{10}$	9.6	~ 0.57	47
	1% Rh-Co/SiO$_2$ ($X_{Co} = 0.5$)	260	18.95	19.79	$\sim 4 \times 10^{12}$	10.2	~ 0.07	74
	Re film	333	16.51	17.51	$\sim 9 \times 10^{10}$	9.4	~ 0.45	37
	Re–Au films ($X_{Au} = 0.22$)	380	16.51	17.51	$\sim 3 \times 10^{11}$	8.9	~ 0.22	37
	($X_{Au} = 0.47$)	315	16.51	17.51	$\sim 1.4 \times 10^{11}$	9.2	~ 0.95	37
	Ir film	335	16.44	17.44	2×10^{11}	9.0	~ 0.57	69
	Ir–Au film ($X_{Au} = 0.73$)	370	16.44	17.44	$\sim 1 \times 10^{11}$	9.3	~ 0.47	69
	0.3% Pt/Al$_2$O$_3$ (BF) ($D = 0.9$)	471	19.94	20.63	$\sim 2 \times 10^{14}$	9.5	~ 0.3	75
	Pt film (40 Å)	273	17.47	18.47	$\sim 1 \times 10^{14}$	9.8	~ 0.03	68

REFERENCES FOR TABLES 7.1 TO 7.41.

1. J.H. Sinfelt. *Adv. Catal.* **23**:91 (1973); J.H. Sinfelt. *Catal. Rev.* **9**:147 (1974); and references cited therein.
2. J.H. Sinfelt, W.F. Taylor, and D.J.C. Yates. *J. Phys. Chem.* **69**:95 (1965).
3. J.H. Sinfelt. *J. Catal.* **29**:308 (1973).

4. D.J.C. Yates and J.H. Sinfelt. *J. Catal.* **8**:348 (1967).

5. J.H. Sinfelt, J.L. Carter, and D.J.C. Yates. *J. Catal.* **24**:283 (1972).

6. L. Babernics, L. Guczi, K. Matusek, A. Sárkány and P. Tétényi. In: *Proceedings of the 6th International Congress on Catalysis*, London, 1976, p. 456.

7. J. R. Anderson and B. G. Baker. *Proc. R. Soc. London* **A271**:402 (1963).

8. J.H. Sinfelt, Y.L. Lam, J.A. Cusumano, and A.E. Barnett. *J. Catal.* **42**:227 (1976).

9. A. Sárkány, K. Matusek, and P. Tétényi. *Faraday Trans.* **73**:1699 (1977).

10. J.R. Anderson and N.R. Avery. *J. Catal.* **5**:446 (1966).

11. D.C. Tajbl. *Ind. Eng. Chem. Proc. Res. Dev.* **8**:365 (1969)

12. R.S. Hansen, private communication.

13. J. Freel and A.K. Galwey. *J. Catal.* **10**:277 (1968).

14. J.T. Richardson. *J. Catal.* **21**:122 (1971).

15. G. Leclercq, L. Leclercq, and R. Manuel. *J. Catal.* **44**:68 (1976).

16. C.P. Huang and J.T. Richardson. *J. Catal.* **52**:332 (1978).

17. T.J. Plunkett and J.K.A. Clarke. *Faraday Trans.* **68**:500 (1972).

18. Y.I. Yermakov, B.N. Kuznetsov, and Y.A. Ryndin, *J. Catal.* **42**:73 (1976).

19. L. Guczi, A. Sárkány, and P. Tétényi. *Faraday Trans.* **70**:1971 (1974).

20. R.S. Dowie, D.A. Whan, and C. Kemball. *Faraday Trans.* **68**:2150 (1972).

21. C. Naccache, N. Kaufherr, M. Dufaux, J. Pandiera, and B. Imelik. In: J.R. Katzer, editor, *Molecular Sieves*, Volume 2. American Chemical Society, Washington, D.C., 1977, p. 538.

22. D.W. McKee and F.J. Norton. *J. Catal.* **3**:252 (1964).

23. C.M. Machiels and R.B. Anderson, *J. Catal.* **58**:253 (1979).

24. F.E. Shephard. *J. Catal.* **14**:148 (1969).

25. L. Guczi, A. Sárkány, and P. Tétényi. In *Proceedings of the 5th International Congress on Catalysis*, Miami, paper 78, p. 1111 (1972).

26. J.R. Anderson and N.R. Avery. *J. Catal.* **8**:48 (1967).

27. T.S. Sridhar and D.M. Ruthven. *J. Catal.* **24**:153 (1972); A. Verma and D.M. Ruthven. *J. Catal.* **46**:160 (1977); A. Verma and D.M. Ruthven. *J. Catal.* **19**:401 (1970).

28. R.A. Dalla Betta, J.A. Cusumano, and J.H. Sinfelt. *J. Catal.* **19**:343 (1970).

29. D.R. Kahn, E.E. Petterson, and G.A. Somorjai. *J. Catal.* **34**:294 (1974).

30. M. Boudart, A. Aldag, J.E. Benson, N.A. Dougharty, and G. C. Harkins. *J. Catal.* **6**:92 (1966).

31. N.A. Dougharty. Ph.D. thesis, University of California, Berkeley, 1964.

32. P.H. Otero-Schipper, W.A. Wachter, J.B. Butt, R.L. Burwell, Jr., and J.B. Cohen. *J. Catal.* **50**:494 (1977).

33. P.A. Camagnon, C. Hoang-Van, and J.C. Teichner. In *Proceedings of the 6th International Congress on Catalysis*, London, 1976, p. 117.

34. J.C. Kempling and R.B. Anderson. In: *Proceedings of the 5th International Congress of Catalysis*, Miami, 1972, p. 1099; *Ind. Eng. Chem. Proc. Res. Dev.* **11**:146 (1972).

35. C. Bolívar, H. Charcosset, R. Frety, L. Tournayan, C. Betizeau, G. Leclercq, and M. Maurel. *J. Catal.* **45**:179 (1976).

36. C. Leclercq, L. Leclercq, and R. Maurel. *J. Catal.* **50**:87 (1977).

37. J.K.A. Clarke and J.F. Taylor. *Faraday Trans.* **71**:2063 (1975).

38. J.R. Anderson and N.R. Avery. *J. Catal.* **2**:542 (1963).

39. J. Datka, P. Gallezot, J. Massardier, and B. Imelik. In: *Proceedings of the 5th Ibero-American Symposium on Catalysis*, Libson, 1976.

40. P.H. Otero-Schipper, W.A. Wachter, J.B. Butt, R.L. Burwell, Jr., and J.B. Cohen. *J. Catal.* **53**:414 (1978).

41. J.C. Schlatter and M. Boudart. *J. Catal.* **25**:93 (1972).

42. H.C. Yao, Y.F. Yu Yao, and K. Otto. *J. Catal.* **56**:21 (1979).

43. H.C. Yao and M. Shelef. *J. Catal.* **56**:12 (1979).

44. J.P. Brunelle, R.E. Montarnal, and A.A. Sugier. In *Proceedings of the 6th International Congress on Catalysis*, London, 1976, p. 844.

45. E. Kikuchi, M. Tsurumi and Y. Morita. *J. Catal.* **22**:226 (1971).

46. V. Ragaini, L. Forni, and Le Van Mao. *J. Catal.* **37**:339 (1975).

47. A. Péter and J.K.A. Clarke. *Faraday Trans.* **72**:1201 (1976).

48. J.R.H. van Schaik, R.P. Dessing, and V. Ponec. *J. Catal.* **38**:273 (1975).

49. J.P. Brunelle, A. Sugier, and J.F. LePage. *J. Catal.* **43**:273 (1976).

50. M. Boudart and L.D. Ptak. *J. Catal.* **16**:90 (1970).

51. K. Foger and J.R. Anderson. *J. Catal.* **54**:318 (1978).

52. B.N. Kuznetsov, Yu.I. Yermakov, M. Boudart, and J.P. Collman. *J. Mol. Catal.* **4**:49 (1978).

53. Z. Karpíński and T. Kóscielski. *J. Catal.* **56**:430 (1979).

54. C.H. Bartholomew and M. Boudart. *J. Catal.* **25**:173 (1972).

55. M. Boudart, A.W. Aldag, L.D. Ptak, and I.E. Benson. *J. Catal.* **11**:35 (1968).

56. R.A. Dalla Betta and M. Boudart. In: *Proceedings of the 5th International Congress on Catalysis*. Miami, 1972, paper 96, p. 1329.

57. S. Fuentes and F. Figueras. *J. Catal.* **54**:397 (1978).

58. S. Fuentes and F. Figueras. *Faraday Trans.* **74**:174 (1978).

59. J.K.A. Clarke and J.F. Taylor. *Faraday Trans.* **72**:917 (1976).

60. A.N. Mitrofanova, V.S. Boronin, and O.M. Poltorak. *Russ. J. Phys. Chem.* **46**, 32 (1972), and references cited therein.

61. R. Burch. *J. Catal.* **58**:220 (1979).

62. V. Ponec and W.M.H. Sachtler. In: *Proceedings of the 5th Int. Congr. Catalysis*, Miami, 1972, paper 43, p. 645.

63. J.K.A. Clarke, T. Manninger, and T. Baird. *J. Catal.* **54**:230 (1978).

64. A. O'Cinneide and F. G. Gault. *J. Catal.* **37**:311 (1975).

65. T.J. Plunkett and J.K.A. Clarke. *J. Catal.* **35**:330 (1974).

66. E. Santacessaria, D. Gelosa, S. Carra, and T. Adami. *Ind. Eng. Chem. Prod. Res. Dev.* **17**:68 (1978).

67. H.C. DeJongste, F.J. Kuijers, and V. Ponec. In: *Proceedings of the 6th International Congress Catalysis*, London, 1976, p. 915.

68. J.R. Anderson and Y. Shimoyama. In: *Proceedings 5th International Congress Catalysis*, Miami, 1972, paper 47, p. 695.

69. Z. Karpinski and J.K.A. Clarke. *Faraday Trans.* **71**:2310 (1975).

70. A. Sárkány, I. Guczi, and T. Tétényi. *J. Catal.* **39**:181 (1975).

71. Y.L. Lam and J.H. Sinfelt. *J. Catal.* **42**:319 (1976).

72. W.D. Gillespie and G.A. Somorjai. *J. Catal.* (1981)

73. A. Roberti, V. Ponec, and W.M.H. Sachtler. *J. Catal.* **28**:381 (1973).

74. J.R. Anderson and D.E. Mainwaring. *J. Catal.* **35**:162 (1974).

75. J.H. Sinfelt, H. Hurwitz, and J.C. Rohrer. *J. Catal.* **1**:481 (1962).

76. J.G. Brandenberger, W.L. Callender, and W.K. Meerbott. *J. Catal.* **42**:282 (1976).

77. C. Corolleur, F.G. Gault, and L. Beránek. *React. Kinet. Catal. Lett.* **5**:459 (1976).

78. H. Kubicka. *J. Catal.* **12**:223 (1968).

79. J.L. Carter, J.A. Cusumano, and J.H. Sinfelt. *J. Catal.* **20**:223 (1971).

80. A.V. Sklyarov, O.V. Kyrlov, and G. Keulks. *Kinet. Katal.* **18**:1213 (1977).

81. D.G. Grenoble. *J. Catal.* **56**:32 (1979).

82. V.N. Mozhaiko, G.L. Rabinovich, G.N. Maslyanskii, and L.P. Erdyakova. *Neftekhimiya* **15**:95 (1975).

83. K. Kochloefl. In: *Proceedings of the International Congress on Catalysis*, London, 1976, p. 1122.

84. G. Lietz and J. Völter. *J. Catal.* **45**:121 (1976).

85. G. Marie, G. Plouidy, J.C. Prudhomme, and F.G. Gault. *J. Catal.* **4**:556 (1967).

86. O. Beek. *Rev. Mod. Phys.* **17**:61 (1945), and references cited therein.

87. J.S. Campbell and P.H. Emmett. *J. Catal.* **7**:252 (1967).

88. I. Yasumori, H. Shinohara, and Y. Inoue. In *Proceedings of the 5th International Congress on Catalysis*, Miami, 1972, paper 52, p. 771.

89. G. Dalmai-Imelik and J. Massardie. In *Proceedings of the 6th International Congress on Catalysis*, London, 1976, p. 90.

90. J.C. Schlatter and M. Boudart. *J. Catal.* **24**:482 (1972).

91. A. Frackiewicz and Z. Karpinski. In *Proceedings of the 5th International Congress on Catalysis*, Miami, 1972, paper 42, p. 635.

92. R.E. Cunningham and A.T. Gwathmey. *Adv. Catal.* **9**:25 (1957).

93. P. Pareja, A. Amariglio, and H. Amariglio. *J. Catal.* **36**:379 (1975).

94. P. Pareja, A. Amariglio, and H. Amariglio. *React. Kinet. Catal. Lett.* **4**:459 (1976).

95. R.L. Moss, D. Pope, and H.R. Gibbens. *J. Catal.* **46**:204 (1977).

96. V.B. Kazanskii and V.P. Strunin. *Kinet. Katal.* **1**:553 (1960).

97. J.H. Sinfelt and P.J. Lucchesi. *J. Am. Chem. Soc.* **85**:3365 (1963).

98. J.H. Sinfelt. *J. Phys. Chem.* **68**:856 (1964).

99. T.A. Dorling, M.J. Eastlake, and R.L. Moss. *J. Catal.* **14**:23 (1969).

100. S.M. Davis and G. A. Somorjai. *J. Catal.* **65**:78 (1980).

101. P. Chatoransky, Jr. and W.L. Kranich. *J. Catal.* **21**:1 (1971).

102. P.A. Sermon, G.C. Bond, and P.B. Wells. *Faraday Trans.* **75**:2385 (1979).

103. R.A. Ross, G.D. Martin, and W.G. Cook. *Ind. Eng. Chem. Prod. Res. Dev.* **14**:151 (1975).

104. G.D. Lyubarskii, L.I. Ivanovskaya, and G.G. Isaeva. *Kinet. Katal.* **1**:235 (1960).

105. J.W.E. Coenen, R.Z.C. Van Meerten, and H.Th. Rijnten. In: *Proceedings of the 5th International Congress Catalysis*, Miami, 1972, paper 45, p. 671.

106. R.Z.C. Van Meerten, A.C.M. Verhaak, and J.W.E. Coenen. *J. Catal.* **44**:217 (1976).

107. G.D. Lyubarskii, L.I. Ivanovskaya, G.L. Isaeva, D.I. Lanier, and N.M. Kogan. *Kinet. Katal.* **1**:358 (1960).

108. G.M. Dixon and K. Singh. *Faraday Trans.* **65**:1129 (1969).

109. P. Gallezot, V. Datka, J. Massardier, M. Primet, and B. Imelik. In *Proceedings of the 6th International Congress on Catalysis*, London, 1976, p. 696.

110. J.M. Bassett, G. Dalmai-Imelik, M. Primet, and R. Mutin. *J. Catal.* **37**:22 (1975).

111. T.A. Dorling and R.L. Moss. *J. Catal.* **5**:111 (1966).

112. W.F. Taylor. *J. Catal.* **9**:99 (1967).

113. P.C. Aben, J.C. Platteeuw, and B. Southamer. In: *Proceedings of the 4th International Congress on Catalysis*, Moscow, 1968, paper 31.

114. M.S. Borisova, V.A. Dziśko, and Yu.O. Bulgakova. *Kinet. Katal.* **12**, 344 (1971).

115. R.Z.C. Van Meerten and J.W.E. Coenen. *J. Catal.* **37**:37 (1975).

116. W.F. Taylor and H.K. Staffin. *Faraday Trans.* **63**, 2309 (1967).

117. R.L. Moss, D. Pope, B.J. Davis, and D.H. Edwards. *J. Catal.* **58**:206 (1979).

118. A.O. Cinneide and J.K.A. Clarke. *J. Catal.* **26**:233 (1972).

119. A.V. Ramaswamy, P. Ratnasamy, S. Sivasanker, and A.J. Leonard. In: *Proceedings of the 6th International Congress on Catalysis*, London, 1976, p. 855.

120. A. Compero, M. Ruiz, and R. Gómez. *React. Kinet. Catal. Lett.* **5**:177 (1976).

121. Yu. S. Snagovskii, G.D. Lyubarskii, and G.M. Ostrovskii. *Kinet. Katal.* **7**:232 (1964).

122. P. Van Der Plank and W.M.H. Sachtler. *J. Catal.* **12**:35 (1968).

123. B. Coughlan, S. Narayanan, W.A. McCann, and W.M. Carroll. *J. Catal.* **49**:97 (1977).

124. R. Gómez, S. Fuentes, F.J. Fernández del Valle, A. Campero, and J.M. Ferreira. *J. Catal.* **38**:47 (1975).

125. M.A. Vannice and W.C. Neikam. *J. Catal.* **23**:401 (1971).

126. K.M. Sancier. *J. Catal.* **23**:404, (1971).

127. R. Maurel, G. Leclercq, and J. Barbier. *J. Catal.* **37**:324 (1975).

128. R. Ratnasamy. *J. Catal.* **31**:466 (1973).

129. G.N. Maslyanskii, B.B. Zharkov, and A.Z. Rubinov. *Kinet. Katal.* **12**:699 (1971).

130. N.R. Bursian, S.B. Kogan, N.M. Dvorova, L.P. Erdyakova, I.A. Korchagina, and N.K. Volnykhina, *Kinet. Katal.* **18**:197 (1977).

131. Yu.I. Ermakov, B.N. Kuznetsov, Yu.A. Ryndin, and V.K. Duplyakin. *Kinet. Katal.* **15**:978 (1974).

132. C.P. Khulbe and R.S. Mann, In: *Proceedings of the 6th International Congress on Catalysis*, London, 1976, p. 447.

133. H.G. Rushford and D.A. Whan. *Faraday Trans.* **67**:3377 (1971).

134. R.L. Burwell Jr., H.H. Kung, and R.J. Pellet. In *Proceedings of the 6th International Congress on Catalysis*, London, 1976, p. 108.

135. E.E. Gonzo and M. Boudart. *J. Catal.* **52**:462 (1978).

136. V.I. Garanin, U.M. Kurkchi, and Kh.M. Minachev. *Kinet. Katal.* **9**:889 (1968).

137. E. Segal, R.J. Madon, and M. Boudart. *J. Catal.* **52**:45 (1978).

138. R. Maatman, W. Ribbens, and B. Vonk, *J. Catal.* **31**, 384 (1973).

139. J. Haro, R. Gómez, and J.M. Ferreira. *J. Catal.* **45**, 326 (1976).

140. R.W. Maatman, P. Mahaffy, P. Hoekstra, and C. Addink. *J. Catal.* **23**, 105 (1971).

141. M. Kraft and H. Spindler. In *Proceedings of the 4th International Congress on Catalysis*, Moscow, paper 69 (1968).

142. N.M. Zaidman, V.A. Dzis'ko, A.P. Karnaukhov, N.P. Krasilenko, N.G. Koroleva, and G.P. Vishnyakova. *Kinet. Katal.* **9**:709 (1968).

143. J.A. Cusumano, G.W. Debinski, and J.H. Sinfelt. *J. Catal.* **5**:471 (1966).

144. I. Horescu and A.P. Rudenko. Russ. *J. Phys. Chem.* **44**:1601 (1970).

145. F. Figueras, B. Mencier, L. DeMorgues, C. Naccache, and Y. Trambouze. *J. Catal.* **19**:315 (1970).

146. P. Biloen, F.M. Dautzenberg, and W.M.H. Sachtler. *J. Catal.* **50**:77 (1977).

147. N. Nakamura, M. Yamoda, and A. Amano. *J. Catal.* **39**:125 (1975).

148. F.M. Dautzenberg and J.C. Platteeuw. *J. Catal.* **19**:41 (1970).

149. J.K.A. Clarke, E. McMahon, and A.D. O'Cinneide. In *Proceedings of the 5th International Congress on Catalysis*, Miami, 1972, paper 46, p. 685.

150. U.M. Kurkchi, V.I. Garanin, and Kh.M. Minachev. *Kinet. Katal.* **9**:472 (1968).

TABLE 7.42. Structure-Sensitive and Structure-Insensitive Catalytic Reactions

Structure-Sensitive	Structure-Insensitive

Hydrogenolysis
  Ethane: Ni
  Methylcylopentane: Pt
	Ring opening
	  Cyclopropane: Pt

Isomerization
  Isobutane: Pt
  Hexane: Pt
	Hydrogenation
	  Benzene: Pt
	  Ethylene: Pt, Rh
	  Carbon Monoxide: Ni, Rh, Ru, Mo, Fe

Cyclization
  Hexane: Pt
  Heptane: Pt
	Dehydrogenation
	  Cyclohexane: Pt

Ammonia synthesis
  Fe, Re
	Hydrodesulfurization
	  Tiophene: Mo

Hydrodesulfurization Tiophene: Re	

TABLE 7.43. Chemical Processes That Are the Largest Users of Heterogeneous Catalysts at Present and the Catalysts That Are Utilized Most Frequently

Reactions	Catalysts
CO, HC oxidation in car exhaust	Pt, Pd on alumina
NO$_x$ reduction in car exhaust	Rh on alumina, V-oxide
Cracking of crude oil	Zeolites
Hydrotreating of crude oil	Co–Mo, Ni–Mo, W–Mo
Re-forming of crude oil	Pt, Pt–Re, and other bimetallics on alumina
Hydrocracking	Metals on zeolites or alumina
Alkylation	Sulfuric acid, hydrofluoric acid, soild acids
Steam reforming	Ni on support
Water–gas shift reaction	Fe–Cr, CuO, ZnO, alumina
Methanation	Ni on support
Ammonia synthesis	Fe
Ethylene oxidation	Ag on support
Nitric acid from ammonia	Pt, Rh, Pd
Sulfuric acid	V-oxide
Acrylonitrile from propylene	Bi, Mo-oxides
Vinyl chloride from ethylene	Cu-chloride
Hydrogenation of oils	Ni
Polyethylene	Cr, Cr-oxide on silica

TABLE 7.44. Chemical Processes That Are the Largest Users of Homogeneous Catalysts at Present and the Catalysts That Are Utilized Most Frequently

Reactions	Catalysts
Hydroformylation	Cobalt, rhodium compounds
Aldehydes and alcohols from olefins	
Carbonylation	Rhodium complexes and methyl iodide
Acetic acid from methanol	
Oxidation	
Adipic acid from cyclohexane	Cu(II), V(V) salts, nitric acid
Terephthalic acid from p-xylene	Co(II), Mn(II) salts, bromide ion
Acetic acid from butane or acetaldehyde	Co(II), Cu(II), Mn(II) salts
Olefin polymerization	
Polyethylene	$TiCl_4$, alkylaluminum, dialkyl magnesium compounds
Ethylene–propylene–diene copolymers	$VOCl_3$, VCl_4, alkylaluminum
Ethylene–butadiene copolymers	$RhCl_3$
Polypropylene, polystyrene	Peroxides
Poly(vinyl chloride)	Percarbonates
Urethane	Amines

TABLE 7.45. Comparison of Initial Specific Rate Data for the Cyclopropane Ring Opening on Platinum Catalysts

Data source	Type of catalyst	Calculated specific reaction rate at $P^0_{CP} = 135$ torr and $T = 75°C$		Comments
		$\dfrac{\text{moles } C_3H_8}{\text{min} \cdot \text{cm}^2 \text{ Pt}}$	$\dfrac{\text{molecules } C_3H_8}{\text{min} \cdot \text{Pt site}}$	
Kahn et al. (1)	Stepped platinum crystal surfaces	2.1×10^{-6} 1.8×10^{-6} 1.8×10^{-6} 2.1×10^{-6}		Rate on Pt(s)-[6(111) × (100)] single crystal based on $\Delta E^* = 12.2$ kcal/mole[a]
	Average	1.95×10^{-6}	812	Value of 812 based upon 87% (111) orientation and 13% polycrystalline orientation
Hegedus and Petersen (2)	0.04 wt% Pt on η-Al$_2$O$_3$	7.7×10^{17} based on 100% Pt dispersion	410	Value of 410 based upon average Pt site density of 1.12×10^{15} atoms/Pt site; this value would be nearly equal to average of values above if dispersion was approximately 50%
Boudart et al. (3)	0.3% and 2.0% Pt on η-Al$_2$O$_3$	8.9×10^{-7}	480	$\eta_{CP} = 0.2$, $E = 8.5$ kcal/mole[a]
Dougharty (4)	0.3% and 0.6% Pt on γ-Al$_2$O$_3$	2.5×10^{-6}	1340	$\eta_{CP} = 0.6$, $E = 8.5$ kcal/mole[a]

[a]Dougharty reports that $E^* = 8$–9 kcal/mol and $\eta = 0.2$–0.6.

REFERENCES

1. D.R. Kahn, E.E. Petersen, and G.A. Somorjai. *J. of Catal.* **34**:294 (1974).
2. L.L. Hegedus and E.E. Petersen, *J. Catal.* **28**:150 (1973).
3. M. Boudart, A. Aldag, N.A. Dougharty, and C.G. Harkins, *J. Catal.* **6**:92 (1966).
4. N.A. Dougharty, Ph.D. thesis, University of California at Berkeley, 1964.

TABLE 7.46. Comparison of Polycrystalline Rh Foil with a 1% Rh/Al$_2$O$_3$ Catalyst in the CO–H$_2$ Reaction at Atmospheric Pressure

	Polycrystalline Rh[a] Foil	Supported 1% Rh/Al$_2$O$_3$[b]
Reaction conditions	300°C, 3:1 H$_2$/CO, 700 torr	300°C[c] 3:1 H$_2$/CO, 760 torr
Type of reactor	Batch	Flow
Conversion	<0.1%	<5%
Product distribution	90% CH$_4$ ± 3	90% CH$_4$
	5% C$_2$H$_4$ ± 1	8% C$_2$H$_6$
	2% C$_2$H$_6$ ± 1	2% C$_3$
	3% C$_3$H$_8$ ± 1	<1% C$_1^+$
	<1% C$_4^1$	
Absolute methanation rate at 300°C (turnover no.)	0.13 ± 0.03 molecule/site/sec	0.034[c] molecule/site/sec
Activation energy (kcal)	24.0 ± 2	24.0

[a] The values is this column are from B.A. Sexton and G.A. Somorjai, *J. Catal.* **46**:167 (1977).
[b] The values in this column are from M.A. Vannice, *J. Catal.* **37**:462 (1975).
[c] Data adjusted from 275°C.

Table 7.47. Comparison of the Kinetic Parameters (TN = A exp ($-\Delta E^*/RT$) for CO and CO$_2$ Hydrogenation Over Polycrystalline Rhodium Foils

Reaction Conditions	Surface Pretreatment	R(CH$_4$) at 600 K (molecules/site/sec)	A (molecules/site/sec)	ΔE^* (kcal/mole)
3 H$_2$:1 CO, 0.92 atm	Clean	0.13 ± 0.03	10^8	24 ± 3
3 H$_2$:1 CO$_2$, 0.92 atm	Clean	0.33 ± 0.05	10^5	16 ± 2
3 H$_2$:1 CO, 6 atm	Preoxidized	1.7 ± 0.4	10^5	12 ± 3

TABLE 7.48. Comparison of Ethylene Hydrogenation Kinetic Parameters for Different Platinum Catalysts

Catalyst	Log Rate[a]	a[b]	b[b]	ΔE^* (kcal/mole)	Ref.
Platinized foil	1.9	−0.8	1.3	10	[41]
Platinum evaporated film	2.7	0	1.0	10.7	[23]
1% Pt/Al$_2$O$_3$	—	−0.5	1.2	9.9	[39]
Platinum wire	0.6	−0.5	1.2	10	[42]
3% Pt/SiO$_2$	1.0	—	—	10.5	[43]
0.05% Pt/SiO$_2$	1.0	0	—	9.1	[44]
Pt(111)	1.4	−0.6	1.3	10.8	Our work

[a] Rate in molecules/Pt atom · sec, corrected for the following conditions: T= 323 K, $P_{C_2H_4}$ = 20 torr, P_{H_2} = 100 torr.
[b] Orders in ethylene (a) and hydrogen (b) partial pressures.

8

MECHANICAL PROPERTIES OF SURFACES

8.1   INTRODUCTION

When we glue a broken chair, light a match, walk on a street, or ski on snow, we make use of the mechanical properties of surfaces. These include (a) static properties such as hardness or adhesion and (b) dynamic properties such as slide, friction, lubrication, or fracture. The study of the mechanical properties of surfaces in relative motion is often called *tribology*. It is the purpose of surface scientists to describe and explore many of the macroscopic mechanical properties on the molecular level in order to provide fundamental answers to some simple questions: Why are materials hard or soft? How do the mechanical properties of surfaces enable us to walk? What occurs when we repeatedly move surfaces relative to each other at variable speeds (such as the piston rod against the piston wall of the internal combustion engine)?

596

8.2 HISTORICAL PERSPECTIVE

There is evidence that in Egypt a lubricant was poured in front of the sledge during the transport of statues and stones [1, 2]. Romans were using iron nails in the sole of their *caligae* (shoes) to reduce wear. They used metal sleeves in olive-crushing and corn grinding mills, indicating an appreciation of the ease of rolling over sliding.

The use of wheel and axle prompted extensive application of animal fat and vegetable oil for lubrication (e.g., see reference [3]). In the Middle Ages, stone studs were employed to reduce the wear of wooden plows. During the Renaissance there was a burst of development of machinery with moving parts. Leonardo da Vinci gained an understanding of the laws of friction, and he designed rolling element bearings. Amontons reported the first comprehensive study of friction in 1699. In the same period, the role of surface texture in determining frictional resistance was realized and utilized; and the concept of deformation during sliding and adhesion was suggested. During the industrial revolution, mineral oil was discovered to be an excellent lubricant. Since that time the formulation of lubricating oils has made the development of fast, steam-turbine-driven ships, automobile, and aircraft engines possible. Because of the importance of mechanical properties of surfaces, centers of tribology research have been established in many countries. From the development of synthetic lubricants (used between moving parts at high speeds and at high temperatures), to the design of steel-belted radial tires and modern skis (and from solid lubricants for applications in ultrahigh vacuum and space research, to lubricants for disc drives of computers) [4], the control of friction between moving surfaces is one of the most challenging problems in the design of modern devices and machinery.

While the macroscopic concepts of hardness, adhesion, friction, and slide have evolved over the last two centuries, atomic level understanding of the mechanical properties of surfaces eluded researchers. The discovery of the atomic force microscope in recent years promises to change this state of affairs. Being able to measure forces as small as 10^{-9} newton or as large as 10^{-2} newton [5] over a very small surface area (few atoms) and by simultaneously providing atomic spatial resolution, this technique permits the study of deformation (elastic and plastic), hardness, and friction on the atomic scale. The buried interface between moving solid surfaces can be studied with spectroscopic techniques on the molecular level. Study of the mechanical properties of interfaces is, again, a frontier research area of surface chemistry.

8.3 HARDNESS

The hardness of a material typically represents its ability to resist indentation. Macroscopic hardness is usually measured by pressing a diamond tip (of known area A) into a solid, with a given load (force). The hardness H, often called Vicker's or Knoop hardness (which differ by the diamond tip geometry), is defined as $H = \mathcal{W}/A$, where $\mathcal{W}$ is the load and A is the indented area. This is a destructive test: The permanent indentation caused by the diamond is produced by breaking bonds and displacing atoms. A typical indentation profile is shown in Figure 8.1, and the Vicker's and Knoop hardnesses of several materials are listed in Tables 8.1 and 8.2.

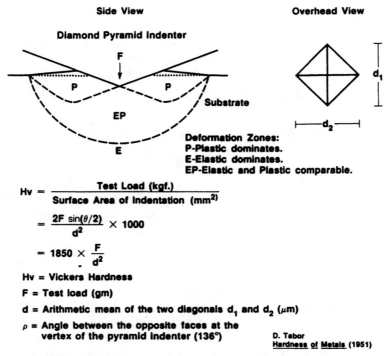

Figure 8.1. Vickers hardness test shown schematically.

An evaluation of actual contact areas in hardness measurements clearly indicates the relevance of microscopic studies of mechanical properties. From the known values of H, we can compute the real area of contact between the solid and the indenter at various loads. These are shown in Table 8.3. The real contact area is a very small fraction of the geometric area, especially at smaller loads. The real area of contact

TABLE 8.1. Hardness Values of Various Materials on the Vickers Scale

Material	Vickers Hardness (kg/mm^2)
Lead	4.2
Gold	22
Aluminum	24
Zinc	35
Silver	36
Platinum	40
Copper	50
Nickel	130
Chromium	200
Tungsten	330

Source: L.E. Samuals. *Metallographic Polishing.* American Society for Metals, Metals Park, OH (1982).

TABLE 8.2. Hardness Values of Various Materials on the Knoop Scale

Substance	Formula	Knoop Value
Gypsum	$CaSO_4 \cdot 2H_2O$	32
Cadmium	Cd	37
Silver	Ag	60
Zinc	Zn	119
Calcite	$CaCO_3$	135
Fluorite	CaF_2	163
Copper	Cu	163
Magnesia	MgO	370
Apatite	$CaF_2 \cdot 3Ca_3(PO4)_2$	430
Nickel	Ni	557
Glass (soda lime)	—	530
Quartz	SiO_2	820
Chromium	Cr	935
Zirconia	ZrO_2	1160
Beryllia	BeO	1250
Topaz	$(AIF)_2SiO_4$	1340
Tungsten carbide alloy	WC, Co	1400–1800
Zirconium boride	ZrB_2	1550
Titanium nitride	TiN	1800
Tungsten carbide	WC	1880
Tantalum carbide	TaC	2000
Zirconium carbide	ZrC	2100
Alumina	Al_2O_3	2100
Beryllium carbide	Be_2C	2410
Titanium carbide	TiC	2470
Silicon carbide	SiC	2480
Aluminum boride	AIB	2500
Boron carbide	B_4C	2750
Diamond	C	7000

Source: CRC Handbook of Chemistry and Physics, 64th edition, CRC press, Boca Raton, FL, p. 20.

TABLE 8.3. Real Area of Contact at Different Loads for Four Metals of Different Hardness

Metal	Hardness (kg/mm^2)	Real Area of Contact (mm^2)[a]		
		1-g load	100-g load	1000-g load
Gold	22	4.5×10^{-5}	4.5×10^{-3}	4.5×10^{-2}
Aluminum	24	4.2×10^{-5}	4.2×10^{-3}	4.2×10^{-2}
Copper	50	2.0×10^{-5}	2.0×10^{-3}	2.0×10^{-2}
Tungsten	330	3.0×10^{-6}	3.0×10^{-4}	3.0×10^{-3}

[a] Real area of contact = Load (kg)/Hardness (kg/m^2).

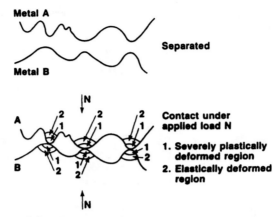

Figure 8.2. Model of solids in contact on the atomic scale.

is 10^{-6}–10^{-4} times the geometric area of the indenter. Thus, the surfaces in contact may be viewed on the microscopic scale as shown in Figure 8.2, where real contact is made at only a few points of a rough interface. Because the pressure is defined as the force per area ($P = \mathcal{F}/A$), there must be very high local pressures, on the order of 10^3–10^5 atm (10^8–10^{10} Pa), at the points of interface contact at typical loads (on the order of several newtons). At such pressures, local sites on the surface are drastically deformed and displaced from their original equilibrium positions, producing an atomically disordered interface.

8.4 MECHANICAL FORCES REQUIRED TO BREAK A CHEMICAL BOND

Let us consider the surface contact on an atomic scale. A surface atom is bound to its neighbors by strong chemical bonds that add up to ≈ 150 kcal/mole (600 kJ/mole). We would like to estimate the mechanical force necessary to break these bonds. If the atom behaves as a harmonic oscillator, stretching its bonds by 0.1 Å (10^{-2} nm) about its equilibrium position—five times its root-mean square displacement of ≈ 0.02 Å (2×10^{-3} nm)—would certainly break the bonds. Thus, the force needed to stretch the bond by 0.1 Å (10^{-2} nm) is given by

$$\mathcal{F} = E/\Delta x = 1 \times 10^{-18}\, \text{J}/1 \times 10^{-11}\, \text{m} = 1 \times 10^{-7}\, \text{N} \qquad (8.1)$$

Therefore, we need about 10 millidynes of force per atom to break the ≈ 150-kcal (600 kJ/mole $= 1 \times 10^{-18}$ J/atom) surface-atom bond.

The atomic-force microscope (AFM) can explore contact and hardness on the atomic scale. Analogous to the STM, the AFM uses a feedback loop to control the distance between sample and a probe tip at the end of an cantilever arm. As opposed to tunneling current however, the AFM monitors an optical signal as feedback to measure lever deflection. Thus both attractive and repulsive interactions of tip and sample can be monitored. As the microscope tip approaches the surface, attractive forces are first exerted on the tip by the surface and can be measured to as small a value as 10^{-9} N. Upon contact with the surface, further motion of the tip results in repulsive forces between tip and sample. This procedure is capable of producing loads which overlap the forces encountered in macroscopic mechanical measurements. After a force of known magnitude is applied, the area of tip contact is scanned for signs of permanent damage. For a smooth gold surface, permanent damage has only been detected above 5×10^{-5} N, as shown in Figure 8.3. This result suggests that the metal surface responds to the approaching metal tip by elastically "bending" in the range of 10^{-9} to 5×10^{-5} N. This range of forces is the regime of elastic deformation. Only on application of a force greater than 5×10^{-5} newton did plastic deformation (the irreversible breaking of metal-metal bonds) begin. Measurements of this type permit one to determine the forces needed for elastic or plastic deformation on the atomic scale and to correlate the results with those obtained by macroscopic studies.

The surface force apparatus (SFA) [6–9] is another type of instrument that can measure the forces on molecules sandwiched between solid interfaces. Two mica sheets with large atomically smooth areas are brought together so that they are separated only by a few molecular diameters. The mica sheets are bent slightly, and there is a relatively large surface area that can be maintained at these very small distances (as compared to the atomic force microscope that has a small tip). The forces between the mica sheets can be measured as a function of the distance of separation, and the orientation of molecules that are sandwiched between the mica sheets can be monitored by laser spectroscopy (SHG).

The behavior of surfaces on the atomic scale is similar to the macroscopic stress–strain behavior of solids, as is demonstrated for a metal in Figure 8.4. Up to a certain applied force, the elongation of the solid is elastic; that is, it is reversible. In the elastic regime one can "store" a great deal of elastic energy in the solid by inducing such a reversible deformation. Perhaps the simple analogy of a long bow demonstrates this most vividly. Above a certain applied load, however, the solid becomes

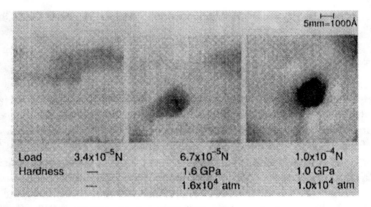

Figure 8.3 Microhardness of gold determined by the atomic force microscope. At 3.4×10^{-5} N load there is no sign of plastic deformation (i.e., permanent damage). At 6.7×10^{-5} N plastic deformation occurs. The measurements were carried out on a gold film deposited on a mica substrate to produce atomically smooth metal surfaces. (Courtesy of C.M. Mate and G.S. Blackman.)

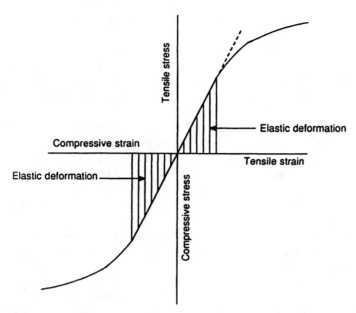

Figure 8.4. Macroscopic stress–strain behavior of metals shown schematically.

plastically, irreversibly deformed and the energy is consumed by the breaking of bonds and rearrangement of atoms. In macroscopic terms, hard materials such as diamond or tungsten carbide possess a much smaller regime of elastic deformation than do soft materials such as gold or polymers.

8.5 ADHESION

The work of adhesion W_{ad} needed to separate two phases A and B (see Chapter 3) is given by

$$W_{ad} = \gamma_A + \gamma_B - \gamma_{AB} \tag{8.2}$$

where γ_A and γ_B are the interfacial energies of the freshly separated surfaces and γ_{AB} is the interfacial energy of the joined A–B phases. In general, solids and liquids that have large interfacial energies form strong adhesive bonds. The work of adhesion ranges from 40 ergs/cm^2 (4×10^{-2} J/m^2) to 140 ergs/cm^2 (14×10^{-2} J/m^2).

The work of adhesion is also related to the force needed to break chemical bonds at the interface on the atomic scale. When bonds form at the interface of two solids, restructuring can occur at both sides of the interface, thereby optimizing the strength of the interfacial chemical bond. This restructuring is related to the adsorbate-induced restructuring processes and to the phenomenon of epitaxy that was discussed in Chapters 2 and 5. It involves the movement of atoms perpendicular, as well as parallel, to the interface. We may call this movement the work of interface restructuring W_r. It involves surface atoms as well as atoms two to three layers away from the topmost exposed atomic layer. As force is applied in order to pull or shear the interface apart by breaking the surface chemical bonds, elastic deformation followed by bond breaking occurs. This process may be called the work of separation W_s. The work of adhesion W_{ad} is therefore $W_{ad} = (W_r + W_s)$.

Strong adhesive bonds are expected to form between solids with atoms of similar size, similar interatomic distances, and chemical bonding that would maximize W_r. Strong adhesive bonds would also be formed for systems that exhibit large elastic deformations before bond breaking occurs, because of the large values of W_s for these systems. Polymers and macromolecules with chains that can rearrange in response to mechanical forces are examples of this type of system. Because of the flexibility of the molecular chains, the work of adhesion also increases due to increased area of contact $\mathcal{A}$ at the interface, since the total adhesive force is $W_{ad}\mathcal{A}$. In some systems, the work of adhesion is often so large that decohesion occurs away from the interface in one of the solids making up the adhesive joint. Composites, where the interface area is maximized as in carbon or glass fibers in a polymer matrix, capitalize on large values of the work of adhesion. These materials often have mechanical strengths that are far superior to those of the individual components.

8.6 SURFACES IN RELATIVE MOTION. TRIBOLOGY

Surfaces in contact and moving parallel to each other exhibit phenomena that are encountered incessantly in our everyday life. These phenomena include friction, slide, wear, and lubrication (the process used to modify friction). We could not walk without controlled friction, nor could the internal combustion engine operate without the lubrication of its moving parts. The name *tribology* (from the Greek *tribein*—to rub) is given to phenomena that involve surfaces in relative motion.

8.6.1 Friction and Sliding

The coefficient of friction μ is defined as the force $\mathcal{F}$ required to initiate and maintain sliding divided by the load $\mathcal{W}$ applied perpendicular to the surface: $\mu = \mathcal{F}/\mathcal{W}$. The coefficient of friction is found to be independent of the area of contact over a wide range of loads that cause plastic deformation. The reason for this is that both force and load are proportional to the same area of interface contact. For plastic deformation, we have already shown that $\mathcal{W} = Ap$, where p is the average pressure over the contacts and A is the area of contact. At the contacts, chemical bonds are made that must be sheared or broken as one solid surface slides over the other. If the shear strength s is constant, the force $\mathcal{S}$ required to shear these contacts equals the area of contact times the shear strength ($\mathcal{S} = As$). If materials of comparable hardness are brought into contact, the sliding force is equal to the shear force and proportional to the area of contact. Therefore we can write $\mu = \mathcal{F}/\mathcal{W} = s/p$; this observation that the frictional force is independent of contact area is known as *Amontons' law*.

An additional phenomenon can occur at the sliding interface of two materials of different hardness. In this case, the chemical bonds that are continuously formed and broken at the interface compete with chemical bonds within the softer material. As a result, parts of the softer of the two sliding partners are transferred to the harder-surface side. This phenomenon is known as *plowing* and is usually associated with the presence of wear tracks, material transport, and the accumulation of debris at the interface. Here, the total frictional force is a sum of the shear force and the force required to displace the softer material.

The frictional forces that operate when a wheel or a ball bearing rolls over a surface are different from the friction between two flat sliding surfaces. When in contact with a surface, the wheel is temporarily deformed elastically and also causes deformation on the other side of the interface. The energy stored in producing the deformation is partially recovered as the wheel moves. However, the deformation is not completely elastic, and friction also occurs due to adhesion, which tends to increase with increasing velocity. Rolling friction plays an all-important role in permitting us to walk or run. It has been found that the sliding force needed to initiate motion is usually greater than the sliding force needed to maintain a given sliding speed. Thus, there is both a static and a kinetic friction coefficient, μ_s and μ_k, respectively, with $\mu_s > \mu_k$. The friction coefficients (μ_s) of several surfaces in contact are listed in Table 8.4. The μ values are usually in the range of 0.4–1.0. The anomalously low friction coefficients over snow or ice will be discussed later.

Although atomic-scale studies of friction have not been carried out until recently—because of the lack of techniques such as the AFM—several mechanisms for friction and slide that depend on the physical and chemical properties of the materials in contact have been suggested. For example, friction and wear of metal surfaces are the highest for ductile materials. This can be understood in terms of the plastic flow that occurs at the interface for these materials under normal loads. With this plastic flow, adhesion proportional to the load also occurs, leading to increased friction and wear. In addition, the oxidation of a metal at the interface can also affect the degree of the friction. If the oxide of a metal has a higher hardness value than the metal itself, measured friction coefficients will be lowered upon oxidation. If,

TABLE 8.4. Static Friction Coefficients for Different Material Contacts

Material	μ_s
Glass on glass, clean	0.9–1.0
Diamond on diamond, clean	0.1
Hard carbon on steel	0.14
Tungsten carbide on steel	0.4–0.6
Polystyrene on steel	0.3–0.35
Brick on wood	0.6
Brake material on cast iron, clean	0.4
Brake material on cast iron, wet	0.2
Copper on copper, in air	2.8
Iron on iron, in air	1.2
Molybdenum on molybdenum, in air	0.8
Ice on ice, 0°C	0.05–0.15
Aluminum on snow, 0°C	0.35
Ski wax on snow, 0°C	0.04

Source: F.P. Bowden and D. Tabor. *The Friction and Lubrication of Solids.* Oxford University Press, New York, 1971.

however, there is little change in hardness at the interface, little change in friction will be seen. (Adhesion of an oxide film to the parent substrate has also been seen to affect friction measurements.) A final example is that of lamellar solids such as graphite or molybdenum disulfide [10]. These materials have low (0.1–0.4) friction coefficients because the crystalline layers that are held together by only weak chemical bonds may slide over each other.

8.6.2 Heating by Friction

Lighting a match by rapidly moving the match head against a rough surface is a good example of how heat is generated by friction. The temperature rise in this circumstance is large enough to ignite the coating material, which has a low flashpoint. Considering the same geometry, if all the frictional work is converted to heat, the amount of heat q is

$$q = \mu \mathcal{W} v \qquad\qquad (8.3)$$

where μ is the kinetic friction coefficient, $\mathcal{W}$ is the load, and v is the sliding velocity. High temperatures, in the range of 200–600°C, can be reached before the heat is dissipated by thermal conduction.

Melting at a sliding interface is a mechanism that can be invoked in some instances to rationalize measured friction coefficients. In the case of a sliding ski on snow, for example, the formation of a film of water by local heating at the interface is thought to be responsible for the low friction coefficients on snow or ice. Thus it is the interaction with water, and not ice, that determines the degree of friction. In looking at the range of μ on ice for different materials, those that cannot be wet by water, such as fluorocarbons, show the lowest friction coefficients. Snow skis are accordingly coated with such materials to enhance performance.

The rapid passage of a weighted wire through ice was first reported in 1872 [11]. This interesting phenomenon has been used to cut large blocks of ice to smaller pieces. Both Faraday [12] and Thomson [13] suggested the presence of a liquid-like layer on the surface of ice below the freezing point to account for the rapid motion of the wire through the ice. Recent studies indicate that this is indeed the correct explanation as compared to melting by pressure or by friction-heating. The liquid-like surface layer on ice appears to exist down to $-30°C$, with its thickness decreasing with decreasing temperature [14, 15]. Thus, wire cutting of ice would not be possible below this temperature.

8.6.3 Applications of Friction

The reduction of friction is essential in machines with rolling and sliding elements to save energy [16–26]. However, the use of brakes and clutches require high levels of steady friction force. The energy liberated by friction can have both harmful and beneficial effects; a large rise in the temperature of the brake can damage it, and the braking efficiency will be reduced dramatically. However, the heat generated by friction can be used for friction cutting and welding. In friction welding, one part is firmly gripped while the other is rotated rapidly and pressed against the static part by axial pressure. Most engineering alloys of copper, aluminum, or titanium (brass and bronze, for example), as well as polymers, can be friction welded [26–29]. Friction-induced deformation at the interface breaks up any contamination layer and the heating causes seizure. Dissimilar materials can be joined this way.

For low-carbon steels, typical sliding speeds, pressures, and friction welding process times are on the order of 1–5 m/sec, 50–140 MPa, and few seconds, respectively [30, 31]. In these conditions the contact interface can reach 1500°C, higher than the melting point of the bulk material. The sliding speeds and pressures needed to friction weld materials varies, but it appears that temperatures close to the melting point of the lower-melting-point component are necessary for the process.

Polytetrafluoroethylene (PTFE, or Teflon) is reported to resist friction welding to other polymers even at high speeds and pressures [24]. PTFE has a much lower coefficient of friction and relatively higher softening point among the polymers. To reach the necessary conditions to weld it, sliding speed and pressure should be increased by a factor of 50 and 20 in comparison with high-density polyethylene [19].

8.6.4 Lubrication

Materials that lower the coefficient of friction are commonly used in tribology at all conditions of relative motion. These materials are called *lubricants*. Long-chain organic molecules with polar groups at one end of the chain are commonly employed to lubricate moving parts with excellent results—often an order of magnitude decrease in μ. Table 8.5 lists well-known and frequently used lubricants. It is thought that the polar groups improve binding and adhesion to the solid while the long hydrocarbon chain provides solubility in the lubricant liquid. In addition, the coefficient of friction decreases with increasing molecular weight of the hydrocarbon chain. One of the most successful lubricant systems uses the dialkyl-dithio zinc phosphates

TABLE 8.5. Frequently Used Lubricants

Name	Structure
Diethyl phosphonate	C_2H_5-O and O double bond; P center; C_2H_5-O and $O-H$
Cyclohexyl amine salt of dibutyl phosphate	C_4H_9-O and O double bond; P center; C_4H_9-O and $O^- \ ^+H_3N \ C_6H_{11}$
Diphenyl disulfide	$C_6H_5-S-S-C_6H_5$
Di-[methyl acetate] disulfide	$C_2H_5\overset{O}{\overset{\|}{C}}-OCH_2-S-S-CH_2O-\overset{O}{\overset{\|}{C}}C_2H_5$
Zinc O,O-diisopropylphosphorodithioate	$Zn\,[S-\overset{S}{\overset{\|}{P}}-(O\ CH\ (CH_3)_2)_2]_2$

as additives,

$$
\begin{array}{c}
\text{R}-\text{O} \diagdown \quad \text{S} \qquad \text{S} \quad \diagup \text{O}-\text{R} \\
\quad\; \diagdown \;\; / \;\; \qquad \diagdown\diagdown \;/ \\
\qquad \text{P} \qquad\qquad \text{P} \\
\quad\; \diagup \; \diagdown \qquad \diagup \; \diagdown \\
\text{R}-\text{O} \qquad \text{S}-\text{Zn}-\text{S} \qquad \text{O}-\text{R}
\end{array}
$$

where R may be either alkyl or aryl groups. The reasons for its lubricating ability are being studied on the molecular level in several laboratories.

The process of lubrication varies with the degree of interaction of the two surfaces. At low load levels (fluid lubrication), there are many layers of organic molecules sandwiched between the solid surfaces. In this circumstance, the friction depends on these molecules sliding by each other—that is, the viscosity of the liquid. When the load is high and/or the speed is low, this pressure may not be sufficient and surfaces come into close contact, resulting in the elastic deformation of the asperities.* We are then in the "elastohydrodynamic" lubrication regime. If the load and/or speed are further increased, the asperities deform plastically and the fluid thickness is reduced to molecular dimensions. This regime is called "boundary lubrication." In this regime the lubricant molecular layer is often not continuous, because high pressures at the asperities or narrow contact points squeeze out the lubricant from that area. A recent model of a lubricated interface is shown in Figure 8.5.

The high temperatures generated by friction are likely to alter the composition of a metal–metal oxide–organic lubricant system. The lubricant may decompose, and its active constituents may alter the surface composition. A graphitic, sulfide, or phosphide layer can form that may act as a solid lubricant in addition to the lubricant in the liquid phase. In this circumstance the lubrication also involves the replenishing of the lubricating coating as it is worn away during sliding. The unique properties of pressurized films have been suggested as contributing in a major way to the production of low friction during boundary-layer lubrication.

As a molecular model of lubrication, let us consider one or two molecular layers adsorbed at a solid–solid interface. The molecules are sandwiched between the solid surfaces and form strong chemical bonds with them. We would like to know the molecular binding configuration in this circumstance, since surfaces are always covered by molecular layers that reduce their surface energy. Because of atomic roughness, which produces uneven pressures at various surface sites when two surfaces are brought into contact, the molecules may assume different orientations or bonding arrangements at areas of high pressure or may even be squeezed out from these sites altogether. One model of molecular orientation at a smooth buried interface is shown in Figure 8.6. Unfortunately, experimental information is lacking in this area of interface science, which is one of the frontiers for the near future.

Until 70 years ago, most of the lubricants were crude-oil-derived mineral oils. Their performance, however, was not suitable for modern machines. The viscosity of the crude mineral oils is strongly affected by the load, temperature, and shear rates developed in the machines. In addition, chemical breakdown of the lubricant molecules is observed. To improve the performance of lubricants, chemists design additives.

*That is, surface protrusions or irregularities.

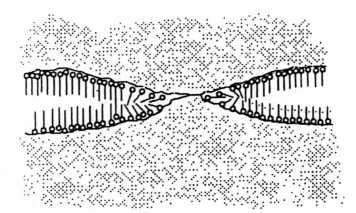

Figure 8.5. Model of a lubricated interface under high loads. The lubricant may be squeezed out at the narrow contact points.

A multigrade oil for engines contains, in addition to mineral oil, a wide variety of additives (e.g., see reference [32]). Polymers such as polymethacrylates and polyisobutylenes increase the viscosity at high temperatures. Rust inhibitors prevent the corrosion of the engines by the oxidation products of the oil. Oxidation inhibitors or antioxidants, mainly compounds of sulfur, phosphorus, oil-soluble amines, and phenols, act on the organic peroxides formed at an intermediate stage of the degradation of the lubricant. The formation of sludge and deposition of products of combustion are prevented by the addition of detergents (oil-soluble and surface-active materials) and by polymeric dispersant additives. Other additives aim to reduce the viscosity at low temperatures.

In spite of the presence of additives, mineral oil lubricants have disadvantages for applications in more severe conditions: They oxidize more rapidly at temperatures higher than 90–120°C and have great reluctance to flow at temperatures below −20°C. For these applications, synthetic lubricants are prepared. Silicones have viscosity varying slightly with temperature, but their chemical decomposition under severe conditions gives SiO_2 that has a harmful abrasive action. Fluorocarbons, diesters, polyglycols, phosphate esters, polybutenes, polyphenyl esters, silicone esters,

Figure 8.6. Scheme of molecules at the buried interface when they are sandwiched between two solids. Studies of their molecular structure and orientation by electron emission is possible as long as one of the solids is a thin layer.

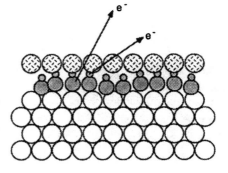

and so on, are among the synthetic lubricants that can work in severe conditions such as $-70°C$ and $300°C$.

For extreme pressures, additives that react with metal surfaces are used. They contain sulfur and phosphorus such as dialkyl dithiophosphates.

Modern lubricants combine (a) components favorable for fluid lubrication and (b) chemical compounds giving good boundary layer lubrication.

8.7   SOLID LUBRICANTS. COATINGS

Liquids are the most common lubricants used to increase the performance of moving mechanical components over long periods. However, in certain environments, in vacuum or at high temperatures, liquids cannot survive. In these cases, we must rely on solid lubricants [33–38].

To be an efficient lubricant, a solid must produce friction coefficients less than 0.2, must adhere to the surfaces of the lubricated pieces in thin film form, and must be durable in harsh environments. Some of the solids in Table 8.6 reduce friction because of their structure. They have low shear strength. Others are softer than the substrate material, and the shear occurs in the film. As an exception to this rule, hard materials such as DLC (diamond-like carbon) and TiN do not possess low shear strength but still possess lubricating properties. For these cases, it is likely that shear occurs in an interfacial layer formed at the contact where adhesion is low. Solid lubricating films are usually less than 1 mm thick. To be durable, these films must have a very low wear rate, even in the presence of corrosive vapors or gases. Applications at high temperatures are also very demanding: The friction coefficient must be low, but must also remain low over the range of required temperatures. At these high temperatures, losses by oxidation or flow become important, and very thin films are less convenient and reliable.

Let us look closely at molybdenum disulfide, MoS_2 [35, 36, 39], an excellent solid lubricant. MoS_2 is a lamellar compound in which each layer is composed of two planes of sulfur atoms and an intermediate plane of molybdenum atoms tightly bound by covalent bonds. These layers are held together by van der Waals bonds. The weakness of these bonds allows the easy shear at this interlayer. Sputtered MoS_2 films are adherent, with long endurance life and low friction coefficients (<0.1). They are used for precision triboelements and show extremely high endurance and low friction in high vacuum (spacecraft). The durability and adherence of MoS_2 films in air can be increased by modifying the synthesis process.

TABLE 8.6.   Materials Used as Solid Lubricants [15]

Lamellar:	MoS_2, graphite
Inorganic:	TiO_2, CaF_2, glasses
Soft metals:	Pd, Ag, Au
Carbon Compounds	
Lamellar	Graphite, graphite fluoride
Diamond-like carbon (DLC)	i-C, a-C y: H
Fats (soap) wax	Stearic acid
Polymers	PTFE (Teflon)

Graphite is probably the most widely used lamellar solid lubricant. Unlike MoS_2, graphite has a lower friction and lower wear in the presence of moisture than in vacuum. Therefore, graphite is not recommended for vacuum or high-temperature applications. But another form of carbon, amorphous hydrogenated carbon films (also called diamond-like carbon), has the reverse behavior: It works extremely well in vacuum, but its friction coefficient is increased by the presence of moisture [37].

The combination of different materials in composite coatings give rise to remarkable properties: Some of these coatings are lubricious from room temperature to 900°C in oxidizing or reducing atmospheres [34]. Promising applications in the field of high-temperature aerospace and advanced heat engines can be foreseen for them.

PTFE (e.g., Teflon) is widely used in the form of composites as self-lubricating cages for ball bearings. A thin film of polymer is transferred from the cage to the rolling/rubbing surface providing continuous lubrication for the system.

In the past decade, durability and integrity of traditional lubricating films have been improved; simultaneously to this improvement, new forms of materials (amorphous MoS_2 and DLC) and composites with greatly improved performance have been developed for the applications in extreme environments [38].

8.7.1 Coatings: Mechanical Protection

The surface of a solid can be changed to modify the friction and wear properties. Hard coatings are used to control friction and wear in such a way as to extend the life of a part. Ideally, a coated part should fail by a mechanism other than wear. If two metallic surfaces are sliding on each other, the dissipation of friction energy can lead to frictional welding at points of real metal-to-metal contact, and wear will follow by detachment of material from welded junctions. If one of the surfaces is coated with a hard coating, exhibiting high levels of covalent bonding, the probability of metal-to-metal welding is lowered, as is the subsequent wear. Cubic carbides and nitrides of transition metals are commonly used to lower this type of wear. We can estimate the hardness level necessary to obtain wear resistance. To prevent important wear of a part sliding on a hard steel piece (maximum Vicker's hardness = 900), it is sufficient to coat this part with a material having a Vicker's hardness in the range of 1500–2000. Many oxides, nitrides, carbides, silicides, and borides possess a Vicker's hardness greater than 1500 [34]. Because the wear rate in presence of the coating is remarkably low, thick hard layers are not necessary. Thin films will be sufficient for most applications as long as their adherence to the substrate is strong enough.

The adhesion strength of the films is the main problem in anti-wear protection. In some cases, this problem can be overcome by forming an interface between the substrate and the coating, with gradually changing composition over several atomic layers.

Among all the hard materials, one is outstanding: diamond [5, 40–46]. Its combination of mechanical, physical, and chemical properties is nearly unique. Its thermal conductivity is five times better than that of copper, and its friction coefficient is close to that of PTFE (Teflon). Therefore, it is very desirable to produce thin diamond coatings. At present, the limiting factor to the development of diamond coatings is the poor adhesion of diamond crystallites to many metals. Studies show that the formation of intermediate carbide at the interface is favorable to the formation of adherent diamond films.

8.7.2 Coatings: Chemical Protection

Thin coatings not only are used for mechanical protection and lubrication, but are widely applied to protect against corrosion and chemical reaction [47–49]. The steel body of a car is first covered by a zinc layer, which is further treated in a phosphate bath, to promote the adhesion of the paint. Damaging this sandwich structure, by breaking one of these layers, results in quick corrosion of the car body: rust. The improvement in the quality of this protective coating during the last 20 years is shown by the increase of the duration of the corrosion warranty proposed by car manufacturers (from 2 years less than 15 years ago, to 8 years at the present time).

Coatings can form naturally by reaction with the surrounding atmosphere: aluminum is quickly covered by an aluminum oxide and hydroxide layer in the presence of oxygen and water. This inert and protective layer can be formed artificially, as is done for window and door frames. A porous layer of alumina can be formed by reaction with oxygen. Pigments can be included in the pores for decorative purposes. After inclusion of the pigments, a suitable treatment of the layer transforms its porous structure to a continuous nonporous one (sealing). These coated frames can withstand relatively harsh conditions for years.

The same type of reaction, of aluminum with oxygen, is disastrous for aluminum mirrors used in space applications. The alumina film absorbs strongly infrared frequencies, limiting the applications of these mirrors to a relatively narrow band of frequencies. This problem can be overcome by coating the aluminum mirror—prior to any exposure of the mirror to oxygen or water—with a very thin film of boron nitride that is transparent to those interesting frequencies. In addition, this hard coating of boron nitride protects the mirror against mechanical wear [50–52].

Boron nitride, when crystallized in its cubic form, has interesting electronic applications [53]. In addition to its chemical inertness, it has high thermal conductivity. This is important in microelectronic applications where the dissipation of heat is one of the main problems. An increase in temperature can induce the diffusion and reaction between the different layers of the electronic device and reduce its performance. The boron nitride coating in this case is protective in at least three ways: mechanical protection due to its hardness, chemical protection, and thermal protection.

Coatings are also very common in food packaging. An example is tomato products that are corrosive to canning materials (such as iron) which must be protected by polymer coatings. Otherwise the food will be contaminated by metallic ions, and rupture of the can can occur.

8.8 CRACK FORMATION AND PROPAGATION. FRACTURE

Consider a small crack, as shown in Figure 8.7. The application of loads that aid in the opening, sliding, or tearing of the crack will cause the crack to grow with time. Environmental attack can also cause the crack to grow. The longer the crack, the higher the stress concentration induced by it, implying that the rate of propagation

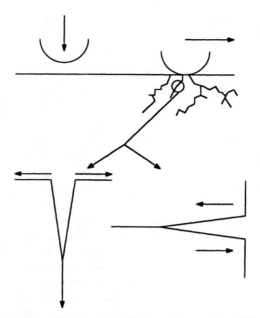

Figure 8.7. Schematic representation of crack formation under sliding and tearing loads.

may increase with time, leading to catastrophic fracture of structural materials. Most often these fractures are brittle and are accompanied by very little plastic deformation. Brittle fracture of steel usually occurs at low temperatures and is initiated by a sharp notch or crack. Cleavage fracture of metal occurs by direct separation along crystallographic planes by breaking of atomic bonds. The atomic structure at steps, as determined by low-energy electron diffraction (LEED) surface crystallography, may be related to crack formation and propagation. One step structure is shown in Figure 8.8. The atoms at the step edge relax inward, resulting in a clustering of atoms near the edge. This clustering opens atomic channels into the bulk. These

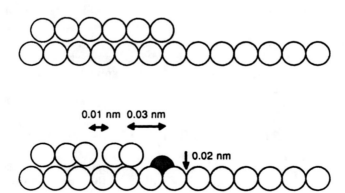

Figure 8.8. Atomic step structure at surfaces. Relaxation and clustering of atoms at the step edge open atomic channels that could be nucleation sites for crack formation.

channels may be surface sites where crack formation could occur when an external stress is applied.

We have previously described how a solid can store elastic energy under stress. Griffith [54] states that crack propagation will occur if the elastic energy released upon crack growth is sufficient to provide all the energy required for the growth of the crack. The elastic energy release varies linearly with the length of the crack and is proportional to the square of the applied stress. Thus, lowering the applied stress can usually stop the crack growth.

8.9 SUMMARY AND CONCEPTS

- The mechanical forces needed to break surface chemical bonds are in the range of $10-10^2$ millidynes per bond.
- Elastic deformation and bond breaking at surfaces under load can be measured by the atomic-force microscope. Elastic deformation can increase the force required to break bonds by two orders of magnitude for soft materials such as gold.
- Macroscopic hardness and adhesive strength are important materials properties that can be controlled by altering the surface bonds.
- Friction, slide, and lubrication are the properties of surfaces in relative motion (tribological properties).
- The formation of debris by plowing, heating, and storage of elastic energy are all observed to occur at moving interfaces.
- A monolayer of adsorbates sandwiched between interfaces of solids can markedly modify adhesive and tribological properties. These monolayers are being investigated increasingly on the molecular level.

8.10 PROBLEMS

8.1 Calculate the local pressure under a diamond tip pressed down on aluminum using a 1000-dyne load if the contact area is (a) 10^{-2} cm^2 and (b) 10^{-4} cm^2. How do these numbers compare with typical values for yield points [55]?

*8.2 A layer of lubricating oil reduces the friction coefficient of sliding steel surfaces from $\mu = 1$ to $\mu \approx 0.001$ when sliding occurs at low speeds (hydrodynamic regime). At high sliding speeds, boundary lubrication occurs and the friction coefficient is again altered. Discuss these regimes and give examples of the types and characteristics of lubricants that work well in this circumstance [21].

*8.3 The surface force apparatus (SFA) operates in a manner similar to that of the atomic force microscope (AFM). However, it provides an atomically smooth interface over a large (≈ 100 μm^2) area that permits the spectroscopic scrutiny of molecules adsorbed at the buried interface. Discuss one study that was recently reported in the literature [56].

*8.4 Additives in lubricants that extend the temperature range of their utility are antioxidants and control the acidity that would build up during use are important ingredients in most motor oils. Review three additives that are utilized [57–59] and explain the reasons for their beneficial properties.

*8.5 Discuss the results of a recent paper that measured friction coefficients using the atomic force microscope (AFM) [60–62].

**8.6 Diamond thin films can be produced from microwave-discharge-heated methane or from carbon that is sputtered by ion bombardment [63–65]. Discuss the methods of preparation and the possible reasons for the formation of metastable phases by these methods.

REFERENCES

[1] D. Dowson. *History of Tribology*. Longman Group, London, 1979.

[2] K.J. Anderson. A History of Lubricants. *MRS Bull.* **16**(10):69 (1991).

[3] E.D. Brown, R.S. Owens, and E.R. Booser. Friction of Dry Surfaces. In: F.F. Ling, E.E. Klaus, and R.S. Fein, editors, *Boundary Lubrication: An Appraisal of World Literature*. American Society of Mechanical Engineers, New York, 1969.

[4] B. Bhushan. *Tribology and Mechanics of Magnetic Storage Devices*. Springer-Verlag, New York, 1990.

[5] S. Ramalingam. Surface Modification for Tribology with PVD Processes: Problems and Prospects. In: L.E. Pope, L.L. Fehrenbacher, and W.O. Winer, editors, *New Materials Approaches to Tribology: Theory and Application. Materials Research Society Symposium Proceedings*, Volume 40, Materials Research Society, Pittsburgh, 1989, p. 465.

[6] J.N. Israelachvili. Adhesion Forces between Surfaces in Liquids and Condensable Vapours. *Surf. Sci. Rep.* **14**:109 (1992).

[7] J.N. Israelachvili and P.M. McGuiggan. Adhesion and Short-Range Forces between Surfaces. Part I: New Apparatus for Surface Force Measurements. *J. Mater. Res.* **5**:2223 (1990).

[8] J.L. Parker, H.K. Christenson, and B.W. Ninham. Device for Measuring the Force and Separation between Two Surfaces Down to Molecular Separations. *Rev. Sci. Instrum.* **60**:3135 (1989).

[9] J.N. Israelachvili. *Intermolecular and Surface Forces*. Academic Press, London, 1985.

[10] P. Sutor. Solid Lubricants: Overview and Recent Developments. *MRS Bull.* **May**:24 (1991).

[11] J.T. Bottomley. Melting and Regelation of Ice. *Nature* **5**:185 (1872).

[12] M. Faraday. On Regelation, and on the Conservation of Force. *London Edinburgh Dublin Philos. Mag. J. Sci. (Ser. 4)* **17**:162 (1859).

[13] W. Thomson. Remarks on the Interior Melting of Ice. *Proc. R. Soc. London* **A9**:141 (1858).

[14] J.G. Davy and G.A. Somorjai. Studies of the Vaporization Mechanism of Ice Single Crystals. *J. Chem. Phys.* **55**:3624 (1971).

[15] R.R. Gilpin. Wire Regelation at Low Temperatures. *J. Colloid Interface Sci.* **77**(2):435–448 (1980).

[16] I.M. Hutchings. *Tribology: Friction and Wear of Engineering Materials*. CRC Press, Boca Raton, FL: 1992.

[17] T.F.J. Quinn. *Physical Analysis for Tribology*. Cambridge University Press, Cambridge, 1991.

[18] D. Landheer and A.W.J. de Gee. Adhesion, Friction, and Wear. *MRS Bull.* **16**(10):36 (1991).

[19] H.S. Kong and M.F. Ashby. Friction-Heating Maps and Their Applications. *MRS Bull.* **16**(10):41 (1991).

[20] A. Erdemir, F.A. Nichols, G.R. Fenske, and J.-H. Hsieh. Sliding Friction and Wear of Ceramics with and without Soft Metallic Films. *MRS Bull.* **16**(10):49 (1991).

[21] B.J. Hamrock. *Lubrication of Machine Elements*. Technical report, NASA, 1984. NASA-RP-1126.

[22] F.P. Bowden and D. Tabor. *Friction—An Introduction to Tribology*. Heinemann, London, 1973.

[23] F.P. Bowden and D. Tabor. *Friction and Lubrication*. Methuen, London, 1967.

[24] F.P. Bowden and D. Tabor. *The Friction and Lubrication of Solids*. Clarendon Press, New York, 1950.

[25] E. Rabinowicz. *Friction and Wear of Materials*. John Wiley & Sons, New York, 1965.

[26] V.I. Vill'. *Friction Welding of Metals*. American Welding Society, New York, 1962, pp. 42–51.

[27] J. Ruge, K. Thomas, C. Eckel, and S. Sundaresan. Joining of Copper to Titanium by Friction Welding. *Weld. J.* **August**:28 (1986).

[28] E.D. Nicholas and W.M. Thomas. Metal Deposition by Friction Welding. *Weld. J.* **August**:17 (1986).

[29] E.D. Nicholas. Radial Friction Welding. *Weld. J.* **July**:17 (1983).

[30] A.L. Phillips, S.T. Walter, L. Griffing, C. Weisman, editors. *Welding Handbook*, 6th edition. American Welding Society, New York, 1968.

[31] T. Lyman, editor. *Metals Handbook*, 8th edition. *Welding and Brazing*, Volume 6. American Society for Metals, Metals Park, OH , 1971.

[32] R.C. Gunter. *Lubrication*. Chilton Book Company, Philadelphia, 1971.

[33] S.M. Hsu. Boundary Lubrication of Materials. *MRS Bull.* **16**(10):54 (1991).

[34] I.L. Singer. Solid Lubricating Films for Extreme Environments. In: L.E. Pope, L.L. Fehrenbacher, and W.O. Winer, editors, *New Materials Approaches to Tribology: Theory and Application. Materials Research Society Symposium Proceedings*, Volume 140, Pittsburgh, 1989, p. 215.

[35] M.R. Hilton and P.D. Fleischauer. Structural Studies of Sputter-Deposited MoS_2 Solid Lubricant Films. In: L.E. Pope, L.L. Fehrenbacher, and W.O. Winer, editors, *New Materials Approaches to Tribology: Theory and Application. Materials Research Society Symposium Proceedings*, Volume 140, Materials Research Society, Pittsburgh, 1989, p. 227.

[36] N.J. Mikkelsen and G. Sorensen. Modifications of Molybdenum-Disulphide Films by Ion Bombardment Techniques. In: L.E. Pope, L.L. Fehrenbacher, and W.O. Winer, editors, *New Materials Approaches to Tribology: Theory and Application. Materials Research Society Symposium Proceedings*, Volume 140, Materials Research Society, Pittsburgh, 1989, p. 271.

[37] A. Grill, V. Patel, and B.F. Meyerson. Abstract 10.1. In: *MRS Spring Meeting '92*, San Francisco, 1992, p. 288.

[38] D.H. Buckley. *Surface Effects in Adhesion, Friction, Wear, and Lubrication*. Elsevier, New York, 1981.

[39] W.E. Campbell. Solid Lubricants. In: F.F. Ling, E.E. Klaus, and R.S. Fein, editors, *Boundary Lubrication: An Appraisal of World Literature*. American Society of Mechanical Engineers, New York, 1969.

[40] M.S. Wong, R. Meilunas, T.P. Ong, and R.P.H. Chang. Thin Diamond Films for Tribological Applications. In: L.E. Pope, L.L. Fehrenbacher, and W.O. Winer, editors, *New Materials Approaches to Tribology: Theory and Application. Materials Research Society Symposium Proceedings*, Volume 140, Materials Research Society, Pittsburgh, 1989, p. 483.

[41] C.J. McHargue. Mechanical Properties of Diamond and Diamond-Like Films. In: Y. Tzeng, M. Yoshikawa, M. Murakawa, and A. Feldman, editors, *Applications of Diamond Films and Related Materials. Materials Science Monographs*, Volume 73. Elsevier, New York, 1991, p. 113.

[42] A. Grill, V. Patel, and B.F. Meyerson. Applications of Diamond-Like Carbon in Computer Technology. In: Y. Tzeng, M. Yoshikawa, M. Murakawa, and A. Feldman, editors, *Applications of Diamond Films and Related Materials. Materials Science Monographs*, Volume 73. Elsevier, New York, 1991, p. 683.

[43] S. Miyake and R. Kaneko. Micro-Tribological Properties and Applications of SuperHard and Lubricating Coatings. In: Y. Tzeng, M. Yoshikawa, M. Murakawa, and A. Feldman, editors, *Applications of Diamond Films and Related Materials. Materials Science Monographs*, Volume 73. Elsevier, New York, 1991, p. 691.

[44] K. Miyoshi. Tribological Studies of Amorphous Hydrogenated Carbon Films in a Vacuum, Spacelike Environment. In: Y. Tzeng, M. Yoshikawa, M. Murakawa, and A. Feldman, editors, *Applications of Diamond Films and Related Materials. Materials Science Monographs*, Volume 73. Elsevier, New York, 1991, p. 699.

[45] B. Lux and R. Haubner. Low-Pressure Diamond Coatings for Tools and Wear Parts. In: R. Messier, J.T. Glass, J.E. Butler, and R. Roy, editors, *New Diamond Science and Technology*. Materials Research Society, Pittsburgh, 1991, p. 805.

[46] T. Ivanova, D. Latev, D. Parvanova, and D. Dimitrov. Diamondlike Thin Films as Protective Layers. In: Y. Tzeng, M. Yoshikawa, M. Murakawa, and A. Feldman, editors, *Applications of Diamond Films and Related Materials. Materials Science Monographs*, Volume 73. Elsevier, New York, 1991, p. 731.

[47] G. Wranglen. *An Introduction to Corrosion and Protection of Metals*. Institut for Metallskydd, Stockholm, 1972.

[48] K.R. Trethewey and J. Chamberlain. *Corrosion for Students of Science and Engineering*. Longman, New York, 1988.

[49] P.A. Schweitzer, editor. *Corrosion and Corrosion Protection Handbook*, 2nd edition. Marcel Dekker, New York, 1988.

[50] K. Miyoshi. *Studies of Mechano-Chemical Interactions in the Tribological Behavior of Materials*. Technical report, NASA, 1990. NASA-TM-102545.

[51] S. Watanabe, S. Miyake, and M. Murakawa. Tribological Properties of Cubic, Amorphous and Hexagonal Boron Nitride Films. *Surf. Coatings Technol.* **49**:406 (1991).

[52] S. Miyake, S. Watanabe, M. Murakawa, R. Kaneko, and T. Miyamoto. A Tribological Study of Cubic Boron Nitride Film. In: Y. Tzeng, M. Yoshikawa, M. Murakawa, and A. Feldman, editors, *Applications of Diamond Films and Related Materials. Materials Science Monographs*, Volume 73. Elsevier, New York, 1991, p. 669.

[53] E. Yamaguchi. Synthesis and Properties of Boron Nitride. *Mater. Sci. Forum* **54/55**:329 (1990).

[54] A.A. Griffith. The Phenomena of Rupture and Flow in Solids. *Philos. Trans. R. Soc. London* **A221**:163 (1920).

[55] R.W. Fitzgerald. *Mechanics of Materials*, 2nd edition. Addison–Wesley, Reading, MA, 1982.

[56] P.M. McGuiggan and J.N. Israelachvili. Adhesion and Short-Range Forces between Surfaces. Part II: Effects of Surface Lattice Mismatch. *J. Mater. Res.* **5**:2232 (1990).

[57] R.M. Mortier and S.T. Orszulik, editors. *Chemistry and Technology of Lubricants*. VCH Publishers, New York, 1992.

[58] W.R. Loomis, editor. *New Directions in Lubrication, Materials, Wear, and Surface Interactions*. Noyes Publications, Park Ridge, NJ, 1985.

[59] W.S. Robertson, editor. *Lubrication in Practice. Mechanical Engineering*, Volume 27. Marcel Dekker, New York, 1984.

[60] C.M. Mate, G.M. McClelland, R. Erlandsson, and S. Chiang. Atomic-Scale Friction of a Tungsten Tip on a Graphite Surface. *Phys. Rev. Lett.* **59**:1942 (1987).

[61] G.M. McClelland and S.R. Cohen. Tribology at the Atomic Scale. In: R. Vanselow and R. Howe, editors, *Chemistry and Physics of Solid Surfaces VIII. Springer Series in Surface Sciences*, Volume 22. Springer-Verlag, Berlin, 1990, p. 419.

[62] E. Meyer, H. Heinzelmann, P. Grütter, T. Jung, H.R. Hidber, R. Rudin, and H.J. Güntherodt. Atomic Force Microscopy for the Study of Tribology and Adhesion. *Thin Solid Films* **181**:527 (1989).

[63] J.E. Field, editor. *The Properties of Natural and Synthetic Diamond*. Academic Press, London, 1992.

[64] K.E. Spear. Diamond—Ceramic Coating of the Future. *J. Am. Ceram. Soc.* **72**:171 (1989).

[65] P.K. Bachmann and R. Messier. Emerging Technology of Diamond Thin Films. *Chem. Eng. News*, **May**:24 (1989).

ANSWERS TO THE PROBLEMS

CHAPTER 1

1.1 We have, using the approximation $\sigma = \rho^{2/3}$,

	$M(\mathrm{g/mole})$	$\rho(\mathrm{g/cm^3})$	$\sigma(\mathrm{cm^{-2}})$
Hg	20	13.6	1.2×10^{14}
Cu	63.5	8.9	1.9×10^{15}
C_6H_6	78	0.879	3.6×10^{14}

1.2 (a) Using the equation

$$F(\mathrm{atoms/cm^2 \cdot sec}) = 3.51 \times 10^{22} \frac{P(\mathrm{torr})}{\sqrt{M(\mathrm{g/mole})T}}$$

we obtain $F \approx 2.9 \times 10^{23}$ molecules/$\mathrm{cm^2}$/sec, assuming $M_{\mathrm{air}} \approx 29$ g/mole.
(b) Assuming that "clean" means 0.01 monolayer, and that $\sigma_{\mathrm{Cu}} = 1.9 \times 10^{15}$ atoms/$\mathrm{cm^2}$, we need a pressure lower than 1.4×10^{-11} torr.

1.3 Assuming $\sigma_{\mathrm{Cu}} = 1.9 \times 10^{15}$ atoms/$\mathrm{cm^2}$, we obtain

	Time		
Face	1 sec	1 hr	8 hr
001	5.3×10^{-6}	1.5×10^{-9}	1.8×10^{-10}
011	2.7×10^{-7}	7.4×10^{-11}	9.2×10^{-12}
111	5.3×10^{-5}	1.5×10^{-8}	1.8×10^{-9}

where the pressures are in torr.

1.4   Assuming a surface area of 16.2 $Å^2$ per molecule of nitrogen, we estimate that for a one monolayer thickness of N_2, 2.5×10^{22} molecules of nitrogen will be adsorbed on the 400 m^2 of zeolite surface area. This is ≈ 1.1 g of N_2.

1.5   Using the equation

$$\tau = \tau_0 \exp\left(\frac{\Delta H_{ads}}{RT}\right) = 1 \times 10^{-12} \times \exp\left(\frac{15000}{8.315 \times T}\right)$$

we estimate $\tau \approx 0.41$ nsec at 300 K and $\tau \approx 0.015$ sec at 77 K.

CHAPTER 2

2.1   STM: Electrons tunnel through the potential energy barrier between an atom of a sharp tip—brought within atomic dimensions of a surface—and that surface. If a small voltage is applied between the tip and the sample, a net current flows across the gap. This tunneling current depends exponentially on the sample-to-tip distance. An atomic scale image of the surface can be obtained, for example, by plotting the tunneling current I as a function of the tip position (x, y) kept at a constant distance from the surface.

FIM: A very high positive potential is applied to a tip of radius $\approx 10^{-4}$ cm in the presence of a gas (usually helium, but hydrogen and neon are also used). The gas atoms are ionized in the high electric field at a critical distance (4–8 Å) from the surface. The positively charged ions are repelled by the positive tip and accelerated onto a fluorescent screen, where a greatly magnified image of the tip is displayed.

2.2   LEED: A monoenergetic beam of electrons (10–500 eV) is directed at a single-crystal surface, and the pattern caused by interference of electrons elastically back-scattered from the surface is observed as bright spots on a fluorescent screen. The pattern has the same symmetry as the surface, but the structural parameters such as the interatomic distances and angles cannot be determined from one pattern. The intensity I of the diffraction spots changes with the energy V of the incoming electrons. Theoretical I–V curves can be calculated for proposed surface structures and compared with the experimental ones. The structure corresponding to the best fit is considered to be the "solved surface structure."

2.3   Letting the open circles represent the spots from the underlying bulk lattice, and letting the smaller, solid circles represent the surface lattice spots, we

have

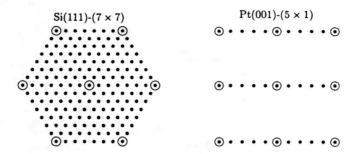

2.4

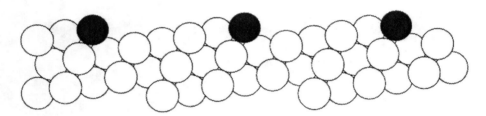

Edge atoms are shown in black.

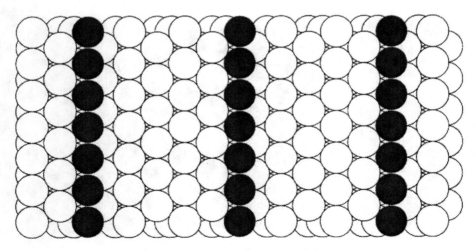

Edge atoms are shown in black.

fcc(332)

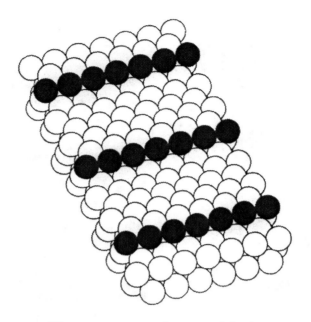

Edge atoms are shown in black.

fcc(430)

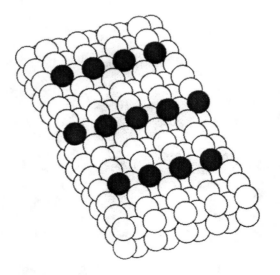

Edge atoms are shown in black.

fcc(430)

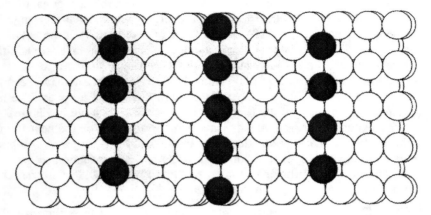

Edge atoms are shown in black.

fcc(430)

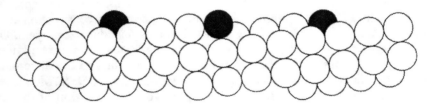

Edge atoms are shown in black.

2.5 From the figure for a fcc(311) crystal face, the first layer relaxation predicted by the theory is $\approx 12\%$, which agrees rather well with the experimentally observed 12–15% relaxation. For the (997) face, which consists of ≈ 8-atom-wide terraces, the surface roughness defined by Jona and Marcus would be very high. But when one looks at the surface, it does not appear "rougher" than a (111) surface. Thus the use of 1/packing density as surface roughness would be inadequate, as would be the extrapolation.

2.6 Ir(100) undergoes a (1×5) reconstruction. Pt(100) shows a variety of LEED patterns such as $\begin{pmatrix} 14 & 1 \\ -1 & 5 \end{pmatrix}$, $\begin{pmatrix} 14 & 1 \\ 0 & 5 \end{pmatrix}$, $\begin{pmatrix} 27 & 2 \\ -3 & 6 \end{pmatrix}$, and $\begin{pmatrix} 12 & 1 \\ -3 & 5 \end{pmatrix}$. Some of these patterns can be observed on stepped Pt(100) surfaces as well. Au(100) exhibits a $c(26 \times 68)$ reconstruction, and on a stepped Au(100) surface a $\begin{pmatrix} 14 & 1 \\ -1 & 5 \end{pmatrix}$ reconstruction has been observed. It was found that a hexagonal arrangement of the top atomic layer can explain the observed LEED patterns. The most likely reason for this reconstruction is the energetically favorable bond length contraction parallel and perpendicular to the surface. The structure is the result of a delicate balance of forces. The hexagonal (111) structure has the lowest surface energy of the possible fcc crystal faces, so there is a considerable driving force for the surface energy of the possible fcc crystal faces, so there is a considerable driving force for the surface to assume this geometry. On the other hand, due to the mismatch between the hexagonal top layer and the fourfold second layer, the strain energy opposes such rearrangement of the top layer. On the (100) crystal faces of Ir, Pt, and Au the balance of the two opposing forces produces a buckled hexagonal top layer. In addition, the rehybridization of the bonding orbitals may also play a role in the rearrangement of the top layer. Dynamical LEED calculations have been carried out for the Ir(100)–(1 × 5) and the Pt(100)–($_{-1}^{14}$ $_5^1$) surfaces. The most probable arrangement of the Ir atoms on the surface is shown in Figure 2.14. The hexagonal layer has a registry involving bridge sites on the next fourfold unit cell layer, and it is buckled. For Pt(100)–($_{-1}^{14}$ $_5^1$), a $\frac{1}{2}$ and two-bridge registry hexagonal layer is favored, showing a 6.3% reduction in the backbond length. The top layer is rotated by approximately 0.7° with respect to the perfect alignment with the substrate.

2.7 Si(100)–(2 × 1): The surface Si atoms on the (100) surface dimerize along the [110] direction. The pairing of surface atoms eliminates 50% of the energetically unfavorable unpaired electrons (dangling bonds). The dimerization takes place despite an increase of strain energy associated with the distortion of the tetrahedral Si–Si bonds near the surface. Recent STM results and total energy calculations show that the dimer is likely symmetric.

Si(111): The (7 × 7) reconstruction is the most complex reconstruction known to date, and it involves not only the top layer, but four surface layers. On a (7 × 7) surface, which would have 49 dangling bonds if it were bulk terminated, the reconstructed surface has only 19, according to the dimer-atom-stacking fault (DAS) model, in agreement with LEED, TED, STM, and RBS results. The energy gain is partially "spent" on the distortion of the tetrahedral bond angles.

2.8 At less than full coverage, the W(100)–c(2 × 2) structure of clean tungsten remains unchanged. On the unreconstructed (1 × 1) surface, hydrogen adsorption generates the c(2 × 2) structure when the sample is cooled. At full coverage the c(2 × 2) reconstruction is removed and there are two hydrogen atoms per unit cell present on the surface, showing a (1 × 1) pattern.

2.9 The carbon atoms occupy hollow sites having local fourfold geometry. The top nickel layer is distorted as shown in Figure 2.17. It appears that the carbon atoms are trying to maximize their bonding with the nickel atoms, and they are forcing their way into the fourfold holes. The position of the C atoms is only slightly higher (0.1 Å) than that of the top nickel layer. The Ni–Ni interlayer spacing also increases by 0.35 Å. The Ni–Ni bond strength is about 17 kcal/mole. This is overshadowed by the Ni–C bond energy of about 52 kcal/mole.

2.10 The preferred adsorption site is closely related to the work function of the substrate. On Pd(111), CO occupies threefold hollow sites up to one-third monolayer coverage. Upon increasing the coverage, CO switches its adsorption sites from threefold to twofold, and if the temperature is now decreased, it occupies top sites also. The transition is smooth as is shown in the figure. CO molecules on adsorption sites that are transition in nature between three-

and twofold are not perfectly perpendicular to the (111) plane but are tilted in the direction of the threefold site.

2.11 CO produces two ordered structures on the (111) surfaces of Pt, Rh, and Pd: a $(\sqrt{3} \times \sqrt{3})R30°$ at one-third monolayer coverage and a c(4 × 2) at one-half monolayer coverage, but benzene is disordered on these surfaces. Coadsorption of these two molecules produces a $(\sqrt{3} \times \sqrt{3})$ structure on Pd(111) and on Rh(111) and a c($2\sqrt{3} \times 4$) on Rh(111) and Pt(111). Based on work function measurements and theoretical studies, it is believed that benzene is an electron donor to Rh(111) and Pt(111), while there is a charge transfer from the metals to the CO molecule. On palladium the benzene skeleton is indistinguishable from gas-phase benzene molecules. On Rh and Pt there is a strong metal–benzene interaction. A significant in-plane distortion of the molecule is observed: The C_6 ring expands by 0.3 and 0.2 Å and assumes a distorted cyclohexatriene structure. HREELS results imply that benzene adsorbs on bridge sites of Pd(111), but when coadsorbed with CO it switches position to the fcc-hollow site. Similarly, on Rh(111), a bridge → hollow site shift is probable, but here the hcp-hollow site is the more likely.

2.12 Just below the xenon's melting point the overlayers form an incommensurate solid. The axes of this structure are aligned 30° relative to those of the graphite surface. On cooling, the layer undergoes a first-order transition and forms a rotated incommensurate phase. Upon further cooling, another first-order transition occurs and a commensurate $(\sqrt{3} \times \sqrt{3})R30°$ structure is formed. When the commensurate phase is heated, first the aligned hexagonal superlight domain wall phase appears, but, at higher temperatures, instead of going through the rotated incommensurate phase, the surface assumes an aligned incommensurate structure. Thus, in the heating cycle, there is only one phase transition. In the ordered phase, regardless of the substrate orientation, the Xe atoms maintain a close-packed hexagonal structure. It appears that the attractive adsorbate–adsorbate interactions control the surface morphology. This is in contrast to molecular adsorbate/single-crystal systems, where the adsorbate–substrate interactions determine the surface structure.

2.13 The bonding of pyridine to metal surfaces can involve either the π ring of electrons, in which case the molecule prefers to lie flat on a metal surface, or it can involve the lone pair of electrons on the nitrogen site. Such a transition as a function of coverage has been observed on Ag(111), Ni(100), and Pt(110). Temperature also changes the bonding, as is seen on Pd(111). Tilted pyridine has been observed on Ir(111) and Cu(110). A third form of bonding has been seen on Ni(100) and Pt(111). In this case, one of the α-hydrogens is lost, resulting in an α-pyridyl species.

CHAPTER 3

3.1 From Eq. 3.9:

$$dG_{T,P} = \gamma \, d\mathcal{Q} = (1670 \text{ ergs}/cm^2) \times (20 \text{ cm}^2)$$
$$= 3.34 \times 10^4 \text{ ergs} = 3.34 \text{ mJ}$$

3.2   We have, from Eq. 3.35:

$$\xi = \frac{\text{bulk heat capacity}}{\text{surface heat capacity}}$$

$$\approx 3.6 \times 10^8 \, L \left(\frac{\Theta^2_{\text{surface}}}{\Theta^3_{\text{bulk}}} \right) \left(\frac{\rho}{M} \right)^{1/3} T$$

We have $\rho_{\text{Ni}} = 8.91$ g/cm^3, $M_{\text{Ni}} = 56.7$ g/mole, $\Theta_{\text{bulk}} = 390$ K, $\Theta_{\text{surface}} \approx 220$ K. Therefore, $\xi \approx 48$, or the surface contributes $\approx 2.1\%$ of the total heat capacity.

3.3   According to Equation 3.51:

$$\frac{x_2^s}{x_1^s} = \frac{x_2^b}{x_1^b} \exp \left\{ \frac{0.16(\Delta H_{\text{subl}_1} - \Delta H_{\text{subl}_2})}{RT} \right\}$$

Subscripts 1 and 2 refer to Ag and Au, respectively. We have:

$$x_1^b = 0.95, \qquad \Delta H_{\text{subl}_1} = 77 \text{ kcal/mole}$$

$$x_2^b = 0.05, \qquad \Delta H_{\text{subl}_2} = 64 \text{ kcal/mole}$$

$$\frac{x_1^s}{x_1^s} = \frac{0.05}{0.95} \exp \left\{ \frac{0.16(77 - 64)}{0.001987 \times T} \right\}$$

Given that $x_1 + x_2 = 1$, we have:
At 300 K:

$$\frac{x_2^s}{x_1^s} = 1.72, \qquad x_1^s = 0.37, \qquad x_2^s = 0.63$$

At 700 K:

$$\frac{x_2^s}{x_1^s} = 0.235, \qquad x_1^s = 0.81, \qquad x_2^s = 0.19$$

3.4   We have, from Eq. 3.70:

$$\ln \left(\frac{P}{P_0} \right) = \frac{2\gamma \overline{V}}{RTr}$$

Using $\gamma = 73.4$ ergs/cm^2, $R = 8.315 \times 10^7$ ergs/K $\cdot$ mole, $\overline{V} = 18$ cm^3/mole, and $P_0 = 17.54$ torr, we obtain

$$\ln \left(\frac{P}{17.54} \right) = \frac{2 \times 73.4 \times 18}{8.315 \times 10^7 \times 298 \times r}$$

$$= 1.0664 \times 10^{-7}/r \text{ (cm)}$$

Thus:

r (cm):	5×10^{-8}	5×10^{-7}	5×10^{-6}
P (torr):	148	21.7	17.9

3.5   We have, from Eq. 3.81,

$$r_{\text{critical}} = \frac{2\gamma \overline{V}}{RT \ln (P/P_{\text{eq}})}$$

and, from Eq. 3.82,

$$\Delta G_{\text{max}} = \tfrac{4}{3} \pi \gamma r_{\text{critical}}^2$$

Therefore,

γ (ergs/cm^2)	V (cm^3/mole)	P_{eq} (torr)	r_{critical} (Å)	ΔG_{max} (ergs)
22.8	58.3	50	48.9	2.28×10^{-11}

3.6   We have, from Eq. 3.72,

$$\cos \psi = \frac{\gamma_{\text{Hg-air}} - \gamma_{\text{Hg-H}_2\text{O}}}{\gamma_{\text{H}_2\text{O-air}}}$$

or

$$\cos 33.3° = \frac{0.436 - \gamma_{\text{Hg-H}_2\text{O}}}{0.073}$$

Therefore $\gamma_{\text{Hg-H}_2\text{O}} = 0.375 \text{ J/m}^2$.

3.7   We have, from Eq. 3.62,

$$\Delta P = \Delta \rho g h = \frac{2\gamma}{r}$$

We have $\gamma = 0.0728 \text{ J/m}^2$, $g = 9.81 \text{ m/sec}^2$, $r = 5 \times 10^{-5}$ cm, $\Delta \rho \approx \rho_{\text{liquid}} = 1 \text{ g/cc} = 10^3 \text{ kg/m}^3$. Therefore, we obtain $\Delta P = 2.9 \times 10^5$ Pa = 2.87 atm and $H = 29.7$ m.

3.8 Analyzing the data, we obtain:

P/P_o	$\dfrac{P}{\sigma(P_o-P)}$
0.055	0.000444
0.061	0.000484
0.077	0.000596
0.094	0.000697
0.120	0.000888
0.158	0.001144
0.177	0.001270
0.209	0.001494
0.240	0.001712
0.270	0.001923
0.300	0.002143
0.330	0.002371
0.352	0.002554

Slope $= 0.00703$ cm^3/g, intercept $= 4.26 \times 10^{-5}$ cm^3/g. Therefore, $c = 166$ and $\sigma_0 = 141$ cm^3/g. Furthermore,

$$\sigma_0 = (141 \text{ cm}^3/\text{g}) \times \left(\frac{1}{22414 \text{ cm}^3/\text{mole}}\right)$$
$$\times (6.022 \times 10^{23} \text{ molecules/mole})$$
$$= 3.8 \times 10^{21} \text{ molecules}$$

Assuming 16.2 Å^2 per molecule, we obtain a surface area of the silica of 614 m^2/g.

3.9 (a) From the article cited in the problem, we have, for platinum,

$$\log_{10} P_0(\text{torr}) = 10.362 - \frac{29,100}{T}$$

$$T = 200°C = 473 \text{ K} \rightarrow P_0 = 6.92 \times 10^{-52} \text{ torr}$$

We have, from Eq. 3.70,

$$\ln\left(\frac{P}{P_0}\right) = \frac{2\gamma \overline{V}_{\text{in}}}{RTr}$$

$$\gamma_{\text{Pt}} = 2340 \text{ erg/cm}^2 \ 1132°C$$

$$\overline{V} = \frac{\text{atomic weight}}{\text{density}} = \frac{195.09}{21.4} = 9.116 \text{ cm}^3$$

$$R = 8.314 \times 10^7 \text{ ergs/K/mole}$$

$$r = 10 \text{ Å} = 10^{-7} \text{ cm}$$

Therefore,

$$\ln P/P_0 = \frac{2 \times 2340 \times 9.116}{8.314 \times 10^7 \times 473 \times 10^{-7}} = 10.85$$

$$P = P_0 e^{10.85}$$

$$= 3.56 \times 10^{-47} \text{ torr}$$

$$= \frac{3.56 \times 10^{-47}}{760} \times 1.013 \times 10^6 \text{ dyne/cm}^2$$

$$= 4.74 \times 10^{-44} \text{ dyne/cm}^2$$

From the same article, the rate of evaporation R in g/cm^2/sec is given by

$$R \approx P\sqrt{\frac{M}{2\pi RT}}$$

where M is the atomic weight of Pt. Therefore, ΔM, the total amount of material lost is

$$\Delta M = (\text{surface area}) \times (\text{evaporation rate}) \times (\text{time})$$

$$= (4\pi \times (10^{-7})^2) \times \left(4.74 \times 10^{-44}\sqrt{\frac{195.09}{2\pi \times 8.315 \times 10^7 \times 473}}\right)$$

$$\times (60 \times 60 \times 24 \times 365 \times 3)$$

$$= 5 \times 10^{-54} \text{ g}$$

$$< 1 \text{ atom}$$

Repeating the calculation at 800°C = 1073 K:

$$T = 800°C = 1073 \text{ K} \rightarrow P_0 = 1.744 \times 10^{-17} \text{ torr}$$

$$\ln P/P_0 = \frac{2 \times 2340 \times 9.116}{8.314 \times 10^7 \times 1073 \times 10^{-7}} = 4.782$$

$$P = P_0 e^{4.782}$$

$$= 2.08 \times 10^{-15} \text{ torr}$$

$$= \frac{2.08 \times 10^{-15}}{760} \times 1.013 \times 10^6 \text{ dyne/cm}^2$$

$$= 2.77 \times 10^{-12} \text{ dyne/cm}^2$$

and

$$\Delta M = (\text{surface area}) \times (\text{evaporation rate}) \times (\text{time})$$

$$= (4\,\pi \times (10^{-7})^2) \times \left(2.77 \times 10^{-12} \right.$$

$$\left. \times \sqrt{\frac{195.09}{2\pi \times 8.315 \times 10^7 \times 1073}} \right)$$

$$\times (60 \times 60 \times 24 \times 365 \times 3)$$

$$= 6 \times 10^{-22}\ \text{g}$$

$$\approx 0.67\%$$

given that the particle's original mass was $\approx 9 \times 10^{-20}$ g. Therefore, these particles may be used continuously for 3 years at both temperatures.

3.10 Most theories of surface segregation in alloys deal with one or more of the following points:

- Surface energy
- Mixing enthalpy
- Mixing entropy
- Atomic size misfit (strain)

According to Eq. 3.57,

$$E_s = 24\pi\, \frac{K_{sm} G_{sm}}{3K_{sm} r_2 + 4G_{sm} r_1}\, r_1 r_2\, (r_2 - r_1)^2$$

Note that this expression is always positive, regardless of whether the solute atoms are smaller or larger than the solvent atoms. Thus, this equation predicts that surface segregation will always occur. Some of the references given in the problem point out, however, that solvent lattices are more likely to "tolerate" a "small" impurity that a "large" one. More precisely, large impurities are expelled over a wider range of conditions than are small ones. This error alone shows that Eq. 3.57 oversimplifies surface segregation.

3.11 The specific evaporation resistance r for the system described in the problem is about 200 sec/m:

$$r \approx \delta_w \Delta \frac{1}{\nu}$$

where δ_w is the equilibrium density of water vapor and $\Delta(1/\nu)$ is the difference between the rate of loss of water with and without the n-octadecanol monolayer. The equilibrium density of water vapor at 25°C is 0.023 kg/m^3. Thus $\Delta(1/\nu) = 8.7 \times 10^3$ (sec · m^2)/kg. During a 24-hr period, the difference in loss of water over a 1-km^2 area is

$$\frac{24 \times 3600 \, \frac{sec}{hour} \times 10^6 m^2}{8.7 \times 10^3} \approx 10^7 \, kg$$

3.12 The trough is a shallow rectangular tray filled with the liquid substrate. The film is prepared by carefully pipetting a dilute solution of the film-forming material onto the substrate. After evaporation of the solvent the sweep is moved towards the float and the force necessary to maintain the position of the float is measured by a torsion balance. Three types of behavior can be described:

- *Gaseous films*: In films with very large surface area/molecule ($> 10^2$ Å^2/molecule), the film behaves like an ideal gas. The films can be expanded or contracted without phase transitions. The gaseous film can be regarded as a dilute two-dimensional solution of the film-forming material and the substrate.

- *Liquid films*: There exists a certain degree of cooperative interaction between the film-forming molecules. Two types have been observed: liquid expanded and liquid condensed films. The first type can be characterized by high compressibility, the absence of islands, and they show a first-order "liquid–gas" phase transition. Condensed films are formed by compressing expanded films.

- *Solid films*: Closest packing of the film-forming material is realized if the surface pressure is further increased.

3.13 High-surface-area aluminosilicates are crystallized from basic aqueous silica-alumina gels, usually at high temperatures. The structure of the material depends on several parameters: silica/alumina ratio and concentration; type and concentration of the cation (Na$^+$, K$^+$, etc.); pH; temperature; and pressure. After crystallization the alkali-metal cations are exchanged with NH$_4^+$ ions, and finally the ammonium salt is decomposed to form highly acidic materials. The surface area of such crystalline aluminosilicates (zeolites) can be as high as 4–600 m^2/g. The primary building blocks of zeolites are SiO$_4$ and AlO$_4$ tetrahedra. These are interconnected to form secondary building units. The difference between the various zeolite structures is the sequence in which these primary and secondary structures are connected to each other. The distribution of the AlO$_4$ units in the zeolite framework is most likely random, but there are some indications that there may be some ordering. It has been observed that, even at low Si/Al ratios, no Al–O–Al bridges are present in the structure (Loewenstein's rule) and that the number of Al–O–Si–O–Al sequences is minimized (Dempsey's rule). The Si/Al ratio of a particular sample can vary and can even be changed after preparation by "dealumination" processes, leaving the structure otherwise intact.

3.14 At low coverages there is no significant interaction between CO molecules because they are far apart from each other. With increasing coverage, however, the repulsive CO–CO interactions influence the metal–CO bonding such that the heat of adsorption decreases.

3.15   The bond energy of the hydrogen molecule is about 104 kcal/mole. When H_2 adsorbs dissociatively on Ni, two Ni-H bonds form. The Ni—H bond strength is about 63 kcal/mole; that is, the formation of two Ni—H bonds is energetically more favorable than maintaining a H—H bond—once the system has overcome the activation energy barrier.

The Au—H bond strength is low (≈ 46 kcal/mole). Thus the energy released during the formation of two Au—H bonds is insufficient to make up for the energy required to dissociate the H_2 molecule.

CHAPTER 4

4.1   We have, from Eq. 4.9,

$$\langle x^2 \rangle^{1/2} = \sqrt{\frac{3N\hbar^2 T}{mk_B\Theta_D^2}}$$

N	6.022×10^{22} mole^{-1}
$\hbar$	1.055×10^{-27} erg · sec
m	58.71 g/mole
k_B	1.38×10^{-16} erg/K
Θ_D	390 K

$$\langle x^2 \rangle^{1/2}|_{300K} = 0.022 \text{ Å}, \quad \langle x^2 \rangle^{1/2}|_{1728K} = 0.053 \text{ Å}$$

Fractional displacement: Ni—Ni distance $= 1.246$ Å (determined from lattice constant). Therefore,

$$\frac{\langle x^2 \rangle^{1/2}|_{300K}}{1.246 \text{ Å}} = 0.018$$

$$\frac{\langle x^2 \rangle^{1/2}|_{1728K}}{1.246 \text{ Å}} = 0.043$$

4.2   We have, from Eq. 4.42,

$$D = 0.014 \exp\left(-\frac{Q_2 T_m}{RT}\right), \qquad \frac{T}{T_m} < 0.77$$

$$D = 0.014 \exp\left(-\frac{54 \cdot 1358}{8.315 \cdot 300}\right)$$

$$\approx 2.41 \times 10^{-15} \text{ cm}^2/\text{sec}$$

$$x = \sqrt{4Dt} = \sqrt{4 \cdot 2.4 \times 10^{-15} \cdot 3600}$$

$$\approx 588 \text{ Å}$$

4.3   Using Eq. 4.48 we obtain

$$\frac{E_1}{RT_p^2} = \frac{\nu_1}{\alpha} \exp\left(-\frac{E_1}{RT_p}\right)$$

$$\frac{E_1}{8.315 \times 480^2} = \frac{10^{12} \text{ sec}^{-1}}{30 \text{ K/sec}} \exp\left(-\frac{E_1}{8.315 \times 480}\right)$$

$$E_1 = 6.4 \times 10^{16} \exp(-E_1 \times 2.5 \times 10^{-4})$$

The activation energy for desorption is 108.4 kJ/mole.

4.4   The bonding of CO to metal atoms occurs via charge donation from the 5σ orbital and back-donation from the metal into the $2\pi^*$ antibonding orbital. This back-donation weakens the $C \equiv O$ bond. At higher coordination, there is an increased back-donation of charge from more than one atom. As a result, the $C \equiv O$ bond becomes weaker and the force constant decreases accordingly. Thus the ν_{CO} stretching frequency shifts to lower wavenumbers. It is generally accepted that $\nu_{CO} \approx 2000$ cm^{-1} corresponds to linear bonding, $1850 < \nu_{CO} < 2000$ cm^{-1} indicates bridge bonding (twofold symmetry), and higher coordination can be assigned to frequencies below 1850 cm^{-1}.

4.5   The Debye–Waller factor quantitatively describes the decrease of the diffracted beam intensity due to thermal vibrations of the atoms in the lattice. As the temperature increases, the root-mean-square displacements of the atoms in the lattice become larger and the diffracted beam intensity decreases. The maximum intensity should be observed at 0 K. According to Eq. 4.11, the Debye–Waller factor for LEED and XRD is

$$\exp\left\{-\left[\left(\frac{12Nh^2}{mk_B}\right)\left(\frac{\cos\psi}{\lambda}\right)^2 \frac{T}{\Theta_D^2}\right]\right\}$$

For He diffraction, the expression has to be modified (see the references suggested) and becomes:

$$\exp\left\{-\frac{24\, mT}{mk_B\Theta_D^2}(D + E\cos^2\psi)\right\}$$

where m is the mass of the gas atom, E its kinetic energy, and D is the depth of the attractive surface well. D is on the order of 13 meV. For any given solid, the Debye–Waller factors for LEED, He atom diffraction, and X-ray scattering are different because of the different wavelengths involved and because of the different surface depths probed. Using nickel as an example ($M_{Ni} = 58.7$ g/mole, $\Theta_D^{bulk} = 390$ K, $\Theta_D^{surface} = 220$ K) and assuming direct backscattering ($\cos^2\psi = 1$) at room temperature ($t = 290$ K), we have the following:
For X-rays:

$$\text{Debye–Waller factor} = \exp\{-[0.746/\lambda(\text{Å})^2]\}$$

For LEED:

$$\text{Debye-Waller factor} = \exp\left\{-\left[2.35/\lambda(\text{Å})^2\right]\right\}$$

For He D:

$$\text{Debye-Waller factor} = \exp\left\{-114\left[0.013 + 0.2/\lambda(\text{Å})^2\right]\right\}$$

Thus, for these techniques under usual operating conditions, we have

Method	Typical Energy Used (eV)	Corresponding Wavelength (Å)	Debye-Waller Factor
LEED	100	1.22	0.207
He D	0.025	0.89	0.013
XRD	10000	1.24	0.616

4.6 According to Eq. 4.11, the Debye-Waller factor is

$$\exp\left\{-\left[\left(\frac{12Nh^2}{mk_B}\right)\left(\frac{\cos\psi}{\lambda}\right)^2\frac{T}{\theta_D^2}\right]\right\}$$

Thus, the intensity of the back-diffracted electron beam depends exponentially on the temperature of the surface. Expressed another way, the logarithm of the beam intensity varies linearly (and inversely) with temperature. If the sample temperature is taken high enough, there will be a sudden, strong downward deviation (a "knee") in the plot of the logarithm of the diffracted beam intensity. For the experiment given in the problem, this transition occurred at about 600 K.

4.7 For the experiment suggested in the problem, the following results were noted:

- $H_2/Ni(111)$
  —Adsorption probability increases with gas temperature, indicating an activation barrier
  —desorption versus angle measurements indicates that the H_2 probably does not adsorb via a precursor state
- $H_2/Ni(110)$
  —measurements indicate that adsorption proceeds via two different channels: direct physisorption at low gas temperatures and via direct chemisorption, with a small activation barrier, at higher gas temperatures.
- $H_2/Ni(100)$: here, as for Ni(110), two adsorption mechanisms are at work.
- The general trends observed are:
  —increasing surface roughness yields increased sticking probability
  —increasing surface roughness yields decreased activation energy for adsorption

4.8 So far there have been at least three different mechanisms observed for diffusion on transition metal surfaces:

- *Hopping*: As stated in the first article, "atom migration takes place by a series of displacements over the minimum in the potential barrier between adjacent binding sites." This simple mechanism explains many of the results for diffusion of various atom/substrate combinations.
- *Cross-channel displacement*: Seen in fcc(110) surfaces. An atom moving crosswise to one of the channels on the (110) surface oes not necessarily "hop" over the top; rather, it displaces one of the atoms in the row. This leads to the apparent "jump over the barrier."
- *Exchange-mediated*: A variant of the cross-channel mechanism. Instead of directly displacing an atom, the "diffusing" atom pushes down into the surface, causing one of its neighbors to "pop up." This leads to an apparent shift in the diffusing atom's "position."

4.9   The technique in question is called "laser-induced thermal desorption." This technique is used to study diffusion in systems where a local concentration gradient would be difficult to prepare. The method is straightforward:

- Prepare the sample, with a uniform surface concentration at a desired value.
- Use a focused laser pulse (or, for thoroughness, a train of several pulses) to "burn a hole" in the adsorbate layer (more precisely, use the focused (e.g., spot size of 1.2 mm × 0.4 mm) laser to cause rapid, local heating of the substrate surface, causing complete thermal desorption of the adsorbate).
- Wait a suitable time interval.
- Using the laser, re-"burn" the same spot; and by noting the rise in chamber pressure, determine how much adsorbate has diffused in from the surrounding portion of the sample.
- Repeat the above steps at different waiting times, sample temperatures, and initial adsorbate coverages to obtain the diffusion activation energy, diffusion prefactor, and effects of adsorbate–adsorbate interactions.

In the case of the *n*-alkane study mentioned in the problem, in addition to measuring the diffusion parameters, several interesting points were noticed. First, the diffusion processes were coverage-independent, indicating that the adsorbate–substrate interactions were dominant over the adsorbate–adsorbate interactions (see, for comparison, the other article mentioned). The other point was that the activation energies for diffusion, for these alkanes on this surface, scaled linearly with the chain length. This was interpreted to mean that the molecules move by "jumping" as a single unit. Diffusion by moving part of the molecule at a time—"snake-like wriggling"—would have led to a $D \propto 1/n^2$ behavior.

4.10   Let us consider the example of $CH_4/Ni(111)$. It was found that the dissociative adsorption had an activation energy of about 12–17 kcal/mole. This is close to the energy needed to distort the CH_4 molecule into a pyramidal shape (base plane containing three hydrogen atoms and the carbon atom). Thus it is argued that the collision serves to deform the molecule sufficiently that the C atom can approach the Ni surface closely enough to form a bond; the normal spherical shape of the methane molecule would otherwise forbid this. Experiments with CD_4 shown an isotope effect, which is taken to mean that tunneling of the hydrogen atom away from the methyl radical also plays a role in the reaction.

4.11 This experiment used ESDIAD to observe rotation of adsorbed PF_3 on Ni(111) about the P—Ni bond. The experiments were performed at two coverages: θ = 0.04 and θ = 0.25 monolayer. At the low coverage, the PF_3 molecules are far enough apart that they do not interact. At low temperature (85 K), F^+ ions were seen coming from the surface along well-defined directions, indicating that for practical purposes, the PF_3 molecules were unable to rotate freely. When the sample was warmed to 275 K, the angular anisotropy in the emission pattern disappeared, indicating that the molecules were free to rotate. Upon cooling the sample back to 85 K, the anisotropy returned, indicating that the rotation was again hindered. When the experiment was performed at θ = 0.25, angular anisotropy was observed at both the high and low temperatures. This is consistent with the molecules' being close enough to each other that they "lock." The molecules would then stay oriented at all temperatures. It was noted that the angular distribution of the ions was somewhat broadened at the higher temperatures.

4.12 Adsorbates on transition metals appear to desorb as a result of thermal desorption: The molecular species has a Boltzmann distribution in its translational energy distribution, and its energy is on the same order as the surface temperature. In this experiment, a laser was used to desorb CH_3Br from LiF. It was concluded that photodissociation, not photodesorption, was taking place. There were several experimental grounds supporting this conclusion:

- Radicals were only detected when the laser energy was tuned to be close to the maximum in the gas-phase absorption spectrum of CH_3Br.
- The energies of the desorbing methyl radicals were independent of the photon flux. Desorption via local heating would be expected to show some dependence.
- When the laser was tuned to a different energy, no radicals were observed. Again, desorption due to local heating would not be expected to show such a strong effect.
- The kinetic energy of the desorbing radicals was consistent with photodissociation, but inconsistent with thermal desorption.

4.13 If the atom has a high enough translational energy (E_T), it typically will elastically scatter from the surface of the crystal. On the other hand, if the translational energy is very small compared to the surface energy, depending on the strength of the surface and atom interaction, the atom either "rolls" into the potential energy well and assumes an adsorbed state or it gains energy and "bounces back" from the surface with a higher kinetic energy.

4.14 The mechanism involves the following steps:

- Electron excitation
- Electron redistribution
- Displacement/desorption of surface atoms
- Structural or electronic modification of the desorbing species

At lower electron kinetic energies (10–10^2 eV) the electron impact excites one electron from a bonding orbital to an antibonding orbital. As a result, the adsorbate-substrate bond weakens and the adsorbate desorbs. At higher kinetic energies, the mechanism more likely involves the formation of a core hole and an interatomic Auger transition precedes the desorption. For polyatomic molecules, after the initial electronic transition, the energy is distributed among the vibrational degrees of freedom and when the substrate–surface bond accumulates sufficient energy, it breaks. The details of the processes are not fully understood. From the angular distribution of the desorbing species, the bonding geometry of the adsorbate may be deduced. For example, the angular distribution of H^+ ions desorbing from $NH_3/Ni(111)$, at low temperature, shows a threefold pattern on a fluorescent screen. This indicates the absence of free rotation. At higher temperatures, the free rotation of the NH_3 molecule about the $N-Ni$ axis gives rise to a ring pattern. The cross section of the process varies from 10^{-23} cm^2 to 10^{-16} cm^2. The cross section depends on the kinetic energy of the incident electron beam and may be very different for different binding sites. The cross section versus kinetic energy curve has a threshold at around 10 eV and a maximum between 100 and 1000 eV.

4.15   Sputtering is a typical multiple collision process that involves a cascade of moving target atoms. The ions penetrate into the solid and lose energy in inelastic collisions with the atoms in their paths. Thus the energy of the incoming ion (500–3000 eV) is deposited in the vicinity of the initial impact, and if a near-surface atom in these multiple collision gains enough energy to escape from the target, it leaves the solid. The most important parameters that determine the sputtering yields are the following: kinetic energy; mass; electronic configuration; angle of incidence of the bombarding ion and the mass; electronic structure; crystal structure; orientation; binding energy of surface atoms; and roughness of the target sample. In the case of a two-component system, preferential sputtering of one component has been observed and this phenomenon can be explained in terms of relative atomic weights to the sputtering ion and the type of bonding between surface atoms. Ion beam sputtering is frequently used to clean solid surfaces in ultrahigh vacuum for surface-science studies, to study the composition of materials as a function of depth, and to deposit materials onto various substrates.

4.16   This study used helium scattering to locate energy levels of bound states of $He/LiF(001)$. If the angle and kinetic energy of the helium beam are varied, "dips" in the surface reflectivity will occur whenever the atom's kinetic energy perpendicular to the surface equals one of the discrete bound-particle energies of the surface-potential well. In the case in question, four bound levels were observed; this information could compare to several different theoretical models.

CHAPTER 5

5.1 The reason for their stability is the electric double layer. On the surface of the colloid particles, adsorbed molecules are oriented such that the outermost layers bear the same positive or negative charge. This prevents the close approach of these particles and stabilizes the colloid.

5.2 For Rh(111), one monolayer $= 1.60 \times 10^{15}$ cm$^{-2} = 1.60 \times 10^{19}$ m^{-2}. From Eq. 5.11, we have

$$\Delta\phi = -\frac{e\sigma\mu}{\epsilon_0}$$

$$= -\frac{-1.8 \text{ eV} \times 1.609 \times 10^{-19} \text{ J/eV}}{8.854 \times 10^{-12} \text{ C}^2/\text{J/m}}$$

$$\mu = 4.99 \times 10^{-30} \text{ C} \cdot \text{m} = q \times l$$

$$q = \mu/l = 4.16 \times 10^{-20} \text{ C} = 0.26 \ e^-$$

5.3 From Eq. 5.22, we have

$$j\left(\frac{\text{A}}{\text{cm}^2}\right) = AT^2 \exp\left(-\frac{\phi}{k_B T}\right)$$

where $k_B = 8.62 \times 10^{-5}$ eV/K. Therefore,

	T(K)	A (A/cm$^2 \cdot$ K^2)	ϕ (eV)	j (mA/cm^2)
W	2200	80	4.54	15.5
BaO	1030	0.0035	1.1	15.5

5.4 *Dipole scattering*: Consider a molecule adsorbed on a metal surface. The image dipole in the solid gives rise to a nonvanishing dipole moment normal to the surface, whereas the parallel component of the dipole moment and the image potential cancel each other. An electron approaching the surface induces an image charge as well. The interaction between the electric field and the molecule is such that only vibrations which show a nonzero dipole moment normal to the surface will be excited. The cross section of the vibrational excitation process has a maximum in the specular direction.

Impact scattering: The incoming electron "hits" the adsorbed atom (molecule) to produce a molecular ion, from which the electron is then re-emitted; this mechanism results in a different angular distribution of the scattering amplitude. It does not have a sharp angular dependence, but it changes rather smoothly.

H on W(100): In the specular direction there is only one energy loss peak. It is present at all coverages, but in off-specular directions three peaks are present. The single peak in the specular direction corresponds to the symmetric W—H—W stretching mode. The other peaks seen only in off specular directions are assigned to asymmetric stretch and wagging modes. At all coverages the H atom occupies bridge sites between two W atoms.

CO on Ni(111): At low coverages, CO forms a c(4 × 2) structure, but at higher coverages a c($\sqrt{7}$ × $\sqrt{7}$)R19° structure has been observed. The HREELS spectra indicate that on the former surface, one ν_{CO} can be observed at 1810–1910 cm^{-1}, indicating the presence of twofold bridge coordination. At >3.5-L CO exposures, terminal CO appears at 2050 cm^{-1}.

CHAPTER 7

7.1 Assuming that at 10^{-6} torr there is one collision/site/sec, at 1 atm there are about 10^9 collisions/site/sec. Thus, this turnover rate corresponds to a reaction probability of 10^{-12} per collision.

7.2 In the derivations which follow, we will use k to denote a rate constant, K to denote an equilibrium constant, and θ_v to denote the concentration of vacant surface sites. Let us do part (b) first. We shall make the following assumptions:

- Step (4) is irreversible.
- Steps (1), (2), and (3) are at equilibrium.
- All species occupy the same type of site.

Because step (4) is rate-limiting, we have

$$R_{CH_4} = r_4 = k_4\theta_{H_2CO}\theta_v \tag{i}$$

Equilibrium in steps (1), (2), and (3) gives us

$$\theta_{CO} = K_1P_{CO}\theta_v \tag{ii}$$

$$\theta_H = (K_2P_{H_2})^{1/2}\theta_v \tag{iii}$$

$$\theta_{H_2CO} = K_3\theta_{CO}\theta_H^2/\theta_v^2 \tag{iv}$$

Substituting (ii) and (iii) into (iv), we have

$$\theta_{H_2CO} = K_1K_2K_3P_{H_2}P_{CO}\theta_v \tag{v}$$

Substituting (v) into (i), we obtain

$$R_{CH_4CO} = k_4K_1K_2K_3P_{H_2}P_{CO}\theta_v^2 \tag{vi}$$

If we assume that $H_{(a)}$ and $CO_{(a)}$ are the dominant surface species, we also have

$$\theta_v + \theta_H + \theta_{CO} = 1 \tag{vii}$$

Substituting (b) and (c) into (g) and solving for θ_v, yields

$$\theta_v = [1 + (K_2 P_{H_2})^{1/2} + K_1 P_{CO}]^{-1} \tag{viii}$$

Substituting (h) into (f), we obtain

$$R_{CH_4} = \frac{k_4 K_1 K_2 K_3}{[1 + K_2^{1/2} P_{H_2}^{1/2} + K_1 P_{CO}]^2} P_{H_2} P_{CO} \tag{ix}$$

Now for part (a) of the problem. If step (3) is the limiting step, we have

$$R_{CH_4} = k_3 \theta_{CO} \theta_H^2 = k_3 \times K_1 \theta_v P_{CO} \times K_2 \theta_v^2 P_{H_2}$$

$$= k_3 K_1 K_2 P_{CO} P_{H_2} \theta_v^3$$

$$= \frac{k_3 K_1 K_2}{[1 + K_2^{1/2} P_{H_2}^{1/2} + K_1 P_{CO}]^3} P_{CO} P_{H_2}$$

CHAPTER 8

8.1

$$\text{Pressure} = \frac{\text{Force}}{\text{Area}}$$

Case (a):

$$P = \frac{1000}{0.01} = 10^5 \text{ dyne/cm}^2 = 10^4 Pa$$

Case (b):

$$P = \frac{1000}{0.0001} = 10^7 \text{ dyne/cm}^2 = 10^6 Pa$$

Yield points are on the order of 100 MPa (e.g., 2.4×10^8 Pa for aluminum and 1×10^8 Pa for brass).

8.2 The two lubrication regimes correspond to the thickness of the lubricating layer. In the hydrodynamic regime, the lubricant film thickness is many times greater than the roughness of the surface(s) in question. The bulk properties of the lubricant (shear resistance, viscosity) determine the friction. Coefficients of friction are usually quite low. In the boundary lubrication regime, the thickness of the lubricating layer is smaller than the surface roughness and the friction is determined by the interactions of asperities. In this case, long-chain polar molecules are frequently used as lubricants. The polar ends bond to the metal surfaces and the nonpolar tails form a layer which prevents contact between opposing metal surfaces.

8.3   The suggested article deals with the attractive force between two mica sheets—in particular, what effect, if any, is caused by the relative orientations of the two mica sheets. Also considered are what effect, if any, is caused by liquids between the two mica sheets. With regard to the first question, it was found that there was little orientation preference when the experiment was performed in air. This was attributed to residual dirt on the mica surfaces. When the experiment was performed in water, which also removes adsorbed dirt, a strong orientational dependence was found. Sharp peaks (width $< \pm 1°$) were seen in the plots of adhesion force versus orientation angle. This implies a strong sensitivity to lattice mismatch in adhesion, and also that the lattices must have been rigid. The experiments also showed that the adhesion had a periodic distance dependence, corresponding to the number of water molecule layers between the mica sheets.

8.4   There are many classes of engine oil additives or ingredients. Three particular examples are:

- *Alkaline detergents*: These are used to neutralize sulfur acids in the oil. Sulfur compounds are present in crude oils, and they form acids as by-products in the presence of the combustion in the engine. While most of the sulfur compounds are expelled from the engine along with the other combustion exhaust gases, any remaining acids tend to attack the engine parts.
- *Long-chain Esters*: These are used as lubricants, given that some of their properties are better than those of normal oil:
  —They burn more cleanly than ''normal'' oil.
  —They do not dissolve in gasoline as readily as ''normal'' oil, so less is burned during any one cycle.
  —Being nonaromatic to start with, fewer polyaromatic hydrocarbons (PAHs) are exhausted into the atmosphere as a result.
  —They have a low pour temperature, so cold engine starts result in less wear.
  —They are somewhat more biodegradable and less toxic than other lubricants.
- *Long-chain carboxylic acids*: The polar ends bond strongly to the metal surfaces, leaving the hydrocarbon ''tails'' sticking outwards. These layers have a high compressibility, but shear past other such layers easily. This protects metal pieces which might otherwise rub together.

8.5   The study by Mate et al. on graphite essentially shows that friction is studiable by AFM. A coefficient of friction of $\mu \approx 0.012$ was seen over a range of loads from 0 to 20×10^{-6} N. In addition, slipping of the tip was observed as it followed the hills and valleys in the graphite surface.

INDEX

Printed in the United States
92328LV00001B/15-22/A